Relations:
Concrete, Abstract, and Applied
An Introduction

Relations:

Concrete, Abstract, and Applied

An Introduction

Herbert Toth

World Scientific

NEW JERSEY • LONDON • SINGAPORE • BEIJING • SHANGHAI • HONG KONG • TAIPEI • CHENNAI • TOKYO

Published by

World Scientific Publishing Co. Pte. Ltd.
5 Toh Tuck Link, Singapore 596224
USA office: 27 Warren Street, Suite 401-402, Hackensack, NJ 07601
UK office: 57 Shelton Street, Covent Garden, London WC2H 9HE

Library of Congress Control Number: 2020025420

British Library Cataloguing-in-Publication Data
A catalogue record for this book is available from the British Library.

RELATIONS: CONCRETE, ABSTRACT, AND APPLIED
An Introduction

ISBN 978-981-122-034-0 (hardcover)
ISBN 978-981-122-035-7 (ebook for institutions)
ISBN 978-981-122-036-4 (ebook for individuals)

For any available supplementary material, please visit
https://www.worldscientific.com/worldscibooks/10.1142/11829#t=suppl

Printed in Singapore

Contents

Preface xi

Acknowledgments xv

Index of Notations xvii

PART 1 : Relations - Historical and Mathematical Preliminaries 1

1. A Short Look into (Relations') History 3

1.1 Syllogisms 4
1.1.1 Categorical Propositions and their Interpretation 5
1.1.2 The Syllogistic Basics 7
1.1.3 Syllogistic Validity 10
1.2 Syllogisms and Relations 12
1.3 From the History of Relations 16
1.3.1 The Groundwork: DE MORGAN and PEIRCE 16
1.3.2 The Followers: SCHRÖDER and TARSKI 21
References 25

2. Mathematical Prerequisites 27

2.1 Sets and their Operations 27
2.1.1 Some Properties of Set Difference 30
2.1.2 Cartesian Products, Relations, and Functions 31
2.1.3 Closures and Interiors 36
2.2 Basics from Predicate Logic 38
2.3 Algebraic Structures 42
2.3.1 A Simple Example: Semigroups 43
2.3.2 Algebras and Homomorphisms 44
2.3.3 Some more Examples of Algebraic Structures 46
2.3.4 Partial Orders and Lattices 48
2.3.5 Boolean Algebras 54

2.3.6 Galois Connections and Residuated Mappings 61
2.3.7 Partially Ordered Semigroups 64
Additional reading 65
References 65

PART 2: Relations - Concrete, two-valued **67**

3. Binary Relations: Basic Properties and Operations 69

3.1 Set Theoretical Properties 70
3.1.1 Images of Sets under a Relation 72
3.2 Basic Relation Operations 73
3.2.1 Representing Relations 79
3.3 Basic Relation Properties 80
3.3.1 Relations and Syllogisms Again 80
3.3.2 (A Lot of) Properties and some of their Interrelationships 81
3.4 Some Results from the Calculus of Relations 85
3.5 Dual Relations, Residuals and Symmetric Quotients 89
3.5.1 Dual Relations 89
3.5.2 Residuals 90
3.5.3 Symmetric Quotients and Bidirectional Residuals 93
3.6 Operations and Properties revisited 97
3.6.1 Point-free Representation of Relation Properties 97
3.6.2 Properties of Compound Relations 101
3.6.3 Powers and Closures of Relations 103
3.7 Homomorphisms and Isomorphisms 108
3.7.1 The Heterogeneous Case 108
3.7.2 The Homogeneous Case 110
References 110

4. Some Basic Kinds of Binary Relations 113

4.1 Relations and Functions 113
4.2 Order Relations 116
4.3 Tolerance and Equivalence Relations 118
4.3.1 Equivalence Relations 119
4.3.2 Tolerance Relations 123
4.3.3 Partial Equivalence Relations 128
References 129

5. Relations, Matrices, and Graphs 131

5.1 Recalling Basic Facts on Boolean Matrices 132
5.2 Graphs and Matrices 135
5.3 Relations and Matrices 137

5.4 Powers of Relations on a Finite Set 139
5.4.1 Indices of Semigroups 139
5.4.2 Inclines, Permanents and Adjoints 141
5.4.3 Properties of Incline Matrices and Relation Powers 145
References 151

6. Some Special Topics around Relations 153

6.1 Bandler-Kohout Products and Images 153
6.1.1 The Basic Concepts 153
6.1.2 On Properties of BK images and products 156
6.2 Some Further Kinds of Relations 163
6.2.1 Subsets, Vectors, and Rectangles 163
6.2.2 Inverses and Regular Relations 172
6.2.3 Commuting Relations 180
References 195

PART 3: Relations - Concrete, many-valued **199**

7. Many-valued Extensions: Preliminaries 201

7.1 Graded Truth Values and their Structures 202
7.2 Order and Lattice Theory: Supplements 204
7.2.1 More on Orders and Residuation 204
7.2.2 More on Lattices 210
References 217

8. Fuzzy Sets and Fuzzy Relations 221

8.1 Fuzzy Sets: Basic Definitions and Properties 222
8.2 Many-valued Connectives on Complete lattices 225
8.3 Fuzzy Relations: Basic Operations and Properties 229
8.3.1 Basic Operations on Fuzzy Relations 233
8.3.2 Basic Properties of Fuzzy Relations 238
8.4 Fuzzy Equivalence Relations 240
8.5 Fuzzy Relation Equations and Inequalities 249
8.5.1 Types of Fuzzy Relation Equations 250
8.5.2 The Lattice Based Approach 251
8.6 Applications of Fuzzy Relations 259
8.6.1 Inference Rules 260
8.6.2 Fuzzy Rule Based Systems 262
References 265

PART 4 : Relations - Abstract **269**

9. Semigroups and Binary Relations 271

9.1 Some Basic Concepts of Semigroup Theory 271
9.1.1 Various Kinds of Semigroups 271
9.1.2 Ideals and Green's Relations 273
9.1.3 Natural Partial Order 279
9.2 Algebraic Properties of Relations on Finite Sets 281
9.2.1 The Semigroup of Binary Relations $\mathcal{B}_X$ 281
9.2.2 The Green's Relations on $\mathcal{B}_X$ 285
9.2.3 The Natural Partial Order on $\mathcal{B}_X$ 291
References 293

10. Relation Algebras 295

10.1 Algebras of relations 295
10.2 Some Basic Concepts from Universal Algebra 297
10.2.1 Subalgebras and Homomorphic Images 297
10.2.2 Congruences and Quotient Algebras 298
10.2.3 Direct Products and Factor Congruences 300
10.2.4 Varieties and Equational Definability 302
10.2.5 Boolean Algebras with Operators 306
10.3 Relation Algebras 309
10.3.1 Arithmetic in Relation Algebras 313
10.3.2 Special Kinds of Elements 319
10.3.3 Representability and other Algebraic Properties of RA 327
References 333

11. Relations and Categories 335

11.1 Categories: Basic Concepts 336
11.1.1 Special Arrows and Structures 342
11.2 Relations and Regular Categories 359
11.2.1 Regular Categories and Factorizations 359
11.2.2 Relations in Regular Categories 368
11.3 Some Related Concepts 376
11.3.1 Using Elements in the Categorical Setting 376
11.3.2 2-categories 378
11.4 Relations and Allegories 379
11.4.1 Allegories: The Basics 379
11.4.2 Special Properties of Arrows 381
11.4.3 Regular Categories and Allegories 383
11.4.4 Special Types of Allegories 384
References 390

PART 5: Relations - Applied 393

12. Semiotics and Semantics 395

12.1 The Meaning Triangle 395
12.2 Ambiguity, Vagueness, Uncertainty and the like 398
12.3 Some relational Clarifications 400
References 405

13. Information Systems and Rough Sets 407

13.1 Information Systems 407
13.2 Rough Set Theory (RST) 410
13.2.1 Basic Ideas of Rough Set Theory 410
13.2.2 Generalized Relation Based Rough Sets 413
References 417

14. Formal Concept Analysis 419

14.1 The Basic Notions and Algebra of FCA 419
14.2 Attribute Implications 423
14.3 FCA and Relations 425
References 426

15. Mereology and Qualitative Spatial and Temporal Reasoning 427

15.1 Mereology 428
15.1.1 Basic Notions of Mereology 428
15.1.2 Ground Mereology 430
15.1.3 Minimal and Extensional Mereology, Atoms 432
15.1.4 Closure and General Mereologies 433
15.2 Introducing Qualitative Models and Inferences 434
15.2.1 Some Formal Background 436
15.3 Qualitative Temporal Calculi 438
15.3.1 Point Algebra 438
15.3.2 Interval Algebra 440
15.4 Qualitative Spatial Calculi: Mereotopology 441
15.4.1 Background and Ontological Aspects 441
15.4.2 Ground Mereotopology 442
15.4.3 The Region-Connection Calculus 445
15.5 Some Logical and Algebraic Considerations 446
15.5.1 Composition Tables 448
15.5.2 Constraint-based Qualitative Reasoning 451
Additional reading 452
References 453

16. Argumentation Frameworks 457

16.1 Introduction 457
16.2 Key Concepts of Argumentation Frameworks 458
16.3 DAFs from a Relational Viewpoint 463
16.4 Generalizations and Variants 465
16.4.1 Values and Preferences 465
16.4.2 Bipolar Frameworks 466
16.4.3 Abstract Dialectical Frameworks 467
Additional reading 468
References 468

17. Describing Graphs and Transition Systems 471

17.1 Graphs and Relations 471
17.1.1 Basic Concepts 471
17.1.2 Describing a Graph by its Relations 474
17.1.3 Paths and Reachability 479
17.1.4 Homomorphisms and Isomorphisms 482
17.2 Transition Systems and Behavioral Equivalence 484
17.2.1 Trace Equivalences 486
17.2.2 Simulation Equivalences 487
References 491

18. Order Relations and Preference Modelling 493

18.1 Ferrers Relations 493
18.2 More on Order Relations 495
18.3 Preferences and their Measurement 501
18.3.1 A Short Sketch of Measurement Theory 501
18.3.2 Semiorders and Interval Orders 506
18.4 Preference Structures 509
References 514

19. Relations and Syllogisms Again 517

19.1 Relational Analysis of Syllogisms 517
19.2 Relations and the Square of Opposition 525
19.2.1 The Square of Opposition 525
19.2.2 Opposition Structures Induced by Sets and Relations 528
References 534

Index 537

Preface

> *The importance of relations is almost self-evident. Science is, in a sense, the discovery of relations between observables.* (GOGUEN, J.: *L-fuzzy sets, J. Math. Anal. Appl.* ***18****(1967), p 161.)*

The above quotation is only one side of the medal; on the other side we have to recognize that relations are so omnipresent in our daily life that in many cases we do not even become aware of them, at least not in the sense of a mathematical concept. Who reasons about what follows from '*She is a relative of mine*'? E.g. that I am also a relative of her, and that her relatives are also my relatives, some of them probably not as near as she is. Compare this with '*is a friend of*'. Or that a cooler is part of your car, but, quite naturally, your car is not part of its cooler. Since any reader can easily find a lot more of such everyday examples, it seems legitimate to reinforce Goguen's quotation from above by saying that without relations we would not be able to manage daily routine and life as a whole (and, of course, one could not do mathematics). So why not give relations the attention they deserve, and do so in a way that corresponds to their role they play in the two areas just mentioned?

Very likely, everyone noticing the existence of this book will be astonished: Why a new book on such a simple mathematical concept? Let me try to give a partial and personal answer to this question.

- Simple concepts are very attractive in principle since they are easy to understand. (But as we all know, they may become dangerous when used in a demagogic or disguised way to manipulate the uncritical masses.) Thus, simple mathematical concepts are much more likely to be looked at and applied than e.g. the category theoretical concept of a *topos*.
- If a simple mathematical concept additionally is very powerful in modeling both real facts and abstract concepts, then a lot of people will enhance the basic ideas and apply them to a wide variety of problems.
- I claim that binary relations are of just that kind, in everyday discourse as well as

in pure and applied mathematics, and also in various engineering disciplines.

- This book is the result of a 'private project' that I have started after some time of inspecting existing literature, since I felt myself in an unsatisfactory situation. I had to realize a big gap: Either I got presented the most basic notions only (but in countless repetitions and with differing notation in both books and papers), or the highly advanced and sophisticated concepts in great detail, but restricted to very narrow contexts (again in both books and papers). In disrespectful words: either broad and shallow, or deep and small. But nowhere I found an overview offering relational concepts in both reasonable depth and broadness, and also providing an insight into the vast diversity of theoretical aspects and application possibilities beyond the commonly known examples.
- This book does not contain new, but only is a new compilation of known results, which tries to give a rather self-contained overview on the huge amount of existing knowledge largely scattered over both time and place. I hope that some anchors for starting out into diverse directions in the field will be provided. In addition to this, the book offers some first glimpses onto the elegance and beauty of the abstract worlds of universal algebra and category theory.

So much about my motivation for undergoing the work of writing this book. According to what I just stated to be missing, its aim is to immerse a little bit deeper, and to get a little broader in choosing the themes presented than is usually done in introductory books. I will try to raise the reader's appetite by providing a reasonable selection of both theoretical results and examples from less familiar application areas. In sum, the material included should enable readers to proceed to the more advanced parts of the theory and applications of binary relations.

If I have succeeded in delivering a text that does not only fulfill my personal needs and interests must be judged by its readers, if any. But independent from their evaluation I enjoyed the process of writing itself, the necessity to both refresh and enhance my LaTeX abilities after some years of interruption, and, above all, the discovery of so many opportunities to learn about new and interesting things.

Let me finish this preface with two technical remarks and a general consideration:

(1) Some of you may feel that this book should be smaller in size because of its subtitle '*An Introduction*'. But though being a mathematically simple concept, relations have very powerful modeling abilities and also a rich formal background, so a lot of results have been accumulated in the course of time. Thus, even an introductory overview needs some space, and therefore I think that the subtitle still applies. It is equally important to the main one and serves to indicate that, for each of the theoretical aspects and application areas described, there exist a lot of much more elaborated and advanced results in books and journals. In addition to the references given, modern technology can certainly support interested readers.

(2) I am fully aware that the text contains another repetition of stuff that has already often been repeated in other books and papers. The reason for doing it this way is my personal experience: I always felt it much more convenient to have collected into a single whole all (or most of) the information that was necessary for enabling me to comprehend a mathematical presentation. This way there was no need to handle several books at a time or to ask the world wide digital resources for mainstream information missing in both my knowledge and in the book of current interest to me.

(3) Because it still applies, it is too tempting to quote the last paragraph from the seminal work of TARSKI, A.: *On the Calculus of Relations*, J. Symbolic Logic **6** (1941), p 89. So please read the final remarks, but not in my own words, because I could not do it better and, of course, not with nearly as much authority: The aim of this book is

> "not so much to present new results, as to awaken interest in a certain neglected logical theory, ... I do believe that the calculus of relations deserves much more attention than it receives. For, aside from the fact that the concepts occurring in this calculus possess an objective importance and are in these times almost indispensable in any scientific discussion, the calculus of relations has an intrinsic charm and beauty which makes it a source of intellectual delight to all who become acquainted with it."

Vienna, October 2019 *Herbert Toth*

Acknowledgments

General Thanks

My sincere thanks go to all the many pioneers, in both mathematics and technology: to the famous and well known, but also to the nameless and unknown ones!

Without their visions, and without the concepts they have invented, the tools they have programmed, the machines they have built, and the data they have collected, writing this book would have been altogether impossible for me.

Special Thanks

I am greatly indebted to the reviewer whose incredibly meticulous work prevented my book from containing still more typos!

Of course, I also owe many thanks to all the people at World Scientific involved in the publishing process, especially to Ms. Rok Ting Tan, for their always smooth and friendly cooperation and support, and for their help in improving the appearance of this book by pointing out some improper formattings.

And I will also be grateful to any reader who will be so kind to give me hints where to correct and improve the text.

Index of Notations

$\mathcal{P}(A)$	29	power set of A
$\mathbf{2}^A$	29	power set of A
Π_A	30	set of the partitions of A
$[a]_\pi$.	30	block of a in partition π
$\langle x, y\rangle$	31	the ordered pair
$x/R, [x]_R$	32	equivalence class of x in R
X/R	32	quotient set of X modulo R
$\mathsf{dom}(f)$	33	domain of function f
$\mathsf{rng}(f)$	33	range of function f
$\mathsf{ker}(f)$	33	kernel of function f
$\exists!$	33	"there is exactly one"
$f\restriction_S$	33	restriction of f to S
$cl_{\mathscr{C}}(S)$	36	closure of S
E_S	47	set of idempotents of semigroup S
$a \asymp b$	48	$a, b \in P$ are comparable in poset P
$a \prec b$	49	b covers a in poset P
C_L	53	set of complemented elements in L
$\mathfrak{B}^*$	56	Boolean ring induced by $\mathfrak{B}$
$\mathsf{At}(B)$	58	set of atoms of B
$\mathsf{dom}(\rho)$	71	domain of relation ρ (a set)
$\mathsf{rng}(\rho)$	71	range of relation ρ (a set)
$\mathsf{fld}(\rho)$	71	field of relation ρ
$\rho^{\smile}$	74	converse relation
$a\rho, \rho b$	74	afterset of $a \in A$, foreset of $b \in B$
ρ_{dom}	75	domain of relation ρ (a relation)
ρ_{rng}	75	range of relation ρ (a relation)

ρ^d	89	dual relation
σ/ρ	90	right residual of σ by ρ
$\sigma\backslash\rho$	90	left residual of σ by ρ
$\mathsf{syQ}(\rho,\sigma)$	93	symmetric quotient of ρ and σ
$\mathsf{biR}(\rho)$	96	bidirectional residual
ρ^+	103	
ρ^*	103	star of ρ
$[x]_\rho$, x/ρ, $[x]$	119	equivalence class of x induced by ρ
S/ρ	119	quotient set of S modulo ρ
Eq_S	120	equivalence relations on S
Π_S	120	the set of partitions of the set S
ρ_π	120	equivalence relation induced by $\pi \in \Pi_A$
π_ρ	120	partition induced by $\rho \in \mathsf{Eq}_S$
PBl_ρ	124	pre-blocks of relation ρ
Bl_ρ	124	blocks of relation ρ
Tol_S	124	tolerance relations on S
PEq_S	129	partial equivalences on S
$\boldsymbol{B}_n$	131	Boolean matrices on any set X with $\|X\| = n$
O, J, I	132	zero, one, and identity matrix
δ_{ij}	132	Kronecker symbol
$\underline{n}$	134	the set $\{1,2,\ldots,n\}$
$\langle A\rangle$	139	intersection of all semigroups $B \supseteq A$
$A(i\|j)$	142	submatrix of A without i-th row and j-th column
$\mathsf{per}(A)$	142	permanent of A
$\mathsf{adj}(A)$	142	adjoint of A
$G_n(L), P_n(L), H_n(L)$	145	invertible, permutation, and Hall matrices
$[n]$	145	least common multiple of the numbers in $\underline{n}$
$\dot{p}$	164	point (relation) $\dot{p} \in \mathcal{B}_V$
$\mathsf{fringe}(\rho)$	172	fringe of relation ρ: $\rho \cap \overline{\rho\rho^d\rho}$
$V(\alpha)$	174	row space of α: $\{A\alpha : A \subseteq X\}$
Idm_S	191	the set of idempotent binary relations on S
χ_B	201	characteristic function of B
$x^{\downarrow}$	205	principal ideal (down-set) generated by x
$x^{\uparrow}$	205	principal filter (up-set) generated by x
$a \frown b$	210	relative pseudocomplement of a in b

$x \rightleftharpoons y$	216	biresiduation of a and b
$\mathcal{F}^L(X)$	222	the set of all L-fuzzy sets over X
$\mathsf{supp}(A)$	224	the support of fuzzy set A
$\Lambda(A)$	224	the level set of fuzzy set A
A_α	224	the α-cut of fuzzy set A
$A_{\vert\alpha}$	224	the α-restriction of fuzzy set A
$\mathsf{hgt}(A)$	224	the height of fuzzy set A
$\mathsf{core}(A)$	224	the core of fuzzy set A
A_L	224	$\{A_\alpha : \alpha \in L\} \subseteq \mathbf{2}^X$
$\mathcal{F}^L_{X\times Y}$, $\mathcal{F}^L_X$	231	set of L-fuzzy relations on $X \times Y$
$A \otimes B$	234	product of L-fuzzy sets A, B
$\mathsf{Eq}^L{}_S$	241	the set of L-fuzzy equivalences on S
E_C	244	
S_C	244	
$\mathsf{cmp}(\mu,\nu)$	246	degree of compatibility of μ and ν
$\mathsf{equ}(\mu,\nu)$	246	degree of equality of μ and ν
$\rho@\sigma$	252	@-composition of relations ρ and σ
$\mathscr{S}(\rho,\tau)$, $\mathscr{S}(\sigma,\tau)$	251	$\{\rho : \rho\sigma = \tau\}$, $\{\sigma : \rho\sigma = \tau\}$
$\rho@\sigma$	252	@-composition of relations ρ and σ
S_ρ, S_σ	253	$\mathscr{S}(\rho,\tau)$, $\mathscr{S}(\sigma,\tau)$
$\check{S}_\rho$	254	minimal elements in S_ρ
$a \check{\sqcap} b$	256	
$a \hat{\sqcap} b$	256	
V_a, $V_{S,a}$	279	inverse elements of a
E_S	279	idempotents of semigroup S
$\mathcal{P}_X$	282	set of all partial maps of X
$\mathcal{T}_X$	282	full transformation semigroup on X
$\mathcal{S}_X$	282	symmetric group on X
$\mathcal{I}_X$	282	symmetric inverse semigroup
BRA	296	algebras of binary relations
$\mathfrak{Re}(S)$	297	an algebra of binary relation with base S
$\mathsf{Con}(\mathfrak{A})$	298	set of all congruences on $\mathfrak{A}$
$\mathfrak{A}/\theta$	299	quotient algebra of $\mathfrak{A}$ by θ
RA	310, 312	class of relation algebras
NA	312	class of nonassociative relation algebras

WA	312	class of weakly associative relation algebras
SA	312	class of semiassociative relation algebras
$\mathsf{Idl}(\mathfrak{A})$	319	set of ideals of relation algebra $\mathfrak{A}$
$\mathsf{Sy}(\mathfrak{A})$	319	symmetric elements of relation algebra $\mathfrak{A}$
$\mathsf{Tr}(\mathfrak{A})$	319	transitive elements of relation algebra $\mathfrak{A}$
$\mathsf{Eq}(\mathfrak{A})$	319	equivalence elements of relation algebra $\mathfrak{A}$
$A(p)$	325	principal ideal generated by p
RRA	328	representable relation algebras
$f : A \rightarrowtail B$	343	monomorphism, monic arrow
$f : A \twoheadrightarrow B$	343	epimorphism, epic arrow
$f : A \rightarrowtail\!\!\!\!\twoheadrightarrow B$	343	isomorphism, iso arrow
$f : A \hookrightarrow B$	343	inclusion
$\mathsf{epi}_{\mathcal{C}}$	343	the epimorphisms of $\mathcal{C}$
$\mathsf{iso}_{\mathcal{C}}$	343	the isomorphisms of $\mathcal{C}$
$\mathsf{mono}_{\mathcal{C}}$	343	the monomorphisms of $\mathcal{C}$
$\mathsf{mor}_{\mathcal{C}}$	343	the morphisms of $\mathcal{C}$
$\langle f, g\rangle$	350	product arrow of f and g
$f \times g$	352	arrow product of f and g
Rel	359	category of sets and relations
$c : A \rightarrowtriangle B$	360	cover, extremal epimorphism
$\rho : A \looparrowright B$	370	isomorphism class of relations from A to B
$\rho \rightarrowtail A \times B$	370	relation from A to B
$[A f \rho g B]$	370	relation from A to B
$\rho \leq \sigma$	371	ρ subrelation of σ (in a category)
Rel($\mathcal{C}$)	376	the relations in a regular category $\mathcal{C}$
L-**Rel**	385	cat. of sets and L-fuzzy relations
1_{XY}	386	greatest L-fuzzy relation from X to Y
def_{ρ}^{U}	410	set of ρ-definable sets
$V_{\rho}^{\perp}$	411	the lower ρ-approximation of V
$V_{\rho}^{\top}$	411	the upper ρ-approximation of V
S_{ρ}	414	set of all solitary elements in ρ
Δ, Δ_X	510	$\{\langle x, x\rangle : x \in X\}$

PART 1

Relations - Historical and Mathematical Preliminaries

Chapter 1

A Short Look into (Relations') History

This chapter gives a short overview on the main historical steps in the mathematical development of relational concepts from their beginnings up to modern times. This also includes, perhaps surprisingly for many, a look at traditional and modern syllogistic logic. (But be aware, that relations are such a common thing in everyday life that their usage usually goes unnoticed. That is to say: Long before relations have become an object of mathematical interest, they already had an important role in dealing with everyday situations. But associated epistemological considerations are not our topic here.)

We can be sure of the fact that what we call *relations* nowadays played a significant, yet unconscious role in everyday life already very early in the development of mankind: It has always been important to be stronger, to associate with a companion more durable than most of the others, to own a larger area of land, to have available richer stocks for the winter period, to raise a greater number of cattle, to have more money at one's disposal than ... .

Nevertheless, compared to e.g. geometry (EUCLID, about 300 BCE), or calculus (GOTTFRIED WILHELM LEIBNIZ (1646-1716), ISAAC NEWTON (1643-1727); for more details see e.g chapter 9 of [22]), it took a rather long time until relations have become a major object of mathematical research beginning in the middle of the 19th century, most notably by the work of AUGUSTUS DE MORGAN (1806-1871) and CHARLES SANDERS PEIRCE (1839-1914). Grattan-Guinness [4], p. 32, calls it "a curious feature of the history of philosophy that ... nobody seems to have taken seriously the fact that relational propositions, such as 'John is taller than Jeremy', cannot fall within the compass of syllogistic logic".

So we have seen that the concept of relation is not a new idea, in both everyday and mathematical discourse. Concerning its mathematical history, the development proceeded in three major steps for which I will give a short account below. One cannot properly appreciate the work of these 'fathers of relations' if one does not have a look at what one of the 'forefathers' had to contribute to our topic. And as in so many other cases of the development of philosophical and scientific ideas of Western civilization, it is inevitable to go back to ARISTOTLE (384-322 BCE) himself, this time to his theory of syllogisms. So the next section turns back in time and is devoted to a short overview on *Aristotelian logic*, also named *syllogistic logic.*

1.1 Syllogisms

Following [8] we call a text-book example which, from "*Some men are philosophers*" and "*All philosophers are wise*" concludes that "*Some men are wise*" a formal logical inference, because the validity of this inference depends only on the form of the given sentences - in particular it does not depend on the truth or falsity of the sentences involved. The *form of a sentence* is that which remains preserved when its predicates, here "*men*" and "*philosophers*", are replaced by *variables.* Variables are signs devoid of meaning, which serve to indicate the places where meaningful predicates are to be inserted.

In order to describe the object of formal logic more precisely, it must be specified which parts of a sentence must not be replaced by variables during the extraction of its form. These are the logical particles, such as *all, some, if - then, and, or* and *not.* We then can consider the investigation of the conditions that allow to infer from sentence forms - consisting of variables and logical particles - other such sentence forms as the fundamental problem of formal logic. If a form B can be inferred from another form A, we say that B is (logically) implied by A, or simply that "*A implies B*". Formal logic can thus be defined as the science of the implications of sentential forms.

This problem was first attacked by ARISTOTLE in his systems of syllogisms, which gives the solution to a part of it, in which he limits the sentence forms to the four patterns given in Table 1.1, leading to what will be called *categorical syllogisms.* There are good reasons for the fact that categorical syllogism lie at the start of classical logic, and that syllogisms are still of interest to logicians: Although, from the viewpoint of modern symbolic logic, syllogisms are but a very small part of what first order logic can handle, and also need only 'simple' formal constructs, they represent some of the most fundamental deductive reasoning principles. And what else to begin with, if not with the simple and fundamental cases, the concepts that are at the basis of large parts of both everyday and scientific reasoning?

I am sure, there are not so few people who are like me: Of course, I have heard about the most elementary properties of syllogisms quite a couple of years ago. But never before I had to really dwell on this subject, and nearly all of this humble knowledge has gone lost. This is but one of the reasons why I decided to include the basic information on syllogistic logic, being completely aware of the fact that this means adding another replication[1]. Two more reasons are (1) my aim to keep things self-contained to a reason-

[1]There are really a lot of books and websites that provide very good exposures of traditional logic; so if you want, I plead guilty to repeating. Hints to books are given in the text; here I want to mention the following three websites: (1) the *History of logic* by Jaakko J. Hintikka on `http://www.britannica.com/EBchecked/topic/346217/history-of-logic/`, (2) Wikipedias syllogism page `http://en.wikipedia.org/wiki/Syllogism` containing Venn diagrams for all 24 valid syllogisms, and, last but not least, (3) `http://philosophy.lander.edu/logic/index.html`.

able degree, and (2) to give a short presentation on when, where, and how the concept of *relation* became a topic of mathematical interest. Moreover, that ARISTOTLE has retained his influences to this day can easily be recognized from a short look into various mathematical journals, mainly on logic and computer science.

To be precise: Although we will limit ourselves almost always to categorical syllogisms, what will be described below is not the Aristotelian theory of the syllogism in its original form, but what ŁUKASIEWICZ [7, §8] calls the "traditional syllogism". Using the word *syllogism* we always mean syllogistic logic in this sense.

The presentation in this chapter heavily rests upon [3] and [5], two excellent books that I strongly recommend not only because they give a much richer treatment of the current subject (including e.g. the traditional Square of Opposition and the usage of Venn diagrams), but, first of all, because their presentation makes us aware of the fact that there has been a long evolution yielding many great insights before modern formal logic could enter the stage.

1.1.1 *Categorical Propositions and their Interpretation*

ARISTOTLE was the first to develop a formal system of logic, conceiving it as a tool for the scientific and philosophical investigation of reality. For him, reality was organized in categories or classes having members, and he thought knowledge of reality to consist of true propositions (assertions) about these categories and their members. What makes this system of logic formal is that, according to Aristotle, every statement, regardless of its content, can be expressed in one of a small number of basic propositional forms. He was aware of the fact that if we could reduce all propositions to a set of few basic forms, then it would be much easier to manage the task of comprehending the ways in which these propositions are related in arguments.

ARISTOTLE realized that the conclusion in some argument types is related to its premises formally, that is, is independent from what the propositions are about, and that all arguments of this form are equally valid. His ground-breaking insight in the study of logic then was the following: What makes the valid arguments valid has nothing to do with their content, but is simply a matter of the formal relation that exists between its premises and its conclusion. And if there are only a few forms of valid arguments, it will become much easier to distinguish between good reasoning and bad reasoning. To honor ARISTOTLE's contribution, this system of logic is often referred to as **Aristotelian Logic**. However brilliant ARISTOTLE's analysis has been, he did overlook one important thing: What he did not consider is that classes can be empty, but instead assumed that every class has at least one member. Recognizing that this assumption is not necessarily true has enormous consequences for categorical logic.

A **categorical statement** (**categorical proposition**) expresses how two sets of things relate to each other. For example, the statement *"All triangles are geometric figures."* tells us that every member of the set of triangles does also belong to the set of geometric figures. Such a categorical statement is made up of four components:

All	triangles	are	geometric figures.
quantifier	subject term	copula	predicate term

The **subject term** and the **predicate term** stand for the two sets whose logical relation is characterized by the statement. Thus, categorical propositions are regarded as being about the classes of objects denoted by their subject and predicate term, respectively. Given two sets, there are four possibilities how they can be related to each other:

(1) Every member of the first set is also a member of the second set.
(2) No member of the first set belongs to the second set.
(3) At least one member of the first set is also a member of the second set.
(4) At least one member of the first set is not a member of the second set.

Since the early Middle Ages these four kinds of categorical propositions have commonly been designated by letter names corresponding to the first four vowels of the Roman alphabet: *A, E, I,* and *O.* (These letters come from the Latin verbs ***affirmo***, meaning "*I affirm*", and ***nego***, meaning "*I deny*").

Categorical propositions are constructed from (1) a grammatical subject, and (2) a grammatical predicate, and have two properties, viz. (a) quality (*affirmative, negative*) and (b) quantity (*universal, particular*). The resulting kinds of propositions (for illustration purposes together with a contemporary transcription in logical and set theoretical terms) are shown in Table 1.1.

TABLE 1.1: ARISTOTLE's proposition types.

Code	Proposition	Quantity	Quality	Logic	Set theory
A	*All S are P*	universal	affirmative	$\forall x\,[S(x) \rightarrow P(x)]$	$S \subseteq P$
E	*No S is P*	universal	negative	$\forall x\,[S(x) \rightarrow \neg P(x)]$	$S \cap P = \emptyset$
I	*Some S are P*	particular	affirmative	$\exists x\,[S(x) \wedge P(x)]$	$S \cap P \neq \emptyset$
O	*Some S are not P*	particular	negative	$\exists x\,[S(x) \wedge \neg P(x)]$	$S \not\subseteq P$

The **quality** of a categorical proposition is either affirmative or negative depending on whether it affirms or denies class membership. Accordingly, '*All S are P*' and '*Some S are P*' have affirmative quality, and '*No S are P*' and '*Some S are not P*' have negative quality. The **quantity** of a categorical proposition is either universal or particular, depending on whether the statement makes a claim about every member or just some member of the class denoted by the subject term. '*All S are P*' and '*No S are P*' each assert something about every member of the S class and thus are universal, whereas '*Some S are P*' and '*Some S are not P*' assert something about one or more members of the S class and hence are particular.

Categorical propositions are building blocks of a certain kind of arguments, but before we can analyze them properly we have to discuss the philosophical problem of *ex-*

istential import and to provide a resolution on which we subsequently will base our analysis of syllogisms. (For a very detailed discussion see [3], Section 5.7.) A proposition is said to have **existential import** if its truth depends on the existence of members of the subject class.

There are two rival views on the question whether a universal categorical sentence like "*All S are P*" implies that there are any S. The origin of this question lies in the *sub-alternation* relation of Aristotle's Square of Opposition, i.e. in the inference: "*All S are P*", therefore "*Some S are P*". The traditional view is called **Aristotelian interpretation**, and the modern **Boolean interpretation** (named after GEORGE BOOLE (1815-1864), an English mathematician). Both views agree on particular propositions: *I* and *O* propositions have existential import; they assert that the classes designated by their subject terms are not empty. But the Aristotelian and the Boolean interpretation differ on universal statements: in Aristotelian logic, *I* and *O* propositions follow validly from *A* and *E* propositions, respectively, by sub-alternation. As a result, Aristotelian logic requires *A* and *E* propositions to have existential import too, because a proposition which has existential import cannot validly be derived from a proposition without it.

Thus, according to ARISTOTLE, if we assert that "*All unicorns have one horn*" then we are committed to assert that unicorns exist. For the modern logician or mathematician, this is an unacceptable result because modern mathematics and logic often deal with empty sets or with imaginary objects. So, in the Boolean interpretation, a universal proposition is understood to assert only that "*If a thing is an S, then it is also a P*". This is where the logical transcriptions given in Table 1.1 above do come from. (Note that the conditional in $\forall x\,[S(x) \rightarrow P(x)]$ will always be true, if there is no thing that is an S, because conditionals with false antecedent are true, and thus a universal conditional whose subject class is empty evaluates to true.)

1.1.2 *The Syllogistic Basics*

With a background of only basic knowledge on modern logic, the following condensed description of syllogisms can be given: In logic, an **argument** is a series of statements (or propositions) typically used to persuade someone of something or to present reasons for accepting another statement. An argument is structured in two parts: the **premise**s and the **conclusion**. In a typical **deductive argument** the premises are meant to provide a guarantee of the truth of the conclusion. Deductive arguments may be either valid or invalid. A **valid argument** links its premises with its conclusions in a way that, if the former are true, the latter must also necessarily be true: A valid argument cannot have true premises and a false conclusion.

An argument as described above then is what modern logicians call a (single-conclusion) sequent: a (finite) set Γ of sentences, the premises, followed by a single sentence φ, the conclusion. Sequents are often represented in the form $\Gamma : \varphi$ or $\Gamma \vdash \varphi$. For our current needs we can simply read the symbol $\vdash$ as "*If all premises listed in* Γ *on the left of* $\vdash$ *are true, then so is the conclusion* φ *on the right side.*". If Γ is empty one writes $\vdash \varphi$, indicating that φ is true on its own as e.g. "*The empty set* $\emptyset$ *has no members.*".

What has come to be called an **Aristotelian syllogism** is a valid single-conclusion argument with two premises and can be written in modern notation as $P_1, P_2 \vdash C$. Interpreting $\vdash$ in the sense of a logical calculus as e.g. natural deduction, then it can be shown that $A_1, A_2 \vdash B$ is equivalent to $\vdash (A_1 \wedge A_2) \rightarrow B$ in the strong sense that a proof of either one may be extended to a proof of the other. And it was indeed this latter form that has originally been used by ARISTOTLE: "…, no syllogism is formulated by Aristotle primarily as an inference, but they are all implications having the conjunction of the premises as the antecedent and the conclusion as the consequent." ([7], p.2).

The sentences occurring in an Aristotelian syllogism are restricted to the few particular forms presented in Table 1.1 above. Each of them involves two non-logical terms (predicates) from a trio of such terms, along with a quantifying expression, which is one of *A, E, I,* or *O.* In addition, the occurrences of the non-logical terms of the trio within the argument have to exhibit one of four certain patterns, called figures.

If this were all what can be said about Aristotelian logic, then why did so many people think and write about it for more than two thousand years? The considerations following below try to give an answer to this question.

ARISTOTLE was the first logician to use variables. We quote from the bottom of page 8 in Łukasiewicz [7]:

> In Aristotle's systematic exposition of his syllogistic no examples are given of syllogisms with concrete terms. Only non-valid combinations of premisses are exemplified through such terms, which are of course universal, like 'animal', 'man', 'horse'. In valid syllogisms all terms are represented by letters, i.e. by variables, e.g. 'If *R* belongs to all *S* and *P* belongs to some *S*, then *P* belongs to some *R*'.
>
> The introduction of variables into logic is one of Aristotle's greatest inventions. It is almost incredible that till now, as far as I know, no one philosopher or philologist has drawn attention to this most important fact.

Using variables instead of concrete terms was indeed a step of utmost importance, since without it the level of generality and abstraction of his logical system would have remained unreachable for ARISTOTLE and his ancient followers.

Although ARISTOTLE defined a syllogism broadly as any inference scheme in which the conclusion follows necessarily from the premises, his own analysis centers on a very specific type of argument that uses only the four types of categorical statements. Therefore, in Aristotelian logic there are no singular propositions, whence the often used example

$$
\text{(S1)} \qquad \begin{array}{l} \text{All men are mortal} \\ \text{Socrates is a man} \\ \hline \text{Socrates is mortal} \quad \therefore \end{array}
$$

is not an Aristotelian syllogism proper, but a traditional one and is first brought into syllogistic logic by the scholastics.

A **categorical syllogism** is a syllogism that consists of three categorical sentences usually written without connectives one below the other, and with a line beneath the first two propositions, the **premises** of the syllogism; the third proposition is called the **conclusion**. Depending on whether the propositions contain concrete terms or variables only, such a compound is either a *proof of the conclusion* (as in (S1) above) or a *rule of proof* (as below in (S2)).

A syllogism involves three terms, each of them appearing exactly twice: S, M, and P, represent the syllogism's "Subject", "Middle", and "Predicate", respectively. The following is one of the most simple syllogistic schemes (later on called '*Barbara*'):

$$
\text{(S2)} \qquad \begin{array}{ll}
\text{All S are M} & (\textit{minor premise}) \\
\text{All M are P} & (\textit{major premise}) \\
\hline
\text{All S are P} & (\textit{conclusion}) \quad \therefore
\end{array}
$$

(To give a more comprehensive impression of the differences between the Aristotelian and the traditional syllogism we present how (S2) would look like in ARISTOTLE's original language: "*If the S belongs to all M and the M belongs to all P, then the S belongs to all P.*". For more details on differences see [16], chapter I.)

The predicate of the conclusion is called the **major term** and its subject **minor term**. The premise that contains the major term is called the **major premise**, and the one containing the minor term the **minor premise**. The third term of the syllogism, which does not occur in the conclusion, but in both premises, is called **middle term**. Although, concerning the order of the two premises one can often find the claim that the major premise has to precede the minor one, this is only a convention and is of no logical significance. However, when the premises of a syllogism are arranged in this order, we say that the syllogism is in **standard form**, or that it is a **standard categorical syllogism**. This is how it looks like in an example (taken from [3], chapter 6):

$$
\begin{array}{c}
\text{No } \overbrace{\text{heroes}}^{\textit{major term}} \text{ are } \overbrace{\text{cowards}}^{\textit{middle term}}. \\
\text{Some } \overbrace{\text{soldiers}}^{\textit{minor term}} \text{ are } \overbrace{\text{cowards}}^{\textit{middle term}}. \\
\hline
\text{Some } \overbrace{\text{soldiers}}^{\textit{minor term}} \text{ are not } \overbrace{\text{heroes}}^{\textit{major term}}.
\end{array}
$$

Given that a syllogism is in standard form, we can then determine its mood and figure: The **mood** of a syllogism depends upon the type of the propositions used (*A, E, I,* or *O*) and is a list of these types, beginning with the major premise and ending with the conclusion. Both the premises of a syllogism and its conclusion take one of the four propositional types given in Table 1.1 above, whence we get $4^3 = 64$ different patterns for syllogisms. The minor premise links S with M, and the major premise links M with P. However, the middle term can be either the subject or the predicate of each premise; its position determines the **figure of a syllogism**. Given that the conclusion always is of the form S - P, the four configurations shown in Table 1.2 can occur. (There is no doubt that ARISTOTLE investigated the first three of them; concerning the history of the fourth see e.g. footnote 6 on page 124 of [24].)

Putting it all together we thus get $64 \times 4 = 256$ possible syllogism patterns: Each of the premises and the conclusion can be of one of the four statement types *A, E, I,* or *O* as given in Table 1.1, and the premises can fall under one of the four figures shown in Table 1.2.

TABLE 1.2: Syllogism Figures.

	Figure 1	**Figure 2**	**Figure 3**	**Figure 4**
Minor premise	S Ⓜ	S Ⓜ	Ⓜ S	Ⓜ S
Major premise	Ⓜ P	P Ⓜ	Ⓜ P	P Ⓜ

Now, at the latest, it is time to switch over to the compact notations for both propositions and syllogisms that have been established in the course of time. As you will easily recognize, all of these abbreviations are self-explanatory: (S2) from above becomes (with notation changed for better readability)

$$\text{(S3)} \qquad \begin{array}{l} \text{SaM} \\ \text{MaP} \\ \hline \text{SaP} \quad \therefore \end{array}$$

or, as an instantiation of Figure 1, *AAA*-1, or with its mnemonic name '*Barbara*' together with the implicit knowledge that (S2) belongs to the Figure 1 syllogisms. And our hero example from above is an apparently valid standard categorical syllogism of the kind *EIO*-2 with the mnemonic '*Festino*'.

1.1.3 *Syllogistic Validity*

A syllogism is a deductive argument in which a conclusion is inferred from two premises. By purely combinatorial considerations we have just learned that there are 256 possible forms of syllogisms, and, of course, like in all deductive logic we aim to discriminate valid arguments from invalid ones. Therefore, the following question comes to mind immediately: Are all these deductive schemes valid inference rules?

For analyzing this question it is reasonable to assume that the constituent propositions of a syllogism are all contingent, that is, neither logically true (as e.g., '*All cheap books are books*') nor logically false (as e.g., '*Some red cars are not cars*'). (The reason for this assumption is that if the argument contained either a logically false premise or a logically true conclusion, then it would be valid regardless of its syllogistic form: it would be logically impossible for its premises to be true and its conclusion to be false.)

Under this assumption, both validity and invalidity of any syllogism depend entirely on its form and are completely independent of the specific content of the argument or its subject matter.

If any syllogism is (in)valid in virtue of its form alone, any other syllogism having that same form will also be (in)valid. The common recognition of this fact is the basis for the frequent use of logical analogy in argumentation: Any misguided argument can be proved invalid by finding another argument with exactly the same form and which is known to be invalid by the fact that its premises are true while its conclusion is false. (Remember that an invalid argument may very well have a true conclusion; that an argument is invalid simply means that its conclusion is not logically implied or necessitated by its premises.)

However, not being able to find an invalid analogy does not prove an argument valid, for this may merely reflect the limitations of our thinking. Moreover, such a search is also cumbersome in principle: there are far too many invalid forms of syllogistic argument, viz. well over 200. So we are in need of more effective methods for establishing the formal validity or invalidity of syllogisms.

One such method is the usage of Venn diagrams to represent the premises of a syllogism by drawing three overlapping circles, one for each of its three different terms (minor, major, and middle term, in a way that result in eight parts for the eight classes. You can find an excellent and detailed account of this method (and of traditional logic in general) in the first half of [3].

Logical analogies and Venn diagrams are two of the three most common methods to test syllogisms. A third method for determining the validity or invalidity of syllogisms is to use rules stating properties that all valid syllogisms must have. If any one of these rules is violated, a specific formal fallacy is committed and the syllogism is invalid. Conversely, if none of the rules is broken, the syllogism is valid. This approach is much older than the usage of Venn diagrams: many such rules are already discussed by ARISTOTLE himself. Details on this approach (which is definitely beyond our current scope) can be found in both Copi et al. [3] and Hurley [5]. Unfortunately, we have to note that the rules given in these two books and also given elsewhere differ slightly in number and wording.

1.1.3.1 *The Valid Syllogisms*

Having given a discussion about a syllogism's validity and how to check it, we finally present the valid syllogism of traditional logic. The vast majority of the 256 possible forms of syllogism are invalid, i.e. the conclusion does not follow logically from the premises. Traditionally it is claimed that there are 24 valid Aristotelian syllogisms, but according to the Boolean view, this is true only with the additional assumption that terms used in the syllogisms refer to non-empty classes. It is common to divide the valid syllogism in two groups according to whether they need this additional assumption or not as shown in Table 1.3 (denoting the valid syllogisms by their mnemonic names).

There are 15 combinations of mood and figure that are valid in both the Aristotelian and Boolean interpretation of categorical propositions; following [5], they are called the "**unconditionally valid**" argument forms. A syllogistic pattern is called "**conditionally valid**" if (1) it can be valid only from the Aristotelian standpoint, where universal propositions have existential import, and (2) the condition in the column "required" is fulfilled, i.e. elements in the classes mentioned actually exist. There are 9 such argument forms for categorical syllogisms in addition to the ones that have been classified as unconditionally valid.

Note that all of the conditionally valid arguments have universal premises and a particular conclusion, which is invalid from the Boolean standpoint because in the Boolean model, universal statements make no claims about existence while particular ones do. Thus, if a syllogism has universal premises, nothing is said about existence. Yet if the conclusion is particular, then it does say something about existence, whence the conclusion contains more information than the premises do, thereby making the argument invalid.

TABLE 1.3: Valid Syllogisms.

	Figure 1	**Figure 2**	**Figure 3**	**Figure 4**	**required**
Major premise Minor premise	Ⓜ P S Ⓜ	P Ⓜ S Ⓜ	Ⓜ P Ⓜ S	P Ⓜ Ⓜ S	
Unconditionally valid forms	Barbara Celarent Darii Ferio	Camestres Cesare Baroco Festino	Datisi Disamis Ferison Bocardo	Calemes Dimaris Fresison	
Conditionally valid forms	Barbari Celaront	Cesaro Camestros		Calemos	$\mathfrak{S} \neq \emptyset$
			Felapton Darapti	Fesapo	$\mathfrak{M} \neq \emptyset$
				Bamalip	$\mathfrak{P} \neq \emptyset$

For example, '*Barbari*', i.e. *AAI*-1, is valid from the Aristotelian standpoint if the subject of the conclusion denotes actually existing things. Thus, an *AAI*-1 syllogism with minor term "*cats*" is valid from the Aristotelian standpoint. But if the minor term is "*unicorns*" then the syllogism is invalid. On the other hand, if the minor term is "*students who failed the exam*" and we are not certain if there are any such students, then the syllogism is conditionally valid.

1.2 Syllogisms and Relations

After our digression into the world of syllogisms we will now return to where we started at the very beginning of this chapter: Around the middle of the nineteenth century two crucial development steps in the history of modern logic took place: GEORGE BOOLE initiated the algebraic treatment of logic and AUGUSTUS DE MORGAN started with the formulation of the logic of relations.

And from ALFRED TARSKI (1901-1983) ([23], p. 73) we learn that

> The first beginnings of the contemporary theory of relations are to be found in the writings of A. De Morgan, who carried out extensive investigations in this domain in the fifties of the Nineteenth Century. De Morgan clearly realized the inadequacy of traditional logic for the expression and justification, not merely of the more intricate arguments of mathematics and the sciences, but even of simple arguments occurring in every-day life; witness his famous aphorism, that all the logic of Aristotle does not permit us, from the fact that a horse is an animal, to conclude that the head of a horse is the head of an animal. In his effort to break the bonds of traditional logic and to expand the limits of logical inquiry, he directed his attention to the general concept of relations and fully recognized its significance. Nevertheless, De Morgan cannot be regarded as the creator of the modern theory of relations, ...
>
> The title of creator of the theory of relations was reserved for C. S. Peirce. In several papers published between 1870 and 1882, he introduced and made precise all the fundamental concepts of the theory of relations and formulated and established its fundamental laws. Thus Peirce laid the foundation for the theory of relations as a deductive discipline; ...

The above quotation not only gives a kind of justification for the preceding section on syllogisms but also directs us to the next stations on our historical path. Without going into details (see [6], Chapter I, Section IV), we have to be aware that around 1850 the main object of logical investigations have still been ARISTOTLE's syllogisms. The main questions being: How to formalize? How to enrich? What are the mathematical properties and structure obtained?

As we come closer to the modern era of logic, we will adapt our notation to the current habits used in modern papers and books on logic. In a way that is both self-explaining and comfortable we can simply write

$$MaP, SaM \vdash SaP$$

which is, of course, the already known '*Barbara*' or AAA-1 syllogism.

According to the preface of [11], DE MORGAN's greatest achievement in logic was his discovery of the logic of relations and that he was the first to sense the pervasiveness of relational arguments and to see their systematic importance for logic. We cite from pages vii and viii:

> This belief in the importance of relational arguments led De Morgan to create much of the logic of relations as we know it today. He made the fundamental discovery that relations could be compounded so that, for instance, one can compound the relations brother and parent to get their relative product, brother of a parent - i.e., uncle. ... De Morgan was also the first person to study the properties of relations, such as transitivity and symmetry (which he called convertibility).

The origins of the calculus of relations in the sense of modern mathematics as we know and use it today is said to start in 1860 with De Morgan's "On the Syllogism, no. IV, and on the Logic of Relations", (published in 1864 as [15], but read April 23, 1860). We start, however, with a short passage from [13], p. 46:

> 162. If two relations combine⋆ into what is contained in a third relation, then the *converse* of either of the two combined with the *contrary* of the third, in the same order, is contained in the *contrary* of the other of the two. Thus the following three assertions are *identically the same*, superior and inferior being taken as contraries, that is, absolute equality not existing. Let the combination be 'master of parent' and the third relation 'superior'.
>
> Every master of a parent is a superior
> Every servant of an inferior is a non-parent
> Every inferior of a child is a non-master.
>
> From either of these the other two follow. This may be generally proved: at present it will be sufficient to deduce one of the assertions before us from another. Assume the second; from it follows that every parent is not any servant of an inferior, and therefore, if servant at all, only servant of superior, whence master of parent must be superior.
>
> ⋆ This theorem ought to be called theorem K, ...

Let xMy mean that the individual x is a master of the individual y and, similarly, xPy that x is a parent of y, and xSy that x is superior to y. Furthermore, noting that "y *is a servant of* x" is equivalent to "x *is a master of* y", i.e. the relation "*servant of*" is the converse of M, denoted by $\breve{M}$, and that $x\breve{P}y$ analogously means "x *is a child of* y", we then get

1. If xMy and yPz, then xSz.
2. If $y\breve{M}x$ and $x\bar{S}z$, then $y\bar{P}z$.
3. If $x\bar{S}z$ and $z\breve{P}y$, then $x\bar{M}y$.

where $\bar{R}$ denotes the negation of relation R. Note that it is assumed that for any two individuals, one of them is the superior of the other, and "x *is inferior to* y" simply translates to $x\bar{S}y$. The above yields the following pattern:

1. If A and B, then C
2. If A and not C, then not B
3. If not C and B, then not A

Where we have arrived at now is, in fact, the original *form* of an Aristotelian syllogism: If A and B, then C. But looking at the *contents* (or matter), we use three different relations in "If xMy and yPz, then xSz" instead of only one copula ("is", or membership in a class).

For taking the last step, let now $M;P$ denote the relational multiplication of the two relations M and P, defined as: $xM;Pz$ means "xMy and yPz" for some y. Leaving away the variables, and letting the set theoretical inclusion $\subseteq$ replace the logical "if ..., then ..." we can achieve a purely algebraic look which, maybe, is already familiar to some of you. (I have attached De Morgan's original notation on the right in order to make you appreciate the kind of notation we have finally arrived at nowadays. But one should also keep in mind the restrictions in typographical possibilities in the middle of the nineteenth century.)

(1) $M;P \subseteq S$	$MP))S$
(2) $\breve{M};\bar{S} \subseteq \bar{P}$	$M^{-1}s))p$
(3) $\bar{S};\breve{P} \subseteq \bar{M}$	$sP^{-1}))m$

The equivalence of these three relational inequalities has been called "**Theorem K**" by De Morgan; see also the last of the three notation examples in the Note at the begin of Section 1.3.1.

To be sure that the great change and enhancement resulting from De Morgan's investigation cannot go unnoticed (because with things that we have got accustomed to as a matter of course, it can very easily happen that we do no longer become aware of them), I like to emphasize:

(1) In ordinary syllogisms, the copula is assumed to be the same in each premise and the conclusion, whatever else it is (the ancient classics being "*is*", or "*are*", respectively).
(2) Considering the two copulas "*is*" and "*is equal to*" one immediately sees that they are formally similar in the sense that the laws of the syllogism apply to each. And in 1847 De Morgan in [12] recognized that any relational term that is symmetric and transitive will satisfy the syllogistic laws. The relation "*is tied to*" is as good as "*is equal to*". But there is no reason for these limitations: we may use any copulas we please. (Called the "doctrine of the abstract copula" in [11, p. 48].)
(3) One may even use different *relations* in the two premises, which results in the composition of these two relations yielding a third one in the conclusion. Note that even such simple everyday relations as "*taller than*" or "*part of*" outrun Aristotle's categorical syllogistic which is based on classes of things: Recall that it is not possible to deduce syllogistically from "*Every man is a mammal*" that "*every leg of a man is the leg of a mammal*".

Aristotle's system is unable to deal with relations between two objects like e.g. the relation "*belongs to*". De Morgan criticized this limitation in the Aristotelian syllogism and sought a framework for reasoning about these higher order relations. A *unary* relation, like "*red*" or "*old*", describes a single object: "*red*"(x) is true just when x is red. A *binary* relation such as "*father of*" or "*older than*" describes a pair of objects: "*older than*"(x, y) is true when x is older than y. *Higher-order* relations such as e.g."*between*" can describe triples, and in general, sequences of objects of any fixed length.

De Morgan can thus be said to have initiated what (following [20], p. 506) we can call "the relational view of syllogisms" : "When two objects, qualities, classes, or attributes, viewed together by the mind, are seen under some connexion, that connexion is called a *relation*. ... The distinction of *subject* and *predicate* is the distinction between the *notion in relation* and the *notion to which it is in relation*." ([14], p. 208). And on page 217 of this paper we find: “The *syllogism* is inference of the relation which exists between two terms, as a necessary consequence of their relations to the same third, or *middle*, term.”

We have now learned about the motivation why, and the way how relations became a topic of mathematical interest via the studies of syllogistic logic by AUGUSTUS DE MORGAN. You can meet syllogisms again as one of the application examples in Part IV of this book in Chapter 19 (together with a discussion of the so-called square of opposition) - either now, if you have already a rudimentary knowledge on relations, or somewhere in the middle or at the end of Part II.

1.3 From the History of Relations

In this section we will give a short sketch of the development of the mathematical theory of relations; as it is almost always a good thing to understand and to know how we have come to where we just are, also a look into the history of relations reveals interesting, and sometimes surprising, details.

Criticizing the limitations of ARISTOTLE's syllogisms, DE MORGAN turned to the logic of binary relations. With the works of PEIRCE and SCHRÖDER, these investigations focused on algebras of binary relations, rather than on Boolean algebras. The third volume of SCHRÖDER's series of books on algebraic logic [21] deals exclusively with the calculus of relations. The subject then more or less fell into neglect until its modern stage began forty years later with TARSKI's [23] in 1941.

1.3.1 *The Groundwork:* DE MORGAN *and* PEIRCE

[**Note.** Before we start to use notational conventions of the modern era, I add some examples for those of you who really want to know more about DE MORGAN's notation; the others are right in trusting that history has taken a good decision to adhere to a notation similar to the one shown in Theorem K on page 15.

The first two examples are taken from Merrill [11], Chapter V.2; and the third is the original formulation concerning theorem K from the top of p. 344 in [15], enhanced with the bracketed explanations in italics as given in Maddux [10], p. 309 (DE MORGAN uses “))” for $\leq$, i.e. $\subseteq$):

- “... “X..LY” is the affirmative judgment that X stands in the relation L to Y, while “X.LY” is the negative judgment that X does not stand in the relation L to Y”.
- “X..LMY” means “X is one of the Ls of one of the Ms of Y (Composition)”, or that X has the relation L to some Z which has the relation M to Y.
- “If a compound relation [$x; y$] be contained in [$\leq$] another relation [z], by the nature of the relations and not by casualty of the predicate [$x; y \leq z$], the same may be said [*that same compound relation is contained in another*] when either component [x *or* y] is converted [*into* $\breve{x}$ *or* $\breve{y}$, *respectively*], and the contrary of the other component [$\overline{y}$ *or* $\overline{x}$, *respectively*] and of the compound [$\overline{z}$] change

places [*yielding* $\breve{x};\overline{z} \leq \overline{y}$ *or* $\overline{z};\breve{y} \leq \overline{x}$, *respectively*]. That is if, be Z what it may, every L of M of Z be an N of Z, say LM))N, then L^{-1}n))m, and nM^{-1}))l. If LM))N, then n))lM' and nM^{-1}))lM'M^{-1}. But an l of *every* M of an M^{-1} of Z must be an l of Z: hence nM^{-1}))l. Again, if LM))N, then n))L,m, whence L^{-1}n))L^{-1}L,m. But an L^{-1} of an L of none but ms of Z must be an m of Z; whence L^{-1}n))m.
I shall call this result theorem K, ..."

]

De Morgan's most mature treatment of the logic of relations is given in [15] where he developed further the technical vocabulary and some of the ideas of previous papers. The quotation in the last example gives a flavor of how development and argumentation has been done at that time. Moreover, it is interesting to see which of the concepts and results have already been known and used by De Morgan; for these purposes we can follow [15], pp. 341ff. Table 1.4 presents a short overview on the most relevant definitions, providing De Morgan's original notation, together with his verbal explanations and a modern notation.

It can be seen from this table that De Morgan considers the complement and converse as well as three kinds of compound relations, and states e.g. the following laws governing combinations of some of these operations (in [15], pp. 342f, before stating theorem K on page 344):

- "Contraries of converses are converses."
- "Converses of contraries are contraries."
- "The contrary of a converse is the converse of the contrary:"
  $\bar{\breve{\lambda}} = \breve{\bar{\lambda}}$, respectively $\overline{\lambda^{\smallsmile}} = \overline{\lambda}^{\smallsmile}$
- "If a first relation be contained in a second, then the converse of the first is contained in the converse of the second : but the contrary of the *second* in the contrary of the *first*."
  If $\lambda \subseteq \mu$, then $\breve{\lambda} \subseteq \breve{\mu}$ and $\bar{\mu} \subseteq \bar{\lambda}$
- "The conversion of a compound relation converts both components, and inverts their order."
  $(\lambda\mu)^{\smallsmile} = \mu^{\smallsmile}\lambda^{\smallsmile}$

The notation I will use from now on is primarily that of Schröder [21] which, hopefully, helps in mastering the changeover from the early to the modern and to the notation still in use in relation algebra literature. Using this contemporary notation, De Morgan's Theorem K (as presented in [15] and quoted above) reads as follows:

$$x;y \leq z \;\rightarrow\; \breve{x};\overline{z} \leq \overline{y} \wedge \overline{z};\breve{y} \leq \overline{x} \tag{1.1}$$

From the quotation of [13] given above on page 14 we know that De Morgan was well aware that these three formulas are in fact equivalent.

Charles Sanders Peirce, according to the Tarski-quotation from above the true "creator of the theory of relations", turned to the subject of relations with the publi-

TABLE 1.4: Relational concepts used by DE MORGAN ([15], pp. 341ff)

operation	**notation**	**explanation** (original and modern)
relation	X..LY	"X is some one of the objects of thought which stand to Y in the relation L, or is one of the Ls of Y" $x\lambda y$
converse	L^{-1}	"The *converse* relation of L, L^{-1}, is defined as usual: if X..LY, Y..L^{-1}X: if X be one of the Ls of Y, Y is one of the L^{-1}s of X." $x\lambda^{\smile} y := x\breve{\lambda} y := y\lambda x$ (Def. 3.10)
contrary	X.LY, X..lY	"Relations are assumed to exist between any two terms whatsoever. If X be not any L of Y, X is to Y in some not-L relation: let this *contrary* relation be signified by l; thus X.LY gives and is given by X..lY." $x\overline{\lambda} y := \neg x\lambda y$ (Def. 3.3)
composition	X..LMY	"if X be one of the Ls of one of the Ms of Y, we may think of X as an ' L of M' of Y, expressed by X..(LM)Y, or simply by X..LMY" $x\lambda\mu y := \exists z[x\lambda z \wedge z\mu y]$ (Def. 3.11)
	LM′	"let LM′ signify an L of *every* M" $x\lambda\mu' y := \forall z[z\mu y \rightarrow x\lambda z]$ or $x\lambda\mu' y := \mu y \subseteq x\lambda$ (see Def. 6.1(c))
	L,M	"let L,M signify ... an L of none but Ms" $x\lambda{,}\mu y := \forall z[x\lambda z \rightarrow z\mu y]$ or $x\lambda{,}\mu y := x\lambda \subseteq \mu y$ (see Def. 6.1(b))

cation of [17] in 1870[2] and contributed several papers and a good many unpublished manuscripts during the next twenty years. But the short outline given below follows [9] which adheres to [19], a later paper from 1883, where readers can find more details and also a comparison of DE MORGAN's and PEIRCE's contributions. We begin with a quotation from [19]:

> Let A, B, C, D, etc., be all the individual objects in the universe; then all the individual pairs may be arrayed in a block, thus:—
>
> | | | | | |
> |---|---|---|---|---|
> | A : A | A : B | A : C | A : D | etc. |
> | B : A | B : B | B : C | B : D | etc. |
> | C : A | C : B | C : C | C : D | etc. |
> | D : A | D : B | D : C | D : D | etc. |
> | etc. | etc. | etc. | etc. | etc. |
>
> A general relative may be conceived as a logical aggregate of a number of such individual relatives. Let l denote "lover;" then we may write
>
> $$l = \Sigma_i \Sigma_j (l)_{ij} (I : J)$$
>
> where $(l)_{ij}$ is a numerical coefficient, whose value is 1 in case I is a lover of J,

[2] In January 1870 PEIRCE presented a 60-page paper on logic to the Academy of Arts and Sciences which was printed in time for him to take it on his European trip in the summer; the paper appeared officially in 1873 as [17] in Vol. 9 of the Memoirs. See [4], p. 145.

> and 0 in the opposite case, and where the sums are to be taken for all individuals in the universe.

PEIRCE's characterization of $(l)_{ij}$ as a numerical coefficient must not mislead us. Since it cannot denote any other numbers than either 0 or 1, we can regard it as a characteristic function.

For the unary operations of converse and contrary (complement) PEIRCE [19] notes the following properties (the numbering is the one used in [9]):

(B1) $\overline{\overline{x}} = x$,
(B2) $\breve{\breve{x}} = x$,
(B3) $\overline{\breve{x}} = \breve{\overline{x}}$,
(B4) if $x \leq y$ then $\overline{y} \leq \overline{x}$,
(B5) if $x \leq y$ then $\breve{x} \leq \breve{y}$.

Besides the operations defined in Table 1.4, some further operations on relations x, y are their *intersection* $x \cdot y$ and their *union* $x + y$:

$$\begin{aligned} x \cdot y &= \{\langle a,b\rangle : \langle a,b\rangle \in x \text{ and } \langle a,b\rangle \in y\}, \\ x + y &= \{\langle a,b\rangle : \langle a,b\rangle \in x \text{ or } \langle a,b\rangle \in y\}. \end{aligned}$$

In PEIRCE's terms ([19, p. 188]):

> Relative terms can be aggregated and compounded like others. Using + for the sign of logical aggregation, and the comma for the sign of logical composition (Boole's multiplication, here to be called non-relative or internal multiplication), we have the definitions
>
> $$(l+b)_{ij} = (l)_{ij} + (b)_{ij},$$
>
> $$(l,b)_{ij} = (l)_{ij} \times (b)_{ij}.$$
>
> The first of these equations, however, is to be understood in a peculiar way: namely, the + in the second member is not strictly addition, but an operation by which
>
> $$0+0=0 \qquad 0+1=1+0=1+1=1.$$

Whereas these are "logical", i.e. set-theoretical, operations with corresponding properties, the following two are relation specific operations, called the relative product $x;y$ and the **relative sum** $x \dagger y$ of two relations x and y is:

$$\begin{aligned} x;y &= \{\langle a,b\rangle : \text{for some } c \in U,\ \langle a,c\rangle \in x \text{ and } \langle c,b\rangle \in y\}, \\ x \dagger y &= \{\langle a,b\rangle : \text{for every } c \in U, \text{ either } \langle a,c\rangle \in x \text{ or } \langle b,c\rangle \in y\}. \end{aligned}$$

Peirce [19, p. 189] explains relative multiplication and addition as follows:

> We now come to the combination of relatives. Of these, we denote two by special symbols; namely, we write
>
> lb for lover of a benefactor,
>
> and
>
> $l \dagger b$ for lover of everything but benefactors.

> The former is called a particular combination, because it implies the *existence* of something *loved by* its relate and a *benefactor of* its correlate. The second combination is said to be *universal,* because it implies the *non-existence* of anything except what is either loved by its relate or a benefactor of its correlate.
> ...
> The two combinations are defined by the equations
> $$(lb)_{ij} = \Sigma_x (l)_{ix} (b)_{xj}$$
> $$(l \dagger b)_{ij} = \Pi_x \{(l)_{ix} + (b)_{xj}\}$$
> The sign of addition in the last formula has the same signification as in the equation defining non-relative multiplication.

The last formula shows that $l \dagger b$ is perhaps more suitably verbalized by "*lover of all non-benefactors*". Among others, the following relationships are given for relative multiplication and addition:

(B6) if $x \le z$ then $x; y \le z; y$,
(B7) if $x \le z$ then $x \dagger y \le z \dagger y$,
(B8) if $y \le z$ then $x; y \le x; z$,
(B9) if $y \le z$ then $x \dagger y \le x \dagger z$,
(B10) $x \dagger (y \dagger z) = (x \dagger y) \dagger z$,
(B11) $x; (y; z) = (x; y); z$,
(B12) $x; (y \dagger z) \le (x; y) \dagger z$,
(B13) $(x \dagger y); z \le x \dagger (y; z)$,
(B14) $(x + y); z = x; z + y; z$ and $x; (y + z) = x; y + y : z$,
(B15) $(x \cdot y) \dagger z = (x \dagger z) \cdot (y \dagger z)$,
(B20) $\overline{x \dagger y} = \overline{x}; \overline{y}$,
(B21) $\overline{x; y} = \overline{x} \dagger \overline{y}$,
(B22) $(x + y)^{\smile} = \breve{x} + \breve{y}$,
(B23) $(x \cdot y)^{\smile} = \breve{x} \cdot \breve{y}$,
(B24) $(x \dagger y)^{\smile} = \breve{y} \dagger \breve{x}$,
(B25) $(x; y)^{\smile} = \breve{y}; \breve{x}$

Note that, by (B20) and (B21), relative sum and relative multiplication are dual to each other, and while relative multiplication is rather familiar, relative addition is not. As can be seen from Table 1.4, DE MORGAN introduced three binary operations, where his *composition* coincides with PEIRCE's relative multiplication. LM$'$ and L,M of Table 1.4 are definable as follows (see [9], p. 430):

$$x{,}y \;=\; \overline{x} \dagger y \;=\; \overline{x; \overline{y}} \tag{1.2}$$

$$xy' \;=\; x \dagger \overline{y} \;=\; \overline{\overline{x}; y} \tag{1.3}$$

PEIRCE also introduced some special relations and gave some laws concerning them. Let 1 denote the *Cartesian square* of a set U, i.e. the set consisting of all ordered pairs

$\langle a, b \rangle$, where a and b are elements of U, and 0 be the *empty relation,* i.e. the relation which never holds between two members of the universe of discourse. Furthermore, he considered the *identity relation* $1'$ and the *diversity relation* $0'$ on U. DE MORGAN did not deal with these distinguished relations. (Be aware that all these relations are always to be seen in relation to the underlying universe of discourse, whence a suffix $_U$ should be appended to them. However, we keep the commonly used notation when appropriate.) We then have

$$\begin{aligned}
1 &= U \times U = \{\langle a, b\rangle : a, b \in U\} \\
0 &= \overline{1} \\
1' &= \{\langle a, a\rangle : a \in U\} \\
0' &= \{\langle a, b\rangle : a, b \in U \text{ and } a \neq b\} = \overline{1'}
\end{aligned}$$

Among the laws concerning these relations and listed in [19] are the following:

(B26) $x \dagger 1 = 1 = 1 \dagger x$
(B27) $x;0 = 0 = 0;x$
(B28) $x \dagger 0' = x - 0' \mid x$
(B29) $x;1' = x = 1';x$
(B30) $1' \leq x \dagger \overline{\breve{x}}$
(B31) $x \dagger \breve{x} \leq 0'$

1.3.2 *The Followers:* SCHRÖDER *and* TARSKI

In Section 1.2 we have outlined the way how the concept of a relation was given birth in the world of logicians and mathematicians. The current section has started with a short sketch of the mathematical development of binary relations during the second half of the 19th, and will now be continued with the progress during the first half of the 20th century. At the end of this way we will find the notion of a *relation algebra* as an abstract algebraic structure which will be the topic of Chapter 10 in *Part IV.*

In 1895, the german mathematician ERNST SCHRÖDER(1841-1902) published [21], the third volume of his *Vorlesungen über die Algebra der Logik,* which is completely devoted to the calculus of relations.

From Maddux [10], p 310 we learn the following: Given DE MORGAN's emphasis on his Theorem K, and the fact that PEIRCE read [15], it is astonishing that this equivalences do not appear in any of the papers that PEIRCE published during his lifetime. He was, however, aware of these equivalences as can be seen from the unpublished paper [18], p. 341 (with the formulas in modern notation, and Theorem K shaded)

Theorem 1.1. *(PEIRCE's equivalences) "Hence the rule is that having a formula of the form $x; y \leq z$, the three letters may be cyclically advanced one place in the order of writing, those which are carried from one side of the copula to the other being both negatived and converted. We have, then, the following twelve propositions, all equivalent*

$$
\begin{array}{llll}
x;y \le z & x \le z \dagger \breve{\overline{y}} & \breve{y};\breve{x} \le \breve{z} & \breve{x} \le \overline{y} \dagger \breve{z} \\
\breve{\overline{z}};x \le \breve{\overline{y}} & y \le \breve{\overline{x}} \dagger z & \overline{z};\breve{y} \le \overline{x} & \overline{z} \le \overline{x} \dagger \overline{y} \\
y;\breve{\overline{z}} \le \breve{\overline{x}} & \breve{\overline{z}} \le \breve{\overline{y}} \dagger \breve{\overline{x}} & \breve{x};\overline{z} \le \overline{y} & \breve{y} \le \breve{z} \dagger \overline{x}
\end{array}
$$

There are in all 64 such sets of 12 equivalent propositions."

(So we proceed from $x;y \le z$ to $\breve{\overline{z}};x \le \breve{\overline{y}}$, to $y;\breve{\overline{z}} \le \breve{\overline{x}}$, where $\breve{\overline{\breve{\overline{x}}}} = \overline{\breve{\breve{\overline{x}}}} = \overline{\overline{x}} = x$ due to (B1) - (B3) in column 1, and similarly for the other columns.)

Extending the results of PEIRCE, SCHRÖDER ([21], p. 242f) presented four of the 64 groups, but with no mention of DE MORGAN. In addition to the one above the following three groups are shown: This probably justifies that nowadays you encounter DE MORGAN's theorem under the name **SCHRÖDER equivalences** ; see Th. 3.26.

Theorem 1.2. (SCHRÖDER*'s equivalences)*

$$
\begin{array}{llll}
y;z \le \breve{\overline{x}} & z \le \breve{\overline{y}} \dagger \breve{\overline{x}} & \breve{z};\breve{y} \le \overline{x} & \breve{z} \le \overline{x} \dagger \overline{y} \\
z;x \le \breve{\overline{y}} & y \le \breve{\overline{x}} \dagger \breve{\overline{z}} & \breve{x};\breve{z} \le \overline{y} & \breve{y} \le \overline{z} \dagger \overline{x} \\
x;y \le \breve{\overline{z}} & x \le \breve{\overline{z}} \dagger \breve{\overline{y}} & \breve{y};\breve{x} \le \overline{z} & \breve{x} \le \overline{y} \dagger \overline{z} \\
 & & & \\
\breve{\overline{x}} \le y \dagger z & \breve{\overline{y}};\breve{\overline{x}} \le z & \overline{x} \le \breve{z} \dagger \breve{y} & \overline{x};\overline{y} \le \breve{z} \\
\breve{\overline{y}} \le z \dagger x & \breve{\overline{x}};\breve{\overline{z}} \le y & \overline{y} \le \breve{x} \dagger \breve{z} & \overline{z};\overline{x} \le \breve{y} \\
\breve{\overline{z}} \le x \dagger y & \breve{\overline{z}};\breve{\overline{y}} \le x & \overline{z} \le \breve{y} \dagger \breve{x} & \overline{y};\overline{z} \le \breve{x} \\
 & & & \\
x \le y \dagger z & \breve{\overline{y}};x \le z & \breve{x} \le \breve{z} \dagger \breve{y} & \breve{x};\overline{y} \le \breve{z} \\
\breve{\overline{y}} \le z \dagger \breve{\overline{x}} & x;\breve{\overline{z}} \le y & \overline{y} \le \overline{x} \dagger \breve{z} & \overline{z};\breve{x} \le \breve{y} \\
\breve{\overline{z}} \le \breve{\overline{x}} \dagger y & \breve{\overline{z}};\breve{\overline{y}} \le \breve{\overline{x}} & \overline{z} \le \breve{y} \dagger \overline{x} & \overline{y};\overline{z} \le \overline{x}
\end{array}
$$

Furthermore, on top of page 244 in [21], SCHRÖDER notes that each of these sets of 12 can be expanded to a set of 60 equivalent formulas. Each formula in a group of 12 is equivalent to 4 others: for e.g. the first formula in PEIRCE's list we have

$$x;y \le z \;\leftrightarrow\; 1 \le (\overline{x} \dagger \overline{y}) + z \;\leftrightarrow\; (x;y) \cdot \overline{z} \;\leftrightarrow\; (x;y) \cdot z = x;y \;\leftrightarrow\; (x;y) + z = z.$$

The works of DE MORGAN, PEIRCE, and SCHRÖDER resulted in a great amount of laws for relations, and at least SCHRÖDER has been aware of this aspect. On top of page 244 of [21] he writes: "...und erkennt man wie ungeheuer vielgestaltig unsre Disziplin ist. Ist das ein Vorteil? Jedenfalls ist es eine Thatsache mit der wir uns abzufinden und vertraut zu machen haben!" [3] But all three of them were not particularly interested in the axiomatic approach to the calculus of relations.

Nearly five decades later, TARSKI [23] (a seminal paper with the text of an invited address delivered at a meeting in Philadelphia on December 28, 1940) took a different view and proposed axiomatizations for the fragment of the calculus of relations consisting of Boolean combinations of equations.

The first approach in [23] consists in constructing the calculus of relations within first-order predicate calculus. It uses two kinds of variables viz. individual variables

[3] [my translation] "...and one recognizes how highly multiform our discipline is. Does this mean being an advantage? Anyway, it is a fact that we have to accept and make ourselves familiar with."

$x, y, z, \ldots$ and relation variables $R, S, T, \ldots$. As constants we have the connectives of the sentential calculus: $\wedge, \vee, \neg, \rightarrow$, and $\leftrightarrow$; and the existential and universal quantifiers $\exists, \forall$. All the formulas (Bn) from PEIRCE's Note B [19] mentioned above belong to this fragment.

This theory is extended by introducing eleven constants which are specific to the calculus of relations. We have four relation constants: 1, 0, 1', and 0' (see page 21); six operations signs: $-, \breve{}, +, \cdot, \dagger$, and ;, all of which we have already met in former sections; and, finally, the identity sign =, which denotes identity between relations.

Twelve axioms are added which are intended to explain the meanings of the new constants, e.g. with the original numbering used in TARSKI [23]

1. $\forall x \forall y [x 1 y]$,
7. $\forall x \forall y [x R^{\smile} y \leftrightarrow yRx]$, or
10. $\forall x \forall y [x R \dagger S y \leftrightarrow \forall z [xRz \vee zSy]]$.

The result is described as follows:

> The theory thus outlined may be called the *elementary theory of (binary) relations*. If we confine ourselves to those sentences and theorems which contain no individual variables, we obtain the fragment of this theory in which we are here interested, namely the calculus of relations. ([23], p. 76)

TARSKI then proceeds with 15 examples of *theorems* provable in the 'elementary theory' from which I-VII characterize the formal structure of a Boolean algebra; the remaining theorems are specific to the calculus of relations and are selected from laws already noted by DE MORGAN, PEIRCE, and SCHRÖDER:

VIII. $R^{\smile\smile} = R$.

IX. $(R;S)^{\smile} = S^{\smile};R^{\smile}$.

X. $R;(S;T) = (R;S);T$.

XI. $R;1' = R$.

XII. $R;1 = 1 \vee 1;\overline{R} = 1$.

XIII. $(R;S) \cdot T^{\smile} = 0 \rightarrow (S;T) \cdot R^{\smile} = 0$.

XIV. $0' = \overline{1'}$.

XV. $R \dagger S = \overline{\overline{R};\overline{S}}$.

(Note that the collection of all binary relations on an arbitrary nonempty set U, together with the appropriate operations and distinguished relations, is a model of theorems I - XV.)

To continue with TARSKI's own words:

> The above way of constructing the elementary theory of relations will probably seem quite natural to any one who is familiar with modern mathematical logic. If, however, we are interested not in the whole theory of relations but merely in the calculus of relations, we must admit that this method has certain defects from the point of view of simplicity and elegance. We obtain the

> calculus of relations in a very roundabout way, and in proving theorems of this calculus we are forced to make use of concepts and statements which are outside the calculus. It is for this reason that I am going to outline another method of developing this calculus. ([23], p. 77)

This other method consists in setting the *theorems* I - XV as *axioms*. This means that only one kind of variables, namely relation variables, are used, and that quantifiers are dismissed from the set of constants. The only rules of inference are the rule of substituting equals for equals and the rule of detachment (*modus ponens*).

The next step then is to check the suitability of this axiom system by trying to (re)prove known results from the elementary theory of relations using only I - XV. So TARSKI continues with proving sixteen theorems in this system, among them e.g.

XVIII. $R = S \rightarrow R^{\smile} = S^{\smile}$.

XIX. $(R + S)^{\smile} = R^{\smile} + S^{\smile}$.

XX. $0^{\smile} = 0 \wedge 1^{\smile} = 1$.

XXIII. $R;(S + T) = (R;S) + (R;T)$.

XXVIII. $0;R = 0$.

XXXI. $(1;S) \cdot T = 0 \rightarrow (1;T) \cdot S = 0$.

He then proves that every sentence (i.e. Boolean combination of equations) of the calculus of relations can be transformed into an equivalent sentence of the form '$R = S$', and even of the form '$T = 1$', generalizing a result of SCHRÖDER. Axioms I-XV also suffice to prove all the formulas (Bn) from PEIRCE's Note B [19] mentioned above and SCHRÖDER's version of DE MORGAN's Theorem K. This gives some evidence that the axioms for the second method have been suitably chosen. A natural question then is whether TARSKI's axiomatization is complete. We finish our excursion into the historical background on relations by letting TARSKI have the final words on some metalogical problems regarding his calculus of relations.

> The first problem concerns the relation between our two methods of constructing the calculus. It is obvious that every theorem which can be proved by the second method can also be proved by the first method (since, as mentioned before, the axioms adopted under the second method can be proved as theorems under the first method, and all the rules of inference used in the second method were also assumed in the first method). It is, however, by no means obvious that the converse is also true, and that consequently our two methods are entirely equivalent. Since it follows from a result of K. Gödel[6] that every sentence which is true in every domain of individuals is provable by our first method, the problem can also be put in the following way: Is it the case that every sentence of the calculus of relations which is true in every domain of individuals is derivable from the axioms adopted under the second method? This problem presents some difficulties and still remains open. I can only say that I am practically sure that I can prove with the help of the second method all of the hundreds of theorems to be found in Schröder's *Algebra und Logik der Relative*. ([23], p. 87)
>
> [6] *Die Vollständigkeit der Axiome des logischen Funktionenkalküls*, ***Monatshefte für Mathematik und Physik***, vol. 37 (1930), pp. 349-360.

Additional Reading

Apart from the relational topics dealt with in Section 1.3 there is a lot of very interesting research in the history of logic at that time, i.e. on how each of DE MORGAN, PEIRCE, and SCHRÖDER contributed to the development of logic, and how they have been influenced by each other or influenced others, e.g. BERTRAND RUSSELL (1872-1970) or GOTTLOB FREGE (1848-1925). Readers interested in these aspects are referred to e.g. [2, 4] or [1]. A short overview on this topic is given in chapter 1.8 of Maddux [10]; this is an excellent starting point that emphasizes the 'Axiomatization of the calculus of relations' (its title) but also sheds light on where and how quantifiers entered the scene.

References

[1] Anellis, I.H.: Some Views of Russell and Russell's Logic by His Early Contemporaries, The Rev Modern Log **10**, 67-98 (2005).
[2] Brady, G.: *From Peirce to Skolem: A Neglected Chapter in the History of Logic*, Elsevier Science B.V., Amsterdam (2000).
[3] Copi, I.M., Cohen, C., McMahon, K.: *Introduction to Logic*, 14th ed., Pearson Education Limited, Harlow, Essex (2014).
[4] Grattan-Guinness, I.: *The Search for Mathematical Roots, 1870-1940*, Princeton University Press, Princeton, New Jersey (2000).
[5] Hurley, P.J.: *A Concise Introduction to Logic*, 11th ed., Wadsworth, Belmont (2012).
[6] Lewis, C.I.: *A survey of symbolic logic*, University of California Press, Berkeley (1918).
[7] Łukasiewicz, J.: *Aristotle's Syllogistic From the Standpoint of Modern Formal Logic*, 2nd ed. Oxford Univ. Press (1957).
[8] Lorenzen, P.: *Formal Logic*, D. Reidel Publishing Company, Dordrecht (1965).
[9] Maddux, R.D.: The Origin of Relation Algebras in the Development and Axiomatization of the Calculus of Relations, Studia Logica **50**, 421-455 (1991).
[10] Maddux, R.D.: *Relation Algebras*, Elsevier Science B.V., Amsterdam (2006).
[11] Merrill, D.D.: *Augustus De Morgan and the Logic of Relations*. Kluwer Academic Publishers, Dordrecht (1990).
[12] De Morgan, A.: *Formal Logic: or,The Calculus of Inference, Necessary and Probable*, Taylor and Walton, London (1847).
[13] De Morgan, A.: *Syllabus of a proposed system of logic*, Walton and Maberly, London (1860).
[14] De Morgan, A.: On the syllogism no. III, and on logic in general, Trans. Cambridge Philos. Soc. **10**, 173-230 (1864).
[15] De Morgan, A.: On the syllogism no. IV, and on the logic of relations, Trans. Cambridge Philos. Soc. **10**, 331-358 (1864).
[16] Patzig, G.: *Aristotle's Theory of the Syllogism*. Springer, Dordrecht (1968).
[17] Peirce, C.S.: Description of a notation for the logic of relatives, resulting from an amplification of the conceptions of Boole's calculus of logic, Memoirs of the American Academy of Arts and Sciences, New Series **9**, 317-378 (1873). Reprinted in *Writings of Charles S. Peirce, A Chronological Edition, Volume 2, 1867-1871*, Indiana University Press, Bloomington and Indianapolis, 359-429 (1984).

[18] Peirce, C.S.: On the Logic of Relatives. Reprinted in *Writings of Charles S. Peirce, A Chronological Edition, Volume 4, 1879-1884*, Indiana University Press, Bloomington and Indianapolis, 336-341 (1989).

[19] Peirce, C.S.: Note B. The Logic of Relatives. In Peirce, C.S.(Ed), *Studies in Logic by Members of the Johns Hopkins University*, Boston: Little, Brown, and Company, 187-203 (1883). Reprinted in *Writings of Charles S. Peirce, A Chronological Edition, Volume 4, 1879-1884*, Indiana University Press, Bloomington and Indianapolis, 453-466 (1989).

[20] Sanchez Valencia, V.: The algebra of logic. In Dov M. Gabbay and John Woods (Eds.), *Handbook of the History of Logic Vol. 3, The Rise of Modern Logic: From Leibniz to Frege.* Elsevier, 389-544 (2004).

[21] Schröder, E.: *Vorlesungen über die Algebra der Logik (Exakte Logik). Dritter Band: Algebra und Logik der Relative.* B.G. Teubner, Leipzig (1895). [Reprinted by Chelsea Publishing Company, New York, 1966]

[22] Stillwell, J.: *Mathematics and Its History*, 3rd edition. Springer, New York (2010).

[23] Tarski, A.: On the Calculus of Relations, J. Symbolic Logic **6**, 73-89 (1941).

[24] Tennant, N.: Aristotle's Syllogistic and Core Logic, History and Philosophy of Logic **35**, 120-147 (2014).

Chapter 2

Mathematical Prerequisites

In order to achieve the goal of getting this book self-contained to a reasonable degree, concerning both the concepts to be introduced later on and the notation used, most of the mathematical prerequisites are gathered in this chapter. You will find standard knowledge on sets and their operations, as well as some information about special collections of sets. Also some theorems of both propositional and predicate logic will be listed. At the end of the chapter some basic general results on algebraic structures will be presented together with familiar, or examples that will become essential later in this book.

We assume that the reader is familiar with the most elementary concepts of set theory and first order logic and their standard notations. Thus, many of the definitions and results are repeated or presented here without any further comment, and to the extent necessary for the understanding of subsequent chapters. Since most of the contents of this chapter can be said to be common folklore of introductory mathematics, as a rule we will not give special credits.

Most of the concepts will not be introduced in their full algebraic generality, because our scope is not abstract algebra, but to use algebra as a tool to the extent that is necessary for our purposes. So I hope to avoid overloading of the text with formalisms that simply are not needed here. Readers interested in more details, may consult one of the books mentioned in the chapter's references for further information. Nevertheless, we will not only refer to secondary literature, but will include some proofs of key results in the text in order to give readers a feeling about the techniques used and to eventually whet their appetite for more. The compilation of this chapter will, however, not save us the task to introduce concepts of a more specialized character at the appropriate places, as e.g. the basics of semigroup and category theory in part IV.

2.1 Sets and their Operations

The notion of a *set* is one of the basic concepts of mathematics, if not *the* basic concept at all. From the naive viewpoint needed here, we might think of a **set** S as any collection of objects whatsoever, called the **elements** of S. As usually, membership of x in S is denoted by $x \in S$, and we write $x \notin S$ if x does not belong to S. If S is finite, then $|S|$ denotes the **cardinality** of S.

Important note: *If not obvious from their meaning, as for e.g. the set of natural numbers $\mathbb{N}$, or if not explicitly stated otherwise, the sets in this book are assumed to be finite.*

There is a special and uniquely determined set $\emptyset$, containing no element, whence it is called the **empty set**; it may formally be defined as $\emptyset := \{x : x \neq x\}$. Thus, $|\emptyset| = 0$, but $|\{\emptyset\}| = |\{\{1,\{a,b,c\}\}\}| = 1$, $|\{1,\{a,b,c\}\}| = 2$, and $|\{a,b,c\}| = 3$. Taking two more simple examples we let $D_6 = \{1,2,3,6\}$ contain the number 6 and all its divisors. In a similar way we can define another set $C_6 = \{\{1,2\}, \{1,3\}, \{1,2,3,6\}\}$ providing analogous information about the divisors of the divisors of 6. Note that there is also a 'syntactical' difference: contrary to D_6, C_6 contains only sets as elements. Sets with this property will be called **collections** and denoted by calligraphic letters like $\mathscr{C}, \mathscr{D}$.

Definition 2.1.

(a) *A set S is called a* **subset** *of a set T, denoted $S \subseteq T$, if every element of S is also an element of T, i.e. if $\forall x[x \in S \rightarrow x \in T]$.*
(b) *Two sets S and T are equal, denoted by $S = T$, if $S \subseteq T$ and $T \subseteq S$.*
(c) *S is a* **proper subset** *of T, $S \subset T$, if $S \subseteq T$ but $S \neq T$.*
(d) *The* **intersection** *of two sets is $S \cap T := \{x : x \in S \wedge x \in T\}$.*
(e) *The* **union** *of two sets is $S \cup T := \{x : x \in S \vee x \in T\}$.*
(f) *The* **difference** *of two sets S and T is $S - T := \{x : x \in S \wedge x \notin T\} = S \cap \overline{T}$.*
(g) *The* **complement** *of a set S is $\overline{S} := \boldsymbol{U} - S$, where $\boldsymbol{U}$ is the universe for the current context under consideration.*

For sake of completeness, some simple properties of these set operations are listed in the following propositions. Complements are meant to be constructed with respect to the same universe. In practice, this set is obvious from the context and therefore not mentioned in many cases.

Proposition 2.2. *For all sets S, T, U we have:*

(1) $S \cup (T \cup U) = (S \cup T) \cup U$ *(associativity)*
(2) $S \cup T = T \cup S$ *(commutativity)*
(3) $S \cup S = S$ *(idempotency)*
(4) $S \cap (T \cap U) = (S \cap T) \cap U$ *(associativity)*
(5) $S \cap T = T \cap S$ *(commutativity)*
(6) $S \cap S = S$ *(idempotency)*
(7) $S \cup \emptyset = S$
(8) $S \cap \emptyset = \emptyset$

Proposition 2.3. *For all sets S, T, U and V we have:*

(1) $S \subseteq S \cup T$;
(2) $S \cap T \subseteq S$;
(3) $S \subseteq U \wedge T \subseteq U \rightarrow S \cup T \subseteq U$;
(4) $U \subseteq S \wedge U \subseteq T \rightarrow U \subseteq S \cap T$;
(5) $S \subseteq T \leftrightarrow S \cap T = S \leftrightarrow S \cup T = T$.
(6) $S \subseteq U \wedge T \subseteq V \rightarrow S \cap T \subseteq U \cap V$

Proposition 2.4. *Let T, U be subsets of S; then the following holds:*

(1) $\overline{\overline{U}} = U$
(2) $\overline{T \cup U} = \overline{T} \cap \overline{U}$ *(de Morgan's law)*
(3) $\overline{T \cap U} = \overline{T} \cup \overline{U}$ *(de Morgan's law)*
(4) $T \subseteq U \leftrightarrow \overline{U} \subseteq \overline{T}$
(5) $T \cap U = \emptyset \leftrightarrow T \subseteq \overline{U} \leftrightarrow U \subseteq \overline{T}$

Proposition 2.5. *For all sets S, T, U the following* **distributivity** *properties hold:*

(1) $(S \cup T) \cap U = (S \cap U) \cup (T \cap U)$
(2) $(S \cap T) \cup U = (S \cup U) \cap (T \cup U)$

Definition 2.6.

(a) *Two non-empty sets S and T are said to be* **disjoint** *if* $S \cap T = \emptyset$.
(b) $\mathscr{C}$ *is called a collection of* **pairwise disjoint** *sets if* $\forall A, B \in \mathscr{C}[A \neq B \rightarrow A \cap B = \emptyset]$.
(c) *The collection* $\mathscr{P} = \{B : B \subseteq A\}$ *of all subsets of some set A is called the* **power set** *of A, denoted by either* $\mathcal{P}(A)$ *or* $\mathbf{2}^A$.
(d) *A collection* $\mathscr{A} \subseteq \mathcal{P}(A)$ *is called a* **system over** *A.*
(e) *A system* $\mathscr{A} = \{A_i : i \in I\} \neq \emptyset$ *is called a* **partial partition** *of A if* $A_i \neq \emptyset$ *for every* $i \in I$, *and* A_i *and* A_j *are pairwise disjoint sets.*

Since $\{\emptyset, A\} \subseteq \mathcal{P}(A)$ we always have $\mathcal{P}(A) \neq \emptyset$. How the power set is related to subset, intersection, and union is shown in

Proposition 2.7. *For all sets A, B we have:*

(1) $A \subseteq B \leftrightarrow \mathcal{P}(A) \subseteq \mathcal{P}(B)$.
(2) $\mathcal{P}(A) \cap \mathcal{P}(B) = \mathcal{P}(A \cap B)$.
(3) $\mathcal{P}(A) \cup \mathcal{P}(B) \subseteq \mathcal{P}(A \cup B)$.

Two important kinds of systems over a set A are introduced in

Definition 2.8.

(a) *A system* $\mathscr{A} = \{A_i, i \in I\} \neq \emptyset$ *over a set* $A \neq \emptyset$ *is called a* **covering** *of A iff*

$$(1) \forall i \in I[A_i \neq \emptyset], \quad \text{and} \quad (2) \bigcup_{i \in I} A_i = A.$$

(b) *A system $\mathscr{A} = \{A_i, i \in I\}$ over a set $A \neq \emptyset$ is called a* **partition** *of A iff it is a partial partition that is also a covering of A. A set $A_i \in \mathscr{A}$ is called a* **block** *of the partition. The set of the partitions of A is denoted by Π_A. If $\pi \in \Pi_A$ then we denote the block of a in π by $[a]_\pi$.*

Definition 2.9. *Let A be a set, and $\pi, \pi' \in \Pi_A$ be two partitions of A. Then π is called a* **refinement** *of (or* **finer** *than) π', denoted by $\pi \leq \pi'$, if every block of π' is a union of blocks of π, i.e. π is a further fragmentation of π'.*

Proposition 2.10. *Let $\pi = \{B_i | i \in I\}$ and $\pi' = \{C_j | j \in J\}$ be two partitions in Π_A. Then $\pi \leq \pi' \leftrightarrow \forall i \in I \, \exists j \in J [B_i \subseteq C_j]$.*

Proof. Assume $\pi \leq \pi'$; then, since π is a covering of A, each $a \in \bigcup_{i \in I} B_i$ must occur in a $C_j \in \pi'$, and since distinct blocks of π' are disjoint, $B_i \cap C_j \neq \emptyset \rightarrow B_i \subseteq C_j$. Thus, for each $B_i \in \pi$ there is a $C_j \in \pi'$ with $B_i \subseteq C_j$.
Conversely, suppose $\forall i \in I \, \exists j \in J [B_i \subseteq C_j]$. We set $B = \bigcup \{B_i \in \pi | B_i \subseteq C_j\}$; then $B \subseteq C_j$. Assume that there is an $x \in C_j - B$; then there is a block $B_k \in \pi$ with $x \in B_k \cap C_j$, whence $B_k \subseteq C_j$. Thus, $x \in B_k \subseteq C_j$, contradicting the assumption we have made about x. So $C_j - B = \emptyset$, and therefore $B = C_j$. □

There are two special partitions in Π_A: The partition of A consisting of all singletons: $\alpha_A = \{\{a\} | a \in A\}$; and the partition that contains A itself: $\omega_A = \{A\}$. Obviously, we have $\alpha_A \leq \pi \leq \omega_A$ for all $\pi \in \Pi_A$.

2.1.1 *Some Properties of Set Difference*

Before turning to the Cartesian products of sets, we continue with a short glance at the properties of the set theoretic difference:

Proposition 2.11. *Let $T, U \subseteq S$; then the following holds:*

(1) $S - (S - T) = S$.
(2) $T \cup (S - T) = S$.
(3) $T \subseteq U \leftrightarrow (S - U) \subseteq (S - T)$.
(4) $S - (T \cup U) = (S - T) \cap (S - U)$.
(5) $S - (T \cap U) = (S - T) \cup (S - U)$.

Proposition 2.12. *For sets S and T the following holds:*

(1) $S - T = S - (S \cap T)$.
(2) $S - \emptyset = S$.
(3) $S - S = \emptyset$.
(4) $S \cap (T - S) = \emptyset$.

Definition 2.13. *The* **symmetric difference** *of two sets S and T is $S \triangle T := (S - T) \cup (T - S)$.*

Proposition 2.14. *For all sets S, T, U we have:*

(1) $S \triangle S = \emptyset$
(2) $S \triangle \emptyset = S$
(3) $S \triangle T = T \triangle S$ *(commutativity)*
(4) $S \triangle (T \triangle U) = (S \triangle T) \triangle U$ *(associativity)*
(5) $S \triangle T = (S \cup T) - (S \cap T)$
(6) $S \triangle U = T \triangle U \rightarrow S = T$

2.1.2 *Cartesian Products, Relations, and Functions*

Definition 2.15. *A set $A := \{\{x\}, \{x, y\}\}$ is called an* **ordered pair** *with first component x and second component y and will be denoted by $\langle x, y \rangle$.*

Theorem 2.16. *Let $A = \langle x, y \rangle$ and $B = \langle u, v \rangle$; then $A = B$ if, and only if, $x = u$ and $y = v$.*

Proof. We prove the non-trivial direction by distinction in two cases: *Case 1.* $x = y$ yields $A = \{\{x\}\}$, whence B must also consist of a single set, i.e. $\{u, v\} = \{u\}$. This implies $u = v$, and from our assumption $A = B$ we get $\{x\} = \{u\}$, and finally $x = u$ and trivially also $y = v$. *Case 2.* If $x \neq y$ then neither A nor B are singletons themselves (i.e. $|A| > 1$), but each of them contains exactly one singleton; thus we again get $\{x\} = \{u\}$. Since $A = B$ we must also have $\{x, y\} = \{u, v\}$. Since $v \in \{x, y\}$ and $v \neq u = x$ we get $v = y$ as desired. □

Th. 2.16 implies that for an ordered pair its first and second component are uniquely determined. This justifies the definition given above.

Definition 2.17.

(a) *The* **Cartesian product** *of sets $A_1, \ldots, A_n$ is defined as*

$$\prod_{i \in I} A_i := A_1 \times \ldots, \times A_n := \{\langle a_1, \ldots, a_n \rangle : a_i \in A_i\}.$$

The vectors $\langle a_1, \ldots, a_n \rangle$ are called n-tuples. In case $A_i = A_j$ for all i, j, the product is written as A^n.

(b) *The mapping $\pi_i : \prod_{i \in I} A_i \longrightarrow A_i, i \in I$, defined by $\pi_i(\langle a_1, \ldots, a_n \rangle) = a_i$, is called the i-***th projection** *of $\prod_{i \in I} A_i$.*

Def. 2.17 describes a concept in a general set-theoretical formalism. By taking $n = 2$ we get ordered pairs as described in Def. 2.15 above: If R designates a concept that applies to pairs of elements from a certain set S, we usually speak of '*the binary relation R on S*'. It is also common practice to tacitly identify the property R with the relation it defines, so that $R(x, y)$ means the same as $\langle x, y \rangle \in R$. And it will be binary relations that are almost exclusively in the focus of this book.

The Cartesian product is (left and right) distributive over union, intersection, and difference of sets. To keep notation simple, we will usually restrict our considerations to the case $n = 2$.

Theorem 2.18. *Let $*$ be one of the oprations $\cup, \cap$, or $-$; then for any sets R, S, and T the following holds:*

(1) $(R * S) \times T = (R \times T) * (S \times T)$
(2) $T \times (R * S) = (T \times R) * (T \times S)$

Proof. As an example we give a proof for $(R - S) \times T = (R \times T) - (S \times T)$. From $\langle x, y\rangle \in (R - S) \times T$ we get $x \in R - S$ and $y \in T$. Thus $\langle x, y\rangle \in R \times T$ and $\langle x, y\rangle \notin S \times T$, whence $\langle x, y\rangle \in (R \times T) - (S \times T)$. The other way round, $\langle x, y\rangle \in (R \times T) - (S \times T)$ implies $x \in R$, $y \in T$, and $\langle x, y\rangle \notin S \times T$. Thus, $x \notin S$, whence $\langle x, y\rangle \in (R - S) \times T$. □

Proposition 2.19. *For all sets S_1, S_2, T_1, T_2 we have: If $S_1 \subseteq S_2$ and $T_1 \subseteq T_2$, then $S_1 \times T_1 \subseteq S_2 \times T_2$.*

In the first paragraph of the preface it has been said "that without relations we ... could not do mathematics". This means, that without a rudimentary knowledge of relations on the meta-language level we cannot introduce and analyze the notion of a relation and its properties on the object level. As a consequence, we have to include the definitions of some basic notions around the concepts of relation and function to our preliminaries, and (following [11]) we will now do it without many comments. Functions are a special kind of relations, and we will return to them later in Section 4.1. We will then notice that function composition fg will correspond to relation composition $g \circ f$. Though both can be taken as the standard notation, do always assure which is the one used in your current text, and whether it is about functions or relations: *left* properties of functions will become *right* properties of relations.

Definition 2.20. *Let X and Y denote two sets.*

(a) $R \subseteq X \times Y$ *is called a (***binary***)* **relation** *from X to Y.*
(b) A binary relation R on X, i.e. $R \subseteq X \times X$, is called
- **reflexive** *iff* $\langle x, x\rangle \in R$, *for all* $x \in X$;
- **symmetric** *iff* $\langle x, y\rangle \in R \rightarrow \langle y, x\rangle \in R$, *for all* $x, y \in X$;
- **transitive** *iff* $\langle x, y\rangle \in R \wedge \langle y, z\rangle \in R \rightarrow \langle x, z\rangle \in R$, *for all* $x, y, z \in X$.

(c) $R \subseteq X \times X$, *is called* **equivalence relation** *if it is reflexive, symmetric, and transitive. The set* $x/R = [x]_R := \{y \in X : \langle x, y\rangle \in R\}$ *is called the* **equivalence class** *of x in R, and the set of set of all equivalence classes is called the* **quotient set** *of X modulo R, denoted as X/R.*

Definition 2.21. *Let X and Y denote two sets.*

(a) $f \subseteq X \times Y$ *is called a (***total***)* **function** *or* **mapping** *from X into Y iff*

$$|f \cap (\{x\} \times Y)| = 1 \text{ for all } x \in X,$$

and is denoted by $f : X \longrightarrow Y$. $f(x)$ is the second component of the unique element in $f \cap (\{x\} \times Y)$. The set of all functions from X into Y is denoted by Y^X.
f is called a **partial function** *iff* $|f \cap (\{x\} \times Y)| \leq 1$ *for all* $x \in X$.

(b) *A function* f *is called*

- **injective** (*or* **one-to-one**) *iff* $|f \cap (X \times \{y\})| \leq 1$ *for all* $y \in Y$;
- **surjective** (*or* **onto**) *iff* $|f \cap (X \times \{y\})| \geq 1$ *for all* $y \in Y$;
- **bijective** *iff* $|f \cap (X \times \{y\})| = 1$ *for all* $y \in Y$.

(c) *The* **domain** *of a function* $f : X \longrightarrow Y$ *is*

$$\mathsf{dom}(f) = \{x \in X : \exists y \in Y[f(x) = y]\},$$

and the **range** *of* f *is*

$$\mathsf{rng}(f) = \{y \in Y : \exists x \in X[f(x) = y]\}.$$

(d) *The* **composition** *of two functions* $f : X \longrightarrow Y$ *and* $g : Y \longrightarrow Z$, *denoted as* $g(f(.))$ *or simply as* $gf(.)$, *is the function* $h : X \longrightarrow Z$, *where* $h(x) = g(f(x))$ *is the uniquely determined element* $z \in Z$ *such that* $f(x) = y$ *and* $g(y) = z$.

(e) *Let* $f : X \longrightarrow Y$ *be a function. The relation*

$$\mathsf{ker}(f) = \{\langle x, x' \rangle \in X \times X : f(x) = f(x')\}$$

on X *is called the* **kernel** *of* f.

A more familiar, but less concise, characterization of the notions defined in item *(b)* above is as follows: A function $f : X \longrightarrow Y$ is called **injective** if

$$\forall x_1, x_2 \in X\,[f(x_1) = f(x_2) \rightarrow x_1 = x_2], \tag{2.1}$$

i.e. elements of X with the same image must be equal. The contrapositive is equally useful:

$$\forall x_1, x_2 \in X\,[x_1 \neq x_2 \rightarrow f(x_1) \neq f(x_2)], \tag{2.2}$$

i.e. different elements of X yield different images. f is called **surjective** if

$$\forall y \in Y\, \exists x \in X\,[f(x) = y], \tag{2.3}$$

i.e. each element of Y has a pre-image. By definition, a function is **bijective**, if it is injective and surjective, i.e.

$$\forall y \in Y\, \exists! x \in X\,[f(x) = y],$$

where $\exists!$ means "*there is exactly one*".

If $S \subseteq X$ then the **restriction of** f **to** S, denoted by $f|_S$, is a function $f|_S : S \longrightarrow Y$ defined by $f|_S(x) = f(x)$, for all $x \in S$. id_X denotes the bijection $id_X : X \longrightarrow X$ with $id_X(x) = x$ and is called the **identity function** on X.

A function $f : X \longrightarrow Y$ gives rise to two induced functions between the powersets of X and Y. Since one of them is also denoted by f, the mapping $f : X \longrightarrow Y$ will be written as $f(x) = y$ which is different from $f[S] = T$ for the induced map $f : \mathcal{P}(X) \longrightarrow \mathcal{P}(Y)$. The latter one being defined as

$$f : \mathcal{P}(X) \longrightarrow \mathcal{P}(Y),\ f[S] = \{f(x) : x \in S\} \subseteq Y \text{ for all } S \subseteq X,$$

where $f[S]$ is called **image of** S **under** f. The **pre-image of** T **under** f is

$$f^{-1} : \mathcal{P}(Y) \longrightarrow \mathcal{P}(X),\ f^{-1}[T] = \{x : f(x) \in T\} \subseteq X, \text{ for all } T \subseteq Y.$$

f **maps** S **into** T if $f[S] \subseteq T$ and **maps** S **onto** T if $f[S] = T$.

2.1.2.1 *Some Basic Function Properties*

In this section we list some important and often used properties of functions. The next proposition reveals the surprising fact that from the two induced functions introduced above the pre-image function is "better behaved" with respect to the usual set operations, since it gives equality for both union and intersection.

Proposition 2.22. *Suppose $f : X \longrightarrow Y$ is a function, and $S, T \subseteq X$ and $U, V \subseteq Y$. Then*

(1) $f[S \cup T] = f[S] \cup f[T]$
(2) $f[S \cap T] \subseteq f[S] \cap f[T]$
(3) $f^{-1}[U \cup V] = f^{-1}[U] \cup f^{-1}[V]$
(4) $f^{-1}[U \cap V] = f^{-1}[U] \cap f^{-1}[V]$

Proof. (2): $b \in f[S \cap T] \rightarrow \exists a \in S \cap T[b = f(a)]$. For this a we have $a \in S \wedge a \in T$, so $b \in f[S] \wedge b \in f[T]$.

(4): For $x \in X$ we have, $x \in f^{-1}[U \cap V] \leftrightarrow f(x) \in U \cap V \leftrightarrow f(x) \in U \wedge f(x) \in V \leftrightarrow x \in f^{-1}[U] \wedge x \in f^{-1}[V]$. □

Proposition 2.23.

(1) Let $f : X \longrightarrow Y$ be a mapping. Then f is surjective iff there exists a mapping $g : Y \longrightarrow X$, called a **right inverse** *of f, such that $fg = id_Y$.*
(2) Let $f : X \longrightarrow Y$ be a surjective mapping, and let $g, h : Y \longrightarrow Z$ be two mappings from Y into Z. If $gf = hf$ then $g = h$. Thus, surjective mappings are **right cancellative**.
(3) Let $f : X \longrightarrow Y$ be a mapping. Then f is injective iff there exists a mapping $g : \mathrm{rng}(f) \longrightarrow X$, called a **left inverse** *of f, such that $gf = id_X$.*
(4) Let $f : X \longrightarrow Y$ be a injective mapping, and let $g, h : Y \longrightarrow Z$ be two mappings from Y into Z. If $fg = fh$ then $g = h$. Thus, injective mappings are **left cancellative**.

Proof. (1): If there exists a mapping g such that $f(g(y)) = y$ for all $y \in Y$, then f clearly is surjective. Conversely, suppose f to be surjective; then, for each $y \in Y$, select $g(y)$ in the non-empty set $f^{-1}(y)$. This defines a mapping $g : Y \longrightarrow X$ such that $f(g(y)) = y$ for all $y \in Y$.

(2): By (1) there exists a mapping $\tilde{f} : Y \longrightarrow X$ such that $f\tilde{f} = id_Y$. Therefore $gf = hf$ implies $gf\tilde{f} = hf\tilde{f}$, whence $g = h$. □

Proposition 2.24. *For functions $f : X \longrightarrow Y$ and $g : Y \longrightarrow Z$ the following holds:*

(1) If f and g are injective then also $gf : X \longrightarrow Z$.
(2) If f and g are surjective then also $gf : X \longrightarrow Z$.

Proof. (1): Since g is injective, $g(f(a)) = g(f(a'))$ implies $f(a) = f(a')$, and since f is injective we get $a = a'$. So gf is injective.

(2): Since g is surjective, $\forall c \in Z \, \exists b \in Y \, [g(b) = c]$, and similarly $\forall b \in Y \, \exists a \in X \, [f(a) = b]$. Together this implies $\forall c \in Z \, \exists a \in X \, [g(f(a)) = c]$. So gf is surjective. □

Proposition 2.25. *For functions $f : X \longrightarrow Y$ and $g : Y \longrightarrow X$ the following holds:*

(1) If gf is injective then also f.
(2) If gf is surjective then also g.

Proof. (1): Assume $x, x' \in X$ and $f(x) = f(x')$. Then $g(f(x)) = g(f(x'))$, and since gf is injective we get $x = x'$, whence f is injective too. □

Proposition 2.26. *Let $f : X \longrightarrow Y$ be a mapping. Then the following statements are equivalent:*

(1) f is bijective;
(2) there exists a (unique and bijective) mapping $g : Y \longrightarrow X$ that is both a right and left inverse of f;
(3) f has both a right and left inverse.

Proof. (1) $\to$ (2): Let f be bijective; then for each $y \in Y$ there is a unique $x \in X$ such that $f(x) = y$. By Prop. 2.23(1) there exists a function $g : Y \longrightarrow X$ such that $fg = id_Y$, defined by $g(y) = x$, where x is the unique element such that $f(x) = y$. This mapping g also satisfies $gf = id_X$, whence it is also a left inverse of f.

Moreover, g is bijective by Prop. 2.25, since fg is injective and gf is surjective. Finally assume that both g_1 and g_2 are two right (hence, by what we have just shown, also left) inverses of f; then $g_1 = g_1 id_Y = g_1(fg_2) = (g_1 f)g_2 = id_X g_2 = g_2$. This unique mapping g is called the **inverse** of f, and is denoted by f^{-1}. (In the proof we have tacitly used that function composition is associative; this is shown in Th. 3.16(3) below for general relations.)

(2) $\to$ (3): Trivial.

(3) $\to$ (1): If f has both a right and left inverse, say r and l respectively, then $fr = id_Y$ and $lf = id_X$. So f is bijective by Prop. 2.25. □

Instead of proving that $f : X \longrightarrow Y$ is a bijection by directly showing that it is both one-to-one and onto, Prop. 2.26 provides an alternative way: If we can define a function $g : Y \longrightarrow X$ which is both a left and right inverse of f, then f is bijective and $g = f^{-1}$.

The next two theorems show some remarkable properties of mappings, involving the concept of the kernel of a function (recall Def. 2.21(e)). Th. 2.28 is a decomposition theorem for mappings. Note that, for any mapping f, $\mathsf{ker}(f)$ clearly is an equivalence relation.

Proposition 2.27. [22, p. 29, Theorem 1.115] *For any mapping $f : X \longrightarrow Y$ there is a bijection $h : X/\mathsf{ker}(f) \longrightarrow f[X]$.*

Proof. We set $[x] := [x]_{\mathsf{ker}(f)}$ and define $h([x]) = f(x)$. The mapping h is well defined, since $x' \in [x]$ implies $\langle x, x' \rangle \in \mathsf{ker}(f)$, hence $f(x) = f(x')$. h is onto since if $y \in f[X]$, then there is an $x \in X$ with $f(x) = y$, so $y \in h([x])$. Finally, to prove the injectivity of h, assume $h([x]) = h([x'])$. This implies $f(x) = f(x')$; thus, $\langle x, x' \rangle \in \mathsf{ker}(f)$, whence $[x] = [x']$. Therefore, h is injective by (2.1). □

Theorem 2.28. [22, p. 30, Theorem 1.116] *Each mapping $f : X \longrightarrow Y$ can be decomposed into three mappings as follows: a surjection $g : X \longrightarrow X/\mathsf{ker}(f)$, a bijection $h : X/\mathsf{ker}(f) \longrightarrow f[X]$, and an injection $k : f[X] \longrightarrow Y$.*

Proof. Defining $g : x \mapsto [x]$ for $x \in X$ and k as the inclusion mapping $k : y \mapsto y$ for all $y \in f[X]$, we get $k(h(g(x))) = k(h([x])) = k(f(x)) = f(x)$, for all $x \in X$, using Prop. 2.27 for h. □

A well known result in set theory is that any function can be decomposed into a surjection and an injection. A generalization of this result we will meet again later in Chapter 11 under the name *epi-mono factorization.*

Corollary 2.29. *For any function $h : X \longrightarrow Z$ there exist a surjection $f : X \longrightarrow Y$ and an injection $g : Y \longrightarrow Z$ such that $h = g \cdot f$.*

Proof. Since by Prop. 2.24(2) the composition of two surjections is again a surjection, the assertion then follows by Th. 2.28. □

2.1.3 *Closures and Interiors*

Definition 2.30.

(a) *A system $\mathscr{C}$ over A is called a* **closure system** *(***Moore family***) over A, if $A \in \mathscr{C}$ and for any subsystem $\mathscr{C}' \subseteq \mathscr{C}$ also $\bigcap \mathscr{C}' \in \mathscr{C}$.*

(b) *Let $\mathscr{C}$ be a closure system over A; the sets $C \in \mathscr{C}$ are called* **closed sets***. Given $S \subseteq A$, the least closed set containing S is called the* **closure of** S*, denoted by $cl_{\mathscr{C}}(S)$.*

By definition, a closure system $\mathscr{C}$ over A is closed under arbitrary intersections, and in particular $A = \bigcap \emptyset \in \mathscr{C}$. Some examples for closure systems are:

- The power set of any set is a closure system.
- For $S \in \mathscr{C}$ we have $cl_{\mathscr{C}}(S) = S$.
- The system of all subgroups of a group G is a closure system.
- The set of all relations from A to B is a closure system over $A \times B$.
- The set of all reflexive relations on A is a closure system over $A \times A$.
- Let T be a topological space, and $\mathcal{T}$ be its closed subsets. Then $\mathcal{T}$ is a closure system with the additional property that: $A, B \in \mathcal{T} \rightarrow A \cup B \in \mathcal{T}$. A closure system with this property is called **topological**.

A second notion which is of interest in this context (and turns out to be equivalent to that of a closure system) is the notion of a *closure operator* on a set:

Definition 2.31. *A* **closure operator** *on a set A is a function $\gamma : \mathcal{P}(A) \longrightarrow \mathcal{P}(A)$ satisfying for all $B, C \in \mathcal{P}(A)$ the following conditions:*

C1. $B \subseteq \gamma(B)$	*(extensive),*
C2. $B \subseteq C \rightarrow \gamma(B) \subseteq \gamma(C)$	*(isotone), and*
C3. $\gamma(\gamma(B)) = \gamma(B)$	*(idempotent).*

$X \subseteq A$ is called a **closed set** *if $\gamma(X) = X$.*

Theorem 2.32. *(Main theorem on closure systems and closure operators)*

(1) Let $\mathscr{C}$ be a closure system over A, and for all $X \subseteq A$ set

$$\gamma_{\mathscr{C}}(X) := \bigcap\{C \in \mathscr{C} | X \subseteq C\}. \tag{2.4}$$

Then $\gamma_{\mathscr{C}} : \mathcal{P}(A) \longrightarrow \mathcal{P}(A)$, $X \mapsto \gamma_{\mathscr{C}}(X)$, is a closure operator on A, and the closed sets of $\gamma_{\mathscr{C}}$ are just the closed sets in $\mathscr{C}$.

(2) Let γ be a closure operator on a set A. Then

$$\mathscr{C}_{\gamma} := \{\gamma(X) | X \subseteq A\}. \tag{2.5}$$

is a closure system over A, and the closed sets of $\mathscr{C}_{\gamma}$ are just the closed sets of A.

(3) For each closure system $\mathscr{C}$ over A, and each closure operator γ on A the following holds:

$$\mathscr{C}_{(\gamma_{\mathscr{C}})} = \mathscr{C}, \quad and \quad \gamma_{(\mathscr{C}_{\gamma})} = \gamma \tag{2.6}$$

Proof. (1): Clearly, $\gamma_{\mathscr{C}}$ is extensive by definition. Let X, Y be two subsets of A with $X \subseteq Y$; then $\{C \in \mathscr{C} | Y \subseteq C\} \subseteq \{C \in \mathscr{C} | X \subseteq C\}$ whence $\gamma_{\mathscr{C}}(X) \subseteq \gamma_{\mathscr{C}}(Y)$. To prove idempotency we first show

$$Y \subseteq \gamma_{\mathscr{C}}(Z) \rightarrow \gamma_{\mathscr{C}}(Y) \subseteq \gamma_{\mathscr{C}}(Z) \tag{2.7}$$

From $Y \subseteq \gamma_{\mathscr{C}}(Z)$ we get $\{C \in \mathscr{C} | Z \subseteq C\} \subseteq \{C \in \mathscr{C} | Y \subseteq C\}$, so $\gamma_{\mathscr{C}}(Y) \subseteq \gamma_{\mathscr{C}}(Z)$. Choosing $Y = \gamma_{\mathscr{C}}(Z)$ in (2.7) yields $\gamma_{\mathscr{C}}(\gamma_{\mathscr{C}}(Z)) \subseteq \gamma_{\mathscr{C}}(Z)$, and since $\gamma_{\mathscr{C}}$ is extensive we finally get $\gamma_{\mathscr{C}}(\gamma_{\mathscr{C}}(Z)) = \gamma_{\mathscr{C}}(Z)$. Thus $\gamma_{\mathscr{C}}$ is a closure operator on A.

It remains to show that

$$C \in \mathscr{C} \leftrightarrow \gamma_{\mathscr{C}}(C) = C. \tag{2.8}$$

Writing $\mathscr{C}_X := \{D \in \mathscr{C} | X \subseteq D\}$ we see that $\mathscr{C}_X \subseteq \mathscr{C}$, so $\gamma_{\mathscr{C}}(C) = \bigcap \mathscr{C}_C \in \mathscr{C}$ since $\mathscr{C}$ is a closure system. Assuming $C \in \mathscr{C}$ gives $\gamma_{\mathscr{C}}(C) = \bigcap\{D \in \mathscr{C} | C \subseteq D\} = \bigcap\{C, D_1, ...\} = C$, so the implication from left to right holds. Conversely assume $\gamma_{\mathscr{C}}(C) = C$; but $\mathscr{C}_C \subseteq \mathscr{C}$ implies $\bigcap \mathscr{C}_C = \gamma_{\mathscr{C}}(C) \in \mathscr{C}$, so $C \in \mathscr{C}$.

(2) Let $\gamma : \mathcal{P}(A) \longrightarrow \mathcal{P}(A)$ be a closure operator, and $\mathscr{C}_{\gamma}$ be defined by (2.5). Then we have

$$\forall X \subseteq A [X \in \mathscr{C}_{\gamma} \leftrightarrow \gamma(X) = X]. \tag{2.9}$$

Let $\mathscr{B} \subseteq \mathscr{C}_\gamma$; then $B \in \mathscr{B}$ implies $B = \gamma(B)$ and $\bigcap_{X\in\mathscr{B}} X \subseteq B$, whence by isotony of γ we get $\gamma(\bigcap_{X\in\mathscr{B}} X) \subseteq \gamma(B) = B$ for each $B \in \mathscr{B}$, and from this $\gamma(\bigcap_{X\in\mathscr{B}} X) \subseteq \bigcap_{B\in\mathscr{B}} B$. On the other hand we also have $\bigcap_{X\in\mathscr{B}} X \subseteq \gamma(\bigcap_{X\in\mathscr{B}} X)$. Thus $\gamma(\bigcap\mathscr{B}) = \bigcap\mathscr{B} \in \mathscr{C}_\gamma$.

(3) Using (2.8) and (2.9) we get $X \in \mathscr{C}_{(\gamma_\mathscr{C})} \;\leftrightarrow\; \gamma_\mathscr{C}(X) = X \;\leftrightarrow\; X \in \mathscr{C}$. Now we choose an arbitrary $Y \subseteq A$ and will show that $\gamma_{(\mathscr{C}_\gamma)}(Y) = \gamma(Y)$. By definition we have $\gamma_{(\mathscr{C}_\gamma)}(Y) = \bigcap\{Z|Z \in \mathscr{C}_\gamma \wedge Y \subseteq Z\}$ which yields $\gamma_{(\mathscr{C}_\gamma)}(Y) = \bigcap\{Z|\gamma(Z) = Z \wedge Y \subseteq Z\}$ by (2.9). Conversely, from $Y \subseteq Z = \gamma(Z)$ we get $\gamma(Y) \subseteq \gamma(Z) = Z$, whence $\gamma(Y) \subseteq \bigcap\{Z|\gamma(Z) = Z \wedge Y \subseteq Z\}$. Thus we finally have $\gamma(Y) = \bigcap\{Z|\gamma(Z) = Z \wedge Y \subseteq Z\}$. □

For more details on closure systems see e.g. the books of Cohn [5] or Lau [15]. The proof given above follows the latter one.

2.2 Basics from Predicate Logic

In this section we will list some basic facts and theorems from both propositional (also: sentential) and first order (also: predicate) logic without much further discussion or justification. They will be used later on in order to facilitate the comprehension of proofs. These proofs will be formulated in a free mixture of the usual "colloquial mathematical speech" and a pure first order logic style. In the following we assume acquaintance with the most basic concepts and notions of logic only as e.g. axioms, connectives, quantifiers, well formed formulas (wffs), inference rules; the rest should become obvious from the text.

We will not focus on mathematical elegance of axiom systems and the like; that is, we avoid 'technical' details which are not necessary for our current needs. Instead we try to provide a toolbox that helps mathematicians or computer scientists master the problems in their daily proof work. The theorems and rules given below should appear in any standard introduction to logic like e.g. Copi et al. [6] or Hurley [13].

Typically, the development of a formal calculus for classical propositional logic, say $\mathscr{L}$, starts with the definition of what counts as a *formula* in $\mathscr{L}$, followed by the proof theory (around the notions of *rule of inference, proof,* and *theorem*) and the semantics of $\mathscr{L}$ (*validity* and *tautology*). There is a set of wffs, called the set $A_\mathscr{L}$ of *axioms of* $\mathscr{L}$ which are taken as given, and a finite set of *inference rules* which tell us when a conclusion can be derived from a set of premises.

A *proof* in $\mathscr{L}$ is a sequence Φ_i such that, for each i, $\Phi_i \in A_\mathscr{L}$ or Φ_i is a direct consequence of some of the preceding wffs in the sequence by application of one of the rules of inference of $\mathscr{L}$. A *theorem* of $\mathscr{L}$ is a wff Φ such that it is the last wff in a proof in $\mathscr{L}$. In a second step the system for $\mathscr{L}$ becomes extended with further axioms and rules to properly deal with quantification.

In our situation, viz. as users of predicate logic, we do not need to take care what are the axioms of a certain logical calculus, and what are theorems derived from them; we also need not differentiate between given and derived rules of inference. We know that we simply can trust the work of the many logicians and their results, and are, therefore, only interested in what are legal steps in the search for a proof of a certain statement.

(A typographical note: $\equiv$ denotes the logical equivalence of the two formulas on its right and left side, i.e. $A \equiv B$ means $(A \rightarrow B) \wedge (B \rightarrow A)$. $\equiv$ thus means the same as $\leftrightarrow$, but it gives better readability of longer formulas, especially if implications are involved.)

Theorem 2.33. (Some useful laws from propositional logic)

(1)	$A \wedge B \rightarrow A$	$A \wedge B \rightarrow B$
(2)	$A \rightarrow A \vee B$	$B \rightarrow A \vee B$
(3)	$\mathsf{true} \wedge A \equiv A$	$\mathsf{false} \vee A \equiv A$
(4)	$A \rightarrow \mathsf{true} \equiv \mathsf{true}$	$\mathsf{true} \rightarrow A \equiv A$
(5)	$A \rightarrow \mathsf{false} \equiv \neg A$	$\mathsf{false} \rightarrow A \equiv \mathsf{true}$
(6)	$\neg A \equiv (A \rightarrow \neg A)$	$A \equiv (\neg A \rightarrow A)$
(7)	$A \rightarrow B \equiv \neg A \vee B$	
(8)	$A \rightarrow B \equiv \neg B \rightarrow \neg A$	
(9)	$A \rightarrow B \equiv \neg (B \wedge \neg A)$	
(10)	$(A \rightarrow B) \wedge (B \rightarrow C) \rightarrow (A \rightarrow C)$	
(11)	$A \wedge (A \rightarrow B) \equiv A$	
(12)	$A \wedge B \rightarrow C \equiv A \rightarrow (B \rightarrow C)$	
(13)	$A \rightarrow (B \rightarrow C) \equiv B \rightarrow (A \rightarrow C)$	
(14)	$(A \rightarrow B) \wedge (A \rightarrow C) \equiv A \rightarrow (B \wedge C)$	
(15)	$(A \rightarrow C) \wedge (B \rightarrow C) \equiv (A \vee B) \rightarrow C$	
(16)	$B \rightarrow (A \rightarrow B)$	
(17)	$(A \rightarrow C) \rightarrow (A \wedge B \rightarrow C)$	
(18)	$(A \rightarrow C) \wedge (B \rightarrow D) \rightarrow [(A \wedge B) \rightarrow (C \wedge D)]$	

Do not forget all the laws concerning commutativity and distributivity of conjunction $\wedge$ and disjunction $\vee$ not mentioned above, as well as the laws concerning negation. All of them should be familiar from the analogies to the operations in set theory.

Before we can continue with some theorems from first order logic (FOL), a few basic notions must be introduced and illustrated by examples. Reading mathematical texts you are likely to encounter strings of the form $\forall x$ and $\exists x$ (to be read as "*for all x*" and "*there is an x*") called *prefixes* as part of some logical formula. Formulas in which $\forall$ and $\exists$ do not occur are termed *quantifier-free* or *open*.

Intuitively, the formulas *(a)* $\forall x \exists y[x + y = 0]$ and *(b)* $\exists y[x + y = 0]$ are different in that in every context with a given meaning for + and 0, the former is either true or false, whereas the latter is not: in formula *(b)* the variable x is waiting to be assigned a value before something can be said about its truth. One says that all variables in *(a)* are bound, while *(b)* contains the free variable x. The *free variables* of a formula φ are those variables occurring in φ that are not quantified, and the *bound variables* of φ are the ones that do have a quantifier referring to them. Formulas without free variables are called *sentences*, or *closed* formulas. For a fully-fledged formal exposition of the topics of free and bound occurrences of variables, substitution etc. see e.g. [8,21].

In the following, let P denote a predicate with at most x and y occurring as free variables, Q a predicate with at most x as a free variable, and A a predicate with no free occurrence of variable x.

Theorem 2.34. (Laws of Quantifier (In)Dependence)

(1) $\forall x \forall y P(x,y) \equiv \forall y \forall x P(x,y)$
(2) $\exists x \exists y P(x,y) \equiv \exists y \exists x P(x,y)$
(3) $\exists x \forall y P(x,y) \rightarrow \forall y \exists x P(x,y)$

Concerning the last implication it is not too hard to give some examples showing the converse direction to fail: From "*To each natural number y there exists another one, x, which is greater than y.*" we cannot conclude "*There is a natural number x that is greater than all other natural numbers y.*". Likewise, from "*Every child has a mother.*" and "*To every event there is a cause.*" we cannot infer "*There is a mother of all children.*" or "*There is one cause for all events.*"

Theorem 2.35. (Duality Laws)

(1) $\forall x Q(x) \equiv \neg\exists x[\neg Q(x)]$ $\quad \forall x \neg Q(x) \equiv \neg\exists x Q(x)$
(2) $\neg\forall x Q(x) \equiv \exists x[\neg Q(x)]$ $\quad \neg\forall x \neg Q(x) \equiv \exists x Q(x)$
(3) $\forall x[P(x) \rightarrow Q(x)] \equiv \neg\exists x[P(x) \wedge \neg Q(x)]$
(4) $\forall x[P(x) \rightarrow \neg Q(x)] \equiv \neg\exists x[P(x) \wedge Q(x)]$
(5) $\forall x[\neg P(x) \rightarrow Q(x)] \equiv \neg\exists x[\neg P(x) \wedge \neg Q(x)]$

Theorem 2.36. (Laws of Quantifier Movement)

(1) (a) $\exists x[A \wedge Q(x)] \equiv A \wedge \exists x Q(x)$
(b) $\forall x[A \wedge Q(x)] \equiv A \wedge \forall x Q(x)$
(2) (a) $\exists x[A \vee Q(x)] \equiv A \vee \exists x Q(x)$
(b) $\forall x[A \vee Q(x)] \equiv A \vee \forall x Q(x)$
(3) (a) $\forall x[Q(x) \rightarrow A] \equiv \exists x Q(x) \rightarrow A$
(b) $\exists x[Q(x) \rightarrow A] \equiv \forall x Q(x) \rightarrow A$
(4) (a) $\exists x[A \rightarrow Q(x)] \equiv A \rightarrow \exists x Q(x)$
(b) $\forall x[A \rightarrow Q(x)] \equiv A \rightarrow \forall x Q(x)$

Theorem 2.37. (Laws of Quantifier Distribution)

(1) (a) $\exists x[P(x) \vee Q(x)] \equiv \exists x P(x) \vee \exists x Q(x)$
(b) $\forall x[P(x) \wedge Q(x)] \equiv \forall x P(x) \wedge \forall x Q(x)$
(2) (a) $\forall x P(x) \vee \forall x Q(x) \rightarrow \forall x[P(x) \vee Q(x)]$
(b) $\exists x[P(x) \wedge Q(x)] \rightarrow \exists x P(x) \wedge \exists x Q(x)$
(3) (a) $\forall x P(x) \wedge \exists x Q(x) \rightarrow \exists x[P(x) \wedge Q(x)]$
(b) $\forall x[P(x) \vee Q(x)] \rightarrow \forall x P(x) \vee \exists x Q(x)$
(4) (a) $\forall x[P(x) \rightarrow Q(x)] \rightarrow (\forall x P(x) \rightarrow \forall x Q(x))$
(b) $\forall x[P(x) \leftrightarrow Q(x)] \rightarrow (\forall x P(x) \leftrightarrow \forall x Q(x))$
(5) $\exists x[P(x) \rightarrow Q(x)] \equiv (\forall x P(x) \rightarrow \exists x Q(x))$

All the logical theorems shown above tell us how to manipulate existential and universal quantifiers if they are already there. However, none of them allows us to prove the following kind of everyday arguments: "All cats are mammals, Luna is a cat, so Luna is a mammal.", in symbols $\forall x\,[C(x) \rightarrow M(x)], C(L) \vdash M(L)$. But if the universal quantifier could be eliminated, then we could use *modus ponens* $p,\ p \rightarrow q \vdash q$ to get the desired result.

This example indicates that we have to think about the questions of where and when quantifiers can be added to or eliminated from a formula. And as you will see immediately, this is a more subtle thing than listing a few theorems. Our presentation will be rather informal but must, of course, be sufficiently precise, so it may become a little bit longer than many would probably expect. It is based upon [6, Section 10.5], [13, Section 8.2], and the detailed and intelligible discussion in [26, Chapter 6]. For the technical formalities we refer to one of the books mentioned at the end of this chapter. We will follow the usual way of presentations as derivation rules.

Theorem 2.38. (Rules for Quantifier Instantiation and Generalization)
In FOL the following are valid inference rules for quantifiers, provided that the respective restrictions are fulfilled:

(1) **Universal instantiation**, (UI): $\forall x[P(x)] \vdash P(b)$,
where b is an individual constant symbol of the underlying language used.
(2) **Universal generalization**, (UG): $P(a) \vdash \forall x[P(x)]$,
where a stands for "any individual arbitrarily chosen" from some domain S, and x is not free in P.
(3) **Existential instantiation**, (EI): $\exists x[P(x)] \vdash P(b)$,
where b is any individual constant symbol that does not occur somewhere earlier in the derivation.
(4) **Existential generalization**, (EG): $P(b) \vdash \exists x[P(x)]$,
where b is is any individual constant symbol.

Instead of *universal/existential instantiation* also the names *universal/existential elimination* can be found, and instead of *universal/existential generalization* the names *universal/existential introduction*. Sometimes these rules are formulated as axiom schemes, e,g, for (UI) as $\forall x[P(x)] \Rightarrow P(t/x)$ for some term t where $P(t/x)$ is the result of substituting t for all free occurrences of x in P.

The *rule (UI)* means: If we have shown somewhere in a derivation that a property P holds for each member of some class C of individuals, then we can conclude later on that P is true for an arbitrary particular individual of that class. Applications of (UI) are obviously and intuitively valid quite generally.

However, for the *rule (UG)* one must take care to ensure that it is truth-preserving: (UG) allows to generalize universally from saying something about some particular individual element of the domain under consideration to saying something about *every* element of this domain. In doing so we must be cautious: the crucial point is what has been formulated above as "*any individual arbitrarily chosen*", i.e. all we should know about that individual, say n, is that it is an element of the domain and nothing else; we *must not* associate any particular property with it. And this must hold for all derivation steps before the one in which we plan to use (UG). In the practice of proof-construction we can always look back at previous lines of proof, checking carefully that there is no line which contains a formula telling us something special about that named individual. Assume that we do already have ... $\vdash x = 0$ somewhere in our proof, then $\forall y[x = 0]$ would be a valid, but $\forall x[x = 0]$ an invalid application of (UG).

(EG), in contrast to (UG), allows us to make inferences whose validity is perfectly obvious, whence there is no need to impose any sort of special restriction upon its use: If we know that $P(a)$, with a any individual constant, then we clearly can infer that there is an individual in the domain with property P, i.e. $\exists x[P(x)]$.

Concerning the fourth rule, viz. existential instantiation (EI), we know that the existential quantification of a propositional function $P(x)$ is true if and only if there is at least one individual a in the domain such that $P(a)$ holds. If an individual constant appears nowhere earlier in the derivation, we may use it to denote an individual that has property P.

(To tell the truth, I have to admit that affairs are still more sophisticated, but the comments just given should give a feeling about the kind of problems one has to consider. (i) Assume you have been able to derive $\forall x[P(x)]$ with $P(x) := \exists y[x \neq y]$; then it is, of course, invalid to conclude $P(y/x) = \exists y[y \neq y]$ by usage of (UI). Such situations need additional restrictions, for which you may consult e.g. [27, 4.1.28 Definition. (Substitution of Terms into Variables)]. (ii) In case you really have to work with more complex formal derivations you cannot save the labor of thoroughly studying one of the books on logic mentioned at the end of the chapter.)

2.3 Algebraic Structures

The notion of an *operation on a set* is needed for the definition of various algebraic structures. Binary operations on a set S are rules for combining two elements of S to get another element of the same set. They occur in a lot of contexts and are a rather familiar thing: Union and intersection of two subsets of a set S are binary operations on $\mathcal{P}(S)$; other examples are the addition of two reals, the multiplication of matrices, etc.

Definition 2.39. *Let $S \neq \emptyset$; a* **binary operation on S** *is a function $f : S \times S \longrightarrow S$ mapping the ordered pair $\langle x, y\rangle$ onto $f(x, y)$.*

Commonly used notations for the image $f(a, b) \in S$ are e.g. $a + b$, $a \cdot b$, or also $a * b$ or $a \circ b$ as more neutral and generic versions, and sometimes only ab. We will not be very careful on which sign to use: you simply will have to live with diverse notations if you look through various books and papers. Thus, there is no harm, if you get accustomed to them.

Definition 2.40. *Let S be a set and $\cdot$ be a binary operation on S; $\cdot$ is called*

(a) **associative** *if* $(a \cdot b) \cdot c = a \cdot (b \cdot c)$ *for all* $a, b, c \in S$*;*
(b) **commutative** *if* $a \cdot b = b \cdot a$ *for all* $a, b \in S$*;*
(c) **idempotent** *if* $a \cdot a = a$ *for all* $a \in S$*.*

Set union and intersection are both associative, commutative, and idempotent operations. The addition of real numbers is associative and commutative on $\mathbb{R}$.

A binary operation $\cdot$ on a finite set $S = \{x_1, \dots, x_n\}$ can conveniently be defined by using an $n \times n$ matrix, called the **Cayley table**[1] for this operation. It is customary to write elements x_i in front of the i-th line (yielding the 0-column), and x_j at the top of the j-th column (yielding the 0-row). Following this convention for the set $\{x_1, x_2, \dots, x_n\}$ and the binary operation $\cdot$ on it, we might define $\cdot$ as e.g. in Table 2.1. If x occupies i-th position in the 0-column and y occupies j-th position in the 0-row, then we get $\cdot : S \times S \longrightarrow S, \langle x, y \rangle \mapsto x \cdot y$, the element at the (i, j)-th position in the table.

TABLE 2.1: A Cayley table example.

$\cdot$	x_1	x_2	$\cdots$	x_j	$\cdots$	x_n
x_1	$x_1 \cdot x_1$	$x_1 \cdot x_2$	$\cdots$	$x_1 \cdot x_j$	$\cdots$	$x_1 \cdot x_n$
x_2	$x_2 \cdot x_1$	$x_2 \cdot x_2$	$\cdots$	$x_2 \cdot x_j$	$\cdots$	$x_2 \cdot x_n$
$\vdots$	$\vdots$	$\vdots$		$\vdots$		$\vdots$
x_i	$x_i \cdot x_1$	$x_i \cdot x_2$	$\cdots$	$x_i \cdot x_j$	$\cdots$	$x_i \cdot x_n$
$\vdots$	$\vdots$	$\vdots$		$\vdots$		$\vdots$
x_n	$x_n \cdot x_1$	$x_n \cdot x_2$	$\cdots$	$x_n \cdot x_j$	$\cdots$	$x_n \cdot x_n$

Definition 2.41. *Let* S *be a set and* $\cdot$ *be a binary operation on* S*.*

(a) $l \in S$ *is called a* **left identity** *of* S *if* $l \cdot a = a$ *for all* $a \in S$*.*
(b) $r \in S$ *is called a* **right identity** *of* S *if* $a \cdot r = a$ *for all* $a \in S$*.*
(c) $e \in S$ *is called an* **identity** *(or* **neutral element***) of* S *if it is both a left and a right identity.*
(d) $l \in S$ *is called a* **left inverse** *for* a *if* $l \cdot a = e$
(e) $r \in S$ *is called a* **right inverse** *for* a *if* $a \cdot r = e$*.*
(f) $b \in S$ *is called an* **inverse** *for* a *if it is both a left and a right inverse for* a*.*

2.3.1 *A Simple Example: Semigroups*

Definition 2.42. *Let* S *be a set and* $\circ$ *be a binary operation on* S*.*

(a) $(S, \circ)$ *is called* **groupoid***.*
(b) *A* **semigroup** *is a groupoid with* $\circ$ *an associative operation. A semigroup is called* **Abelian** *(or* **commutative***) if* $\circ$ *is a commutative operation.*

[1] ARTHUR CAYLEY(1821-1895), a British mathematician, was the first to define the concept of a group in the modern way, i.e. as a set with a binary operation satisfying certain laws.

Proposition 2.43. *If a semigroup S has a left identity l and a right identity r, then $l = r$ is a neutral element. There is only one neutral element in S.*

Proof. By definition of a left identity we have $lr = r$, and also $lr = l$ since r is right identity. Therefore, $l = r$ is a neutral element, say e. Now suppose, there exists another neutral element e'; then e' is both left and right identity and therefore $e = ee' = e'$. □

This proposition shows that a semigroup has at most one identity element which we will usually denote by 1, e or I, dependent on the context. As a consequence, exactly one of the following statements must hold for a groupoid S:

(1) S has no left and no right identity element.
(2) S has one or more left, but no right identity elements.
(3) S has one or more right, but no left identity elements.
(4) S has a unique two-sided, and no other right or left identity element.

Proposition 2.44. *If, in a semigroup S, x has a left inverse l and a right inverse r, then $l = r$ is an inverse x' for x. x' is the unique inverse for x in S.*

Proof. We obviuosly have $l = le = l(xr) = (lx)r = er = r$. Assume that both y and z are inverses of x; then y is a left inverse, and z is a right inverse, whence $y = z$ as we have just shown. □

2.3.2 *Algebras and Homomorphisms*

Semigroups are but a simple example of a general concept which will be introduced in this section. Let A be a nonempty set and $n \in \mathbb{N}$ a natural number, $n \geq 0$. Then A^n denotes the set of all n-tuples of elements of A; we set $A^0 = \{\emptyset\}$. A function $f : A^n \longrightarrow A$ is called an **n-ary operation on** A, and n is called the **arity** or **rank of an operation** of f. Operations of arity 0,1, and 2 are called *nullary, unary, and binary operations,* respectively. A nullary operation on A is a mapping $f : \{\emptyset\} \longrightarrow A$, which is completely determined by $f(\emptyset) \in A$, the image of the only element $\emptyset \in A^0$. Nullary operations may be identified with elements $a \in A$, and are often called **constants**: $\emptyset \mapsto f(\emptyset) = a_f$.

Definition 2.45.

(a) *A* **type** *(also* **similarity type** *or* **signature***) is a pair (Ω, θ), with $\Omega = \{f_i | i \in I\}$ a set of operation symbols and a mapping $\theta : I \longrightarrow \mathbb{N}$ indicating their ranks. Usually we will speak of θ as the type, leaving Ω implicit. Two algebras are said to be* **similar** *if and only if they have the same type.*
(b) *An* **algebra of type** θ *(also* **Ω-algebra***) is a pair $\mathfrak{A} = (A, \Omega^A)$ such that (1) A is a non-void set, and (2) $\Omega^A = \{f_i^A | f_i \in \Omega\}$ is a set of operations on A, and $f_i^A \in \Omega^A$ has arity $\theta(i)$ for every $i \in I$.*

If the type θ is clear from the context we will shortly speak of an algebra. Let $\mathfrak{A} = (A, \Omega^A)$ be an algebra; A is called the **carrier** of $\mathfrak{A}$, and $\mathfrak{A}$ is called a **finite algebra** if its carrier is a finite, and **trivial** (or **degenerate**) if it is a singleton set. If the indexing set

I is finite, we say that the type θ is finite and that $\mathfrak{A}$ is an **algebra of finite type**. For finite types the indexing set I we will usually be assumed to have the form $\{0,1,\dots,n-1\}$ and θ will be written as $\langle\theta(0),\theta(1),\dots,\theta(n-1)\rangle \in \mathbb{N}^n$. Algebras of finite type are usually denoted in a simplified manner as $(A, f_1, f_2, \dots, f_k)$, with the operations f_i usually ordered by decreasing rank or by structural aspects.

Roughly speaking, an algebra is to be thought of as a set with certain operations defined on it; thus, an algebra can also be said to be an **operational structure**. But note that, given any operation f on a set A with arity $n-1$, there is a naturally corresponding n-ary relation ρ on A consisting of those n-tuples $a_1, a_2, \dots, a_n$ for which $f(a_1, a_2, \dots, a_{n-1}) = a_n$.

The concept of an algebra as defined above includes most of the algebraic structures encountered in practice as you can see from the examples given later on. But there are some important exceptions like an ordered group, which is a structure that needs a relation for its description in addition to the usual group operations. For a convenient handling one needs to have relations as well as operations, or relations alone, since operations may be obtained as a special case as we have just argued. For sake of convenience we vote for the first possibility. The corresponding algebraic concept is that of a *relational structure*, which generalizes the notion of an algebra:

Definition 2.46. *Let $\Omega = \{f_i | i \in I\}$ and $\Sigma = \{r_j | j \in J\}$ be disjoint sets of operation and relation symbols, respectively. A* **type** *is a mapping $\theta : I \cup J \longrightarrow \mathbb{N}$. A* **relational structure of type** *θ is a triple $\mathfrak{R} = (A, \Omega^A, \Sigma^A)$ such that*

(1) A is a non-void set,
(2) $\Omega^A = \{f_i^A | f_i \in \Omega\}$ is a set of operations on A, and $f_i^A \in \Omega^A$ has arity $\theta(i)$ for every $i \in I$, and
(3) $\Sigma^A = \{r_i^A | r_i \in \Sigma\}$ is a set of relations on A such that for every relation $r_j^A \in \Sigma^A$ we have $r_j^A \subseteq A^{\theta(j)}$.

Depending on which of Ω^A or Σ^A, if any, is empty, we can distinguish three types of algebraic structures: we speak of *algebras* or *operational structures* if $\Sigma^A = \emptyset$, and of *relational structures* if $\Sigma^A \neq \emptyset$. In the latter case we say that $\mathfrak{R} = (A, \Omega^A, \Sigma^A)$ is a **hybrid structure** if neither $\Omega^A = \emptyset$ nor $\Sigma^A = \emptyset$. If the indexing sets I and J are finite, we say that the type θ is finite and that $\mathfrak{R}$ is an **relational structure of finite type**. For finite types the indexing sets I and J we will usually be assumed to have the forms $\{0,1,\dots,n-1\}$ and $\{n, n+1, \dots, n+k-1\}$, respectively. In this case θ will be written as $\langle\theta(0),\theta(1),\dots,\theta(n-1);\theta(n),\dots,\theta(n+k-1)\rangle \in \mathbb{N}^{n+k}$. Algebras of finite type are usually denoted in a simplified manner as $(A, f_1, f_2, \dots, f_n; r_1, r_2, \dots, r_k)$, with the f_i and r_j ordered by decreasing rank.

Definition 2.47. *Let $\mathfrak{A} = (A, \Omega^A)$ and $\mathfrak{B} = (B, \Omega^B)$ be two similar algebras. A function $h : A \longrightarrow B$ is called a* **homomorphism** *from $\mathfrak{A}$ to $\mathfrak{B}$, and is denoted by $h : \mathfrak{A} \longrightarrow \mathfrak{B}$, if*

$$h(f_i^A(a_1, a_2, \dots, a_n)) = f_i^B(h(a_1), \dots, h(a_n))$$

for any $n \in \mathbb{N}$, $i \in I$, and $(a_1, \dots, a_n) \in A^n$.

Lemma 2.48. [3, Theorem 6.5] *Let $\mathfrak{A}$, $\mathfrak{B}$, and $\mathfrak{C}$ be algebras; furthermore, suppose that $g : \mathfrak{A} \longrightarrow \mathfrak{B}$ and $h : \mathfrak{B} \longrightarrow \mathfrak{C}$ are homomorphisms. Then the composition $hg : \mathfrak{A} \longrightarrow \mathfrak{C}$ is a homomorphism too.*

A homomorphism $h : \mathfrak{A} \longrightarrow \mathfrak{B}$ which is bijective is called an **isomorphism**, and $\mathfrak{A}$ and $\mathfrak{B}$ are said to be *isomorphic* algebras, denoted as $\mathfrak{A} \cong \mathfrak{B}$. In mathematics, the notion of an isomorphism between two structures is fundamental, since it allows to prove that objects of different nature do obey identical laws, i.e. that they have the same properties. In other words, two isomorphic structures may be regarded identical up to the naming of their elements.

There are some other special cases of homomorphisms: a homomorphism is called a **monomorphism** if it is injective, and an **epimorphism** if it is surjective. A homomorphism $h : \mathfrak{A} \longrightarrow \mathfrak{A}$ is called **endomorphism**, and an **automorphism** if h is an isomorphism. Given two Ω-algebras $\mathfrak{A}$ and $\mathfrak{B}$, a monomorphism $h : \mathfrak{A} \longrightarrow \mathfrak{B}$, $\mathfrak{A}$ is called an **embedding** of $\mathfrak{A}$ into $\mathfrak{B}$. If there is an epimorphism from $\mathfrak{A}$ to $\mathfrak{B}$, then $\mathfrak{B}$ is called a **homomorphic image** of $\mathfrak{A}$.

Usually, compared to Def. 2.47, we can simplify notation because it is clear from the context, if the operation belongs to $\mathfrak{A}$ or to $\mathfrak{B}$. Look for the following

Example: Let h be a homomorphism between two groups $\mathfrak{A} = (A, \cdot, ^{-1}, e)$ and $\mathfrak{B} = (B, \cdot, ^{-1}, e)$, i.e. a function $h : A \longrightarrow B$ that satisfies:

(1) $h(e) = e$
(2) $h(x_1 \cdot x_2) = h(x_1) \cdot h(x_2)$, and
(3) $h(x_1^{-1}) = (h(x_1))^{-1}$.

(Note the simplification from $\cdot^A$ and $\cdot^B$ to $\cdot$; analogous for e^A and ${^{-1}}^A$.) Now, choose $x_2 = e$; then $h(x_1 \cdot x_2) = h(x_1) \cdot h(x_2)$ gives $h(x_1) = h(x_1) \cdot h(e)$. Multiplying both sides from the left with $h(x_1)^{-1}$ yields $h(x_1)^{-1} \cdot h(x_1) = e = h(x_1)^{-1} \cdot h(x_1) \cdot h(e) = h(e)$, thus $e = h(e)$. Choosing $x_2 = x_1^{-1}$ we get $e = h(e) = h(x_1) \cdot h(x_1^{-1})$, which implies $h(x_1^{-1}) = (h(x_1))^{-1}$. Thus, in case of groups, homomorphisms can be characterized by imposing the condition on the binary operation alone.

2.3.3 *Some more Examples of Algebraic Structures*

With the new concepts at hand, taking a short look back to Section 2.3.1 we see that a *groupoid* on a set A is an algebra of type $\langle 2 \rangle$, $\mathfrak{A} = (A, \cdot)$. Groupoids with an associative operation are called *semigroups*, and *monoids* are semigroups with an identity, i.e. algebras $\mathfrak{A} = (A, \cdot, e)$ of type $\langle 2, 0 \rangle$. Some familiar kinds of algebras, repeating semigroup definition for sake of completeness, are presented in

Definition 2.49.

(a) *A* **semigroup** *is an algebra $(A, \cdot)$ of type $\langle 2 \rangle$ such that for all $x, y, z \in A$*
  (G1) $x \cdot (y \cdot z) = (x \cdot y) \cdot z$ *(associativity).*

(b) *A* **monoid** *is a semigroup with identity* e*, i.e. an algebra* $(A, \cdot, e)$ *of type* $\langle 2, 0\rangle$ *such that for all* $x \in A$

(G2) $x \cdot e = e \cdot x = x$ *(identity, neutral element).*

(c) *A* **group** *is an algebra* $(A, \cdot, ^{-1}, e)$ *of type* $\langle 2, 1, 0\rangle$ *where* $(A, \cdot, e)$ *is a monoid in which for every* $x \in A$ *there is an element* x^{-1} *such that:*

(G3) $x \cdot x^{-1} = x^{-1} \cdot x = e$ *(inverse).*

(d) *A (semi)group is called* **Abelian** *(or* **commutative***) if for all* $x, y \in A$

(G4) $x \cdot y = y \cdot x$ *(commutativity).*

(e) *An element* $e \in A$ *is called* **idempotent** *if* $e^2 = ee = e$*. The subset of idempotents of a semigroup* A *is often denoted* E_A*.* A *is called an* **idempotent semigroup** *or* **band** *if* $E_A = A$*, and an* **orthodox semigroup** *if* E_A *is a subsemigroup of* A*.*

Examples: Let $\mathbb{N}^+$ denote the set of positive integers; then both $(\mathbb{N}^+, +)$ and $(\mathbb{N}^+, \cdot)$ are commutative semigroups. The set $\mathbb{N}$ of nonnegative integers is a commutative monoid for both addition and multiplication, with 0, respectively 1, as neutral element. Also $(\mathcal{P}(A), \cup)$ is a commutative monoid with $\emptyset$ as neutral element. And the set $\mathbb{Z}$ of integers is a commutative group for addition, with 0 as identity. In the foregoing examples the semigroup operation is denoted as addition '+' or as multiplication '·'; in the first case, an identity is called the **zero**, and in the second case the **one** or **multiplicative unit** of the semigroup.

Note that any semigroup can be extended to a monoid by what is called the "the adjunction of an identity element to S" ([4, p. 4]): Let S be any semigroup, and let 1 be a symbol not in S. In $S \cup \{1\}$ we define multiplication by $11 = 1$ and $1a = a1 = a$ for every $a \in S$. Then $S \cup \{1\}$ becomes a monoid. It is common to use the following notation:

$$S^1 = \begin{cases} S & \text{if } S \text{ has an identity element,} \\ S \cup \{1\} & \text{otherwise.} \end{cases}$$

Definition 2.50.

(a) *A* **semiring** *is an algebra* $(R, +, \cdot, 0, 1)$ *of type* $\langle 2, 2, 0, 0\rangle$ *such that*

(SR1) $(R, +, 0)$ *is an Abelian monoid with neutral element* 0*,*

(SR2) $(R, \cdot, 1)$ *is a monoid with neutral element* 1*,*

(SR3) $x \cdot (y + z) = (x \cdot y) + (x \cdot z)$*,*
$(x + y) \cdot z = (x \cdot z) + (y \cdot z)$*, for all* $x, y, z \in R$*.*

(SR4) $x0 = 0x = 0$*, for all* $x \in R$

(b) *A* **ring** *is an algebra* $(R, +, \cdot, -, 0, 1)$ *of type* $\langle 2, 2, 1, 0, 0\rangle$ *such that*

(R1) $(R, +, -, 0)$ *is an Abelian group,*

(R2) $(R, \cdot, 1)$ *is a monoid with neutral element* 1*,*

(R3) $x \cdot (y + z) = (x \cdot y) + (x \cdot z)$*,*
$(x + y) \cdot z = (x \cdot z) + (y \cdot z)$*, for all* $x, y, z \in R$*.*

A semiring is called **commutative** if $(R, \cdot)$ is an Abelian semigroup, and **additively idempotent** if $x + x = x$ for all $x \in R$. If $x + y = 0$ implies $x = y = 0$ for all $x, y \in R$ then S is called an **antiring** or **zerosumfree**. A ring is a semiring where $(R,+)$ is an Abelian group.

Unfortunately, terminology is not uniform: Sometimes, the identity elements are excluded from the definition of a semiring, i.e. only semigroup properties are required in (SR1) and (SR2). So one has to be careful with respect to such subtleties. Note that in a semiring as defined above, i.e. implicitly stating that $0 \neq 1$, (SR4) follows from the other requirements: we have $0x + x = (0+1)x = 1x = x$, whence $0x = 0$ by Prop. 2.43, for all $x \in R$.

Examples: Clearly, any ring is a semiring. $(\mathbb{N},+,\cdot)$ and $(\mathbb{R}^+,+,\cdot)$ are semirings, both with zero element 0 and multiplicative unit 1. Any Boolean algebra, and also the fuzzy algebra $([0,1],+,\cdot)$ under the operations $a+b = \sup\{a,b\}, ab = \inf\{a,b\}$ are semirings with $0,1$. The simplest example of a semiring which is not a ring is the Boolean semiring $\mathbf{B} = (\{0,1\},+,\cdot)$ with addition defined as $1+0 = 0+1 = 1+1 = 1$ and $0+0 = 0$; and with multiplication given by $1\cdot 0 = 0\cdot 1 = 0\cdot 0 = 0$ and $1\cdot 1 = 1$. Boolean and fuzzy algebras are commutative, additively idempotent, and antirings. The prototype example for a ring is $(\mathbb{Z},+,\cdot,-,0)$, where $-: \mathbb{Z} \longrightarrow \mathbb{Z},\ z \mapsto -z$.

These examples are simple and familiar ones. Of course, in each case there are many other and more sophisticated examples, as you can easily find out by looking at pertinent sources. To name three: (1) The set of all ideals of a given ring forms a semiring under addition and multiplication of ideals. (2) The square $n \times n$ matrices with non-negative entries form a semiring under ordinary addition and multiplication of matrices. (3) The set of all $n \times n$ matrices over an arbitrary ring R form a ring which is usually referred to as the "*full ring of n-by-n matrices*"

2.3.4 *Partial Orders and Lattices*

This section is devoted to the presentation of basic notions and results around the concept of an order relation. In this chapter, providing the mathematical preliminaries, this has to be done on the metalanguage level. As you will see later on, the stuff included here is indispensable for the further development of the subject matter, especially in part III. Moreover, the contents of this section are usually regarded to belong to lattice and order theory rather than to a theory of relations. You only have to look at e.g. the books by Blyth [1] or Schröder [23] which are recommended for further reading.

Definition 2.51.

(a) *A* **partially ordered set** *(or* **poset***) is a relational structure* $\mathfrak{P} = (A,\leq)$ *of type* $\langle 2\rangle$ *such that for all* $a,b,c \in A$*:*

(PO1)	$a \leq a$	*(reflexive)*
(PO2)	$a \leq b \wedge b \leq a \rightarrow a = b$	*(antisymmetric)*
(PO3)	$a \leq b \wedge b \leq c \rightarrow a \leq c$	*(transitive)*

$\leq$ *is called a* **partial order** *on* A.

Two elements $a,b \in A$ *are called* **comparable** *if* $a \leq b$ *or* $b \leq a$*, denoted by* $a \asymp b$*, and* **incomparable** *otherwise, denoted by* $a \not\asymp b$.

(b) *A mapping* $f : A \longrightarrow B$ *is said to be*

- **isotone** *iff* $\forall x, y \in A[x \leq y \rightarrow f(x) \leq f(y)]$;
- **antitone** *iff* $\forall x, y \in A[x \leq y \rightarrow f(x) \geq f(y)]$.

(c) *A poset A is called a* **chain** *(or* **linearly ordered set***) iff* $\forall a, b \in A[a \asymp b]$. *A subset* $B \subseteq A$ *of a poset A with pairwise incomparable elements, i.e.* $\forall a, b \in B[a \not\asymp b]$, *is called an* **antichain**.

(A warning: Terminology in literature is not consistent. The unspecified term '*order*' is sometimes used for our notion of a partial order and sometimes for the notion of a total order.) For poset examples see Section 4.2.

Some more notions need to be mentioned in connection with partial orders: Let $(A, \leq)$ be a poset. We shall use $x < y$ as abbreviation for $x \leq y \wedge x \neq y$. Note that the relation $<$ thus defined is transitive but is not an order since it fails to be reflexive; moreover, $x < y$ and $y < x$ are incompatible, i.e. $x \neq y$. Nonetheless, $<$ is called a **strict partial order**. We say that y **covers** x in $(A, \leq)$, denoted as $x \prec y$, if $x < y$ and $x \leq z \leq y \rightarrow z = x \vee z = y$, i.e. there is no z in between x and y.

Definition 2.52. *Let* $(A, \leq)$ *be a poset, and* $B \subseteq A$.

(a) $a \in A$ *is called a* **lower bound** *of B in A if* $\forall b \in B[a \leq b]$, *and an* **upper bound** *of B in A if* $\forall b \in B[b \leq a]$.

(b) *B is said to be* **bounded from below** *if there exists a lower bound, and* **bounded from above** *if there exists a upper bound of B in A. If B is bounded from both below and above, it is called* **bounded**.

(c) *If for every two-element set* $\{a, b\} \subseteq B$ *there exists a lower (respectively upper) bound of* $\{a, b\}$, *then* $(A, \leq)$ *is called* **directed from below** *(respectively* **directed from above***).* $(A, \leq)$ *is* **directed** *if it is directed from below and from above.*

(d) *The* **infimum**, *or* **greatest lower bound** *(g.l.b.), of B (when it exists) is a lower bound* $a \in A$ *with* $a \geq x$ *for all lower bounds x of B; it is denoted by* $\inf B$. *Dually, if it exists, the* **supremum**, *or* **least upper bound** *(l.u.b.) of B is an upper bound* $a \in A$ *such that* $a \leq x$ *for all upper bounds x of B; it is denoted by* $\sup B$.

(e) *If* $a = \inf B \in B$ *then a is called the* **least element** *or* **minimum** *of B; dually,* $a \in B$ *is the* **greatest element** *or* **maximum** *of B if* $\forall x \in B[a \geq x]$.

(f) *An element* $m \in A$ *is said to be* **minimal** *if there is no* $a \in A$ *with* $a < m$, *i.e.* $\forall a \in A[a \leq m \rightarrow a = m]$, *and* **maximal** *if* $\forall a \in A[m \leq a \rightarrow a = m]$.

Obviously, a maximum, if it exists, is uniquely determined: Suppose that both a and b are a maximum of A, then both $a \leq b$ and $b \leq a$, so $a = b$ by the antisymmetry of $\leq$. An analogous result holds for the minimum. It is customary to denote maximum and minimum by 1 and 0, respectively.

Having defined the notion of a poset, our next step is to introduce semilattices and lattices. Both of them have two forms of existence which, however, stand in close relationship as will be shown below: The first form arises from a pure structural view as a special algebra, the other as a special kind of poset as described in the two definitions below.

Definition 2.53.

(a) *An algebra* $(A, \cdot)$ *with a binary operation* $\cdot$ *on* A *is called* **semilattice** *iff* $\cdot$ *is idempotent, commutative, and associative.*

(b) *An algebraic structure* $(L, +, \cdot)$ *of type* $\langle 2, 2\rangle$ *is called* **lattice** *iff*

(AL1) *both* $(L, +)$ *and* $(L, \cdot)$ *are semilattices,*

(AL2) *the* **absorption laws** $a + (a \cdot b) = a = a \cdot (a + b)$ *hold.*

Definition 2.54.

(a) *A poset* $(L, \leq)$ *in which* $\inf\{a, b\}$ *exists for all* $a, b \in L$ *is called* **meet semilattice***; this element is denoted by* $a \wedge b$ *or* $a \cdot b$ *and called the meet of* a *and* b.

(b) *A poset* $(L, \leq)$ *in which* $\sup\{a, b\}$ *exists for all* $a, b \in L$ *is called* **join semilattice***; this element is denoted by* $a \vee b$ *or* $a + b$ *and called the join of* a *and* b.

(c) *A* **lattice** *is a poset* $(L, \leq)$ *which is both a meet and join semilattice, i.e. an algebra* $(L, \vee, \wedge)$ *of type* $\langle 2, 2\rangle$.

(d) *A* **distributive lattice** *is a lattice which satisfies*

(D1) $x \wedge (y \vee z) = (x \wedge y) \vee (x \wedge z)$

(D2) $x \vee (y \wedge z) = (x \vee y) \wedge (x \vee z)$

(e) L *is called* **complete lattice** *if every* $S \subseteq L$ *has both a supremum and an infimum.*

Note that the duality principle for posets implies a duality principle for lattice theory, which implies the exchange of $\wedge$ by $\vee$ and conversely; see Section 7.2.1.1.

Examples: A familiar example of a meet semilattice is $(\mathcal{P}(X), \cap)$, or more generally any collection of subsets of X closed under intersection, and the prototype example for a (distributive) lattice is $(\mathcal{P}(X), \cup, \cap)$.

Between semilattices and partial orders there exist important relationships, which are presented below.

Theorem 2.55.

(1) *Let* $(S, \cdot)$ *be a semilattice; the relation* $\leq$ *on* S*, defined by* $x \leq y$ *iff* $x \cdot y = x$ *for* $x, y \in S$*, is a partial order on* S.

(2) *Moreover, in* $(S, \leq)$ $\inf\{s, t\}$ *exists for all* $s, t \in S$ *and* $s \cdot t = \inf\{s, t\}$.

Proof. (1): Since $\cdot$ is idempotent we get $x = x \cdot x$, thus $x \leq x$. And if $x \leq y \wedge y \leq x$, then $x \cdot y = x \wedge y \cdot x = y$, whence $x = y$ since $\cdot$ is commutative. If $x \leq y \wedge y \leq z$ then $x \cdot y = x \wedge y \cdot z = y$, so we have $x \cdot z = (x \cdot y) \cdot z = x \cdot (y \cdot z) = x \cdot y = x$, i.e. $x \leq z$.

(2) Let $s, t \in S$. First we show that $s \cdot t \leq s$ and $s \cdot t \leq t$: We have $(s \cdot t) \cdot s = s \cdot (s \cdot t) = (s \cdot s) \cdot t = s \cdot t$ and $(s \cdot t) \cdot t = s \cdot (t \cdot t) = s \cdot t$. Thus $s \cdot t$ is a lower bound of $\{s, t\}$. Now suppose that b is an arbitrary lower bound of $\{s, t\}$, i.e. $b \cdot s = b$ and $b \cdot t = b$. We have $b \cdot (s \cdot t) = (b \cdot s) \cdot t = b \cdot t = b$, whence $b \leq s \cdot t$. □

Proposition 2.56.

(1) *Let* $(S, \leq)$ *be a partially ordered set such that* $\inf\{s, t\}$ *exists for all* $s, t \in S$. *If* $\cdot$ *is the operation defined by* $s \cdot t = \inf\{s, t\}$ *for* $s, t \in S$, *then* $(S, \cdot)$ *is a semilattice.*
(2) *Let* $(S, \leq)$ *be a partially ordered set such that* $\sup\{s, t\}$ *exists for all* $s, t \in S$. *If* $+$ *is the operation defined by* $s + t = \sup\{s, t\}$ *for* $s, t \in S$, *then* $(S, +)$ *is a semilattice.*

Since the infimum and supremum are uniquely determined if they exist, we can define two pairs of functions $\wedge, \vee : S \times S \to S$ or $\cdot, + : S \times S \to S$ such that for all $a, b \in S$: $\wedge(a, b) = \cdot(a, b) = \inf\{a, b\}$ and $\vee(a, b) = +(a, b) = \sup\{a, b\}$. Using an infix notation for these functions we arrive at the familiar forms $a \wedge b$ and $a \vee b$, or $a \cdot b$ and $a + b$, respectively.

(There are several reasons for introducing two different notations for the same thing: You can find both of them in existing literature and should therefore get accustomed to both; the $\cdot/+$-combination is the one commonly used in the context of relation algebras. And sometimes there is the practical need to typographically distinguish between the logical and '$\wedge$' and the order theoretical meet '$\wedge$'. Nonetheless, both notations will still be used in a generic sense also, but it should always be clear from the context which is the case. If appropriate for better readability, we will alternatively write $\&$ for the logical and, and $\|$ for the logical or in the context of lattices and orders.)

Proposition 2.57. *A meet semilattice* $(S, \wedge)$ *is a commutative and idempotent semigroup, hence a semilattice i.e. the following identities hold:*

(SL1) $x \wedge (y \wedge z) = (x \wedge y) \wedge z$ *(associative).*
(SL2) $x \wedge x = x$ *(idempotent)*
(SL3) $x \wedge y = y \wedge x$ *(commutative)*

Proof. We give only some parts of the proofs, since they are easy applications of the definitions and properties of logical operations:
(SL1): Let $p = a \wedge (b \wedge c)$ & $q = (a \wedge b) \wedge c$; then by definition of $\wedge$

$$p \leq a \;\&\; p \leq b \wedge c.$$

Thus

$$p \leq a \;\&\; (p \leq b \;\&\; p \leq c),$$

and by the associativity of logical conjunction we get

$$(p \leq a \;\&\; p \leq b) \;\&\; p \leq c,$$

i.e. $$p \leq a \;\&\; b \wedge p \leq c, \text{ whence } p \leq (a \wedge b) \wedge c = q.$$

A similar argument shows $q \leq p$.
(SL2): Clearly $a \wedge a = \inf\{a, a\} = a$.
(SL3): Let $p = a \wedge b$ & $q = b \wedge a$; then by definition of $\wedge$ we have $p \leq a$ & $p \leq b$, whence $p \leq b$ & $p \leq a$, i.e. $p \leq b$ & $a = q$. A similar argument shows $q \leq p$. □

Theorem 2.58. *Let $(L, \vee, \wedge)$ be a lattice according to Def. 2.54(c); then it is also lattice according to Def. 2.53(b).*

Proof. From Prop. 2.57 we get that $(L, \wedge)$ and $(L, \vee)$ are semilattices according to Def. 2.53(a). So we have to show that the absorption laws $a \vee (a \wedge b) = a = a \wedge (a \vee b)$ hold. By definition of $\wedge$ we have $a \wedge (a \vee b) \le a$. On the other hand, $a \le a$ and $a \le a \vee b$, whence $a \le a \wedge (a \vee b)$. □

Proposition 2.59. *In any lattice $(L, +, \cdot)$ the following holds:*

(L1) $a + (a \cdot b) = a \wedge a \cdot (a + b) = a.$ *(absorption)*
(L2) $a \le b \leftrightarrow a \cdot b = a \leftrightarrow a + b = b.$
(L3) $a \cdot 0 = 0,\ a + 0 = a,\ a \cdot 1 = a,$ *and* $a + 1 = 1.$
(L4) $a \le c \wedge b \le d \rightarrow a \cdot b \le c \cdot d \wedge a + b \le c + d.$

Proof. (L1): By Th. 2.58.

(L2): If $a \le b$ then a fulfills exactly the constraints defining $a \cdot b$: $a \le a \wedge a \le b$, and $c \le a \wedge c \le b \rightarrow c \le a$; thus $a \cdot b = a$. Conversely, $a = a \cdot b \le b \rightarrow a \le b$. For the second equivalence we have: $a + b = b$ and (L1) imply $a \cdot b = a$, and assuming $a \cdot b = a$ we get $b = b + (b \cdot a) = b + (a \cdot b) = b + a$ by (L1).

(L3): Apply (L2).

(L4): By (L2) we get $a + c = c \wedge b + d = d \wedge a \cdot c = a \wedge b \cdot d = b$; then

$$(a \cdot b)(c \cdot d) = (a \cdot c)(b \cdot d) = a \cdot b \quad \text{and}$$
$$(a + b) + (c + d) = (a + c) + (b + d) = c + d.$$

□

Proposition 2.60. *In any lattice $(L, \wedge, \vee)$ the following holds for all $a, b, c \in L$:*

(L5) $a \wedge (b \vee c) \ge (a \wedge b) \vee (a \wedge c)$ *(distributive inequalities)*
(L5') $a \vee (b \wedge c) \le (a \vee b) \wedge (a \vee c).$

Proof. (L5): Clearly $a \wedge b \le a$ and $a \wedge b \le b \le b \vee c$; hence $a \wedge b \le a \wedge (b \vee c)$. We also have $a \wedge c \le a$ and $a \wedge c \le c \le b \vee c$; hence $a \wedge c \le a \wedge (b \vee c)$. Thus, $a \wedge (b \vee c)$ is an upper bound of $a \wedge b$ and $a \wedge c$, and we have shown (L5). (L5') follows from (L5) by duality. □

Proposition 2.61. *In any lattice $(L, \wedge, \vee)$ the following holds for all $a, b, c \in L$:*

(L6) $a \wedge (b \vee c) = (a \wedge b) \vee (a \wedge c)$ *iff*
(L6') $a \vee (b \wedge c) = (a \vee b) \wedge (a \vee c).$

Proof. We show that (L6) implies (L6'): For any $a, b, c \in L$ we have

$$\begin{aligned}
(a \vee b) \wedge (a \vee c) &= [(a \vee b) \wedge a] \vee [(a \vee b) \wedge c] && \text{by (L6)}\\
&= a \vee [c \wedge (a \vee b)] && \text{by absorption, commutativity}\\
&= a \vee [(c \wedge a) \vee (c \wedge b)] && \text{by (L6)}\\
&= [a \vee (c \wedge a)] \vee (c \wedge b) && \text{by associativity}\\
&= a \vee (b \wedge c) && \text{by absorption, commutativity.}
\end{aligned}$$

□

Definition 2.62. *Let L be lattice.*

(a) *A* **bounded lattice** *is a structure* $(L, \vee, \wedge, 0, 1)$ *of type* $\langle 2, 2, 0, 0\rangle$ *such that* 0 *is the least and* 1 *the greatest element in L, i.e.* $\forall x \in L[0 \leq x \leq 1]$.

(b) *Let L be bounded lattice.* $x \in L$ *is called* complemented *if there is an element* $y \in L$ *such that* $x \wedge y = 0$ *and* $x \vee y = 1$. *Every element y with this property is called a* **complement** *of x. The set of complemented elements in L is denoted by* C_L. *L is called* **complemented** *if* $C_L = L$, *i.e. every* $x \in L$ *is complemented.*

(c) *An* **interval** *of a lattice L is a subset of L of the form* $[a, b] = \{x : a \leq x \leq b\}$, $a, b \in L$ *with* $a \leq b$. *Let* $a \in [b, c]$; *x is a* **relative complement** *of a in* $[b, c]$ *if* $a \wedge x = b$ *and* $a \vee x = c$. *L is* **relatively complemented** *if every interval* $[x, y]$ *of L is complemented.*

Note that if y is a complement of x then x is also a complement of y. An element may have several complements, a single one, or none.

Proposition 2.63. *For a bounded distributive lattice* $(L, \vee, \wedge, 0, 1)$ *the following holds:*

(1) *Every* $x \in L$ *has at most one complement denoted by* $\overline{x}$.
(2) $\{0, 1\} \subseteq C_L$.
(3) $\{x, y\} \subseteq C_L \rightarrow \{x \wedge y, x \vee y, \overline{x}\} \subseteq C_L$: *We have* $\overline{x \wedge y} = \overline{x} \vee \overline{y}$, $\overline{x \vee y} = \overline{x} \wedge \overline{y}$, *and* $\overline{\overline{x}} = x$.

Proof. (1) Let both y and y' be complements of x. Then

$$\begin{aligned} y &= y \wedge 1 = y \wedge (x \vee y') = (y \wedge x) \vee (y \wedge y') \\ &= 0 \vee (y \wedge y') = (x \wedge y') \vee (y \wedge y') = (x \vee y) \wedge y' = 1 \wedge y' \\ &= y'. \end{aligned}$$

(3) Using distributivity we get

$$\begin{aligned} (x \wedge y) \wedge (\overline{x} \vee \overline{y}) &= (x \wedge y \wedge \overline{x}) \vee (x \wedge y \wedge \overline{y}) \\ &= (0 \wedge y) \vee (x \wedge 0) = (0 \vee 0) = 0, \text{ and} \end{aligned}$$

$$\begin{aligned} (x \wedge y) \vee (\overline{x} \vee \overline{y}) &= (x \vee \overline{x} \vee \overline{y}) \wedge (y \vee \overline{x} \vee \overline{y}) \\ &= (1 \vee \overline{y}) \wedge (1 \vee \overline{x}) = (1 \wedge 1) = 1. \end{aligned}$$

This implies $\overline{x \wedge y} = \overline{x} \vee \overline{y}$; the second equation follows by duality. And since $\overline{x} \wedge x = 0$ and $\overline{x} \vee x = 1$, x is the complement of $\overline{x}$, i.e. $\overline{\overline{x}} = x$. □

Proposition 2.64. *In a bounded distributive lattice* $(L, \vee, \wedge, 0, 1)$ *the following properties hold for every* $x \in C_L$ *and* $y \in L$:

(1) $x \wedge (\overline{x} \vee y) = x \wedge y$,
(2) $x \vee (\overline{x} \wedge y) = x \vee y$,
(3) $x \geq y \leftrightarrow \overline{x} \wedge y = 0$.
(4) $x \leq y \leftrightarrow \overline{x} \vee y = 1$.

Proof. (1) $x \wedge (\overline{x} \vee y) = (x \wedge \overline{x}) \vee (x \wedge y) = x \wedge y$.

(3) $x \geq y \to 0 = \overline{x} \wedge x \geq \overline{x} \wedge y$. Conversely, $\overline{x} \wedge y = 0 \to y = (x \vee \overline{x}) \wedge y = (x \wedge y) \vee (\overline{x} \wedge y) = x \wedge y$. □

2.3.5 *Boolean Algebras*

Boolean algebras will play a major role later on in part IV dealing with the abstract algebraic aspects of relations: *Relation algebras* are Boolean algebras with additional operators and properties. So there is need to provide some relevant concepts and results (see e.g. [9, 16]).

Definition 2.65.

(a) *An algebra* $(L, \vee, \wedge, \eta, 0, 1)$ *of type* $\langle 2, 2, 1, 0, 0 \rangle$ *is called a* **De Morgan algebra** *if* $(L, \vee, \wedge, 0, 1)$ *is a bounded distributive lattice and* η *fulfills*

(DM1) $\eta(\eta(x)) = x$,
(DM2) $\eta(x \wedge y) = \eta x \vee \eta y$,
(DM3) $\eta(x \vee y) = \eta x \wedge \eta y$.

(b) *A* **Boolean algebra** *is an algebra* $\mathfrak{B} = (B, \vee, \wedge, \overline{}, 0, 1)$ *such that*

(B1) $\mathfrak{B}$ *is a De Morgan algebra, and*
(B2) $L = C_L$, *i.e. every element is complemented.*

Examples: The most common example for a Boolean algebra, and the motivating source for the algebraic abstraction, is of course $(\mathcal{P}(X), \cup, \cap, \overline{}_X, \emptyset, X)$, with $\overline{}_X$ denoting the set theoretical complement of $A \subseteq X$ with respect to X. Another well-known example is $\mathbf{2} = (\{0,1\}, \vee, \wedge, \neg, 0, 1)$,with $\vee, \wedge$ denoting max, min, $\neg 1 = 0$ and $\neg 0 = 1$. **2** is called the **binary Boolean algebra**, and it is the simplest non-trivial Boolean algebra, i.e. having more than one element. By Prop. 2.63, the set C_L of complemented elements in a bounded distributive lattice is a Boolean algebra.

Another example of a Boolean algebra is a tuple $\mathscr{F} = \langle F, \emptyset, X, \cup, \overline{} \rangle$, where $\emptyset \subset F \subseteq \mathbf{2}^X$ such that $S, T \in F \to (S \cup T) \in F$ and $\overline{S} \in F$, called a **field of sets**. A field of sets is thus closed under finite unions and under complementation relative to X. It also follows that $\emptyset, X \in F$ and that F is also closed under finite intersections.

Proposition 2.66. *The following identities are satisfied in every Boolean algebra* $\mathfrak{B} = (B, +, \cdot, \overline{}, 0, 1)$ *for arbitrary elements* $x, y \in B$:

(1) $x = \overline{\overline{x}}$.
(2) $\overline{x+y} = \overline{x} \cdot \overline{y}$ *and* $\overline{x \cdot y} = \overline{x} + \overline{y}$. *(De Morgan Laws)*
(3) $\overline{1} = 0$, *and* $\overline{0} = 1$.
(4) $x \cdot y = \overline{\overline{x} + \overline{y}}$.
(5) $x = \overline{\overline{x} + y} + \overline{\overline{x} + \overline{y}}$. *(Huntington's axiom)*

Proof. (1)-(3): By Prop. 2.63.
(4): From Prop. 2.63 we get $\overline{x \cdot y} = \overline{x} + \overline{y}$ and $x \cdot y = \overline{\overline{x \cdot y}} = \overline{\overline{x} + \overline{y}}$.

(5): $\overline{\overline{x}+y}+\overline{\overline{x}+\overline{y}} = x\cdot\overline{y}+x\cdot y = x\cdot(y+\overline{y}) = x$, using (4) and distributivity. □

Proposition 2.67. *Let $\mathfrak{B}$ be a Boolean algebra; then the following holds:*

(1) *The following statements are equivalent for all $x, y \in B$:*
  (a) $x \le y$,
  (b) $\overline{y} \le \overline{x}$,
  (c) $x + y = y$,
  (d) $x \cdot y = x$,
  (e) $\overline{x} + y = 1$,
  (f) $x \cdot \overline{y} = 0$.

(2) $\forall x, y, z \in B[x\cdot y \le z \;\leftrightarrow\; y \le \overline{x}+z]$.

Proof. (1) (a), (c) and (d) are equivalent by Prop. 2.59(L2). Using Prop. 2.64(3) we get (f) equivalent to (b): $x\cdot\overline{y} = 0 \leftrightarrow \overline{x} \ge \overline{y} \leftrightarrow \overline{y} \le \overline{x}$; and (f) equivalent to (a): $0 = x\cdot\overline{y} = \overline{y}\cdot x \leftrightarrow y \ge x \leftrightarrow x \le y$. Finally, by Prop. 2.64(4) (a) is equivalent to (e).

(2)		
	$x\cdot y \le z$	
$\leftrightarrow$	$x\cdot y\cdot\overline{z} = 0$	by Prop. 2.64(3)
$\leftrightarrow$	$y\cdot(x\cdot\overline{z}) = 0$	by associativity of $\cdot$
$\leftrightarrow$	$y\cdot(\overline{\overline{x}+\overline{\overline{z}}}) = 0$	by Prop. 2.66(4)
$\leftrightarrow$	$y\cdot(\overline{\overline{x}+z}) = 0$	by Prop. 2.66(1)
$\leftrightarrow$	$y \le \overline{x}+z$	by Prop. 2.64(3)

□

2.3.5.1 *Boolean Rings*

(This and the following section will not become relevant until we deal with relation algebras in Section 3.6. So they can be skipped at first reading.)

In Section 10.3.2.4, we will need the concept of a **Boolean ring**, which is, by definition, a ring in which every element is idempotent, i.e. $bb = b$. So writing down all the constraints explicitly we get

Definition 2.68. *A* **Boolean ring** *is an algebra $(B, +, \cdot, -, 0, 1)$ of type $\langle 2,2,1,0,0\rangle$ such that for all $a, b, c \in B$*

(1) $a + b = b + a$
(2) $(a+b)+c = a+(b+c)$
(3) $\exists 0 \in B\, \forall a \in B[a+0 = a]$
(4) $\forall a \in B\, \exists -a \in B\, [a+(-a) = 0]$
(5) $aa = a$
(6) $(ab)c = a(bc)$
(7) $(a+b)c = ac+bc$ *and* $c(a+b) = ca+cb$
(8) $\exists 1 \in B\, \forall a \in B[1a = a1 = a]$

Likewise, we list the properties of a Boolean algebra, which has been introduced above in Section 2.3.4 by building it on top of more simpler structures by gradually extending the set of requirements. So, for sake of convenience, it makes sense to list them as a whole in

Definition 2.69. *A* **Boolean algebra** *is an algebra* $(B, \vee, \wedge, \bar{\ }, 0, 1)$ *of type* $\langle 2,2,1,0,0\rangle$ *such that for all* $a, b, c \in B$

(1) $\overline{0} = 1, \quad \overline{1} = 0,$
(2) $a \wedge 0 = 0, \quad a \vee 1 = 1,$
(3) $a \wedge 1 = a, \quad a \vee 0 = a,$
(4) $a \wedge \overline{a} = 0, \quad a \vee \overline{a} = 1,$
(5) $\overline{\overline{a}} = a$
(6) $a \wedge a = a, \quad a \vee a = a,$
(7) $\overline{(a \wedge b)} = \overline{a} \vee \overline{b}, \quad \overline{(a \vee b)} = \overline{a} \wedge \overline{b},$
(8) $a \wedge b = b \wedge a \quad a \vee b = b \vee a$
(9) $(a \wedge b) \wedge c = a \wedge (b \wedge c) \quad (a \vee b) \vee c = a \vee (b \vee c)$
(10) $a \wedge (b \vee c) = (a \wedge b) \vee (a \wedge c), \quad a \vee (b \wedge c) = (a \vee b) \wedge (a \vee c).$

The above set of axioms is larger than necessary, but fine for doing calculations. A much more parsimonious characterization is given by the requirements (R1) – (R3) of Def. 10.29.

Lemma 2.70. *In a Boolean ring,* $a + a = 0$ *and* $ab = ba$*, for all* $a, b \in B$.

Proof. We have $a + b = (a + b)^2 = a^2 + ab + ba + b^2 = a + b + ab + ba$, whence $ab + ba = 0$ (since $(R, +, -, 0)$ is an Abelian group by Def. 2.50(b), and in a group G there is a unique $c \in G$ such that $a + c = b$ for all $a, b \in G$). Putting $b = a$ in the last equation, we get $a + a = 0$. Furthermore we have $ab + ab = 0$; thus $ab + ab = ab + ba$, whence $ab = ba$. □

In the next theorem the precise relationship between Boolean algebras and Boolean rings is explained. The results go back to [24], a 1936 paper of Stone, our presentation below follows [2, 3]. The proofs basically are carried out by calculations, for which we include some typical examples.

Theorem 2.71. [3, p. 123, Theorem 2.3]

(1) *Let* $\mathfrak{B} = (B, \vee, \wedge, \bar{\ }, 0, 1)$ *be a Boolean algebra. Then* $\mathfrak{B}^* = (B, +, \cdot, -, 0, 1)$ *with*

$$\begin{aligned} a + b &= (a \wedge \overline{b}) \vee (\overline{a} \wedge b) \\ a \cdot b &= a \wedge b \\ -a &= a \end{aligned}$$

is a Boolean ring.

(2) *Let* $\mathfrak{R} = (R, +, \cdot, -, 0, 1)$ *be a Boolean ring. Then* $\mathfrak{R}^* = (R, \vee, \wedge, \bar{\ }, 0, 1)$ *with*

$$\begin{aligned} a \vee b &= a + b + a \cdot b \\ a \wedge b &= a \cdot b \\ \overline{a} &= 1 + a \end{aligned}$$

is a Boolean algebra.

(3) With $\mathfrak{B}$ *and* $\mathfrak{R}$ *as defined above,* $(\mathfrak{B}^*)^* = \mathfrak{B}$ *and* $(\mathfrak{R}^*)^* = \mathfrak{R}$.

Proof. (1): We have to show that conditions (a) - (h) of Def. 2.68 are valid in $\mathfrak{B}^*$.

(a):
$$\begin{aligned} a+b &= (a \wedge \overline{b}) \vee (\overline{a} \wedge b) \\ &= (b \wedge \overline{a}) \vee (\overline{b} \wedge a) = b+a \end{aligned}$$

(b):
$$\begin{aligned} a+(b+c) &= \left[a \wedge \overline{(b+c)}\right] \vee \left[\overline{a} \wedge (b+c)\right] \\ &= \left[a \wedge \overline{\left((b \wedge \overline{c}) \vee (\overline{b} \wedge c)\right)}\right] \vee \left[(\overline{a} \wedge \left((b \wedge \overline{c}) \vee (\overline{b} \wedge c)\right)\right] \\ &= \left[a \wedge \left((\overline{b} \vee c) \wedge (b \vee \overline{c})\right)\right] \vee \left[(\overline{a} \wedge b \wedge \overline{c}) \vee (\overline{a} \wedge \overline{b} \wedge c)\right] \\ &= \left[a \wedge \left((\overline{b} \wedge \overline{c}) \vee (c \wedge b)\right)\right] \vee \left[(\overline{a} \wedge b \wedge \overline{c}) \vee (\overline{a} \wedge \overline{b} \wedge c)\right] \\ &= (a \wedge \overline{b} \wedge \overline{c}) \vee (a \wedge b \wedge c) \vee (\overline{a} \wedge b \wedge \overline{c}) \vee (\overline{a} \wedge \overline{b} \wedge c) \\ &= (a \wedge b \wedge c) \vee (a \wedge \overline{b} \wedge \overline{c}) \vee (b \wedge \overline{c} \wedge \overline{a}) \vee (c \wedge \overline{a} \wedge \overline{b}) \end{aligned}$$

Since the last expression is symmetric in a, b, and c, i.e. it has the same value however we permute them, we get $c+(a+b) = a+(b+c)$. Therefore, by (a), $a+(b+c) = (a+b)+c$.

(c):
$$a+0 = (a \wedge 1) \vee (0 \wedge \overline{a}) = a \vee 0 = a$$

(g):
$$\begin{aligned} a \cdot (b+c) &= a \wedge [(b \wedge \overline{c}) \vee (\overline{b} \wedge c)] \\ &= (a \wedge b \wedge \overline{c}) \vee (a \wedge \overline{b} \wedge c) \\ &= [(a \wedge b) \wedge (\overline{a} \vee \overline{c})] \vee [(\overline{a} \vee \overline{b}) \wedge (a \wedge c)] \\ &= [(a \wedge b) \wedge \overline{(a \wedge c)}] \vee [\overline{(a \wedge b)} \wedge (a \wedge c)] \\ &= (a \wedge b) + (a \wedge c) \\ &= (a \cdot b) + (a \cdot c) \end{aligned}$$

(2): By easy calculations we can show that the conditions (a) - (j) of Def. 2.69 are valid in $\mathfrak{R}^*$, using Def. 2.68 and Lemma 2.70.

(a): $\overline{0} = 1+0 = 1$

(b): $a \vee 1 = a+1+a \cdot 1 = 1+a+1 \cdot a = 1 \vee a$

(d): $a \wedge \overline{a} = a \cdot (1+a) = a+a \cdot a = a+a = 0$

(g):
$$\begin{aligned} \overline{a \wedge b} &= 1+(a \wedge b) = 1+a \cdot b, \text{ and} \\ \overline{a} \vee \overline{b} &= (1+a) \vee (1+b) \\ &= 1+a+1+b+(1+a) \cdot (1+b) \\ &= a+b+1+a+b+a \cdot b \\ &= 1+a \cdot b \end{aligned}$$

(h): $a \vee b = a+b+a \cdot b = b+a+b \cdot a = b \vee a$

(j):
$$\begin{aligned} a \wedge (b \vee c) &= a \cdot (b+c+b \cdot c) \\ &= a \cdot b+a \cdot c+a \cdot b \cdot c \\ &= a \cdot b+a \cdot c+a \cdot b \cdot a \cdot c \\ &= (a \wedge b) \vee (a \wedge c) \end{aligned}$$

(3): Let $\mathfrak{B} = (B, \vee, \wedge, \overline{\ }, 0, 1)$ be a Boolean algebra, and $a, b \in B$. Then with $\mathfrak{B}^* = (B, +, \cdot, -, 0, 1)$ we get

(◇) $a \cdot b = a \wedge b$

(◇) $1 + a = (1 \wedge \overline{a}) \vee (\overline{1} \wedge a) = \overline{a}$

(◇)
$$\begin{aligned} a + b + a \cdot b &= a + (b + a \cdot b) \\ &= a + b \cdot (1 + a) \\ &= a + b \cdot \overline{a} \\ &= [a \wedge \overline{(b \wedge \overline{a})}] \vee [\overline{a} \wedge (b \wedge \overline{a})] \\ &= [a \wedge (\overline{b} \vee a)] \vee [\overline{a} \wedge b] \\ &= [a \wedge (\overline{a} \wedge b) \\ &= (a \vee \overline{a}) \wedge (a \vee b) \\ &= a \vee b \end{aligned}$$

So $(\mathfrak{B}^*)^* = \mathfrak{B}$. Starting from the other side, assume $\mathfrak{R} = (R, +, \cdot, -, 0, 1)$ to be a Boolean ring, and $a, b \in R$. Then with $\mathfrak{R}^* = (R, \vee, \wedge, \overline{\ }, 0, 1)$ we get

(◇) $a \wedge b = a \cdot b$

(◇)
$$\begin{aligned} (a \wedge \overline{b}) \vee (\overline{a} \wedge b) &= (a \wedge \overline{b}) + (\overline{a} \wedge b) + (a \wedge \overline{b}) \cdot (\overline{a} \wedge b) \\ &= [a \cdot (1 + b)] + [(1 + a) \cdot b] + [a \cdot (1 + b) \cdot (1 + a) \cdot b] \\ &= [a + a \cdot b] + [b + a \cdot b] + [(a + a^2) \cdot (b + b^2)] \\ &= a + b \end{aligned}$$

Therefore, $(\mathfrak{R}^*)^* = \mathfrak{R}$. □

This theorem shows that, with additional operations suitably defined, every Boolean algebra becomes a Boolean ring, and conversely, every Boolean ring can be turned into a Boolean algebra. Moreover, starting with either a Boolean algebra or a Boolean ring, one can go forth and back again, arriving at the original structure. This is usually summarized by saying that the theories of Boolean algebras and Boolean rings are *definitionally equivalent*. So, on each occasion, you can select from these two concepts the one that seems intuitively more appropriate for your current considerations.

2.3.5.2 *Atomic Boolean Algebras*

Definition 2.72. *Let $\mathfrak{B} = (B, +, \cdot, \overline{\ }, 0, 1)$ be a Boolean algebra, and let $B^+ := B - \{0\}$ denote the set of the non-zero elements of B.*

(a) An **atom** *of a Boolean algebra B is a minimal element $a \in B^+$, i.e. $0 < a$. The set of atoms of B is denoted by* $\mathrm{At}\, B$.
(b) $\mathfrak{B}$ is called **atomless** *if it has no atoms.*
(c) B is called **atomic** *iff every non-zero element dominates at least one atom, i.e.* $\forall b \in B^+ \exists a \in \mathrm{At}\, B [a \leq b]$.

The power set algebra $(\mathcal{P}(X), \cup, \cap, \overline{\ }_X, \emptyset, X)$ is atomic, with $\mathrm{At}\, \mathcal{P}(X) = \{\{x\} : x \in X\}$. Accordingly, in the Boolean algebra $\{0, 1\}^X$, the atoms are the characteristic functions of singletons (functions that map a single element $x \in X$ to 1 and the remaining elements of $y \in X$ and $y \neq x$, to 0). A few characterizations of atoms are given in

Corollary 2.73. *The following conditions on an element* a *in a Boolean algebra* $\mathfrak{B} = (B, +, \cdot, \overline{\ }, 0, 1)$ *are equivalent:*

(1) *a is an atom;*
(2) *for each $b \in B$, $a \leq b$ or $a \leq \overline{b}$ but not both;*
(3) $0 < a$ *and,* $\forall b, c \in B\, [a \leq b + c \leftrightarrow a \leq b \vee a \leq c]$;
(4) *for each $b \in B$, $a \leq b$ or $a \cdot b = 0$ but not both;*
(5) $\forall b \in B [b \leq a \rightarrow (b = 0 \vee b = a)]$;
(6) *a is a minimal element in B^+.*

Proof.
(1) $\rightarrow$ (2): Let a be an atom, and $b \in B$. From $a \leq b \wedge a \leq \overline{b}$ we get $a \leq b \cdot \overline{b} = 0$, a contradiction. By Prop. 2.67(1) we have $x \leq y \leftrightarrow x \cdot \overline{y} = 0$; so if If $\neg a \leq b$ then $0 < a \cdot \overline{b} \leq a$, and since a is an atom, $a \cdot \overline{b} = a$, whence $a \leq \overline{b}$. A similar argument holds for $\neg a \leq \overline{b}$.

(2) $\rightarrow$ (3): Obviously, $a \leq b \vee a \leq c$ implies $a \leq b + c$. Conversely, assume $a \leq b + c$, but $\neg a \leq b$. Then $a \leq \overline{b}$ by (2), hence $a \leq \overline{b} \cdot (b + c) = \overline{b} \cdot c \leq c$. We also have $0 < a$, since for $a = 0$ we would get $a \leq b \wedge a \leq \overline{b}$ for each $b \in B$, contradicting (2).

(3) $\rightarrow$ (1): Assume that $0 \leq b < a$ for some $b \in B$. Then $a = a \cdot 1 = a \cdot b + a \cdot \overline{b} = b + a \cdot \overline{b}$, and since $\neg a \leq b$, we get $a \leq a \cdot \overline{b}$ by (3). Thus, $0 = a \cdot b = b$, using Prop. 2.67(1),(d) and (f). □

Proposition 2.74. *A finite Boolean algebra $\mathfrak{B}$ is atomic.*

Proof. Assume $\mathfrak{B}$ to be finite, but non-atomic. Then there is an element $0 < b \in B$ dominating no atom, i.e. $\neg \exists a \in \mathrm{At}\, B\, [a \leq b]$. So b itself is no atom, whence there must exist a $b_1 \in B$ with $0 \leq b_1 < b$ and $a \neq b_1$ for each $a \in \mathrm{At}\, B$. Thus, $0 < b_1 < b$. But also b_1 must not dominate any atom a, since we would then get $a \leq b$, contrary to our assumption. So there exists a $b_2 \in B$ such that $0 < b_2 < b_1 < b$. Since B has been assumed to be finite, this step can be repeated a finite number of times only, so that finally we arrive at an atom dominated by b, contradicting our hypothesis. Hence $\mathfrak{B}$ must be atomic. □

From. chapters 14 and 15 of [9]) we record some further well known results on Boolean algebras:

Lemma 2.75. *The following conditions are equivalent for a Boolean algebra* $\mathfrak{B} = (B, \vee, \wedge, \overline{\ }, 0, 1)$*:*

(1) *$\mathfrak{B}$ is atomic.*
(2) *Every element is the supremum of the atoms it dominates.*
(3) *1 is the supremum of* $\mathrm{At}\, B$.

Proof.
(1) $\rightarrow$ (2): Clearly each element $b \in B$ is an upper bound of the set of atoms, say E, it dominates. Assuming $c \in B$ to be another upper bound of E, we have to show that $b \leq c$, i.e. $b \wedge \overline{c} = 0$ by Prop. 2.67(1). Let $b \wedge \overline{c} > 0$; then, since $\mathfrak{B}$ is assumed atomic,

$\exists a \in \mathrm{At}\,B\,[a \le b \wedge \overline{c}]$. Since $b \wedge \overline{c} \le b$, we get $a \in E$, so $a \le c$. Therefore $a \le b \wedge \overline{c} \wedge c = 0$, contradicting $a \in \mathrm{At}\,B \subseteq B^+$.

(2) → (3): Obvious from (2).

(3) → (1): Let $0 < b \in B$. Then $0 < b = b \wedge 1 = b \wedge \bigvee \mathrm{At}\,B = \bigvee\{b \wedge a : a \in \mathrm{At}\,B\}$. Therefore, for at least one atom $a \in \mathrm{At}\,B$ we must have $b \wedge a \ne 0$, whence $a \le b$ by Cor. 2.73(4). □

Theorem 2.76. *Let $\mathfrak{B} = (B, \vee, \wedge, \overline{}, 0, 1)$ be a finite Boolean algebra, $b \in B^+$, and $E_b :=$ $\{a \in \mathrm{At}\,B : a \le b\}$. Then the following holds:*

(1) $b = \bigvee E_b$.
(2) Let $a, a_1, a_2, \ldots, a_k$ be any atoms in $\mathfrak{B}$, and $b = a_1 \vee a_2 \vee \ldots \vee a_k$. If $a \le b$ then for some $1 \le i \le k$ we have $a = a_i$.
(3) b can be uniquely expressed in the form $b = \bigvee E_b$.

Proof.
(1): By Prop. 2.74 $\mathfrak{B}$ is atomic, whence the assertion follows by Lemma 2.75(2).

(2): Since a is an atom, by Cor. 2.73(4) we have $a \le b$ or $a \wedge b = 0$. Assuming $a \wedge b = 0$ gives $a \le \overline{b}$, so $a \le b \wedge \overline{b} = 0$, a contradiction to $a \in \mathrm{At}\,B$. So we have $a \le b$, i.e $a \wedge b = a$. Thus $a \wedge (a_1 \vee a_2 \vee \ldots \vee a_k) = \bigvee_{1 \le i \le k}(a \wedge a_i) = a > 0$, whence, for some $1 \le i \le k$, $a \wedge a_i > 0$. Therefore, by definition of atoms, $a = a \wedge a_i = a_i$.

(3): Follows from (1) and (2). □

Theorem 2.77. *Let $\mathfrak{B} = (B, \vee, \wedge, \overline{}, 0, 1)$ be a Boolean algebra. Then $h : B \longrightarrow \mathcal{P}(\mathrm{At}\,B)$, $b \mapsto \{a \in \mathrm{At}\,B : a \le b\}$, is a homomorphism. It is injective if $\mathfrak{B}$ is atomic, and surjective if $\mathfrak{B}$ is complete.*

Proof. We have $h(0) = \emptyset$ and $h(1) = \mathrm{At}\,B$. From Cor. 2.73(2) we get

$$h(\overline{b}) = \{a \in \mathrm{At}\,B : a \le \overline{b}\} = \overline{\{a \in \mathrm{At}\,B : a \le b\}} = \overline{h(b)},$$

and similarly $h(a \vee b) = h(a) \cup h(b)$ follows from Cor. 2.73(3). Since, for every $c \in B$, $c \le a \wedge b$ iff $c \le a \wedge c \le b$, also $h(a \wedge b) = h(a) \cap h(b)$. Thus h is a homomorphism.

Now let $\mathfrak{B}$ be atomic, and $b \ne c$ be two elements in B. Without loss of generality we assume $b \not\le c$. Then $b \wedge \overline{c} > 0$ by Prop. 2.67(1), so there is an $a \in \mathrm{At}\,B$ with $a \le b \wedge \overline{c}$. Thus we get $a \in h(b)$ and $a \notin h(c)$, whence $h(b) \ne h(c)$ and h is injective.

Let $\mathfrak{B}$ be complete, $C = \{a_1, a_2, \ldots, a_k\} \subseteq \mathrm{At}\,B$, and $s := \bigvee C$. We prove that $h(s) = C$ which shows that h maps B onto $\mathrm{At}\,B = h(1)$: Clearly, $C \subseteq h(s)$ and, as already stated above, $h(0) = \emptyset$. Since for any atom $a \le s$, we have $a = a_j$, for some $1 \le j \le k$, by Th. 2.76(2), we get $h(s) \subseteq C$. □

The last theorem has important consequences as it describes the complete atomic and the finite Boolean algebras:

Corollary 2.78.

(1) Every atomic Boolean algebra is isomorphic to an algebra of sets.
(2) Every complete and atomic Boolean algebra is isomorphic to a power set algebra.

(3) (Representation theorem for finite Boolean algebras) *Every finite Boolean algebra is, up to isomorphism, exactly the power set algebra of a finite set.*

Proof. (3): If $\mathfrak{B}$ is finite, then also $\mathrm{At}\,B$, and $\mathfrak{B}$ is both complete and, by Prop. 2.74, also atomic. Thus, by Th. 2.77, it is isomorphic to $\mathcal{P}(\mathrm{At}\,B)$. □

Lemma 2.79. *Let X and Y be two sets such that $|X| = |Y|$. For any bijection $\varphi : X \longrightarrow Y$, let h be the mapping $h : \mathcal{P}(Y) \longrightarrow \mathcal{P}(X)$ be defined as $h[Q] = \varphi^{-1}[Q] = \{a \in X : \varphi(a) \in Q\}$. Then h is an isomorphism from $\mathcal{P}(X)$ to $\mathcal{P}(Y)$.*

Proof. Routine calculations reveal that h is an isomorphism: Obviously, $h(\emptyset) = \{a \in X : \varphi(a) \in \emptyset\} = \emptyset$, and similarly $h[Y] = X$. It can be easily checked that $h[A \cap B] = h[A] \cap h[B]$ and $h[A \cup B] = h[A] \cup h[B]$. Finally we have $a \in h[\overline{A}] \leftrightarrow \varphi(a) \in \overline{A} \leftrightarrow \varphi(a) \notin A \leftrightarrow a \notin h[A] \leftrightarrow a \in \overline{h[A]}$. So we have shown that h is a homomorphism.

From the definition of h we get $\varphi[h[Q]] = \varphi[\varphi^{-1}[Q]] = Q$, so $\varphi h = id_Y$. And $h[\varphi[P]] = [\varphi^{-1}[\varphi[P]] = P$, i.e. $h\varphi = id_X$. Thus, h has both a right and a left inverse, hence it is bijective by Prop. 2.26. □

Theorem 2.80. *Two complete, atomic Boolean algebras with the same number of atoms are isomorphic.*

Proof. Let $\mathfrak{A}$ and $\mathfrak{B}$ be complete and atomic Boolean algebras with sets of atoms X and Y respectively, and let f and g be the isomorphisms $f : A \longrightarrow \mathcal{P}(X)$ and $g : B \longrightarrow \mathcal{P}(Y)$ according to Th. 2.77. By assumption we have $|X| = |Y|$. Then, by Lemma 2.79, for any bijection $\varphi : X \longrightarrow Y$, the mapping $h : \mathcal{P}(Y) \longrightarrow \mathcal{P}(X)$, $Q \mapsto \varphi^{-1}[Q] = \{a \in X : \varphi(a) \in Q\}$ is an isomorphism.

The composition $k = g^{-1}h^{-1}f$ is an isomorphism from $\mathfrak{A}$ to $\mathfrak{B}$, by Lemma 2.48 and Prop. 2.24:

$$\mathfrak{A} \xrightarrow{f} \mathcal{P}(X) \xleftarrow{h} \mathcal{P}(Y) \xleftarrow{g} \mathfrak{B}.$$

□

2.3.6 *Galois Connections and Residuated Mappings*

Galois connections consist of a pair of maps with certain properties. Such maps arise in a great many number in various contexts; it is therefore desirable to give a short exposition of their main properties and interpretations. As we will see shortly, there are two different kinds of these pairs of maps that appear in literature, unfortunately often under the same name (see e.g. Novak [19] for the monotone, and Ore [20] for the other variant).

Definition 2.81. *Let $(P, \leq)$ and $(Q, \leq)$ be posets, and $f : P \longrightarrow Q$ and $g : Q \longrightarrow P$ mappings between P and Q.*

(a) *The pair (f, g) is called a* **Galois connection** *between P and Q if f, g are antitone and the following conditions hold for all $p \in P$ and $q \in Q$:*

$$p \leq g(f(p)) \quad (i.e.\ id_P \leq gf), \tag{2.10}$$

$$q \leq f(g(q)) \quad (i.e.\ id_Q \leq fg). \tag{2.11}$$

(b) *The pair (f,g) is called a* **monotone Galois connection** *between P and Q if f,g are isotone and the following conditions hold for all $p \in P$ and $q \in Q$:*

$$p \leq g(f(p)) \quad (i.e.\ id_P \leq gf), \tag{2.12}$$

$$f(g(q)) \leq q \quad (i.e.\ fg \leq id_Q). \tag{2.13}$$

Later in this book we will meet both kinds of Galois connections: The original form plays an important role in Formal Concept Analysis dealt with in Chapter 14; more details on the monotone variant are given in Section 7.2.1.2 as a prerequisite for our remarks on many-valued relations. See the overview in Table 7.1.

If (f,g) is a monotone Galois connection between P and Q, then f is also called a **residuated mapping**, and g its **residual**. f is also called the **lower** (or **left**) **adjoint** of g, and g the **upper** (or **right**) **adjoint** of f. The two maps are called **dually adjoint** to each other. (Note that it does not make sense to talk about the lower and upper adjoint for (antitone) Galois connections, since the situation is completely symmetrical.)

Before giving some basic of their properties, we continue with a few examples of Galois connections (for examples of monotone Galois connections see Section 7.2.1.2):

(1) The motivating example comes from Galois theory, whence the name; details are beyond our scope, so the reader may revert to e.g. chapter VI of [14].

(2) A semigroup S with at least two elements containing an element 0 such that, for all $x \in S, 0x = x0 = 0$, is called a **semigroup with zero**, and 0 is called a **zero element** of S. As for identity elements it can be shown that a semigroup can have at most one zero, so properties analogous to those given in Section 2.3.1 for identities also hold for zero elements.

For every $A \subseteq S$ define respectively the **left** and **right annihilators** of A by

$$L(A) = \{x \in S : \forall a \in A[xa = 0]\} \text{ and } R(A) = \{x \in S : \forall a \in A[ax = 0]\}.$$

Then the maps $L, R : \mathbf{2}^S \longrightarrow \mathbf{2}^S$ form a Galois connection. (Both L and R are obviously antitone: Let $A \subseteq B \subseteq S$, and set $P_A(x) := \forall a \in A[xa = 0]$. Then we get $x \in L(B) \rightarrow P_A(x) \wedge P_{B-A}(x) \rightarrow P_A(x) \rightarrow x \in L(A)$, whence $L(B) \subseteq L(A)$. Using the shorthand $L(x)$ for singletons $\{x\}$ we get $a \in L(x) \rightarrow ax = 0 \rightarrow x \in R(a)$, whence $x \in R(L(x))$, i.e. $id_{\mathbf{2}^S} \leq RL$. The other half of the proof runs in a similar way.)

(3) Let a relation $\rho \subseteq X \times Y$, and subsets $S \subseteq X$ and $T \subseteq Y$ be given. Defining $f(S) = \{y \in Y : \forall s \in S[sRy]\}$ and $g(T) = \{x \in X : \forall t \in T[xRt]\}$ then $f,g : \mathbf{2}^X \longrightarrow \mathbf{2}^Y$ form a Galois connection where the ordering is given by set inclusion $\subseteq$.

Proposition 2.82. *A pair (f,g) of maps $f : P \longrightarrow Q$ and $g : Q \longrightarrow P$ is*

(1) *a Galois connection if and only if for all $p \in P$ and $q \in Q$*

$$p \leq g(q) \leftrightarrow q \leq f(p); \tag{2.14}$$

(2) *a monotone Galois connection if and only if for all $p \in P$ and $q \in Q$*

$$p \leq g(q) \leftrightarrow f(p) \leq q. \tag{2.15}$$

Proof. We only prove assertion (1): [$\Rightarrow$] Let (f, g) be a Galois connection, and assume $p \leq g(q)$. Then $f(p) \geq f(g(q))$ since f is antitone, whence $f(p) \geq q$ by (2.11). The other direction can be shown in a similar way.
[$\Leftarrow$] Suppose that $p \leq g(q) \leftrightarrow q \leq f(p)$ holds for all $p \in P$ and $q \in Q$. Applied to $f(p) \leq f(p)$ this gives $p \leq g(f(p))$, i.e. (2.10); and by the same argument we get (2.11). Let $p_1 \leq p_2$; then, using (2.10), we get $p_1 \leq p_2 \leq g(f(p_2))$. By assumption this is equivalent to $f(p_2) \leq f(p_1)$, i.e. f is antitone. □

Proposition 2.83. *Let (f, g) be a Galois connection of either kind. Then*

$$f = f \circ g \circ f \quad \text{and} \quad g = g \circ f \circ g.$$

Proof. By (2.11) we get $f(p) \leq f(g(f(p)))$. And from (2.10), i.e. $p \leq g(f(p))$, we get $f(p) \geq f(g(f(p)))$ since f is antitone.. □

Proposition 2.84. *A map $f : (P, \leq) \longrightarrow (Q, \leq)$ between complete lattices has a dual adjoint if and only if*

$$f\left(\bigvee_{i \in I} x_i\right) = \bigwedge_{i \in I} f(x_i) \tag{2.16}$$

holds for all families $(x_i)_{i \in I} \subseteq V$.

Proof. [$\Rightarrow$] If g is a dual adjoint to f, then by Prop. 2.82

$$\begin{aligned} y \leq \bigwedge_{i \in I} f(x_i) &\iff y \leq f(x_i) \text{ for all } i \in I \\ &\iff x_i \leq g(y) \text{ for all } i \in I \\ &\iff \bigvee_{i \in I} x_i \leq g(y) \\ &\iff y \leq f\left(\bigvee_{i \in I} x_i\right), \end{aligned}$$

whence (2.16) holds.
[$\Leftarrow$] Conversely, suppose that $f\left(\bigvee_{i \in I} x_i\right) = \bigwedge_{i \in I} f(x_i)$ for holds for all families $(x_i)_{i \in I} \subseteq V$; then

$$\begin{aligned} x_1 \leq x_2 &\iff x_2 = x_1 \vee x_2 \\ &\Longrightarrow f(x_2) = f(x_1 \vee x_2) = f(x_1) \wedge f(x_2) \\ &\iff f(x_1) \geq f(x_2), \end{aligned}$$

i.e. f is antitone. The map $g : Q \longrightarrow P$ defined by $g(y) := \bigvee\{x \in P \mid y \leq f(x)\}$ is antitone. And since f is antitone, for $z \in P$ we have

$$g(f(z)) = \bigvee\{x \in V \mid f(z) \leq f(x)\} \geq \bigvee\{x \in V \mid z \geq x\} = z,$$

so we get (2.10), and similar for (2.11) we have for $y \in Q$

$$f(g(y) = f(\bigvee\{x \in V \mid y \leq f(x)\}) = \bigwedge\{f(x) \mid y \leq f(x)\} \geq y.$$

Thus we have shown that (f, g) is a Galois connection. □

2.3.7 *Partially Ordered Semigroups*

Definition 2.85. *Let $(S,\cdot;\leq)$ be an algebraic structure of type $\langle 2;2\rangle$. S is called a* **partially ordered semigroup** *(also* **posemigroup***) if*

(POSG1) $(S,\cdot)$ *is a semigroup,*
(POSG2) $(S,\leq)$ *is a partially ordered set,*
(POSG3) $\forall a,b,c \in S[a \leq b \rightarrow ac \leq bc \wedge ca \leq cb]$

If the partial order on S is total then S is called an **ordered semigroup**.

Examples:

- $(\mathbb{N}_0,\cdot)$, ordered by divisibility: $m \leq n$ iff $m|n$ iff $\exists k \in \mathbb{N}_0\,[mk = n]$, is a partially ordered monoid, **pomonoid** for short, with top element 0 and bottom element 1, since for all $n \in \mathbb{N}_0$ we have $1|n$ and $n|0$. Reflexivity and transitivity of $|$ are obvious, and for antisymmetry we have: $mk = n \wedge nl = m$ implies $nlk = n$, so $lk = 1$. Thus $l = k = 1$ and $m = n$.
- For a semigroup S $(\mathbf{2}^S,\cdot;\subseteq)$ is a posemigroup under product $XY = \{xy : x \in X, y \in Y\}$.
- The additive groups $(\mathbb{Z},+)$, $(\mathbb{Q},+)$, and $(\mathbb{R},+)$ are all ordered Abelian groups under the natural ordering. Note that $(\mathbb{Z},+)$ is the usual shorthand for $(\mathbb{Z},+,-,0)$ (where $-$ denotes the unary operator $- : z \mapsto -z$), so the 'official' notation would be $(\mathbb{Z},+,-,0;\leq)$.
- In an ordered group G the subset $P_G = \{x : x \geq 1_G\}$, called the *positive cone* of G, is a partially ordered monoid.

Definition 2.86. *Let $(S,\cdot)$ be a semigroup.*

(a) Let $x,y \in S$. The **left** *(resp.* **right***)* **translation** *on S determined by x is defined as $\lambda_x(y) = x\cdot y$ (resp. $\rho_x(y) = y\cdot x$).*
(b) An ordered semigroup $(S,\cdot;\leq)$ is **left (right) residuated** *if each right (left) translation is a residuated mapping, and* **residuated** *if it is both left and right residuated.*
(c) Let $(S,\cdot;\leq)$ be an ordered semigroup and $a,b \in S$; the following notation is used

$$\langle a.\!\cdot b\rangle = \{x \in S : bx \leq a\}, \quad \langle a\cdot\!. b\rangle = \{x \in S : xb \leq a\}.$$

Note the 'directional cross relationship' between the residuation of mappings and semigroup, and be careful to not confuse them: If a *left* translation of S, say $\lambda_a : S \longrightarrow S : x \mapsto ax$ is residuated then, according to Prop. 7.10 (4), the set $\{x \in S : ax \leq b\}$ has a maximal element $\lambda_a^\sharp(b)$. Considering what is said about residuation in the context of Table 3.1, $\lambda_a^\sharp(b)$ is the *right* residual $b.\!\cdot a$ of b by a. Therefore, $b.\!\cdot a$ is characterized by the properties

(1) $a(b.\!\cdot a) \leq b$,
(2) $ac \leq b \rightarrow c \leq b.\!\cdot a$.

Analogously, we have for the left residual

(3) $(b \cdot\!. a)a \leq b$,
(4) $ca \leq b \rightarrow c \leq b \cdot\!. a$.

Additional Reading

For set theory you may wish to have a look at one of the introductory, but rigorous texts by Devlin [7], Monk [18], or Suppes [25].

Readers interested in more details and results, or in the foundations of mathematical logic might want to consult e.g. one of the following books: Hinman [12], Mendelson [17], Rautenberg [21], or Tourlakis [27].

Concerning general algebraic concepts, in addition to [3] a good source for information are the opening chapters of Grätzer [] or Wechler [28].

References

[1] Blyth, T.S.: *Lattices and Ordered Algebraic Structures*, Springer, London (2005).
[2] Birkhoff, G.: *Lattice Theory*, 3rd ed., American Mathematical Society, Providence, Rhode Island (1973).
[3] Burris, S., Sankappanavar, H.P.: *A Course in Universal Algebra*, The Millenium Edition, Update 2012; available online at `https://www.math.uwaterloo.ca/~snburris/htdocs/UALG/univ-algebra2012.pdf` (2012).
[4] Clifford, A.H., Preston, G.B.: *The Algebraic Theory of Semigroups, Vol. 1*, 2nd ed., reprinted with corrections, American Mathematical Society, Providence, Rhode Island (1977).
[5] Cohn, P.M.: *Universal Algebra*, D. Reidel Publishing Company, Dordrecht, Holland (1981).
[6] Copi, I.M., Cohen, C., McMahon, K.: *Introduction to Logic*, 14th ed. Pearson Education Limited, Harlow, Essex (2014).
[7] Devlin, K.: *The Joy of Sets*, 2nd ed., Springer, New York (1993).
[8] Ebbinghaus, H.-D., Flum, J., Thomas, W.: *Mathematical Logic*, 2nd ed., Springer Science+Business Media, New York (1994).
[9] Givant, S., Halmos, P.: *Introduction to Boolean Algebras*, Springer Science+Business Media (2009).
[10] Grätzer, G.: *Universal Algebra*, 2nd ed., Springer Science+Business Media (2008).
[11] Graver, J.E., Watkins, M.E.: *Combinatorics With Emphasis on the Theory of Graphs*, Springer, New York(1977).
[12] Hinman, P.G.: *Fundamentals of Mathematical Logic*, A K Peters, Wellesley, Massachusetts (2005).
[13] Hurley, P.J.: *A Concise Introduction to Logic*, 12th ed. Wadsworth, Belmont (2015).
[14] Lang, S.: *Algebra*, 3rd ed., Springer-Verlag New York, Inc. (2002).
[15] Lau, D.: *Algebra und Diskrete Mathematik 2*, Springer-Verlag Berlin Heidelberg (2004).
[16] Maddux, R.D.: Relation-algebraic semantics, Theoretical Computer Science **160**, 1-85 (1996).
[17] Mendelson, E.: *Introduction to Mathematical Logic*, 5th ed., Taylor & Francis Group (2010).
[18] Monk, J.D.: *Introduction to Set Theory*, McGraw-Hill, New York (1969).

[19] Novak, D.: Generalization of continuous posets, Trans. Amer. Math. Soc. **272**, 645-667 (1982).

[20] Ore, O.: Galoios Connexions, Trans. Amer. Math. Soc. **55**, 493-513 (1944).

[21] Rautenberg, W.: *A Concise Introduction to Mathematical Logic*, 3rd ed., Springer Science+Business Media, New York (2010).

[22] Simovici, D.A., Djeraba, C.: *Mathematical Tools for Data Mining*, 2nd edition. Springer, London (2014).

[23] Schröder, B.S.W.: *Ordered Sets - An Introduction*, Springer Science+Business Media, New York (2003).

[24] Stone, M.H.: The Theory of Representations of Boolean Algebras, Trans. Amer. Math. Soc. **40**, 37-111 (1936).

[25] Suppes, P.: *Axiomatic Set Theory*, Dover Publications, New York (1972).

[26] Tomassi,P.: *Logic*, Routledge, London and New York (1999).

[27] Tourlakis, G.: *Mathematical Logic*, John Wiley & Sons, Inc., New Jersey (2008).

[28] Wechler, W.: *Universal Algebra for Computer Scientists*, Springer-Verlag, Berlin Heidelberg (1992).

PART 2

Relations - Concrete, two-valued

Chapter 3

Binary Relations: Basic Properties and Operations

In this chapter we start to deal with the concept of a *relation* on the object level, i.e. we make it the central point of our considerations. After relations have been formally defined, both their set-theoretical (in Section 3.1) and their genuine relational operations and properties, i.e. the characteristics beyond their mere existence as sets (in Section 3.2 and Section 3.3), will be discussed. We use the perhaps most popular non-mathematical example of family relations for illustration purposes.

Of course, everyone reading these lines knows what (binary) relations are from many and rather diverse everyday situations. When we refer to such a relation, say R, in ordinary contexts we use some intuitive description of the connection existing between items which stand in relation R. To add some examples to the ones already listed at the beginnings of the Preface and of Chapter 1: *owner of, actor in, overlaps, drives, visits, earlier than, heavier than, ...* . Indeed every transitive verb and every comparison may serve as an example.

Before we present relations in the way as it is usually done in texts not dealing with axiomatic set theory, let us have a short look at how relations are introduced in just this mathematical area.

Definition 3.1.

(a) *A set A is a* **relation** $:\Leftrightarrow \forall x[x \in A \rightarrow \exists y, z[x = \langle y, z\rangle]]$.

(b) $xAy :\Leftrightarrow \langle x, y\rangle \in A$, *and we say that x is in relation A to y.*

At the beginning of Section 2.1 we used $C_6 = \{\{1,2\}, \{1,3\}, \{1,2,3,6\}\}$ to encode some information about the number 6 and its divisors. We can represent the more or less same information in a more concise form as the relation $R_6 = \{\langle 2,6\rangle, \langle 3,6\rangle\}$ leaving away all "trivial" (i.e. redundant) pairs like $\langle 1,3\rangle$ or $\langle 6,6\rangle$. Some elementary properties immediately following from Def. 3.1 are given in

Proposition 3.2.

(1) *$\emptyset$ is a relation.*

(2) *If R is a relation and $S \subseteq R$, then S is a relation.*

(3) *If R and S are relations, $R \cap S$, $R \cup S$, and $R - S$ are relations too.*

In axiomatic set theory proofs of statements like these have to be given using the axioms of the underlying theory (e.g in ZFC, the Zermelo-Fraenkel set theory with the axiom of choice), and especially the existence of sets with the desired properties has to be shown to follow from the axioms (e.g. the axiom of union, or the axiom-schema of replacement etc.). We refer readers interested in these aspects to a standard text on set theory. For a detailed discussion of relations we recommend [22, Chapter 3] and [23, Chapter 6]. From the intuitive and so-called "naive" set-theoretical point of view taken in this book it is "obvious" that the assertions in Prop. 3.2 hold.

3.1 Set Theoretical Properties

Before focusing on the proper topic of this book, we start off with the presentation of the general case: Let $X_1, X_2, \ldots, X_n$ be sets; an **n-ary relation** ρ is a subset of the Cartesian product of the X_i, i.e. $\rho \subseteq X_1 \times X_2 \times \cdots \times X_n$. And it is not an accident that the special case $n = 2$ of binary relations has already been considered (on the metalevel) in Section 2.1.2: They occur most often in practice and theory, and as already stated in the first paragraph of the Preface, they are indispensable for managing daily routine, or for doing mathematics. So, for convenience, we make official our habit of speech: the term "*relation*" always means "*binary relation*".

But also the cases $n = 1$, giving an **unary relation**, and $n = 3$, called a **ternary relation**, are of interest. Examples for the latter are "*x lies between y and z*", or "*a likes b more than c*". For $n = 1$ we get $\rho \subseteq X$, i.e. ρ is a subset of X, often also called a *property*. Let $X = \{$*Jack, Kate, Leonardo, Rose*$\}$ be a set of names; then each of $\{$*Kate, Rose*$\}$ and $\{$*Leonardo, Jack*$\}$ represents a certain property, but also each of $\{$*Kate, Leonardo*$\}$ and $\{$*Rose, Jack*$\}$.

Definition 3.3.

(a) *A (**binary**) **relation** ρ is a set of ordered pairs.*
(b) *If $\rho \subseteq S \times T$ for two sets S and T, then ρ is said to be a **relation from** S **to** T.*
(c) *A relation from S to S is called a **relation on** S.*
(d) *If $\langle x, y\rangle \in \rho$ we say that x is in relation ρ to y and usually write this as $x\rho y$. For $\langle x, y\rangle \notin \rho$ we also use $\neg x\rho y$, or $x\overline{\rho}y$. $\overline{\rho}$ is called the **complement of** ρ.*
(e) *The **restriction of a relation** $\rho \subseteq S \times T$ to a subset $A \subseteq S$ is the relation $\rho|_A \subseteq S \times T$ defined by $\rho|_A = \rho \cap (A \times T) = \{\langle x, y\rangle | \langle x, y\rangle \in \rho \wedge x \in A\}$.*

Two special relations are the **empty relation** $\emptyset$ and the **universal relation** $S \times T$, also denoted as $U_{S\times T}$, and as U_S if $S = T$. The **identity relation** on a set S is $I_S := \{\langle x, x\rangle : x \in S\}$. Relations *on* a set are also called **homogeneous**, and those between two different sets **heterogeneous**.

Examples of relations are such diverse sets as $\{\langle Adam, Eve\rangle, \langle Bonny, Clyde\rangle\}$ and $\{\langle x, y\rangle : x, y \in S \wedge \exists z \in \mathbb{R}_{\geq 0}(y = x + z)\}$, the latter one representing "*less than or equal*" on S, with $\mathbb{R}_{\geq 0}$ denoting the non-negative reals. A relation between different kinds of objects is "*capital city of*" which is a relation from cities to countries.

From Def. 3.3 we get $\mathcal{P}(S \times T)$ as the set of all relations from S to T. Following [2], an element $\rho \in \mathcal{P}(S \times T)$ is said to be a relation of type $S \times T$, and to emphasize our focus on binary relations we will write $\mathcal{B}_{S\times T}$ for this set of relations, and $\mathcal{B}_S$ if $S = T$. Now, $\mathcal{B}_{S\times T}$ is defined to be a power set, viz. that of $S \times T$, which means that it bears already the following lattice theoretical structure:

(1) Existence of the null relation: $\emptyset \in \mathcal{B}_{S\times T}$;
(2) Existence of the full (or universal) relation: $U_{S\times T} \in \mathcal{B}_{S\times T}$;
(3) Three set theoretical operations are defined as follows:
   (a) $\cup : \mathcal{B}_{S\times T} \times \mathcal{B}_{S\times T} \mapsto \mathcal{B}_{S\times T}$ (Union)
   (b) $\cap : \mathcal{B}_{S\times T} \times \mathcal{B}_{S\times T} \mapsto \mathcal{B}_{S\times T}$ (Intersection)
   (c) $\bar{\ } : \mathcal{B}_{S\times T} \mapsto \mathcal{B}_{S\times T}$ (Complement)

Theorem 3.4. *The structure* $(\mathcal{B}_{S\times T}, \cup, \cap, \bar{\ })$ *is a complete atomic Boolean algebra with least element* $\emptyset$ *and greatest element* $U_{S\times T}$*, with its set of atoms given by* $\mathsf{At}(\mathcal{B}_{S\times T}) = \{\{\langle x, y\rangle\} : x \in S, y \in T\}$.

Definition 3.5. *Let* $\rho \subseteq S \times T$ *be a relation from* S *to* T. *Then*

(a) $\mathsf{dom}(\rho) = \{x \in S : \exists y \in T[x\rho y]\}$ *is called the* **domain** *of* ρ;
(b) $\mathsf{rng}(\rho) = \{y \in T : \exists x \in S[x\rho y]\}$ *is called the* **range** *of* ρ;
(c) $\mathsf{fld}(\rho) = \mathsf{dom}(\rho) \cup \mathsf{rng}(\rho)$ *is called the* **field** *of* ρ.

Proposition 3.6. *Let* ρ *and* σ *be two relations.*

(1) $\rho \subseteq S \times T$ *if and only if* $\mathsf{dom}(\rho) \subseteq S$ *and* $\mathsf{rng}(\rho) \subseteq T$. *Obviously,* ρ *is always a relation from* $\mathsf{dom}(\rho)$ *to* $\mathsf{rng}(\rho)$.
(2) If $\rho \subseteq \sigma$, *then* $\mathsf{dom}(\rho) \subseteq \mathsf{dom}(\sigma)$ *and* $\mathsf{rng}(\rho) \subseteq \mathsf{rng}(\sigma)$.
(3) $\mathsf{dom}(\rho \cup \sigma) = \mathsf{dom}(\rho) \cup \mathsf{dom}(\sigma)$.
(4) $\mathsf{dom}(\rho \cap \sigma) \subseteq \mathsf{dom}(\rho) \cap \mathsf{dom}(\sigma)$.
(5) $\mathsf{dom}(\rho - \sigma) \supseteq \mathsf{dom}(\rho) - \mathsf{dom}(\sigma)$.
(6) $\mathsf{rng}(\rho \cup \sigma) = \mathsf{rng}(\rho) \cup \mathsf{rng}(\sigma)$.
(7) $\mathsf{rng}(\rho \cap \sigma) \subseteq \mathsf{rng}(\rho) \cap \mathsf{rng}(\sigma)$.
(8) $\mathsf{rng}(\rho - \sigma) \subseteq \mathsf{rng}(\rho) - \mathsf{rng}(\sigma)$.

Proof. As examples for the relevance of correct usage of the quantifier distribution laws we give the straightforward proofs of (3) and (4).

$$
\begin{aligned}
(3)\!: x \in \mathsf{dom}(\rho \cup \sigma) &\leftrightarrow \exists y[x(\rho \cup \sigma)y] \\
&\leftrightarrow \exists y[x\rho y \vee x\sigma y] \\
&\leftrightarrow \exists y[x\rho y] \vee \exists y[x\sigma y] && \text{by Th. 2.37(1)} \\
&\leftrightarrow x \in \mathsf{dom}(\rho) \vee x \in \mathsf{dom}(\sigma) \\
&\leftrightarrow x \in \mathsf{dom}(\rho) \cup \mathsf{dom}(\sigma).
\end{aligned}
$$

$$(4)\!: x \in \mathsf{dom}(\rho \cap \sigma) \leftrightarrow \exists y[x(\rho \cap \sigma)y]$$

$$
\begin{aligned}
&\leftrightarrow \exists y[x\rho y \wedge x\sigma y] \\
&\rightarrow \exists y[x\rho y] \wedge \exists y[x\sigma y] && \text{by Th. 2.37(2)} \\
&\leftrightarrow x \in \mathrm{dom}(\rho) \wedge x \in \mathrm{dom}(\sigma) \\
&\leftrightarrow x \in \mathrm{dom}(\rho) \cap \mathrm{dom}(\sigma).
\end{aligned}
$$

□

(**Note.** The proofs for the propositions and theorems of this chapter are not difficult in general, and will therefore be omitted in most cases. Those presented serve to give a flavor of the techniques to be used for demonstrating the corresponding assertions.)

Proposition 3.7.

(1) fld *has properties analogous to Prop. 3.6 (1)-(vi).*
(2) $\mathrm{fld}(\rho^{\smile}) = \mathrm{fld}(\rho)$
(3) $\mathrm{fld}(\rho\sigma) \subseteq \mathrm{dom}(\rho) \cup \mathrm{rng}(\sigma)$.

3.1.1 *Images of Sets under a Relation*

In this section, we consider one more notion concerning relations, together with some easy to prove properties, which plays an important role in e.g. the discussion of functions, and also in the definitions of approximations, neighborhoods and various other concepts.

Definition 3.8. *Let* $\rho \subseteq S \times T$, $A \subseteq S$ *and* $B \subseteq T$.

(a) The **image of** A **under** ρ *is the set* $\rho[A] = \{y \in T : \exists x \in A[x\rho y]\}$.
(b) The **pre-image of** B **under** ρ *(also* **inverse image of** B **under** ρ*) is the set* $\rho^{-1}(B) = \{x \in S : \exists y \in B[x\rho y]\}$.

Proposition 3.9. *Let* $\rho \subseteq S \times T$, *and* $A, B \subseteq S$. *Then the following holds:*

(1) $\emptyset[A] = \emptyset$.
(2) $\rho[\emptyset] = \emptyset$.
(3) $I[A] = A$.
(4) $A \subseteq B \rightarrow \rho[A] \subseteq \rho[B]$.
(5) $\rho[A \cup B] = \rho[A] \cup \rho[B]$.
(6) $\rho[A \cap B] \subseteq \rho[A] \cap \rho[B]$.
(7) $\rho[A \cap B] = \rho[A] \cap \rho[B]$, *if* ρ *is one-to-one.*
(8) $\rho[A] - \rho[B] \subseteq \rho[A - B]$.
(9) $\rho[A] = \mathrm{rng}(\rho|_A)$.

Proof. The following proofs are the only ones that will be given in nearly full length (i.e. including laws from set theory and propositional as well as predicate logic), thus illustrating the machinery at work. Having acquired some routine, one can considerably shorten the arguments.

$$\begin{aligned}
(5)\colon y \in \rho[A \cup B] &\leftrightarrow \exists x[x\rho y \wedge x \in A \cup B] \\
&\leftrightarrow \exists x[x\rho y \wedge (x \in A \vee x \in B)] \\
&\leftrightarrow \exists x[(x\rho y \wedge x \in A) \vee (x\rho y \wedge x \in B)] \\
&\leftrightarrow \exists x[x\rho y \wedge x \in A] \vee \exists x[x\rho y \wedge x \in B] && \text{Th. 2.37(1)} \\
&\leftrightarrow y \in \rho[A] \vee y \in \rho[B] \\
&\leftrightarrow y \in \rho[A] \cup \rho[B]
\end{aligned}$$

$$\begin{aligned}
(7)\colon y \in \rho[A] \cap \rho[B] &\to \exists z_A[z_A\rho y \wedge z_A \in A] \wedge \exists z_B[z_B\rho y \wedge z_B \in B] \\
&\to \exists z[z = z_A = z_B \wedge z\rho y \wedge z \in A \cap B] && \text{since } \rho \text{ is one-to-one} \\
&\to y \in \rho[A \cap B] \\
&\to \rho[A] \cap \rho[B] \subseteq \rho[A \cap B] \\
&\to \rho[A] \cap \rho[B] = \rho[A \cap B] && \text{by (6)}
\end{aligned}$$

$$\begin{aligned}
(8)\colon y \in \rho[A] - \rho[B] &\leftrightarrow y \in \rho[A] \wedge y \notin \rho[B] \\
&\leftrightarrow \exists x[x\rho y \wedge x \in A] \wedge \neg\exists x[x\rho y \wedge x \in B] \\
&\leftrightarrow \exists x[x\rho y \wedge x \in A] \wedge \forall x \neg[x\rho y \wedge x \in B] && \text{Th. 2.35(1)} \\
&\leftrightarrow \exists x[x\rho y \wedge x \in A] \wedge \forall x[\neg x\rho y \vee x \notin B] && \text{de Morgan's law} \\
&\to \exists x[x\rho y \wedge x \in A \wedge (\neg x\rho y \vee x \notin B)] && \text{Th. 2.37(3)} \\
&\leftrightarrow \exists x[(x\rho y \wedge x \in A \wedge \neg x\rho y) \vee (x\rho y \wedge x \in A \wedge x \notin B)] \\
&\leftrightarrow \exists x[x\rho y \wedge x \in A \wedge x \notin B] && \text{Th. 2.33(3)} \\
&\leftrightarrow \exists x[x\rho y \wedge x \in A - B] \\
&\leftrightarrow y \in \rho[A - B]
\end{aligned}$$

That the last inclusion cannot be strengthened to equality is shown by the following example (from [22], p. 65): Let $\rho = \{\langle 1,3\rangle, \langle 2,3\rangle\}$, $A = \{1\}$, and $B = \{2\}$. Then $\rho[A] - \rho[B] = \emptyset$, but $\rho[A - B] = \rho[A] = \{3\}$. □

3.2 Basic Relation Operations

Before proceeding to the general formal presentation, let us take a short glance at some well known examples of relations that we encounter in our everyday life (already used by DE MORGAN in [16] , on e.g. p. 344): Let P be the set of all people (dead or alive), and let us consider e.g the relations $\mathsf{sonOf}, \mathsf{daughterOf} \in \mathcal{B}_P$. Then, with the set-theoretical operations at hand, we can easily build the relations like e.g. $\mathsf{childOf}$ or $\mathsf{spouseOf} \in \mathcal{B}_P$ as follows:

$$\begin{aligned}
\mathsf{childOf} &= \mathsf{sonOf} \cup \mathsf{daughterOf}, \\
\mathsf{spouseOf} &= \mathsf{husbandOf} \cup \mathsf{wifeOf}.
\end{aligned}$$

But what can we do about such simple concepts as parentOf or auntOf? Nothing, if we do not define some additional operations for relations: For parentOf we have to reverse childOf in a suitable way, and for auntOf we need something like a concatenation of parentOf and sisterOf, the latter one still to be constructed with what we have already at hand.

The definitions of converse and composition, the two basic operations on relations, are given below, as well as some of their fundamental properties.

Definition 3.10. *The relation* $\rho^{\smile} = \{\langle y, x\rangle : \langle x, y\rangle \in \rho\}$ *is called the* **converse** *of* ρ.

It has turned out convenient to use the following abbreviations: Let $\rho \subseteq A \times B$. The **afterset** of $a \in A$ is $a\rho = \rho[\{a\}] = \{b \in B : a\rho b\}$, and the **foreset** of $b \in B$ is $\rho b = \rho^{\smile}[\{b\}] = \{a \in A : a\rho b\}$. Due to their meaning in Boolean matrix representations (see Def. 3.19) aftersets (foresets) are also called *rows* (*columns*) of a relation. Sometimes we will also use the following generalized notation: we write e.g. $A\rho := \bigcup_{x\in A} x\rho = \rho[A]$. Note that our notation from Def. 3.8 follows the common habit: Of course $\rho^{-1}(B) = \rho^{\smile}[B]$. Using all these notations one gets, and can also find in the pertinent literature:

$$\mathsf{rng}(\rho) = \rho[A] = A\rho \quad \text{and} \quad \mathsf{dom}(\rho) = \rho^{-1}(B) = \rho B. \tag{3.1}$$

Definition 3.11. *Let* ρ *and* σ *be two relations. The* **product** *or* **composition** *of* ρ *and* σ *is the relation* $\rho\sigma = \{\langle x, z\rangle : \exists y[x\rho y \wedge y\sigma z]\} = \{\langle x, z\rangle : \exists y[x\rho y\sigma z]\}$.

For $\langle x, z\rangle \in \rho\sigma$ we may also write $x(\rho\sigma)z$; occasionally $\rho\cdot\sigma$ or $\rho\circ\sigma$ will be used for typographical reasons, e.g. in the product $\mathsf{childOf}\cdot\mathsf{sonOf}$. And sometimes, e.g. in historical contexts or when we will consider (abstract) relation algebras later on, we will adhere to the commonly used $R;S$ or $x;y$ instead of $\rho\sigma$ to be nearer to the original literature.

Proposition 3.12. *Let* $\rho, \sigma \subseteq A \times B$ *and* $\tau \subseteq B \times C$*; then the following holds:*

(1) $x\overline{\rho} = \overline{x\rho}$ *and* $\overline{\rho}y = \overline{\rho y}$.
(2) $x(\rho\cup\sigma) = x\rho\cup x\sigma$ *and* $(\rho\cup\sigma)y = \rho y\cup\sigma y$.
(3) $x(\rho\cap\sigma) = x\rho\cap x\sigma$ *and* $(\rho\cap\sigma)y = \rho y\cap\sigma y$.
(4) $x(\rho\tau) = (x\rho)\tau$.

Proof. (4): $y \in x(\rho\tau) \leftrightarrow \exists z[x\rho z\tau y] \leftrightarrow \exists z[z \in x\rho \wedge z \in \tau y] \leftrightarrow y \in (x\rho)\tau$. □

To keep notation simple in complex cases, we fix the following priority for operations: negation, conversion, composition, intersection, and union. We keep, however, freedom in inserting parentheses and the product operator. Thus, $(\overline{\rho})^{\smile}$ is the same as $\overline{\rho}^{\smile}$, and the short version of $(\rho\cdot\sigma)\cap(\sigma\cdot\tau)$ is $\rho\sigma\cap\sigma\tau$.

Proposition 3.13. *For any relations* ρ *and* σ *we have*

(1) $\mathsf{dom}(\rho^{\smile}) = \mathsf{rng}(\rho)$.
(2) $\mathsf{rng}(\rho^{\smile}) = \mathsf{dom}(\rho)$.
(3) *If* $\rho \subseteq S \times T$*, then* $\rho^{\smile} \subseteq T \times S$.
(4) $\rho^{\smile\smile} = \rho$.
(5) $\overline{\rho}^{\smile} = \overline{\rho^{\smile}}$.
(6) $\rho \subseteq \sigma$ *iff* $\rho^{\smile} \subseteq \sigma^{\smile}$.
(7) $(\rho\cap\sigma)^{\smile} = \rho^{\smile}\cap\sigma^{\smile}$.
(8) $(\rho\cup\sigma)^{\smile} = \rho^{\smile}\cup\sigma^{\smile}$.

(9) $(\rho - \sigma)^{\smile} = \rho^{\smile} - \sigma^{\smile}$.
(10) $x\rho = \rho^{\smile} x$ *and* $\rho y = y\rho^{\smile}$.

Proof. We give two example proofs, both easy applications of the definitions:
(5): $x(\overline{\rho})^{\smile} y \leftrightarrow y\overline{\rho}x \leftrightarrow \neg y\rho x \leftrightarrow \neg x\rho^{\smile} y \leftrightarrow x\overline{\rho^{\smile}} y$.
(7): $x(\rho \cap \sigma)^{\smile} y \leftrightarrow y(\rho \cap \sigma)x \leftrightarrow y\rho x \wedge y\sigma x \leftrightarrow x\rho^{\smile} y \wedge x\sigma^{\smile} y \leftrightarrow x(\rho^{\smile} \cap \sigma^{\smile})y$. □

Some further remarks referring to notation are advisable at this place: Whenever we write a relational expression using operations like e.g. $\cap, \cup, \circ$, we assume the relations involved to be of appropriate type, without stating this explicitly. We will also refrain from always stating all technical details if they become obvious from the context. This simplifies formulas and makes them concise and clear.

Thus, the assertion $UU = U$ is written as a concise and parsimonious (but maybe somewhat sloppy) shorthand for a whole set of equations like $U_S U_S = U_S$, ..., $U_{S\times T} U_{T\times V} = U_{S\times V}$. Take a look at the Tarski rule in Prop. 3.55: $\rho \neq \emptyset \rightarrow U\rho U = U$. Here it is e.g. assumed that the domain of the first and third universal relation are the same, and that the range of the first coincides with the domain of ρ. Similar remarks hold for the identity relations I which look alike with respect to comprising only diagonal elements, but which may differ in size for different sets S and T. But for a relation $\rho \subseteq S \times T$ we always have $I_S\rho = \rho = \rho I_T$. Henceforth we will, with some exceptions, use the generic shorthand notation for identity and universal relations.

It should be mentioned that there exists a second pair of related concepts also bearing the names *domain* and *range* (or *codomain*) of a relation which are themselves relations again and not sets; see e.g. [20, 6.1.1 Definition] or [3].

Definition 3.14. *Let* $\rho \subseteq X \times Y$ *be a (possibly heterogeneous) relation.*

(a) $\rho_{\mathrm{dom}} = I \cap \rho\rho^{\smile}$ *is called the* **domain** *of* ρ.
(b) $\rho_{\mathrm{rng}} = I \cap \rho^{\smile}\rho$ *is called the* **range** *of* ρ.

By definition we have that both ρ_{dom} and ρ_{rng} are partial identities, i.e. relations included in the identity relation. ρ_{dom} obviously identifies those elements in X which are in relation ρ to at least one element of Y. (More on how sets can be characterized by relations can be found in Section 6.2.1.2.) Some straightforward consequences of the definition are the following:

Proposition 3.15. *For the domain and range of a relation* ρ *the following holds:*

(1) $\rho_{\mathrm{dom}} = \{\langle x, x\rangle : \exists y \in Y[x\rho y]\} = I \cap \rho U$.
(2) $\rho_{\mathrm{rng}} = \{\langle y, y\rangle : \exists x \in X[x\rho y]\} = I \cap U\rho$.
(3) $\pi_1(\rho_{\mathrm{dom}}) = \pi_2(\rho_{\mathrm{dom}}) = \{x \in X : \exists y \in Y[x\rho y]\} = \mathrm{dom}(\rho)$.
(4) $\pi_1(\rho_{\mathrm{rng}}) = \pi_2(\rho_{\mathrm{rng}}) = \{y \in Y : \exists x \in X[x\rho y]\} = \mathrm{rng}(\rho)$.
(5) $\rho_{\mathrm{rng}} = \rho^{\smile}_{\mathrm{dom}}$.
(6) $\rho_{\mathrm{dom}}\rho = \rho = \rho\rho_{\mathrm{rng}}$.

Proof. (1): Clearly, $\rho_{\text{dom}} \subseteq I$. Thus, $x(I \cap \rho\rho^{\smile})x \leftrightarrow \exists y[x\rho y \wedge y\rho^{\smile}x] \leftrightarrow \exists y[x\rho y \wedge x\rho y] \leftrightarrow \exists y[x\rho y] \leftrightarrow \exists y[x\rho y \wedge yUx] \leftrightarrow x(I \cap \rho U)x$.

(6): By Prop. 3.54. □

Theorem 3.16. *Let ρ, σ, and τ be relations such that the various compositions are well defined. Then the following holds:*

(1) $I_S\rho = \rho = \rho I_T$.
(2) $\emptyset\rho = \emptyset = \rho\emptyset$.
(3) *Associativity of relation product:* $\rho(\sigma\tau) = (\rho\sigma)\tau$.
(4) *Subdistributivity of relation product over intersection:*
$$\rho(\sigma \cap \tau) \subseteq (\rho\sigma) \cap (\rho\tau) \quad \textit{and} \quad (\rho \cap \sigma)\tau \subseteq (\rho\tau) \cap (\sigma\tau).$$
(5) *Monotonicity of relation product: If* $\sigma \subseteq \tau$, *then*
$$\rho\sigma \subseteq \rho\tau \quad \textit{and} \quad \sigma\rho \subseteq \tau\rho.$$
(6) *Distributivity of relation product over union:*
$$\rho(\sigma \cup \tau) = (\rho\sigma) \cup (\rho\tau) \quad \textit{and} \quad (\rho \cup \sigma)\tau = (\rho\tau) \cup (\sigma\tau).$$
(7) *If* $\rho \subseteq \sigma$ *and* $\tau \subseteq \omega$, *then* $\rho\tau \subseteq \sigma\omega$.
(8) $(\rho\sigma)^{\smile} = \sigma^{\smile}\rho^{\smile}$.
(9) $\rho\sigma = \bigcup_{y \in Y}(\rho y \times y\sigma)$.

Proof. For illustration purposes we give some proofs, most of them being easy routine applications of FOL.

$$\begin{aligned}
(3):\quad x\rho(\sigma\tau)w &\leftrightarrow \exists y[x\rho y \wedge y(\sigma\tau)w] \\
&\leftrightarrow \exists y[x\rho y \wedge \exists z(y\sigma z \wedge z\tau w)] \\
&\leftrightarrow \exists y\exists z[x\rho y \wedge (y\sigma z \wedge z\tau w)] && \text{Th. 2.36(1)(a)} \\
&\leftrightarrow \exists z[\exists y(x\rho y \wedge y\sigma z) \wedge z\tau w)] && \text{Th. 2.34(2)} \\
&\leftrightarrow \exists z[x\rho\sigma z \wedge z\tau w] \\
&\leftrightarrow x(\rho\sigma)\tau w
\end{aligned}$$

(5): Assume $x\rho y\sigma z$; by hypothesis we have $\sigma \subseteq \tau$, whence $y\tau z$. Thus $x\rho y\tau z$, i.e. $\rho\sigma \subseteq \rho\tau$.

(6): (This proof makes use of the SCHRÖDER rule, which is given only later in Th. 3.26(1) as a direct companion of the DEDEKIND rule. But if you look at the proof of the former, you will see that it is a pure FOL proof that does not need any additional assumptions besides theorems from set theory and logic, so we can safely use it without danger of a vicious circle in the proof.)

By (5) we get $\rho\sigma \subseteq \rho(\sigma \cup \tau)$ and $\rho\tau \subseteq \rho(\sigma \cup \tau)$, whence $\rho\sigma \cup \rho\tau \subseteq \rho(\sigma \cup \tau)$. For the other direction we start from $\rho\sigma \subseteq \rho\sigma$ and $\rho\tau \subseteq \rho\tau$, from which, by the first equivalence of Th. 3.26(1), we get $\rho^{\smile}\overline{\rho\sigma} \subseteq \overline{\sigma}$ and $\rho^{\smile}\overline{\rho\tau} \subseteq \overline{\tau}$. Furthermore, using (4) we have $\rho^{\smile}(\overline{\rho\sigma \cup \rho\tau}) = \rho^{\smile}(\overline{\rho\sigma} \cap \overline{\rho\tau}) \subseteq \rho^{\smile}\overline{\rho\sigma} \cap \rho^{\smile}\overline{\rho\tau} \subseteq \overline{\sigma} \cap \overline{\tau} = \overline{\sigma \cup \tau}$, i.e. $\rho^{\smile}(\overline{\rho\sigma} \cap \overline{\rho\tau}) \subseteq \overline{\sigma \cup \tau}$. Again applying Th. 3.26(1) we finally get $\rho(\sigma \cup \tau) \subseteq \rho\sigma \cup \rho\tau$.

(7): By (5) $\rho \subseteq \sigma \wedge \tau \subseteq \omega \rightarrow \rho\tau \subseteq \rho\omega \wedge \rho\omega \subseteq \sigma\omega$, whence $\rho\tau \subseteq \sigma\omega$.

(8): $x(\rho\sigma)^{\smile}y \leftrightarrow y(\rho\sigma)x \leftrightarrow \exists z[y\rho z \wedge z\sigma x] \leftrightarrow \exists z[x\sigma^{\smile}z \wedge z\rho^{\smile}y] \leftrightarrow x(\sigma^{\smile}\rho^{\smile})y$.

(9): $\langle x,z\rangle \in \bigcup_{y\in Y}(\rho y \times y\sigma) \leftrightarrow \exists y \in Y[x \in \rho y \wedge z \in y\sigma] \leftrightarrow \exists y \in Y[y \in x\rho \wedge y \in \sigma z] \leftrightarrow x\rho \cap \sigma z \neq \emptyset \leftrightarrow \langle x,z\rangle \in \rho\sigma$ □

Theorem 3.17. *Let* ρ *and* σ *be relations. The following holds:*

(1) $\mathsf{dom}(\rho\cup\sigma) = \mathsf{dom}(\rho)\cup\mathsf{dom}(\sigma)$ *and* $\mathsf{rng}(\rho\cup\sigma) = \mathsf{rng}(\rho)\cup\mathsf{rng}(\sigma)$.
(2) $\mathsf{dom}(\rho\cap\sigma) \subseteq \mathsf{dom}(\rho)\cap\mathsf{dom}(\sigma)$ *and* $\mathsf{rng}(\rho\cap\sigma) \subseteq \mathsf{rng}(\rho)\cap\mathsf{rng}(\sigma)$.
(3) $\mathsf{dom}(\rho) - \mathsf{dom}(\sigma) \subseteq \mathsf{dom}(\rho-\sigma)$ *and* $\mathsf{rng}(\rho) - \mathsf{rng}(\sigma) \subseteq \mathsf{rng}(\rho-\sigma)$.
(4) *If* $\rho\subseteq\sigma$ *then* $\mathsf{dom}(\rho)\subseteq\mathsf{dom}(\sigma)$ *and* $\mathsf{rng}(\rho)\subseteq\mathsf{rng}(\sigma)$.
(5) $\mathsf{dom}(\emptyset) = \emptyset = \mathsf{rng}(\emptyset)$.
(6) $\mathsf{dom}(I_S) = S = \mathsf{rng}(I_S)$.
(7) $\mathsf{dom}(\rho\sigma) \subseteq \mathsf{dom}(\rho)$ *and* $\mathsf{rng}(\rho\sigma) \subseteq \mathsf{rng}(\sigma)$.
(8) $(\rho\sigma)[A] = \sigma[\rho[A]]$.
(9) $\mathsf{dom}(\rho\sigma) = \rho\breve{\ }[\mathsf{dom}(\sigma)]$ *and* $\mathsf{rng}(\rho\sigma) = \sigma[\mathsf{rng}(\rho)]$.
(10) $\mathsf{dom}(\rho\sigma) = [\mathsf{rng}(\rho)\cap\mathsf{dom}(\sigma)]\rho\breve{\ }$.
(11) $\mathsf{rng}(\rho\sigma) = [\mathsf{rng}(\rho)\cap\mathsf{dom}(\sigma)]\sigma$.

(The contents of assertions (10) and (11) are illustrated in Fig. 3.1. Clearly, if $\mathsf{rng}(\rho)\cap\mathsf{dom}(\sigma) = \emptyset$ then $\rho\sigma = \emptyset$.)

Proof. (7): $\mathsf{dom}(\rho\sigma) = \{x : \exists z[x\rho\sigma z]\} = \{x : \exists y,z[x\rho y\sigma z]\} = \{x : \exists y[x\rho y \wedge \exists z[y\sigma z]]\} \subseteq \{x : \exists y[x\rho y]\} = \mathsf{dom}(\rho)$.

(8): $y \in (\rho\sigma)[A] \rightarrow \exists a \in A[a(\rho\sigma)y]$; let $a\rho x \wedge x\sigma y$; then, by definition, $x \in \rho[A]$, whence $y \in \sigma[\rho[A]]$. Conversely, assume $z \in \sigma[\rho[A]]$, and let $t \in \rho[A]$ such that $t\sigma z$. Then choose $b \in A$ with $b\rho t$; we get $b(\rho\sigma)z$, i.e. $z \in (\rho\sigma)[A]$.

(9): Let $x \in \mathsf{dom}(\rho\sigma)$, and choose a z with $x \in (\rho\sigma)z$. Then $\exists y[y \in \mathsf{dom}(\sigma) \wedge x\rho y\sigma z]$; thus $y\rho\breve{\ }x$, whence $x \in \rho\breve{\ }[\mathsf{dom}(\sigma)]$. To demonstrate the other direction, suppose $x \in \rho\breve{\ }[\mathsf{dom}(\sigma)]$. Then choose a $y \in \mathsf{dom}(\sigma)$ with $y\rho\breve{\ }x$; thus $x\rho y$, and taking z such that $y\sigma z$ we get $x(\rho\sigma)z$, so that finally $x \in \mathsf{dom}(\rho\sigma)$.

(10): Assume that $x \in \mathsf{dom}(\rho\sigma)$. Then there exist $y,z \in X$ such that $x\rho z\sigma y$. Clearly, $z \in \mathsf{rng}(\rho)\cap\mathsf{dom}(\sigma)$ and $z\rho\breve{\ }x$. Hence $x \in z\rho\breve{\ } \subseteq [\mathsf{rng}(\rho)\cap\mathsf{dom}(\sigma)]\rho\breve{\ }$. Conversely, suppose that $x \in [\mathsf{rng}(\rho)\cap\mathsf{dom}(\sigma)]\rho\breve{\ }$. Then there exists a $z \in \mathsf{rng}(\rho)\cap\mathsf{dom}(\sigma)$ such that $x \in z\rho\breve{\ }$, i.e. such that $x\rho z$. Since $z \in \mathsf{dom}(\sigma)$, there exists a y with $z\sigma y$. Therefore, $x\rho\sigma y$ and so $x \in \mathsf{dom}(\rho\sigma)$. Thus $[\mathsf{rng}(\rho)\cap\mathsf{dom}(\sigma)]\rho\breve{\ } \subseteq \mathsf{dom}(\rho\sigma)$. □

In Th. 3.4 above the set-theoretical properties of $\mathcal{B}_{S\times T}$, the set of all relations from S to T, have been described. Now we turn to the question which algebraic structure is induced on $\mathcal{B}_{S\times T}$ by the operations of product and conversion. For every three sets S, T, and U we have

(1) For every set S there is the identity relation I_S on S.
(2) The two operations of product and conversion are defined as follows (where we symbolize the product with a $\cdot$):
   - $\cdot : \mathcal{B}_{S\times T} \times \mathcal{B}_{T\times U} \mapsto \mathcal{B}_{S\times U}$ (Composition)

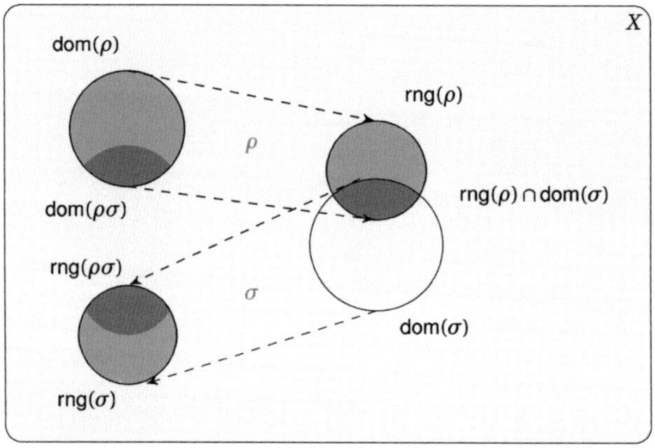

FIG. 3.1: Composition of relations.

- $\breve{}: \mathcal{B}_{S\times T} \mapsto \mathcal{B}_{T\times S}$ (Conversion)

The product of relations is associative, and for all $\rho \in \mathcal{B}_{S\times T}$ we have $\rho I_T = \rho$ and $I_S\rho = \rho$. So the first algebraic result is (for more of them see Section 9.2):

Proposition 3.18. *The structure $(\mathcal{B}_S, \cdot, I_S)$ is a monoid.*

After the basic mathematics on conversion and composition of relations we return to family configurations for first application examples. The new operations allow for a lot more definitions in this context:

- $\mathsf{grandsonOf} = \mathsf{sonOf} \cdot \mathsf{childOf}$
- $\mathsf{parentOf} = \mathsf{childOf}\breve{} = \mathsf{sonOf}\breve{} \cup \mathsf{daughterOf}\breve{}$
- $\mathsf{grandfatherOf} = \mathsf{fatherOf} \cdot (\mathsf{fatherOf} \cup \mathsf{motherOf})$
- $\mathsf{auntOf} = \mathsf{sisterOf} \cdot \mathsf{parentOf}$

The new possibilities demonstrate the usefulness of the two additional operations. However, what we are still missing, is the definition for sisterOf. Taking

$$\mathsf{sisterOf} = \mathsf{daughterOf} \cdot \mathsf{parentOf}$$

yields a problem for girls:

$$s\,\mathsf{sisterOf}\,me \leftrightarrow \exists p[s\,\mathsf{daughterOf}\,p \wedge p\,\mathsf{parentOf}\,me]$$

But, of course, $me\,\mathsf{daughterOf}\,p \wedge p\,\mathsf{parentOf}\,me$, thus $me\,\mathsf{sisterOf}\,me$! So, we must be more carefully. The problem can easily be solved as follows: At the beginning of Section 3.1 we have defined the identity relation on a set as $I_S := \{\langle x,x\rangle : x \in S\}$. We simply can use its complement, often called **diversity**, to exclude ourselves from the set of siblings:

$$\mathsf{sisterOf} = (\mathsf{daughterOf} \cdot \mathsf{parentOf}) \cap \overline{I_P}.$$

3.2.1 *Representing Relations*

A convenient way to represent relations, particularly relations between finite sets, is by using Boolean matrices. We do already know from Section 2.3.3 that $\mathbf{2} = (\{0,1\}, \vee, \wedge, 1 - ., 0, 1)$ is a Boolean algebra.

Definition 3.19.

(a) *A $m \times n$ matrix over* **2** *is called a* **Boolean matrix** *(also* **binary matrix***). The set of all $m \times n$ Boolean matrices is denoted by $\mathfrak{M}^{\mathbf{2}}_{m\times n}$; we let $\mathfrak{M}^{\mathbf{2}}_{n} := \mathfrak{M}^{\mathbf{2}}_{n\times n}$.*
(b) *Let $\rho \subseteq A \times B$ be a relation from $A = \{a_1, a_2, \ldots, a_m\}$ to $B = \{b_1, b_2, \ldots, b_n\}$. $M_\rho = (m_{ij}) \in \mathfrak{M}^{\mathbf{2}}_{m\times n}$ is the* **representation of** *ρ in case $m_{ij} = 1$ iff $a_i \rho b_j$.*
(c) *Let $M = (m_{ij}) \in \mathfrak{M}^{\mathbf{2}}_{m\times p}$ and $P = (p_{ij}) \in \mathfrak{M}^{\mathbf{2}}_{p\times n}$. The* **product of Boolean matrices** *M and P, denoted by MP is the matrix $Q = (q_{ij}) \in \mathfrak{M}^{\mathbf{2}}_{m\times n}$ with*

$$q_{ij} = \bigvee_{1\le k\le p} (m_{ik} \wedge p_{kj}). \tag{3.2}$$

Proposition 3.20. *Let $\rho \subseteq A \times B$ and $\sigma \subseteq B \times C$ be relations from A to B, and from B to C, respectively, with $A = \{a_1, a_2, \ldots, a_m\}$, $B = \{b_1, b_2, \ldots, b_p\}$, and $C = \{c_1, c_2, \ldots, c_n\}$. Furthermore, let $R = (r_{ij}) \in \mathfrak{M}^{\mathbf{2}}_{m\times p}$ and $S = (s_{ij}) \in \mathfrak{M}^{\mathbf{2}}_{p\times n}$ be representations of ρ and σ. Then $P = RS \in \mathfrak{M}^{\mathbf{2}}_{m\times n}$ is a representation of $\rho\sigma$.*

Proof. We have to show that $a_i \rho\sigma c_j$ if and only if $p_{ij} = 1$ (see Def. 3.11). Assuming $p_{ij} = 1$, we get from

$$p_{ij} = \bigvee_{1\le k\le p} (r_{ik} \wedge s_{kj}) = 1$$

that at least one of the expressions $r_{ik} \wedge s_{kj}$ must be equal to 1, for $1 \le i \le m$ and $1 \le j \le n$. Thus, $\exists l[r_{il} = 1 \wedge s_{lj} = 1]$, which implies $a_i \rho b_l$ and $b_l \sigma c_j$ for a $b_l \in B$, and therefore $a_i \rho\sigma c_j$. The proof of the other direction is just as easy. □

In a similar way we can use Boolean matrices for the representation of relation union and intersection by defining $M \vee P$ and $M \wedge P$ for two matrices $M, P \in \mathfrak{M}^{\mathbf{2}}_{m\times n}$ by

$$(m \vee p)_{ij} = m_{ij} \vee p_{ij} \text{ and } (m \wedge p)_{ij} = m_{ij} \wedge p_{ij} \tag{3.3}$$

Then, by assuming R and S to be representations of the relations ρ and σ as above in Prop. 3.20, it can easily be seen that $R \vee S$ and $R \wedge S$ are representations of $\rho \cup \sigma$ and $\rho \cap \sigma$. Moreover, the operations $\wedge$ and $\vee$ are associative, commutative, and idempotent, and also distributivity holds:

$$M(P \vee Q) = MP \vee MQ,$$
$$(M \vee P)Q = MQ \vee PQ.$$

Another often used way to visualize relations are directed graphs. Viewed informally, a *directed graph* consists of a set of *vertices*, depicted as points, and of a set of *arcs* represented by arrows joining one vertex to another. The position of the vertices in a drawing of the directed graph is irrelevant: only which vertices are joined by arcs is significant.

3.3 Basic Relation Properties

3.3.1 *Relations and Syllogisms Again*

Have a look back to Table 1.2 and remember the information contained in it: The core structure of the Aristotelian syllogism is a compound of three categorical propositions, two premises and a conclusion (which is assumed to be of the form $S - P$ in all four cases and has therefore been omitted). A syllogism involves three terms, each of them appearing exactly twice: S, M, and P, represent the syllogism's "Subject", "Middle", and "Predicate", respectively. The term which does not occur in the conclusion, but in both premises, is called the middle term. According to the positions of M in the premises, four different patterns can occur, called "figures" of the syllogism. It is the occurrence of the middle term that eventually allows us to draw an inference about the subject S and predicate P of the conclusion.

ARISTOTLE's investigation have been restricted to four kinds of categorical statements which where allowed to occur in a syllogism in each of the available places, thus yielding 256 possible inference patterns, of which less then ten percent could be identified as valid in Section 1.1.3.1. If we now switch over to the viewpoint of a relation, we are in an analogous situation: Only if there is an element, say y, which relates somehow - or not - to two other elements x and z, something can eventually be inferred about how the last two stand to each other. Following [10, Kap. II, § 2], we can take the four syllogistic figures as the basis for the subsequent considerations, but allow only one relation to occur in all the positions. So our starting point is this:

I.	$x\rho y \wedge y\rho z \to x\rho z$	(*transitive*)
II.	$x\rho y \wedge z\rho y \to x\rho z$	(*left Euclidean*)
III.	$y\rho x \wedge y\rho z \to x\rho z$	(*[right] Euclidean*)
IV.	$y\rho x \wedge z\rho y \to x\rho z$	(*circular*)

(II. and III. are generalizations of the first axiom from EUCLID's Elements: "Things which are equal to the same thing are also equal to one another." ([14], p. 95.) Though the naming for IV. is evident, this property is rarely listed in literature.)

Let ρ be a relation from S to T. Without loss of generality we can take

V. $\forall x \in S \exists y \in T[x\rho y]$ (*total*)

as an additional and convenient property. It simply means that we consider relations as being defined everywhere on the underlying universe, which is $\mathsf{dom}(\rho)$ if we follow this convention.

Though this may seem somewhat surprising, we regard I. - V. as a plausible and natural starting point for the definition of relation properties, for eventually deriving other properties, and for the investigation how these properties relate to each other. Therefore, as I have already done above, I will use what may aptly be called the *intuitive* or *point-based* kind of description for introducing a relation property (in contrast to the *abstract* or *algebraic* one to be presented later in Section 3.6.1). But to give an early example, let us describe *symmetry* in both ways:

$$\text{point-based: } \forall x, y[x\rho y \rightarrow y\rho x], \text{ and algebraic: } \rho = \rho^{\smile}.$$

Due to their formal characteristic as formulas of predicate (or first order) logic, pointwise descriptions will occasionally also be called *FOL-descriptions*.

3.3.2 *(A Lot of) Properties and some of their Interrelationships*

As the heading of this section suggests, the number of genuine relation properties that can be found in book and journal literature is considerably large. There are two principle strategies for mastering this list that come to mind immediately: Order the properties according to their importance or popularity, or simply alphabetically. Since the first method is disputable, whereas the latter is not, I decided to chose this one. Moreover, an alphabetical order is like an index: it makes look up in a list easier.

The contents of the list itself are compiled from what I found in all the papers and books I consulted during the process of writing, and also from

`https://proofwiki.org/wiki/Category:Definitions/Relation_Theory.`

To get this list comprehensive, the few properties introduced already above are repeated. Looking at Def. 3.21 you will notice:

- that a few common synonyms have been included;
- that, in rare cases, alternative equivalent definitions are given;
- that no underlying sets for the quantifiers are usually mentioned. This serves to simplify the list of definitions without damaging the presentation: the characteristic features of the property to be defined remain clearly visible. Take as an example for a generally formulated definition the following long version: $\rho \in \mathcal{B}_{S\times T}$ is called *functional* if $\forall x \in S \forall y, z \in T[x\rho y \wedge x\rho z \rightarrow y = z]$.

Definition 3.21. *This definition is given in form of the subsequent alphabetically ordered list of properties a binary relation ρ may have. The definitions are to be read as: A relation ρ is* "Name" *iff the related* "FOL-description" *holds.*

Name	**FOL-description**
acyclic (1)	$\neg\exists x_1,\ldots,x_k[x_1\rho x_2\rho\ldots x_{k-1}\rho x_k\rho x_1]$
acyclic (2)	$\forall n \geq 2\forall x_1,\ldots,x_n[x_1\rho x_2\rho\ldots x_{n-1}\rho x_n \rightarrow x_1 \neq x_n]$
antisymmetric (1)	$\forall x, y[x\rho y \wedge y\rho x \rightarrow x = y]$
antisymmetric (2)	$\forall x, y[x\rho y \wedge x \neq y \rightarrow \neg y\rho x]$
asymmetric	$\forall x, y[x\rho y \rightarrow \neg y\rho x]$
bijective	*injective* ∧ *surjective*
biorder (1)	$\forall u, v, x, y[u\rho v \wedge \neg x\rho v \wedge x\rho y \rightarrow u\rho y]$
biorder (2)	see *Ferrers*
circular	$\forall x, y, z[x\rho y \wedge y\rho z \rightarrow z\rho x]$
complete	$\forall x, y[x \neq y \rightarrow x\rho y \vee y\rho x]$
connected	see *complete*

coreflexive	$\forall x, y[x\rho y \to x = y]$
cotransitive	$\forall x, y, z[x\rho z \to x\rho y \vee y\rho z]$
dense	$\forall x, y[x\rho y \wedge x \neq y \to \exists z \notin \{x, y\}[x\rho z \wedge z\rho y]]$
deterministic	see *functional*
difunctional	$\forall x, y, u, v[x\rho u \wedge y\rho u \wedge y\rho v \to x\rho v]$
Euclidean	$\forall x, y, z[y\rho x \wedge y\rho z \to x\rho z]$
extensional	$\forall x, y[x\rho = y\rho \to x = y]$
Ferrers (1)	$\forall u, v, x, y[u\rho v \wedge x\rho y \to u\rho y \vee x\rho v]$
Ferrers (2)	see *biorder*
functional	$\forall x, y, z[x\rho y \wedge x\rho z \to y = z]$
injective	$\forall x, y, z[x\rho z \wedge y\rho z \to x = y]$
intransitive	$\forall x, y, z[x\rho y \wedge y\rho z \to \neg x\rho z]$
irreflexive	$\forall x[\neg x\rho x]$
left Euclidean	$\forall x, y, z[x\rho y \wedge z\rho y \to x\rho z]$
left total	see *total*
left unique	see *injective*
locally reflexive	$\forall x\big[\exists y[x\rho y \vee y\rho x] \to x\rho x\big]$
many-to-one	see *functional*
mapping	*total* $\wedge$ *functional*
mediate	see *dense*
negatively transitive (1)	$\forall x, y, z[x\rho z \to x\rho y \vee y\rho z]$
negatively transitive (2)	$\forall x, y, z[\neg x\rho y \wedge \neg y\rho z \to \neg x\rho z]$
onto	$\forall y \exists x[x\rho y]$
reflexive	$\forall x[x\rho x]$
regular	$\forall x, y[x\rho y \to \exists u, v[x\rho v \wedge u\rho y \wedge \forall s, t[s\rho v \wedge r\rho t \to s\rho t]]$
right Euclidean	see *Euclidean*
right unique	see *functional*
semitransitive	$\forall x, y, z, w[x\rho y \wedge y\rho z \to x\rho w \vee w\rho z]$
separating	see *injective*
serial	see *total*
strongly complete	$\forall x, y[x\rho y \vee y\rho x]$
strongly connected	see *strongly complete*
a subidentity	see *coreflexive*
surjective	see *onto*
symmetric	$\forall x, y[x\rho y \to y\rho x]$
total	$\forall x \exists y[x\rho y]$
totally connected	see *strongly complete*
transitive	$\forall x, y, z[x\rho y \wedge y\rho z \to x\rho z]$
univalent	see *functional*

We begin with picking out some prominent properties and will investigate their interrelationships. In doing so we will keep formulations as concise as possible, bypassing grammatical rules for sake of brevity.

Proposition 3.22. *let ρ be a binary relation on some set S. The following relationships hold for properties of ρ:*

(1) *left Euclidean $\wedge$ total $\to$ reflexive*
(2) *Euclidean $\wedge$ surjective $\to$ reflexive*
(3) *left Euclidean $\wedge$ reflexive $\to$ symmetric*
(4) *Euclidean $\wedge$ reflexive $\to$ symmetric*
(5) *left Euclidean $\wedge$ symmetric $\to$ transitive $\wedge$ Euclidean $\wedge$ circular*
(6) *Euclidean $\wedge$ symmetric $\to$ transitive $\wedge$ left Euclidean $\wedge$ circular*
(7) *left Euclidean $\wedge$ total $\to$ reflexive $\wedge$ symmetric $\wedge$ transitive*
(8) *ρ symmetric $\wedge$ ρ transitive $\to$ $\mathsf{dom}(\rho) = \mathsf{rng}(\rho) = \mathsf{fld}(\rho)$*
(9) *ρ symmetric $\wedge$ ρ transitive $\to$ ρ reflexive on $\mathsf{fld}(\rho)$*
(10) *symmetric $\wedge$ transitive $\to$ Euclidean*
(11) *reflexive $\wedge$ symmetric $\wedge$ transitive $\leftrightarrow$ symmetric $\wedge$ transitive $\wedge$ total $\leftrightarrow$ reflexive $\wedge$ Euclidean*
(12) *strongly complete $\leftrightarrow$ complete $\wedge$ reflexive*
(13) *reflexive $\wedge$ circular $\to$ equivalence*

Proof. (1): For a given x we choose a y such that $x\rho y$; thus $x\rho y \wedge x\rho y$, whence $x\rho x$ since ρ is assumed to be left Euclidean.

(2): $x\rho y \to y\rho x \to x\rho x$.

(3): $x\rho y \to y\rho y \wedge x\rho y \to y\rho x$.

(5): Each of the three properties: transitive, Euclidean, and circular has the same conclusion as left Euclidean: $x\rho z$; and using symmetry, each of the three premises can be transformed into the premise of left Euclidean, so we are done.

(7): By (1) we get reflexivity, by (3) symmetry, and by (5) transitivity.

(8): ([15, p. 21]) If $x \in \mathsf{dom}(\rho)$ then $\exists y[x\rho y$, whence $\exists y[y\rho x$ by symmetry. So $x \in \mathsf{rng}(\rho)$, i.e. $\mathsf{dom}(\rho) \subseteq \mathsf{rng}(\rho)$. The other direction follows by a similar argument. Therefore, $\mathsf{dom}(\rho) = \mathsf{rng}(\rho)$, yielding the assertion.

(9): ([15, p. 21]) If $x \in \mathsf{fld}(\rho)$ then by (8) $x \in \mathsf{dom}(\rho)$, so that $\exists y[x\rho y$. Since by symmetry $\exists y[y\rho x$ we get $x\rho x$ by transitivity.

(11): The assertion has the logical structure $(A) \leftrightarrow (B) \leftrightarrow (C)$. $(A) \to (B)$ follows from (9); so we need to show $(B) \to (C)$: Let $x \in S$; then $\exists y[x\rho y]$. For each such y we have $y\rho x$ by symmetry, whence $x\rho x$ by transitivity. Since $y\rho x \wedge y\rho z \to x\rho y \wedge y\rho z \to x\rho z, \rho$ is also Euclidean. $(C) \to (A)$: By (4) ρ is symmetric, and by (6) transitive.

(13): Since ρ is circular, $x\rho x \wedge x\rho y \to y\rho x$, so ρ is symmetric; and $x\rho y \wedge y\rho z \to z\rho x$, whence $x\rho z$ by symmetry, i.e. ρ is transitive. □

Note that by Prop. 3.22(7), and using only two of the five syllogistically induced properties of binary relations that we have discussed in Section 3.3.1 above, we have arrived at the properties that characterize (1) *preorders*: reflexive, transitive; (2) *tolerance relations*: reflexive, symmetric; and (3) *equivalence relations*: reflexive, symmetric, transitive. For more details see Section 4.3.1.

Proposition 3.23. ([1], Example 1.2.5) *The properties: reflexive, symmetric (or antisymmetric) and transitive are independent of each other.*

Proof. (1): Example for a relation which is reflexive and symmetric, but not transitive: Let ρ on $\mathbb{Z}$ be given as $x\rho y :\Leftrightarrow x-y \in \{-5,0,5\}$. Symmetry and reflexivity obviously hold. However, $12\rho 7 \wedge 7\rho 2$, but $\neg 12\rho 2$.
(2): Example for a relation which is reflexive and transitive, but neither symmetric nor antisymmetric: Let $\mathbb{Z}^* = \mathbb{Z} - \{0\}$, and ρ be the following relation on $\mathbb{Z}^*$ $a\rho b :\Leftrightarrow a|b$ (a is a factor of b). Then ρ clearly is reflexive and transitive, but symmetry fails by e.g. $3|6 \wedge \neg 6|3$, and antisymmetry since, for all $a \in \mathbb{Z}^*$, $-a|a \wedge a|-a \wedge a \neq -a$.
(3) Example for a relation which is symmetric and transitive, but not reflexive: Let ρ be given on $\mathbb{R}$, the set of real numbers, by $x\rho y :\Leftrightarrow xy > 0$. Clearly, ρ is symmetric and transitive, but $\neg 0\rho 0$. □

Proposition 3.24. *Some basic relationships involving irreflexivity, antisymmetry, and asymmetry are:*

(1) *irreflexive* $\wedge$ *transitive* $\rightarrow$ *asymmetric.*
(2) *irreflexive* $\wedge$ *transitive* $\rightarrow$ *acyclic.*
(3) *asymmetric* $\leftrightarrow$ *(irreflexive* $\wedge$ *antisymmetric).*
(4) *asymmetric* $\wedge$ *antisymmetric* $\rightarrow$ *transitive.*
(5) *transitive* $\rightarrow$ *(irreflexive* $\leftrightarrow$ *symmetric).*
(6) *asymmetric* $\wedge$ *negatively transitive* $\rightarrow$ *transitive.*
(7) *asymmetric* $\wedge$ *transitive* $\rightarrow$ *negatively transitive.*

Proof. (1): $a\rho b \wedge b\rho a \rightarrow a\rho a$ by the transitivity of ρ, a contradiction to its irreflexivity. So ρ must be asymmetric.

(3): Assume $x\rho x$; then by asymmetry we get $x\rho x \rightarrow \neg x\rho x$ and, from Th. 2.33(6), $\neg x\rho x$. Thus, ρ is irreflexive. By asymmetry we have $x\rho y \rightarrow \neg y\rho x$, whence $x\rho y \wedge x \neq y \rightarrow \neg y\rho x$ by Th. 2.33(17). Thus, ρ is also antisymmetric. To prove the other direction, assume $x\rho y$; then $x \neq y$ by the irreflexivity of ρ, whence $\neg y\rho x$ by antisymmetry. Thus ρ is asymmetric.

(4): Let $x\rho y \wedge y\rho z$; then $y\rho x$ by symmetry, But by the antisymmetry of ρ we get $x\rho y \wedge y\rho x \rightarrow x = y$, so $y\rho z$ becomes $x\rho z$, whence ρ is transitive.

(5): By (3) asymmetric relations are irreflexive. So it remains to show that a transitive and irreflexive relation is asymmetric: Assume that ρ is not asymmetric, i.e. $x\rho y \wedge y\rho x$ for some x, y; then transitivity implies $x\rho x$, a contradiction to irreflexivity.

(6): From $x\rho y \wedge y\rho z$, by definition of negative transitivity, we get $(x\rho z \vee z\rho y) \wedge (y\rho x \vee x\rho z)$. Since $\neg z\rho y$ and $\neg y\rho x$ by asymmetry, $x\rho z$ follows.

(7): From $x\rho y \wedge y\rho z$ we get $y\overline{\rho}x \wedge z\overline{\rho}y$ by asymmetry and $x\rho z$ by transitivity; so $z\overline{\rho}x$ by asymmetry. □

Proposition 3.25. *Properties of converse and complement relations* (see e.g. [17])

(1) ρ reflexive $\rightarrow$ $\rho^{\smile}$ reflexive $\wedge$ $\overline{\rho}$ irreflexive.
(2) ρ irreflexive $\rightarrow$ $\rho^{\smile}$ irreflexive $\wedge$ $\overline{\rho}$ reflexive.
(3) ρ symmetric $\rightarrow$ $\rho^{\smile}$ symmetric $\wedge$ $\overline{\rho}$ symmetric.
(4) ρ asymmetric $\rightarrow$ $\rho^{\smile}$ asymmetric $\wedge$ $\overline{\rho}$ not symmetric.
(5) ρ antisymmetric $\rightarrow$ $\rho^{\smile}$ antisymmetric.
(6) ρ transitive $\rightarrow$ $\rho^{\smile}$ transitive.
(7) ρ intransitive $\rightarrow$ $\rho^{\smile}$ intransitive.
(8) ρ connected $\rightarrow$ $\rho^{\smile}$ connected.

Proof. (3): Assuming that $\overline{\rho}$ is not symmetric we get $\exists x, y[x\overline{\rho}y \wedge \neg y\overline{\rho}x]$. Thus $y\rho x$, and the symmetry of ρ gives $x\rho y$, a contradiction to $x\overline{\rho}y$, whence $\overline{\rho}$ must be symmetric.

(6): Assume $z\rho^{\smile}y \wedge y\rho^{\smile}x$; then $x\rho y \wedge y\rho z$, whence $x\rho z$ by transitivity of ρ. Thus, also $z\rho^{\smile}x$. (We anticipate the algebraic version of transitivity from Prop. 3.50, viz. $\rho\rho \subseteq \rho$, to give an illustration of the different character of the proof methods: We have $\rho\rho \subseteq \rho$, whence $(\rho\rho)^{\smile} = \rho^{\smile}\rho^{\smile} \subseteq \rho^{\smile}$ by Prop. 3.13(6) and Th. 3.16(8).)

(7): Assume that $z\rho^{\smile}y \wedge y\rho^{\smile}x$; then we have $x\rho y \wedge y\rho z$, and by intransitivity of ρ we get $\neg x\rho z$. This implies $\neg z\rho^{\smile}x$. □

3.4 Some Results from the Calculus of Relations

According to what we have learned up to here, it comes at no surprise that the calculus of relations [11, 13, 18, 24] comprises five operations, viz. *union, intersection, complementation, composition* and *converse*, together with four standard constant relations: the *empty* and the *universal* relations $\emptyset$ and U, the *identity* I and the *diversity* $\overline{I}$. Applying these nine elements to given binary relations allows for the creation of a lot of new concepts useful in many applications some of which will be considered in the last part of this book.

The most important results in the calculus of relations, and their interrelationships, are presented in the next two theorems.

Theorem 3.26. *Let ρ, σ, and τ be relations, and, as usual, assume that the constructs are well formed. Then the following holds:*

(1) **SCHRÖDER *equivalences:***
(a) $\rho\sigma \subseteq \tau \;\leftrightarrow\; \rho^{\smile}\overline{\tau} \subseteq \overline{\sigma}$,
(b) $\rho\sigma \subseteq \tau \;\leftrightarrow\; \overline{\tau}\sigma^{\smile} \subseteq \overline{\rho}$.
(c) $\rho^{\smile}\overline{\tau} \subseteq \overline{\sigma} \;\leftrightarrow\; \overline{\tau}\sigma^{\smile} \subseteq \overline{\rho}$.

(2) **Modular laws***:*

(a) $\rho\sigma \cap \tau \subseteq (\rho \cap \tau\sigma^{\smile})\sigma$,

(b) $\rho\sigma \cap \tau \subseteq \rho(\sigma \cap \rho^{\smile}\tau)$.

(3) **DEDEKIND *rule*:** $\rho\sigma \cap \tau \subseteq (\rho \cap \tau\sigma^{\smile})(\sigma \cap \rho^{\smile}\tau)$.

Proof. (1)(a) *Schröder rule*, first equivalence:

$$\begin{aligned}
\rho\sigma \subseteq \tau &\leftrightarrow \forall x,y[x\rho\sigma y \rightarrow x\tau y] \\
&\leftrightarrow \forall x,y[\exists z[x\rho z \wedge z\sigma y] \rightarrow x\tau y] \\
&\leftrightarrow \forall x,y[\neg\exists z[x\rho z \wedge z\sigma y] \vee x\tau y] \\
&\leftrightarrow \forall x,y,z[\neg z\rho^{\smile}x \vee x\tau y \vee \neg z\sigma y] && \text{Th. 2.37(1)} \\
&\leftrightarrow \forall y,z[\neg\exists x[z\rho^{\smile}x \wedge \neg x\tau y] \vee \neg z\sigma y] \\
&\leftrightarrow \forall y,z[\exists x[z\rho^{\smile}x \wedge \neg x\tau y] \rightarrow \neg z\sigma y] && \text{Th. 2.33(7)} \\
&\leftrightarrow \forall y,z[z\rho^{\smile}\overline{\tau}y \rightarrow \neg z\sigma y] \\
&\leftrightarrow \rho^{\smile}\overline{\tau} \subseteq \overline{\sigma}
\end{aligned}$$

(1)(b) *Schröder rule*, second equivalence:

$$\begin{aligned}
\rho\sigma \subseteq \tau &\leftrightarrow \forall x,y[x\rho\sigma y \rightarrow x\tau y] \\
&\leftrightarrow \forall x,y[\exists z[x\rho z \wedge z\sigma y] \rightarrow x\tau y] \\
&\leftrightarrow \forall x,y[\neg\exists z[x\rho z \wedge z\sigma y] \vee x\tau y] \\
&\leftrightarrow \forall x,y,z[x\tau y \vee \neg x\rho z \vee \neg z\sigma y] \\
&\leftrightarrow \forall x,y,z[\neg x\overline{\tau}y \vee \neg y\sigma^{\smile}z \vee \neg x\rho z] \\
&\leftrightarrow \forall x,z[\neg\exists y[x\overline{\tau}y \wedge y\sigma^{\smile}z] \vee \neg x\rho z] \\
&\leftrightarrow \forall x,z[x\overline{\tau}\sigma^{\smile}z \rightarrow \neg x\rho z] \\
&\leftrightarrow \overline{\tau}\sigma^{\smile} \subseteq \overline{\rho}
\end{aligned}$$

(2)(a) *Modular laws*, relational proof:

$$\begin{aligned}
\rho\sigma \cap \tau &= [\rho \cap (\tau\sigma^{\smile} \cup \overline{\tau\sigma^{\smile}})]\sigma \cap \tau && \text{since } \rho \cap U = \rho \\
&= [(\rho \cap \tau\sigma^{\smile})\sigma \cup (\rho \cap \overline{\tau\sigma^{\smile}})\sigma] \cap \tau && \text{distributivity, twice} \\
&= [(\rho \cap \tau\sigma^{\smile})\sigma \cap \tau] \cup [(\rho \cap \overline{\tau\sigma^{\smile}})\sigma \cap \tau] && \text{distributivity} \\
&= [(\rho \cap \tau\sigma^{\smile})\sigma \cap \tau] \\
&\subseteq (\rho \cap \tau\sigma^{\smile})\sigma
\end{aligned}$$

(The last equation follows, because $(\rho \cap \overline{\tau\sigma^{\smile}})\sigma \cap \tau = \emptyset$ as shown below:

$$\begin{aligned}
\tau\sigma^{\smile} \subseteq \tau\sigma^{\smile} &\leftrightarrow \overline{\tau\sigma^{\smile}}\sigma \subseteq \overline{\tau} && \text{Schröder rule} \\
&\rightarrow (\rho \cap \overline{\tau\sigma^{\smile}})\sigma \subseteq \overline{\tau} && \text{monotonicity} \\
&\leftrightarrow (\rho \cap \overline{\tau\sigma^{\smile}})\sigma \cap \tau = \emptyset && \text{set theory)}
\end{aligned}$$

(3) *Dedekind rule*, relational proof:

$$\begin{aligned}
\rho\sigma \cap \tau &= [\rho\sigma \cap \tau] \cap \tau \\
&\subseteq [(\rho \cap \tau\sigma^{\smile})\sigma] \cap \tau && \text{by (2)(a), monotonicity} \\
&= [(\rho \cap \tau\sigma^{\smile})\sigma \cap \tau] \cap \tau \\
&\subseteq [(\rho \cap \tau\sigma^{\smile})((\rho \cap \tau\sigma^{\smile})^{\smile}\tau \cap \sigma)] \cap \tau && \text{by (2)(b), with } \rho := \rho \cap \tau\sigma^{\smile} \\
&\subseteq [(\rho \cap \tau\sigma^{\smile})(\rho^{\smile}\tau \cap \sigma)] \cap \tau && \text{monotonicity} \\
&\subseteq (\rho \cap \tau\sigma^{\smile})(\sigma \cap \rho^{\smile}\tau)
\end{aligned}$$

(3) *Dedekind rule*, FOL proof for direct comparison:

$$x(\rho\sigma \cap \tau)y \leftrightarrow \exists z[x\rho z \wedge z\sigma y] \wedge x\tau y$$

$$
\begin{array}{lll}
\leftrightarrow & \exists z[x\rho z \wedge x\tau y \wedge y\sigma\breve{}\, z \wedge z\sigma y \wedge z\rho\breve{}\, x \wedge x\tau y] & \text{by Th. 2.36(1)} \\
\rightarrow & \exists z[x\rho z \wedge \exists s[x\tau s \wedge s\sigma\breve{}\, z] \wedge z\sigma y \wedge \exists t[z\rho\breve{}\, t \wedge t\tau y]] & \text{by Th. 2.38(EG)} \\
\leftrightarrow & \exists z[x(\rho \cap \tau\sigma\breve{}\,)z \wedge z(\sigma \cap \rho\breve{}\,\tau)y] & \\
\leftrightarrow & x(\rho \cap \tau\sigma\breve{}\,)(\sigma \cap \rho\breve{}\,\tau)y &
\end{array}
$$

□

(Note that comparing the formulation of the so-called Schröder equivalences from above with the discussion of DE MORGAN's **Theorem K** on p. 15, one might conclude that the wrong name has been chosen for them.)

Theorem 3.27. *For relations ρ, σ and τ the following holds:*

(1) *The two modular rules together are equivalent to the* DEDEKIND *rule.*
(2) *The* SCHRÖDER *equivalences are equivalent to the* DEDEKIND *rule.*

Proof. (1) The modular rules follow directly from the DEDEKIND rule by virtue of the monotonicity properties of relation composition:

$$\rho\sigma \cap \tau \subseteq (\rho \cap \tau\sigma\breve{}\,)(\sigma \cap \rho\breve{}\,\tau) \subseteq \begin{cases} (\rho \cap \tau\sigma\breve{}\,)\sigma \\ \rho(\sigma \cap \rho\breve{}\,\tau). \end{cases}$$

Conversely, assume that both modular rules hold; then we have

$$
\begin{array}{lll}
\rho\sigma \cap \tau & \subseteq \rho(\sigma \cap \rho\breve{}\,\tau) & \text{Th. 3.26(2)(b)} \\
\rho\sigma \cap \tau \cap \tau & \subseteq \rho(\sigma \cap \rho\breve{}\,\tau) \cap \tau & \text{Prop. 2.3(6)} \\
& \subseteq (\rho \cap \tau(\sigma \cap \rho\breve{}\,\tau)\breve{}\,)(\sigma \cap \rho\breve{}\,\tau) & \text{Th. 3.26(2)(a)} \\
& = (\rho \cap \tau(\sigma\breve{}\, \cap \tau\breve{}\,\rho))(\sigma \cap \rho\breve{}\,\tau) & \\
& \subseteq (\rho \cap \tau\sigma\breve{}\,)(\sigma \cap \rho\breve{}\,\tau) & \text{Monotonicity of product.}
\end{array}
$$

(2) We show that the DEDEKIND rule implies $\rho\sigma \subseteq \tau \leftrightarrow \rho\breve{}\,\overline{\tau} \subseteq \overline{\sigma}$. Now assume that the DEDEKIND rule holds; then by (1) also the modular laws are valid. Further assume $\rho\sigma \subseteq \tau$. We have $\rho\sigma \subseteq \tau \leftrightarrow \rho\sigma \cap \overline{\tau} = \emptyset$ by set theory, and $\rho\breve{}\,\overline{\tau} \cap \sigma \subseteq \rho\breve{}\,(\overline{\tau} \cap \rho\sigma)$ by Th. 3.26(2)(b). But $\rho\breve{}\,(\overline{\tau} \cap \rho\sigma) = \rho\breve{}\,\emptyset = \emptyset$, whence $\rho\breve{}\,\overline{\tau} \cap \sigma = \emptyset$, and therefore $\rho\breve{}\,\overline{\tau} \subseteq \overline{\sigma}$. Conversely, let $\rho\breve{}\,\overline{\tau} \subseteq \overline{\sigma}$. Then $\rho\breve{}\,\overline{\tau} \cap \sigma = \emptyset$. Again using Th. 3.26(2)(b) we get $\rho\sigma \cap \overline{\tau} \subseteq \rho(\rho\breve{}\,\overline{\tau} \cap \sigma) = \rho\emptyset = \emptyset$. So $\rho\sigma \subseteq \tau$.

For the other direction we assume that the Schröder equivalences hold and get

$$
\begin{array}{ll}
[\rho][\sigma] & \\
= [(\rho \cap \tau\sigma\breve{}\,) \cup (\rho \cap \overline{\tau\sigma\breve{}}\,)][(\sigma \cap \rho\breve{}\,\tau) \cup (\sigma \cap \overline{\rho\breve{}\,\tau})] & \text{using } (A \cap B) \cup (A \cap \overline{B}) = A \\
= (\rho \cap \tau\sigma\breve{}\,)(\sigma \cap \rho\breve{}\,\tau) \cup (\rho \cap \tau\sigma\breve{}\,)(\sigma \cap \overline{\rho\breve{}\,\tau}) \cup & \\
\quad (\rho \cap \overline{\tau\sigma\breve{}}\,)(\sigma \cap \rho\breve{}\,\tau) \cup (\rho \cap \overline{\tau\sigma\breve{}}\,)(\sigma \cap \overline{\rho\breve{}\,\tau}) &
\end{array}
$$

The last three products are bounded by $\overline{\tau}$ (for a proof see below); therefore

$$\rho\sigma \cap \tau \subseteq [(\rho \cap \tau\sigma\breve{}\,)(\sigma \cap \rho\breve{}\,\tau) \cup \overline{\tau}] \cap \tau \subseteq (\rho \cap \tau\sigma\breve{}\,)(\sigma \cap \rho\breve{}\,\tau).$$

By the first Schröder equivalence we have $\rho\breve{}\,\tau \subseteq \rho\breve{}\,\tau \leftrightarrow \rho\overline{\rho\breve{}\,\tau} \subseteq \overline{\tau}$ whence for the second product we get $(\rho \cap \tau\sigma\breve{}\,)(\sigma \cap \overline{\rho\breve{}\,\tau}) \subseteq \rho\overline{\rho\breve{}\,\tau} \subseteq \overline{\tau}$ by monotonicity of composition. Similarly, the second Schröder equivalence gives $\tau\sigma\breve{}\, \subseteq \tau\sigma\breve{}\, \leftrightarrow \overline{\tau\sigma\breve{}}\,\sigma \subseteq \overline{\tau}$. From this it follows for the product term $(\rho \cap \overline{\tau\sigma^{\smile}})(\sigma \cap \rho^{\smile}\tau) \subseteq \overline{\tau\sigma^{\smile}}\sigma \subseteq \overline{\tau}$, and for the product term $(\rho \cap \overline{\tau\sigma^{\smile}})(\sigma \cap \overline{\rho^{\smile}\tau}) \subseteq \overline{\tau\sigma^{\smile}}\sigma \subseteq \overline{\tau}$. □

Proposition 3.28. ([9, p. 250]) *For relations ρ, σ and τ the following holds:*

(1) Restriction identities:

$$\begin{aligned} (\rho \cap \sigma U)\tau &= \rho\tau \cap \sigma U, \\ \rho(\sigma \cap U\tau) &= \rho\sigma \cap U\tau, \\ (\rho \cap U\sigma^{\smile})\tau &= \rho(\tau \cap \sigma U). \end{aligned}$$

(2) Diagonal identities:

$$\begin{aligned} I \cap U\rho &= I \cap \rho^{\smile}\rho, \\ I \cap \rho U &= I \cap \rho\rho^{\smile}. \end{aligned}$$

Proof.

(1)
$$\begin{aligned} (\rho \cap \sigma U)\tau &\subseteq \rho\tau \cap \sigma U\tau \\ &\subseteq \rho\tau \cap \sigma U && U\tau \subseteq U\text{, monotonicity of } \circ \\ &\subseteq (\rho \cap \sigma U\tau^{\smile})(\tau \cap \rho^{\smile}\sigma U) && \text{Dedekind rule 3.26(3)} \\ &\subseteq (\rho \cap \sigma U)\tau && U\tau^{\smile} \subseteq U\text{, monotonicity of } \circ \end{aligned}$$

The second equation can be proved in a similar way, and the third as follows:

$$\begin{aligned} (\rho \cap U\sigma^{\smile})\tau &= (\rho \cap U\sigma^{\smile})\tau \cap U \\ &\subseteq [\rho \cap U\sigma^{\smile} \cap U\tau^{\smile}][\tau \cap (\rho \cap U\sigma^{\smile})^{\smile}U] && \text{Dedekind rule 3.26(3)} \\ &\subseteq \rho[\tau \cap (\rho^{\smile} \cap \sigma U)U] && \text{monotonicity of } \circ \\ &\subseteq \rho(\tau \cap \rho^{\smile}U \cap \sigma UU) && \text{Th. 3.16(4)} \\ &\subseteq \rho(\tau \cap \sigma U) && \text{monotonicity of } \circ\text{, } UU = U \\ &= \rho(\tau \cap \sigma U) \cap U \\ &\subseteq (\rho \cap U(\tau \cap \sigma U)^{\smile})(\tau \cap \sigma U \cap \rho^{\smile}U) && \text{Dedekind rule 3.26(3)} \\ &\subseteq (\rho \cap U(\sigma U)^{\smile})\tau && \text{monotonicity of } \circ \text{ and } {}^{\smile} \\ &\subseteq (\rho \cap U\sigma^{\smile})\tau && \text{Th. 3.16(8), } UU^{\smile} = U \end{aligned}$$

(2) By applying the Dedekind rule 3.26(3) we get $U\rho \cap I \subseteq (U \cap I\rho^{\smile})(\rho \cap U^{\smile}I) = \rho^{\smile}(\rho \cap U) = \rho^{\smile}\rho$; thus $U\rho \cap I \subseteq \rho^{\smile}\rho \cap I$. The converse inclusion follows by $\rho^{\smile}\rho \subseteq U\rho$, whence $\rho^{\smile}\rho \cap I \subseteq U\rho \cap I$. (Compare to Prop. 3.15(1) and (2).) □

Proposition 3.29. *Let ρ, σ, τ, and ω be relations. The following holds:*

(1) $\rho\sigma \cap \tau = \emptyset \;\leftrightarrow\; \rho^{\smile}\tau \cap \sigma = \emptyset$.

(2) $\rho\sigma \cap \tau\omega \subseteq \rho(\rho^{\smile}\tau \cap \sigma\omega^{\smile})\omega$

Proof. (1): Using the Schröder equivalence Th. 3.26(1)(a) we get $\rho\sigma \cap \tau = \emptyset \leftrightarrow \rho\sigma \subseteq \overline{\tau} \leftrightarrow \rho^{\smile}\tau \subseteq \overline{\sigma} \leftrightarrow \rho^{\smile}\tau \cap \sigma = \emptyset$.

(2): Assume $x(\rho\sigma \cap \tau\omega)z$; choose s, t such that $x\rho s\sigma z$ and $x\tau t\omega z$. Then $s\rho^{\smile}x\tau t$, whence $s(\rho^{\smile}\tau)t$. In the same way we get $s(\sigma\omega^{\smile})t$. Thus $s(\rho^{\smile}\tau \cap \sigma\omega^{\smile})t, x\rho s$, and $t\omega z$, and finally $x\,\rho(\rho^{\smile}\tau \cap \sigma\omega^{\smile})\omega\,z$. □

3.5 Dual Relations, Residuals and Symmetric Quotients

This section introduces some operations that arise by a concatenation of elementary operations defined so far: complement, conversion, and product. It often turns out more comfortable to use such abbreviations instead of aggregates built from the basic ones.

3.5.1 *Dual Relations*

Definition 3.30. $\rho^d = \{\langle x, y\rangle : \langle y, x\rangle \notin \rho\}$ *is called the* **dual relation** *for* ρ.

Note that by Prop. 3.13(5) we have $\rho^d = \overline{\rho^\smile} = \overline{\rho}^\smile$. For dual relations the following easy to prove relationships hold:

Corollary 3.31. *Let* $\rho, \sigma \subseteq X \times Y$ *be binary relations. Then*

(1) $\rho \subseteq \sigma$ *iff* $\sigma^d \subseteq \rho^d$;
(2) $(\rho^d)^d = \rho$.
(3) $(\rho^d)^\smile = \overline{\rho} = (\rho^\smile)^d$.
(4) $\overline{\rho}^d = \rho^\smile$.
(5) $\rho\sigma^\smile\rho \subseteq \rho \;\leftrightarrow\; \sigma \subseteq \overline{\rho\rho^d\rho}$.

Proof. (1): By Prop. 3.13(6), $\rho \subseteq \sigma$ iff $\rho^\smile \subseteq \sigma^\smile$. But $\rho^\smile \subseteq \upsilon^\smile$ iff $\overline{\sigma^\smile} \subseteq \overline{\rho^\smile}$.
(2): $(\rho^d)^d = \overline{(\overline{\rho^\smile})^\smile} = (\overline{\overline{\rho^\smile}})^\smile = \rho^{\smile\smile} = \rho$.
(3): $(\rho^d)^\smile = (\overline{\rho}^\smile)^\smile = \overline{\rho} = \overline{\rho^{\smile\smile}} = (\rho^\smile)^d$.
(4): By (2) and (3).

(5):

$$\begin{array}{lll} & \rho\sigma^\smile\rho \subseteq \rho & \\ \leftrightarrow & \sigma^\smile\rho \subseteq \overline{\rho^\smile\overline{\rho}} & \text{by (1)} \\ \leftrightarrow & \sigma^\smile \subseteq \overline{\rho^\smile\overline{\rho}\rho^\smile} & \text{by (1)} \\ \leftrightarrow & \sigma \subseteq \overline{\rho\rho^d\rho} & \text{by Prop. 3.13(5),(6)} \end{array}$$

□

Proposition 3.32. ([8, Lemma 7.10]) *Let* ρ *be a binary relation on* X. *Then:*

(1) *For all* $x, y \in X$ *we have*
- $x\rho\rho^d y$ *iff* $x\rho \not\subseteq y\rho$;
- $x\rho^d\rho y$ *iff* $\rho y \not\subseteq \rho x$;
- $x(\rho\rho^d)^d y$ *iff* $x\rho \supseteq y\rho$;
- $x(\rho^d\rho)^d y$ *iff* $\rho y \supseteq \rho x$.

(2) *Both* $\rho\rho^d$ *and* $\rho^d\rho$ *are negatively transitive.*
(3) *The following properties are equivalent:*
   (a) $\rho\rho^d$ *is asymmetric;*
   (b) $\rho^d\rho$ *is asymmetric;*

(c) $\rho\rho^d$ *is a weak order (i.e. asymmetric and negatively transitive);*
(d) $\rho^d\rho$ *is a weak order.*

Proof. (1): Assume $x\rho\rho^d y$; then $x\rho z$ and $y\overline{\rho}z$ for some $z \in X$, which is equivalent to $x\rho \not\subseteq y\rho$. For proving the third assertion, first note that $(\rho\rho^d)^d = \overline{(\rho\overline{\rho}^{\smile})}^{\smile} = \overline{(\rho\overline{\rho}^{\smile})^{\smile}} = \overline{\overline{\rho}\rho^{\smile}}$. Then we get

$$\begin{aligned} x\overline{\overline{\rho}\rho^{\smile}}\,y &\leftrightarrow \neg\exists z \in X[x\overline{\rho}z \wedge z\rho^{\smile}y] & \\ &\leftrightarrow \forall z \in X[x\rho z \vee z\overline{\rho^{\smile}}y] & \text{by Th. 2.35(1)} \\ &\leftrightarrow \forall z \in X[y\rho z \to x\rho z] & \text{by Th. 2.33(7).} \end{aligned}$$

(2): Let $\tau := \rho\rho^d$; then from the first equivalence in (1) we get $x\overline{\tau}y$ iff $x\rho \subseteq y\rho$. Now assume that also $y\overline{\tau}z$, so $y\rho \subseteq z\rho$. Thus, $x\rho \subseteq z\rho$, i.e. $x\overline{\tau}z$.

(3): Assuming that $\rho\rho^d$ is asymmetric, but $\rho^d\rho$ is not, implies $x\rho^d\rho\,y$ and $y\rho^d\rho\,x$ for two elements $x \neq y$. So there are two distinct elements z and t with $z\overline{\rho}x \wedge z\rho y$ from the first, and $t\overline{\rho}y \wedge t\rho x$ from the second relationship, whence we have $t\,\rho\rho^d\,z$. But from $z\rho y$ and $t\overline{\rho}y$ we also get $z\,\rho\rho^d\,t$, a contradiction to the asymmetry of $\rho\rho^d$. In a similar way one can show that the asymmetry of $\rho^d\rho$ implies that of $\rho\rho^d$; thus (a) is equivalent to (b).

By (2) $\rho^d\rho$ is negatively transitive, so (b) implies (c). Since (c) trivially implies (b) by the definition of a weak order, we get that (b) is equivalent to (c). And using (2), (a), and (b) we get that (c) is equivalent to (d). □

3.5.2 *Residuals*

(A warning: Unfortunately, mainly notation but also definition of residuals is a little bit confusing. So you will have to be very careful in order to assure to always work with the correct translations between them. Some examples are listed in Table 3.1.)

Since we have defined an operation called *product*, a natural question is, if we can also define an operation that gives something like a *quotient* of two relations. To eventually find an answer to this question, the concept of a (left and right, since product of relations is not commutative) **residual** is introduced: Residuals are the greatest solutions χ of the relational inequalities $\rho\chi \subseteq \sigma$ and $\chi\rho \subseteq \sigma$, respectively.

Remembering that we look for a kind of division, the modern notation as used in Boyd & Everett(1999) [7] seems to be suitable: in the expressions $\rho(\sigma/\rho) \subseteq \sigma$ and $(\sigma\backslash\rho)\rho \subseteq \sigma$, the ρ's "cancel out". Moreover, this scheme of notation follows the majority of the sample given in Table 3.1 and is also used in e.g. [6] and many others: the upper end of the / or \ symbol points in the direction indicated by its name, which is the place of the unknown factor χ in the product.

Definition 3.33.

(a) Let $\rho \in \mathcal{B}_{C\times A}$ *and* $\sigma \in \mathcal{B}_{C\times B}$ *be two relations; then the* **right residual** *of* σ *by* ρ *is* $\sigma/\rho \in \mathcal{B}_{A\times B}$ *defined as* $\sigma/\rho := \overline{\rho^{\smile}\overline{\sigma}}$.

(b) *Let* $\rho \in \mathcal{B}_{A\times C}$ *and* $\sigma \in \mathcal{B}_{B\times C}$ *be two relations; then the* **left residual** *of* σ *by* ρ *is* $\sigma\backslash\rho \in \mathcal{B}_{B\times A}$ *defined as* $\sigma\backslash\rho := \overline{\overline{\sigma}\,\rho^{\smile}}$.

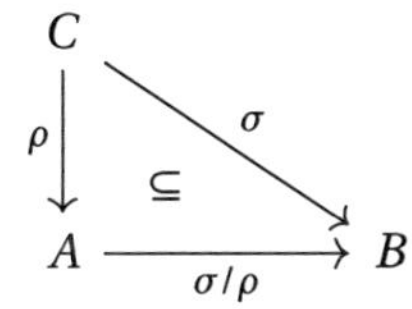

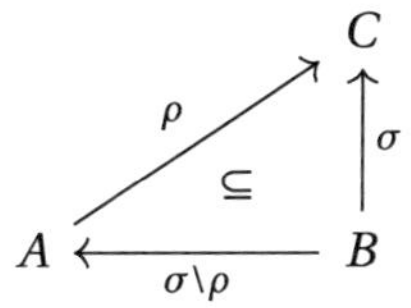

Note that the definitions of the right and left residual involve relations with common source and target, respectively,

TABLE 3.1: Residuals: Terminology and Notation.

Notation	Name	Definition	Source
$a\,.\!\cdot b$	*right residual* of a by b	largest x such that $bx \le a$	Birkhoff(1973)
$a\cdot\!.\,b$	*left residual* of a by b	largest y such that $yb \le a$	[5, p. 325]
$\sigma\,.\!\cdot\rho$	*right residue* of σ by ρ	greatest rel. χ with $\rho\chi \subseteq \sigma$; $\sigma\,.\!\cdot\rho := \overline{\rho^{\smile}\overline{\sigma}}$	Schmidt & Ströhlein (1993) [21, p. 19]
$\sigma\cdot\!.\,\rho$	*left residue* of σ by ρ	greatest rel. χ with $\chi\rho \subseteq \sigma$; $\sigma\cdot\!.\,\rho := \overline{\overline{\sigma}\rho^{\smile}}$	
$\rho\backslash\sigma$	*right residual* of σ by ρ	$\rho\backslash\sigma := \overline{\rho^{\smile}\overline{\sigma}}$	Berghammer(2012)
σ/ρ	*left residual* of σ by ρ	$\sigma/\rho := \overline{\overline{\sigma}\rho^{\smile}}$	[2, p. 223]
σ/ρ	*right residual* of σ by ρ	$\sigma/\rho := \overline{\rho^{\smile}\overline{\sigma}}$	Boyd & Everett
$\sigma\backslash\rho$	*left residual* of σ by ρ	$\sigma\backslash\rho := \overline{\overline{\sigma}\rho^{\smile}}$	(1999) [7, p. 151]
$\rho\backslash\sigma$	***left** residual*	$\rho\backslash\sigma := \overline{\rho^{\smile}\overline{\sigma}}$	Schmidt(2011)
σ/ρ	***right** residual*	$\sigma/\rho := \overline{\overline{\sigma}\rho^{\smile}}$	[20, p. 43]

Proposition 3.34. *Left and right residuals have the following properties:*

(1) $\rho\chi \subseteq \sigma \leftrightarrow \chi \subseteq \sigma/\rho$
(2) $\chi\rho \subseteq \sigma \leftrightarrow \chi \subseteq \sigma\backslash\rho$
(3) $x(\sigma/\rho)y \leftrightarrow \rho x \subseteq \sigma y$, *i.e.* $\forall z[z\rho x \to z\sigma y]$
(4) $x(\sigma\backslash\rho)y \leftrightarrow x\rho \subseteq y\sigma$, *i.e.* $\forall z[x\rho z \to y\sigma z]$
(5) $\rho(\sigma/\rho) \subseteq \sigma$, *i.e.* σ/ρ *is the greatest solution* χ *of* $\rho\chi \subseteq \sigma$
(6) $(\sigma\backslash\rho)\rho \subseteq \sigma$, *i.e.* $\sigma\backslash\rho$ *is the greatest solution* χ *of* $\chi\rho \subseteq \sigma$
(7) $\rho\tau \subseteq \sigma \leftrightarrow \tau \subseteq \sigma/\rho \leftrightarrow \rho \subseteq \sigma\backslash\tau$
(8) $\rho\tau \subseteq \sigma \leftrightarrow \rho \subseteq \overline{\overline{\sigma}\tau^{\smile}} \leftrightarrow \tau \subseteq \overline{\rho^{\smile}\overline{\sigma}}$.

Proof. (1): Using the first equivalence of the Schröder rule, we have $\rho\chi \subseteq \sigma \leftrightarrow \rho^{\smile}\overline{\sigma} \subseteq \overline{\chi}$, which is equivalent to $\chi \subseteq \overline{\rho^{\smile}\overline{\sigma}}$ by Prop. 2.4(4).

(2): $\chi\rho \subseteq \sigma \leftrightarrow \overline{\sigma}\rho^{\smile} \subseteq \overline{\chi} \leftrightarrow \chi \subseteq \overline{\overline{\sigma}\rho^{\smile}} = \sigma\backslash\rho$

(3): $x(\sigma/\rho)y \leftrightarrow x(\overline{\rho^{\smile}\,\overline{\sigma}})y \leftrightarrow \neg\exists z[x\rho^{\smile}z \wedge z\overline{\sigma}y]$ by definition of composition

$\leftrightarrow \forall z[x\overline{\rho^{\smile}}z \vee z\sigma y]$ by Th. 2.35(1)

$\leftrightarrow \forall z[x\rho^{\smile}z \rightarrow z\sigma y]$ by Th. 2.33(7)

$\leftrightarrow \forall z[z\rho x \rightarrow z\sigma y]$

$\leftrightarrow \rho x \subseteq \sigma y$

(5): We have $\rho^{\smile}\overline{\sigma} \subseteq \rho^{\smile}\overline{\sigma} \leftrightarrow \rho^{\smile}\overline{\sigma} \subseteq \overline{\overline{\rho^{\smile}\overline{\sigma}}} \leftrightarrow \rho^{\smile}\overline{\sigma} \subseteq \overline{\sigma/\rho}$, which is equivalent to $\rho(\sigma/\rho) \subseteq \sigma$ by the Schröder rule, and by (1) σ/ρ is the greatest solution.

(7): By (1) and (2).

(8): Using Th. 3.26(1)(a),(b) we get $\rho\tau \subseteq \sigma \leftrightarrow \rho^{\smile}\overline{\sigma} \subseteq \overline{\tau} \leftrightarrow \tau \subseteq \overline{\rho^{\smile}\overline{\sigma}}$, and similarly Th. 3.26(1)(b) implies $\rho\tau \subseteq \sigma \leftrightarrow \overline{\sigma}\tau^{\smile} \subseteq \overline{\rho} \leftrightarrow \rho \subseteq \overline{\overline{\sigma}\tau^{\smile}}$. □

By (3) of the above proposition, x is (σ/ρ)-related to y iff the x-column of ρ is is a subset of the y-column of σ; and by (4) $\sigma\backslash\rho$ is characterized by the fact that the y-row of ρ is a subset of the x-row of σ. Do also note that by (1) and (2) of the above proposition the following holds:

Corollary 3.35. *Let ρ be a transitive relation; then $\rho \subseteq \rho/\rho = \overline{\rho^{\smile}\overline{\rho}}$ and $\rho \subseteq \rho\backslash\rho = \overline{\overline{\rho}\rho^{\smile}}$.*

Proposition 3.36. ([4, Lemma 2.3]) *Let $\rho, \rho' \subseteq Y \times Z$, $\sigma, \sigma' \subseteq X \times Y$, and $\tau, \tau' \subseteq X \times Z$ be relations. Then the following properties hold:*

(1) If $\tau \subseteq \tau'$ and $\rho' \subseteq \rho$ then $\tau\backslash\rho \subseteq \tau'\backslash\rho'$.
(2) If $\tau' \subseteq \tau$ and $\sigma \subseteq \sigma'$ then $\tau/\sigma \subseteq \tau'/\sigma'$.
(3) $\rho\backslash I = \rho$, $I/\rho = \rho$, and $\overline{I}\backslash\overline{\rho} = \rho^{\smile}$.
(4) $I \subseteq \rho/\rho$ and $I \subseteq \rho\backslash\rho$.
(5) Both ρ/ρ and $\rho\backslash\rho$ are transitive.
(6) $\rho(\rho/\rho) = \rho$ and $(\rho\backslash\rho)\rho = \rho$
(7) $(\rho\backslash\sigma)(\sigma\backslash\tau) \subseteq \rho\backslash\tau$ and $(\sigma/\rho)(\tau/\sigma) \subseteq \tau/\rho$.

Proof. (1): By hypothesis $\overline{\tau'} \subseteq \overline{\tau}$ and $\rho'^{\smile} \subseteq \rho^{\smile}$. So $\overline{\tau'}\rho'^{\smile} \subseteq \overline{\tau}\rho^{\smile}$ which implies $\tau\backslash\rho \subseteq \tau'\backslash\rho'$.

(3): Trivially, $\overline{I}\backslash\overline{\rho} = \overline{I\overline{\rho}^{\smile}} = \overline{\overline{\rho}^{\smile}} = \rho^{\smile}$, using Prop. 3.13(5).

(4): By Prop. 3.34(6) $\rho\backslash\rho$ is the greatest solution χ of $\chi\rho \subseteq \rho$, so $I \subseteq \rho\backslash\rho$.

(5): By Prop. 3.34(5) we have $\rho(\rho/\rho) \subseteq \rho$, whence $\rho(\rho/\rho)^2 \subseteq \rho(\rho/\rho) \subseteq \rho$. And since ρ/ρ is the greatest solution χ of $\rho\chi \subseteq \rho$, $(\rho/\rho)^2 \subseteq \rho/\rho$ follows.

(6): By Prop. 3.34(5) we have $\rho(\rho/\rho) \subseteq \rho$, and by (4) $\rho = \rho I \subseteq \rho(\rho/\rho)$

(7): Using Prop. 3.34(5),(6) we get $(\rho\backslash\sigma)(\sigma\backslash\tau)\tau \subseteq (\rho\backslash\sigma)\sigma \subseteq \rho$ and $\rho(\sigma/\rho)(\tau/\sigma) \subseteq \sigma(\tau/\sigma) \subseteq \tau$, from which the assertions follow by Prop. 3.34(1),(2) □

There are still more properties for residuals, which exhibit a further regularity of this concept:

Proposition 3.37. *Using conversion and complement, the following interrelationships hold for residuals:*

(1) $\sigma / \rho = (\sigma^{\smile} \backslash \rho^{\smile})^{\smile}$
(2) $\sigma \backslash \rho = (\sigma^{\smile} / \rho^{\smile})^{\smile}$
(3) $\overline{\sigma} / \overline{\rho} = \rho^{\smile} \backslash \sigma^{\smile}$
(4) $\overline{\rho} / \overline{\sigma} = \rho^{\smile} / \sigma^{\smile}$

Proof.

(1): $\sigma / \rho = \overline{\rho^{\smile}\,\overline{\sigma}}$ by definition of right residual
$= \overline{(\overline{\sigma}^{\smile}\,\rho)^{\smile}}$ by Th. 3.16(8)
$= \overline{(\overline{\sigma^{\smile}}\,\rho^{\smile\smile})}^{\smile}$ using Prop. 3.13(5) twice
$= (\sigma^{\smile} \backslash \rho^{\smile})^{\smile}$ by definition of left residual

(3): $\overline{\sigma} / \overline{\rho} = \overline{\overline{\rho}^{\smile}\,\sigma} = \rho^{\smile} \backslash \sigma^{\smile}$.

□

3.5.3 *Symmetric Quotients and Bidirectional Residuals*

There are two more notions that occur in connections with residuals, viz. the *bidirectional residual* $\mathsf{biR}(\rho)$ and the *symmetric quotient* $\mathsf{syQ}(\rho, \sigma)$ that are useful in various considerations on relations. We begin with the latter one by providing its definition and afterwards listing some remarkable properties of symmetric quotients. The material of this section has been compiled from [2, Section 7.5], [20, Section 8.5], and [21, Section 4.4].

Definition 3.38. *Let $\rho \subseteq X \times Y$ and $\sigma \subseteq X \times Z$ be two relations with coinciding source; then the* **symmetric quotient** *of ρ and σ is defined as*

$$\mathsf{syQ}(\rho, \sigma) := \overline{\rho^{\smile}\,\overline{\sigma}} \cap \overline{\overline{\rho}^{\smile}\,\sigma} \subseteq Y \times Z.$$

Corollary 3.39.

(1) $\mathsf{syQ}(\rho, \sigma) = (\sigma / \rho) \cap (\rho^{\smile} \backslash \sigma^{\smile}) = (\sigma / \rho) \cap (\overline{\sigma} / \overline{\rho})$.
(2) $\langle y, z \rangle \in \mathsf{syQ}(\rho, \sigma)$ *iff* $\rho y = \sigma z$.
(3) $\tau \subseteq \mathsf{syQ}(\rho, \sigma)$ *iff* $\rho\tau \subseteq \sigma \wedge \tau\sigma^{\smile} \subseteq \rho^{\smile}$.

Proof. (1): By definition and Prop. 3.37(3).

(2): By Prop. 3.34(3),(4) we get $x(\sigma / \rho)y \leftrightarrow \rho x \subseteq \sigma y$ and $y(\rho^{\smile} \backslash \sigma^{\smile})z \leftrightarrow y\sigma^{\smile} \subseteq x\rho^{\smile} \leftrightarrow \sigma y \subseteq \rho x$.

(3): By Prop. 3.34(7) we get $\tau \subseteq \sigma / \rho \leftrightarrow \rho\tau \subseteq \sigma$ and $\tau \subseteq \rho^{\smile} \backslash \sigma^{\smile} \leftrightarrow \tau\sigma^{\smile} \subseteq \rho^{\smile}$. Now apply (1). □

Thus, given two relations $\rho \subseteq X \times Y$ and $\sigma \subseteq X \times Z$, the symmetric quotient $\mathsf{syQ}(\rho, \sigma)$ relates a $y \in Y$ to $z \in Z$ iff y and z have identical foresets. In terms of the representing matrices for ρ and σ this means that the columns corresponding to y and z, respectively, are equal.

Proposition 3.40. *For the symmetric quotient of two relations ρ,σ the following holds:*

(1) $\mathsf{syQ}(\rho,\sigma) = \mathsf{syQ}(\sigma,\rho)^{\smile}$.
(2) $\mathsf{syQ}(\rho,\sigma) = \mathsf{syQ}(\overline{\rho},\overline{\sigma})$.
(3) $I \subseteq \mathsf{syQ}(\rho,\rho)$.

Proof. (1): Cor. 3.39(1) implies $\mathsf{syQ}(\sigma,\rho)^{\smile} = \left[(\rho/\sigma)\cap(\sigma^{\smile}\backslash\rho^{\smile}\right]^{\smile} = \left[(\rho^{\smile}\backslash\sigma^{\smile})^{\smile}\cap(\sigma^{\smile}\backslash\rho^{\smile})\right]^{\smile} = (\rho^{\smile}\backslash\sigma^{\smile})\cap(\sigma^{\smile}\backslash\rho^{\smile})^{\smile} = (\rho^{\smile}\backslash\sigma^{\smile})\cap(\sigma/\rho) = \mathsf{syQ}(\rho,\sigma)$ by Prop. 3.37.

(2): From Cor. 3.39(1) we get $\mathsf{syQ}(\rho,\sigma) = (\sigma/\rho)\cap(\overline{\sigma}/\overline{\rho}) = (\overline{\sigma}/\overline{\rho})\cap(\overline{\overline{\sigma}}/\overline{\overline{\rho}}) = \mathsf{syQ}(\overline{\rho},\overline{\sigma})$.

(3): By Prop. 3.36(4) we have $I \subseteq \rho/\rho$, and from Prop. 3.37(1) we have $\rho/\rho = (\sigma^{\smile}\backslash\rho^{\smile})^{\smile}$, so $I^{\smile} \subseteq (\rho/\rho)^{\smile} = (\sigma^{\smile}\backslash\rho^{\smile})^{\smile\smile}$. Thus $I \subseteq (\rho/\rho)\cap(\sigma^{\smile}\backslash\rho^{\smile}) = \mathsf{syQ}(\rho,\rho)$. □

Corollary 3.41. *For the symmetric quotient of two relations ρ,σ the following holds:*

(1) $\rho\cdot\mathsf{syQ}(\rho,\sigma) = \sigma\cap U\cdot\mathsf{syQ}(\rho,\sigma)$
(2) $\rho\cdot\mathsf{syQ}(\rho,\sigma) \subseteq \sigma$.
(3) *If* $\mathsf{syQ}(\rho,\sigma)$ *is surjective, then* $\rho\cdot\mathsf{syQ}(\rho,\sigma) = \sigma$.

Proof. (1): We have

$$\rho\cdot\mathsf{syQ}(\rho,\sigma) = \rho(\overline{\rho^{\smile}\overline{\sigma}}\cap\overline{\overline{\rho}^{\smile}\sigma}) \subseteq \rho\overline{\rho^{\smile}\overline{\sigma}}\cap\rho\overline{\overline{\rho}^{\smile}\sigma} \subseteq \rho\overline{\rho^{\smile}\overline{\sigma}},$$

by subdistributivity and monotony. And from $\rho^{\smile}\overline{\sigma} \subseteq \rho^{\smile}\overline{\sigma}$ we get $\rho\overline{\rho^{\smile}\overline{\sigma}} \subseteq \sigma$ by application of the Schröder rule Th. 3.26(1)(a). So finally $\rho\cdot\mathsf{syQ}(\rho,\sigma) \subseteq \sigma$. Obviously $\rho\cdot\mathsf{syQ}(\rho,\sigma) \subseteq U\cdot\mathsf{syQ}(\rho,\sigma)$, whence we have proved the inclusion $\subseteq$.
To show the converse inclusion note that we have (∗) $\overline{\rho}\cdot\mathsf{syQ}(\rho,\sigma) \subseteq \overline{\sigma}$, which follows from the $\subseteq$-inclusion just shown with $\overline{\rho},\overline{\sigma}$ instead of ρ,σ and Prop. 3.40(2). Then

$$\begin{aligned}
\sigma\cap U\cdot\mathsf{syQ}(\rho,\sigma) &= \sigma\cap(\rho\cup\overline{\rho})\cdot\mathsf{syQ}(\rho,\sigma)\\
&= \sigma\cap[\rho\cdot\mathsf{syQ}(\rho,\sigma)\cup\overline{\rho}\cdot\mathsf{syQ}(\rho,\sigma)]\\
&= [\sigma\cap\rho\cdot\mathsf{syQ}(\rho,\sigma)]\cup[\sigma\cap\overline{\rho}\cdot\mathsf{syQ}(\rho,\sigma)]\\
&\subseteq [\rho\cdot\mathsf{syQ}(\rho,\sigma)]\cup[\sigma\cap\overline{\sigma}] \qquad \text{by monotonicity, } (*)\\
&= \rho\cdot\mathsf{syQ}(\rho,\sigma).
\end{aligned}$$

(2): Trivial by (1).

(3): An easy consequence from (1), since $U\cdot\mathsf{syQ}(\rho,\sigma) = U$ by Prop. 4.2(4). □

Proposition 3.42. *Let ρ,σ and τ be relations; then* $\mathsf{syQ}(\rho,\sigma)\cdot\mathsf{syQ}(\sigma,\tau) \subseteq \mathsf{syQ}(\rho,\tau)$.

Proof. Due to the Schröder rule Th. 3.26(1)(b) the following equivalence holds:

$$\mathsf{syQ}(\rho,\sigma)\cdot\mathsf{syQ}(\sigma,\tau) \subseteq \mathsf{syQ}(\rho,\tau) \leftrightarrow \overline{\mathsf{syQ}(\rho,\tau)}\cdot\mathsf{syQ}(\sigma,\tau)^{\smile} \subseteq \overline{\mathsf{syQ}(\rho,\sigma)}$$

Thus, to prove the validity of the left inclusion, we have to show that of the right one:

$$\begin{aligned}
&\overline{\mathsf{syQ}(\rho,\tau)}\cdot\mathsf{syQ}(\sigma,\tau)^{\smile}\\
=\ &\overline{\mathsf{syQ}(\rho,\tau)}\cdot\mathsf{syQ}(\tau,\sigma) && \text{Prop. 3.40(1)}\\
=\ &(\rho^{\smile}\overline{\tau}\cup\overline{\rho}^{\smile}\tau)\cdot\mathsf{syQ}(\tau,\sigma) && \text{definition of syQ(.,.)}\\
=\ &\rho^{\smile}\overline{\tau}\cdot\mathsf{syQ}(\tau,\sigma)\cup\overline{\rho}^{\smile}\tau\cdot\mathsf{syQ}(\tau,\sigma)\\
=\ &\rho^{\smile}\overline{\tau}\cdot\mathsf{syQ}(\overline{\tau},\overline{\sigma})\cup\overline{\rho}^{\smile}\tau\cdot\mathsf{syQ}(\tau,\sigma) && \text{Prop. 3.40(2)}
\end{aligned}$$

$$\begin{array}{rll} \subseteq & \rho^{\smile}\overline{\sigma}\cup\overline{\rho}^{\smile}\sigma & \text{Cor. 3.41(2)} \\ = & \overline{\mathsf{syQ}(\rho,\sigma)}. & \end{array}$$

□

Proposition 3.43. *For the symmetric quotient of two relations we have*

$$\mathsf{syQ}(\rho,\sigma)\cdot\mathsf{syQ}(\sigma,\tau)=\mathsf{syQ}(\rho,\tau)\cap U\cdot\mathsf{syQ}(\sigma,\tau)=\mathsf{syQ}(\rho,\tau)\cap\mathsf{syQ}(\rho,\sigma)\cdot U.$$

Proof. We prove the first equality and begin with the $\subseteq$-inclusion: By Prop. 3.42 we have $\mathsf{syQ}(\rho,\sigma)\cdot\mathsf{syQ}(\sigma,\tau)\subseteq\mathsf{syQ}(\rho,\tau)$, and from $\mathsf{syQ}(\rho,\sigma)\subseteq U$ we get $\mathsf{syQ}(\rho,\sigma)\cdot\mathsf{syQ}(\sigma,\tau)\subseteq U\cdot\mathsf{syQ}(\sigma,\tau)$, whence $\mathsf{syQ}(\rho,\sigma)\cdot\mathsf{syQ}(\sigma,\tau)\subseteq\mathsf{syQ}(\rho,\tau)\cap U\cdot\mathsf{syQ}(\sigma,\tau)$. The converse inclusion is verified by

$$\begin{array}{rll} & \mathsf{syQ}(\rho,\tau)\cap U\cdot\mathsf{syQ}(\sigma,\tau) & \\ = & U\cdot\mathsf{syQ}(\sigma,\tau)\cap\mathsf{syQ}(\rho,\tau) & \\ \subseteq & [U\cap\mathsf{syQ}(\rho,\tau)\cdot\mathsf{syQ}(\sigma,\tau)^{\smile}]\cdot[\mathsf{syQ}(\sigma,\tau)\cap U^{\smile}\cdot\mathsf{syQ}(\rho,\tau)] & \\ & & \text{by the Dedekind rule Th. 3.26(3)} \\ \subseteq & [\mathsf{syQ}(\rho,\tau)\cdot\mathsf{syQ}(\sigma,\tau)^{\smile}]\cdot[\mathsf{syQ}(\sigma,\tau)] & \\ = & \mathsf{syQ}(\rho,\tau)\cdot\mathsf{syQ}(\tau,\sigma)\cdot\mathsf{syQ}(\sigma,\tau) & \text{Prop. 3.40(1)} \\ \subseteq & \mathsf{syQ}(\rho,\sigma)\cdot\mathsf{syQ}(\sigma,\tau) & \text{Prop. 3.42} \end{array}$$

□

Corollary 3.44.

(1) *If* $\mathsf{syQ}(\rho,\sigma)$ *is total, or if* $\mathsf{syQ}(\sigma,\tau)$ *is surjective, then* $\mathsf{syQ}(\rho,\sigma)\cdot\mathsf{syQ}(\sigma,\tau)=\mathsf{syQ}(\rho,\tau)$.
(2) $\mathsf{syQ}(\rho,\sigma)\cdot\mathsf{syQ}(\rho,\sigma)^{\smile}\subseteq\mathsf{syQ}(\rho,\rho)$.
(3) $\mathsf{syQ}(\rho,\rho)\cdot\mathsf{syQ}(\rho,\sigma)=\mathsf{syQ}(\rho,\sigma)$.
(4) $\mathsf{syQ}(\rho,\rho)$ *is an equivalence relation.*

Proof. (1): Similar to Cor. 3.41(1), by Prop. 3.43(2) and Prop. 4.2(4).

(2): By Prop. 3.42, using Prop. 3.40(1).

(3): We have $I\subseteq\mathsf{syQ}(\rho,\rho)$ by Prop. 3.40(3), so $\mathsf{syQ}(\rho,\sigma)\subseteq\mathsf{syQ}(\rho,\rho)\cdot\mathsf{syQ}(\rho,\sigma)\subseteq\mathsf{syQ}(\rho,\sigma)$ by Prop. 3.42.

(4): $\mathsf{syQ}(\rho,\rho)$ is reflexive by Prop. 3.40(3), symmetric by construction, and transitive by (3). □

Proposition 3.45.

(1) ρ *is an equivalence relation iff* $\rho=\mathsf{syQ}(\rho,\rho)$.
(2) *If* ρ *is functional and surjective then* $I=\mathsf{syQ}(\rho,\rho)$.

Proof. (1): $\mathsf{syQ}(\rho,\rho)$ is an equivalence by Cor. 3.44(4). Assume that ρ is an equivalence; then from $I\subseteq\rho$ we get $\mathsf{syQ}(\rho,\rho)\subseteq\rho\cdot\mathsf{syQ}(\rho,\rho)\subseteq\rho$ by Cor. 3.41(2). To obtain the converse inclusion note that $\rho\rho\subseteq\rho\leftrightarrow\rho^{\smile}\overline{\rho}\subseteq\overline{\rho}$ using Th. 3.26(1)(a), so $\rho\subseteq\overline{\rho^{\smile}\overline{\rho}}=\mathsf{syQ}(\rho,\rho)$.

(2): Since ρ is functional and surjective, we have $\rho^{\smile}\rho = I$ by Prop. 3.50. Thus, $\mathsf{syQ}(\rho,\rho) = I \cdot \mathsf{syQ}(\rho,\rho) = \rho^{\smile}\rho \cdot \mathsf{syQ}(\rho,\rho) \subseteq \rho^{\smile}\rho = I$, where the inequality follows from Cor. 3.41(2). Together with Prop. 3.40(3) the assertion follows. □

Proposition 3.46. *For injective relations ρ,σ we have*

(1) $\rho^{\smile}\sigma \subseteq \mathsf{syQ}(\rho,\sigma)$,
(2) $\rho \cdot \mathsf{syQ}(\rho,\sigma) = \rho U \cap \sigma$.
(3) $\sigma \subseteq \mathsf{syQ}(\rho^{\smile}, \mathsf{syQ}(\rho,\sigma))$.

Proof. (1): Since ρ is injective, $\rho\rho^{\smile} \subseteq I$, whence $\rho\rho^{\smile}\sigma \subseteq \sigma$, which is equivalent to $\rho^{\smile}\sigma \subseteq \sigma/\rho$ by Prop. 3.34(7). Since σ is injective, $\sigma^{\smile}$ is functional, i.e. $\sigma\sigma^{\smile} \subseteq I$, whence $\rho^{\smile}\sigma\sigma^{\smile} \subseteq \rho^{\smile}$, which is equivalent to $\rho^{\smile}\sigma \subseteq \rho^{\smile}\backslash\sigma^{\smile}$ by Prop. 3.34(7). Therefore, $\rho^{\smile}\sigma \subseteq \sigma/\rho \cap \rho^{\smile}\backslash\sigma^{\smile} = \mathsf{syQ}(\rho,\sigma)$ using Cor. 3.39(1).

(2): We get $\rho U \cap \sigma \subseteq (\rho \cap \sigma U^{\smile})(U \cap \rho^{\smile}\sigma) \subseteq \rho\rho^{\smile}\sigma \subseteq \rho \cdot \mathsf{syQ}(\rho,\sigma)$ by the Dedekind rule Th. 3.26(3) and (1). The opposite inclusion follows using Cor. 3.41(2).

(3): Using Prop. 3.34(7), from (1) we get $\sigma \subseteq \mathsf{syQ}(\rho,\sigma)/\rho^{\smile}$. By Prop. 3.40(1) we have $\mathsf{syQ}(\rho,\sigma)^{\smile} = \mathsf{syQ}(\sigma,\rho)$ and by Cor. 3.41(2) $\sigma \cdot \mathsf{syQ}(\sigma,\rho) \subseteq \rho$. This implies $\sigma \subseteq \rho\backslash\mathsf{syQ}(\sigma,\rho) = \rho\backslash\mathsf{syQ}(\rho,\sigma)^{\smile}$, so $\sigma \subseteq \mathsf{syQ}(\rho^{\smile}, \mathsf{syQ}(\rho,\sigma))$ by Cor. 3.39(1). □

We now turn to the second notion that we have mentioned at the begin of this section in connection with residuals, that of the *bidirectional* residual with

Definition 3.47. [4, p. 43] *Let $\rho \subseteq X \times Y$; then the* **bidirectional residual** $\mathsf{biR}(\rho)$ *is defined as the greatest relation $\sigma \subseteq Y \times X$ satisfying $\rho\sigma\rho \subseteq \rho$, i.e.*

$$\sigma \subseteq \mathsf{biR}(\rho) \;\leftrightarrow\; \rho\sigma\rho \subseteq \rho,$$

for all $\sigma \subseteq Y \times X$.

Corollary 3.48.

(1) $\mathsf{biR}(\rho) = \rho/(\rho\backslash\rho) = (\rho/\rho)\backslash\rho = \overline{\rho^{\smile}\overline{\rho}\rho^{\smile}}$.
(2) $\rho\sigma^{\smile}\rho \subseteq \rho \;\leftrightarrow\; \sigma \subseteq \overline{\rho\rho^{d}\rho}$.

Proof. (1): A simple application of Def. 3.33 gives $\rho\sigma\rho \subseteq \rho \;\leftrightarrow\; \rho\sigma \subseteq \rho\backslash\rho \;\leftrightarrow\; \sigma \subseteq \rho/(\rho\backslash\rho) = \overline{\rho^{\smile}\overline{(\rho\backslash\rho)}} = \overline{\rho^{\smile}\overline{\overline{(\overline{\rho}\rho^{\smile})}}} = \overline{\rho^{\smile}\overline{\rho}\rho^{\smile}}$.

(2): Straightforward using the definition of $\mathsf{biR}(\rho)$ and (1). □

Proposition 3.49. [4, Lemmas 2.6 and 3.1] *For relations ρ,σ the following properties hold*

(1) $\mathsf{biR}(\rho)^{\smile} = \mathsf{biR}(\rho^{\smile})$.
(2) $\rho \cdot \mathsf{biR}(\rho) \subseteq \rho/\rho$ *and* $\mathsf{biR}(\rho) \cdot \rho \subseteq \rho\backslash\rho$.
(3) $\mathsf{biR}(\rho) \cdot (\rho\backslash\rho) \subseteq \mathsf{biR}(\rho)$ *and* $(\rho/\rho) \cdot \mathsf{biR}(\rho) \subseteq \mathsf{biR}(\rho)$.

Proof. (1): By Prop. 3.37(1),(2) we have $(\rho/\sigma)^{\smile} = \sigma^{\smile}\backslash\rho^{\smile}$ and $(\sigma\backslash\rho)^{\smile} = \sigma^{\smile}/\rho^{\smile}$. We then get $\mathsf{biR}(\rho)^{\smile} = \big(\rho/(\rho\backslash\rho)\big)^{\smile} = (\rho\backslash\rho)^{\smile}\backslash\rho^{\smile} = (\rho^{\smile}/\rho^{\smile})\backslash\rho^{\smile} = \mathsf{biR}(\rho^{\smile})$ by using Cor. 3.48(1) twice.

(2): By Prop. 3.34(5) $\mathsf{biR}(\rho)$ is the greatest relation χ such that $\rho\cdot\mathsf{biR}(\rho)\subseteq\rho\backslash\rho$.

(3): With Prop. 3.42 we get $\mathsf{biR}(\rho)\cdot(\rho\backslash\rho) = ((\rho/\rho)\backslash\rho)\cdot(\rho\backslash\rho)$. □

3.6 Operations and Properties revisited

If we take the set theoretical properties of relations summarized in Section 3.1 as a sine qua non, then the relations specific news we learned about up to here essentially are:

- Three special relations: $\emptyset, I_S$ and U_S, in Section 3.1;
- two operations: Conversion $^{\smile}$ and composition $\cdot$, in Section 3.2;
- the many properties listed in Section 3.3, Def. 3.21.

Using the modest tools mentioned in the first and second item, this section will address two problems: (1) the algebraic representation of the properties using the set-theoretical and relational operations at hand; and (2) the logical relationships of the properties among each other.

3.6.1 *Point-free Representation of Relation Properties*

As a mathematician you might perhaps wonder about what might have been the reason for using the cumbersome FOL-descriptions, if we can do it much more compact and elegant in many cases, as shown below. In my view, this question has a simple and obvious answer: because we usually think in an element-based, or set theoretical, way, or at least start to think so. It is in line with the fact that syllogisms preceded the emergence of formal predicate calculus quite a couple of centuries, or with the naming of the *natural* numbers. And you surely remember the question, if it is a pure accident that we typically use the decimal number system, or if the fact that we have then fingers has something to do with it.

We present the algebraic representations using the structure of Def. 3.21, and we will also give some of the proofs. It should be clear from the context if ρ denotes a relation on S or a relation from S to T without overloading the list with this detail. Using the relational, or algebraic, instead of the set-theoretical definitions leads to equational proofs that are more compact and easier to verify. Working with this representation makes up what is usually called the *calculus of relations* and which is the starting point for the axiomatic abstraction that is named *relation algebra* (see Section 3.6).

Proposition 3.50. *This proposition is given in form of the subsequent, alphabetically ordered list of properties a binary relation ρ may have; the list corresponds to the one given in Def. 3.21, but here the properties are described algebraically, i.e. point-free. The assertions are to be read as: A relation ρ is* "Name" *iff the related* "Algebraic Property" *holds.*

Name	**Algebraic Property**
acyclic	$\forall n \geq 2[\rho^n \cap I_S = \emptyset]$
antisymmetric	$\rho \cap \rho^{\smile} \subseteq I_S \leftrightarrow \rho^{\smile} \subseteq \overline{\rho} \cup I$
asymmetric	$\rho \cap \rho^{\smile} = \emptyset \leftrightarrow \rho^{\smile} \subseteq \overline{\rho}$
bijective	*injective* $\wedge$ *surjective*
biorder	$\rho \cdot \rho^d \cdot \rho \subseteq \rho$
circular	$\rho \cdot \rho \subseteq \rho^{\smile}$
complete	see *connected*
connected	$\overline{I_S} \subseteq \rho \cup \rho^{\smile}$
coreflexive	$\rho \subseteq I$
cotransitive	$\rho \subseteq \rho \cdot \rho$
dense	$\rho \cap \overline{I} \subseteq (\rho \cap \overline{I})(\rho \cap \overline{I})$
deterministic	see *functional*
difunctional	$\rho \rho^{\smile} \rho \subseteq \rho$
Euclidean	$\rho^{\smile} \cdot \rho \subseteq \rho$
extensional	$\overline{(\rho \overline{\rho^{\smile}})} \cap \overline{(\rho^{\smile} \overline{\rho})} \subseteq I$
Ferrers	(see *biorder*)
functional	$\rho^{\smile} \cdot \rho \subseteq I_T$
injective	$\rho \cdot \rho^{\smile} \subseteq I_S \leftrightarrow \rho^{\smile}$ is *functional*
intransitive	$\rho \cdot \rho \subseteq \overline{\rho}$
irreflexive	$I_S \cap \rho = \emptyset \leftrightarrow I_S \subseteq \overline{\rho}$
left Euclidean	$\rho \cdot \rho^{\smile} \subseteq \rho$
left total	see *total*
left unique	see *injective*
locally reflexive	$(\rho \cup \rho^{\smile})_{\mathsf{dom}} \subseteq \rho$
many-to-one	see *functional*
mapping	$\rho \cdot \overline{I} = \overline{R} \leftrightarrow \rho$ is *functional* and *total*
mediate	see *dense*
negatively transitive	$\overline{\rho} \cdot \overline{\rho} \subseteq \overline{\rho}$
one-to-many	see *injective*
onto	$I_T \subseteq \rho^{\smile} \cdot \rho$
reflexive	$I_S \subseteq \rho$
regular	$\exists \sigma \in \mathcal{B}_S[\rho \cdot \sigma \cdot \rho = \rho]$
right Euclidean	see *Euclidean*
right unique	see *functional*
semitransitive	$\rho \cdot \rho \cdot \rho^d \subseteq \rho$
separating	$\rho \cdot \rho^{\smile} = I_S$
serial	see *total*
strongly complete	see *strongly connected*
strongly connected	$\rho \cup \rho^{\smile} = U_{S \times S}$
a subidentity	see *coreflexive*
surjective	$I_S \subseteq \rho^{\smile} \cdot \rho$

symmetric	$\rho^{\smile} = \rho \;\leftrightarrow\; \rho^{\smile} \cap \overline{\rho} = \emptyset$
total	$I_S \subseteq \rho \cdot \rho^{\smile}$
totally connected	see *strongly connected*
transitive	$\rho \cdot \rho \subseteq \rho$
univalent	see *functional*

Proof. (For illustration purposes we include some of the proofs for the equivalence of the corresponding FOL and algebraic description of a property. And for the convenience of the reader, we will repeat the point-based definition.)

***(connected)**:* $\forall x, y[x \neq y \to x\rho y \vee y\rho x] \iff \overline{I_S} \subseteq \rho \cup \rho^{\smile}$
[⇒] Assume

$$x\overline{I_S}y\,, \tag{3.4}$$

then

$$\{x, y\} \subseteq \mathsf{fld}(\rho) \wedge x \neq y\,, \tag{3.5}$$

which, by connectedness of ρ, gives

$$x\rho y \vee y\rho x. \tag{3.6}$$

Thus

$$x(\rho \cup \rho^{\smile})y. \tag{3.7}$$

[⇐] We have to demonstrate that (3.6) follows from (3.5) based on the hypothesis

$$\overline{I_S} \subseteq \rho \cup \rho^{\smile}\,. \tag{3.8}$$

But by (3.8) we obviously have (3.4) $\to$ (3.7), and, since (3.4) $\leftrightarrow$ (3.5) and (3.6) $\leftrightarrow$ (3.7), we get the desired result.

***(difunctional)**:* $\forall x, y, u, v[x\rho u \wedge y\rho u \wedge y\rho v \to x\rho v] \iff \rho\rho^{\smile}\rho \subseteq \rho$
We have

$$\begin{aligned} & \forall x, y, u, v[x\rho u \wedge y\rho u \wedge y\rho v \to x\rho v] \\ \leftrightarrow\; & \forall x, y, u, v[x\rho u \wedge u\rho^{\smile} y \wedge y\rho v \to x\rho v] \\ \leftrightarrow\; & \exists u, y[x\rho u \wedge u\rho^{\smile} y \wedge y\rho v] \to x\rho v && \text{by Th. 2.36(3)(a)} \\ \leftrightarrow\; & x\rho\rho^{\smile}\rho v \to x\rho v \\ \leftrightarrow\; & \rho\rho^{\smile}\rho \subseteq \rho. \end{aligned}$$

***(extensional)**:* $\forall x, y[x\rho = y\rho \to x = y] \iff \overline{(\rho\overline{\rho}^{\smile})} \cap \overline{(\rho^{\smile}\overline{\rho})} \subseteq I$
[⇒] Assume that ρ is extensional and $x[\overline{(\rho\overline{\rho}^{\smile})} \cap \overline{(\rho^{\smile}\overline{\rho})}]y$; it must be shown that $x = y$. Suppose that $x\rho \subset y\rho$; then $\exists z[x\rho z \wedge \neg y\rho z]$, i.e. $x(\rho\overline{\rho}^{\smile})y$, contradicting the hypothesis $x\overline{(\rho\overline{\rho}^{\smile})}y$. Thus we have $x\rho = y\rho$, and by extensionality $x = y$, i.e. xIy.
[⇐] Suppose that the right hand side holds, and that $x\rho = y\rho$. Assuming $x \neq y$ we get $x(\rho\overline{\rho}^{\smile})y$ or $x(\rho^{\smile}\overline{\rho})y$. In the first case, $\exists z[x\rho z \wedge \neg z\rho^{\smile} y]$, i.e. $\exists z[x\rho z \wedge y\overline{\rho}z]$, which contradicts $x\rho = y\rho$. Similarly, the second case cannot hold, whence $x = y$.

***(Ferrers)**:* $\forall u, v, x, y[u\rho v \wedge x\rho y \to u\rho y \vee x\rho v] \iff \rho\rho^{d}\rho \subseteq \rho$

Note that the left hand side is equivalent to $\forall u, v, x, y[u\rho v \wedge \neg x\rho v \wedge x\rho y \rightarrow u\rho y]$.

(functional): $\forall x, y, z[x\rho y \wedge x\rho z \rightarrow y = z] \iff \rho^{\smile} \cdot \rho \subseteq I_T$

$[\Rightarrow]$ $\forall x, y, z[x\rho y \wedge x\rho z \rightarrow y = z] \leftrightarrow \forall y, z\exists x[y\rho^{\smile} x \wedge x\rho z] \rightarrow y = z$ by definition of conversion and Th. 2.36(3)(a).

$[\Leftarrow]$ Assume that $\rho^{\smile} \cdot \rho \subseteq I_T$, and let $x\rho^{\smile} y$ and $y\rho z$. This gives $x\rho^{\smile} \cdot \rho z$ and, by assumption, $x = z$.

(injective): $\forall x, y, z[x\rho z \wedge y\rho z \rightarrow x = y] \iff \rho \cdot \rho^{\smile} \subseteq I_S$

$[\Rightarrow]$ $x\rho \cdot \rho^{\smile} z \rightarrow \exists y[x\rho y \wedge z\rho y]$. Thus $x = z$ since ρ is assumed to be injective, whence $\langle x, z\rangle = \langle x, x\rangle$, and finally $\rho \cdot \rho^{\smile} \subseteq I_S$.

$[\Leftarrow]$ Now assume that $\rho \cdot \rho^{\smile} \subseteq I_S$, and let $x\rho y$ and $y\rho^{\smile} z$. This gives $x\rho \cdot \rho^{\smile} z$ and, by the assumption, $x = z$.

(total): $\forall x\exists y[x\rho y] \iff I_S \subseteq \rho \cdot \rho^{\smile}$

$[\Rightarrow]$ $\forall x\exists y[x\rho y]$ implies for each $x \in S$ there is a y such that: $x\rho y$, whence $y\rho^{\smile} x$, and thus $x\rho\rho^{\smile} x$ for each $x \in S$. So $I_S \subseteq \rho \cdot \rho^{\smile}$.

$$
\begin{aligned}
[\Leftarrow]\ I_S \subseteq \rho \cdot \rho^{\smile} &\rightarrow \forall x\exists y[x\rho y \wedge y\rho^{\smile} x]] \\
&\rightarrow \forall x[\exists y[x\rho y] \wedge \exists y[y\rho^{\smile} x]] && \text{by Th. 2.37(2)(b)} \\
&\rightarrow \forall x\exists y[x\rho y] \wedge \forall x\exists y[y\rho^{\smile} x]. && \text{by Th. 2.37(1)(b)}
\end{aligned}
$$

(transitive): $\forall x, y, z[x\rho z \wedge z\rho y \rightarrow x\rho y] \iff \rho \cdot \rho \subseteq \rho$

$$
\begin{aligned}
\rho \cdot \rho \subseteq \rho &\leftrightarrow \forall x, z[\exists y[x\rho y \wedge y\rho z] \rightarrow x\rho z] \\
&\leftrightarrow \forall x, z[\forall y[\neg x\rho y \vee \neg y\rho z] \vee x\rho z] && \text{by Th. 2.35(1), Th. 2.33(7)} \\
&\leftrightarrow \forall x, y, z[\neg x\rho y \vee \neg y\rho z \vee x\rho z] \\
&\leftrightarrow \forall x, y, z[x\rho y \wedge y\rho z \rightarrow x\rho z].
\end{aligned}
$$

□

To get a feeling for how to work with this algebraic machinery we reformulate two proofs whose point-based version has been given above in Prop. 3.22. Prop. 3.51(2) shows that the algebraic need not necessarily be shorter or simpler than the point-based proof ($x\rho y \rightarrow y\rho y \wedge x\rho y \rightarrow y\rho x$).

Proposition 3.51.

(1) total $\wedge$ left Euclidean $\rightarrow$ reflexive
(2) left Euclidean $\wedge$ reflexive $\rightarrow$ symmetric

Proof. (1): $I_S \subseteq \rho \cdot \rho^{\smile} \wedge \rho \cdot \rho^{\smile} \subseteq \rho \rightarrow I_S \subseteq \rho$

(2): By assumption we have $\rho \cdot \rho^{\smile} \subseteq \rho$ and $I_S \subseteq \rho$. Thus, by monotonicity of composition, $I_S \cdot \rho^{\smile} \subseteq \rho \cdot \rho^{\smile}$, and by Th. 3.16(1) $I_S \cdot \rho^{\smile} = \rho^{\smile}$, whence $\rho^{\smile} \subseteq \rho \cdot \rho^{\smile} \subseteq \rho$. To demonstrate the other inequality we apply Prop. 3.13(6) and Th. 3.16(8): From $\rho \cdot \rho^{\smile} \subseteq \rho$, we get $(\rho \cdot \rho^{\smile})^{\smile} = \rho \cdot \rho^{\smile} \subseteq \rho^{\smile}$; and $I_S \subseteq \rho$ gives $I_S^{\smile} = I_S \subseteq \rho^{\smile}$. Thus $\rho = \rho \cdot I_S \subseteq \rho \cdot \rho^{\smile} \subseteq \rho^{\smile}$. □

3.6.2 *Properties of Compound Relations*

We have seen earlier in this chapter that relations can have a lot of properties that have arisen in the course of time by using relations as a tool in various domains of pure and applied science. The same, of course holds true, for relations that emerge by one ore more applications of one ore more of the operations defined so far. Relations of this kind will be called **compound**. Thus, the most simple compound relations that can be built out of a relation ρ are $\overline{\rho}$ and $\rho^{\smile}$. Some of the most basic results for these, and also for intersection and union have already been presented in sections 3.1 and 3.2; some more are shown here, but it is hard to structure them in a systematic way.

As already stated above, we work with the general assumption that the relations involved have suitable types for the operations to be applied. Thus, we need not to be too explicit, so that formulas can become concise. Remember that e.g. IU can denote one of $I_S U_S$ or $I_S U_{S\times T}$ etc..

Proposition 3.52. *Some basic relationships concerning reflexivity, symmetry, and transitivity:*

(1) ρ reflexive $\to$ $\rho \cup \sigma$ reflexive.
(2) ρ, σ reflexive $\to$ $\rho \cap \sigma$ reflexive.
(3) ρ, σ irreflexive $\to$ $\rho \cap \sigma$ irreflexive $\wedge$ $\rho \cup \sigma$ irreflexive.
(4) ρ, σ symmetric $\to$ $\rho \cap \sigma$ symmetric $\wedge$ $\rho \cup \sigma$ symmetric.
(5) $\rho \cup \rho^{\smile}, \rho \cap \rho^{\smile}, \rho\rho^{\smile}$, and $\rho^{\smile}\rho$ are symmetric.
(6) ρ, σ transitive $\to$ $\rho \cap \sigma$ transitive.

Proof. (1): $I_S \subseteq \rho \to I_S \subseteq \rho \cup \sigma$.

(2): $I_S \subseteq \rho \wedge I_S \subseteq \sigma \to I_S \subseteq \rho \cap \sigma$.

(3): We have $I_S \cap (\rho \cap \sigma) \subseteq I_S \cap \rho = \emptyset$, and $\emptyset = (I \cap \rho) \cup (I \cap \sigma) = I \cap (\rho \cup \sigma)$.

(4): $\rho^{\smile} = \rho \wedge \sigma^{\smile} = \sigma \to (\rho \cap \sigma)^{\smile} = \rho^{\smile} \cap \sigma^{\smile} = \rho \cap \sigma$, by Prop. 3.13(7).

(5): For union we have: $(\rho \cup \rho^{\smile})^{\smile} = \rho^{\smile\smile} \cup \rho^{\smile} = \rho \cup \rho^{\smile}$; and for composition: $(\rho\rho^{\smile})^{\smile} = \rho^{\smile\smile}\rho^{\smile} = \rho\rho^{\smile}$.

(6): We have $\rho \cap \sigma \subseteq \rho$ and $\rho \cap \sigma \subseteq \sigma$, and by hypothesis $\rho^2 \subseteq \rho$ and $\sigma^2 \subseteq \sigma$. From this it follows that $(\rho \cap \sigma)^2 \subseteq \rho(\rho \cap \sigma) \subseteq \rho^2 \cap \rho\sigma \subseteq \rho^2 \subseteq \rho$ by Th. 3.16(4) and (5). An analogous argument gives $(\rho \cap \sigma)^2 \subseteq \sigma$, whence $(\rho \cap \sigma)^2 \subseteq \rho \cap \sigma$. □

Proposition 3.53.

(1) Let ρ and σ be reflexive; then also $\rho\sigma$ and $\sigma\rho$ are reflexive.
(2) Let ρ and σ be reflexive; then $\rho \cup \sigma \subseteq \rho\sigma \cap \sigma\rho$.
(3) Let ρ be reflexive and transitive; then ρ is idempotent, i.e. $\rho^2 = \rho$.
(4) Let ρ and σ be symmetric, and $\rho\sigma \subseteq \sigma\rho$; then $\rho\sigma = \sigma\rho$.
(5) Let ρ be coreflexive; then ρ is symmetric, transitive, and $\rho^2 = \rho$.
(6) Let ρ and σ be coreflexive relations; then $\rho\sigma = \rho \cap \sigma$.
(7) Let ρ and σ be coreflexive relations; then $\rho\sigma = \sigma\rho$.

Proof. (1): From $I \subseteq \rho \wedge I \subseteq \sigma$ we get $I = II \subseteq \rho\sigma$ by Th. 3.16(7).

(2): From $I \subseteq \rho \wedge I \subseteq \sigma$ we get $\rho = \rho I \subseteq \rho\sigma$ and $\sigma = I\sigma \subseteq \rho\sigma$. The remaining assertions follow in a similar way.

(3): By assumption $I \subseteq \rho \wedge \rho^2 \subseteq \rho$; then $\rho = \rho I \subseteq \rho^2$ by monotonicity of composition.

(4): We have $(\rho\sigma)\breve{} \subseteq (\sigma\rho)\breve{}$ by Prop. 3.13(6); furthermore $(\rho\sigma)\breve{} = \sigma\breve{}\rho\breve{} = \sigma\rho$, and $(\sigma\rho)\breve{} = \rho\breve{}\sigma\breve{} = \rho\sigma$, i.e. $\sigma\rho \subseteq \rho\sigma$.

(5): Symmetry is obvious, and transitivity follows by monotonicity: $\rho \subseteq I \rightarrow \rho\rho \subseteq I\rho = \rho$. Now assume that $\rho^2 \subset \rho$, i.e. $\exists x[x\rho x \wedge \neg x\rho^2 x]$; but then it follows that $t\rho t \wedge \neg\exists y[t\rho y \wedge y\rho t]$ for some t, a contradiction, whence $\rho^2 = \rho$.

(6): $\rho\sigma \subseteq \rho I = \rho \wedge \rho\sigma \subseteq I\sigma = \sigma$, whence $\rho\sigma \subseteq \rho \cap \sigma$. On the other hand, $\rho \cap \sigma$ is coreflexive; thus $(\rho \cap \sigma) = (\rho \cap \sigma)^2$ by (5), and $(\rho \cap \sigma)^2 \subseteq \rho\sigma$.

(7): Follows from (6). □

Some more properties of compound relations are given later on in Section 4.1 on relations and functions.

Proposition 3.54. *For an arbitrary relation ρ the following relationships hold:*

(1) $\rho \subseteq \rho\rho\breve{}\rho$.
(2) $\rho = \rho(\rho\breve{}\rho \cap I) = \rho\,\mathsf{rng}(\rho)$.
(3) $\rho = (\rho\rho\breve{} \cap I)\rho = \mathsf{dom}(\rho)\rho$.

Proof. (1) Let ρ be reflexive; thus $I \subseteq \rho \rightarrow I \subseteq \rho\breve{} \rightarrow \rho = I\rho \subseteq \rho\breve{}\rho$; and from $I \subseteq \rho \subseteq \rho\breve{}\rho$ we finally get $\rho = I\rho \subseteq \rho\rho\breve{}\rho$. And if ρ is coreflexive, then $\rho = \rho\breve{} = \rho^2$ by Prop. 3.53(5); thus $\rho\rho\breve{}\rho = \rho^3 = \rho$.

(2): $x\rho y \rightarrow x\rho y \wedge y\rho\breve{}x \rightarrow y\rho\breve{}\rho y \rightarrow y(\rho\breve{}\rho \cap I)y \rightarrow x[\rho(\rho\breve{}\rho \cap I)]y$, whence $\rho \subseteq \rho(\rho\breve{}\rho \cap I)$. To show the other direction, note that by Prop. 3.52(5) $\rho\breve{}\rho$ is symmetric; thus $x[\rho(\rho\breve{}\rho \cap I)]y \rightarrow x\rho y \wedge y(\rho\breve{}\rho \cap I)y$. □

The next proposition is formulated in a way that makes the types of the identity and universal relations explicit.

Proposition 3.55. *Let $\rho \subseteq S \times T$. Then for universal relations U the following relationships hold:*

(1) $U_{S\times T}U_{T\times V} = U_{S\times V}$.
(2) $\rho \subseteq \rho U_T \subseteq U_{S\times T}$ *and* $\rho \subseteq U_S\rho \subseteq U_{S\times T}$
(3) *If* $\rho \subseteq S \times S$ *is reflexive, then* $\rho U_S = U_S = U_S\rho$.
(4) $\rho \neq \emptyset \leftrightarrow U_{X\times S}\,\rho\,U_{T\times Y} = U_{X\times Y}$ (**Tarski rule**).

Proof. (1): For homogeneous instantiations U_S we can use Prop. 3.53(3); for the general, i.e. heterogeneous, case the assertion follows from the fact that for each $s \in S$, $t \in T$, and $v \in V$ we have $\langle s,t\rangle \in U_{S\times T}$, $\langle t,v\rangle \in U_{T\times V}$. Thus, $\langle s,v\rangle \in U_{S\times V}$ for each $s \in S$ and $v \in V$.

(2): We have $I_T \subset U_T$ and $\rho \subseteq U_{S\times T}$; thus $\rho = \rho I_T \subseteq \rho U_T \subseteq U_{S\times T} U_T = U_{S\times T}$, using monotonicity, and (1) for the last equation.

(3): $I_S \subseteq \rho \rightarrow U_S = I_S U_S \subseteq \rho U_S \subseteq U_S$.

(4): For the aftersets and foresets induced by U we have

$$\forall x \in X[xU_{X\times S} = S] \quad \text{and} \quad \forall y \in Y[T = U_{T\times Y} y].$$

To show the non-trivial direction assume $\rho \neq \emptyset$ and $s\rho t$; then we get $xUs\rho tUy$ for all x, y, whence $U_{X\times S}\, \rho\, U_{T\times Y} = U_{X\times Y}$. □

3.6.3 *Powers and Closures of Relations*

As it is customary for various kinds of multiplication arising in mathematics, we define the powers of a relation ρ on a set S defined recursively as follows:

$$\rho^0 = I_S \text{ and } \rho^{n+1} = \rho^n \rho \text{ for } n \geq 1.$$

Proposition 3.56. *Let ρ be relation relation on a set S; for powers of ρ we have:*

(1) $\rho^m \rho^n = \rho^{m+n}$ (for $m, n \geq 0$).
(2) $(\rho^m)^n = \rho^{m\cdot n}$ (for $m, n \geq 0$).

Two unary oprations related to the powers of a relation that are useful for further considerations are the following

Definition 3.57.

(a) $\rho^+ = \bigcup_{n\geq 1} \rho^n$.
(b) $\rho^* = I_S \cup \rho^+ = \bigcup_{n\geq 0} \rho^n$ is called the **star** *of ρ*

Definition 3.58. *A relation ρ' on a set S is the* **closure of a relation** *ρ on S with respect to a property P if*

(1) $\rho \subseteq \rho'$,
(2) *ρ' has property P, and*
(3) *If ρ'' has property P and $\rho \subseteq \rho''$, then $\rho' \subseteq \rho''$.*

We will use $\mathbf{r}(\rho)$, $\mathbf{s}(\rho)$, and $\mathbf{t}(\rho)$ to denote the **reflexive**, **symmetric**, and **transitive closure** of a relation ρ, respectively; composite closures will be written with only the innermost parentheses, e.g. $\mathbf{rt}(\rho)$.

Theorem 3.59. *Let ρ be a relation on a set S, then*

(1) $\mathbf{r}(\rho) = \rho \cup I_S$.
(2) $\mathbf{s}(\rho) = \rho \cup \rho\breve{\ }$.
(3) $\mathbf{t}(\rho) = \rho^+$.

Proof. (1): Trivially we have (a) $\rho \subseteq \mathbf{r}(\rho)$, and (b) $I_S \subseteq \mathbf{r}(\rho)$, whence $\mathbf{r}(\rho)$ is reflexive. And for ρ' a reflexive relation with $\rho \subseteq \rho'$, we get (c) $I_S \subseteq \rho' \wedge \rho \subseteq \rho' \rightarrow I_S \cup \rho \subseteq \rho'$; thus $\mathbf{r}(\rho) \subseteq \rho'$.

(2): Obviously, (a) $\rho \subseteq \mathbf{s}(\rho)$, and (b) $\mathbf{s}(\rho)$ is symmetric by Prop. 3.52(5). (c) If ρ' is symmetric with $\rho \subseteq \rho'$, then $\rho^{\smile} \subseteq \rho'^{\smile} = \rho'$. So $\rho \subseteq \rho' \wedge \rho^{\smile} \subseteq \rho'$, whence $\rho \cup \rho^{\smile} = \mathbf{s}(\rho) \subseteq \rho'$.

(3): As before (a) $\rho \subseteq \rho^+$. (b) ρ^+ is transitive: $x\rho^+ y \wedge y\rho^+ z \rightarrow \exists j,k \in \mathbb{N}[x\rho^k y \wedge y\rho^j z]$. Thus $x\rho^k\rho^j y$; therefore $x\rho^{j+k}y$ and thus $x\rho^+ y$. (c) Let ρ' be transitive with $\rho \subseteq \rho'$. To show that $\rho^+ \subseteq \rho'$ we prove $\rho^n \subseteq \rho'$ for each $n \in \mathbb{N}$: The case $n = 1$ follows from the assumption $\rho \subseteq \rho'$; if $\rho^k \subseteq \rho'$, then $\rho^{k+1} = \rho^k\rho \subseteq \rho'\rho \subseteq \rho'\rho' \subseteq \rho'$ by monotonicity of composition and transitivity of ρ'. Thus $\rho^{k+1} \subseteq \rho'$ for all $k \in \mathbb{N}$, whence $\rho^+ \subseteq \rho'$. So $\rho^+ = \mathbf{t}(\rho)$. □

Proposition 3.60. *For a relation ρ on S the following holds:*

(1) $\rho = \mathbf{r}(\rho)$ iff ρ is reflexive; furthermore, $\mathbf{rr}(\rho) = \mathbf{r}(\rho)$.
(2) $\rho = \mathbf{s}(\rho)$ iff ρ is symmetric; furthermore, $\mathbf{ss}(\rho) = \mathbf{s}(\rho)$.
(3) $\rho = \mathbf{t}(\rho)$ iff ρ is transitive; furthermore, $\mathbf{tt}(\rho) = \mathbf{t}(\rho)$.

Proof. (1): A reflexive relation ρ is clearly the smallest reflexive relation containing itself, thus $\rho = \mathbf{r}(\rho)$. If $\rho = \mathbf{r}(\rho)$, then, since $\mathbf{r}(\rho)$ is reflexive, ρ also is. And clearly $\mathbf{rr}(\rho) = \mathbf{r}(I_S \cup \rho) = I_S \cup I_S \cup \rho = I_S \cup \rho = \mathbf{r}(\rho)$. □

Note that each $\mathbf{o} \in \{\mathbf{r}, \mathbf{s}, \mathbf{t}\}$ is indeed a closure operator (see Def. 2.31): $\mathbf{o}$ is extensive by Th. 3.59, isotone by the properties of set union, and idempotent by Prop. 3.60.

Proposition 3.61. ([19], p. 584) *For a relation ρ on S the following holds:*

(1) If ρ is reflexive, then also $\mathbf{s}(\rho)$ and $\mathbf{t}(\rho)$.
(2) If ρ is symmetric, then also $\mathbf{r}(\rho)$ and $\mathbf{t}(\rho)$.
(3) If ρ is transitive, then also $\mathbf{r}(\rho)$.

Proof. (1): $I_S \subseteq \rho \rightarrow I_S \subseteq \mathbf{s}(\rho) \wedge I_S \subseteq \mathbf{t}(\rho)$.

(2): I_S is symmetric, whence $\mathbf{r}(\rho)$ is symmetric by Prop. 3.52(4). Now we demonstrate that ρ^k is symmetric for each $n \in \mathbb{N}$: ρ^1 is symmetric by assumption; if $\rho^k = (\rho^k)^{\smile}$, then $\rho^{k+1} = \rho^k\rho = (\rho^k)^{\smile}\rho^{\smile} = (\rho\rho^k)^{\smile} = (\rho^{k+1})^{\smile}$. So, again by Prop. 3.52(4), it follows that $\mathbf{t}(\rho)$ is symmetric.

(3): $\mathbf{r}(\rho)^2 = (\rho \cup I_S)(\rho \cup I_S) = (\rho \cup I_S)\rho \cup (\rho \cup I_S)I_S = \rho^2 \cup \rho \cup \rho \cup I_S = \rho \cup I_S$, by the transitivity of ρ. (For sake of direct comparison I include the point-based proof given in [19]: Let $x(\rho \cup I_S)y$ and $y(\rho \cup I_S)z$. If $\langle x,y\rangle \in I_S$, then $x = y$ and $\langle x,z\rangle = \langle y,z\rangle \in \rho \cup I_S$. If $\langle y,z\rangle$ is in I_S, then $y = z$ and $\langle x,z\rangle = \langle x,y\rangle \in \rho \cup I_S$. If none of these two cases holds, then both pairs are in ρ and therefore $x(\rho \cup I_S)z$ by the transitivity of ρ.) □

The following examples are taken from [12]: Let $<$ be taken on e.g. the integers, then $\mathbf{r}(<) = \leq$. For the symmetric closure of parentOf we get $\sigma := \mathbf{s}(\text{parentOf}) = \text{parentOf} \cup \text{childOf}$: clearly, $a\sigma b$ implies $b\sigma a$. And $\mathbf{t}(\text{parentOf}) = \text{ancestorOf}$: whenever $\langle a,b\rangle$ and $\langle b,c\rangle$ are in ancestorOf, then also $\langle a,c\rangle$.

Proposition 3.62. *Let ρ and σ be relations on a set S. Some consequences from the definitions of ρ^+ and ρ^* are:*

(1) $\rho \subseteq \rho^+$, $\quad \rho \subseteq \rho^*$.
(2) $\rho \subseteq \sigma \;\rightarrow\; \rho^+ \subseteq \sigma^+ \wedge \rho^* \subseteq \sigma^*$.
(3) $\rho^*\rho^* = \rho^*$.
(4) $\rho^* \subseteq \rho^*\sigma^*$, $\quad \rho^* \subseteq \sigma^*\rho^*$, $\quad \sigma^* \subseteq \rho^*\sigma^*$, $\quad \sigma^* \subseteq \sigma^*\rho^*$.
(5) $\rho^*\sigma^* \subseteq (\rho \cup \sigma)^*$.
(6) $\rho^* = \mathbf{rt}(\rho)$.
(7) $\rho^{**} = \rho^*$.
(8) $(\rho^+)^{\smile} = (\rho^{\smile})^+$, $\quad (\rho^*)^{\smile} = (\rho^{\smile})^*$.
(9) *Let σ be transitive with $\rho \subseteq \sigma$; then $\mathbf{t}(\rho) \subseteq \sigma$.*

Proof. (3): ρ^+ is transitive, thus $\mathbf{r}(\rho^+) = I_S \cup \rho^+ = \rho^*$ is transitive by Prop. 3.61(3). Idempotency then follows by Prop. 3.53(3).

(4): We have $I \subseteq \sigma^*$, thus $\rho^* = I\rho^* \subseteq \sigma^*\rho^*$.

(5): Applying (2) and (3) we get $\rho, \sigma \subseteq \rho \cup \sigma \;\rightarrow\; \rho^*, \sigma^* \subseteq (\rho \cup \sigma)^* \;\rightarrow\; \rho^*\sigma^* \subseteq (\rho \cup \sigma)^*(\rho \cup \sigma)^* = (\rho \cup \sigma)^*$.

(6): By Th. 3.59(3), $\mathbf{t}(\rho) = \rho^+$, whence $\mathbf{rt}(\rho^+) = \rho^+ \cup I_S = \rho^*$. Moreover, (a) $\rho \subseteq \rho^*$ by (1); (b) ρ^* is transitive by (3) and reflexive by definition; and (c) If σ is a reflexive and transitive relation with $\rho \subseteq \sigma$, then $\sigma = \mathbf{t}(\sigma) = \sigma^+$ by Prop. 3.60. By (2), $\rho^+ \subseteq \sigma^+$, whence $\rho^* \subseteq \sigma^*$ by monotonicity of union.

(7): We will show $\rho^{**} = \mathbf{rtrt}(\rho) = \mathbf{rt}(\rho)$. By construction $\mathbf{rt}(\rho)$ is reflexive, and by Prop. 3.61(1) also $\mathbf{trt}(\rho)$, whence $\mathbf{trt}(\rho) = \mathbf{rtrt}(\rho)$ by Prop. 3.60(1). And $\mathbf{rt}(\rho)$ is transitive by Prop. 3.61(3), whence $\mathbf{rt}(\rho) = \mathbf{trt}(\rho)$ by Prop. 3.60(3).

(8): $(\rho^+)^{\smile} = \left(\bigcup_{i\geq 1} \rho^i\right)^{\smile} = \bigcup_{i\geq 1} (\rho^i)^{\smile} = \bigcup_{i\geq 1} (\rho^{\smile})^i = (\rho^{\smile})^+$, using Th. 3.16(8).

(9): $I_S \cup \sigma$ is reflexive and transitive, hence $\rho^* \subseteq I_S \cup \sigma$. Thus, $\mathbf{t}(\rho) = \rho^+ = \rho\rho^* \subseteq \rho(I_S \cup \sigma) = \rho \cup \rho\sigma \subseteq \sigma \cup \sigma^2 = \sigma$. □

Proposition 3.63. *For relations ρ and σ on S the following holds:*

(1) $\mathbf{rs}(\rho) = \mathbf{sr}(\rho)$.
(2) $\mathbf{rt}(\rho) = \mathbf{tr}(\rho)$.
(3) $\mathbf{st}(\rho) \subseteq \mathbf{ts}(\rho)$.
(4) $\mathbf{rts}(\rho) = \mathbf{trs}(\rho) = \mathbf{tsr}(\rho)$ *and* $\mathbf{rst}(\rho) = \mathbf{srt}(\rho) = \mathbf{str}(\rho)$.

Proof. (1): $\mathbf{rs}(\rho) = I_S \cup \rho \cup \rho^{\smile} = I_S \cup \rho \cup I_S \cup \rho^{\smile} = \mathbf{s}(I_S \cup \rho) = \mathbf{sr}(\rho)$.

(2): It is not hard to show by induction that $(I_S \cup \rho)^k = \bigcup_{0\leq i\leq k} \rho^i$. Then $\mathbf{tr}(\rho) = \mathbf{t}(I_S \cup \rho) = \bigcup_{0\leq n} (I_S \cup \rho)^n = I_S \cup \bigcup_{1\leq n} \bigcup_{0\leq i\leq n} \rho^i = I_S \cup \bigcup_{1\leq n} \rho^n = I_S \cup \rho^+ = \rho^* = \mathbf{rt}(\rho)$ by Prop. 3.62(5).

(3): We have $\mathbf{st}(\rho) = \mathbf{s}(\rho^+) = \rho^+ \cup (\rho^+)^{\smile}$. We show by induction that $\rho^n \cup (\rho^n)^{\smile} \subseteq (\rho \cup \rho^{\smile})^n$ for all $n \in \mathbb{N}$: For $n = 1$ we trivially have $\rho^1 \cup (\rho^1)^{\smile} \subseteq (\rho \cup \rho^{\smile})^1$. Now assume $\rho^k \cup (\rho^k)^{\smile} \subseteq (\rho \cup \rho^{\smile})^k$; then, by monotonicity of composition,

$$(\rho \cup \rho^{\smile})^k(\rho \cup \rho^{\smile})$$
$$\supseteq [\rho^k \cup (\rho^k)^{\smile}](\rho \cup \rho^{\smile})$$

$= \rho^{k+1} \cup (\rho^k)^\smile \rho \cup \rho^k \rho^\smile \cup (\rho^k)^\smile \rho^\smile$
$= \rho^{k+1} \cup (\rho^{k+1})^\smile \cup \ldots,$ by applying Th. 3.16(8) twice to the last term
$\supseteq \rho^{k+1} \cup (\rho^{k+1})^\smile.$

Thus, $\bigcup_{n\geq 1} \rho^i \cup (\rho^i)^\smile \subseteq \bigcup_{n\geq 1} (\rho \cup \rho^\smile)^i$, and therefore $\mathbf{st}(\rho) \subseteq \mathbf{ts}(\rho)$.

(4): By (1) we have $\mathbf{trs}(\rho) = \mathbf{tsr}(\rho)$, and by (2) $\mathbf{rts}(\rho) = \mathbf{trs}(\rho)$. □

Proposition 3.64. *For relations ρ and σ on S the following holds:*

(1) $\mathbf{r}(\rho \cup \sigma) = \mathbf{r}(\rho) \cup \mathbf{r}(\sigma)$.
(2) $\mathbf{s}(\rho \cup \sigma) = \mathbf{s}(\rho) \cup \mathbf{s}(\sigma)$.
(3) $\mathbf{t}(\rho_1 \cup \ldots \cup \rho_K) = \bigcup_{1\leq i_k \leq K,\ n\geq 1} \rho_{i_1}\rho_{i_2}\cdots\rho_{i_n}$, *for all $K \geq 1$.*
(4) $\mathbf{r}(\rho \cap \sigma) = \mathbf{r}(\rho) \cap \mathbf{r}(\sigma)$.
(5) $\mathbf{s}(\rho \cap \sigma) \subseteq \mathbf{s}(\rho) \cap \mathbf{s}(\sigma)$.

Proof. (2): $\mathbf{s}(\rho \cup \sigma) = (\rho \cup \sigma) \cup (\rho \cup \sigma)^\smile = \rho \cup \rho^\smile \cup \sigma \cup \sigma^\smile = \mathbf{s}(\rho) \cup \mathbf{s}(\sigma)$.

(3): Let (TC_K) denote the equation of the assertion. We have to show that (TC_K) holds for all $K \geq 1$. (TC_1) holds by Th. 3.59(3). For sake of ease of notation writing $\rho := \rho_1$ and $\sigma := \rho_2$, we show (TC_2), because it will be needed later on in the proof of the general case:

$$\begin{aligned}\mathbf{t}(\rho \cup \sigma) &= \bigcup_{i\geq 1} (\rho \cup \sigma)^i \\ &= (\rho \cup \sigma) \cup (\rho^2 \cup \sigma\rho \cup \rho\sigma \cup \sigma^2) \cup \\ &\quad (\rho^3 \cup \sigma\rho^2 \cup \rho\sigma\rho \cup \sigma^2\rho \cup \rho^2\sigma \cup \sigma\rho\sigma \cup \sigma^2\rho \cup \rho\sigma^2 \cup \sigma^3) \cup \ldots \\ &= \bigcup_{1\leq i_k \leq 2,\ n\geq 1} \rho_{i_1}\rho_{i_2}\cdots\rho_{i_n}.\end{aligned}$$

To prove the general case, we use mathematical induction, and assume that (TC_K) holds. Then we have $\mathbf{t}(\rho_1 \cup \ldots \cup \rho_K \cup \rho_{K+1}) = \mathbf{t}((\rho_1 \cup \ldots \cup \rho_K) \cup \rho_{K+1})$. Setting $\rho := (\rho_1 \cup \ldots \cup \rho_K)$ and $\sigma := \rho_{K+1}$ we now apply (TC_2) to get

$$\begin{aligned}\bigcup_{i\geq 1} (\rho \cup \sigma)^i &= \bigcup_{i\geq 1} \rho^i \cup (\rho\sigma \cup \sigma\rho \cup \sigma^2) \cup \\ &\quad (\sigma\rho^2 \cup \rho\sigma\rho \cup \sigma^2\rho \cup \rho^2\sigma \cup \sigma\rho\sigma \cup \sigma^2\rho \cup \rho\sigma^2 \cup \sigma^3) \cup \ldots \\ &= \bigcup_{1\leq i_k \leq K,\ n\geq 1} \rho_{i_1}\rho_{i_2}\cdots\rho_{i_n} \cup \text{ all products involving } \sigma \\ &= \bigcup_{1\leq i_k \leq K+1,\ n\geq 1} \rho_{i_1}\rho_{i_2}\cdots\rho_{i_n}.\end{aligned}$$

(5): $\mathbf{s}(\rho \cap \sigma) = (\rho \cap \sigma) \cup (\rho^\smile \cap \sigma^\smile) = (\rho \cup \rho^\smile) \cap (\sigma \cup \rho^\smile) \cap (\rho \cup \sigma^\smile) \cap (\sigma \cup \sigma^\smile) \subseteq \mathbf{s}(\rho) \cap \mathbf{s}(\sigma)$. □

The next two propositions are due to Wechler [25, Section 1.3].

Proposition 3.65.

(1) Let ρ be a relation on S; then $\chi = I_S \cup \rho\chi \ \rightarrow\ \rho^* \subseteq \chi$.
(2) Let ρ and σ be relations on a set S; then $(\rho \cup \sigma)^* = \rho^*(\sigma\rho^*)^*$.
(3) Given relations ρ and σ, then $\sigma\rho^* \subseteq \rho^*\sigma^* \ \rightarrow\ (\sigma\rho^*)^* \subseteq \rho^*\sigma^*$.

Proof. (1): We first show that ρ^* is a solution of the equation: By definition, $\rho^* = I_S \cup \rho^+$, whence $\rho\rho^* = \rho(I_S \cup \rho^+) = \rho^+$. Thus $I_S \cup \rho\rho^* = I_S \cup \rho^+ = \rho^*$. For any other solution σ we have $\sigma = I_S \cup \rho\sigma \ \rightarrow\ I_S = \rho^0 \subseteq \sigma$ and $\rho\sigma \subseteq \sigma$. We proceed by induction: Assuming $\rho^n \subseteq \sigma$ we get $\rho\rho^n = \rho^{n+1} \subseteq \rho\sigma \subseteq \sigma$. Thus, $\forall n \in \mathbb{N}[\rho^n \subseteq \sigma]$, therefore $\bigcup_{n\geq 0} \rho^n = \rho^* \subseteq \sigma$.

(2) By (1) we have $\rho^* = I_S \cup \rho\rho^*$. Thus

$$\begin{aligned}\rho^*(\sigma\rho^*)^* &= (I_S \cup \rho\rho^*)(\sigma\rho^*)^* \\ &= (\sigma\rho^*)^* \cup \rho\rho^*(\sigma\rho^*)^* \\ &= I_S \cup \sigma\rho^*(\sigma\rho^*)^* \cup \rho\rho^*(\sigma\rho^*)^* \\ &= I_S \cup (\rho\cup\sigma)\rho^*(\sigma\rho^*)^*.\end{aligned}$$

So, by (1), $(\rho\cup\sigma)^* \subseteq \rho^*(\sigma\rho^*)^*$. To show the converse inclusion we use the monotonicity properties of union and composition, and apply results from Prop. 3.62:

$$\begin{aligned}[\sigma\rho^*]^* &\subseteq [(\rho\cup\sigma)\rho^*]^* \\ &\subseteq [(\rho\cup\sigma)(\rho\cup\sigma)^*]^* \\ &\subseteq [(\rho\cup\sigma)^*(\rho\cup\sigma)^*]^* \\ &\subseteq [(\rho\cup\sigma)^*]^* \\ &= (\rho\cup\sigma)^{**} \\ &= (\rho\cup\sigma)^*.\end{aligned}$$

Since also $\rho^* \subseteq (\rho\cup\sigma)^*$ we get $\rho^*(\sigma\rho^*)^* \subseteq (\rho\cup\sigma)^*(\rho\cup\sigma)^* = (\rho\cup\sigma)^*$.

(3): By assumption we have the base case for $n = 1$: $\sigma\rho^* \subseteq \rho^*\sigma^*$. Given that $(\sigma\rho^*)^k \subseteq \rho^*\sigma^*$, then we get for the inductive step $\sigma\rho^*(\sigma\rho^*)^k \subseteq \sigma\rho^*\rho^*\sigma^* = \sigma\rho^*\sigma^*$. But, applying the base case, it follows that $\sigma\rho^*\sigma^* \subseteq \rho^*\sigma^*\sigma^* = \rho^*\sigma^*$. Thus also $(\sigma\rho^*)^{k+1} \subseteq \rho^*\sigma^*$. □

Proposition 3.66. *For any two relations ρ and σ on a set A, the following inequalities are equivalent:*

(1) $\sigma\rho^* \subseteq \rho^*\sigma^*$,
(2) $\sigma^*\rho^* \subseteq \rho^*\sigma^*$,
(3) $(\rho\cup\sigma)^* \subseteq \rho^*\sigma^*$.

Proof. (2) → (1): $\sigma\rho^* \subseteq \sigma^*\rho^*$ by monotonicity of composition, whence $\sigma\rho^* \subseteq \rho^*\sigma^*$ using (2).

(1) → (3): By Prop. 3.65(3), (1) implies $(\sigma\rho^*)^* \subseteq \rho^*\sigma^*$; thus $\rho^*(\sigma\rho^*)^* \subseteq \rho^*\rho^*\sigma^* = \rho^*\sigma^*$. But the left side of the last inequality equals $(\rho\cup\sigma)^*$ by 3.65(2), and finally this gives (3).

(3) → (2): We have $\rho^*, \sigma^* \subseteq (\rho\cup\sigma)^*$. Thus $\sigma^*\rho^* \subseteq (\rho\cup\sigma)^*(\rho\cup\sigma)^* = (\rho\cup\sigma)^* \subseteq \rho^*\sigma^*$, for the last inequality applying the assumption (3). □

Proposition 3.67. ([26]) *Let ρ and σ be reflexive relations; then $(\rho\cup\sigma)^* = (\rho^*\sigma^*)^* = (\sigma^*\rho^*)^*$*

Proof. From $\rho \subseteq \rho^*$ and $I \subseteq \sigma^*$ one has $\rho \subseteq \rho^*\sigma^*$, and similarly $\sigma \subseteq \rho^*\sigma^*$; thus $(\rho\cup\sigma) \subseteq \rho^*\sigma^*$, whence

$$(\rho\cup\sigma)^* \subseteq (\rho^*\sigma^*)^*. \tag{3.9}$$

Furthermore, $\rho^* \subseteq (\rho\cup\sigma)^*$ and $\sigma^* \subseteq (\rho\cup\sigma)^*$ implies $\rho^*\sigma^* \subseteq (\rho\cup\sigma)^*(\rho\cup\sigma)^* = (\rho\cup\sigma)^*$, whence

$$(\rho^*\sigma^*)^* \subseteq ((\rho\cup\sigma)^*)^* = (\rho\cup\sigma)^* \tag{3.10}$$

Combining (3.9) and (3.10) gives the desired result. □

3.7 Homomorphisms and Isomorphisms

Concerning the definition of homomorphisms between two relations we have to differentiate between the homogeneous ($\rho \in \mathcal{B}_X$) and heterogeneous case ($\rho \in \mathcal{B}_{X\times Y}, X \neq Y$). Though the former clearly is a special case of the latter with $X = Y$, it is worthwhile to dedicate it a separate discussion. The subsequent presentation roughly follows the ones in [2, Chapter 8] and [20, Section 5.7]. We start with the general case and will specialize it afterwards.

We begin with shortly recalling the notion of a homomorphism for algebras from Def. 2.47, using ring homomorphisms as example: If R and S are rings, then a mapping $f : R \longrightarrow S$ is a *ring homomorphism* if $f(a +_R b) = f(a) +_S f(b)$ and $f(a \cdot_R b) = f(a) \cdot_S f(b)$ for all $a, b \in R$, and $f(1_R) = 1_S$.

Translating this to the context of relations we get two general configurations: one for the homogeneous and the second for the heterogeneous case as shown in the two diagrams below.

$$\begin{array}{ccc} A & \xrightarrow{\rho} & A \\ \varphi\big\downarrow & & \big\downarrow\varphi \\ B & \xrightarrow{\sigma} & B \end{array} \qquad\qquad \begin{array}{ccc} A & \xrightarrow{\rho} & A' \\ \varphi\big\downarrow & & \big\downarrow\psi \\ B & \xrightarrow{\sigma} & B' \end{array}$$

The right diagram means that if one follows relation ρ and then proceeds with ψ to the set B', it is always possible to go to the set B first and then follow σ to reach the same element.

3.7.1 *The Heterogeneous Case*

When we want to compare two relations $\rho \subseteq X_1 \times Y_1$ and $\sigma \subseteq X_2 \times Y_2$, we must be able to relate them in a reasonable way. Typically this means that two mappings $\varphi : X_1 \longrightarrow X_2$ and $\psi : Y_1 \longrightarrow Y_2$ are provided. The pair (φ, ψ) is said to form a homomorphism from ρ into σ if whenever two elements x, y are related by ρ then their images $\varphi(x), \psi(y)$ are related by σ. This gives us

Definition 3.68. *Let $\rho \subseteq X_1 \times Y_1$ and $\sigma \subseteq X_2 \times Y_2$ be two relations. A pair (φ, ψ) of mappings $\varphi : X_1 \longrightarrow X_2$ and $\psi : Y_1 \longrightarrow Y_2$ is called a* **homomorphism from ρ to σ** *if $\forall x \in X_1, \forall y \in Y_1 [x\rho y \rightarrow \varphi(x)\sigma\psi(y)]$.*

The proposition below gives a compact form of this definition using the relations involved:

Proposition 3.69. ([20, p. 86]) *A pair (φ, ψ) of mappings $\varphi : X_1 \longrightarrow X_2$ and $\psi : Y_1 \longrightarrow Y_2$ is a homomorphism from $\rho \subseteq X_1 \times Y_1$ into $\sigma \subseteq X_2 \times Y_2$ iff $\rho\psi \subseteq \varphi\sigma$.*

Proof. (To get it short, throughout the proof we use e.g $\forall x_i$ as a shorthand for $\forall x_i \in X_i$ etc., for $i = 1, 2$. For convenience, justifications are given by the logical theorems themselves instead of references to Section 2.2.)

$\forall x_1 \forall y_1 \forall x_2 \forall y_2 [x_1 \rho y_1 \wedge x_1 \varphi x_2 \wedge y_1 \psi y_2 \rightarrow x_2 \sigma y_2]$	
$\leftrightarrow \forall x_1 \forall y_1 \forall x_2 \forall y_2 [\neg x_1 \rho y_1 \vee \neg y_1 \psi y_2 \vee \neg x_1 \varphi x_2 \vee x_2 \sigma y_2]$	$A \rightarrow B \equiv \neg A \vee B$
$\leftrightarrow \forall x_1 \forall y_1 \forall y_2 \big[\neg x_1 \rho y_1 \vee \neg y_1 \psi y_2 \vee \forall x_2 [\neg x_1 \varphi x_2 \vee x_2 \sigma y_2]\big]$	$\forall x[A \vee Q(x)] \equiv A \vee \forall x Q(x)$
$\leftrightarrow \forall x_1 \forall y_2 \forall y_1 \big[\neg x_1 \rho y_1 \vee \neg y_1 \psi y_2 \vee \neg \exists x_2 [x_1 \varphi x_2 \wedge \neg x_2 \sigma y_2]\big]$	$\forall x Q(x) \equiv \neg \exists x [\neg Q(x)]$
$\leftrightarrow \forall x_1 \forall y_2 \big[\neg \exists y_1 [x_1 \rho y_1 \wedge y_1 \psi y_2] \vee \neg [x_1 \varphi \overline{\sigma} y_2]\big]$	
$\leftrightarrow \forall x_1 \forall y_2 \big[\neg [x_1 \rho \psi y_2] \vee [x_1 \varphi \sigma y_2]\big]$	Cor. 4.6(1)
$\leftrightarrow \forall x_1 \forall y_2 \big[x_1 \rho \psi y_2 \rightarrow x_1 \varphi \sigma y_2\big]$	$A \rightarrow B \equiv \neg A \vee B$
$\leftrightarrow \rho \psi \subseteq \varphi \sigma$	

□

The relational homomorphism condition proved in the preceding theorem has three variants:

Lemma 3.70. *Let φ and ψ be mappings, and ρ, σ be relations; then the following equivalences hold:*

$$\rho\psi \subseteq \varphi\sigma \quad \leftrightarrow \quad \rho \subseteq \varphi\sigma\psi^{\smile} \quad \leftrightarrow \quad \varphi^{\smile}\rho \subseteq \sigma\psi^{\smile} \quad \leftrightarrow \quad \varphi^{\smile}\rho\psi \subseteq \sigma$$

Proof. Immediate from suitable multiplications using the mapping properties $\varphi^{\smile}\varphi \subseteq I$ etc. □

Comparing the definition of a homomorphism given above with the one in Def. 2.47 for algebraic structures, the question arises why here we have containment only instead of equality. Remember that e.g. in group theory $\varphi(a +_1 b) = \varphi(a) +_2 \varphi(b)$ would be required, where $+_1, +_2$ are binary operations, i.e. functions. Here we have a more general context in which we use arbitrary relations instead and do therefore need a more general condition for homomorphy, viz. $\rho\psi \subseteq \varphi\sigma$. But if ρ and σ are mappings, this condition will become an equality, whence it encompasses both algebraic as well as relational structures.

Definition 3.71. *Let $\rho \subseteq X_1 \times Y_1$ and $\sigma \subseteq X_2 \times Y_2$ be two relations. A pair (φ, ψ) of mappings $\varphi : X_1 \longrightarrow X_2$ and $\psi : Y_1 \longrightarrow Y_2$ is called a* **isomorphism between** ρ **and** σ *if it is a homomorphism from ρ to σ and $(\varphi^{\smile}, \psi^{\smile})$ is a homomorphism from σ to ρ.*

Corollary 3.72. *Let ρ, σ be relations together with a homomorphism (φ, ψ) from ρ to σ such that φ, ψ are bijective and $\rho\psi = \varphi\sigma$ then (φ, ψ) is an isomorphism.*

Proof. In order that $(\varphi^{\smile}, \psi^{\smile})$ becomes a homomorphism from σ to ρ we have to show $\sigma\psi^{\smile} \subseteq \varphi^{\smile}\rho$. We get $\sigma\psi^{\smile} = I\sigma\psi^{\smile} = \varphi^{\smile}\varphi\sigma\psi^{\smile} = \varphi^{\smile}\rho\psi\psi^{\smile} = \varphi^{\smile}\rho$. □

We now turn to composition of homomorphisms, which we define as $(\varphi_1, \psi_1) \cdot (\varphi_2, \psi_2) := (\varphi_1\varphi_2, \psi_1\psi_2)$. That homomorphisms are closed with respect to composition is shown in

Proposition 3.73. *Let (φ_1, ψ_1) and (φ_2, ψ_2) be two homomorphisms from ρ_1 to σ_1 and from σ_1 to σ_2, respectively; then their composition $(\varphi_1\varphi_2, \psi_1\psi_2)$ is a homomorphism from ρ_1 to σ_2.*

Proof. We have to show $\rho_1\psi_1\psi_2 \subseteq \varphi_1\varphi_2\sigma_2$. But this easily follows from the hypotheses $\rho_1\psi_1 \subseteq \varphi_1\sigma_1$ and $\sigma_1\psi_2 \subseteq \varphi_2\sigma_2$ by suitable multiplications; we get $\rho_1\psi_1\psi_2 \subseteq \varphi_1\sigma_1\psi_2$ and $\varphi_1\sigma_1\psi_2 \subseteq \varphi_1\varphi_2\sigma_2$. □

3.7.2 *The Homogeneous Case*

Specializing Def. 3.68 to the homogeneous case, the following results: Let $\rho \subseteq A \times A$ and $\sigma \subseteq B \times B$ be two relations. A mapping $\varphi : A \longrightarrow B$ is called a **homomorphism from ρ to σ** if $\forall a, a' \in A[a\rho a' \to \varphi(a)\sigma\varphi(a')]$, equivalently $\rho\varphi \subseteq \varphi\sigma$.

Isotone mappings between posets (see Def. 2.51(b)) are a well known example of homomorphisms between homogeneous relations: With fully explicit notation, let $(A, \leq_A)$ and $(B, \leq_B)$ be two posets, and $\varphi : A \longrightarrow B$ an isotone mapping, i.e.

$$\forall a, a' \in A[a \leq_A a' \to \varphi(a) \leq_B \varphi(a')].$$

This is equivalent to the much more compact and less error-prone formulation

$$\leq_A \cdot \varphi \subseteq \varphi \cdot \leq_B,$$

as shown in the following diagram.

$$\begin{array}{ccc} a & \xrightarrow{\leq_A} & a' \\ \varphi\big\downarrow & & \big\downarrow\varphi \\ b & \xrightarrow{\leq_B} & b' \end{array}$$

Another example can be found in Section 17.1.4 where homomorphisms and isomorphisms for 1-graphs are discussed.

References

[1] Adhikari, M.R., Adhikari, A.: *Basic Modern Algebra with Applications*, Springer India (2014).

[2] Berghammer R.: *Ordnungen, Verbände und Relationen mit Anwendungen*, 2nd ed., Springer Vieweg, Wiesbaden (2012).

[3] Berghammer, R., Winter, M.: Embedding mappings and splittings with applications. Acta Informatica **47**, 77-110 (2010).

[4] Berghammer, R., Winter, M.: Decomposition of relations and concept lattices. Fundamenta Informaticae **126**, 37-82 (2013).

[5] Birkhoff, G.: *Lattice Theory*, American Math. Soc., Providence, Rhode Island(1973).

[6] Blyth, T.S., Janowitz, M.F.: *Residuation Theory*, Pergamon Press, Oxford (1972).

[7] Boyd, J.P., Everett, M.G.: Relations, residuals, regular interiors, and relative regular equivalence, Social Networks **21**, 147-165 (1999).

[8] Caspard, N., Leclerc, B., Monjardet, B.: *Finite Ordered Sets-Concepts, Results and Uses*, Cambridge University Press, Cambridge, UK (2012).

[9] Desharnais, J., Jaoua, A.: A relational division operator: the conjugate kernel, Theoretical Computer Science **114**, 247-272 (1993).

[10] Gericke,H.: *Theorie der Verbände*, 2nd ed., Bibliographisches Institut, Mannheim (1967).

[11] Givant, S.: The Calculus of Relations as a Foundation for Mathematics, Journal of Automated Reasoning **37**, 277-322 (2006).

[12] Gries, D., Schneider, F.B.: *A Logical Approach to Discrete Math*, Springer-Verlag, New York (1993).

[13] Maddux, R.D.: The Origin of Relation Algebras in the Development and Axiomatization of the Calculus of Relations, Studia Logica **50**, 421-455 (1991).

[14] Merzbach, U.C., Boyer, C.B.: *A History of Mathematics*, 3rd ed., John Wiley & Sons, Hoboken (2011).

[15] Mirin, A.C.: Relation Algebras, Master of Arts in Mathematics Thesis, Mills College (2013).

[16] De Morgan, A.: On the syllogism no. IV, and on the logic of relations, Trans. Cambridge Philos. Soc. **10**, 331-358 (1864).

[17] Ono, K.: On some properties of binary relations, Nagoya Math. J. **12**, 161-170 (1957).

[18] Pratt, V.R.: Origins of the calculus of binary relations. In Proc. 7th Annual IEEE Symp. on Logic in Computer Science, LICS'92, 248–254, (1992).

[19] Ross, K.A., Wright, C.R.B.: *Discrete Mathematics*, 4th ed., Prentice-Hall, New Jersey (1999).

[20] Schmidt, G.: *Relational mathematics*, Cambridge University Press, (2011).

[21] Schmidt, G., Ströhlein, T.: *Relations and Graphs. Discrete Mathematics for Computer Scientists*, Springer-Verlag, Berlin Heidelberg (1993).

[22] Suppes, P.: *Axiomatic Set Theory*, Dover Publications, Mineola, N.Y. (1972).

[23] Takeuti, G., Zaring, W.M.: *Introduction to Axiomatic set theory*, Springer-Verlag Berlin Heidelberg (1971).

[24] Tarski, A.: On the Calculus of Relations, J. Symbolic Logic **6**, 73-89 (1941).

[25] Wechler, W.: *Universal Algebra for Computer Scientists*, Springer-Verlag, Berlin Heidelberg (1992).

[26] Yoeli, M.: A Note on a Generalization of Boolean Matrix Theory, The American Mathematical Monthly **68** 552-557 (1961).

Chapter 4

Some Basic Kinds of Binary Relations

Chapter 4 continues with the presentation of the most popular and most often used kinds of relations and some of their variants: We begin with considering functions (in contrast to Chapter 2 now on the object level). Then we have a short look on order relations before we give a detailed account of tolerance and equivalence relations in the final section.

There are two main kinds of relations which are of foremost importance in pure and applied mathematics, in computer science, and also, of course, in daily life (remember the examples given at the beginning of Chapter 1): partial orders and equivalence relations. Therefore, these two classes of relations will be in the center of the current chapter.

4.1 Relations and Functions

We do already know about the concept of a function from Def. 2.21 in Chapter 2 on preliminaries, and also from Def. 3.21. But since functions play such a prominent role in all kinds of mathematical activities, we devote a separate section to the presentation of their main properties. To avoid cumbersome look-ups to former parts we shortly recall the relevant notions in

Definition 4.1. *Let $\rho \subseteq S \times T$ be a relation. ρ is called*

(a) **partial map** *or* **functional** *if $\forall x \in S, \forall y, z \in T[x\rho y \wedge x\rho z \rightarrow y = z]$;*
(b) **total** *if $\forall x \in S\, \exists y \in T[x\rho y]$, i.e. $\mathrm{dom}(\rho) = S$;*
(c) a **mapping** *or* **function** *if it is functional and total;*
(d) **injective** *or* **one-to-one** *if $\forall x, y \in S, \forall z \in T[x\rho z \wedge y\rho z \rightarrow x = y]$;*
(e) **surjective** *or* **onto** *if $\forall y \in T\, \exists x \in S[x\rho y]$;*
(f) **bijective** *if it is injective and surjective.*

We continue with listing some easy consequences from the definitions above. Some characterizations of functional, injective and surjective relations have already been shown in Prop. 3.50. We will use these results to give some more relationships of that kind in

Proposition 4.2. *Let $\rho \subseteq S \times T$ be a relation.*

(1) *ρ is functional $\iff \rho\,\breve{}\,\rho \subseteq I_T \iff \rho\overline{I_T} \subseteq \overline{\rho} \iff \rho\,\breve{}$ is injective;*
(2) *ρ is total $\iff I_S \subseteq \rho\rho\,\breve{} \iff \overline{\rho} \subseteq \rho\overline{I_T} \iff U_{S\times T} = \rho U_T$;*
(3) *ρ is injective $\iff \rho\rho\,\breve{} \subseteq I_S \iff \overline{I_S}\rho \subseteq \overline{\rho} \iff \rho\,\breve{}$ is functional;*
(4) *ρ is surjective $\iff I_T \subseteq \rho\,\breve{}\,\rho \iff \overline{\rho} \subseteq \overline{I_T}\rho \iff U_{S\times T} = U_S\rho$
$\iff \rho\,\breve{}$ is total;*
(5) *ρ is a mapping $\iff \rho\overline{I_T} = \overline{\rho}$.*
(6) *If both ρ and $\rho\,\breve{}$ are mappings, then both are bijective.*

Proof. Note that the first equivalences in (1) - (4) are due to Prop. 3.50.

(1): From $\rho\breve{}\rho \subseteq I_T$ we get $\rho\overline{I_T} \subseteq \overline{\rho}$ by applying the SCHRÖDER equivalences Th. 3.26(1)(a). Now assume ρ to be functional, and let $\{\langle y_1, x\rangle, \langle y_2, x\rangle\} \subseteq \rho\breve{}$. Then $\{\langle x, y_1\rangle, \langle x, y_2\rangle\} \subseteq \rho$; thus $y_1 = y_2$, i.e. $\rho\breve{}$ is injective. The proof for the converse direction is similar.

(2): $I_S \subseteq \rho\rho\breve{}$ and $\rho\breve{}U_{S\times X} \subseteq U_{T\times X}$ imply $U_{S\times X} = I_S U_{S\times X} \subseteq \rho\rho\breve{}U_{S\times X} \subseteq \rho U_{T\times X} \subseteq U_{S\times X}$, i.e. $U_{S\times X} = \rho U_{T\times X}$, for any set X, which is more general than we need to show. For the proof of the other direction assume $U_{S\times T} = \rho U_T$. Then $I_S = \rho U_{T\times S} \cap I_S$ and, by applying the Dedekind rule Th. 3.26(3), $\rho U_{T\times S} \cap I_S \subseteq (\rho \cap I_S U_{S\times T})(U_{T\times S} \cap \rho\breve{}I_S) = \rho\rho\breve{}I_S = \rho\rho\breve{}$, so $I_S \subseteq \rho\rho\breve{}$. (Note that $U_{T\times S}\breve{} = U_{S\times T}$.)

Furthermore, we have $\overline{\rho} \subseteq S \times T$ and $\rho\overline{I_T} \subseteq S \times T$, whence $\overline{\rho} \subseteq \rho\overline{I_T} \iff \rho \cup \rho\overline{I_T} = U_{S\times T}$. But $\rho \cup \rho\overline{I_T} = \rho I_T \cup \rho\overline{I_T} = \rho(I_T \cup \overline{I_T}) = \rho U_T$

(3): $\rho\rho\breve{} \subseteq I_S \iff \overline{I_S}\rho \subseteq \overline{\rho}$ holds due to the SCHRÖDER equivalence Th. 3.26(1)(b). By (1) we have $\rho\breve{}$ is functional iff $\rho\breve{}\breve{} = \rho$ is injective.

(4): Similar to (2).

(5): Follows directly from the inequalities shown in (1) and (2).

(6): By hypothesis ρ is functional and total, i.e. $\rho\breve{}\rho \subseteq I_T$ and $I_S \subseteq \rho\rho\breve{}$; for $\rho\breve{}$ we similarly get $\rho\rho\breve{} \subseteq I_S$ and $I_T \subseteq \rho\breve{}\rho$. From the last two inequalities ρ is injective and surjective, according to (3) and (4), and from the first two the same follows for $\rho\breve{}$. □

Corollary 4.3. *Let $\rho \subseteq X \times Y$ and $\sigma \subseteq Y \times Z$ be relations. If both ρ and σ are*

(1) *functional, then also $\rho\sigma$ and $\sigma\rho$;*
(2) *total, then also $\rho\sigma$ and $\sigma\rho$;*
(3) *injective, then also $\rho\sigma$ and $\sigma\rho$;*
(4) *surjective, then also $\rho\sigma$ and $\sigma\rho$;*
(5) *mappings, then also $\rho\sigma$ and $\sigma\rho$;*
(6) *bijective, then also $\rho\sigma$ and $\sigma\rho$.*

Proof. (1): $(\rho\sigma)\breve{}\rho\sigma = \sigma\breve{}\rho\breve{}\rho\sigma \subseteq \sigma\breve{}I_Y\sigma = \sigma\breve{}\sigma \subseteq I_Z$.

(2): $I_X \subseteq \rho\rho\breve{} \subseteq \rho I_Y\rho\breve{} \subseteq \rho\sigma\sigma\breve{}\rho\breve{} = \rho\sigma(\rho\sigma)\breve{}$.

(3): $(\rho\sigma)(\rho\sigma)\breve{} = \rho\sigma\sigma\breve{}\rho\breve{} \subseteq \rho I_Y\rho\breve{} = \rho\rho\breve{} \subseteq I_X$.

(4): $I_Z \subseteq \sigma\breve{}\sigma \subseteq \sigma\breve{}I_Y\sigma \subseteq \sigma\breve{}\rho\breve{}\rho\sigma = (\rho\sigma)\breve{}\rho\sigma$.

(5): By (1) and (2).

(6): By (3) and (4). □

Corollary 4.4. *For all relations ρ, σ and τ the following holds:*

(1) τ *is functional* $\Longleftrightarrow \tau(\rho \cap \sigma) = \tau\rho \cap \tau\sigma$;
(2) τ *is functional* $\Longleftrightarrow \rho\sigma \cap \tau = (\rho \cap \tau\sigma^{\smile})\sigma$;
(3) τ *is functional* $\Longleftrightarrow \tau\overline{\rho} = \tau U \cap \overline{\tau\rho}$;
(4) τ *is functional* $\wedge\ \rho \subseteq \tau \wedge \rho U \supseteq \tau U \Longrightarrow \rho = \tau$.

Proof. (1) [⇒] By Th. 3.16(4) we have $\tau(\rho \cap \sigma) \subseteq \tau\rho \cap \tau\sigma$. Using the Dedekind rule Th. 3.26(3), monotonicity, and $\tau^{\smile}\tau \subseteq I$ gives $\tau\rho\cap\tau\sigma \subseteq (\tau\cap\tau\sigma\rho^{\smile})(\rho\cap\tau^{\smile}\tau\sigma) \subseteq \tau(\rho\cap\tau^{\smile}\tau\sigma) \subseteq \tau(\rho \cap I\sigma)$.
[⇐] We take $\rho := \overline{I}$ and $\sigma := I$; then $\tau\overline{I} \cap \tau = \tau\overline{I} \cap \tau I = \tau(\overline{I} \cap I) = \tau\emptyset = \emptyset$. But $\tau\overline{I} \cap \tau = \emptyset \rightarrow \tau\overline{I} \subseteq \overline{\tau}$, whence τ is functional by Prop. 4.2(1).

(2) [⇒] $(\rho \cap \tau\sigma^{\smile})\sigma \subseteq \rho\sigma \cap \tau\sigma^{\smile}\sigma \subseteq \rho\sigma \cap \tau \subseteq (\rho \cap \tau\sigma^{\smile})(\sigma \cap \rho^{\smile}\tau) \subseteq (\rho \cap \tau\sigma^{\smile})\sigma$.
[⇐] Taking $\rho := \sigma^{\smile}$ and $\sigma := \overline{I}$ we get $\sigma^{\smile}\sigma \cap \overline{I} = (\sigma^{\smile} \cap \overline{I}\sigma^{\smile})\sigma = (I\sigma^{\smile} \cap \overline{I}\sigma^{\smile})\sigma$. Now $I\sigma^{\smile} \cap \overline{I}\sigma^{\smile} = \{\langle x, y\rangle : xIx \wedge x\sigma^{\smile}y \wedge \neg xIx \wedge x\sigma^{\smile}y\} = \emptyset$. So $\sigma^{\smile}\sigma \cap \overline{I} = \emptyset$, whence $\sigma^{\smile}\sigma \subseteq I$.

(3) [⇒] We have $\rho \subseteq \rho \rightarrow \tau^{\smile}\tau\rho \subseteq I\rho \leftrightarrow \tau\overline{\rho} \subseteq \overline{\tau\rho}$ by the functionality of τ and Th. 3.26(1). Since $\tau\overline{\rho} \subseteq \tau U$, we get $\tau\overline{\rho} \subseteq \overline{\tau\rho} \cap \tau U$. For the other inequality, using set theory and distributivity, we conclude as follows: $U = U \leftrightarrow \tau U \cup \overline{\tau U} = U \leftrightarrow \tau(\rho \cup \overline{\rho}) \cup \overline{\tau U} = U \leftrightarrow \tau\rho \cup \tau\overline{\rho} \cup \overline{\tau U} = U \leftrightarrow \overline{\overline{\tau\rho} \cap \tau U} \cup \tau\overline{\rho} = U \leftrightarrow \overline{\tau\rho} \cap \tau U \subseteq \tau\overline{\rho}$.
[⇐] Taking $\rho := I$, from the assumption we get $\tau\overline{I} = \tau U \cap \overline{\tau} = \tau(I \cup \overline{I}) \cap \overline{\tau} = (\tau I \cup \tau\overline{I}) \cap \overline{\tau} = \tau\overline{I} \cap \overline{\tau}$. So $\tau\overline{I} \subseteq \overline{\tau}$, whence τ is functional by Prop. 4.2(1).

(4): By Prop. 3.55(2) and our hypothesis we have $\tau \subseteq \tau U \subseteq \rho U$, whence $\tau = \tau U \cap \tau \subseteq \rho U \cap \tau$. Applying the Dedekind rule Th. 3.26(3) this gives

$$\tau \subseteq \rho U \cap \tau \subseteq (\rho \cap \tau U^{\smile})(U \cap \rho^{\smile}\tau) \subseteq \rho\rho^{\smile}\tau \subseteq \rho\tau^{\smile}\tau,$$

with the last inequality following from $\rho \subseteq \tau \rightarrow \rho^{\smile} \subseteq \tau^{\smile}$. And by the functionality of τ we finally get $\tau \subseteq \rho\tau^{\smile}\tau \subseteq \rho I = \rho$. □

Proposition 4.5. *For all relations ρ and τ the following holds:*

(1) τ *is total* $\Longrightarrow \overline{\tau\rho} \subseteq \tau\overline{\rho}$;
(2) τ *is total* $\Longrightarrow (\rho\tau = \emptyset \leftrightarrow \rho = \emptyset)$;

Proof. Note that for τ total we have $U = \tau U$ by Prop. 4.2.
(1): Then $U = \tau U = \tau(\rho \cup \overline{\rho}) = \tau\rho \cup \tau\overline{\rho}$. So we have $\tau\rho \supseteq \overline{\tau\overline{\rho}}$, whence $\overline{\tau\rho} \subseteq \tau\overline{\rho}$.

(2): Using the SCHRÖDER equivalence Th. 3.26(1)(b) we get, since $U = U^{\smile}$,

$$\rho\tau \subseteq \emptyset \leftrightarrow U\tau^{\smile} \subseteq \overline{\rho} \leftrightarrow U^{\smile}\tau^{\smile} \subseteq \overline{\rho} \leftrightarrow (\tau U)^{\smile} \subseteq \overline{\rho} \leftrightarrow U \subseteq \overline{\rho} \leftrightarrow \rho \subseteq \emptyset.$$

□

Corollary 4.6. *Let τ be a mapping; then for all relations ρ and σ the following holds:*

(1) $\tau\overline{\rho} = \overline{\tau\rho}$;

(2) $\rho\tau \subseteq \sigma \iff \rho \subseteq \sigma\tau^{\smile}$.

Proof. (1): From Cor. 4.4(3), since $\tau U = U$ by Prop. 4.2(2).
(2): [$\Rightarrow$] Since τ is total and $U = U^{\smile}$ we get: $U = \tau U = U^{\smile} = U\tau^{\smile} = (\overline{\sigma} \cup \sigma)\tau^{\smile} = \overline{\sigma}\tau^{\smile} \cup \sigma\tau^{\smile} \subseteq \overline{\rho} \cup \sigma\tau^{\smile}$ (for the last inequality we used that , by Th. 3.26(1)(b), $\rho\tau \subseteq \sigma \rightarrow \overline{\sigma}\tau^{\smile} \subseteq \overline{\rho}$). Thus $U \subseteq \overline{\rho} \cup \sigma\tau^{\smile}$, whence $\rho \cap \overline{\sigma\tau^{\smile}} \subseteq \emptyset$, i.e. $\rho \subseteq \sigma\tau^{\smile}$.
[$\Leftarrow$] $\rho \subseteq \sigma\tau^{\smile} \rightarrow \rho\tau \subseteq \sigma\tau^{\smile}\tau \subseteq \sigma$, since τ is functional by hypothesis. □

We close this section with the following two remarks:

- Given two relations ρ and σ, then their composition $\rho\sigma$ is commonly denoted as $\sigma\rho$ in case ρ and σ are functions: $x(\rho\sigma) := \sigma(\rho(x))$.
- A relation $\rho \subseteq S \times T$ from S to T can also be viewed as a function

$$f_\rho : S \longrightarrow \mathbf{2}^T \text{ with } f_\rho(x) := x\rho \tag{4.1}$$

for each $x \in S$. Similarly, $\rho^{\smile}$ con be considered as a function

$$f_{\rho^{\smile}} : T \longrightarrow \mathbf{2}^S \text{ with } f_{\rho^{\smile}}(y) := \rho y. \tag{4.2}$$

4.2 Order Relations

Here we give only a short introduction to the main notions and results concerning order relations, focusing on algebraic properties. Some information about the role of order relations in the context of preference modeling can be found in Section 18.3.

Definition 4.7. *Let ρ be a relation on a set S. ρ is called a*

(a) **preorder** *(or* **quasi-order***), if ρ is reflexive and transitive;*
(b) **partial order***, if it is a preorder that is additionally antisymmetric, usually denoted in a generic sense by $\preceq$;*
(c) **strict partial order***, if it is irreflexive and transitive;*
(d) **total order** *if it is a partial order that is additionally complete;*
(e) **linear order** *if it is a strict partial order that is additionally complete;*
(f) If $\preceq$ is a partial order on a set S, then the pair $\langle S, \preceq\rangle$ is called a **partially ordered set** *or* **poset** *for short.*

A partial order is thus characterized by the following relationships: $I \subseteq \rho, \rho^2 \subseteq \rho$, and $\rho \cap \rho^{\smile} \subseteq I$ (see Prop. 3.50); and for a total order $\rho \cup \rho^{\smile} = U$ additionally holds.

Before we proceed with more details on partial orders, we present some elementary properties and an example of a preorder.

Proposition 4.8. *Let ρ and σ be two preorders on a set A. Then*

(1) $\rho^{\smile}$ is a preorder.
(2) $\rho \cap \sigma$ is a preorder.
(3) $\rho^2 \subseteq \rho^{\smile}\rho$.

Proof. (1): By Prop. 3.25(1) and (6).
(2): By Prop. 3.52.
(3): By (1) $\rho^\smile$ is a preorder, so $I \subseteq \rho^\smile$. Therefore, $\rho^2 = (I \cap \rho^\smile)\rho^2 \subseteq \rho^\smile\rho^2 \subseteq \rho^\smile\rho$ by transitivity of ρ. □

Proposition 4.9. *Let ρ and σ be two preorders on a set S. Then ρ and σ commute iff both $\rho\sigma$ and $\sigma\rho$ are preorders.*

Proof. Assume $\rho\sigma = \sigma\rho$. $\rho\sigma$ is reflexive by Prop. 3.53(1). Furthermore, using Prop. 3.53(3), we have $(\rho\sigma)^2 = \rho\sigma\sigma\rho = \rho\sigma\rho = \rho^2\sigma = \rho\sigma$. Thus $\rho\sigma$ and $\sigma\rho$ are preorders. Conversely, assume both $\rho\sigma$ and $\sigma\rho$ to be preorders, i.e. each of $\rho, \sigma, \rho\sigma$ and $\sigma\rho$ are reflexive, transitive and idempotent. Then, using $I \subseteq \sigma$ and $I \subseteq \rho$, we get $\rho\sigma \subseteq \sigma(\rho\sigma) \subseteq \sigma\rho\sigma\rho = \sigma\rho$. Similarly one shows $\sigma\rho \subseteq \rho\sigma$. □

A characterization of preorders in terms of BK products is given in Prop. 6.24. From Prop. 4.8(2) it follows that, given a set $R \subseteq \mathcal{B}_X$, there is a smallest preorder containing all the relations of R, called the *preorder generated by R*.

Proposition 4.10. *Let ρ and σ be two preorders on a set S. If they commute, the preorder generated by ρ and σ is equal to $\rho\sigma$.*

Proof. Suppose that ρ and σ commute; then by Prop. 4.9 both $\rho\sigma$ and $\sigma\rho$ are preorders. Then, using $I \subseteq \sigma$ and $I \subseteq \rho$, we have $\rho \subseteq \rho\sigma$ and $\sigma \subseteq \rho\sigma$. Now let τ be another preorder with $\rho \subseteq \tau$ and $\sigma \subseteq \tau$. Then $\rho\sigma \subseteq \tau\tau = \tau$, since preorders are idempotent by Prop. 3.53(3). □

As an example of a preorder consider the following (from [10, Section 1.4]): Let S denote the set of all sequences $(a_n)_{n\in\mathbb{N}}$ of real numbers. For $\boldsymbol{a}, \boldsymbol{b} \in S$ we define $\boldsymbol{a} \leq \boldsymbol{b}$ if $\boldsymbol{a}$ is a subsequence of $\boldsymbol{b}$. Then $\leq$ is reflexive and transitive, and thus a preorder, but not antisymmetric since $(0,1,0,1,\ldots) \leq (1,0,1,0,\ldots) \leq (0,1,0,1,\ldots)$ without the last two sequences being equal.

Proposition 4.11. *Let ρ and σ be two partial orders on a set A. Then*

(1) $\rho^\smile$ is a partial order.
(2) $\rho \cap \sigma$ is a partial order.

Proof. (1): By Prop. 4.8 and Prop. 3.25(5).
(2): From the antisymmetry of ρ and σ we get $\rho \cap \rho^\smile \subseteq I$ and $\sigma \cap \sigma^\smile \subseteq I$; thus $\rho \cap \rho^\smile \cap \sigma \cap \sigma^\smile = \rho \cap \sigma \cap (\rho \cap \sigma)^\smile \subseteq I$. □

The strict order $\prec$ associated with a partial order $\preceq$ is the relation defined as $x \prec y := x \preceq y \wedge x \neq y$. Note that if $\prec$ is the strict order associated with $\preceq$, then $\prec$ is irreflexive. Conversely, if $\prec$ is an irreflexive and transitive relation on a set S, then $\preceq$ defined by $x \preceq y := x \prec y \vee x = y$ is a partial order.

As a shorthand we will often refer to $\langle S, \preceq\rangle$ as the poset S, leaving $\preceq$ implicit. And if it is necessary to distinguish it from others, we will write $\preceq_S$ for a partial order on S. If there is already a commonly used sign for a certain partial order, like $\leq$ for numbers or $\subseteq$ for sets, we will of course use these signs instead of $\preceq$. A few examples of partial orders are (see [8]):

(1) The subset relation $\subseteq$: For any set S, $\langle \mathbf{2}^S, \subseteq\rangle$ is a partial order.
(2) The natural orders on $\mathbb{N}, \mathbb{Q}$ or $\mathbb{Z}$.
(3) Also $(\mathbb{N}, |)$ is a poset, where $m|n$ means $\exists k \in \mathbb{N}[k \geq 1 \wedge n = km]$.
(4) *Orders on strings*: Let $\Sigma = \{a_1, \ldots, a_n\}$ be a finite alphabet; the a_i are called the *symbols* of Σ . *Strings* over Σ are finite sequences $w = x_1 x_2 \ldots x_m$ such that each $x_j \in \Sigma$. The length of a string $w = x_1 x_2 \ldots x_m$ is $|w| = m$. The special string with length 0 is called the *empty string* and is denoted by ϵ. The set of all strings over Σ (including ϵ) is usually denoted by Σ^*. On Σ^* we can define the binary operation of **concatenation** as follows: If $u = u_1 u_2 \ldots u_n$ and $w = w_1 w_2 \ldots w_m$ are two strings in Σ^*, then their concatenation is $u \cdot w = u_1 u_2 \ldots u_n w_1 w_2 \ldots w_m$. We set $w \cdot \epsilon = \epsilon \cdot w = w$. Obviously, $(\Sigma^*, \cdot, \epsilon)$ is a monoid.
Let $u, v \in \Sigma^*$ be two strings over an alphabet Σ; then
– *u is a* **prefix** *of v*, denoted by $u\pi v$, iff there is a $y \in \Sigma^*$ such that $v = uy$;
– *u is a* **suffix** *of v*, denoted by $u\sigma v$, iff there is a $x \in \Sigma^*$ such that $v = xu$.
It is not hard to show, that both π and σ are partial orders: E.g. $v = v\epsilon$, for all $v \in \Sigma^*$, thus $u\pi u$, or if $u\sigma v \wedge v\sigma w$ then there are $x, y \in \Sigma^*$ with $v = xu \wedge w = yv$, whence $w = yxu$; thus $u\sigma w$. Clearly, $u\pi v \wedge v\pi u \rightarrow v = ux \wedge u = vy \rightarrow v = vyx \wedge u = uxy$; therefore $x = y = \epsilon$ and $u = v$. These orderings are, however, not total; but sometimes we need just this property, e.g. for building the index of a book. It can be shown, that the **lexicographic order** $\preceq$ has the desired property. We assume that Σ is totally ordered, say $a_1 < a_2 < \ldots < a_n$. Then for any two strings $u, v \in \Sigma^*$ we define

$$u \preceq v := \begin{cases} v = uy \text{ for some } y \in \Sigma^* \text{ or} \\ u = xa_i y, v = xa_j z, \text{ and } a_i < a_j \text{ for some } x, y, z \in \Sigma^*. \end{cases}$$

In words: either u is a prefix of v or else u and v share a common prefix x followed by differing symbols, a_i in u and a_j in v, with $a_i < a_j$.

4.3 Tolerance and Equivalence Relations

From a formal point of view, this section is about a few basic properties of three variants of symmetric relations, as given in

Definition 4.12. *A symmetric binary relation ρ on a set A is said to be a*

(a) **tolerance relation** *iff ρ is also reflexive;*
(b) **partial equivalence relation** *(short: PER), iff ρ is also transitive;*
(c) **equivalence relation**, *iff ρ is also transitive and reflexive.*

An equivalence relation thus is a tolerance that is also transitive or, equivalently, a preorder that is also symmetric (see Def. 4.7(a)), or a partial equivalence relation that is reflexive too; therefore an equivalence relation is a special case of each of the other three mentioned concepts. Because of their prevalence (and popularity) we begin our exposition with this special case, followed by the presentation of the two generalizations mentioned in Def. 4.12. This way, most of the material in Section 4.3.1 on equivalence relations can be read independently from the contents of the other two sections.

4.3.1 *Equivalence Relations*

Equivalence relations are the most radical notion to model similarity in the sense that they partition a collection of things into pairwise disjoint sub-collections. Note that we do already know two trivial examples for an equivalence relation on a set S: I_S which fragments S into single-element sets, and U_S which does not fragment S at all.

A result that is inseparably connected to equivalence relations concerns their relationship to partitions. The fundamental significance of equivalence relations is that they motivate a rather general principle of abstraction: If we want to ignore some inessential attributes of the objects in a certain collection, then all objects with the same essential attributes can be identified. And analysis of such equivalence classes of objects rather than of the objects themselves often simplifies a problem significantly. Before we turn to the details, we have to introduce some new notation that has become standard for equivalences:

Definition 4.13.

(a) *Let ρ be an equivalence relation on a set S, then it is customary to denote the aftersets $x\rho$ by $[x]_\rho$, x/ρ, or simply by $[x]$ if no confusion about the equivalence involved can occur: $[x]_\rho = \{y \in S : x\rho y\}$ is called the* **equivalence class** *of x induced by ρ.*

(b) *The set of all equivalence classes is called the* **quotient set** *of S modulo ρ, denoted by S/ρ.*

Proposition 4.14. *Let ρ be an equivalence relation on a set S. Then the following three statements are equivalent:*

(1) $x\rho y$;
(2) $[x] = [y]$;
(3) $[x] \cap [y] \neq \emptyset$.

Proof. (1) $\to$ (2): Suppose $x\rho y \wedge z \in [x]$; then $x\rho z$ and, by symmetry, $y\rho x$. Thus, $y\rho z$ by transitivity, whence $z \in [y]$. So $[x] \subseteq [y]$. Conversely, from $z \in [y]$ we get $y\rho z$ and, by transitivity, $x\rho z$. Thus, $[y] \subseteq [x]$.
(2) $\to$ (3): Obvious, since $x \in [x]$.
(3) $\to$ (1): $[x] \cap [y] \neq \emptyset$ implies $\exists z[x\rho z \wedge y\rho z]$. Thus, $z\rho y$, which gives $x\rho y$ by the transitivity of ρ. □

The following two theorems contain the presumably most well known and important results on equivalence relations:

Theorem 4.15. *Let ρ be an equivalence relation on a set S. Then the following holds:*

(1) $[x] \neq \emptyset$
(2) *For $x, y \in S$, either $[x] = [y]$ or $[x] \cap [y] = \emptyset$;*
(3) $\bigcup\{[x] : x \in S\} = S$;
(4) *S/ρ is a partition of S.*

Proof. (1): Since ρ is reflexive, for each $x \in S : x \in [x]$.
(2): By Prop. 4.14 $[x] = [y] \leftrightarrow [x] \cap [y] \neq \emptyset$.
(3): Since for each $x \in S$ we have $x \in [x]$, we get $S \subseteq \bigcup\{[x] : x \in S\}$. And $[x] \subseteq S$ gives $\bigcup\{[x] : x \in S\} \subseteq S$.
(4): An immediate consequence from (1) - (3). □

For the following we need some new notation: The set of all equivalence relations on a set S will be denoted by Eq_S. If $\pi = \{A_i | i \in I\} \in \Pi_A$, the set of partitions of the set A, we define the relation ρ_π by: $x\rho_\pi y$ if there is an $A_i \in \pi$ such that $\{x, y\} \subseteq A_i$. Conversely, for $\rho \in \mathsf{Eq}_A$ we define the partition π_ρ of A by: $\pi_\rho = S/\rho = \{[x]_\rho | x \in A\}$.

Theorem 4.16. *Every partition π of a set A induces a corresponding equivalence relation ρ_π on A.*

Proof. Let $x \in A_i$; then $x\rho_\pi x$ since $\{x\} \subseteq A_i$. Obviously, ρ_π is symmetric. Assume $x\rho_\pi y$ and $y\rho_\pi z$; then there are blocks $A_i, A_j \in \pi$ with $\{x, y\} \subseteq A_i$ and $\{y, z\} \subseteq A_j$, and, since $y \in A_i \cap A_j$, we get $A_i = A_j$ by the definition of partitions. Thus, $x\rho_\pi z$. □

Theorem 4.17. *There is a one-to-one correspondence between partitions and equivalence relations on a set. With the notation introduced above we have*

(1) $\pi_{\rho_\pi} = \pi$;
(2) $\rho_{\pi_\rho} = \rho$.

Proof. (1): For all $y \in [x]_\pi$ we have $[x\rho_\pi y]$, whence $[x]_\pi \subseteq [x]_{\rho_\pi}$. Let $z \in [x]_{\rho_\pi}$, then there exists some $B \in \pi$ with $\{x, z\} \subseteq B$. By assumption $x \in [x]_\pi$, so $x \in [x]_\pi \cap B$, whence $[x]_\pi = B$ by definition of a partition, and $z \in [x]_\pi$. Therefore $[x]_{\rho_\pi} \subseteq [x]_\pi$, whence we have shown $[x]_{\rho_\pi} = [x]_\pi$. Since $[x]_{\rho_\pi} \in \pi_{\rho_\pi}$, we have $\pi \subseteq \pi_{\rho_\pi}$.

We now want to show the converse inequality, $\pi_{\rho_\pi} \subseteq \pi$, i.e. that for $x \in S$, $[x]_{\rho_\pi} \in \pi$. By definition of ρ_π we have $[x]_{\rho_\pi} = \{y \in S : \{x, y\} \subseteq B\}$ for some $B \in \pi$. Since π is a covering of S, there is a $B_0 \in \pi$ with $x \in B_0$, whence we are done if $[x]_{\rho_\pi} = B_0$. For $y \in S$ assume $\{x, y\} \subseteq B \in \pi$, i.e. $y \in [x]_{\rho_\pi}$. Then $x \in B \cap B_0 \neq \emptyset$, so $B = B_0$ since π is a partition; so $y \in B_0$, whence $[x]_{\rho_\pi} \subseteq B_0$. And if $y \in B_0$, then obviously $\{x, y\} \subseteq B_0 \in \pi$, so $B_0 \subseteq [x]_{\rho_\pi}$.

(2): Let $x\rho y$; then $y \in [x]_\rho$ and $x\pi_{\rho_\pi} y$, whence $\rho \subseteq \rho_{\pi_\rho}$. Again, $x\pi_{\rho_\pi} y$ implies that there is an equivalence class $[z]_\rho$ such that $\{x, y\} \subseteq [z]_\rho$. Thus $z\rho x \wedge z\rho y \to x\rho z \wedge z\rho y \to x\rho y$, by symmetry and transitivity of ρ. Therefore, $\rho_{\pi_\rho} \subseteq \rho$. There is a bijection $\phi : \mathsf{Eq}_A \longrightarrow \Pi_A, \phi(\rho) \mapsto \pi_\rho$, with the inverse $\psi : \Pi_A \longrightarrow \mathsf{Eq}_A, \psi(\pi) \mapsto \rho_\pi$. □

In what follows we will consider some properties of equivalence relations besides their correspondence to partitions, beginning with

Proposition 4.18. *Let $\rho, \sigma \in \mathsf{Eq}_S$. Then the following holds:*

(1) *$\rho \cap \sigma$ is an equivalence relation.*
(2) *$\rho\rho = \rho$.*
(3) *$\rho\sigma$ is an equivalence iff $\rho\sigma = \sigma\rho$.*
(4) *$\rho\rho\breve{}\rho = \rho$.*
(5) *$\rho \subseteq \sigma$ implies $\rho\sigma = \sigma$.*
(6) *If $\rho \cup \sigma$ is an equivalence then $\rho \cup \sigma = \rho\sigma$.*
(7) *If $\rho \cup \sigma$ is an equivalence then ρ and σ commute.*

Proof. (1): By Prop. 3.52.

(2): By Prop. 3.53(3).

(3): If $\rho\sigma$ is an equivalence, then $\rho\sigma = (\rho\sigma)\breve{} = \sigma\breve{}\rho\breve{} = \sigma\rho$, by the symmetry of each of $\rho\sigma, \rho$ and σ. Conversely, assume $\rho\sigma = \sigma\rho$. By Prop. 3.53(1) $\rho\sigma$ is reflexive; by the symmetry of ρ and σ we get $\rho\sigma = \sigma\rho = \sigma\breve{}\rho\breve{} = (\rho\sigma)\breve{}$, whence $\rho\sigma$ is symmetric; and, finally, $\rho\sigma$ is transitive: $\rho\sigma\rho\sigma = \rho\sigma\sigma\rho = \rho\sigma\rho = \rho\rho\sigma = \rho\sigma$ by (2).

(4): Using symmetry and (2) we get $\rho\rho\breve{}\rho = \rho^2 = \rho$.

(5): Obviously, $\sigma = I\sigma \subseteq \rho\sigma$; by assumption we also have $\rho\sigma \subseteq \sigma\sigma = \sigma$ by monotonicity of composition and (2).

(6): By Prop. 3.53(2) we have $\rho \cup \sigma \subseteq \rho\sigma$. Moreover, $\rho\sigma \subseteq (\rho \cup \sigma)^2 \subseteq \rho \cup \sigma$.

(7): By (6) and (3). □

The natural question about the structure of Eq_S is answered in the first part of

Proposition 4.19. *Let $\rho, \sigma \in \mathsf{Eq}_S$. Then the following holds:*

(1) *$(\mathsf{Eq}_S, \subseteq)$ is a complete lattice.*
(2) *$\rho \vee \sigma = \mathbf{t}(\rho \cup \sigma)$.*
(3) *$\rho \vee \sigma = \rho \cup \rho\sigma \cup \rho\sigma\rho \cup \ldots \cup \sigma \cup \sigma\rho \cup \sigma\rho\sigma \ldots$.*
(4) *ρ and σ commute iff $\rho \vee \sigma = \rho\sigma$ iff $\rho\sigma \in \mathsf{Eq}_S$.*

Proof. (1): A first trivial property is that, for every $S \neq \emptyset$, $|\mathsf{Eq}_S| \geq 2$, since $\{I_S, U_S\} \subseteq \mathsf{Eq}_S$. Furthermore, we have $I_S \subseteq \rho \subseteq U_S$ for each $\rho \in \mathsf{Eq}_S$, whence Eq_S is a bounded poset. We have to show that any non-empty family $\{\rho_i : i \in I\} \subseteq \mathsf{Eq}_S$ has a supremum and infimum. From Prop. 4.18(1) we get that $\bigwedge\{\rho_i : i \in I\} = \bigcap_{i \in I} \rho_i \in \mathsf{Eq}_S$. The supremum $\bigvee\{\rho_i : i \in I\}$ of a family of equivalence relations on S is $\bigcap\{\rho \in \mathsf{Eq}_S : \rho \supseteq \rho_i$ for all $i \in I\}$. But $U_S \in \mathsf{Eq}_S$ and $\rho_i \subseteq U_S$ for each $i \in I$, whence this intersection exists and is an equivalence on S due to what has just been shown. Moreover, it is the least such equivalence and therefore the supremum of $\{\rho_i : i \in I\} \subseteq \mathsf{Eq}_S$.

(2): By Prop. 3.52(1),(4) $\rho \cup \sigma$ is reflexive and symmetric, and by Prop. 3.61 $\mathbf{t}(\rho \cup \sigma)$ is also reflexive and symmetric. So by Def. 3.58 $\mathbf{t}(\rho \cup \sigma)$, is the smallest equivalence relation with $\rho \subseteq \mathbf{t}(\rho \cup \sigma)$ and $\sigma \subseteq \mathbf{t}(\rho \cup \sigma)$.

$$\begin{aligned}
(3):\ \rho \vee \sigma &= \textstyle\bigcup_{i\geq 1}(\rho \cup \sigma)^i \\
&= (\rho \cup \sigma) \cup (\rho^2 \cup \sigma\rho \cup \rho\sigma \cup \sigma^2) \cup (\rho \cup \sigma)^3 \ldots \\
&= (\rho \cup \sigma) \cup (\sigma\rho \cup \rho\sigma) \cup (\sigma\rho \cup \rho\sigma)(\rho \cup \sigma) \ldots \\
&= (\rho \cup \sigma) \cup (\sigma\rho \cup \rho\sigma) \cup (\sigma\rho^2 \cup \rho\sigma\rho \cup \sigma\rho\sigma \cup \rho\sigma^2) \ldots \\
&= (\rho \cup \sigma) \cup (\sigma\rho \cup \rho\sigma) \cup (\rho\sigma\rho \cup \sigma\rho\sigma) \ldots \\
&= \rho \cup \rho\sigma \cup \rho\sigma\rho \cup \ldots \cup \sigma \cup \sigma\rho \cup \sigma\rho\sigma \ldots
\end{aligned}$$

(Note that by transitivity $\rho^2 \subseteq \rho$, whence $\rho^2\chi \subseteq \rho\chi$ and thus $\rho^2\chi \cup \rho\chi = \rho\chi$.)

Assume $\rho\sigma = \sigma\rho$; then $\rho \vee \sigma$ reduces to $\rho \cup \rho\sigma \cup \sigma$. But by reflexivity we get $\rho = \rho I_S \subseteq \rho\sigma$ and $\sigma = I_S\sigma \subseteq \rho\sigma$, so $\rho \vee \sigma = \rho\sigma$. By (2), $\rho \vee \sigma \in \mathsf{Eq}_S$; thus $\rho\sigma \in \mathsf{Eq}_S$ and by Prop. 4.18(3) ρ and σ commute. □

Now that we have learned in Th. 4.17 about the one-to-one correspondence between equivalence relations and partitions, and just above about $(\mathsf{Eq}_S, \subseteq)$ being a complete lattice, we will have a look at how this latter result can be reformulated from the viewpoint of partitions (following[, Lemma 403]) in a more constructive way.

It is useful to be familiar with a few tantamount notations when scanning the pertinent literature: If $\rho \in \mathsf{Eq}_S$, then $a\rho b$ is often denoted by $a \equiv b \pmod{\rho}$; the equivalence class of ρ, often also called a *block*, containing a is denoted by $a\rho$, $[\rho]_a$ and sometimes as a/ρ. If $a \equiv b \pmod{\rho}$ then $a, b \in [\rho]_a$, also denoted as $a \equiv b \pmod{\pi_\rho}$. Note that $a \equiv b \pmod{\rho}$ and $a \equiv b \pmod{\pi_\rho}$ express the same fact, viz. $a\rho b$, whence each of the three notations can be used at will.

In Def. 2.9 the refinement relation has been defined on Π_A as follows: $\pi \leq \pi'$, if every block of π' is a union of blocks of π, i.e. π is a further fragmentation of π'. And in Prop. 2.10 $\leq$ has been shown to hold between two partitions π and π' iff every block of π is a subset of some block of π'.

Proposition 4.20. *Let $\rho, \sigma \in \mathsf{Eq}_A$, and $\pi_i, i \in I$, partitions of A. Then the following holds:*

(1) *$\pi_\rho \leq \pi_\sigma$ iff $\rho \subseteq \sigma$.*

(2) *Π_A is a complete lattice with*

 (a) *$a \equiv b \pmod{\bigwedge_{i\in I}\pi_i}$ iff $a \equiv b \pmod{\pi_i}$, for all $i \in I$;*

 (b) *$a \equiv b \pmod{\bigvee_{i\in I}\pi_i}$ iff there are $i_0, \ldots, i_n \in I$ and $x_0, \ldots, x_{n+1} \in A$, for some $n \in \mathbb{N}$, such that $a = x_0$, $b = x_{n+1}$ and $x_j \equiv x_{j+1}$ mod(π_{i_j}) for all $0 \leq j \leq n$.*

The zero of Π_A is $A/I_A = \{\{a\} : a \in A\}$, and the one is $A/U_A = \{A\}$.

Proof. (For sake of notational simplicity the arguments of the proof will be given for the case of two relations.)

(1): Obvious: the more pairs $\langle a, b\rangle$ are related by an equivalence, the bigger the blocks, and conversely.

(2)(a): By Prop. 4.18(1), $\rho \cap \sigma \in \mathsf{Eq}_A$, and it is the infimum of ρ and σ, whence $\rho \wedge \sigma = \rho \cap \sigma$. So $a \equiv b \pmod{\pi_{\rho\wedge\sigma}}$ iff $a \equiv b \pmod{\pi_\rho}$ and $a \equiv b \pmod{\pi_\sigma}$.

(2)(b): Restricting the above assertion to two relations, their join is described as follows: $a \equiv b \pmod{\pi_{\rho\vee\sigma}}$ iff there is a sequence $x_0, \ldots, x_{n+1} \in A$ such that

$$a = x_0, b = x_{n+1} \text{ and } x_i \equiv x_{i+1} \operatorname{mod}(\pi_\rho) \text{ or } x_i \equiv x_{i+1} \operatorname{mod}(\pi_\sigma), 0 \le i \le n. \qquad (*)$$

Let $a\gamma b$ the relation described by (*) above; then clearly $\rho \subseteq \gamma$ and $\sigma \subseteq \gamma$. If δ is an equivalence with $\rho \subseteq \delta$ and $\sigma \subseteq \delta$, and $a\gamma b$, then either $x_i \rho x_{i+1}$ or $x_i \sigma x_{i+1}$ and thus $x_i \delta x_{i+1}$ for each i. And by the transitivity of δ, $a\delta b$; so $\gamma \subseteq \delta$. Therefore, if γ is an equivalence, then $\gamma = \rho \vee \sigma$, i.e. $a \equiv b \pmod{\pi_{\rho\vee\sigma}}$.

Since $I \subseteq \rho \subseteq \gamma$, γ clearly is reflexive. By concatenation of two sequences corresponding to $x\gamma y$ and $y\gamma z$, we get $x\gamma z$ for arbitrary x, y, z; thus γ is transitive. Supposing $a\gamma b$, there is a sequence $x_0, \ldots, x_{n+1} \in A$ such that (*); but then, by the symmetry of ρ and σ, one has $x_{i+1}\rho x_i$ or $x_{i+1}\sigma x_i$, $0 \le i \le n$, whence $b\gamma a$. Thus γ is an equivalence relation on A.

The other direction follows from Prop. 4.19(2),(3): $\rho \vee \sigma = \mathbf{t}(\rho \cup \sigma) = \rho \cup \rho\sigma \cup \rho\sigma\rho \cdots \cup \sigma \cup \sigma\rho \cup \sigma\rho\sigma \cdots$. □

4.3.2 *Tolerance Relations*

Among the three generalizations of equivalences (recall the other two are *preorders* and *partial equivalence relations*) *tolerances* are presumably the most popular because of their appealing intuitive interpretation: Abandoning transitivity, they offer a less radical notion of similarity than equivalence does and provide a general tool for studying indiscernibility phenomena in the investigation of resemblance between perceptual objects, occurring with close-by sensations. With this purpose in mind it has been introduced by Poincaré [13] and Carnap [2], who called them *similarity*. The name "*tolerance*" itself is due to [16]; for more details see [12].

A paradigmatic example was given in [13]: Consider a *metric* d[1] on a set X, and define $x\tau y := d(x, y) < \Delta$ for some small positive value Δ. Since $2\Delta > \Delta$, we do not necessarily get $x\tau z$ from $x\tau y$ and $y\tau z$, i.e. τ is, in general, not transitive. So $x\tau_\epsilon y := |x - y| < \epsilon$ for some fixed real $\epsilon > 0$ can be regarded as the "classical" tolerance.

We continue with some basic concepts and properties around tolerances, using as main sources [1, 12] as well as some early papers (given in their logical order) of Chajda and his colleagues [3–5]. Before we are going to have a closer look at what makes equivalences so special compared with other kinds of relations, we list some elementary properties of tolerance relations in

---

[1]A **metric** on a set X is a function $d : X \times X \longrightarrow \mathbb{R}_0$, called *distance*, such that
(1) $d(x, y) \ge 0$ (non-negativity),
(2) $d(x, y) = 0$ iff $x = y$ (identity of indiscernibles),
(3) $d(x, y) = d(y, x)$ (symmetry), and
(4) $d(x, z) \le d(x, y) + d(y, z)$ (triangle inequality).

Proposition 4.21. *Let ρ and σ be two tolerances on a set S. Then*

(1) $\rho^{\smile} \in \mathsf{Tol}_S$.
(2) $\rho \cap \sigma \in \mathsf{Tol}_S$.
(3) $\rho\sigma \in \mathsf{Tol}_S$ *iff* $\rho\sigma = \sigma\rho$.
(4) $\rho^2 \in \mathsf{Tol}_S$.

Proof. (1): By Prop. 3.25.
(2): By Prop. 3.52.
(3): By Prop. 4.18(3), using the parts for reflexivity and symmetry.
(4): By (3). □

The concept of an equivalence class can be generalized in a natural way for arbitrary relations as done in Def. 4.22 below. Some elementary facts will then be formulated on the most general level possible. Later on these facts will be used when focusing on tolerance relations in this, and partial equivalence relations in the next section.

Recall that an equivalence relation ρ on a set S defines a partition of S via its aftersets $x\rho, x \in S$. For arbitrary relations the notion of an equivalence class splits into different concepts:

Definition 4.22. ([1,4]) *Let ρ be an arbitrary, and τ be a tolerance relation on a set S.*

(a) A **pre-block** *of ρ is a non-void subset $B \subseteq S$, such that $B \times B \subseteq \rho$. The set of pre-blocks of ρ is denoted by* PBl_ρ.
(b) A **block** *of ρ is a maximal pre-block: if $B \subseteq C$ and $C \times C \subseteq \rho$, then $B = C$. The set of blocks of ρ is denoted by Bl_ρ. We say that ρ has blocks provided $\mathsf{Bl}_\rho \neq \emptyset$*
(c) The **tolerance class** *of an element $x \in S$ is the afterset $x\tau$, also denoted as x/τ, and sometimes also called the τ-***neighborhood of** *x. The set of all tolerance classes is called the* **quotient set** *of S modulo τ, denoted by S/τ.*
(d) A set S together with a tolerance relation τ is called a **tolerance space**, *denoted (S,τ); and Tol_S denotes the set of all tolerances on S.*

A block of ρ thus is a maximal non-void subset of $B \subseteq S$ such that the restriction of ρ to B is the universal relation U_B. Looking at a tolerance as an undirected graph, its blocks correspond to its maximal cliques and the tolerance class of a node corresponds to the set of its neighbors, together with the node itself. As for equivalences, due to reflexivity, the blocks and the classes of a tolerance on a set S form coverings of S, but they do not have to be mutually disjoint, and so these coverings need not be partitions. If, however, one of these coverings is a partition, then the tolerance is an equivalence and the coverings by blocks and by classes coincide.

Lemma 4.23. ([4]) *Let ρ be a binary relation on a set S and $\emptyset \neq C \subseteq S$; then C is a pre-block of ρ iff there is a $B \in \mathsf{Bl}_\rho$ with $C \subseteq B$.*

Proof. Obviously, every subset of a block of ρ is a pre-block. Conversely, every pre-block is contained in a block: For finite S this follows from the existence of maximal elements in every finite non-empty poset; and in the general case it is a consequence of Zorn's Lemma[2] since the union of an ascending chain of pre-blocks is again a pre-block. □

Corollary 4.24. ([4])

(1) Let ρ be a binary relation on S; then ρ has blocks iff $I \cap \rho \neq \emptyset$

(2) Let ρ, σ be a binary relations on S. If $\rho \subseteq \sigma$ and ρ has blocks, then also σ has blocks, and every block of ρ is contained in some block of σ.

Proof. (1): Take an x with $x(I \cap \rho)x$, and apply Lemma 4.23 to $C = \{x\}$.
(2): By Lemma 4.23 and (1). □

Another consequence of Lemma 4.23 is

Corollary 4.25. ([1, p. 196]) *A tolerance τ on a set S is completely determined by its blocks. If $\tau, \tau' \in \mathsf{Tol}_S$ such that $\mathsf{Bl}_\tau = \mathsf{Bl}_{\tau'}$ then $\tau = \tau'$.*

Proof. $\{x, y\} \subseteq S$ is a pre-block iff $x\tau y$, since this implies $y\tau x, x\tau x$, and $y\tau y$. So $x\tau y$ iff there exists a $B \in \mathsf{Bl}_\tau$ with $\{x, y\} \subseteq B$. □

Not every covering of a set S is of the form Bl_τ for some tolerance τ. The next proposition characterizes those that are.

Proposition 4.26. [1, Theorem 2] *Let $\mathcal{C}$ be a covering of a set S. Then $\mathcal{C} = \mathsf{Bl}_\tau$ for some $\tau \in \mathsf{Tol}_S$ iff it satisfies the following two conditions:*

(i) For every $B, C \in \mathcal{C}$, $B \subseteq C$ implies $B = C$.

(ii) For every $T \subseteq S$, if for every two elements $x, y \in T$ there is a $B \in \mathcal{C}$ such that $x, y \in B$, then $T \subseteq C$ for some $C \in \mathcal{C}$.

Proof. Let $\mathcal{C} = \mathsf{Bl}_\tau$ for some $\tau \in \mathsf{Tol}_S$. Then (i) is straightforward by the maximality of blocks. Concerning (ii), if for every pair $x, y \in T \subseteq S$ we have $\{x, y\} \subseteq B$ for some $B \in \mathcal{C}$, then T is a pre-block and, by Lemma 4.23, contained in some block $C \in \mathcal{C}$.

Conversely, let $\mathcal{C}$ be a covering of S satisfying conditions (i) and (ii). Obviously, the relation $\tau = \bigcup_{C \in \mathcal{C}} C \times C = \{\langle x, y\rangle \in S^2 : x, y \in C \text{ for some } C \in \mathcal{C}\}$ is a tolerance on S such that every $C \in \mathcal{C}$ is a pre-block of τ. By (ii) for every such pre-block T there is a $C \in \mathcal{C}$ with $T \subseteq C$. So, for every pre-block T of τ, we have $T \in \mathcal{C}$, whence every $C \in \mathcal{C}$ is a block of τ by (i). □

Proposition 4.27. ([4]) *Let ρ be a binary relation on S.*

(1) ρ is reflexive iff Bl_ρ is a covering of S.

---

[2] **Zorn's Lemma** is a result in set theory that appears in proofs of some non-constructive existence theorems throughout mathematics: *If $(P, \leq)$ is a partially ordered set such that every chain in P has an upper bound in P, then P has a maximal element m.* For details see e.g. [6, 7]. (The situation resembles the one we have mentioned concerning the SCHRÖDER rules and DE MORGAN's Theorem K just before Th. 3.27: Though this theorem was proved in 1922 by KAZIMIERZ KURATOWSKI and independently by MAX ZORN in 1935, in essentially all published sources the name ZORN is associated to it.)

(2) *ρ is a tolerance on S iff* Bl_ρ *is a covering of S and* $\bigcup_{B\in\mathsf{Bl}_\rho} B^2 = \rho$.
(3) *If ρ is transitive and has blocks, then they are pairwise disjoint.*
(4) *If ρ is antisymmetric then* $\mathsf{Bl}_\rho \subseteq \{\{a\} : a \in S\}$.

Proof. (1): If ρ is reflexive then for all $x \in S$ we have $\{x\} \times \{x\} \subseteq \rho$. So by Lemma 4.23 there is a block B with $\{x\} \subseteq B$, and Bl_ρ is therefore a covering of S. Conversely, if Bl_ρ is a covering of S then for all $x \in S$ there is a block B with $\{x\} \subseteq B$, whence $\{x\} \times \{x\} \subseteq \rho$ again by Lemma 4.23, i.e. $x\rho x$ for all x.

(2): Let ρ be a tolerance; then by (1) Bl_ρ is a covering. For the remainder it is sufficient to prove $\rho \subseteq \bigcup_{B\in\mathsf{Bl}_\rho} B^2$. Assume $x\rho y$; then also $y\rho x$, $x\rho x$, and $y\rho y$. Therefore $\{x, y\}^2 \subseteq \rho$, whence by Lemma 4.23 we get $\langle x, y\rangle \in \bigcup_{B\in\mathsf{Bl}_\rho} B^2$. Conversely, if Bl_ρ is a covering then ρ is reflexive by (1), and symmetry follows from the fact that ρ equals the union of squares of blocks.

(3): Let B_1, B_2 be blocks of a transitive relation ρ, and assume $x \in B_1 \cap B_2$. Then, for each $y \in B_1$ and $z \in B_2$ we have $y\rho x$ and $x\rho z$, and similarly also $z\rho x$ and $x\rho y$. So, by the transitivity of ρ, we get $y\rho z$ and $z\rho y$. Since, obviously, $y\rho y$ and $z\rho z$, it follows that $\{y, z\}^2 \subseteq \rho$. Thus $(B_1 \cup B_2)^2 \subseteq \rho$, and since B_1 and B_2 have been assumed to be blocks of ρ, we finally have $B_1 = B_1 \cup B_2 = B_2$.

(4): To prove the non-trivial case assume that B is a block of an antisymmetric relation ρ with $x, y \in B$. But then we have $x\rho y$ and $y\rho x$; thus $x = y$ and $B = \{x\}$. □

Definition 4.28. ([5]) *Let $\mathcal{C} = \{M_i : i \in I\} \subseteq \mathbf{2}^S$ be a covering of a non-void set S. $\mathcal{C}$ is called a t-***covering** *if*

(1) *$i_0 \in I$ and $I_0 \subseteq I$ then $M_{i_0} \subseteq \bigcup_{i\in I_0} M_i \to \bigcap_{i\in I_0} M_i \subseteq M_{i_0}$, and*
(2) *$T \subseteq S$ such that $T \nsubseteq M_i$ for $i \in I$, then there are two elements $x, y \in T$ such that $\{x, y\} \nsubseteq M_i$ for $i \in I$.*

Proposition 4.29. ([5, Theorem 1]) *For any non-empty set S there exists a one-to-one correspondence between tolerance relations on S and t-coverings of S such that if $\tau \in \mathsf{Tol}_S$ and $\mathcal{C}_\tau$ is the t-covering of S corresponding to τ, then for each $x, y \in S$ we have $x\tau y$ iff $\{x, y\} \subseteq M$ for some $M \in \mathcal{C}_\tau$.*

Proof. Let $\tau \in \mathsf{Tol}_S$, then Bl_τ is a covering of S by Prop. 4.27(1). Let $\mathsf{Bl}_\tau = \{B_i : i \in I\}$, $i_0 \in I$, $I_0 \subseteq I$ and $B_{i_0} \subseteq \bigcup_{i\in I_0} B_i$. Furthermore, let $P = \bigcap_{i\in I_0} B_i$, and suppose $B_{i_0} \subset P$, i.e. there is an $x \in P - B_{i_0}$, and $y \in B_{i_0}$. This means that $y \in \bigcup_{i\in I_0} B_i$, so $y \in B_{i_1}$ for some $i_1 \in I_0$; we have $x \in P$ which implies $x \in B_{i_1}$, so we have $x\tau y$. And since y was chosen arbitrarily, $x\tau y$ for each $y \in B_{i_0}$, and therefore $B_{i_0} \cup \{x\} \in \mathsf{Bl}_\tau$. But this means that $B_{i_0} \notin \mathsf{Bl}_\tau$ since it is a proper subset of $B_{i_0} \cup \{x\}$, a contradiction. Thus we get $P = \bigcap_{i\in I_0} B_i \subseteq B_{i_0}$, and the first condition of Def. 4.28 is fulfilled. If $T \subseteq S$ is not a subset of any $B_i \in \mathsf{Bl}_\tau$, then $T \notin \mathsf{PBl}_\tau$ and there exist two elements $x, y \in T$ with $\neg x\tau y$. Thus, there is no $B_i \in \mathsf{Bl}_\tau$ with $\{x, y\} \subseteq B_i$. So the second condition of Def. 4.28 is fulfilled, and we have shown that Bl_τ is a t-covering.

Conversely, let $\mathcal{C} = \{C_i : i \in I\}$ be a t-covering of S, and let τ be a relation on S with $x\tau y$ iff $\{x, y\} \subseteq C_i$ for some $i \in I$. Then τ is reflexive and symmetric by construction, i.e. a tolerance. We have to show that $\mathcal{C} = \mathsf{Bl}_\tau$, i.e. that each $C_i, i \in I$, is maximal in PBl_τ and each maximal element of PBl_τ is in $\mathcal{C}$. Assume that $C_k \in \mathsf{PBl}_\tau$ is not maximal; then there is a P such that $C_k \subset P \in \mathsf{PBl}_\tau$. For $x \in P - C_k$ we then get $x\tau y$ for each $y \in C_k$. But then, for $y \in C_k$, there is an $i(y) \in I$ such that $x, y \in C_{i(y)}$, and $C_k \subseteq \bigcup_{y \in C_k} C_{i(y)}$. We have $x \in C_{i(y)}$ for all $y \in C_k$, thus $x \in \bigcap_{y \in C_k} C_{i(y)}$, and since $\mathcal{C}$ is a t-covering of S, $\bigcap_{y \in C_k} C_{i(y)} \subseteq C_k$, whence $x \in C_k$. This is a contradiction to $x \in P - C_k$, so each $C_i \in \mathsf{PBl}_\tau$ is maximal, since C_k was chosen arbitrarily. Thus, $\mathcal{C} \subseteq \mathsf{Bl}_\tau$.

Suppose there exists a $P \in \mathsf{Bl}_\tau - \mathcal{C}$. Since, as just shown, $\mathcal{C} \subseteq \mathsf{Bl}_\tau$, there is no $C \in \mathcal{C}$ with $P \subseteq C$. Therefore, since $\mathcal{C}$ is a t-covering of S, there exist two elements $x, y \in P$ such that for no $C \in \mathcal{C}$ we have $\{x, y\} \subseteq C$. So $\neg x\tau y$, whence $P \not\in \mathsf{PBl}_\tau$ and also $P \not\in \mathsf{Bl}_\tau$, a contradiction to $P \in \mathsf{Bl}_\tau - \mathcal{C}$. So each maximal element of PBl_τ is in $\mathcal{C}$. Thus, $\mathsf{Bl}_\tau \subseteq \mathcal{C}$. □

Up to here we have considered how coverings $\mathcal{C}$ of a set S are in relation to the blocks Bl_τ of a tolerance relation τ on S: When is it the case that $\mathcal{C} = \mathsf{Bl}_\tau$? We can now proceed to a still more general view and ask when a covering is determined by S/τ, the set of tolerance classes of τ. Coverings by blocks and by classes share few properties, but differ in many other ones: Since τ is reflexive, $x \in x\tau$, for every $x \in S$; thus, like Bl_τ also S/τ is a covering of S. Actually, if one of these coverings is a partition of S then τ is an equivalence relation and the coverings by blocks and by classes are exactly the same. And as it was the case with blocks, not every covering $\mathcal{C}$ of a set S is of the form $\mathcal{C} = S/\tau$ for some $\tau \in \mathsf{Tol}_S$. See [1, Example 5].

A main difference is the following: The covering by blocks completely determines a tolerance, because for a pair of elements x, y we have $x\tau y$ if and only if $x, y \in B$ some block B of τ (see Cor. 4.25). This makes it easier to characterize those coverings of a set that are families of blocks of tolerances; see Prop. 4.26 above. On the other hand, different tolerances may yield the same covering by classes. See [1, Example 6].

Beyond the results mentioned above, the authors of [1] answer the question when a covering of a *finite* set S equals S/τ for some tolerance on S. We must, however, refrain from presenting their results, because the lemmas, propositions and theorems needed for it are dependent to such a high extent, that this would mean to reproduce the whole paper. We close this section with two lemmas which connect tolerance classes and blocks.

Lemma 4.30. [1, Lemma 3] *Let τ be a tolerance on a set S. Then, for every $B \subseteq S$*

(1) B is a pre-block of τ iff $B \subseteq \{x \in S : B \subseteq x\tau\}$.
(2) B is a block of τ iff $B = \{x \in S : B \subseteq x\tau\}$.

Proof. (1): B is a pre-block of τ iff, for every $x \in B$, $y \in B \rightarrow y \in x\tau$, i.e. iff $B \subseteq x\tau$ for every $x \in B$. Thus, B is a pre-block of τ iff $B \subseteq \{x \in S : B \subseteq x\tau\}$.

(2): Let B a block of τ, and $y \in S$ such that $B \subseteq y\tau$; then $B \cup \{y\}$ is a pre-block of τ: By Th. 2.18 $(B \cup \{y\})^2 = B \times B \cup B \times \{y\} \cup \{y\} \times B \cup \{\langle y, y\rangle\}$. We have $B \times B \subseteq \tau$, since B is assumed to be a block of τ and $\{\langle y,y\rangle\}\subseteq\tau$, since τ is reflexive. Moreover, by the symmetry of τ, $B\times\{y\}=\{y\}\times B$. But our assumption $B\subseteq y\tau$ means $b\in B\to\exists z[y\tau z]$, so $\{y\}\times B\subseteq\tau$. Thus we have shown $(B\cup\{y\})^2\subseteq\tau$, i.e. $B\cup\{y\}$ is a pre-block of τ Since, obviously, $B\subseteq B\cup\{y\}$ we get $B=B\cup\{y\}$ because B is maximal, i.e. $y\in B$. Therefore, $B=\{x\in S:B\subseteq x\tau\}$.

Conversely, if $B=\{x\in S:B\subseteq x\tau\}$, then B is a pre-block of τ by (1). For any pre-block B' containing B we get $B\subseteq B'\subseteq x'\tau$ for every $x'\in B'$, again using (1). Thus, $B'\subseteq\{x\in S: B\subseteq x\tau\}=B$, whence B is a block of τ. □

Lemma 4.31. [1, Lemma 4] *Let τ be a tolerance on a set S.*

(1) For every $x\in S$, $x\tau$ is the union of all blocks of τ containing x.
(2) Every block of τ is the intersection of the tolerance classes of its elements.

4.3.3 *Partial Equivalence Relations*

Note that the difference between a PER and an equivalence relation is that the former need not be reflexive. So every equivalence relation is a PER, but not conversely: $\{\langle 0,0\rangle\}$ is a PER on $\{0,1\}$ only, since it is not reflexive. But by Prop. 3.22(9) we know that a PER ρ is reflexive on its field $\mathsf{fld}(\rho)$, so ρ is an equivalence on $\mathsf{fld}(\rho)$. We begin by listing some elementary properties of PERs in

Proposition 4.32. *Let ρ and σ be partial equivalence relations on a set S. Then the following holds:*

(1) $\rho\cap\sigma$ is a partial equivalence relation;
(2) $\rho^2=\rho$;
(3) ρ is a partial equivalence iff it is symmetric and idempotent.
(4) $\rho\sigma$ is a partial equivalence iff $\rho\sigma=\sigma\rho$;
(5) ρ is a partial equivalence iff $\rho\rho\breve{}=\rho$;
(6) $\rho\rho\breve{}\rho=\rho$;
(7) $\rho\subseteq\sigma$ implies $\rho\sigma=\sigma$.

Proof. (1): By Prop. 3.52(4) and (6).

(2): By Prop. 3.54(1) we have $\rho\subseteq\rho\rho\breve{}\rho$. Using symmetry and transitivity of ρ we get $\rho\rho\breve{}\rho=\rho^2\rho\subseteq\rho\rho$, so $\rho\subseteq\rho\rho$, which together with transitivity of ρ gives the desired result.

(3): Obvious from (2) and the definition of transitivity.

(4): If $\rho\sigma$ is a PER, then $\rho\sigma=(\rho\sigma)\breve{}=\sigma\breve{}\rho\breve{}=\sigma\rho$, by the symmetry of each of $\rho\sigma,\rho$ and σ. Conversely, assume $\rho\sigma=\sigma\rho$. By the symmetry of ρ and σ we get $\rho\sigma=\sigma\rho=\sigma\breve{}\rho\breve{}=(\rho\sigma)\breve{}$, whence $\rho\sigma$ is symmetric; and, finally, $\rho\sigma$ is transitive: $\rho\sigma\rho\sigma=\rho\sigma\sigma\rho=\rho\sigma\rho=\rho\rho\sigma=\rho\sigma$ by (2).

(5): Suppose ρ to be a partial equivalence . Then $\rho\rho=\rho$ by (2) and $\rho\breve{}=\rho$, whence $\rho\rho\breve{}=\rho$. To show the other direction, let $\rho\rho\breve{}=\rho$. Then $\rho\breve{}=(\rho\rho\breve{})\breve{}=\rho\breve{}\breve{}\rho\breve{}=\rho\rho\breve{}=\rho$; whence ρ is symmetric. So $\rho\rho=\rho\rho\breve{}=\rho$; thus ρ is transitive.

(6): Applying (5) and (2) we get $\rho\rho\breve{}\rho=\rho^2=\rho$.

(7): Obviously, $\sigma = I\sigma \subseteq \rho\sigma$; by assumption we also have $\rho\sigma \subseteq \sigma\sigma = \sigma$ by monotonicity of composition and (2). □

Concerning the algebraic structure of the set of all partial equivalences on a set, denoted by PEq_S, we have the following results which, as well as their proofs, are very similar to the ones of Prop. 4.19:

Proposition 4.33. *Let $\rho, \sigma \in \mathsf{PEq}_S$. Then the following holds:*

(1) $(\mathsf{PEq}_S, \subseteq)$ *is a complete lattice.*
(2) $\rho \vee \sigma = \mathbf{t}(\rho \cup \sigma)$.
(3) $\rho \vee \sigma = \rho \cup \rho\sigma \cup \rho\sigma\rho \cup \ldots \cup \sigma \cup \sigma\rho \cup \sigma\rho\sigma \ldots$.
(4) If $\rho\sigma = \sigma\rho$, *then* $\rho \vee \sigma = \rho \cup \sigma \cup \rho\sigma$.

Proof. (4): $\rho \vee \sigma = \rho \cup \rho\sigma \cup \rho\sigma\rho \cup \ldots \cup \sigma \cup \sigma\rho \cup \sigma\rho\sigma \ldots$ reduces to $\rho \cup \sigma \cup \rho\sigma$ because the products with more than two factors are smaller than $\rho\sigma = \sigma\rho$ by transitivity. □

The well-known bijection between equivalence relations and partitions (see Th. 4.17) extends to their partial counterparts (recall Def. 2.6(e) for the notion of a partial partition):

Theorem 4.34. ([14, Proposition 4]) *There is a one-to-one correspondence between partial partitions π and partial equivalence relations ρ on a set S via the following connection*

$$\text{for all } x, y \in S\text{: } x\rho_\pi y \text{ iff there is a } B \in \pi \text{ such that } \{x, y\} \subseteq B.$$

Partial equivalences also appear under the names *symmetric idempotents* (especially in connection with *splittings*), *weak congruences* (see [11, 15]), and in connection with *partial partitions.*

References

[1] Bartol, W., Miró, J., Pióro, K., Rosselló, F.: On the coverings by tolerance classes, Information Sciences **166**, 193-211 (2004).
[2] Carnap, R.: *Der Logische Aufbau der Welt,* Weltkreis-Verlag (1928).
[3] Chajda, I.: Partitions, coverings and blocks of binary relations, Glasnik Mat. **14**, 21-26 (1979).
[4] Chajda, I., Duda, J.: Blocks of binary relations, Annales Univ. Sci. Budapest, sectio Math **22**, 3-9 (1979).
[5] Chajda, I., Niederle, J., Zelinka, B.:. On existence conditions for compatible tolerances, Czechoslovak Mathematical Journal **26**, 304-311 (1976).
[6] Ciesielski, K.: *Set Theory for the Working Mathematician,* Cambridge University Press (1997).
[7] Devlin, K.: *The Joy of Sets: Fundamentals of Contemporary Set Theory,* 2nd ed., Springer-Verlag New York (1993).
[8] Gallier, J.: *Discrete Mathematics,* Springer Science+Business Media (2011).
[9] Grätzer, G.: *Lattice Theory: Foundation,* Springer Basel AG (2011).
[10] Harzheim, E.: *Ordered Sets,* Springer (2005).

[11] Madarász, R.: On the partial algebras of weak congruences, Univ Novi Sad J Math **25**, 89-97 (1995).

[12] Peters, J. F., Wasilewski, P.: Tolerance spaces: Origins, theoretical aspects and applications. Information Sciences **195**, 211-225 (2012).

[13] Poincaré, J.H. : *La valeur de la science*, Flammarion (1905).

[14] Ronse, C.: Partial partitions, partial connections and connective segmentation, J. Math. Imaging Vis. **32**, 97-125 (2008).

[15] Vojvodić, G. Šešelja, B.: On the lattice of weak congruence relations, Algebra Universalis **25**, 121-130 (1988).

[16] Zeeman, E.C.: The topology of the brain and the visual perception. In: Fort, K.M. (Ed.): *Topology of 3-manifolds and Selected Topics*, 240-256. Prentice Hall, Englewood Cliffs (1965).

Chapter 5

Relations, Matrices, and Graphs

This chapter provides all the concepts, together with their associated terminology and notation, which are essential for an adequate presentation of the tight interrelationships between the three areas mentioned in its title. There are one-to-one correspondences between relations and certain kinds of matrices and graphs; so one can take the liberty to use that terminology which is regarded the most suitable for a certain context.

There is a set of concepts in the crossing area of matrix theory, relations and graphs that will be needed in topics to come, like e.g. the inverse of a relation, or idempotent relations. This chapter serves to put together all of them in one place. I suggest to eventually skip this chapter on first reading and to follow the references given at the places where these concepts will be actually used.

Boolean matrices play a prominent role in the analysis of relations. In Section 3.2.1 we have already given a few short remarks on the interrelationships between relations and Boolean matrices. Since it is a common habit, we will henceforth use the simple notation $\boldsymbol{B}_n$ for $\mathfrak{M}^{\mathbf{2}}_{n\times n}$, and also for relations $\alpha \in \mathcal{B}_X$ with $|X| = n$. Boolean matrices also appear under the names "*(Boolean) relation matrices*", "*binary Boolean matrices*" or "*(0,1)-matrices*".

Remark: This chapter is interdisciplinary in the sense that it puts together concepts and facts from the three overlapping areas mentioned in its title. This is a first hint that, in case you want to know more, or certain special things, about relations you will also have to look for other notions like e.g. "*Boolean matrix*" or "*graph matrices*". But there is also another direction that deserves attention in this context: Here we focus on the two element Boolean algebra $\mathbf{2} := \{0, 1\}$; if, however, results on matrices over more general algebraic structures are of interest, you will have to look for e.g. "*matrices over distributive lattices / semirings / inclines*", "*lattice-valued matrices*", "*fuzzy relations / matrices*", etc. Since distributive lattices, semirings and inclines are all generalizations of Boolean algebras, you may find information of interest also under these headings. Some examples will be presented in Section 5.4 below, where we will have a look on matrices over inclines.

5.1 Recalling Basic Facts on Boolean Matrices

We shortly present some basic terminology and properties of matrix theory, mainly following [5]. You may also wish to consult [1] or [21] for more details. Let m and n be positive integers; a rectangular array

$$A = [a_{ij}] = [a_{ij}]_{m,n} \qquad (i = 1,2,\dots,m;\ j = 1,2,\dots,n)$$

is said to be a *matrix* of size m by n, and to have m *rows* and n *columns*. In case $m = n$ A is called a square matrix of order n. An m by n matrix contains $m \cdot n$ *entries* or *scalars* out of a certain set, like e.g. the set of rational or real numbers, in the context of engineering problems or in the field of statistics, or the sets of states, or of input or output alphabets in automata theory etc. In case $a_{ij} \in \mathbf{2} := \{0,1\}$, A is called a Boolean matrix; see Def. 3.19(a).

It is customary to denote zero matrices with an O, matrices with each entry equal to 1 with J, and the identity matrix with $I := [\delta_{ij}]$, where δ is the so-called **Kronecker symbol** defined as $\delta_{ij} = 1$, if $i = j$, and $\delta_{ij} = 0$ otherwise. As we did it already for relations, we will eventually use subscripts to emphasize the size of matrices.

Two matrices A and B are called **conformable** for multiplication in the order AB, if A has exactly as many columns as B has rows. (Below it will be tacitly assumed that the required size restrictions are fulfilled for the matrix operations used.) The **product of two matrices** $A = [a_{ij}]_{m,p}$ and $B = [b_{ij}]_{p,n}$, denoted by $A \cdot B$ (or just by juxtaposition AB), is the matrix $C = [c_{ij}]_{m,n}$, with $c_{ij} = \sum_{k=1}^{p} a_{ik} b_{kj}$, for $i = 1,2,\dots,m$; $j = 1,2,\dots,n$. Thus the (i,j)-entry of AB is determined only by the i-th row of A and the j-th column of B. If you compare the definition of matrix product with Def. 3.19(c), you can see that $\vee$ has the role of $+$, and $\wedge$ the one of multiplication. In general, matrix product depends on the operations defined in the underlying algebra from which the entries are taken.

Let $A = [a_{ij}]_{m,n}$; then the **transpose** of A is the matrix $A^T = [a_{ji}]_{n,m}$ which is obtained by interchanging the rows and columns of A. A is called **symmetric** if $A = A^T$. Matrix multiplication and transposition satisfy some properties analogous to ones we have already seen to hold for relations:

Corollary 5.1. *Let A, B and C be conformable matrices, i.e. with sizes as required, and let $+$ and juxtaposition stand for either the usual $(+,\times)$ or the Boolean $(\vee,\wedge)$ operations. Then the following holds:*

(1) $(A^T)^T = A$;
(2) $(A+B)^T = A^T + B^T$, *if A and B have the same size;*
(3) $(AB)^T = B^T A^T$ *(note that if A and B are conformable then B^T and A^T also are);*
(4) $A^T A$ *and* AA^T *are symmetric matrices;*
(5) $A(BC) = (AB)C$;
(6) $A(B+C) = AB + AC$, *and* $(B+C)A = BA + CA$.

It is important to have in mind that in general, even for square matrices, $AB \neq BA$, so $(A+B)^2 = A^2 + BA + AB + B^2$.

Square matrices $P = [p_{ij}]_m$ with exactly one 1 in each row and column are called

permutation matrices. They correspond to permutations in the following way: If $(k_1, k_2, \ldots, k_m)$ is a permutation of $(1,2,\ldots,m)$, define $P = [p_{ij}]_m$ by $p_{ij} = 1$, if $j = k_i$, and $p_{ij} = 0$ otherwise. For an m by n matrix A, PA is obtained by permuting the rows of A so that in PA they appear in the order $k_1, k_2, \ldots, k_m$:

$$\text{If } A = \begin{bmatrix} \boldsymbol{r}_1 \\ \boldsymbol{r}_2 \\ \vdots \\ \boldsymbol{r}_m \end{bmatrix} \quad \text{then} \quad PA = \begin{bmatrix} \boldsymbol{r}_{k_1} \\ \boldsymbol{r}_{k_2} \\ \vdots \\ \boldsymbol{r}_{k_m} \end{bmatrix}.$$

Obviously, the transpose P^T of a permutation matrix P is also a permutation matrix. Note that the set $\boldsymbol{P}_n \subseteq \boldsymbol{B}_n$ of all permutation matrices forms a subgroup of the semigroup $\boldsymbol{B}_n$.

The definition of a permutation matrix treats rows and columns in the same way. So to permute the columns of $A_{m,n}$ so that they appear in the order $l_1, l_2, \ldots, l_n$ one has to multiply it from the right by the permutation matrix Q^T, where Q corresponds to the permutation $(l_1, l_2, \ldots, l_n)$ of $(1,2,\ldots,n)$. If A is a square matrix of order n, then the matrix QAQ^T is obtained from A by putting its rows in the order $l_1, l_2, \ldots, l_n$, and then by putting its columns in the order $l_1, l_2, \ldots, l_n$. We say that the matrix QAQ^T is obtained from A by *simultaneous permutations* of its rows and columns.

A matrix A is said to be **partitioned** if horizontal and vertical lines have been introduced subdividing it into submatrices, called **blocks**. Partitioning allows A to be written as a matrix whose entries are its blocks. As an example consider the permutation matrix

$$A = \begin{bmatrix} O & I_2 \\ I_2 & O \end{bmatrix} = \left[\begin{array}{cc|cc} 0 & 0 & 1 & 0 \\ 0 & 0 & 0 & 1 \\ \hline 1 & 0 & 0 & 0 \\ 0 & 1 & 0 & 0 \end{array}\right].$$

In properly partitioned matrices blocks that appear to the left or right of other blocks have the same number of rows, and blocks that appear above or below other blocks have the same number of columns.

Partitioning of matrices not only frequently reveals underlying structures, but it often speeds up computation, especially when matrices are being multiplied, and when they have many blocks of zeros. If matrices are very large, as it is often the case in real-life applications of matrix theory, the computational overhead and the risk of accumulated numerical effects due to rounding and truncation may become substantial.

To be familiar with some simple facts about operations with partitioned matrices is very useful. It turns out that the multiplication of partitioned matrices is just like ordinary matrix multiplication.To avoid cumbersome notation for the general case, we restrict our presentation of standard matrix operations to block matrices having a few blocks only:

Corollary 5.2. *For scaling, sum, product, and transposition of block matrices the following holds:*

$$(1)\ \lambda\begin{bmatrix}A_{11} & A_{12}\end{bmatrix} = \begin{bmatrix}\lambda A_{11} & \lambda A_{12}\end{bmatrix};$$

$$(2)\ \begin{bmatrix}A_{11} & A_{12}\\ A_{21} & A_{22}\end{bmatrix} + \begin{bmatrix}B_{11} & B_{12}\\ B_{21} & B_{22}\end{bmatrix} = \begin{bmatrix}A_{11}+B_{11} & A_{12}+B_{12}\\ A_{21}+B_{21} & A_{22}+B_{22}\end{bmatrix};$$

$$(3)\ \begin{bmatrix}A_{11} & A_{12}\\ A_{21} & A_{22}\\ A_{31} & A_{32}\end{bmatrix} \cdot \begin{bmatrix}B_{11} & B_{12}\\ B_{21} & B_{22}\end{bmatrix} = \begin{bmatrix}A_{11}B_{11}+A_{12}B_{21} & A_{11}B_{12}+A_{12}B_{22}\\ A_{21}B_{11}+A_{22}B_{21} & A_{21}B_{12}+A_{22}B_{22}\\ A_{31}B_{11}+A_{32}B_{21} & A_{31}B_{12}+A_{32}B_{22}\end{bmatrix};$$

$$(4)\ \begin{bmatrix}A_{11} & A_{12}\\ A_{21} & A_{22}\\ A_{31} & A_{32}\end{bmatrix}^T = \begin{bmatrix}A_{11}^T & A_{21}^T & A_{31}^T\\ A_{12}^T & A_{22}^T & A_{32}^T\end{bmatrix}.$$

This gives us the following rules:

(1) To multiply partitioned matrices, treat its blocks as though they were elements of a matrix, and multiply the corresponding submatrices. Thus if A is an $(m+n)\times(p+q)$ matrix and B a $(p+q)\times(r+s)$ matrix such that A_{11} is of order $m\times p$ and B_{11} is of order $p\times r$, then all submatrices are well-defined and AB is an $(m+n)\times(r+s)$ partitioned matrix,

(2) To transpose a partitioned matrix, treat its submatrices as though they were elements of a matrix, and transpose each submatrix. Thus the transpose of an $m\times n$ partitioned matrix results in an $n\times m$ partitioned form.

Definition 5.3. *Let $A=[a_{ij}]\in \boldsymbol{B}_n$, and $\underline{n} := \{1,2,\ldots,n\}$. A is called*

(a) **diagonal** *if $a_{ij}=0$ for all $i\neq j, i,j\in\underline{n}$.*
(b) **upper triangular** *if $a_{ij}=0$ for all $i>j, i,j\in\underline{n}$.*
(c) **lower triangular** *if A^T is upper triangular.*
(d) **block diagonal** *if it is partitioned into submatrices $A_{ij}\in\boldsymbol{B}_{n_i\times n_j}$ such that $A=[A_{ij}]$, $\sum_{i=1}^k n_i = n$, and $A_{ij}=O$, for all $i\neq j$.*
(e) **block upper triangular** *f it is partitioned into submatrices $A_{ij}\in\boldsymbol{B}_{n_i\times n_j}$ such that $A=[A_{ij}]$, $\sum_{i=1}^k n_i = n$, and $A_{ij}=O$, for all $i>j, 1\le i,j\le k$..*
(f) **block lower triangular** *if A^T is block upper triangular.*

Thus, a square matrix A is *upper triangular* if all its entries below the main diagonal equal zero, and *block-diagonal* if it takes the form

$$\begin{bmatrix}A_{11} & O & \cdots & O\\ O & A_{22} & \cdots & O\\ \vdots & & \ddots & \vdots\\ O & O & \cdots & A_{kk}\end{bmatrix}.$$

$A\in\boldsymbol{B}_n$ is called **reducible** if $n\ge 2$ and for some integer $r, 1\le r<n$, there exists an $(n-r)\times r$ zero submatrix such that

$$PAP^T = \begin{bmatrix} A_1 & B \\ O_{n-r,r} & A_2 \end{bmatrix} \tag{5.1}$$

for some permutation matrix P; otherwise, A is called **irreducible**. Note that both A_1 and A_2 are square matrices with size $r \times r$ and $(n-r) \times (n-r)$, respectively. In particular, both of the 1×1 binary matrices are irreducible.

If A is reducible and in the form (5.1), and if a diagonal block is also reducible, then this block can be reduced further; if this process is continued then finally there exists a suitable permutation matrix P such that PAP^T is block upper triangular

$$PAP^T = \begin{bmatrix} A_{11} & A_{12} & \cdots & A_{1k} \\ O & A_{22} & \cdots & A_{2k} \\ \vdots & & \ddots & \vdots \\ O & O & \cdots & A_{kk} \end{bmatrix}, \tag{5.2}$$

and each block A_{ii} is a square and irreducible submatrix. (5.2) is called the **Frobenius normal form** of A. $A \in \boldsymbol{B}_n$ is called **primitive** if $A^k = J$, and **nilpotent** if $A^k = O$, for some positive integer k.

We continue with a short look at the topic of invertibility of matrices (over $\mathbb{R}$, say, as a source of inspiration) by recalling some well known facts from linear algebra (see e.g. [3, 23, 24, 26]):

(1) A matrix A with determinant $|A| \neq 0$ is called **nonsingular**.
(2) A square matrix A is nonsingular iff there exists a matrix B with $AB = I$, where I is the identity matrix. B is the unique matrix with $AB = BA = I$, and is called the **inverse** of A denoted by A^{-1}, and A is called **invertible**.
(3) If A is an invertible matrix, then the system of linear equations given by $A\boldsymbol{x} = \boldsymbol{b}$ has the unique solution $A^{-1}\boldsymbol{b}$.

There are, however, various fields of applied mathematics where there is need for some kind of partial inverse of a matrix that is singular or even rectangular; and for e.g. solving $\boldsymbol{x}A = \boldsymbol{b}$ it is sufficient for A to have a left inverse, since $\boldsymbol{x} = \boldsymbol{b}A^{-1}$ then is a solution. So a **generalized inverse** of a given matrix A is a matrix X that ([3])

- exists for matrices other than nonsingular square matrices only;
- has some of the properties of the inverse as defined above;
- and reduces to the usual inverse when A is nonsingular.

5.2 Graphs and Matrices

In this section we will have a first and short look at the concept of a graph and present the most basic facts and terminology of graph theory as they can be found in standard textbooks as e.g. [2,4]. More details on relations and graphs can be found in Section 17.1. Moreover, a few important interrelationships between graph theoretical properties and structural properties of the associated adjacency matrix will be presented.

Definition 5.4.

(a) *A* **directed graph** *(or* **digraph** *for short) is a pair* $G = (V, E)$*, where* $E \subseteq V \times V$*. An element* $v \in V = \{v_1, v_2, \dots, v_n\}$ *is called a* **vertex** *(also* **node***), and* $e \in E$ *an* **edge** *(also* **arc***) of* G*. If* $e = \langle a, b\rangle$ *then* a *is called the* **initial vertex***, and* b *the* **target vertex** *of* e*, which is called an* edge from a to b.

(b) *The Boolean matrix representing* E *is called the* **adjacency matrix** *of* G*:* $A_G \in \boldsymbol{B}_n$ *defined by* $A_G = [a_{ij}]$*,* $i, j \in \underline{n}$*, with* $a_{ij} = 1$ *iff* $\langle v_i, v_j\rangle \in E$*.*

Since arcs are defined to be ordered pairs, directed graphs provide a natural means to represent and visualize relations. Conversely, it is clear that to every Boolean matrix $A \in \boldsymbol{B}_n$ there is an associated digraph of order n, denoted by G_A, such that $A_{G_A} = A$.

Definition 5.5. *Let* $G = (V, E)$ *be a digraph.*

(a) *A* **path** *in* G *from vertex* u *to vertex* v *of length* k *is a sequence of* $k+1$ *distinct nodes starting with* u *and ending with* v*:* $u = v_0, v_1, \dots, v_k = v$ *such that* $\langle v_m, v_{m+1}\rangle \in E$*, for* $0 \le m < k$*.*

(b) *If* $u = v_0 = v_k = v$ *and the vertices* $v_1, \dots, v_{k-1}$ *are distinct, then the sequence* $v_0, v_1, \dots, v_k$ *is said to form a* **cycle***.* G *is* **acyclic** *if it has no cycles.*

(c) *Two vertices* u *and* v *are called* **strongly connected** *if there is a path from* u *to* v *and a path from* v *to* u*. A vertex is regarded as trivially strongly connected to itself. G is strongly connected if each two distinct vertices* u *and* v *are.*

Obviously, strong connectivity between nodes is a reflexive, symmetric, and transitive relation, and hence an equivalence on V, say γ; the elements of the quotient set V/γ are called the **strong components** of G. So G is a strongly connected graph if $|V/\gamma| = 1$.

Corollary 5.6. ([5, Theorem 1.2.3(i)]) *Let* $G = (V, E)$ *be a digraph. Then* G *is strongly connected iff there is no partition* $\{U, W\}$ *of* V *such that all edges between* U *and* W *have their initial vertex in* U *and their target vertex in* W*.*

Proposition 5.7. ([6, Theorem 3.2.1]) $A \in \boldsymbol{B}_n$ *is irreducible if and only if its digraph* $D := G_A$ *is strongly connected.*

Proof. If A is reducible, then, by Cor. 5.6, the vertex set V of D can be partitioned into two nonempty sets U and W such that there is no arc from a $u \in U$ to a $w \in W$. So if $a \in U$ and $b \in W$ there is no path from a to b, whence D is not strongly connected.

Conversely assume that D is not strongly connected. Then there are $a, b \in V, a \neq b$, for which there is no path from a to b. Let $W_1 := \{b\} \cup \{$all vertices from which there is a directed walk to $b\}$ and let $W_2 := \{a\} \cup \{$all vertices to which there is a directed walk from $a\}$. Then $W_1 \cap W_2 = \emptyset$ and $W_1, W_2 \neq \emptyset$. Furthermore we set $W_3 := V - (W_1 \cup W_2)$. If we now simultaneously permute the lines of A so that the lines are ordered according to the vertices in W_2 followed by those in W_3 and W_1 we get the following matrix

$$\begin{array}{c} \\ W_2 \\ W_3 \\ W_1 \end{array} \begin{array}{c} \begin{array}{ccc} W_2 & W_3 & W_1 \end{array} \\ \left[\begin{array}{cc|c} X_{11} & X_{12} & O \\ X_{21} & X_{22} & O \\ \hline X_{31} & X_{32} & X_{33} \end{array}\right] \end{array}$$

Since there is no directed path from a to b there is no arc from a vertex in W_2 to a vertex in W_1, whence $X_{13} = O$. There is no arc from a vertex c in W_3 to a vertex in W_1, because such an arc implies that c belongs to W_1; therefore, we also have $X_{23} = O$. And since the blocks on the diagonal are square submatrices, A is reducible. □

If A is irreducible then for each pair of distinct nodes a and b in D there is a path from a to b of length at most equal to $n-1$. Thus it follows from Prop. 5.7 that A is irreducible if and only if $(I + A)^{n-1} = J_n$.

5.3 Relations and Matrices

It is a well established result that $\boldsymbol{B}_n$ with Boolean matrix multiplication and $(\mathcal{B}_X, \cdot)$, with $|X| = n$, are isomorphic semigroups. Therefore we will sometimes denote elements from both $\mathcal{B}_X$ and $\boldsymbol{B}_n$ with small Greek letters, and usually will use the same letter for a relation and its Boolean matrix representation, that is, we will henceforth also use the simple notation $\rho \in \boldsymbol{B}_n$ instead of the more complicated $M_\rho \in \mathfrak{M}_n^{\mathbf{2}}$ (see Def. 3.19(b)).

In connection with matrices it is common to speak of rows and columns, which gives rise to the notion of a *Boolean vector space* in a natural way: The rows of any $\beta \in \boldsymbol{B}_n$ are elements of the set $V_n := \mathbf{2}^n = \{0,1\}^n$, i.e. *Boolean vectors* of dimension n, naturally corresponding to the aftersets $x\beta$. Analogously we can define $V_n^T = \{v^T : v \in V_n\}$ as the set of all Boolean column vectors of length n comprising the columns of any $\beta \in \boldsymbol{B}_n$, corresponding to the foresets βy. In the sequel we will introduce the basic notions and properties in terms of row vectors.

If we define addition and multiplication for $\mathbf{2}$ in the usual arithmetical way with the exception that we set $1 + 1 = 1$, then $+$ is called *Boolean addition*. We thus have $1+1 = 1+0 = 0+1 = 1 = 1 \cdot 1$ and $0 \cdot 0 = 1 \cdot 0 = 0 \cdot 1 = 0 = 0+0$, whence $+$ and $\cdot$ become equivalent to the $\vee$ and $\wedge$ operations on $\mathbf{2}$. As a consequence you will find both kinds of notation in literature (and also in the following). From a computational point of view one can add and multiply as usual, but has to replace all positive entries in the final result by ones. With $\mathbf{0}$ we will denote the zero row vector $(0,0,\ldots,0)$ and also the zero column vector vector $(0,0,\ldots,0)^T$, and with $\mathbf{1}$ the vectors consisting of 1's only; row vectors $(v_1, v_2, \ldots, v_n)$ will be denoted by $\boldsymbol{v}$.

Definition 5.8. ([18, Section 1.1])

(a) *V_n together with the operation of component-wise addition is called the* **Boolean vector space of dimension** n : $(u_1, u_2, \ldots, u_n) + (v_1, v_2, \ldots, v_n) = (u_1 + v_1, u_2 + v_2, \ldots, u_n + v_n)$,

(b) $\emptyset \neq W \subseteq V_n$ *is called a* **subspace** *of* V_n*, if it contains the zero vector and is closed under addition.*

(c) *The* **span** *of* $W \subseteq V_n$*, denoted by* $[W]$*, is the set of all finite sums of elements of* W*,* $[W] = \{\sum_B : B \subseteq W\}$ *with the convention that* $\sum_\emptyset = \mathbf{0}$.

If we define **scalar multiplication** by $a\boldsymbol{u} = (a \cdot u_1, a \cdot u_2, \ldots, a \cdot u_n)$, for $a \in \mathbf{2}$ and $\boldsymbol{u} \in V_n$, then it is easy to see that the following properties hold for $(V_n, +, \cdot)$: For all $s, t \in \mathbf{2}$ and all $\boldsymbol{u}, \boldsymbol{v}, \boldsymbol{w} \in V_n$ we have

(A1) $\boldsymbol{u} + \boldsymbol{v} \in V_n$ (*Closure under addition*).
(A2) $(\boldsymbol{u} + \boldsymbol{v}) + \boldsymbol{w} = \boldsymbol{u} + (\boldsymbol{v} + \boldsymbol{w})$ (*Associativity*).
(A3) $(\boldsymbol{u} + \boldsymbol{v}) = (\boldsymbol{v} + \boldsymbol{u})$ (*Commutativity*).
(A4) V_n has a zero vector $\mathbf{0}$ such that, for every $\boldsymbol{u} \in V_n, \boldsymbol{u} + \mathbf{0} = \boldsymbol{u}$ (*Additive identity*).

(M1) $s\boldsymbol{u} \in V_n$ (*Closure under scalar multiplication*).
(M2) $(s \cdot t)\boldsymbol{u} = s(t\boldsymbol{u})$ (*Associativity*).
(M3) $s(\boldsymbol{u} + \boldsymbol{v}) = s\boldsymbol{u} + s\boldsymbol{v}$ (*Distributivity*).
(M4) $(s + t)\boldsymbol{u} = s\boldsymbol{u} + t\boldsymbol{u}$ (*Distributivity*).
(M5) $1\boldsymbol{u} = \boldsymbol{u}$ (*Scalar identity*).

If we compare this with the definition of a **vector space** as given in e.g [21, p. 160], then the only missing of the constituting properties is

(A5) For each $\boldsymbol{a} \in V_n$ there is an element $(-\boldsymbol{a}) \in V_n$ such that $\boldsymbol{a} + (-\boldsymbol{a}) = \mathbf{0}$,

the existence of an *additive inverse*; but this clearly does not make sense for the binary Boolean algebra **2** and V_n, respectively. So the economical definition of a Boolean vector space becomes comprehensible. As in linear algebra we have

Corollary 5.9. *Let* $\emptyset \neq W \subseteq V_n$*; then* W *is a subspace of* V_n *iff the two closure conditions (A1) and (M1) hold for* W.

Proof. Assume W to be a subspace of V_n, then W is a Boolean vector space and therefore it is closed under addition and scalar multiplication. If $\boldsymbol{u}, \boldsymbol{v}, \boldsymbol{w} \in W$ then $\boldsymbol{u}, \boldsymbol{v}, \boldsymbol{w} \in V_n$, whence all the above mentioned conditions hold in W except (A4). But, since W is closed under scalar multiplication, $0\boldsymbol{u} = \mathbf{0} \in W$, whence (A4) also holds in W. □

Let $\boldsymbol{u}, \boldsymbol{v} \in V_n$; then $\boldsymbol{u} \leq \boldsymbol{v}$ iff for each $1 \leq i \leq n$, $u_i = 1$ implies $v_i = 1$. $\boldsymbol{v} \in V_n$ is said to be **dependent** on $B \subseteq V_n$ if $\boldsymbol{v} \in [B]$. B is called **independent** if, for each $\boldsymbol{v} \in B$, $\boldsymbol{v}$ is not dependent on $B - \boldsymbol{v}$. A **basis** of a subspace W of V_n is an independent set B such that $W = [B]$. (Note that the basis of $\mathbf{0}$ is $\emptyset$.) We also say that B *generates* W.

Corollary 5.10. ([18, Theorem 1.1.1]) *Let* W *be a subspace of* V_n*; then there exists a unique basis* $B \subseteq V_n$ *of* W.

Proof. Let $B = \{\boldsymbol{b}_j\}$ be the set of all vectors of W that are not sums of $\boldsymbol{w}_i \in W$ smaller than themselves: If $\boldsymbol{b}_j = \sum_{i \in I_j} \boldsymbol{w}_i$ then there is a $k \in I_j$ with $\boldsymbol{w}_k \geq \boldsymbol{b}_j$. Then B is an independent set. Suppose that $[B] \subset W$ and that $\boldsymbol{v}$ is a minimal vector in $W - [B]$. Then $\boldsymbol{v}$ is expressible as a sum of vectors $\boldsymbol{w}_i < \boldsymbol{v}$ since $\boldsymbol{v} \notin B$. But these smaller vectors must be in $[B]$ since $\boldsymbol{v}$ was minimal in $W - [B]$. Thus $\boldsymbol{v} \in [B]$, a contradiction. So $[B] = W$ and it is a basis of W.

To prove uniqueness note that by independence $B \subseteq B'$ for every other basis B'. Assume $\boldsymbol{u}$ to be minimal in $B' - B$; then by the above $\boldsymbol{u}$ is dependent, a contradiction. So $B = B'$. □

5.4 Powers of Relations on a Finite Set

This section is a slight deviation from the main path of the book in that we generalize from two-valued to many-valued binary relations, but do so with a syntactical motivation: We simply look what is going on when concepts are lifted from a an ordered set with only two elements to some ordered structure consisting of more than two elements. This is different from the approach in Part III which is driven from epistemological and engineering considerations, which clearly suggest that allowing for only two levels of discrimination is a much too coarse approach for building realistic models of phenomena in the real world. But it must be noted that the investigations on fuzzy relations have been a big stimulation for the work on e.g. incline matrices during the past twenty-five years, and that research efforts have switched from the two-valued context to the general one. So it is worthwhile to have a look on the newer concepts and methods. Examples for classical papers pioneering the newer developments are e.g. [19, 22, 25].

5.4.1 *Indices of Semigroups*

Let S be a semigroup, and $\emptyset \neq A \subseteq S$. A is called a **subsemigroup** of S if it forms a semigroup under the same operation as S. Let $\{A_i : i \in I \neq \emptyset\}$ be a family of subsemigroups of S. Then, if it is non-empty, the intersection $\bigcap_{i \in I} A_i$ is again a subsemigroup of S. For every $\emptyset \neq A \subseteq S$ there is at least one subsemigroup of S containing it, namely S itself; thus the intersection of all semigroups $B \supseteq A$ is a subsemigroup of S containing A, denoted by $\langle A \rangle$. Note that $\langle A \rangle$ is defined by two properties: (a) $A \subseteq \langle A \rangle$, and (b) $\langle A \rangle \subseteq B$ for any subsemigroup B of S containing A.

$\langle A \rangle$ consists of all $s \in S$ that can be expressed as finite products $a_1 a_2 \dots a_k$ with $a_i \in A, i \in \underline{k}$. If $\langle A \rangle = S$, then A is said to be a **set of generators** for S, or to **generate** S. Of special interest is the case where $A = \{a\}$; then we write $\langle a \rangle = \{a, a^2, a^3, \dots\}$. $\langle a \rangle$ is called the **cyclic** subsemigroup of S generated by a.

For a semigroup S, $|S|$ is called the **order** of S, and $|\langle a \rangle|$ is called the **order of the element** $a \in S$. A semigroup is called **periodic** if all its elements are of finite order. Thus, a finite semigroup is necessarily periodic.

Definition 5.11. *Let S be a semigroup, $a \in S$, and assume that $a^m = a^{m+d}$ for some $m, d \geq 1$. The least such positive integers m and d are called the* **index** *(of convergence) and the* **period** *of a, denoted by $k = k(a) \geq 1$ and $p = p(a) \geq 1$, respectively. If $p = 1$, a is said to* **converge**, *and otherwise, to* **oscillate** *with finite period p.*

The following is a well known result from semigroup theory (see e.g. [15, p. 10]):

Theorem 5.12. *Let S be a semigroup and $a \in S$. Then either*

(1) *all the powers of a are distinct, and $\langle a \rangle$ isomorphic to $(\mathbb{N}, +)$, i.e. infinite; or*
(2) *there exist the index k and the period p of a such that the following holds*
   (i) $a^k = a^{k+mp}$, *for each* $m \geq 1$;
   (ii) *for all* $u, v \in \mathbb{N}_0$, $a^{k+u} = a^{k+v}$ *iff* $u \equiv v$ *(mod p);*
   (iii) $\langle a \rangle = \{a, a^2, \dots, a^{k+p-1}\}$, *i.e. finite;*
   (iv) $K_a := \{a^k, \dots, a^{k+p-1}\}$ *is a cyclic subgroup of* $\langle a \rangle$ *of order p.*

Proof. (1): If there are no repetitions in the list $a, a^2, a^3, \dots$, i.e. $a^m = a^n \to m = n$, then all the powers are distinct and, obviously, $(\langle a \rangle, \cdot)$ is isomorphic to $(\mathbb{N}, +)$, since multiplication is carried out according to the rule of addition of exponents.

(2): (i) Let k be the index and p be the period of a; then by successive substitution of a^k by a^{k+p} we get $a^k = a^{k+p} = a^k a^p = a^{k+p} a^p = a^{k+2p} = \dots$, and finally $a^k = a^{k+mp}$, for each $m \geq 1$.

(ii) Without loss of generality we let $i < j$. If $i, j < k$, then clearly $i \equiv j \pmod p$ $\leftrightarrow i = j \leftrightarrow a^i = a^j$. Assume $i, j \geq k$ and $i \equiv j \pmod p$; then $j = i + mp$ with $m > 0$; so $a^j = a^{k+mp+(i-k)} = a^{k+i-k} = a^i$. Conversely, if $a^i = a^j$, then $i \geq k$ by definition of k. If $j - i = mp + r$ with $0 \leq r < p$ and $m \geq 0$, then

$$a^{k+r} = a^{k+mp+r} = a^{k+j-i} = a^{k+ip+j-i} = a^j a^{k+ip-i} = a^i a^{k+ip-i} = a^{k+ip} = a^k$$

By definition of the period p, $r \equiv 0 \pmod p$, whence $i \equiv j \pmod p$.

(iii) From the minimality of k and p it follows that $k + p$ is the least positive integer such that a^{k+p} equals some lower power of a. Thus $a, a^2, \dots, a^{k+p-1}$ are all distinct. It is a elementary fact from number theory that any $n \geq 1$ can be written as $n = mp + r$ with $m \geq 0$ and $0 \leq r < p$. Thus from $a^{k+n} = a^{k+mp+r} = a^{k+r}$ we see that, for each $s > k$, $a^s \in K_a := \{a^k, \dots, a^{k+p-1}\}$ whence $\langle a \rangle = \{a, a^2, \dots, a^{k+p-1}\}$.

(iv) K_a is a cyclic subgroup of $\langle a \rangle$: For $a^{k+u}, a^{k+v} \in K_a$, choosing s such that $s \equiv v - u - k \pmod p$ and $0 \leq s \leq p - 1$, we have $a^{k+u} a^{k+s} = a^{k+v}$. Further note that $k, k+1, \dots, k+p-1$ form a complete set of incongruent residues modulo p. Then there exists an l such that $0 \leq l < p$ and $k + l \equiv 1 \pmod p$, Thus for each $m \in \mathbb{N}$ $m(k+l) \equiv m \pmod p$, whence the powers $(a^{k+l})^m$ exhaust K_a. And for $0 \leq q < p$ with $k + q \equiv 0$ (mod p) a^{k+q} is idempotent, since $a^{k+q} a^{k+q} = a^{k+q} a^{mp} = a^{k+mp+q} = a^{k+q}$, and hence the identity of K_a. Thus K_a is a cyclic group of order p, generated by the element a^{k+l}. □

Proposition 5.13. *Every element of a periodic semigroup has a power which is idempotent. Hence in every periodic (and in particular, in every finite) semigroup there is at least one idempotent.*

Proof. If S is a periodic semigroup then $\langle a \rangle$ is finite; thus, for some k, a^k is the identity of the group K_a from Th. 5.12(2)(iv). □

So, in addition to the two integer numbers that have been introduced in Def. 5.11, viz. the index $k = k(A)$ and the period $p = p(A)$, we can associate the following two numbers to any relation or matrix $A \in \boldsymbol{L}_n$ (see e.g. [25, p. 634] for Boolean relations, and [28, p. 5] or [11] for incline matrices):

- With $r := r(A)$ we denote the least integer $r \geq 1$ such that A^r is idempotent, i.e. $A^{2r} = A^r$; and
- with $t := t(A)$ we denote the least integer $t \geq 1$ such that A^t is transitive, i.e. $A^{2t} \leq A^t$.

Clearly, $t(A) \leq r(A)$.

Corollary 5.14. *Let S be a semigroup and $a \in S$ with index k and period p. Then $a^m = a^{m+d}$ if and only if $k \geq m$ and $p|d$.*

Proof. Assume $a^m = a^{m+d}$. By definition of index and period we have $k \leq m$ and $p \leq d$. Let $d = qp + r$, $0 \leq r < p$ and $q \geq 1$. Then $a^k = a^{k+p}$ implies $a^m = a^{m+p}$, whence $a^k = a^{k+p} = a^{k+2p} = \cdots = a^{k+qp}$. Thus $a^k = a^{k+d} = a^{k+qp+r} = a^{k+qp}a^r = a^k a^r = a^{k+r}$. But since $0 \leq r < p$, we get $r = 0$ by the minimality of p; therefore $p|d$. The converse direction is obvious. □

5.4.2 *Inclines, Permanents and Adjoints*

Before turning to relations themselves we have to add some supplemental information on algebraic structures and semigroups that will be needed later on.

Definition 5.15. ([7]) *An* **incline** *is an algebra $(L, +, \cdot)$ of type $\langle 2, 2\rangle$ such that*

(In1) *$(L, +)$ is a semilattice,*
(In2) *$(L, \cdot)$ is a commutative semigroup,*
(In3) *$a \cdot (b + c) = a \cdot b + a \cdot c$, for all $a, b, c \in L$,*
(In4) *$a + a \cdot b = a$, for all $a, b \in L$.*

If in an incline L we define $a \leq b :\iff a + b = b$; then $\leq$ becomes a partial order on L. If L has an additive identity 0, it is called the *zero* of L, and If L has an multiplicative identity 1, it is called the *identity* of L. From now on L will always denote an incline with zero and identity, also called a 0-1-**incline**.

Corollary 5.16. ([8]) *Let L be an 0-1-incline and $a, b, c \in L$; then*

(1) $0 \leq a \leq 1$,
(2) if $b \leq c$, then $a + b \leq a + c$, $ab \leq ac$, and $ba \leq ca$,
(3) $a \leq a + b$, and $a + b = \sup\{a, b\}$,
(4) $ab \leq a$, $ab \leq b$, i.e. $ab = \inf\{a, b\}$,
(5) $acb \leq ab$,
(6) $a + 0 = 0 + a = a$, and $a + 1 = 1 + a = 1$,
(7) $a0 = 0a = 0$, and $a1 = 1a = a$,
(8) $a + b = 0$ iff $a = b = 0$, and $ab = 1$ iff $a = b = 1$,
(9) $a = a + a$.

Proof. The proofs are immediate consequences of the definition or straightforward; e.g.:
(2): Assume $b \le c$; then $b+c=c$. Thus $ab+ac=a(b+c)=ac$, i.e. $ab \le ac$.
(4): $b \le 1 = b+1$. By (2), (In3), and (In4) $ab \le a(b+1) = ab+a \le a$.
(8): $a=b=1$ implies $ab=a=b=1$ by the identity property of a,b. Conversely, $1=ab \le a \le 1$ by (4).
(9): $a = a1 = a(1+1) = a1 + a1 = a+a$. □

We have the following inclusion relationships: Boolean algebras ⊂ fuzzy algebras ⊂ distributive lattices ⊂ 0-1-inclines ⊂ semirings. (A *fuzzy algebra* is an algebra $([0.1], \vee, T)$, where T is a t-norm; see Def. 8.6.) A commutative semiring R is an incline if $x+1=1$ for all $x \in R$. And due to Cor. 5.16(8), any incline is a commutative antiring. There is another structure that sometimes appears in the context of semiring and incline matrices (see e.g. [30]): A **path algebra** is an additively idempotent semiring, and thus generalizes the notions of Boolean algebras, fuzzy algebras, bounded distributive lattices and 0-1-inclines. Thus literature using path algebras as underlying structure for matrix entries may also contain results pertinent to the topics under consideration.

In analogy to $\boldsymbol{B}_n$, by $\boldsymbol{L}_n$ we will denote the set of all $n \times n$ matrices with entries $l_{ij} \in L$, called incline matrices, and likewise all square relations ρ of size $n \times n$, n finite, such that the valuation of $x\rho y$ is an element $l \in L$. For $A \in \boldsymbol{L}_n$, with $A(i|j)$ we denote the $(n-1)\times(n-1)$ submatrix of A obtained by deleting the ith row and the jth column of A.

For two matrices $A = [a_{ij}] \in \boldsymbol{L}_n$ and $B = [b_{ij}] \in \boldsymbol{L}_n$, their product $A \cdot B \in \boldsymbol{L}_n$ is defined by

$$A \cdot B = [c_{ij}], \quad \text{with} \quad c_{ij} := \sum\nolimits_{1 \le k \le n} a_{ik} b_{kj}.$$

The sum of A and B is defined as $A + B = [a_{ij} + b_{ij}]$. Furthermore we let $A \le B :\leftrightarrow a_{ij} \le b_{ij}$ for all $i, j \in \underline{n}$. $\emptyset, I_n$ and U_n are the same as in the Boolean case. Then it is easy to show that $\boldsymbol{L}_n$ forms a partially ordered monoid (whence Th. 5.12 applies).

Definition 5.17. *Let $A \in \boldsymbol{L}_n$ be a square incline matrix.*

(a) *The* **permanent** *of A is defined by*

$$\mathsf{per}(A) = \sum\nolimits_{\sigma \in S_n} a_{1\sigma(1)}\, a_{2\sigma(2)} \cdots a_{n\sigma(n)},$$

where $\sigma \in S_n$, the set of all permutations of $\{1, 2, \ldots, n\}$.

(b) *The* **adjoint** *of A is the matrix* $\mathsf{adj}(A) = [b_{ij}] \in \boldsymbol{L}_n$, *with* $b_{ij} = \mathsf{per}(A(j|i))$ *for all* $i, j \in \underline{n}$.

Proposition 5.18. ([8, Prop. 2], [12, Lemmas 4.1 and 4.3(3)]) *Permanents of incline matrices $A, B \in \boldsymbol{L}_n$ have the following properties:*

(1) *If $A \le B$, then* $\mathsf{per}(A) \le \mathsf{per}(B)$.
(2) $\mathsf{per}(A) = \mathsf{per}(A^T)$.
(3) *If P and Q are permutation matrices then* $\mathsf{per}(PAQ) = \mathsf{per}(A)$.
(4) $\mathsf{per}(A) = \sum_j a_{ij}\,\mathsf{per}(A(i|j)) = \sum_i a_{ij}\,\mathsf{per}(A(i|j))$, *where* $\mathsf{per}(A(i|j))$ *is called the* **minor** *of the entry a_{ij} (in analogy to determinants in linear algebra).*

Proof. (1): Obvious from definition and Cor. 5.16

(2): Consider any term $a_{1i_1}\, a_{2i_2} \dots a_{ni_n}$ of $\mathsf{per}(A)$. Since the subscripts $i_1, i_2, \dots, i_n$ form an arrangement of the numbers $1, 2, \dots, n$, all factors $a_{1i_1}, a_{2i_2}, \dots, a_{ni_n}$ are in different rows and different columns of A^T, whence $a_{1i_1}\, a_{2i_2} \dots a_{ni_n}$ occurs also as a term of $\mathsf{per}(A^T)$. Since the converse also holds, $\mathsf{per}(A)$ and $\mathsf{per}(A^T)$ consist of the same terms and are therefore equal.

(3): Let $A = [a_{kj}]$ and $P = [p_{ik}]$, and assume $p_{ik_i} = 1$ for any $i \in \underline{n}$, where $k_1, k_2, \dots, k_m$ are all different. Then the (i, j)-entry of PA is $a_{k_i j}$, whence

$$\mathsf{per}(PA) = \sum_{\sigma\in S_n} \prod_{i\in\underline{n}} a_{k_i\sigma(k_i)} = \sum_{\sigma\in S_n} \prod_{k_i\in\underline{n}} a_{k_i\sigma(k_i)} = \mathsf{per}(A).$$

Similarly, we have $\mathsf{per}(AQ) = \mathsf{per}(A)$. Thus, $\mathsf{per}(PAQ) = \mathsf{per}(AQ) = \mathsf{per}(A)$.

(4): We have

$$\begin{aligned}\mathsf{per}(A) &= \sum_{j\in\underline{n}} \sum_{\sigma\in S_n, \sigma(i)=j} a_{1\sigma(1)}\, a_{2\sigma(2)} \cdots a_{n\sigma(n)} \\ &= \sum_{j\in\underline{n}} a_{ij} \sum_{\pi\in S_{i|j}} a_{1\pi(1)} \dots a_{i-1\pi(i-1)}\, a_{i+1\pi(i+1)} \dots a_{n\pi(n)},\end{aligned}$$

where $S_{i|j}$ denotes the set of bijections from $\underline{n} - \{i\}$ to $\underline{n} - \{j\}$. But from the definition of $\mathsf{per}(A)$ we get that the inner sum is just $\mathsf{per}(A(i|j))$. For the second equality note that by (2) $\mathsf{per}(A) = \mathsf{per}(A^T)$ and that the columns of A are the rows in A^T. So the result follows by application of what we have just shown to $\mathsf{per}(A^T)$. □

Proposition 5.19. ([16, Proposition 3.1]) *Adjoints of incline matrices have the following properties:*

(1) *If* $A \le B$, *then* $\mathsf{adj}(A) \le \mathsf{adj}(B)$.
(2) $\mathsf{adj}(A) + \mathsf{adj}(B) \le \mathsf{adj}(A + B)$.
(3) $(\mathsf{adj}(A))^T = \mathsf{adj}(A^T)$.

Proof. (1): By definition of adjoints and Prop. 5.18(1).
(2): Since both $A \le A + B$ and $B \le A + B$, using (1) one gets $\mathsf{adj}(A) \le \mathsf{adj}(A + B)$ and $\mathsf{adj}(B) \le \mathsf{adj}(A + B)$, whence $\mathsf{adj}(A) + \mathsf{adj}(B) \le \mathsf{adj}(A + B)$. □

Theorem 5.20. ([8, Theorems 4,5], [16, Theorem 3.3]) *Let L be an incline and $A \in \mathbf{L}_n$ with $a_{ii}a_{jk} = a_{jk}$ (equivalently: $a_{ii} \ge a_{jk}$), for all $i, j, k \in \underline{n}$ (see Def. 5.33, where this property is named* diagonally dominant*). Then $A^{n-1} = \mathsf{adj}(A)$.*

Proof. The proof proceeds in two steps: We show that (a): $\mathsf{per}(A(j|i)) \le a_{ij}^{(n-1)}$, and that (b:) $\mathsf{per}(A(j|i)) \ge a_{ij}^{(n-1)}$. From this we get $A^{n-1} = \mathsf{adj}(A)$ by Def. 5.17(b).

(a): Assume $i = j$; then

$$\begin{aligned}\mathsf{per}(A(i|i)) &= \sum_{\sigma\in S_n} a_{1\sigma(1)} \dots a_{i-1\sigma(i-1)}\, a_{i+1\sigma(i+1)} \dots a_{n\sigma(n)} \\ &\le \sum_{\sigma\in S_n} a_{11} \dots a_{i-1,i-1}\, a_{i+1,i+1} \dots a_{nn}\end{aligned}$$

$$
\begin{aligned}
&= a_{11} \dots a_{i-1,i-1}\, a_{i+1,i+1} \dots a_{nn} && \text{by Cor. 5.16(4)}\\
&= a_{ii}^{n-1} && \text{by Cor. 5.16(9)}\\
&\le a_{ii}^{(n-1)}.
\end{aligned}
$$

For $i \neq j$ we get

$$\mathsf{per}(A(j|i)) = \sum_{\sigma \in S_n, \sigma(j)=i} a_{1\sigma(1)} \cdots a_{i-1\sigma(i-1)}\, a_{i+1\sigma(i+1)} \cdots a_{n\sigma(n)}.$$

Let $T_\sigma = a_{1\sigma(1)} \cdots a_{i-1\sigma(i-1)}\, a_{i+1\sigma(i+1)} \cdots a_{n\sigma(n)}$ denote any term of $\mathsf{per}(A(j|i))$. Since $\sigma(j) = i$, there is a $u \geq 1$ such that $\sigma^u(i) = j$, $0 \leq u \leq n-1$. Let t be the least such u; then $i, \sigma(i), \cdots, \sigma^{t-1}(i), j$ are all different. Thus

$$T_\sigma \le a_{i\sigma(i)}\, a_{\sigma(i)\sigma^2(i)} \cdots a_{\sigma^{t-1}(i)j}\, a_{jj}^{n-1-t} \le a_{ij}^{(n-1)}.$$

Since this holds for any term T_σ we get

$$\mathsf{per}(A(j|i)) = \sum_{\sigma \in S_n, \sigma(j)=i} T_\sigma \le \sum_{\sigma \in S_n, \sigma(j)=i} a_{ij}^{(n-1)} = a_{ij}^{(n-1)}.$$

(b): Assume $i = j$; then

$$
\begin{aligned}
a_{ii}^{(n-1)} &= \sum_{j_1, j_2, \dots, j_n \in \underline{n}} a_{ij_1}\, a_{j_1 j_2} \cdots a_{j_{n-2} i}\\
&\le \sum_{j_1, j_2, \dots, j_n \in \underline{n}} a_{11} \cdots a_{i-1,i-1}\, a_{i+1,i+1} \cdots a_{nn}\\
&= a_{11} \cdots a_{i-1,i-1}\, a_{i+1,i+1} \cdots a_{nn}\\
&\le \mathsf{per}(A(i|i)).
\end{aligned}
$$

Now we let $i \neq j$, and denote by $T = a_{ij_1}\, a_{j_1 j_2} \cdots a_{j_{n-2} j}$ any summand of $a_{ij}^{(n-1)}$. If the subscripts $i = j_0, j_1, \dots, j_{n-2}, j_{n-1} = j$ are all different, then T is a term of $\mathsf{per}(A(j|i))$; thus $T \le \mathsf{per}(A(j|i))$. If $j_r = j_s$ for some $r, s \in \{0, 1, \dots, n-1\}$, $r < s$, we delete the subproduct $a_{j_r j_{r+1}} \cdots a_{j_{s-1} j_s}$ from T and get $a_{ij_1} \cdots a_{j_{r-1} j_r} a_{j_s j_{s+1}} \cdots a_{j_{n-2} j}$. This process is repeated until we get $a_{il_1}\, a_{l_1 l_2} \cdots a_{l_m j}$, where $0 \le m < n-2$ and $i, l_1, \dots, l_m, j$ are all different. We let $\{p_1, p_2, \dots, p_{n-m-2}\} = \underline{n} - \{i, l_1, \dots, l_m, j\}$; then

$$T \le a_{il_1}\, a_{l_1 l_2} \cdots a_{l_m j} a_{p_1 p_1} a_{p_2 p_2} \cdots a_{p_{n-m-2} p_{n-m-2}} \le \mathsf{per}(A(j|i)).$$

Since T has been arbitrarily chosen, $a_{ij}^{(n-1)} \le \mathsf{per}(A(j|i))$. □

Corollary 5.21. *Let L be a 0-1-incline and $A \in \mathbf{L}_n$ be a reflexive matrix, i.e. $I_n \le A$. Then $A^{n-1} = \mathsf{adj}(A)$.*

The two-valued counterpart for Boolean matrices has been shown already in the early fifties in Lunts [20] as reported in Hammer and Rudeanu [10, p. 15]. As it is to be expected, the method of proof is the same for both Boolean and incline matrices. Some more properties of adjoints are given in Prop. 5.37 below.

5.4.3 *Properties of Incline Matrices and Relation Powers*

When a relation is constructed on an infinite set, then all its powers $\rho^i, i \in \mathbb{N}$, may be different. If, however, the underlying set A is finite, then only a finite number of different relations on A is available for the ρ^i; so the question for eventual patterns in the sequence $\rho, \rho^2, \rho^3, \ldots$ arises in a natural way. In the following assume that ρ is any relation on a finite set $A = \{a_1, a_2, \ldots, a_n\}$. Remember that by $\rho \in \mathcal{B}_A$, $\rho \in \boldsymbol{B}_n$ or $B \in \boldsymbol{B}_n$ (with B denoting a Boolean matrix of size $n \times n$), we mean the same thing. For inclines L we generalize this notation to $\rho \in \mathcal{L}_A$, $\rho \in \boldsymbol{L}_n$ or $B \in \boldsymbol{L}_n$.

For each of the results reported below we will give the source where it is taken from, and we will explicitly state for which kind of underlying structure it holds. Though we will not do it in each case, we will give all the necessary proofs for at least one example derivation. Our presentation of some basic facts on incline matrices (relations) is based on [11–13].

For two matrices $A = [a_{ij}] \in \boldsymbol{L}_n$ and $B = [b_{ij}] \in \boldsymbol{L}_n$, their product $A \cdot B \in \boldsymbol{L}_n$ is defined by For sake of completeness we mention the following

Definition 5.22. (see [13]) *Let $A, B, C, P \in \boldsymbol{L}_n$.*

(a) A is said to be **right invertible***(***left invertible***) if there is a matrix B such that $AB = I$ $(BA = I)$. B is called the* **right inverse** *(***left inverse***) of A. If A is both right and left invertible, then it is called* **invertiblc***, and B is called the* **inverse** *of A denoted by A^{-1}.*

(b) P is called a **permutation matrix** *if there is exactly one 1 in each row and each column, and all other entries are 0.*

(c) A is called **reflexive** *if $I \le A$, and* **transitive** *if $A^2 \le A$.*

(d) A is called **Hall matrix** *if there exists an invertible matrix C with $C \le A$.*

If we denote by $G_n(L), P_n(L)$ and $H_n(L)$ the sets of all invertible, permutation, and Hall matrices, respectively, then the following holds:

- $G_n(L)$ is a subgroup of $\boldsymbol{L}_n$.
- $P_n(L)$ is a subgroup of $G_n(L)$ and is isomorphic to the symmetric group of degree n.
- $H_n(L)$ is a subsemigroup of $\boldsymbol{L}_n$.
- $P_n(L) \subseteq G_n(L) \subseteq H_n(L) \subseteq \boldsymbol{L}_n$. These inclusions are proper in general (see [13, Example 2.1]).

Let $A \in \boldsymbol{L}_n$. In the following we write $A^m = [a_{ij}^{(m)}]$, and let $k = k(\rho)$ (or $k(A)$) denote the index and $p = p(\rho)$ (or $p(A)$) the period of ρ (or A). Furthermore, recall that we use $\underline{n}$ as abbreviation for $\{1, 2, \ldots, n\}$, and we write $[n]$ for the least common multiple of the numbers in $\underline{n}$. Note that for any $i, j \in \underline{n}$ one has

$$a_{ij}^{(m)} = \sum_{i_1, i_2, \ldots, i_{m-1} \in \underline{n}} a_{i i_1} a_{i_1 i_2} \ldots a_{i_{m-1} j}. \tag{5.3}$$

Lemma 5.23. ([11, Lemma 3.1]) *If $A \in \boldsymbol{L}_n$ then $A^{n^2[n]} \le A^{n^2[n]-[n]}$.*

Proof. Let $m = n^2[n]$. Now consider any summand of $a_{ij}^{(m)}$ and divide it into $n[n]$ products of length n, yielding

$$a_{ii_1} \cdots a_{i_{n-1}i_n} \mid a_{i_n i_{n+1}} \cdots a_{i_{2n-1}i_{2n}} \mid \cdots \mid a_{i_{m-n}i_{m-n+1}} \cdots a_{i_{m-1}i_j}$$

In every product of n elements there are $n+1$ subscripts taken out of $\underline{n}$, whence two of them must be equal. Therefore, each of the n-element products contains a sequence $S_p^k := a_{pp_1} \cdots a_{p_k p}$, $k \geq 1$. From each of these $n[n]$ n-element products select exactly one such sequence S_p^k. Then there is an $r \in \underline{n}$ such that at least $[n]$ subproducts S_p^r of r elements are chosen. (Otherwise, for every $r \in \underline{n}$, the number of selected sequences S_p^r is smaller than $[n]$, and hence the total number of selected sequences S_p^r smaller than $n[n]$.)

Now select exactly $[n]/r$ sequences S_p^r and delete them from $a_{ii_1} a_{i_1 i_2} \ldots a_{i_{m-1}i_j}$, yielding a term of $a_{ij}^{(m-[n])}$. Thus $a_{ii_1} a_{i_1 i_2} \ldots a_{i_{m-1}i_j} \leq a_{ij}^{(m-[n])}$. Since the last inequality holds for an arbitrary summand of $a_{ij}^{(m)}$, we have shown that $a_{ij}^{(m)} \leq a_{ij}^{(m-[n])}$. □

Proposition 5.24. ([12, Theorem 3.1]) *If $A \in \mathbf{L}_n$ is right invertible then $A^{[n]} = I$.*

Proof. Since A is right invertible we have $AB = I$ for some $B \in \mathbf{L}_n$. Therefore, for any $m \geq 1$,

$$A^m B^m = \underbrace{A \cdots (A(A}_{m \text{ times}} \underbrace{B)B) \cdots B}_{m \text{ times}} = AB = I$$

Let $m = n^2[n]$ and $d = [n]$; then $A^{m+d} \leq A^m$ by Lemma 5.23. Hence

$$A^d = A^d \cdot I = A^d \cdot AB = A^d \cdot A^m B^m = A^{m+d} B^m \leq A^m B^m = I$$

For the other direction note that $A^d B^d = I$, whence $\sum_{s \in \underline{n}} a_{is}^{(d)} b_{si}^{(d)} = 1$. Since $A^d \leq I$, $a_{ij}^{(d)} = 0$, for $i \neq j$. Thus, for every $\in \underline{n}$, $a_{ii}^{(d)} = \sum_{s \in \underline{n}} a_{is}^{(d)} \geq \sum_{s \in \underline{n}} a_{is}^{(d)} b_{si}^{(d)} = 1$, i.e. $a_{ii}^{(d)} = 1$, so $A^d = I$. □

Proposition 5.25. ([12, Theorem 3.2]) *If $A \in \mathbf{L}_n$ is an invertible matrix, then $A^T A = AA^T = I$.*

Proof. By assumption, $AB = BA = I$ for some $B \in \mathbf{L}_n$. From $AB = I$ we have $\sum_{s \in \underline{n}} a_{is} b_{sj} = 0$ for $i \neq j$. So, for all $s \in \underline{n}$, $a_{is} b_{sj} = 0$. And from $BA = I$ one gets

$$\textstyle\sum_{s \in \underline{n}} a_{si} \geq \sum_{s \in \underline{n}} b_{is} a_{si} = 1 \quad \text{and} \quad \sum_{s \in \underline{n}} b_{is} \geq \sum_{s \in \underline{n}} b_{is} a_{si} = 1,$$

for all $i \in \underline{n}$. Thus

$$\begin{aligned} a_{ij} &= a_{ij} \cdot 1 = a_{ij} \cdot \textstyle\sum_{s \in \underline{n}} b_{js} = \sum_{s \in \underline{n}} a_{ij} b_{js} = a_{ij} b_{ji} + \sum_{s \neq i} a_{ij} b_{js} \\ &= a_{ij} b_{ji} + 0 = a_{ij} b_{ji} + \textstyle\sum_{s \neq i} a_{sj} b_{ji} = \sum_{s \in \underline{n}} a_{sj} b_{ji} \\ &= \left(\textstyle\sum_{s \in \underline{n}} a_{sj}\right) \cdot b_{ji} = 1 \cdot b_{ji} \\ &= b_{ji}, \end{aligned}$$

for all $i, j \in \underline{n}$. Thus $B = A^T$. □

To relate this proposition to some early classical results: It generalizes [9, Theorem 6] and [19, Theorem 4.2].

Theorem 5.26. ([12, Theorem 3.4]) *For $A \in \mathbf{L}_n$ the following statements are equivalent:*

(1) *A is right invertible;*
(2) *A is left invertible;*
(3) *A is invertible;*
(4) *$AA^T = I$;*
(5) *$A^T A = I$;*
(6) *$AA^T = A^T A = I$*
(7) *$A^l = I$ for some $l \geq 1$.*

Proof. We have (1) → (7) by Prop. 5.24, and can show (2) → (7) in an analogous way. We can prove the implications (1) → (7) → (3) → (6) → (4) → (1) by application of Prop. 5.24 and Prop. 5.25; the second implication is clear from $A^l = AA^{l-1} = A^{l-1}A = I$. Similarly we get (2) → (7) → (3) → (6) → (5) → (2). □

There are also the following interrelationships between invertibility and the permanent and adjoint of an incline matrix $A \in \mathbf{L}_n$.

Proposition 5.27. ([12, Theorems 4.1], [27, Lemma 2.10]) *Let $A \in \mathbf{L}_n$ be invertible. Then*

(1) $\mathsf{per}(A) = 1$.
(2) $\mathsf{adj}(A) = A^T$.

Proof. (1): By Th. 5.26(4), $AA^T = I$. Then for any $i \in \underline{n}$, $1 = \sum_{s \in \underline{n}} a_{is} a_{is} \leq \sum_{s \in \underline{n}} a_{is}$, i.e. $\sum_{s \in \underline{n}} a_{is} = 1$. And for $i \neq j$ and $t \in \underline{n}$ we have $a_{it} a_{jt} \leq \sum_{k \in \underline{n}} a_{ik} a_{jk} = 0$, i.e. $a_{it} a_{jt} = 0$. Thus

$$1 = \prod_{i \in \underline{n}} 1 = \prod_{i \in \underline{n}} \Big(\sum_{s \in \underline{n}} a_{is} \Big) = \sum_{\sigma \in S_n} a_{1\sigma(1)}\, a_{2\sigma(2)} \ldots a_{n\sigma(n)} = \mathsf{per}(A),$$

where S_n is the set of all permutations of $\underline{n}$.

(2): By Th. 5.26(6), $AA^T = A^T A = I$. Then, as in (1) above, for any $i \in \underline{n}$ we have $\sum_{s \in \underline{n}} a_{is} = 1$. Since $\sum_{k \in \underline{n}} a_{ik} a_{jk} = 0$ for $i \neq j$ we have $a_{ij} a_{ik} = a_{ji} a_{ki} = 0$ for all $i, j, k \in \underline{n}$, $j \neq k$. From the definition of $\mathsf{per}(A)$ we get

$$\mathsf{per}(A(i|j)) = \textstyle\sum_{\pi \in S_{i|j}} a_{1\pi(1)} \ldots a_{i-1\pi(i-1)}\, a_{i+1\pi(i+1)} \ldots a_{n\pi(n)},$$

where $S_{i|j}$ denotes the set of bijections from $\underline{n} - \{i\}$ to $\underline{n} - \{j\}$. For $t \neq j$, every term of $\mathsf{per}(A(i|t))$ has a factor which is an entry of the jth column of A and is different from a_{ij}, say $a_{ij'}$, $j \neq j'$. Hence, every term in $a_{ij}\mathsf{per}(A(i|t))$ contains a product $a_{ij} a_{ij'} = 0$, whence $a_{ij}\mathsf{per}(A(i|t)) = 0$.

By (1) and Prop. 5.18(4) we obtain

$$1 = \mathsf{per}(A) = \textstyle\sum_{j \in \underline{n}} a_{ij}\, \mathsf{per}(A(i|j)) \leq \sum_{j \in \underline{n}} \mathsf{per}(A(i|j)),$$

i.e. $\sum_{j \in \underline{n}} \mathsf{per}(A(i|j)) = 1$. Thus,

$$\begin{aligned} a_{ij} &= a_{ij} \cdot 1 = a_{ij} \cdot \big(\textstyle\sum_{t \in \underline{n}} \mathsf{per}(A(i|t)) \big) \\ &= a_{ij}\, \mathsf{per}(A(i|j)) + \textstyle\sum_{t \in \underline{n}, t \neq j} a_{ij}\, \mathsf{per}(A(i|t)) \end{aligned}$$

$= a_{ij}\,\mathsf{per}(A(i|j))$
$= a_{ij}\,\mathsf{per}(A(i|j)) + \sum_{t\in \underline{n}, t\neq j} a_{it}\,\mathsf{per}(A(i|j))$
$= \sum_{t\in \underline{n}} a_{it}\,\mathsf{per}(A(i|j))$
$= \left(\sum_{t\in \underline{n}} a_{it}\right)\cdot \mathsf{per}(A(i|j))$
$= 1\cdot \mathsf{per}(A(i|j))$.

Thus $a_{ij} = \mathsf{per}(A(i|j))$ for all $i, j \in \underline{n}$, whence $\mathsf{adj}(A) = \left[\mathsf{per}(A(j|i))\right] = A^T$. □

Generalization of the foregoing results to antirings are given in [29].

Proposition 5.28. *Let L be a path algebra, and $A \in \mathbf{L}_n$. Then A is transitive iff PAP^T is transitive for any permutation matrix $P \in \mathbf{L}_n$.*

Proof. Let A be transitive. We have $(PAP^T)^2 = (PA)(P^TP)(AP^T) = (PA)I_n(AP^T) = P(AI_nA)P^T = PA^2P^T \leq PAP^T$. Conversely, assume $(PAP^T)^2 \leq PAP^T$; then, similar to the above, we get $PA^2P^T \leq PAP^T$. This implies $P^TPA^2P^TP \leq P^TPAP^TP$, i.e. $I_nA^2I_n \leq I_nAI_n$, so $A^2 \leq A$. □

Anticipating the considerations of Chapter 8 we use a slightly different way of formulation in the following proposition: $a\rho b$, or equivalenty $\langle a, b\rangle \in \rho$, has to be read as $M_\rho(a, b) = l > 0, l \in L$, and $\rho \subseteq \sigma$ as $M_\rho(a, b) \leq M_\sigma(a, b)$, for all a, b.

Proposition 5.29. ([8, Lemma 1]) *For $\rho \in \mathbf{L}_n$ and all $s \geq 1$, $\rho^s \subseteq \rho \cup \rho^2 \cup \ldots \cup \rho^n$.*

Proof. The assertion obviously holds for $\rho = \emptyset$ or $s \in \underline{n}$. So we take $s = n+1$; then $a\rho^{n+1}b$ iff there are numbers $i_1, i_2, \ldots, i_{n+1}$ such that

$$a = a_{i_1}, b = a_{i_{n+2}}, \text{ and } a_{i_1}\,\rho\, a_{i_2}\,\rho \ldots \rho\, a_{i_{n+1}}\,\rho\, a_{i_{n+2}}.$$

Since the numbers $i_1, i_2, \ldots, i_{n+1}$ cannot all be different, there exist two integers, say $l < m$, such that $i_l = i_m$, whence in the above product there is a subsequence

$$\ldots a_{i_{l-1}}\,\rho\, a_{i_l}\,\rho\, a_{i_{l+1}}\,\rho \ldots \rho\, a_{i_{m-1}}\,\rho\, a_{i_l}\,\rho\, a_{i_{m+1}} \ldots .$$

Deleting $a_{i_{l+1}}\,\rho \ldots \rho\, a_{i_{m-1}}\,\rho\, a_{i_l}$ from the term $a_{i_1}\,\rho\, a_{i_2}\,\rho \ldots \rho\, a_{i_{n+1}}\,\rho\, a_{i_{n+2}}$ gives

$$a_{i_1}\,\rho\, a_{i_2}\,\rho \ldots \rho\, a_{i_l}\,\rho\, a_{i_{m+1}}\,\rho \ldots \rho\, a_{i_{n+2}}.$$

So we still have a path from a_{i_1} to , $a_{i_{n+2}}$, but now we need at most n steps, whence this path is also contained in $\rho \cup \rho^2 \cup \ldots \cup \rho^n$. Therefore, $\rho^{n+1} \subseteq \rho \cup \rho^2 \cup \ldots \cup \rho^n$.

Furthermore, $\rho^{n+1} \subseteq \rho \cup \rho^2 \cup \ldots \cup \rho^n$ implies $\rho^{n+2} \subseteq \rho^2 \cup \rho^3 \cup \ldots \cup \rho^{n+1} \subseteq \rho \cup \rho^2 \cup \ldots \cup \rho^n$, and repeating this argument we get $\rho^s \subseteq \rho \cup \rho^2 \cup \ldots \cup \rho^n$ for any $s \geq n+1$, and the assertion is proved.. □

Corollary 5.30. *For $\rho \in \mathbf{L}_n$, $\mathbf{t}(\rho) = \rho \cup \rho^2 \cup \ldots \cup \rho^n$.*

Proof. Like in the binary case (see Def. 3.57) we define $\rho^+ = \bigcup_{n\geq 1} \rho^n$. Then, according to Prop. 5.29, $\rho^+ \subseteq \rho \cup \rho^2 \cup \ldots \cup \rho^n$. Since clearly $\rho \cup \rho^2 \cup \ldots \cup \rho^n \subseteq \rho^+$, equality holds: $\rho \cup \rho^2 \cup \ldots \cup \rho^n = \rho^+$. As for Th. 3.59(3) one can show that ρ^+ is the transitive closure of ρ. □

The proposition and the subsequent corollary below refer early classical results on binary Boolean relations, which we present with the original proof. Then we will formulate the corresponding generalization for incline matrices.

Proposition 5.31. ([14, Theorem 3.2.1]) *Let $B \in \mathbf{B}_n$ be reflexive. Then*

$$I_n \subseteq B \subseteq B^2 \subseteq \ldots \subseteq B^{n-1} = B^n = B^{n+1} = \ldots$$

Proof. Due to isotonicity of relation product we get $I_n \subseteq B \subseteq B^2 \subseteq \ldots \subseteq B^{n-1} \subseteq B^n \subseteq B^{n+1} \subseteq \ldots$ .Thus it suffices to prove $B^n \subseteq B^{n-1}$.

Note that for all $h \geq 1, 1 \leq j \leq n$ we have $b_{jj}^{(h)} = 1$ and that

$$b_{ij}^{(h+1)} = \sum_k b_{ik}^{(h)} b_{kj} = b_{ij}^{(h)} + \sum_{k \neq j} b_{jk}^{(h)} b_{kj}. \tag{5.4}$$

Consider a term

$$b_{ii_1} b_{i_1 i_2} \ldots b_{i_{n-2} i_{n-1}} b_{i_{n-1} j} \tag{5.5}$$

of the sum (5.3) for $k = n$. Since there are only n distinct values, at least two of $i = i_0, i_1, \ldots, i_{n-1}, i_n = j$ must be equal. Suppose $j = i_s$ for some $s \leq n-1$; then (5.5) takes the form

$$b_{ii_1} b_{i_1 i_2} \ldots b_{i_{s-1} j} b_{j i_{s+1}} \ldots b_{i_{n-1} j}. \tag{5.6}$$

We clearly have (5.6) $\leq b_{ii_1} b_{i_1 i_2} \ldots b_{i_{s-1} j}$, a term of $b_{ij}^{(s)}$, and hence also (5.6) $\leq b_{ij}^{(n-1)}$ by (5.4). A similar argument works in the case $i = i_s$.

If neither i nor j is equal to any i_k, then there are $1 \leq r < s \leq n-1$ such that $i_r = i_s$, and (5.5) becomes

$$b_{ii_1} \ldots b_{i_{r-1} i_s} b_{i_s i_{r+1}} \ldots b_{i_{s-1} i_s} b_{i_s i_{s+1}} \ldots b_{i_{n-1} j} \leq b_{ii_1} \ldots b_{i_{r-1} i_s} b_{i_s i_{s+1}} \ldots b_{i_{n-1} j}$$

with the last term again being smaller than $b_{ij}^{(n-1)}$. Since the conclusions hold for all possible value sequences $i_1, \ldots, i_{n-1}$ and all $1 \leq i, j \leq n$, $B^n \subseteq B^{n-1}$ has been proved. □

Corollary 5.32. *Let $B \in \mathbf{B}_n$ a reflexive square Boolean matrix.*

(1) *For the index k of B we have $k \leq n-1$.*
(2) *k is the index of B iff $B \subset B^2 \subset \ldots \subset A^{k-1} \subset B^k = B^{k+1} = \ldots = B^n = \ldots$.*
(3) *B^m is idempotent for all $m \geq k$.*

Turning now to the generalization of these results to incline matrices as announced above we will need the following:

Definition 5.33. ([11, Definition 5.1], [17, Definition 2.3]) *Let L be an incline. A matrix $A = [a_{ij}] \in \mathbf{L}_n$ is said to be*

(a) **row diagonally dominant** *if $a_{ij} = a_{ii} a_{ij}$ (equivalently: $a_{ii} \geq a_{ij}$), $i, j \in \underline{n}$;*
(b) **column diagonally dominant** *if $a_{ij} = a_{ij} a_{jj}$ (equivalently: $a_{ii} \geq a_{ji}$), $i, j \in \underline{n}$;*
(c) **weakly diagonally dominant** *if for any $i \in \underline{n}$,*

$$\text{either } \forall j \in \underline{n}\, [a_{ij} = a_{ii} a_{ij}], \quad \text{or} \quad \forall j \in \underline{n}\, [a_{ji} = a_{ji} a_{ii}]$$

(equivalently:

$$\text{either } \forall j \in \underline{n}\, [a_{ii} \geq a_{ij}], \quad \text{or} \quad \forall j \in \underline{n}\, [a_{ii} \geq a_{ji}]).$$

(d) **diagonally dominant** *if* $a_{ii}a_{jk} = a_{jk}$ *(equivalently:* $a_{ii} \geq a_{jk}$*), for all* $i, j, k \in \underline{n}$.

Corollary 5.34. *Let* L *be an incline and* $A \in \mathbf{L}_n$ *be diagonally dominant. Then*

(1) $a_{ii} = a_{jj}$ *for all* $i, j \in \underline{n}$;
(2) A *is weakly diagonally dominant.*

Proof. (1): By assumption, $a_{ii}a_{jk} = a_{jk}$ for all $i, j, k \in \underline{n}$. Thus $a_{ii}a_{jj} = a_{jj}$, i.e. $a_{ii} \geq a_{jj}$; conversely, we also have $a_{jj}a_{ii} = a_{ii}$, i.e. $a_{jj} \geq a_{ii}$. Thus $a_{ii} = a_{jj}$ for all $i, j \in \underline{n}$.

(2): Since $a_{ii}a_{jk} = a_{jk}$ for all $i, j, k \in \underline{n}$, we have, for any $i \in \underline{n}$, $a_{ii}a_{ij} = a_{ij}$ for all $j \in \underline{n}$. And since $A \to A \vee B$ by Th. 2.33(2), the assertion is proved. □

Proposition 5.35. ([11, Theorem 5.3], [17, Proposition 3.8]) *Let* L *be an incline and* $A \in \mathbf{L}_n$ *be weakly diagonally dominant. Then the following holds:*

(1) $A^2 \leq A^3$;
(2) a_{ii} *is idempotent and* $a_{ii} = a_{ii}^{(2)} = \cdots = a_{ii}^{(n)}$, *for all* $i \in \underline{n}$;
(3) A *converges to* $A^{k(A)}$ *with* $k(A) \leq n-1$.

Proof. (1): $a_{ij}^{(2)} = \sum_{k\in\underline{n}} a_{ik}a_{kj}$, for all $i, j, k \in \underline{n}$. Then, assuming for all $j \in \underline{n}$, $a_{ij} = a_{ii}a_{ij}$, we get

$$a_{ij}^{(2)} = \textstyle\sum_{k\in\underline{n}} a_{ik}a_{kj} = \sum_{k\in\underline{n}} a_{ii}a_{ik}a_{kj} \leq \sum_{i_1,i_2\in\underline{n}} a_{ii_1}\, a_{i_1i_2}\, a_{i_2j} = a_{ij}^{(3)},$$

by (5.3). And if for all $j \in \underline{n}$, $a_{ji} = a_{ji}a_{ii}$, then

$$a_{ij}^{(2)} = \textstyle\sum_{k\in\underline{n}} a_{ik}a_{kj} = \sum_{k\in\underline{n}} a_{ik}a_{kk}a_{kj} \leq a_{ij}^{(3)}.$$

So $a_{ij}^{(2)} \leq a_{ij}^{(3)}$ for arbitrary $i, j \in \underline{n}$, whence $A^2 \leq A^3$.

(2): Assume as first case that, for all $j \in \underline{n}$, $a_{ij} = a_{ii}a_{ij}$; then $a_{ii} = a_{ii}a_{ii}$, i.e. a_{ii} is an idempotent element of L. Thus, $a_{ii}\, a_{ii} \cdots a_{ii} = a_{ii}$. Using Cor. 5.16 we then get

$$a_{ii} = a_{ii}^s \leq a_{ii}^{(s)} = \sum_{i_1,\ldots,i_{s-1}\in\underline{n}} a_{ii_1}\, a_{i_1i_2} \cdots a_{i_{s-1}i} \leq \sum_{k\in\underline{n}} a_{ik} = \sum_{k\in\underline{n}} a_{ii}\, a_{ik} \leq \sum_{k\in\underline{n}} a_{ii} = a_{ii}.$$

As an analogous result holds for the second case, for all $j \in \underline{n}$, $a_{ji} = a_{ji}a_{ii}$, $a_{ii} = a_{ii}^{(s)}$ for all $s \in \underline{n}$.

(3): From (1) we get $A^2 \leq A^3 \leq \cdots \leq A^{n-1} \leq A^n$. To prove equality we have to show $A^{n-1} \geq A^n$. Considering (2), it suffices to show $a_{ij}^{(n-1)} \geq a_{ij}^{(n)}$ for $i \neq j$.

If $n = 2$, then product terms must have the form $a_{ii}a_{ij}$ or $a_{ij}a_{jj}$. By Cor. 5.16 we have $a_{ii}a_{ij} \leq a_{ij}$ and $a_{ij}a_{jj} \leq a_{ij}$ for arbitrary i, j. So for $i \neq j$ we get

$$a_{ij}^{(2)} = \textstyle\sum_{i\neq j\in\underline{n}} a_{ik}a_{kj} = \sum_{i\neq j\in\underline{n}} a_{ii}a_{ij} + a_{ij}a_{jj} \leq \sum_{i\neq j\in\underline{n}} a_{ij} + a_{ij} = a_{ij}^{(1)}.$$

Let $n \geq 3$. In any term $t := a_{ii_1}\, a_{i_1i_2} \ldots a_{i_{n-1}j}$ of $a_{ij}^{(n)}$ $n+1$ subscripts occur, so two of them must be equal, say $i_s = i_t$, $0 \leq s < t \leq n$. Proceeding in the usual way we delete the corresponding cycle and get a term t' of $a_{ij}^{(n-(t-s))}$ with $t \leq t'$. Since this holds for each summand, we finally get $a_{ij}^{(n)} \leq a_{ij}^{(n-(t-s))}$. Therefore, $A^n \leq A^m$, $m \leq n-1$; and since, by (1), for any $k \geq 3$, $A^{k-1} \leq A^k$, we get $A^m \leq A^n$, $m \leq n-1$. So $A^{n-1} \leq A^n$, whence finally $A^{n-1} = A^n$. □

Corollary 5.36. *Let L be an incline and $A \in \mathbf{L}_n$.*

(1) If A is row (or column) diagonally dominant, then $k(A) \leq n-1$.
(2) If A is reflexive, then $k(A) \leq n-1$.
(3) If A is reflexive, then $I_n \subseteq A \subseteq A^2 \subseteq \ldots \subseteq A^{n-1} = A^n = A^{n+1} = \ldots$.

Proof. (1): By Prop. 5.35, since a row (or column) diagonally dominant matrix is weakly diagonally dominant.
(2): By (1), since a reflexive matrix is row (and column) diagonally dominant.
(3): Reformulation of (2). □

Note that Cor. 5.36(2),(3) are the general counterparts to Cor. 5.32(1) and Prop. 5.31 announced above.

Proposition 5.37. ([13, Lemma 2.7]) *Let L be a 0-1-incline and $A \in \mathbf{L}_n$ be a reflexive matrix, i.e. $I_n \leq A$. Then the following holds:*

(1) $\mathsf{adj}(A)$ is idempotent.
(2) $A \cdot \mathsf{adj}(A) = \mathsf{adj}(A) \cdot A = \mathsf{adj}(A)$.
(3) $A^2 = A$ implies $\mathsf{adj}(A) = A$.
(4) $\mathsf{adj}(A)$ is reflexive.

Proof. (1): $[\mathsf{adj}(A)]^2 = [A^{n-1}]^2 = A^{2(n-1)} = A^{n-1} = \mathsf{adj}(A)$ by Cor. 5.21 and Cor. 5.36(3).
(2): Again using Cor. 5.21 and Cor. 5.36(3) we get $A \cdot \mathsf{adj}(A) = A \cdot A^{n-1} = A^n = A^{n-1} = \mathsf{adj}(A)$. Similarly, $\mathsf{adj}(A) \cdot A = \mathsf{adj}(A)$.
(3): By Cor. 5.21 $\mathsf{adj}(A) = A^{n-1}$, and since A is assumed idempotent, we get $A^{n-1} = A$.
(4): Take $B := \mathsf{adj}(A)$; then by Def. 5.17

$$b_{ii} = \mathsf{per}(A(i|i)) = \textstyle\sum_{\sigma \in S_{n-1}} a_{1\sigma(1)} \ldots a_{(i-1)\sigma(i-1)}\, a_{(i+1)\sigma(i+1)} \ldots a_{n\sigma(n)}.$$

So, taking the identity permutation $\sigma(i) = i$, we get

$$b_{ii} \geq a_{11} \ldots a_{(i-1)(i-1)}\, a_{(i+1)(i+1)} \ldots a_{nn} = 1.$$

□

References

[1] Abadir, K.M., Magnus, J.R.: *Matrix Algebra*, Cambridge University Press, New York (2005).
[2] Balakrishnan, R., Ranganathan, K.: *A Textbook of Graph Theory*, 2nd ed., Springer Science+Business Media, New York (2012).
[3] Ben-Israel, A., Greville, T.: *Generalized Inverses - Theory and Applications*, 2nd ed., Springer-Verlag New York, Inc. (2003).
[4] Bondy, J.A., Murty, U.S.R.: *Graph Theory*, Springer (2008).
[5] Brualdi, R.A., Cvetković, D.: *A Combinatorial Approach to Matrix Theory and Its Applications*, Taylor & Francis Group, Boca Raton (2009).
[6] Brualdi, R.A., Ryser, H.J.: *Combinatorial Matrix Theory*, Cambridge University Press (1991).

[7] Cao, Z.-Q. Kim, K.H., Roush, F.W.: *Incline Algebra and Applications*, Ellis Horwood Limited, Chichester (1984).
[8] Duan, J.-S.: The transitive closure, convergence of powers and adjoint of generalized fuzzy matrices, Fuzzy Sets and Systems **145**, 301-311 (2004).
[9] Give'on, Y.: Lattice matrices, Information and Control **7**, 477-484 (1964).
[10] Hammer P.L., Rudeanu, S.: *Boolean Methods in Operations Research and Related Areas*, Springer-Verlag Berlin Heidelberg (1968).
[11] Han,S.-C., Li, H.-X.: Indices and periods of incline matrices, Linear Algebra and its Applications **387**, 143-165 (2004).
[12] Han,S.-C., Li, H.-X.: Invertible incline matrices and Cramer's rule over inclines, Linear Algebra and its Applications **389**, 121-138 (2004).
[13] Han,S.-C., Li, H.-X.: The semigroup of incline Hall matrices, Linear Algebra and its Applications **390**, 183-196 (2004).
[14] Hohn, P.E., Schissler, L.R.: Boolean matrices and combinational circuit design, Bell System Tech. J. **34**, 177-202 (1955).
[15] Howie, J.M: *Fundamentals of Semigroup Theory*, Clarendon Press, Oxford (1995).
[16] Huang, Y., Lian, H.: On permanents of matrices over a commutative additively idempotent semiring, `https://arxiv.org/abs/1205.6891` (2012).
[17] Jiang, J., Shu, L. Tian, X.: On Generalized Transitive Matrices, J. Appl. Math. 2011 , Article ID 164371, 16 pages, doi:10.1155/2011/164371 (2011).
[18] Kim, K.H.: *Boolean Matrix Theory and Applications*, Marcel Dekker, Inc., New York and Basel (1982).
[19] Luce, R. D.: A note on Boolean matrix theory, Proc. of the American Mathematical Society **3**, 382-388, (1952).
[20] Lunts, A.G.: Algebraic methods of analysis and synthesis of relay contact networks, Izv. Akad. Nauk SSSR Ser. Mat. **16**, 405-426 (1952) (in Russian).
[21] Meyer, C.D.: *Matrix Analysis and Applied Linear Algebra*, Soc. for Industrial and Applied Mathematics, Philadelphia (2001).
[22] Ore, O.: Theory of equivalence relations, Duke Math J **9**, 573-627 (1942).
[23] Poole, D.: *Linear Algebra - A Modern Introduction*, 3rd ed., Brooks/Cole, Cengage Learning (2011).
[24] Rao, K.P.S. Bhaskara: *Theory of Generalized Inverses Over Commutative Rings*, Taylor & Francis, London and New York (2002).
[25] Schwarz, S.: On the semigroup of binary relations on a finite set, Czech. Math. J. **20**, 632-679 (1970).
[26] Shafarevich, I.R., Remizov, A.O.: *Linear Algebra and Geometry*, Springer-Verlag, Berlin Heidelberg (2013).
[27] Tan, Y.-J.: The Semigroup of Hall Matrices over Distributive Lattices, Semigroup Forum **61**, 303-314 (2000).
[28] Tan, Y.-J.: On the powers of matrices over a distributive lattice, Linear Algebra and its Applications **336**, 1-14 (2001).
[29] Tan, Y.-J.: On invertible matrices over antirings, Linear Algebra and its Applications **423**, 428-444 (2007).
[30] Tan, Y.-J.: On transitivity of generalized fuzzy matrices, Fuzzy Sets and Systems **210**, 69-88 (2013).

Chapter 6

Some Special Topics around Relations

In this chapter we present some concepts around relations that usually cannot be found in the mainstream accounts: The first section is devoted to the Bandler-Kohout products and images, which permit a compact representation of knowledge based on available data. In section 6.2 we introduce various types of relations, among them regular and difunctional, as well as idempotent relations, and consider their algebraic properties.

6.1 Bandler-Kohout Products and Images

6.1.1 *The Basic Concepts*

Given two relations, say ρ and σ, one can create a new relation $\rho * \sigma$ by applying a binary operation that takes ρ and σ as its arguments. The $*$ operation is called a *relational product*, and the resulting product is called a *composition* of ρ and σ. The most common and usual notion of a relational product has been introduced in Def. 3.11 as

$$\rho\sigma := \rho \circ \sigma := \{\langle x, z\rangle : \exists y[x\rho y \wedge y\sigma z]\} = \{\langle x, z\rangle : \exists y[x\rho y\sigma z]\}.$$

This definition can be rewritten in terms of after- and foresets as follows:

$$\rho \circ \sigma = \{\langle x, z\rangle : x\rho \cap \sigma z \neq \emptyset\}.$$

From this point of view three more product constructions come to mind immediately which have been introduced by Bandler and Kohout [2] in 1980, whence they are called **Bandler-Kohout products**, or **BK products** for short. For sake of completeness we present all four notions in Def. 6.1, the BK products not in their original form but in their revised version as given in [13,14] imposing some non-emptiness conditions. (Note that, by pure chance, the acronym "BK" is suitable for whatever you prefer: Bandler-Kohout or de Baets-Kerre.)

Definition 6.1. *Let ρ and σ be two relations from X to Y, and from Y to Z, respectively. The* **circle product** $\rho\circ\sigma$, *the* **subdirect product** $\rho\lhd\sigma$, *the* **superdirect product** $\rho\rhd\sigma$, *and the* **square product** $\rho\,\square\,\sigma$ *of ρ and σ are relations from X to Z defined as follows:*

(a) $\rho\circ\sigma = \{\langle x, z\rangle \in X\times Z : x\rho\cap\sigma z \supset \emptyset\}$,

$$
\begin{aligned}
&(b)\ \rho \lhd \sigma = \{\langle x,z\rangle \in X\times Z : \emptyset \subset x\rho \subseteq \sigma z\}\\
&\quad = \{\langle x,z\rangle \in X\times Z : \emptyset \subset x\rho \wedge \forall y\in Y[x\rho y \rightarrow y\sigma z]\},\\
&(c)\ \rho \rhd \sigma = \{\langle x,z\rangle \in X\times Z : x\rho \supseteq \sigma z \supset \emptyset\}\\
&\quad = \{\langle x,z\rangle \in X\times Z : \emptyset \subset \sigma z \wedge \forall y\in Y[y\sigma z \rightarrow x\rho y]\},\\
&(d)\ \rho \square \sigma = \{\langle x,z\rangle \in X\times Z : \emptyset \subset x\rho = \sigma z\}\\
&\quad = \{\langle x,z\rangle \in X\times Z : \emptyset \subset x\rho \wedge \forall y\in Y[x\rho y \leftrightarrow y\sigma z]\},
\end{aligned}
$$

A visualization of these definitions helps understanding their meaning, so some illustrations are given in Fig. 6.1 and Fig. 6.2.

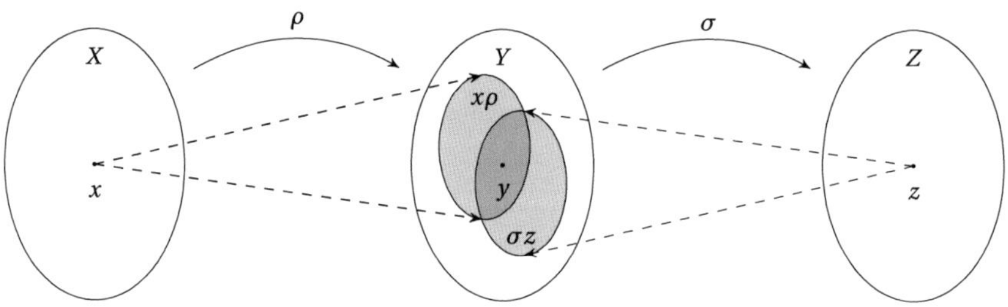

FIG. 6.1: Circle product $\rho \circ \sigma$ of relations ρ and σ.

Note that without the non-emptiness conditions we would have

$$\overline{\mathsf{dom}(\rho)} \times Z \subseteq \rho \lhd \sigma, \text{ and}$$

$$X \times \overline{\mathsf{rng}(\rho)} \subseteq \rho \rhd \sigma.$$

The first expression means that x is in relation $\rho \lhd \sigma$ with all elements of Z, even if there is no element of Y that is in relation $\rho^{\smile}$ with x, and similar for the second expression. The compositions $\rho \lhd \sigma$, $\rho \rhd \sigma$ and $\rho \square \sigma$ can thus contain a lot of unwanted pairs, whereas only those pairs can be accepted for which both components are involved in the relations.

One of the running examples for BK products is from the medical area, and has ω as an observation relation from patients to individual symptoms, and δ as the diagnostic relation from symptoms to diseases. Then the meaning of the products is as follows (see e.g. [14,25]):

- $p(\omega \circ \delta) i$: p has at least one symptom of illness i;
- $p(\omega \lhd \delta) i$: all of p's symptoms are among those which characterize i;
- $p(\omega \rhd \delta) i$: p's symptoms include all those which characterize i;
- $p(\omega \square \delta) i$: p's symptoms are exactly those of illness i.

But one can go further than that. The BK-products can be used to compare relational structures. Consider the generic case where ρ is a relation between a set of objects X and a set of properties such that $o\rho x$ means that object o has property x. Taking the products of ρ and $\rho^{\smile}$ we get relations on X with the following semantics:

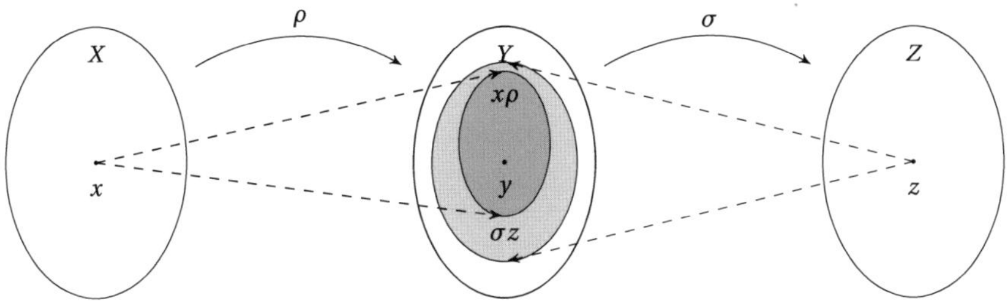

FIG. 6.2: Subdirect product $\rho \lhd \sigma$ of relations ρ and σ.

- $a(\rho \circ \rho^{\smile})b$: a shares at least one property with b;
- $a(\rho \lhd \rho^{\smile})b$: all of a's properties are among those of b;
- $a(\rho \rhd \rho^{\smile})b$: a's properties include all those of b;
- $a(\rho \square \rho^{\smile})b$: a and b have the same properties.

Returning to the medical example from above we get e.g.

- $p_1(\omega \lhd \omega^{\smile})p_2$: the symptoms of patient p_1 are among those of patient p_2;
- $s_1(\omega^{\smile} \lhd \omega)s_2$: whenewer symptom s_1 occurs, so does s_2 (in the patient group observed);
- $s_1(\delta \square \delta^{\smile})s_2$: symptom s_1 characterizes exactly the same diseases as s_2;

In Def. 3.8 we have already defined the notion of the (direct) image of A under a relation $\rho \subseteq X \times Y$ as the set

$$\rho[A] = \{y \in Y : \exists x \in A[x\rho y]\}$$

As for the usual composition of relations this can be equivalently rewritten in terms of after- and foresets as

$$\rho[A] = \{y \in Y : A \cap \rho y \neq \emptyset\}.$$

In the spirit of the Bandler-Kohout extensions for relation products, also for images three more variants can be defined; this has been done in [13, 14]. As for relation products above we present all four kinds of images in Def. 6.2. The direct image $\rho[A]$ is the set of those elements of Y that are in relation $\rho^{\smile}$ with *at least one* element of A. The purpose now is to refine the direct image $\rho[A]$ in order to distinguish those elements of Y that are in relation $\rho^{\smile}$ with *all* elements of A and those elements of Y that are in relation $\rho^{\smile}$ with elements of A *only*.

Definition 6.2. *Let ρ be a relation from X to Y, and $A \subseteq X$. The* **direct image** $\rho[A]$, *the* **subdirect image** $\rho^{\lhd}[A]$, *the* **superdirect image** $\rho^{\rhd}[A]$, *and the* **square image** $\rho^{\square}[A]$ *of A under ρ are the subsets of Y defined as:*

(a) $\rho[A] = \{y \in Y : A \cap \rho y \neq \emptyset\} = \{y : \exists x \in X[x \in A \wedge x\rho y]\}$,
(b) $\rho^{\lhd}[A] = \{y \in Y : \emptyset \subset A \subseteq \rho y\} = \{y : \forall x \in X[x \in A \rightarrow x\rho y]\}$,
(c) $\rho^{\rhd}[A] = \{y \in Y : \emptyset \subset \rho y \subseteq A\} = \{y : \forall x \in X[x\rho y \rightarrow x \in A]\}$,

(d) $\rho^{\square}[A] \;=\{y\in Y:\emptyset\subset A=\rho y\} \;=\{y:\forall x\in X[x\rho y\leftrightarrow x\in A]\}.$

Again note the non-emptiness conditions: Without $\emptyset\subset A$ it would follow that $\rho^{\lhd}[\emptyset]=Y$, which is unacceptable. And without $\emptyset\subset\rho y$ in the definition of the superdirect image it would follow that $\overline{\mathsf{rng}(\rho)}\subseteq\rho^{\rhd}[A]$, which is unacceptable again: The condition $\emptyset\subset\rho y$ ensures that $\rho^{\rhd}[A]$ contains those elements of Y that are in relation $\rho^{\smile}$ with elements of A only. Furthermore, the non-emptiness conditions imply that all images are contained in $\mathsf{rng}(\rho)$ and that the images of $\emptyset$ under a relation all yield $\emptyset$.

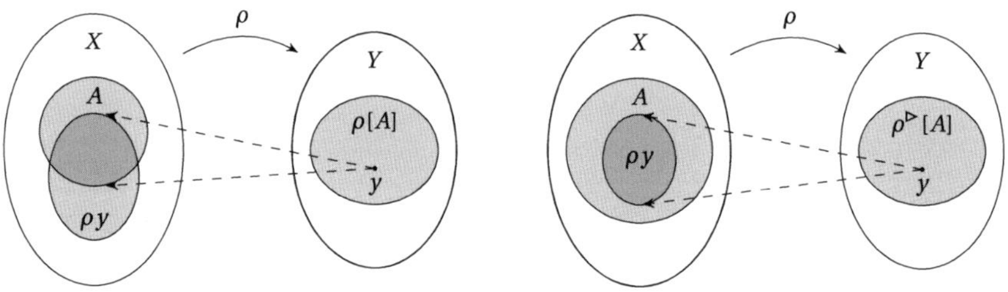

FIG. 6.3: Direct image $\rho[A]$ and superdirect image $\rho^{\rhd}[A]$ of a set A under ρ.

For illustration purposes we use the medical example from above: Let $F\neq\emptyset$ be the non-empty set of female patients; then the images of F under ω are given by

- $\omega[F]$ is the set of symptoms shown by at least one female patient;
- $\omega^{\lhd}[F]$ is the set of symptoms shown by all female patients;
- $\omega^{\rhd}[F]$ is the set of symptoms shown by at least one female patient but no male patient;
- $\omega^{\square}[F]$ is the set of symptoms shown by all female patients but no male patient.

6.1.2 *On Properties of BK images and products*

The material in this section is largely based on the results given in [13, 14, 47]. For direct comparison we will repeat some results from sections 3.1.1 and 3.2.

6.1.2.1 *Image Properties*

Below, let $\rho\subseteq X\times Y$ and $\sigma\subseteq y\times Z$ be two relations, and furthermore A, B, and A_i be subsets of X. We start with some properties of BK images which are an immediate consequence of Def. 6.2:

Proposition 6.3. *(Containment properties)*

(1) $\rho^{\square}[A]=\rho^{\lhd}[A]\cap\rho^{\rhd}[A]$,
(2) $\rho^{\square}[A]\subseteq\rho^{\lhd}[A]\subseteq\rho[A]$,
(3) $\rho^{\square}[A]\subseteq\rho^{\rhd}[A]\subseteq\rho[A]$.

The next three propositions describe how the (new) BK images and the (usual) direct image of a set can be expressed in terms of each other.

Proposition 6.4. *(Relationships between images 1)*

(1) $\rho^{\lhd}[A] = \overline{\overline{\rho}[A]}$, *if* $\emptyset \subset A$;
(2) $\rho^{\rhd}[A] = \overline{\rho[\overline{A}]} \cap \mathrm{rng}(\rho)$;
(3) $\rho^{\square}[A] = \overline{\overline{\rho}[A]} \cap \overline{\rho[\overline{A}]}$, *if* $\emptyset \subset A$.

Proof.

(1) $y \in \overline{\overline{\rho}[A]} \leftrightarrow \neg[A \cap \overline{\rho} y \neq \emptyset)] \leftrightarrow A \cap \overline{\rho} y = \emptyset \leftrightarrow \emptyset \subset A \subseteq \rho y \leftrightarrow y \in \rho^{\lhd}[A]$. □

Proposition 6.5. *(Relationships between images 2)*

(1) $\rho[A] = \overline{\overline{\rho}^{\lhd}[A]}$, *if* $\emptyset \subset A$;
(2) $\rho[A] = \overline{\rho^{\rhd}[\overline{A}]} \cap \mathrm{rng}(\rho)$.

Proof.

$$
\begin{aligned}
(2)\ y \in \overline{\rho^{\rhd}[\overline{A}]} \cap \mathrm{rng}(\rho) &\leftrightarrow \neg[\emptyset \subset \rho y \subseteq \overline{A}] \wedge \rho y \neq \emptyset \\
&\leftrightarrow [\neg(\emptyset \subset \rho y) \vee \neg(\rho y \subseteq \overline{A})] \wedge \rho y \neq \emptyset \\
&\leftrightarrow [\emptyset = \rho y \vee \neg(\rho y \subset \overline{A})] \wedge \rho y \neq \emptyset \\
&\leftrightarrow \neg(\rho y \subseteq \overline{A}) \wedge \rho y \neq \emptyset \\
&\leftrightarrow \neg(\rho y \cap A = \emptyset) \wedge \rho y \neq \emptyset \\
&\leftrightarrow A \cap \rho y \neq \emptyset \\
&\leftrightarrow y \in \rho[A]
\end{aligned}
$$

□

Proposition 6.6. *(Relationships between images 3)*

(1) $\rho^{\lhd}[A] = \overline{\rho}^{\rhd}[\overline{A}] \cup \overline{\mathrm{rng}(\overline{\rho})}$, *if* $\emptyset \subset A$;
(2) $\rho^{\rhd}[A] = \overline{\rho}^{\lhd}[\overline{A}] \cap \mathrm{rng}(\rho)$, *if* $A \subset X$;
(3) $\rho^{\square}[A] = \overline{\rho}^{\square}[\overline{A}]$, *if* $\emptyset \subset A \subset X$.

Proof.

$$
\begin{aligned}
(1)\ y \in \overline{\rho}^{\rhd}[\overline{A}] \cup \overline{\mathrm{rng}(\overline{\rho})} &\leftrightarrow \emptyset \subset \overline{\rho} y \subseteq \overline{A} \vee \neg(\overline{\rho} y \neq \emptyset) \\
&\leftrightarrow [\emptyset \subset \overline{\rho y} \wedge \overline{\rho y} \subseteq \overline{A}] \vee \overline{\rho y} = \emptyset \\
&\leftrightarrow [\rho y \subset X \wedge \emptyset \subset A \subseteq \rho y] \vee \rho y = X \\
&\leftrightarrow [\rho y \subset X \vee \rho y = X] \wedge [\emptyset \subset A \subseteq \rho y \vee \rho y = X] \\
&\leftrightarrow \emptyset \subset A \subseteq \rho y \vee \rho y = X, \quad \text{by Th. 2.33(3): true} \wedge A \equiv A \\
&\leftrightarrow y \in \rho^{\lhd}[A]
\end{aligned}
$$

□

Proposition 6.7. *(Expressions in terms of aftersets)*

(1) $\rho[A] = \bigcup_{x \in A} x\rho$;
(2) $\rho^{\lhd}[A] = \bigcap_{x \in A} x\rho$, *if* $\emptyset \subset A$.

Proof. (1) $y \in \rho[A] \leftrightarrow y \in \{y \in T : \exists x \in A[x\rho y]\} \leftrightarrow y \in \bigcup_{x\in A}\{y \in T : x\rho y\} \leftrightarrow y \in \bigcup_{x\in A} x\rho$.
(2) $\rho^{\triangleleft}[A] = \overline{\overline{\rho}[A]} = \overline{\bigcup_{x\in A} x\overline{\rho}} = \overline{\bigcup_{x\in A} \overline{x\rho}} = \bigcap_{x\in A} \overline{\overline{x\rho}} = \bigcap_{x\in A} x\rho$, using (1) for the second and Prop. 3.12(1) for the third equation. □

Proposition 6.8. *(Monotonicity properties)*

$$
\begin{array}{lrcl}
(1) & A \subseteq B & \to & \rho[A] \subseteq \rho[B];\\
(2) & \emptyset \subset A \subseteq B & \to & \rho^{\triangleleft}[B] \subseteq \rho^{\triangleleft}[A];\\
(3) & A \subseteq B & \to & \rho^{\triangleright}[A] \subseteq \rho^{\triangleright}[B];\\
(4) & \rho \subseteq \sigma & \to & \rho[A] \subseteq \sigma[A];\\
(5) & \rho \subseteq \sigma & \to & \rho^{\triangleleft}[A] \subseteq \sigma^{\triangleleft}[A];\\
(6) & \mathrm{rng}(\rho) = \mathrm{rng}(\sigma) \wedge \rho \subseteq \sigma & \to & \sigma^{\triangleright}[A] \subseteq \rho^{\triangleright}[A].
\end{array}
$$

Proof. (6) $y \in \sigma^{\triangleright}[A] \leftrightarrow \emptyset \subset \sigma y \subseteq A$. From the assumption $\mathrm{rng}(\rho) = \mathrm{rng}(\sigma)$ we get $\emptyset \subset \rho y$, and from $\rho \subseteq \sigma$ we have $\rho y \subseteq \sigma y \subseteq A$, whence $\emptyset \subset \rho y \subseteq A$, i.e. $y \in \rho^{\triangleright}[A]$. □

Proposition 6.9. *(Interaction with union 1)*

$$
\begin{array}{lrcll}
(1) & \rho[\bigcup_{i\in I} A_i] & = & \bigcup_{i\in I} \rho[A_i]; & \\
(2) & \rho^{\triangleleft}[\bigcup_{i\in I} A_i] & = & \bigcap_{i\in I} \rho^{\triangleleft}[A_i], & \textit{if } \forall i \in I[\emptyset \subset A_i];\\
(3) & \bigcap_{i\in I} \rho^{\triangleleft}[A_i] \subseteq \rho^{\triangleleft}[\bigcup_{i\in I} A_i] & \subseteq & \bigcup_{i\in I} \rho^{\triangleleft}[A_i]; & \\
(4) & \rho^{\triangleright}[\bigcup_{i\in I} A_i] & \supseteq & \bigcup_{i\in I} \rho^{\triangleright}[A_i]. &
\end{array}
$$

Proof.

$$
\begin{aligned}
(2)\ y \in \rho^{\triangleleft}[\textstyle\bigcup_{i\in I} A_i] &\leftrightarrow \emptyset \subset (\textstyle\bigcup_{i\in I} A_i) \subseteq \rho y\\
&\leftrightarrow \forall i \in I[\emptyset \subset A_i \subseteq \rho y]\\
&\leftrightarrow y \in \textstyle\bigcap_{i\in I} \rho^{\triangleleft}[A_i]
\end{aligned}
$$

□

Proposition 6.10. *(Interaction with intersection 1)*

$$
\begin{array}{lrcll}
(1) & \rho[\bigcap_{i\in I} A_i] & \subseteq & \bigcap_{i\in I} \rho[A_i]; & \\
(2) & \rho^{\triangleleft}[\bigcap_{i\in I} A_i] & \supseteq & \bigcup_{i\in I} \rho^{\triangleleft}[A_i], & \textit{if} \forall i \in I[\emptyset \subset A_i];\\
(3) & \rho^{\triangleright}[\bigcap_{i\in I} A_i] & = & \bigcap_{i\in I} \rho^{\triangleright}[A_i]. &
\end{array}
$$

Proposition 6.11. *(Interaction with union 2)*

$$
\begin{array}{lrcl}
(1) & (\bigcup_{i\in I} \rho_i)[A] & = & \bigcup_{i\in I} \rho_i[A];\\
(2) & (\bigcup_{i\in I} \rho_i)^{\triangleleft}[A] & \supseteq & \bigcup_{i\in I} \rho_i^{\triangleleft}[A];\\
(3) & \bigcap_{i\in I} \rho_i^{\triangleright}[A] \subseteq (\bigcup_{i\in I} \rho_i)^{\triangleright}[A] & \subseteq & \bigcup_{i\in I} \rho_i^{\triangleright}[A].
\end{array}
$$

Proposition 6.12. *(Interaction with intersection 2)*

$$
\begin{array}{lrcll}
(1) & (\bigcap_{i\in I} \rho_i)[A] & \subseteq & \bigcap_{i\in I} \rho_i[A]; & \\
(2) & (\bigcap_{i\in I} \rho_i)^{\triangleleft}[A] & = & \bigcap_{i\in I} \rho_i^{\triangleleft}[A]; & \\
(3) & (\bigcap_{i\in I} \rho_i)^{\triangleright}[A] & \supseteq & \bigcup_{i\in I} \rho_i^{\triangleright}[A], & \textit{if } \forall y \in Y[\bigcap_{i\in I} \rho_i y \neq \emptyset].
\end{array}
$$

Proof.

$$
\begin{aligned}
(1)\ y \in (\textstyle\bigcap_{i\in I}\rho_i)[A] &\rightarrow A\cap(\textstyle\bigcap_{i\in I}\rho_i)y \neq \emptyset\\
&\rightarrow \textstyle\bigcap_{i\in I}((A\cap\rho_i)y) \neq \emptyset\\
&\rightarrow y\in\textstyle\bigcap_{i\in I}\rho_i[A]
\end{aligned}
$$

□

6.1.2.2 *Product Properties*

After this investigation of the most important properties of BK images, we will now turn to the analogue properties of the BK products of relations. In the following, let ρ, ρ_1, and ρ_2 be relations from X to Y, and σ a relation from Y to Z; and let $(\rho_i)_{i\in I}$ be a family of relations from X to Y.

Proposition 6.13. *(Containment properties)*

(1) $\rho\,\square\,\sigma = (\rho \lhd \sigma)\cap(\rho \rhd \sigma)$,

(2) $\rho\,\square\,\sigma \subseteq \rho \lhd \sigma \subseteq \rho\circ\sigma$,

(3) $\rho\,\square\,\sigma \subseteq \rho \rhd \sigma \subseteq \rho\circ\sigma$.

The new compositions can be expressed in terms of the classical composition as follows:

Proposition 6.14. *(Relationships)*

(1) $\rho \lhd \sigma = \overline{\rho\circ\overline{\sigma}} \,\cap\, (\mathrm{dom}(\rho)\times Z)$;

(2) $\rho \rhd \sigma = \overline{\overline{\rho}\circ\sigma} \,\cap\, (X\times \mathrm{rng}(\sigma))$;

(3) $\rho\,\square\,\sigma = \overline{\rho\circ\overline{\sigma}} \,\cap\, \overline{\overline{\rho}\circ\sigma} \,\cap\, (\mathrm{dom}(\rho)\times Z)$
$\quad = \overline{\rho\circ\overline{\sigma}} \,\cap\, \overline{\overline{\rho}\circ\sigma} \,\cap\, (X\times \mathrm{rng}(\sigma))$.

Proof.

$$
\begin{aligned}
(1){:}\ \langle x,z\rangle \in \overline{\rho\circ\overline{\sigma}} \cap (\mathrm{dom}(\rho)\times Z) &\leftrightarrow \neg(\langle x,z\rangle\in\rho\circ\overline{\sigma}) \wedge x\in\mathrm{dom}(\rho)\\
&\leftrightarrow \neg(x\rho\cap\overline{\sigma}z\neq\emptyset)\wedge x\rho\neq\emptyset\\
&\leftrightarrow x\rho\cap\overline{\sigma}z=\emptyset\wedge x\rho\neq\emptyset\\
&\leftrightarrow \emptyset\subset x\rho\subseteq\sigma z\\
&\leftrightarrow \langle x,z\rangle\in\rho\lhd\sigma
\end{aligned}
$$

□

And also the other way round, the usual composition can be expressed in terms of the new compositions in a convenient way:

Proposition 6.15. *(Relationships)*

(1) $\rho\circ\sigma = \overline{\rho\lhd\overline{\sigma}} \,\cap\, (\mathrm{dom}(\rho)\times Z)$;

(2) $\rho\circ\sigma = \overline{\overline{\rho}\rhd\sigma} \,\cap\, (X\times\mathrm{rng}(\sigma))$.

A simple way of interdefinability for the new compositions holds only for the original version of Bandler and Kohout, but not for the improved version of De Baets and Kerre which is used here. See [13], Properties 4.4.

Proposition 6.16. *(Convertibility properties)*

(1) $(\rho \circ \sigma)\breve{} = \sigma\breve{} \circ \rho\breve{}$;
(2) $(\rho \lhd \sigma)\breve{} = \sigma\breve{} \rhd \rho\breve{}$;
(3) $(\rho \rhd \sigma)\breve{} = \sigma\breve{} \lhd \rho\breve{}$;
(4) $(\rho \Box \sigma)\breve{} = \sigma\breve{} \Box \rho\breve{}$;

Proof.
(2) Note that $\rho \lhd \sigma = (\rho \circ \sigma) \cap \{\langle x,z\rangle : x\rho \subseteq \sigma z\}$. Thus, by Prop. 3.13(7), $(\rho \lhd \sigma)\breve{} = (\rho \circ \sigma)\breve{} \cap \{\langle x,z\rangle : x\rho \subseteq \sigma z\}\breve{} = (\sigma\breve{} \circ \rho\breve{}) \cap \{\langle z,x\rangle : \rho\breve{} x \subseteq z\sigma\breve{}\}$, using Prop. 3.13(10). But the last intersection is nothing else than $\sigma\breve{} \rhd \rho\breve{}$.

$$\begin{aligned}
(3)\text{: } x(\rho \rhd \sigma)z &\leftrightarrow x\rho \supseteq \sigma z \supset \emptyset \\
&\leftrightarrow \emptyset \subset \sigma z \subseteq x\rho \\
&\leftrightarrow \emptyset \subset z\sigma\breve{} \subseteq \rho\breve{} x \\
&\leftrightarrow z(\sigma\breve{} \lhd \rho\breve{})x \\
&\leftrightarrow x(\sigma\breve{} \lhd \rho\breve{})\breve{} z
\end{aligned}$$

□

Proposition 6.17. *(Monotonicity properties)*

$$\begin{array}{lrcl}
(1) & \rho_1 \subseteq \rho_2 & \rightarrow & \rho_1 \circ \sigma \subseteq \rho_2 \circ \sigma; \\
(2) & \mathsf{dom}(\rho_1) = \mathsf{dom}(\rho_2) \wedge \rho_1 \subseteq \rho_2 & \rightarrow & \rho_2 \lhd \sigma \subseteq \rho_1 \lhd \sigma; \\
(3) & \rho_1 \subseteq \rho_2 & \rightarrow & \rho_1 \rhd \sigma \subseteq \rho_2 \rhd \sigma; \\
(4) & \sigma_1 \subseteq \sigma_2 & \rightarrow & \rho \circ \sigma_1 \subseteq \rho \circ \sigma_2; \\
(5) & \sigma_1 \subseteq \sigma_2 & \rightarrow & \rho \lhd \sigma_1 \subseteq \rho \lhd \sigma_2; \\
(6) & \mathsf{rng}(\sigma_1) = \mathsf{rng}(\sigma_2) \wedge \sigma_1 \subseteq \sigma_2 & \rightarrow & \rho \rhd \sigma_2 \subseteq \rho \rhd \sigma_1.
\end{array}$$

Proof. (2): $\langle x,z\rangle \in \rho_2 \lhd \sigma \leftrightarrow \emptyset \subset x\rho_2 \subseteq \sigma z \rightarrow \emptyset \subset x\rho_1 \subseteq \sigma z$, using both of the assumptions. But the last expression is equivalent to $\langle x,z\rangle \in \rho_1 \lhd \sigma$. □

Proposition 6.18. *(Interaction with union 1)*

$$\begin{array}{lrcl}
(1) & (\bigcup_{i\in I} \rho_i) \circ \sigma & = & \bigcup_{i\in I}(\rho_i \circ \sigma); \\
(2) & \bigcap_{i\in I}(\rho_i \lhd \sigma) \subseteq (\bigcup_{i\in I} \rho_i) \lhd \sigma & \subseteq & \bigcup_{i\in I}(\rho_i \lhd \sigma); \\
(3) & (\bigcup_{i\in I} \rho_i) \rhd \sigma & \supseteq & \bigcup_{i\in I}(\rho_i \rhd \sigma).
\end{array}$$

Proposition 6.19. *(Interaction with intersection 1)*

(1) $(\bigcap_{i\in I} \rho_i) \circ \sigma \subseteq \bigcap_{i\in I}(\rho_i \circ \sigma)$;
(2) $(\bigcap_{i\in I} \rho_i) \rhd \sigma = \bigcap_{i\in I}(\rho_i \rhd \sigma)$.

Proposition 6.20. *(Interaction with union 2)*

$$\begin{array}{lrcl}
(1) & \rho \circ (\bigcup_{i\in I} \sigma_i) & = & \bigcup_{i\in I}(\rho \circ \sigma_i); \\
(2) & \rho \lhd (\bigcup_{i\in I} \sigma_i) & \supseteq & \bigcup_{i\in I}(\rho \lhd \sigma_i); \\
(3) & \bigcap_{i\in I}(\rho \rhd \sigma_i) \subseteq \rho \rhd (\bigcup_{i\in I} \sigma_i) & \subseteq & \bigcup_{i\in I}(\rho \rhd \sigma_i).
\end{array}$$

Proposition 6.21. *(Interaction with intersection 2)*

(1) $\rho \circ (\bigcap_{i\in I} \sigma_i) \subseteq \bigcap_{i\in I}(\rho \circ \sigma_i)$;
(2) $\rho \lhd (\bigcap_{i\in I} \sigma_i) = \bigcap_{i\in I}(\rho \lhd \sigma_i)$.

Proof. (2) $\langle x,z\rangle \in \rho \lhd (\bigcap_{i\in I}\sigma_i) \leftrightarrow \emptyset \subset x\rho \subseteq (\bigcap_{i\in I}\sigma_i)z \leftrightarrow \forall i \in I[\emptyset \subset x\rho \subseteq \sigma_i z] \leftrightarrow \forall i \in I[\langle x,z\rangle \in \rho \lhd \sigma_i] \leftrightarrow \langle x,z\rangle \in \bigcap_{i\in I}(\rho \lhd \sigma_i)$ □

In [13], Properties 4.13, the following six mixed pseudo-associativity properties between the sub-, the super- and the classical composition of relations are listed, repeating the well known classical result from Th. 3.16 in (1):

Proposition 6.22. *(Mixed and pseudo associativity laws)*
Let $\rho \subseteq X \times Y$, $\sigma \subseteq Y \times Z$, and $\tau \subseteq Z \times U$ be three relations. Then the following relationships hold:

(1) $\rho \circ (\sigma \circ \tau) = (\rho \circ \sigma) \circ \tau$;
(2) $\rho \circ (\sigma \rhd \tau) \subseteq (\rho \circ \sigma) \rhd \tau$;
(3) $\rho \lhd (\sigma \circ \tau) \supseteq (\rho \lhd \sigma) \circ \tau$;
(4) $\rho \lhd (\sigma \lhd \tau) \subseteq (\rho \circ \sigma) \lhd \tau$;
(5) $\rho \lhd (\sigma \rhd \tau) = (\rho \lhd \sigma) \rhd \tau$;
(6) $\rho \rhd (\upsilon \circ \iota) \supseteq (\rho \rhd \sigma) \rhd \tau$;

Proof. (2): We have

$$\begin{aligned}
\langle x,u\rangle \in \rho \circ (\sigma \rhd \tau) &\leftrightarrow x\rho \cap (\sigma \rhd \tau)u \neq \emptyset \\
&\leftrightarrow \exists y \in x\rho[y \in (\sigma \rhd \tau)u] \\
&\leftrightarrow \exists y \in x\rho[\emptyset \subset \tau u \subseteq y\sigma] \\
&\leftrightarrow \exists y \in x\rho\ \forall z \in \tau u[z \in y\sigma] \ \wedge\ \tau u \neq \emptyset \qquad \text{(A)}
\end{aligned}$$

and

$$\begin{aligned}
\langle x,u\rangle \in (\rho \circ \sigma) \rhd \tau &\leftrightarrow \emptyset \subset \tau u \subseteq x(\rho \circ \sigma) \\
&\leftrightarrow \forall z \in \tau u[z \in x(\rho \circ \sigma)] \ \wedge\ \tau u \neq \emptyset \\
&\leftrightarrow \forall z \in \tau u[x\rho \cap \sigma z \neq \emptyset] \ \wedge\ \tau u \neq \emptyset \\
&\leftrightarrow \forall z \in \tau u\ \exists y \in x\rho[y \in \sigma z] \ \wedge\ \tau u \neq \emptyset \\
&\leftrightarrow \forall z \in \tau u\ \exists y \in x\rho[z \in y\sigma] \ \wedge\ \tau u \neq \emptyset \qquad \text{(B)}
\end{aligned}$$

By Th. 2.34(3), $\exists x \forall y P(x,y) \rightarrow \forall y \exists x P(x,y)$, (B) follows from (A), whence the inclusion holds.

(3): Follows from (2) using convertibility properties from Prop. 6.16:

$$\begin{aligned}
& \tau^{\smile} \circ (\sigma^{\smile} \rhd \rho^{\smile}) && \subseteq (\tau^{\smile} \circ \sigma^{\smile}) \rhd \rho^{\smile} \\
\leftrightarrow\ & (\tau^{\smile} \circ (\sigma^{\smile} \rhd \rho^{\smile}))^{\smile} && \subseteq ((\tau^{\smile} \circ \sigma^{\smile}) \rhd \rho^{\smile})^{\smile} \\
\leftrightarrow\ & (\sigma^{\smile} \rhd \rho^{\smile})^{\smile} \circ \tau && \subseteq \rho \lhd (\tau^{\smile} \circ \sigma^{\smile})^{\smile} \\
\leftrightarrow\ & (\rho \lhd \sigma) \circ \tau && \subseteq \rho \lhd (\sigma \circ \tau)
\end{aligned}$$

□

These results have been considerably enhanced in [47], where the author has systematically considered all the $4^4 = 256$ possibilities to arrange the four signs of compositions. But note that the proofs given in this paper are at least partially based upon the above mentioned results from [13]: E.g. in the numbering of [47], the proof of formula (25) uses classical associativity from (19) (that is our Prop. 6.22,(1)); or that of (29) uses (20) (our Prop. 6.22,(2)) etc.. Since the results of [47] are based upon those of [13], it seems justified to include the latter as the basic associativity-like properties in Prop. 6.22.

6.1.2.3 *Some More Properties of BK Products*

The interplay of relational compositions as given in the next theorem generates powerful inequalities that enrich the computational power of both crisp and fuzzy relational calculi.

Theorem 6.23. *Let $\rho \subseteq X \times Y$, $\sigma \subseteq Y \times Z$, and $\tau \subseteq X \times Z$ be three relations. Then the following equivalences holds:*

$$\rho \circ \sigma \subseteq \tau \;\leftrightarrow\; \rho \subseteq \tau \rhd \sigma^{\smallsmile} \;\leftrightarrow\; \sigma \subseteq \rho^{\smallsmile} \lhd \tau.$$

Proof. The proof is given in two parts:

1)
$$\begin{aligned}
\rho \circ \sigma \subseteq \tau &\leftrightarrow \forall x, z[x\rho\sigma z \to x\tau z] \\
&\leftrightarrow \forall x, z[\exists y[x\rho y \wedge y\sigma z \wedge \emptyset \subset y\rho^{\smallsmile}] \to x\tau z] \\
&\leftrightarrow \forall x, y, z[(y\sigma z \wedge x\rho y \wedge \emptyset \subset y\rho^{\smallsmile}) \to x\tau z] && \text{Th. 2.36(3)(a)} \\
&\leftrightarrow \forall x, y, z[y\sigma z \to (x\rho y \wedge \emptyset \subset y\rho^{\smallsmile} \to x\tau z)] && \text{Th. 2.33(12)} \\
&\leftrightarrow \forall y, z[y\sigma z \to \forall x[\emptyset \subset y\rho^{\smallsmile} \wedge x\rho y \to x\tau z]] && \text{Th. 2.36(4)(b)} \\
&\leftrightarrow \forall y, z[y\sigma z \to (\emptyset \subset y\rho^{\smallsmile} \wedge \forall x[y\rho^{\smallsmile} x \to x\tau z])], && \text{Th. 2.36(1)(b)} \\
&\leftrightarrow \forall y, z[y\sigma z \to (\emptyset \subset y\rho^{\smallsmile} \wedge y\rho^{\smallsmile} \subseteq \tau z)], \\
&\leftrightarrow \sigma \subseteq \rho^{\smallsmile} \lhd \tau, && \text{by Def. 6.1}
\end{aligned}$$

2) We can use a similar proof beginning with

$$\begin{aligned}
\rho \circ \sigma \subseteq \tau &\leftrightarrow \forall x, z[x\rho\sigma z \to x\tau z] \\
&\leftrightarrow \forall x, z[\exists y[x\rho y \wedge y\sigma z \wedge \emptyset \subset \sigma^{\smallsmile} y] \to x\tau z] \\
&\leftrightarrow \cdots \\
&\leftrightarrow \forall x, y[x\rho y \to (\emptyset \subset \sigma^{\smallsmile} y \wedge \forall z[z\sigma^{\smallsmile} y \to x\tau z])] \\
&\leftrightarrow \forall x, y[x\rho y \to (\emptyset \subset \sigma^{\smallsmile} y \wedge \sigma^{\smallsmile} y \subseteq x\tau)] \\
&\leftrightarrow \rho \subseteq \tau \rhd \sigma^{\smallsmile}, && \text{by Def. 6.1}
\end{aligned}$$

□

Using BK products we can give the following characterizations ([3]):

Proposition 6.24.

(1) *ρ is transitive iff $\rho \subseteq \rho \rhd \rho^{\smallsmile}$.*
(2) *ρ is reflexive iff $\rho \rhd \rho^{\smallsmile} \subseteq \rho$.*
(3) *ρ is a preorder iff $\rho = \rho \rhd \rho^{\smallsmile}$.*

(4) *ρ is an equivalence iff $\rho = \rho \square \rho^{\smile}$.*
(5) *If $\rho \rhd \rho^{\smile} = \rho$ then it is idempotent.*

Proof. (1): From the transitivity of ρ and Th. 6.23 we get $\rho \circ \rho \subseteq \rho \leftrightarrow \rho \subseteq \rho \rhd \rho^{\smile}$.

(2): $\rho \rhd \rho^{\smile} = \{\langle x, z\rangle : x\rho \supseteq \rho^{\smile} z\} = \{\langle x, z\rangle : x\rho \supseteq z\rho\}$. But certainly $x\rho \supseteq x\rho$, whence $\langle x, x\rangle \in \rho \rhd \rho^{\smile}$, i.e. $I \subseteq \rho \rhd \rho^{\smile} \subseteq \rho$. Conversely assume $I \subseteq \rho$, and let $x(\rho \rhd \rho^{\smile})y$. Then $x\rho \supseteq y\rho$, and from $y\rho y$ we get $y \in y\rho \subseteq x\rho$ whence $x\rho y$.

(3): Follows immediately from (1) and (2).

(4): Let ρ be an equivalence. Then by (3) we have $\rho = \rho \rhd \rho^{\smile}$. So by Prop. 6.16(3), $\rho^{\smile} = \rho \lhd \rho^{\smile}$. Moreover, $\rho = \rho^{\smile}$ by symmetry, whence using Prop. 6.13(1) we get

$$\rho \rhd \rho^{\smile} = \rho = \rho^{\smile} = \rho \lhd \rho^{\smile} = \rho \rhd \rho^{\smile} \cap \rho \lhd \rho^{\smile} = \rho \square \rho^{\smile}.$$

Conversely, let $\rho = \rho \square \rho^{\smile} = \rho \rhd \rho^{\smile} \cap \rho \lhd \rho^{\smile}$. So $\rho \subseteq \rho \rhd \rho^{\smile}$, i.e. ρ is transitive by (1). It has already been mentioned in the proof of (2) that $I \subseteq \rho \rhd \rho^{\smile}$; similarly we have $I \subseteq \rho \lhd \rho^{\smile}$. Thus $I \subseteq \rho \square \rho^{\smile} = \rho$, i.e. ρ is reflexive. Finally, by Prop. 6.16(4), we have $\rho^{\smile} = (\rho \square \rho^{\smile})^{\smile} = \rho \square \rho^{\smile} = \rho$.

(5): By (1) and (2), $\rho \rhd \rho^{\smile}$ is reflexive and transitive, whence also idempotent by Prop. 3.53(3). □

6.2 Some Further Kinds of Relations

This section presents introductory information on some special kinds of relations, some of their properties and how they can be used in various problem areas.

6.2.1 *Subsets, Vectors, and Rectangles*

6.2.1.1 *Vectors and Points*

In linear algebra vectors are usually denoted by lower case letters, and we will follow this habit; see e.g. [42] and [43, Section 2.4]. Besides these two books, from which most of the material of this section is taken, I have also used [5, 7] as a source.

Definition 6.25. ([42, 5.2.4 Definition]) *Let p, v be relations from X to Y.*

(a) *v is called a **vector** if $v = vU$;*
(b) *A bijective vector p is called a **point**, i.e. $U = pU$, $pp^{\smile} \subseteq I$, and $Up = U$.*

(In [43] a point is defined as an injective and non-empty vector. But using the Tarski rule Prop. 3.55(4) we get $p \neq \emptyset \leftrightarrow U = UpU = Up \leftrightarrow p$ is surjective, by Prop. 4.2(4). So the two variants coincide.)

The condition $v = vU$ means that v can be written as $v = Z \times Y$ with $Z \subseteq X$; v is said to *describe the subset Z of X* (or the respective predicate), since for all $x \in X, y \in Y$ we have $x \in Z \leftrightarrow \langle x, y\rangle \in v$. A point corresponds to a singleton subset, i.e. it has the form $p = \{x\} \times Y$, and we say that it *describes the element $x \in X$*. The example in Fig. 6.4

illustrates the case where $T = \{b, c\} \subseteq \{a, b, c, d\} = V$ is described by a vector τ, p is a point, and t is an abbreviation for τ in form of a Boolean column vector.

In the beginning it will maybe be instructive to make a difference between an element $p \in V$ and its representation as a point (relation) $\dot{p} \in \mathcal{B}_V$ to become accustomed to the new viewpoint without getting confused. However, we will not be consequent with this notation and will often abandon it without further comment, in case it is clear from the context if p denotes an element or a relation.

$$\tau = \begin{array}{c} \\ a \\ b \\ c \\ d \end{array} \begin{array}{c} \begin{array}{cccc} a & b & c & d \end{array} \\ \begin{bmatrix} 0 & 0 & 0 & 0 \\ 1 & 1 & 1 & 1 \\ 1 & 1 & 1 & 1 \\ 0 & 0 & 0 & 0 \end{bmatrix} \end{array}, \qquad \dot{p} = \begin{array}{c} \\ a \\ b \\ c \\ d \end{array} \begin{array}{c} \begin{array}{cccc} a & b & c & d \end{array} \\ \begin{bmatrix} 0 & 0 & 0 & 0 \\ 0 & 0 & 0 & 0 \\ 1 & 1 & 1 & 1 \\ 0 & 0 & 0 & 0 \end{bmatrix} \end{array}, \qquad t = \begin{bmatrix} 0 \\ 1 \\ 1 \\ 0 \end{bmatrix}$$

FIG. 6.4: Subsets, vectors, and points.

For vectors as a means to characterize subsets one would expect some properties as e.g. that complementation, intersection and union of vectors are again vectors. These and some more properties are shown in

Proposition 6.26. *Let $\rho \subseteq X \times Y$ be a relation, and let v, w be vectors such that all compositions are defined.*

(1) ρU is a vector.
(2) ρv is a vector.
(3) $v \cup w$, $v \cap w$ and $\overline{v}$ are vectors.
(4) $uv^{\smile} = uU \cap Uv^{\smile}$.
(5) v has constant rows.
(6) ρ has constant rows iff $\rho = \rho U$.

Proof. (1): U is an equivalence relation and thus idempotent by Prop. 4.18(2), and therefore $\rho U = \rho UU$.

(2): Follows directly from the definition of a vector: $v = vU \;\rightarrow\; \rho v = \rho v U$.

(3): (a) $v \cup w = vU \cup wU = (v \cup w)U$.

(b)
$$\begin{aligned} (v \cap w)U &\subseteq vU \cap wU && \text{Th. 3.16(4), subdistributivity} \\ &\subseteq (v \cap wUU^{\smile})(U \cap v^{\smile} wU) && \text{Th. 3.26(3)} \\ &= (v \cap w)(U \cap v^{\smile} wU) && \\ &\subseteq (v \cap w)U && \text{monotonicity} \end{aligned}$$

But $vU \cap wU = v \cap w$, and the assertion for intersection follows.
(c) Clearly, $\overline{v} = \overline{v}I \subseteq \overline{v}U$; v is a vector, so $v \subseteq vU$ which is equivalent to $\overline{v}U^{\smile} = \overline{v}U \subseteq \overline{v}$ by Th. 3.26((1))(b).

(4): From $u \subseteq U$ and $v^{\smile} \subseteq U$ one gets $uv^{\smile} \subseteq uU \cap Uv^{\smile}$. Using the Dedekind rule Th. 3.26(2), $u = uU$ and $U^{\smile} = U$ we get the other inclusion as follows:

$$\begin{aligned} uU \cap Uv^{\smile} &\subseteq (u \cap Uv^{\smile}U^{\smile}) \cdot (U \cap u^{\smile}Uv^{\smile}) \\ &\subseteq u \cdot u^{\smile}Uv^{\smile} = u \cdot U^{\smile}uv^{\smile} = uUvv^{\smile} = uvv^{\smile} \\ &\subseteq uUv^{\smile} \\ &= uv^{\smile}. \end{aligned}$$

(5): Let $v = X \times Y$ and $z \in xv \neq \emptyset$; then $xvzUy$ for all $y \in Y$. whence $xvU = Y$; and if $xv = \emptyset$, then also $xvU = \emptyset$. So $vU = v$ has constant rows.

(6): By Th. 3.16(9): $\rho\sigma = \bigcup_{y \in Y}(\rho y \times y\sigma)$. Furthermore, note that if ρ is row-constant then either $x\rho = \emptyset$ or $x\rho = Y$. Thus we have

$$\rho U = \bigcup_{y \in Y}(\rho y \times yU) = \bigcup_{y \in Y}(\rho y \times Y).$$

Therefore, assuming that ρ has constant rows, we get $x\rho = \emptyset \rightarrow x\rho U = \emptyset$, since $\neg\exists z \in Y[x\rho z]$, and $x\rho = Y \rightarrow x\rho U = Y$, since $\forall y \in Y[zUy]$, so that $\rho = \rho U$ follows. The converse direction holds by (1) and (5). □

Clearly, points are of the form $\{x\} \times Y$, i.e. they describe singleton sets: $\dot{p}\dot{p}^{\smile} \subseteq I$ implies $\forall x, y[\exists z[\langle x, z\rangle \in \dot{p} \wedge \langle z, y\rangle \in \dot{p}^{\smile}] \rightarrow \langle x, y\rangle \in I]$, or equivalently $\forall x, y[\exists z[\langle x, z\rangle \in \dot{p} \wedge \langle y, z\rangle \in \dot{p}] \rightarrow x = y]$.

Proposition 6.27. *Let $\dot{p}$ and $\dot{q}$ be points; then*

(1) $\rho \subseteq \sigma\dot{p} \leftrightarrow \rho\dot{p}^{\smile} \subseteq \sigma$;
(2) $\dot{q} \subseteq \sigma\dot{p} \leftrightarrow \dot{p} \subseteq \sigma^{\smile}\dot{q}$.

Proof. (1): Assume $\rho \subseteq \sigma p$; then $\rho p^{\smile} \subseteq \sigma p p^{\smile} \subseteq \sigma I = \sigma$. Conversely, note that from $\rho p^{\smile} \subseteq \sigma$ we get $\overline{\sigma}p \subseteq \overline{\rho}$ by the SCHRÖDER equivalence Th. 3.26(1)(b). Using Prop. 3.55(4) then gives $U = UpU = Up = \overline{\sigma}p \cup \sigma p \subseteq \overline{\rho} \cup \sigma p$, whence $\overline{(\overline{\rho} \cup \sigma p)} = \rho \cap \overline{\sigma p} = \emptyset$, so $\rho \subseteq \sigma p$.

(2): From (1) we get $q \subseteq \sigma p \leftrightarrow qp^{\smile} \subseteq \sigma$, so $pq^{\smile} \subseteq \sigma^{\smile}$. Applying (1) again we then get $p \subseteq \sigma^{\smile}q$. □

Proposition 6.28. *Let v be a vector, and p and q be points; then*

(1) $\emptyset \neq v \subseteq p \rightarrow p = v$, *i.e. every point is an atom among the vectors.*
(2) (i) $p \neq q \leftrightarrow p \subseteq \overline{q} \leftrightarrow pq^{\smile} \subseteq \overline{I} \leftrightarrow p^{\smile}q = \emptyset$;
 (ii) $p = q \leftrightarrow p \subseteq q \leftrightarrow pq^{\smile} \subseteq I \leftrightarrow p^{\smile}q = U$.

Proof. (1): By Prop. 3.55(4) we have $Uv = UvU = U$; so using the Dedekind rule Th. 3.26(3) and monotonicity we get $p = Uv \cap p \subseteq (U \cap pv^{\smile})(v \cap U^{\smile}p) \subseteq pv^{\smile}v$; and from the assumptions we have $v^{\smile} \subseteq p^{\smile}$ and $pp^{\smile} \subseteq I$, whence $pv^{\smile}v \subseteq pp^{\smile}v \subseteq v$. Thus $p \subseteq v$.

(2)(i): Assume $p \subset q$; then by (1) we would get the contradiction $p = q$; so $p \cap \overline{q} \neq \emptyset$ by Prop. 2.4(5). By Prop. 6.26(3) $p \cap \overline{q} \subseteq p$ is a vector, whence $p \cap \overline{q} = p$ by (1). Thus $p \subseteq \overline{q}$. Conversely, $p \subseteq \overline{q}$ clearly implies $p \neq q$. The second and third equivalence follow by application of the SCHRÖDER equivalences Th. 3.26(1)(c) and (1)(a) to $p = Ip \subseteq \overline{q}$ and $p = pU \subseteq \overline{q}$, respectively.
(ii): Note that $\overline{I}q = \overline{q}$: $qq^{\smile} \subseteq I$ is equivalent to $\overline{I}q \subseteq \overline{q}$ by Th. 3.26(1)(b). For the reverse inclusion we have, again using Th. 3.26(1)(b), $Iq \subseteq q \leftrightarrow \overline{q}q^{\smile} \subseteq \overline{I}$, and $\overline{q}q^{\smile} \subseteq \overline{I} \leftrightarrow \overline{q} \subseteq \overline{I}q$

by Prop. 6.27(1).
The first equivalence follows from (1). For the second equivalence we use $p \subseteq q \leftrightarrow \overline{q} \subseteq \overline{p}$, whence $\overline{I}q \subseteq \overline{p}$, and this is equivalent to $pq^{\smile} \subseteq I$ by Th. 3.26(1)(b). The third equivalence follows from $p = pU \subseteq q \leftrightarrow U^{\smile}p^{\smile} \subseteq q^{\smile}$ by application of Prop. 6.27(1) to the latter inclusion: $U^{\smile}p^{\smile} \subseteq q^{\smile} \leftrightarrow U^{\smile} \subseteq q^{\smile}p \leftrightarrow U \subseteq p^{\smile}q$. □

Proposition 6.29. *Let X be a set, and $p, q \in X$.*

(1) $\dot{p}\dot{q}^{\smile} = \{\langle p, q\rangle\}$ is an atom in $\mathcal{B}_X$.
(2) (Point axiom) *To every relation $\emptyset \neq \rho \in \mathcal{B}_X$ there are two points $\dot{p}, \dot{q} \in \mathcal{B}_X$ such that $\dot{p}\dot{q}^{\smile} \subseteq \rho$.*
(3) Every vector $v \neq \emptyset$ contains a point.

Proof. (1): Easy calculation, using the relation-algebraic representations as points according to Def. 6.25(b).

(2): From the assumption $\rho \neq \emptyset$ we see that there are elements $p, q \in X$ with $p\rho q$. By (1) it follows that $\dot{p}\dot{q}^{\smile} \subseteq \rho$.

(3): If $\emptyset \neq v = vU$ is a vector, then by (2) there are points p, q such that $pq^{\smile} \subseteq v$. By Prop. 6.28(2)(ii) we then get $p = pU = pq^{\smile}q \subseteq vq \subseteq vU = v$. □

(*Note:* Prop. 6.29(2) cannot be proved in a point-free relation algebraic way, whence it has to be requested as the so-called **point axiom** if it should hold for relation algebras. For more details see e.g. [42, Section 8.6] or [44].)

The result stated in the next proposition is not a new one: $x\rho\sigma y \leftrightarrow \exists z\,[x\rho z \wedge z\sigma y]$. What is new is its proof in a pure relation-algebraic fashion. We will use it later on in Section 17.1, where we analyze what kind of information on graphs can be expressed using certain relations, when proving Prop. 17.16.

Proposition 6.30. ([43, Proposition 2.4.8], Intermediate point theorem) *For points $\dot{p}, \dot{q}$ and suitably chosen relations ρ, σ we have*

$$\dot{p} \subseteq \rho\sigma\dot{q} \quad \textit{iff} \quad \textit{There is a point } \dot{r} \textit{ such that } \dot{p} \subseteq \rho\dot{r} \wedge \dot{r} \subseteq \sigma\dot{q}.$$

Proof. For the proof of the non-trivial direction first note that for two sets S, T we have $(S \cap \overline{T}) \cup (S \cap T) = S$. So, taking $S := \sigma q$ and $T := \overline{\rho}^{\smile} p$, we get

$$\rho\sigma q = \rho[(\sigma q \cap \overline{\overline{\rho}^{\smile} p}) \cup (\sigma q \cap \overline{\rho}^{\smile} p)] = \rho(\sigma q \cap \overline{\overline{\rho}^{\smile} p}) \cup \rho(\sigma q \cap \overline{\rho}^{\smile} p). \tag{6.1}$$

Since $pp^{\smile} \subseteq I$, we get $p(p^{\smile}\overline{\rho}) \subseteq \overline{\rho}$, and by the Schröder rule Th. 3.26(1)(b) the last expression is equivalent to $\rho(p^{\smile}\overline{\rho})^{\smile} \subseteq \overline{p}$, i.e. $\rho\overline{\rho}^{\smile}p \subseteq \overline{p}$. Thus $\rho\sigma q \cap \rho\overline{\rho}^{\smile}p \subseteq \overline{p}$, whence $\rho(\sigma q \cap \overline{\rho}^{\smile}p) \subseteq \overline{p}$ by Th. 3.16(4).

Now assuming $p \subseteq \rho\sigma q$ and writing τ for $\sigma q \cap \overline{\overline{\rho}^{\smile} p}$ we get $p \subseteq \rho\tau$ from (6.1). Since, as a point $p \neq \emptyset$, it follows that $\tau \neq \emptyset$. Since τ is a vector by Prop. 6.26, we can apply Prop. 6.29(3) and get some point $r \subseteq \tau$. Thus $r \subseteq \sigma q$, and also $r \subseteq \overline{\overline{\rho}^{\smile} p}$. The latter inclusion implies $\overline{\rho}^{\smile} p \subseteq \overline{r} \rightarrow (\overline{\rho}^{\smile} p)^{\smile} \subseteq \overline{r}^{\smile} \rightarrow p^{\smile}\overline{\rho} \subseteq \overline{r}^{\smile} = \overline{r^{\smile}}$, with the equality from Prop. 3.13(5). But the last inclusion is equivalent to $pr^{\smile} \subseteq \rho$ by Th. 3.26(1)(a), whence $p \subseteq \rho r$ by Prop. 6.27(1). □

6.2.1.2 *Describing Sets with Relations*

Relations can be used to characterize sets in various ways.The first representation of subsets uses *vectors*, which have been discussed in the preceding section where we have found the following situation: Let $T \subseteq V$, and define $\tau := \{\langle t, v\rangle : t \in T\} \subseteq V \times V$. This definition only restricts the first component of pairs of elements,which means that an element $t \in T$ is either in relation τ to none of the elements of V or to all elements of V. So τ is row-constant, i.e. $\tau = \tau U$ by Prop. 6.26(6). Thus, the Boolean representation M_τ has constant rows of **1**'s for all $x \in T$ and constant rows of **0**'s for all other $x \in V$. Apparently, there is a one-to-one correspondence between subsets $T \subseteq V$ and row-constant relations $\tau \in \mathcal{B}_V$. Such a row-constant matrix can be abbreviated by a vector equal to one of its identical columns.

(Note that from a formal point of view the range of vectors is irrelevant for describing sets. Therefore, we can consider a vector v as a relation $v \subseteq X \times 1$, with $1 = \{\mathsf{u}\}$ a specific singleton set. Then a vector v describes the subset $\{x \in X : xv\mathsf{u}\} \subseteq X$.)

Two other ways for set representation are *injective mappings* and the *set-theoretic membership relation*; our presentation mainly follows the one in [6, 7, 43].

Given an injective mapping $\iota : Y \longrightarrow X$, we may consider Y as a subset of X by identifying it with its image $\iota[Y]$. If $Y \subseteq X$ and $\iota = id_X$, then the vector $\iota^{\smile} U \subseteq X \times 1$ represents Y in the way described above. The transition in the other direction, i.e. the construction of a mapping $\mathsf{inj}(v) \subseteq Y \times X$ from a given non-empty vector $v \subseteq X \times 1$ describing Y such that

$$\mathsf{inj}(v) \text{ is injective} \quad \text{and} \quad \langle y, x\rangle \in \mathsf{inj}(v) \leftrightarrow y = x \tag{6.2}$$

is also possible by removing all pairs $\langle x, x\rangle$ from id_X with $x \notin Y$. Such a relation is called the *injective embedding generated by* v and can be axiomatized by the conditions

$$\mathsf{inj}(v) \text{ is an injective function} \quad \text{and} \quad v = \mathsf{inj}(v)^{\smile} U, \tag{6.3}$$

That (6.3) characterizes injective embeddings generated by vectors up to isomorphism is shown in

Theorem 6.31. ([8, Theorem 1]) *Let $v \subseteq X \times Y$ be a vector, and $\iota_1 \subseteq Z_1 \times X$ and $\iota_2 \subseteq Z_2 \times X$ be injective mappings with $\iota_i^{\smile} U = v$, for $i = 1, 2$. Then ι_1 and ι_2 are isomorphic via the pair (φ, I_X), where $\varphi := \iota_1 \iota_2^{\smile} \subseteq Z_1 \times Z_2$.*

Proof. We first show that φ is functional, using the hypotheses that ι_1 is functional and ι_2 is injective:

$$\varphi^{\smile} \varphi = (\iota_1 \iota_2^{\smile})^{\smile} \iota_1 \iota_2^{\smile} = \iota_2 \iota_1^{\smile} \iota_1 \iota_2^{\smile} \subseteq \iota_2 I_X \iota_2^{\smile} \subseteq I_X.$$

Concerning the totality of φ we use $\iota_1^{\smile} U = v$, $\iota_2^{\smile} U = v$, and $I \subseteq \iota_1 \iota_1^{\smile}$ to get

$$\varphi U = \iota_1 \iota_2^{\smile} U = \iota_1 v = \iota_1 \iota_1^{\smile} U \supset IU = U$$

by Prop. 4.2(2). So we have shown φ to be a mapping. Similarly we get the same result for $\varphi^{\smile}$.

I_X obviously is bijective. Furthermore, since ι_2 is functional, $\iota_1 I_X \supset \iota_1 \iota_2^{\smile} \iota_2 = \varphi \iota_2$. Letting $\sigma := \iota_1 I_X$ and $\rho := \varphi \iota_2$ we have thus shown $\rho \subseteq \sigma$. Applying Cor. 4.3(5), we get that both

ρ and σ are mappings, whence $\rho U = \sigma U = U$ by Prop. 4.2(2). Thus,

$$\sigma = \rho U \cap \sigma \subseteq (\rho \cap \sigma U^{\smile})(U \cap \rho^{\smile}\sigma) \subseteq \rho\rho^{\smile}\sigma \subseteq \rho\sigma^{\smile}\sigma \subseteq \rho,$$

using in that order the Dedekind rule Th. 3.26(3), monotonicity, $\rho^{\smile} \subseteq \sigma^{\smile}$, and the functionality of σ. So $\rho = \sigma$, i.e. $\iota_1 = \varphi\iota_2$. Since both ρ and $\rho^{\smile}$ are mapping as shown above, it follows by Prop. 4.2(6) that they are bijective. □

A third possibility to deal with subsets of a given set X is to work with a relation-level equivalent $\epsilon \subseteq X \times \mathbf{2}^X$ of the meta-level set-theoretic membership relation $\in$, defined as $x\epsilon Y \iff x \in Y, Y \subseteq X$; see [4, 5, 7]. In matrix terminology: each column of ϵ represents a subset of X. ϵ can be characterized by relation-algebraic means as follows:

Definition 6.32. *A relation $\sigma \subseteq X \times \mathbf{2}^X$ is called a* **powerset relation** *if*

(PS1) $\mathsf{syQ}(\sigma,\sigma) \subseteq I$;

(PS2) $U \cdot \mathsf{syQ}(\sigma,\rho) = U$, *i.e.* $\mathsf{syQ}(\sigma,\rho)$ *is surjective, for all relations* ρ.

First note that the powerset relation is uniquely characterized up to isomorphism: If $\sigma' \subseteq Y \times \mathbf{2}^Y$ is another powerset relation, $\phi : X \longrightarrow Y$ is a bijective function, and if one defines the bijective function $\psi : \mathbf{2}^X \longrightarrow \mathbf{2}^Y$ by $\psi := \mathsf{syQ}(\sigma,\sigma')$, then (ϕ,ψ) is an isomorphism between σ and σ'.

We now verify that ϵ is indeed a powerset relation, i.e. that it satisfies the two conditions of Def. 6.32: For (PS1) this is shown by the following calculation:

$$\begin{aligned}
\langle X,Y\rangle \in \mathsf{syQ}(\epsilon,\epsilon) &\iff \epsilon X = \epsilon Y && \text{by Cor. 3.39(2)}\\
&\iff \forall x[x \in X \leftrightarrow x \in Y]\\
&\iff X = Y\\
&\iff X\, I_{\mathbf{2}^X}\, Y.
\end{aligned}$$

And for (PS2), we know from Prop. 3.40(3) that $\mathsf{syQ}(\epsilon,\epsilon)$ is reflexive, whence by Prop. 3.55(3) we get $U \cdot \mathsf{syQ}(\epsilon,\epsilon) = U$, i.e. $\mathsf{syQ}(\epsilon,\epsilon)$ is surjective. Applied to ϵ, Def. 6.32 has the following meaning:

- (PS1) describes the equality of sets, and thus corresponds to the extensionality axiom of set theory.
- And (PS2) says that each vector v has a corresponding point in the powerset: Recall from above that vectors can be considered as a relation $v \subseteq S \times 1$, with $1 = \{\mathsf{u}\}$. Then, if v is a vector describing a subset $T \subseteq S$, we get $\mathsf{syQ}(\epsilon, v) \subseteq \mathbf{2}^S \times 1$. Moreover, $\langle X,\mathsf{u}\rangle \in \mathsf{syQ}(\epsilon, v) \leftrightarrow \forall x[x \in X \leftrightarrow xv\mathsf{u}] \leftrightarrow X = T$, for all $X \in \mathbf{2}^S$. By (PS2) $\mathsf{syQ}(\epsilon, v)$ is surjective; that it is also injective follows from (PS1) and Prop. 3.42 since $\mathsf{syQ}(\epsilon, v) \cdot \mathsf{syQ}(\epsilon, v)^{\smile} = \mathsf{syQ}(\epsilon, v) \cdot \mathsf{syQ}(v,\epsilon) \subseteq \mathsf{syQ}(\epsilon,\epsilon) = I$. So $\mathsf{syQ}(\epsilon, v)$ is the point describing $T \subseteq S$ as member of $\mathbf{2}^S$.

We conclude this section with a few examples of how injective mappings and the relation-level equivalent ϵ of the set-theoretic membership relation $\in$ can be used to represent subsets; for more details see e.g. [5, 7] or [6, Section 2.2]. (To keep things simple, note that we overload the symbol $\emptyset$ and let it denote both the empty set and the empty relation, on both the meta and object level.)

- Let $\mathcal{S} \subseteq \mathbf{2}^S$ be represented by a vector $v \subseteq \mathbf{2}^S \times 1$. Then, from (6.2) and (6.3) we get that the elements of $\mathcal{S}$ are just the columns of $\epsilon \cdot \mathsf{inj}(v)^{\smile} \subseteq X \times \mathcal{S}$.
- The empty set: By what has just been said above, if v is a vector describing a subset $T \subseteq S$, then $\mathsf{syQ}(\epsilon, v)$ is the point describing $T \subseteq S$ as member of $\mathbf{2}^S$. So $\mathsf{syQ}(\epsilon, \emptyset) = (\emptyset / \epsilon) \cap (\epsilon^{\smile} \backslash \emptyset) = \emptyset / \epsilon$, since $\epsilon^{\smile} \backslash \emptyset = \overline{\overline{\epsilon^{\smile}} \emptyset} = U$.
- Complementation, $\langle X, Y \rangle \in \mathsf{C}$ iff $\overline{X} = Y$: $\mathsf{C} = \mathsf{syQ}(\epsilon, \overline{\epsilon})$. By Cor. 3.39(2) we have $\langle y, z \rangle \in \mathsf{syQ}(\rho, \sigma)$ iff $\rho y = \sigma z$, whence

$$\begin{aligned}\langle X, Y \rangle \in \mathsf{syQ}(\epsilon, \overline{\epsilon}) &\iff \forall x [x \epsilon X \leftrightarrow x \overline{\epsilon} Y] \\ &\iff \forall x [x \in X \leftrightarrow x \notin Y] \\ &\iff \overline{X} = Y.\end{aligned}$$

- Set inclusion $\sqsubseteq \; = \epsilon / \epsilon$: By Prop. 3.34(3), $x(\sigma / \rho) y \leftrightarrow \forall z [z \rho x \rightarrow z \sigma y]$; so

$$\begin{aligned}X \sqsubseteq Y &\iff \forall x [x \epsilon X \rightarrow x \epsilon Y] \\ &\iff \forall x [x \in X \rightarrow x \in Y] \\ &\iff X \subseteq Y.\end{aligned}$$

6.2.1.3 *Rectangles and Fringes*

Let $\rho \subseteq S \times T$ be a relation, and $A \subseteq S$ and $B \subseteq T$ be two non-empty sets. A relation $\sigma = A \times B \subseteq \rho$ is called a **rectangle of** ρ. A rectangle $\sigma \subseteq \rho$ is called **non-enlargeable**, also **maximal** or a **formal concept**, if $\sigma \subseteq A' \times B' \subseteq \rho$ implies $A = A'$ and $B = B'$. (For more information on formal concepts see Chapter 14.)

Proposition 6.33. ([26, Proposition 1.13]) *$\sigma = A \times B \subseteq \rho \subseteq S \times T$ is a rectangle of ρ iff $x \sigma y \wedge s \sigma t \rightarrow x \sigma t$.*

Proof. Assume that $\sigma \subseteq A \times B$ is a rectangle , and $x \sigma y \wedge s \sigma t$; then clearly $\langle x, t \rangle \in A \times B = \sigma$. Conversely, assume $x \sigma y \wedge s \sigma t \rightarrow x \sigma t$, and define $A = \sigma Y$ and $B = X \sigma$. Clearly, $\sigma \subseteq A \times B$ from the definition of A and B. And if $\langle x, y \rangle \in A \times B$, then $x \in A = \sigma Y$, so $x \sigma y'$ for some $y' \in Y$. Similarly, $x' \sigma y$ for some $x' \in X$. Thus, applying the assumption, we get $x \sigma y$, so $A \times B \subseteq \sigma$. □

Rectangles are not only interesting in Formal Concept Analysis; just note that an equivalence relation can be represented by a block diagonal Boolean matrix (see Def. 5.3) with all the blocks (which stand for the equivalence classes) being mutually disjoint and square. This immediately raises the question for a relation algebraic characterization of relations made up from disjoint rectangles. Before we turn to it in Section 6.2.2.1 we will dwell on general properties of rectangles. Remember that vectors have been shown to be suitable representations for subsets in Prop. 6.26(3). So it is nearby to represent the Cartesian product of two sets by a product of two vectors as given in the following

$$\begin{bmatrix} 0 & 1 & 1 & 1 & 0 & 0 & 1 & 1 & 0 \end{bmatrix} = v^{\smile}$$

$$u = \begin{bmatrix} 0 \\ 1 \\ 1 \\ 0 \\ 0 \\ 1 \\ 0 \end{bmatrix} \quad \begin{bmatrix} 0 & 1 & 1 & 1 & 0 & 0 & 1 & 1 & 0 \\ 1 & 1 & 1 & 1 & 1 & 1 & 1 & 1 & 1 \\ 1 & 1 & 1 & 1 & 1 & 1 & 1 & 1 & 1 \\ 0 & 1 & 1 & 1 & 0 & 0 & 1 & 1 & 0 \\ 0 & 1 & 1 & 1 & 0 & 0 & 1 & 1 & 0 \\ 1 & 1 & 1 & 1 & 1 & 1 & 1 & 1 & 1 \\ 0 & 1 & 1 & 1 & 0 & 0 & 1 & 1 & 0 \end{bmatrix} = uv^{\smile}$$

(a) vectors

$$\begin{bmatrix} 1 & 1 & 1 & 1 & 1 & 0 & 0 & 0 & 0 \end{bmatrix} = \tilde{v}^{\smile}$$

$$\tilde{u} = \begin{bmatrix} 1 \\ 1 \\ 1 \\ 0 \\ 0 \\ 0 \\ 0 \end{bmatrix} \quad \begin{bmatrix} 1 & 1 & 1 & 1 & 1 & 1 & 1 & 1 & 1 \\ 1 & 1 & 1 & 1 & 1 & 1 & 1 & 1 & 1 \\ 1 & 1 & 1 & 1 & 1 & 1 & 1 & 1 & 1 \\ 1 & 1 & 1 & 1 & 1 & 0 & 0 & 0 & 0 \\ 1 & 1 & 1 & 1 & 1 & 0 & 0 & 0 & 0 \\ 1 & 1 & 1 & 1 & 1 & 0 & 0 & 0 & 0 \\ 1 & 1 & 1 & 1 & 1 & 0 & 0 & 0 & 0 \end{bmatrix} = \tilde{u}\tilde{v}^{\smile}$$

(b) reordered vectors

FIG. 6.5: A rectangular relation.

Definition 6.34. ([42, Definition 6.2.1]) *Given vectors $u \subseteq S$ and $v \subseteq T$, the relation $\rho := uv^{\smile}$ is called* **rectangular** *or simply a* **rectangle***, and u and v are called the* **source** *and* **target vector** *of ρ, respectively.*

Corollary 6.35. *Let $\rho := uv^{\smile}$ be a rectangular relation. Then $uv^{\smile} = uU \cap (vU)^{\smile}$.*

Proof. By Prop. 6.26(4). □

Although this corollary might seem a mere repetition, it is a change of view that should be emphasized: More precisely, let $u \subseteq X \times Z$ and $v \subseteq Y \times Z$ be two row-constant relations, i.e. $u = uU_Z$ and $v = vU_Z$. Then, by Cor. 6.35, the rectangular relation $uv^{\smile}$ arises as the intersection of a row-constant with a column-constant relation as $uv^{\smile} = uU_{Z,Y} \cap (vU_{Z,X})^{\smile}$, illustrated in Fig. 6.5: First note that, if a vector u is interpreted as a subset S of an underlying base set B, then each reordering $\tilde{u}$ of u again represents S (where implicitly the identical reordering has to be supposed for the base set B). So we can always get a pattern with all the **1**'s on consecutive positions in $\tilde{u}$. One out of five possible reorderings of v with this property is shown in part (b).

In order to properly comprehend why a vector product $uv^{\smile}$ can simply be called a *rectangle* you have to read the matrix shown for $\tilde{u}\tilde{v}^{\smile}$ as a three layer representation, where **1**'s in a higher layer hide 0's in a layer below: Its rows represent uU and the columns $(vU)^{\smile}$, and the highlighted rectangle stands, of course, for the Cartesian product of $S \times T$ (no 0's in a hidden layer), i.e. it is what has been called a *rectangle of* $uv^{\smile}$ just at the beginning of this section. And if you view a vector as a subset $S \subseteq B$, then its 0's symbolize that the corresponding elements of the base set B are not member in S, and thus they cannot contribute to a pair $\langle s, t\rangle$ in $S \times T$, i.e. *only* the highlighted **1**'s in Fig. 6.5 actually represent pairs $\langle s, t\rangle \in S \times T$. So, nomenclature may be felt a little bit confusing.

Proposition 6.36. ([42, 6.2.2 Proposition]) *For a binary relation $\rho \subseteq X \times Y$ the following conditions are equivalent:*

(1) ρ is rectangular.

(2) $\rho U \rho \subseteq \rho$.
(3) $\rho U \rho = \rho$.
(4) $\rho \rho^d \rho = \emptyset$
(5) *For any fitting pair of relations α, β, the Dedekind rule becomes an equality:* $\alpha\beta \cap \rho = (\alpha \cap \rho\beta^{\smile})(\beta \cap \alpha^{\smile}\rho)$

Proof. (1) $\Rightarrow$ (2): By assumption there exist two vectors u, v such that $uv^{\smile} = \rho$. Thus, $\rho U \rho = uv^{\smile} U uv^{\smile} \subseteq uUUUv^{\smile} = uv^{\smile} = \rho$.

(2) $\iff$ (3): The Dedekind rule Th. 3.26(2) gives $\rho = \rho I \cap \rho \subseteq (\rho \cap \rho I)(I \cap \rho^{\smile}\rho) \subseteq \rho\rho^{\smile}\rho \subseteq \rho U \rho \subseteq \rho$, with the last inequality following from (2). The direction (3) $\Rightarrow$ (2) is obvious.

(2) $\iff$ (4): By twice applying the Schröder equivalences Th. 3.26(1)(a) and (b) in that order, we get $\rho U \rho \subseteq \rho \leftrightarrow U\rho^{\smile}\overline{\rho} \subseteq \overline{\rho} \leftrightarrow \rho\overline{\rho}^{\smile}\rho \subseteq \emptyset$.

(2) $\Rightarrow$ (5): Since '$\subseteq$'-direction is the content of the Dedekind rule, we only have to show the reverse inequality. Obviously, the right hand side is contained in $\alpha\beta$. Moreover, $(\alpha \cap \rho\beta^{\smile})(\beta \cap \alpha^{\smile}\rho) \subseteq \rho\beta^{\smile}\alpha^{\smile}\rho \subseteq \rho U \rho \subseteq \rho$ by monotonicity and (2).

(5) $\Rightarrow$ (1): We take $\alpha := U_{X,Y}$ and $\beta := U_Y$. Then we get $\alpha\beta := U_{X,Y}$ etc., so

$$\rho = \alpha\beta \cap \rho = (\alpha \cap \rho\beta^{\smile})(\beta \cap \alpha^{\smile}\rho) = \rho\beta^{\smile}\alpha^{\smile}\rho = \rho U_Y (\rho^{\smile} U_{X,Y})^{\smile}.$$

Using our simplified notation, by Prop. 6.26(1) γU is a vector for any γ, whence we have shown ρ to be rectangular with vectors ρU and $\rho^{\smile} U$. □

Rectangles are often studied with reference to a given relation [41, 42]. In an order, e.g., we have that all $u < e$ are related via $<$ to all $v > e$, and, as already mentioned above, an equivalence has squares along the diagonal, i.e. it can be transformed into the so-called block-diagonal form. So we continue with

Definition 6.37. ([42, p. 95, 6.2.3]) *Let ρ be a relation; then we say that two vectors u, v constitute a*

(a) **rectangle inside** ρ $:\iff uv^{\smile} \subseteq \rho \iff \overline{\rho} v \subseteq \overline{u} \iff \overline{\rho}^{\smile} u \subseteq \overline{v}$;
(b) **rectangle containing** ρ $:\iff \rho \subseteq uv^{\smile}$;
(c) **rectangle outside** ρ $:\iff$ *u, v constitute a rectangle inside* $\overline{\rho}$.

From these three types the rectangles inside a relation are most prominent, and among them the maximal ones are especially interesting. This property has the following algebraic characterization:

Proposition 6.38. ([41, 3.3 Proposition], [42, 10.1.2 Proposition])

(1) *Let u, v be two vectors defining a rectangle inside the relation ρ. Then there is no strictly greater rectangle u', v' inside ρ iff*

$$\text{(i)}\quad \overline{u} \subseteq \overline{\rho} v \qquad \text{and} \qquad \text{(ii)}\quad \overline{v} \subseteq \overline{\rho}^{\smile} u. \tag{6.4}$$

(2) *Vectors constitute a maximal rectangle inside ρ iff*

$$\overline{u} = \overline{\rho} v \wedge \overline{v} = \overline{\rho}^{\smile} u \quad \text{iff} \quad u = \rho \backslash v^{\smile} \wedge v^{\smile} = \rho / u.$$

Proof. (1): Assume that the rectangle $uv^{\smile}$ does not satisfy the first inclusion, i.e. $\overline{\rho}v \subset \overline{u}$. Then there exists a point $p \subseteq \overline{u} \cap \overline{\overline{\rho}v}$, and $u' := u \cup p \supset u$ and $v' := v$ is a strictly greater rectangle since $pv^{\smile} \subseteq \rho$.
Conversely, assume that u, v satisfy the two additional conditions (6.4), and that u', v' is another rectangle inside ρ with $u \subseteq u'$ and $v \subseteq v'$. Then, from (6.4)(ii) we have $\overline{v} \subseteq \overline{\rho}^{\smile}u$, and from Def. 6.37 $\overline{\rho}^{\smile}u' \subseteq \overline{v'}$, whence by monotonicity $\overline{v} \subseteq \overline{v'}$. So, $v' \subseteq v$, which together with the hypothesis $v \subseteq v'$ gives $v = v'$. A similar argument produces $u = u'$; therefore, the rectangle u', v' cannot be strictly greater than u, v.

(2): By Def. 6.37, $\overline{\rho}v \subseteq \overline{u}$ and $\overline{\rho}^{\smile}u \subseteq \overline{v}$, and by (1) rectangle u, v is maximal iff $\overline{u} \subseteq \overline{\rho}v$ and $\overline{v} \subseteq \overline{\rho}^{\smile}u$. Recalling Def. 3.33 we see that $\overline{u} = \overline{\rho}v \leftrightarrow u = \overline{\overline{\rho}v} = \rho\backslash v^{\smile}$, and $\overline{v} = \overline{\rho}^{\smile}u \leftrightarrow \overline{v}^{\smile} = \overline{v^{\smile}} = u^{\smile}\overline{\rho} \leftrightarrow v^{\smile} = \overline{u^{\smile}\overline{\rho}} = \rho / u$. □

The notion of a **fringe** of a relation has been introduced in [41] (the concept without a name is already mentioned in [51], denoted as R^d): For a relation ρ we define $\mathsf{fringe}(\rho) := \rho \cap \overline{\rho\rho^d}\rho$. Obviously, $\mathsf{fringe}(\rho^{\smile}) = \mathsf{fringe}(\rho)^{\smile}$.

Proposition 6.39. ([23, Proposition 1]) *Let ρ be an arbitrary binary relation; the greatest relation $\chi \subseteq \rho$ verifying $\rho\chi^{\smile}\rho \subseteq \rho$ is $\chi = \mathsf{fringe}(\rho)$.*

Proof. Assume that $\chi \subseteq \rho \wedge \rho\chi^{\smile}\rho \subseteq \rho$; this implies $\chi \subseteq \rho \wedge \chi \subseteq \overline{\rho\rho^d\rho}$ by Cor. 3.48(2). So we have shown that every relation $\chi \subseteq \rho$ satisfying $\rho\chi^{\smile}\rho \subseteq \rho$ is smaller than $\mathsf{fringe}(\rho)$. It remains to show that $\mathsf{fringe}(\rho)$ itself satisfies the two conditions: Obviously, $\mathsf{fringe}(\rho) \subseteq \rho$ and $\mathsf{fringe}(\rho) \subseteq \overline{\rho\rho^d\rho}$ by definition; and from the second inequality we get $\rho \cdot \mathsf{fringe}(\rho)^{\smile} \cdot \rho \subseteq \rho$, again using Cor. 3.48(2). □

Some remarkable properties of partial orders are listed in:

Lemma 6.40. ([9, Lemma 3.1], [41, p. 7]) *Let ρ be a partial order; then*

(1) $\rho\backslash\rho = \rho/\rho = \rho$
(2) $\mathsf{biR}(\rho) = \rho$
(3) $\mathsf{fringe}(\rho) = I$.

Proof. (1): From transitivity $\rho\rho \subseteq \rho$ we get $\rho \subseteq \rho\backslash\rho$ by Prop. 3.34(2). Since $I \subseteq \rho$, for the converse inclusion we have $\rho\backslash\rho \subseteq \rho\backslash I = \rho$ using Prop. 3.36(1), (3).

(2): We have $\mathsf{biR}(\rho) = \rho/(\rho\backslash\rho)$ from Cor. 3.48(1), and applying (1) yields the assertion.

(3): $\mathsf{fringe}(\rho) = \rho \cap \overline{\rho\rho^d}\rho = \rho \cap \mathsf{biR}(\rho^{\smile})$ by Cor. 3.48(1). By Prop. 4.11(1) $\rho^{\smile}$ is a partial order, so we can apply (2) and get $\mathsf{fringe}(\rho) = \rho \cap \rho^{\smile} = I$. □

6.2.2 *Inverses and Regular Relations*

In analogy to the standard classification of matrix theory using the Moore-Penrose equations (see Def. 6.41(f) below), the following kinds of inverses are of interest in the context of binary relations:

Definition 6.41. *Let α be a given relation; a relation γ is called*

(a) *a* **left** *(***right***)* **inverse** *of α, and α is called* left (right) invertible, *if*
$\gamma\alpha = I\ (\alpha\gamma = I)$;
(b) *an* **inverse** *of α, denoted as α^{-1}, and α is called* invertible, *if*
$\gamma\alpha = I = \alpha\gamma$;
(c) *a* **subinverse** *of α if*
$\alpha\gamma\alpha \subseteq \alpha$;
(d) *a* **generalized inverse** *(or g-***inverse***) of α if*
$\alpha\gamma\alpha = \alpha$;
(e) *a* **pseudoinverse** *of α (sometimes also* **Vagner inverse***) if*
$\alpha\gamma\alpha = \alpha,\ \gamma\alpha\gamma = \gamma$;
(f) *a* **Moore-Penrose inverse** *of α if*
$\alpha\gamma\alpha = \alpha,\ \gamma\alpha\gamma = \gamma,\ (\alpha\gamma)^{\smallsmile} = \alpha\gamma,\ (\gamma\alpha)^{\smallsmile} = \gamma\alpha.$

In the following, let

$$\widetilde{\alpha} := \overline{\alpha^{\smallsmile}\,\overline{\alpha}\alpha^{\smallsmile}} = \overline{\alpha\,\overline{\alpha}^{\smallsmile}\alpha}^{\smallsmile}. \tag{6.5}$$

Clearly, subinverses do always exist, since $\emptyset = \rho\emptyset\rho \subseteq \rho$ for every relation ρ; and since the union of two subinverses is again a subinverse, one may ask for the greatest subinverse of a relation. This question is answered in

Proposition 6.42. ([40, 41]) *$\widetilde{\alpha}$ is the greatest subinverse of α.*

Proof. Let χ be a subinverse of α; then $\alpha\chi\alpha \subseteq \alpha$. Using Prop. 3.13(5) and the SCHRÖDER equivalences Th. 3.26(1) we get $\alpha\chi\alpha \subseteq \alpha \ \leftrightarrow\ (\alpha\chi)^{\smallsmile}\overline{\alpha} = \chi^{\smallsmile}(\alpha^{\smallsmile}\overline{\alpha}) \subseteq \overline{\alpha} \ \leftrightarrow\ \alpha(\alpha^{\smallsmile}\overline{\alpha})^{\smallsmile} = \alpha\overline{\alpha}^{\smallsmile}\alpha \subseteq \overline{\chi^{\smallsmile}}$. But the last inequality is equivalent to $\chi^{\smallsmile} \subseteq \overline{\alpha\overline{\alpha}^{\smallsmile}\alpha}$, whence $\chi \subseteq \overline{\alpha\overline{\alpha}^{\smallsmile}\alpha}^{\smallsmile}$ by Prop. 3.13(6). □

A relation ρ is called **regular**, if it has a generalized inverse; recall Prop. 3.50 and Def. 9.1. g-inverses are not uniquely determined: as an example take U_X, for which both I_X and U_X are generalized inverses. Moreover, the union of two g-inverses is again a g-inverse; a regular relation thus has a greatest one. We will meet regular relations (and elements) again in Chapter 9.

Proposition 6.43. ([40, 41])

(1) *α is regular iff $\alpha \subseteq \alpha\widetilde{\alpha}\alpha$.*
(2) *If α is regular, its maximum pseudoinverse is $\widetilde{\alpha}\alpha\widetilde{\alpha}$.*
(3) *Moore-Penrose inverses are uniquely determined if they exist.*

Proof. (1): If α is regular then there is a relation χ such that $\alpha\chi\alpha = \alpha$, whence χ is a subinverse and $\chi \subseteq \widetilde{\alpha}$ by Prop. 6.42. Then $\alpha = \alpha\chi\alpha \subseteq \alpha\widetilde{\alpha}\alpha$. Conversely, assume $\alpha \subseteq \alpha\widetilde{\alpha}\alpha$; since $\widetilde{\alpha}$ is a subinverse of α, we have $\alpha\widetilde{\alpha}\alpha \subseteq \alpha$ by definition, whence equality follows, and α is regular.
(2): $\alpha' := \widetilde{\alpha}\alpha\widetilde{\alpha}$ is a pseudoinverse of α: $\alpha\alpha'\alpha = \alpha$ and $\alpha'\alpha\alpha' = \alpha'$ can easily be shown

using (1). Let χ be a pseudoinverse of α; then $\chi = \chi\alpha\chi \subseteq \tilde{\alpha}\alpha\tilde{\alpha} = \alpha'$, since $\chi \subseteq \tilde{\alpha}\alpha\tilde{\alpha}$ by Prop. 6.42.

(3): Let χ and γ be two Moore-Penrose inverses of α. Then, using Def. 6.41(4), we have $\gamma = \gamma\alpha\gamma = \gamma(\alpha\gamma)^{\smile} = \gamma\gamma^{\smile}\alpha^{\smile} = \gamma\gamma^{\smile}(\alpha\chi\alpha)^{\smile} = \gamma\gamma^{\smile}\alpha^{\smile}\chi^{\smile}\alpha^{\smile} = \gamma\gamma^{\smile}\alpha^{\smile}\alpha\chi = \gamma\alpha\gamma\alpha\chi = \gamma\alpha\chi = \gamma\alpha\chi\alpha\chi = \gamma\alpha\alpha^{\smile}\chi^{\smile}\chi = \alpha^{\smile}\gamma^{\smile}\alpha^{\smile}\chi^{\smile}\chi = \alpha^{\smile}\chi^{\smile}\chi = \chi\alpha\chi = \chi$. □

Anticipating Def. 7.1(a), an *ideal* of a poset $(A, \leq)$ is a subset $\emptyset \neq D \subseteq A$ such that for $y \in A$ we have $x \in D \wedge y \leq x$ implies $y \in D$. If L is a complete lattice and $a \in L$, let $K(a) := \bigcap\{D : D \text{ ideal in } L \wedge a \leq \bigvee D\}$. Then the following holds:

Proposition 6.44. ([34, Theorem 1]) *Let L be a complete lattice. L is completely distributive iff $a = \bigvee K(a)$, for every $a \in L$.*

Regularity in $\boldsymbol{B}_n$ was first investigated by Zaretskiĭ [52], who showed the remarkable result of Th. 6.46 below, the proof of which uses

Lemma 6.45. ([1], [28, p. 495], [35, proof of Theorem 4]) *A complete lattice L is completely distributive iff $x = \bigvee\{a \in L : x \not\leq \bigvee_{b \in L,\, a \not\leq b} b\}$, for all $x \in L$.*

Proof. Let $i(a) := \{b \in L : a \not\leq b\}$; then, for a non-zero $a \in L$, $i(a)$ is an ideal in L with $a \notin i(a)$. Furthermore, we set $a^* := \bigvee i(a)$; so, due to Prop. 6.44, we have to show $x = \bigvee K(x) = \bigvee\{a \in L : x \not\leq a^*\}$.

Following [35] we give an indirect proof that for $x \in L$, $a \in K(x)$ iff $x \not\leq a^*$: Let $x \leq a^*$; then by definition of a^* we have $x \leq \bigvee i(a) = \bigvee\{b \in L : a \not\leq b\}$. Since $i(a)$ is an ideal in L not containing a, a cannot be in $K(x)$. Therefore, $a \in K(x)$ implies $x \not\leq a^*$. Conversely, if $a \notin K(x)$, then, by definition of $K(x)$, there is an ideal i_a such that $x \leq \bigvee i_a$ and $a \notin i_a$. Then $i_a \subseteq \{b \in L : a \not\leq b\} = i(a)$, whence $\bigvee i_a \leq \bigvee i(a) = a^*$. Therefore, $x \leq a^*$, so $x \not\leq a^*$ implies $a \in K(x)$. □

For $\alpha \in \mathcal{B}_X$ we call $V(\alpha) := \{A\alpha : A \subseteq X\}$ the **row space** of α. Then $V(\alpha) \subseteq V_n$ is the collection consisting of the zero vector $\mathbf{0}$ and all possible unions of (one or more) non-zero rows of α. For every $\alpha \in \boldsymbol{B}_n$, $V(\alpha)$ is a complete lattice with respect to inclusion of Boolean vectors where joins are set-theoretic unions (see [1, p. 123]).

Theorem 6.46. *A relation $\alpha \in \mathcal{B}_X$ is regular iff $V(\alpha)$ forms a completely distributive lattice under inclusion.*

Proof. (The detailed proof given here follows [1], but note that relation composition $\rho \circ \sigma$ used in this paper is our $\sigma\rho$).

Given three relations ρ, σ, τ we have

$$\begin{aligned} x(\rho\sigma\tau)y &\iff \exists s, t \in X\,[x\rho s\sigma t\tau y] \\ &\iff \exists s, t \in X\,[s \in x\rho \wedge s \in t\sigma^{\smile} \wedge y \in t\tau] \\ &\iff \exists t \in X\,[x\rho \not\subseteq \overline{t\sigma^{\smile}} \wedge y \in t\tau]. \end{aligned}$$

Thus $x(\rho\sigma\tau) = \bigcup_{t\in X}\{t\tau : x\rho \not\subseteq \overline{t\sigma^{\smallsmile}}\}$, $x \in X$, whence

$$A(\rho\sigma\tau) := (\rho\sigma\tau)[A] = \bigcup_{B\subseteq X} \{\tau[B] : \rho[A] \not\subseteq \overline{\sigma^{\smallsmile}}[B]\}, \quad A \subseteq X. \tag{6.6}$$

Applying (6.6) to $\rho\tilde{\rho}\rho$ gives

$$\rho\tilde{\rho}\rho[A] = \bigcup_{B\subseteq X} \{\rho[B] : \rho[A] \not\subseteq \overline{\tilde{\rho}^{\smallsmile}}[B]\}, \quad A \subseteq X. \tag{6.7}$$

But using (6.5) we get $\overline{\tilde{\rho}^{\smallsmile}} = \bar{\tilde{\rho}}^{\smallsmile} = \left(\overline{\overline{\rho^{\smallsmile}\overline{\rho}\rho^{\smallsmile}}}\right)^{\smallsmile} = (\rho^{\smallsmile}\overline{\rho}\rho^{\smallsmile})^{\smallsmile} = \rho\overline{\rho}^{\smallsmile}\rho$, which, by applying (6.6) to it, yields

$$\overline{\tilde{\rho}^{\smallsmile}}[B] = \rho\overline{\rho}^{\smallsmile}\rho[B] = \bigcup_{C\subseteq X} \{\rho[C] : \rho[B] \not\subseteq \overline{\overline{\rho}^{\smallsmile\smallsmile}}[C]\}, \quad B \subseteq X. \tag{6.8}$$

Further note that $\overline{\overline{\rho}^{\smallsmile\smallsmile}} = \rho$ and that $\rho\tilde{\rho}\rho = \rho$ if ρ is regular (see proof of Prop. 6.43(1)); so from (6.7) we get by substitution of (6.8)

$$\rho[A] = \bigcup_{B\subseteq X} \Big\{\rho[B] : \rho[A] \not\subseteq \bigcup_{C\subseteq X} \{\rho[C] : \rho[B] \not\subseteq \rho[C]\}\Big\}, \quad A \subseteq X. \tag{6.9}$$

Now apply Lemma 6.45 with $x := \rho[A]$, $a := \rho[B]$, $b := \rho[C]$, and $L := V(\rho)$ to see that $V(\rho)$ is a completely distributive lattice if ρ is regular (noting that $S \subseteq X$ means $\rho[S] \in V(\rho)$).

Conversely, if $V(\rho)$ is a completely distributive lattice then (6.9) holds by Lemma 6.45. Using (6.7) and (6.8), and since $\overline{\tilde{\rho}^{\smallsmile}} = \rho\overline{\rho}^{\smallsmile}\rho$, and $\overline{\overline{\rho}^{\smallsmile\smallsmile}} = \rho$, we get

$$\rho\tilde{\rho}\rho[A] = \bigcup_{B\subseteq X} \{\rho[B] : \rho[A] \not\subseteq \overline{\tilde{\rho}^{\smallsmile}}[B]\}, \quad A \subseteq X,$$

and

$$\overline{\tilde{\rho}^{\smallsmile}}[B] = \rho\overline{\rho}^{\smallsmile}\rho[B] = \bigcup_{C\subseteq X} \{\rho[C] : \rho[B] \not\subseteq \rho[C]\}, \quad B \subseteq X.$$

Therefore

$$\rho\tilde{\rho}\rho[A] = \bigcup_{B\subseteq X} \Big\{\rho[B] : \rho[A] \not\subseteq \bigcup_{C\subseteq X} \{\rho[C] : \rho[B] \not\subseteq \rho[C]\}\Big\}, \quad A \subseteq X, \tag{6.10}$$

and by comparison with (6.9) (for which we did not use the regularity of ρ but Lemma 6.45!) we get $\rho\tilde{\rho}\rho[A] = \rho[A]$, $A \subseteq X$. This equality holds for arbitrary subsets of X, thus for all singleton sets $\{x\}_{x\in X}$. So $x(\rho\tilde{\rho}\rho) = x\rho$ for all $x \in X$, whence $\rho\tilde{\rho}\rho = \rho$, i.e. ρ is regular. □

The notion of an inverse shows up in the context of square boolean matrices as early as 1934 in [50]. Following Luce [30], we proceed with

Definition 6.47. *Let ρ be a given finite square Boolean relation of order n, i.e. $\rho \in \boldsymbol{B}_n$. Then ρ is called*

(a) **orthogonal** *(or* **permutation relation***) if $\rho^{-1} = \rho^{\smallsmile}$, i.e. $I = \rho\rho^{\smallsmile} = \rho^{\smallsmile}\rho$.*
(b) **row (column) consistent** *if $\rho U = U$ ($U\rho = U$).*

Next we present some early classical results on orthogonal relations and inverses. (Note that *row consistent* is what has been defined as *vector* in Def. 6.25 above.)

Proposition 6.48. ([30, 38]) *Let ρ be a given finite square Boolean relation of order n, i.e. $\rho \in \boldsymbol{B}_n$. Then the following holds:*

(1) *ρ is row consistent iff $I \subseteq \rho\rho^{\smallsmile}$, and column consistent iff $I \subseteq \rho^{\smallsmile}\rho$;*
(2) *ρ is orthogonal iff it is both row and column consistent, and*

$$\forall i, j, k[i \neq j \rightarrow \rho_{ik} \wedge \rho_{jk} = 0 = \rho_{ki} \wedge \rho_{kj}];$$

(3) *ρ has an inverse iff it is orthogonal.*

Proof. (1): We show the case of row consistency: By Prop. 3.55(3) we have $I \subseteq \rho \rightarrow \rho U = U = U\rho$, and by Prop. 3.28(2) $I \cap \rho U = I \cap \rho\rho^{\smallsmile}$. So $I \subseteq \rho\rho^{\smallsmile}$ implies $I = I \cap \rho\rho^{\smallsmile} = I \cap \rho U$, whence $I \subseteq \rho U$. Thus $U = IU \subseteq \rho UU = \rho U \subseteq U$, and therefore $U = \rho U$.

Conversely, let $U = \rho U$. Then, using the DEDEKIND rule Th. 3.26(3), we get $I = \rho U \cap I \subseteq (\rho \cap IU^{\smallsmile})(U \cap \rho^{\smallsmile} I) = (\rho \cap U)(U \cap \rho^{\smallsmile}) = \rho\rho^{\smallsmile}$, i.e. $I \subseteq \rho\rho^{\smallsmile}$.

(2): If ρ is orthogonal, then by (1) it is both row and column consistent. From $I = \rho\rho^{\smallsmile}$ we get for every $i, j, i \neq j$: $0 = \bigvee_k(\rho_{ik} \wedge \rho^{\smallsmile}_{kj})$, whence for every k: $\rho_{ik} \wedge \rho_{jk} = 0$. Similarly, for every $i, j, k, i \neq j$ we get $\rho_{ki} \wedge \rho_{kj} = 0$ from $I = \rho^{\smallsmile}\rho$.

For the reverse direction assume that $\rho_{ik} \wedge \rho_{jk} = 0$ for every $i, j, k, i \neq j$; then $\bigvee_k(\rho_{ik} \wedge \rho^{\smallsmile}_{kj}) = 0$. Moreover, since ρ is assumed row consistent, we have $I \subseteq \rho\rho^{\smallsmile}$ by (1), i.e. $\bigvee_k(\rho_{ik} \wedge \rho^{\smallsmile}_{ki}) = 1$. Thus $\rho\rho^{\smallsmile} \subseteq I$, hence, $I = \rho\rho^{\smallsmile}$; the second case, $I = \rho^{\smallsmile}\rho$, can be shown in an analogous way.

(3): (As noted in [30], this result appears already in [50].) We only have to show the direction that is not obvious by definition. So assume that ρ has an inverse, (say $\sigma := \rho^{-1}$); then $I = \rho\sigma = \sigma\rho$ and by Prop. 3.55(3) we get $\rho\sigma U = U\rho\sigma = U$ and $\sigma\rho U = U\sigma\rho = U$. Now $\rho U \subseteq U$ and $\sigma U \subseteq U$ always hold. Thus $U = (\rho\sigma)U = \rho(\sigma U) \subseteq \rho U \subseteq U$, i.e. ρ is row consistent. The other cases, $U = U\rho, \sigma U = U$, and $U = U\sigma$ can be shown analogously, whence by (1) we get $I \subseteq \rho\rho^{\smallsmile}$, $I \subseteq \rho^{\smallsmile}\rho$, $I \subseteq \sigma\sigma^{\smallsmile}$, and $I \subseteq \sigma^{\smallsmile}\sigma$. Thus, both ρ and $\sigma := \rho^{-1}$ are row and column consistent. Using (2) it remains to show that $\forall i, j, k[i \neq j \rightarrow \rho_{ik} \wedge \rho_{jk} = 0 = \rho_{ki} \wedge \rho_{kj}]$. Let $I = \rho\rho^{-1}$; then we have, for every $i, j, i \neq j$, $\bigvee_k(\rho_{ik} \wedge \rho^{-1}_{kj}) = 0$, whence, for every k, $\rho_{ik} \wedge \rho^{-1}_{kj} = 0$. Thus, for k fixed, $\bigvee_{i\neq j}(\rho_{ik} \wedge \rho^{-1}_{kj}) = 0$, and therefore

$$\rho_{ik} \wedge \rho^{-1}_{ki} = \bigvee_{i\neq j}(\rho_{ik} \wedge \rho^{-1}_{kj}) \vee (\rho_{ik} \wedge \rho^{-1}_{ki}) = \bigvee_j(\rho_{ik} \wedge \rho^{-1}_{kj}) = \rho_{ik} \wedge \bigvee_j \rho^{-1}_{kj} = \rho_{ik}.$$

(Note that $\bigvee_j \rho^{-1}_{kj} = 1$ because ρ^{-1} is row consistent.) So, for every $i, j, k, i \neq j$ we have

$$\rho_{ik} \wedge \rho_{jk} = (\rho_{ik} \wedge \rho^{-1}_{ki}) \wedge (\rho_{jk} \wedge \rho^{-1}_{kj}) = (\rho_{ik} \wedge \rho^{-1}_{kj}) \wedge (\rho_{jk} \wedge \rho^{-1}_{ki}) = 0,$$

whence ρ is orthogonal. □

In [38] it has been shown that if a square Boolean matrix α possesses a one-sided inverse, that inverse is also a two-sided inverse. Furthermore such an inverse, if it exists, is unique and is $\alpha^{\smallsmile}$. Permutation matrices in $\boldsymbol{B}_n$ are the only matrices which have inverses. We need the following

Lemma 6.49. *Let α be a square Boolean matrix such that*

$$\bigvee_j a_{ij} = 1, \quad (i = 1, \ldots, n), \tag{6.11}$$

and

$$a_{ij} \wedge a_{kj} = 0, \quad (i \neq k). \tag{6.12}$$

Then equations (6.11) and (6.12) are equivalent to (6.13) and (6.14)

Proof. From (6.11) it follows that $1 = \bigwedge_i(\bigvee_j a_{ij}) = \bigvee_{f\in\Phi}(\bigwedge a_{if(i)})$, the last equation being a well known result from lattice theory, where Φ is the set of functions from $\{1,2,\ldots,n\}$ into $\{1,2,\ldots,n\}$; see [46, p. 124]. A typical intersection $a_{1j_1} \wedge a_{2j_2} \wedge \ldots \wedge a_{nj_n}$ evaluates to 0 according to (6.12), unless all the j_k are distinct. This means that only permutations of $\{1,2,\ldots,n\}$ contribute a 1, whence for the permanent we have $\mathsf{per}(\alpha) = 1$ (see Def. 5.17(a)). Expanding $\mathsf{per}(\alpha)$ according to Prop. 5.18(4) and using monotonicity results in $\bigvee_i a_{ij} \geq \bigvee_i(a_{ij} \wedge \mathsf{per}(\alpha(i|j))) = 1$, and so

$$\bigvee_i a_{ij} = 1, \quad (j = 1,\ldots,n). \tag{6.13}$$

So what we have shown is that there are 1's in each row and each column, by (6.11) and (6.13), and that there is exactly one 1 in each column by (6.12) together with (6.13). Now assume $a_{ij} = a_{ik} = 1, j \neq k$. Then there are $n-1$ rows which must be assigned a 1 such that each column contains exactly one. But, since columns j and k are no longer available, there remain only $n-2$ columns for $n-1$ required occurrences of 1's. Thus there is one line with no 1 in it, a contradiction to (6.11), and therefore

$$a_{ki} \wedge a_{kj} = 0, \quad (i \neq j). \tag{6.14}$$

The other direction follows using the above argumentation for $\alpha^{\smile}$ instead of α. □

Theorem 6.50. ([38, p. 51f]) *If a square relation (square Boolean matrix) β possesses a one-sided inverse, this inverse is also a two-sided inverse. Furthermore, such an inverse, if it exists, is unique and equals $\beta^{\smile}$.*

Proof. Consider the equation $\beta\gamma^{\smile} = I$, with β, γ square Boolean matrices of order n. This equation implies the following relationships:

$$\bigvee_j(b_{ij} \wedge c_{ij}) = 1, \quad (i = 1,\ldots,n), \tag{6.15}$$

$$\bigvee_j(b_{ij} \wedge c_{kj}) = 0, \quad (i \neq k).$$

From the second equation we get

$$b_{ij} \wedge c_{kj} = 0, \quad (i \neq k, j = 1,\ldots,n). \tag{6.16}$$

Setting $a_{ij} := b_{ij} \wedge c_{ij}$ obviously gives

$$a_{ij} \wedge a_{kj} = b_{ij} \wedge c_{ij} \wedge b_{kj} \wedge c_{kj} = 0, \quad (i \neq k). \tag{6.17}$$

From (6.15) and (6.17) we see that $\alpha = [a_{ij}]$ satisfies (6.11) and (6.12), and therefore also (6.13) and (6.14) by Lemma 6.49. Since $a_{ij} \leq b_{ij}$ and $a_{ij} \leq c_{ij}$ we have

$$\bigvee_i b_{ij} = 1, \quad \bigvee_i c_{ij} = 1, \quad (i = 1,\ldots,n),$$

by (6.13). And applying (6.16) we get

$$\begin{aligned} b_{kj} = b_{kj} \wedge \left(\bigvee_i c_{ij}\right) &= \bigvee_i(b_{kj} \wedge c_{ij}) = b_{kj} \wedge c_{kj} \\ &= \bigvee_i(b_{ij} \wedge c_{kj}) = \left(\bigvee_i b_{ij}\right) \wedge c_{kj} = c_{kj}. \end{aligned}$$

So, by the definition of a_{ij} above, we have shown $\alpha = \beta = \gamma$, whence the equation $\beta\gamma^{\smile} = I$ can only be satisfied if $\gamma = \beta$. Thus, if β has a right inverse, this must be $\beta^{\smile}$, and β is orthogonal. Then, (6.15) and (6.16) reduce to (6.11) and (6.12), which, by Lemma 6.49, imply (6.13) and (6.14). These in turn imply $\beta^{\smile}\beta = I$. Moreover, $\beta^{\smile}$ is the only left inverse of β: Let $\chi\beta = I$, then $\beta = \chi^{\smile}$ and so $\chi = \beta^{\smile}$. □

6.2.2.1 *Difunctional Relations*

In Prop. 3.50 regular relations have been characterized by $\exists\sigma \in \mathcal{B}_S[\rho\sigma\rho = \rho]$, and **difunctional relations** by $\rho\rho^{\smile}\rho \subseteq \rho$. Since $\rho \subseteq \rho\rho^{\smile}\rho$ by Prop. 3.54(1) for any relation ρ, difunctional are a special kind of regular relations, whence they got this place.

Following [43, p. 74f] one can verify that the following is characteristic for rows i of a matrix $A = [a_{ij}] \in \boldsymbol{B}_{m\times n}$ that is transformable into block diagonal form: whenever there is a row k and a column j such that if $a_{ij} = 1$ and $a_{kj} = 1$, then, for all columns m with $a_{km} = 1$ we also have $a_{im} = 1$. This is illustrated in the following example (simplified after [43, p. 75]):

$$
\begin{array}{r|cccccccc|}
 & & j & & & & m & & \\
 & & \downarrow & & & & \downarrow & & \\
 & 1 & 2 & 3 & 4 & 5 & 6 & 7 & 8 \\
\hline
1 & \mathbf{1} & 0 & \mathbf{1} & \mathbf{1} & 0 & 0 & 0 & 0 \\
k\to 2 & 0 & \textcircled{1} & 0 & 0 & \mathbf{1} & \textcircled{1} & \mathbf{1} & 0 \\
3 & 0 & 0 & 0 & 0 & 0 & 0 & 0 & \mathbf{1} \\
i\to 4 & 0 & \textcircled{1} & 0 & 0 & \mathbf{1} & \boxed{1} & \mathbf{1} & 0 \\
5 & \mathbf{1} & 0 & \mathbf{1} & \mathbf{1} & 0 & 0 & 0 & 0 \\
6 & 0 & \mathbf{1} & 0 & 0 & \mathbf{1} & \mathbf{1} & \mathbf{1} & 0 \\
\hline
\end{array}
\qquad
\begin{array}{r|cccccccc|}
 & & & & j & & m & & \\
 & & & & \downarrow & & \downarrow & & \\
 & 1 & 3 & 4 & 2 & 5 & 6 & 7 & 8 \\
\hline
1 & \mathbf{1} & \mathbf{1} & \mathbf{1} & 0 & 0 & 0 & 0 & 0 \\
5 & \mathbf{1} & \mathbf{1} & \mathbf{1} & 0 & 0 & 0 & 0 & 0 \\
k\to 2 & 0 & 0 & 0 & \textcircled{1} & \mathbf{1} & \textcircled{1} & \mathbf{1} & 0 \\
i\to 4 & 0 & 0 & 0 & \textcircled{1} & \mathbf{1} & \boxed{1} & \mathbf{1} & 0 \\
6 & 0 & 0 & 0 & \mathbf{1} & \mathbf{1} & \mathbf{1} & \mathbf{1} & 0 \\
3 & 0 & 0 & 0 & 0 & 0 & 0 & 0 & \mathbf{1} \\
\hline
\end{array}
$$

In a more formal way the above condition becomes

$$\forall i \forall j,k\big[(a_{ij} = 1 \wedge a_{kj} = 1) \to \forall m[a_{km} = 1 \to a_{im} = 1]\big]$$

or, equivalently, in relational notation, with A representing ρ, and keeping i, j, k, m for better readability:

$$\forall i \forall j,k\big[(i\rho j \wedge j\rho^{\smile}k) \to \forall m[k\rho m \to i\rho m]\big] \tag{6.18}$$

Looking at the left example matrix, one can easily see that, given the circled 1's, condition (6.18) requires the squared 1. This ensures, with some reordering of the rows and columns involved, the existence of squares of 1's with size 2×2. By iteration this finally results in blocks of 1's as shown in the matrix on the right-hand side. We can proceed as follows:

$$
\begin{array}{lll}
 & \forall i \forall j,k\big[(i\rho j \wedge j\rho^{\smile}k) \to \forall m[k\rho m \to i\rho m]\big] & (6.18) \\
\leftrightarrow & \forall i \forall j,k\big[\forall m[(i\rho j \wedge j\rho^{\smile}k) \to (k\rho m \to i\rho m)]\big] & \text{by Th. 2.36(4)(b)} \\
\leftrightarrow & \forall i,j,k,m[i\rho j \wedge j\rho^{\smile}k \wedge k\rho m \to i\rho m] & \text{by Th. 2.33(12)}
\end{array}
$$

And the last line is just how difunctional relations have been defined in Def. 3.21; so rectangles contribute to difunctional relations in an essential way. Note that the last line is equivalent to

$$\forall i,m \forall j,k[i\rho j \wedge j\rho^{\smile}k \wedge k\rho m \to i\rho m],$$

by Th. 2.34(1), which, by Th. 2.36(3)(a), is equivalent to

$$\forall i,m\big[\exists j,k[i\rho j\rho^{\smile}k\rho m] \to i\rho m\big].$$

Proposition 6.51. *Let $\rho, \sigma \in \mathcal{B}_{X\times Y}$ be difunctional relations. Then the following holds:*

(1) ρ is difunctional iff $\rho^{\smile}$ is.
(2) $\rho \cap \sigma$ is difunctional.

Proof. (1): $\rho\rho^{\smile}\rho = \rho$ implies $(\rho\rho^{\smile}\rho)^{\smile} = \rho^{\smile}\rho\rho^{\smile} = \rho^{\smile}$, so if ρ is difunctional, then also its converse. Using this result we get the proof for the other direction: if $\rho^{\smile}$ is difunctional then also $\rho^{\smile\smile} = \rho$.
(2): $(\rho\cap\sigma)(\rho\cap\sigma)^{\smile}(\rho\cap\sigma) \subseteq \rho\rho^{\smile}\rho \subseteq \rho$ and $(\rho\cap\sigma)(\rho\cap\sigma)^{\smile}(\rho\cap\sigma) \subseteq \sigma\sigma^{\smile}\sigma \subseteq \sigma$. □

Proposition 6.52. ([33])

(1) If ρ is symmetric and circular, then also difunctional.
(2) If $\rho \in \mathcal{B}_{X\times Y}$ is total and difunctional, then $\rho\rho^{\smile} \in \mathsf{Eq}_X$.

Proof. (1): Circularity of ρ means that $\rho\rho \subseteq \rho^{\smile}$, so we get $\rho\rho^{\smile}\rho = \rho^3 \subseteq \rho^{\smile}\rho = \rho\rho \subseteq \rho^{\smile} = \rho$.
(2): $\rho\rho^{\smile}$ is symmetric by Prop. 3.52(5). Moreover, since $(\rho\rho^{\smile})^2 = (\rho\rho^{\smile}\rho)\rho^{\smile} = \rho\rho^{\smile}$, it is also transitive. And by Prop. 3.22(9) it is reflexive, since $\mathsf{fld}(\rho) = X$. □

Proposition 6.53. ([22, Lemma 2.1.11]) *Let ρ be a relation on a set S; then the following holds:*

(1) If ρ is reflexive and difunctional, then also symmetric.
(2) If ρ is transitive and symmetric, then also difunctional.
(3) If ρ is reflexive and transitive, then ρ is difunctional iff it is symmetric.

Proof. (1): $\rho^{\smile} = I\rho^{\smile}I \subseteq \rho\rho^{\smile}\rho \subseteq \rho$.
(2): $\rho\rho^{\smile}\rho \subseteq \rho\rho\rho \subseteq \rho$.
(3): By (1) and (2). □

Note that partial equivalences, hence also equivalences, are always difunctional due to (2) of the last proposition.

Proposition 6.54. ([36]) *A relation $\rho \in \mathcal{B}_{X\times Y}$ is difunctional iff their aftersets are pairwise disjoint, i.e. $\forall a, b \in X[a\rho \cap b\rho \neq \emptyset \rightarrow a\rho = b\rho]$.*

Proof. Let $\rho \in \mathcal{B}_{X\times Y}$ be difunctional and $a, b \in \mathsf{dom}(\rho)$ such that $a\rho \cap b\rho \neq \emptyset$. For $c \in a\rho \cap b\rho$ and $d \in a\rho$ then $b\rho c, c\rho^{\smile} a$, and $a\rho d$, whence $b\rho\rho^{\smile}\rho d$. So, by regularity, $b\rho d$, i.e. $d \in a\rho \rightarrow d \in b\rho$. Since a, b have symmetric roles, we also get $b\rho \subseteq a\rho$. Conversely assume $\forall a, b \in X[a\rho \cap b\rho \neq \emptyset \rightarrow a\rho = b\rho]$ and $a\rho\rho^{\smile}\rho d$. Then there exist b, c such that $a\rho b\rho^{\smile} c\rho d$. Then $b \in a\rho \cap c\rho$, so $a\rho = c\rho$. Since $d \in c\rho$, we thus get $d \in a\rho$, whence $a\rho d$. □

Proposition 6.55. ([33]) *For a reflexive relation ρ the following are equivalent:*

(1) ρ is an equivalence.
(2) ρ is difunctional.

(3) ρ is circular.

Proof. (1)⇒(2): We have $I \subseteq \rho$, $\rho = \rho^{\smile}$, and $\rho\rho \subseteq \rho$. So $\rho\rho^{\smile}\rho = \rho^3 \subseteq \rho\rho \subseteq \rho$.
(2)⇒(3): We have $I \subseteq \rho^{\smile}$, and $\rho^{\smile}$ is difunctional by Prop. 6.51(1). Therefore,

$$\rho\rho = I\rho I\rho I \subseteq \rho^{\smile}(\rho\rho^{\smile}\rho)\rho^{\smile} = \rho^{\smile}\rho\rho^{\smile} = \rho^{\smile}.$$

(3)⇒(1): If ρ is reflexive and circular; then it is an equivalence by Prop. 3.22(14). □

Proposition 6.56. ([23, Proposition 2]) *Let ρ be an arbitrary binary relation; then* fringe(ρ) *is difunctional.*

Proof. Setting $\tau :=$ fringe(ρ), we have to show that $\tau\tau^{\smile}\tau \subseteq \tau$. We have $\tau \subseteq \rho$ and $\tau \subseteq \overline{\rho\rho^d\rho}$ by definition (see page 172), and $\rho\sigma^{\smile}\rho \subseteq \rho \leftrightarrow \sigma \subseteq \overline{\rho\rho^d\rho}$, for σ an arbitrary relation, by Cor. 3.31(5). Thus $\rho\tau^{\smile}\rho \subseteq \rho$, and therefore

$$\rho(\tau\tau^{\smile}\tau)^{\smile}\rho = \rho\tau^{\smile}\tau\cdot\tau^{\smile}\rho \subseteq \rho\tau^{\smile}\rho\cdot\tau^{\smile}\rho \subseteq \rho\cdot\tau^{\smile}\rho \subseteq \rho.$$

Applying Cor. 3.31(5) to $\rho(\tau\tau^{\smile}\tau)^{\smile}\rho \subseteq \rho$ gives $(\tau\tau^{\smile}\tau) \subseteq \overline{\rho\rho^d\rho}$. Moreover, $\tau\tau^{\smile}\tau \subseteq \rho\tau^{\smile}\rho \subseteq \rho$, So, finally $\tau\tau^{\smile}\tau \subseteq \rho \cap \overline{\rho\rho^d\rho} =: \tau$. □

Corollary 6.57. ([42, 10.3.9 Corollary])

(1) *A relation ρ is difunctional iff $\rho =$* fringe(ρ).
(2) *Forming the fringe of a relation is idempotent:* fringe(fringe(ρ)) = fringe(ρ).

Proof. (1): If $\rho =$ fringe(ρ) then ρ is difunctional according to Prop. 6.56. Conversely, if ρ is difunctional then $\rho\rho^{\smile}\rho \subseteq \rho$, so $\rho \subseteq \overline{\rho\rho^d\rho}$ by Cor. 3.31(5). But then fringe(ρ) = $\rho \cap \overline{\rho\rho^d\rho} = \rho$.
(2): Follows from (1) since fringe(ρ) is difunctional by Prop. 6.56. □

6.2.3 *Commuting Relations*

Two relations ρ and σ are called **commuting** in case $\rho\sigma = \sigma\rho$. For sake of completeness, we begin with recalling some results that we have reported earlier in this book:

(1) Let ρ and σ be symmetric, and $\rho\sigma \subseteq \sigma\rho$; then $\rho\sigma = \sigma\rho$. (Prop. 3.53(4))
(2) *Let ρ and σ be two preorder relations on a set S. Then ρ and σ commute iff both $\rho\sigma$ and $\sigma\rho$ are preorders.* (Prop. 4.9)

We continue with some easy to prove properties of commuting relations.

Proposition 6.58.

(1) *If a relation ρ commutes with two relations σ and τ then it also commutes with their union.*
(2) *If both ρ and ρ' commute with a relation σ then also their union does.*

Proof. (1): $\rho(\sigma \cup \tau) = \rho\sigma \cup \rho\tau = \sigma\rho \cup \tau\rho = (\sigma \cup \tau)\rho$.
(2): Just a reformulation of (1). □

Proposition 6.59. *For two commuting relations ρ and σ on a set S the following holds:*

(1) If ρ and σ are symmetric, then also their composition.
(2) If ρ and σ are transitive, then also their composition.

Proof. (1): $(\rho\sigma)^{\smile} = \sigma^{\smile}\rho^{\smile} = \sigma\rho = \rho\sigma$.
(2): $(\rho\sigma)^2 = \rho\sigma\rho\sigma = \rho^2\sigma^2 \subseteq \rho\sigma$. □

6.2.3.1 *Commuting Equivalences*

For sake of completeness we first repeat two results concerning commuting equivalences which have already been mentioned in Prop. 4.18(3) and Prop. 4.19(4):

Proposition 6.60. *Let $\rho, \sigma \in \mathsf{Eq}_S$. Then the following holds:*

(1) $\rho\sigma$ is an equivalence iff ρ and σ commute;
(2) ρ and σ commute iff $\rho \vee \sigma = \rho\sigma$ iff $\rho\sigma \in \mathsf{Eq}_S$.

Commuting equivalence relations occur frequently. In the following we give some short remarks on them in independence concepts and social networks.

(a) Independent equivalences:

Recall from Th. 4.17 that there exists a one-to-one correspondence between partitions and equivalence relations on a set. Accordingly, we will sometimes use terminology for equivalences and their associated partitions interchangeably. The presentation in this section largely follows [10, 12, 17].

Definition 6.61. *Let $\rho, \sigma \in \mathsf{Eq}_S$; ρ and σ (equivalently, their associated partitions π_ρ and π_σ) are called*

(a) **independent** ([17]) *if, for any two blocks $A \in \pi_\rho$ and $B \in \pi_\sigma$, we have $A \cap B \neq \emptyset$.*
(b) **orthogonal** ([10]) *if, for any two blocks $A \in \pi_\rho$ and $B \in \pi_\sigma$, we have $|A \cap B| = 1$.*

Intuitively, the meaning of independence of two equivalence relations is the following: (Recall that $x\rho$ denotes the equivalence class of x induced by ρ.) Given the classification of S according to ρ, since any equivalence class $x\rho$ intersects with all equivalence classes $y\sigma$, you cannot infer anything about the classification of S according to σ. In other words: Knowing the properties of an element $s \in S$ in terms of ρ does not give you any information on which properties may be attributed to s from the viewpoint of σ.

[17, p. 74] illustrates the meaning of the independence of two partitions as follows: A partition $\pi \in \Pi_S$ of a set S can be seen as a phase in the search process of an unknown element $s \in S$, because each $\pi \in \Pi_S$ determines a block containing the element s. A set $P \subseteq \Pi_S$ of partitions of S is a good search process, whenever $\bigcap P$ is the minimum partition $\alpha_S = S/I_S$ where each block has exactly one element.

Proposition 6.62. *Let ρ and σ be independent equivalences on a set S; then*

(1) ρ and σ commute.
(2) $\rho \vee \sigma = \rho\sigma = U_S$.

Proof. (1): We first show that $\rho\sigma \subseteq \sigma\rho$: Let $x\rho\sigma z$; then, by independence of ρ and σ, $x\rho \cap y\sigma \neq \emptyset$ for each $y \in S$. We take an element y_0 such that $y_0\sigma x$ which must exist since one characteristic property of a partitions is to be a covering (see Def. 2.8). By symmetry of σ we get $x\sigma y_0$, and by independence $x\sigma \cap y\rho \neq \emptyset$, for each $y \in S$. We choose y_1 such that $y_1\rho z$, and thus get $x\sigma\rho z$, whence $\rho\sigma \subseteq \sigma\rho$. By Prop. 3.53(4) $\rho\sigma = \sigma\rho$.

(2): Let $x, y \in S$. By reflexivity, there exists a block $A \in \pi_\rho$ with $x \in A$, and a block B in π_σ with $y \in B$. By independence of ρ and σ $A \cap B \neq \emptyset$. For $z \in A \cap B$ we have $x\rho z\sigma y$, and since x, y have been arbitrary elements, $\rho\sigma = U_S$. By (1) and Prop. 6.60(2) $\rho \vee \sigma = \rho\sigma$. □

For two equivalence relations ρ and σ to be independent is more than to commute: I_S commutes with all other equivalences on S but is not independent of any of them with the exception of U_S. But the following (partial) converse of Prop. 6.62(1) holds:

Lemma 6.63. *If ρ and σ are commuting equivalence relations on a set S such that $\rho \vee \sigma = U_S$, then ρ and σ are independent.*

Proof. Let $A \in \pi_\rho$ and $B \in \pi_\sigma$ be two blocks, and let $r \in A$ and $t \in B$ be two elements in S. From $\rho \vee \sigma = U_S$ we get $r(\rho \vee \sigma)t$. Now, ρ and σ commute, whence $\rho \vee \sigma = \rho\sigma$ by Prop. 6.60(2). Thus, there is an element s such that $r\rho s$ and $s\sigma t$. So $s \in A \cap B$, and, as A and B are arbitrary blocks, ρ and σ are independent □

We know that each partition of a set S does uniquely correspond to an equivalence relation on S. So we may ask how independence and commutativity may be characterized in terms of the former. The answer to this question has been given already in 1939 in [16]. For its presentation we need the following concept: If π is a partition of S and $T \subseteq S$, then the **restriction** $\pi|_T := \{B \cap T : B \in \pi \wedge B \cap T \neq \emptyset\}$ is a partition of T. Note that $\pi|_T$ is the partition associated to $\rho_\pi|_T$, and that $\rho|_T$ is an equivalence on $T \subseteq S$ if ρ is an equivalence on S.

Lemma 6.64. *Let π_ρ and π_σ two partitions of a set S and C be a block of $\pi_{\rho\vee\sigma}$. Then $\pi_\rho|_C \vee \pi_\sigma|_C = \pi_{\rho\vee\sigma}|_C$.*

Proof. In Π_C we have $\pi_\rho|_C \vee \pi_\sigma|_C \leq \pi_{\rho\vee\sigma}|_C = \{C\}$ by Prop. 4.20(1). Let $a, b \in C$, then $a\ (\rho \vee \sigma)|_C\ b$. By Prop. 4.20(2)(b) there is a sequence $x_0, \ldots, x_{n+1} \in C$ such that $a = x_0, b = x_{n+1}$ and $x_i \equiv x_{i+1} \bmod(\pi_\rho|_C)$ or $x_i \equiv x_{i+1} \bmod(\pi_\sigma|_C)$, i.e. $\langle x_i, x_i + 1\rangle \in (\rho \cup \sigma)|_C \subseteq (\rho \vee \sigma)|_C$, $0 \leq i \leq n$. For each of those i, $a\ (\rho \vee \sigma)|_C\ x_i$ due to the transitivity of $(\rho \vee \sigma)|_C$, whence $\langle x_i, x_{i+1}\rangle \in \rho|_C \cup \sigma|_C$. Therefore, $a(\rho|_C \vee \sigma|_C)b$, and so we have $\pi_{\rho\vee\sigma}|_C \leq \pi_\rho|_C \vee \pi_\sigma|_C$. □

Theorem 6.65. (Dubreil-Jacotin) *Two equivalence relations ρ and σ on a set S commute iff for each block C of the partition $\pi_{\rho\vee\sigma}$, the restrictions $\pi_\rho|_C$ and $\pi_\sigma|_C$ are independent.*

Proof. Let ρ and σ be commuting equivalences on S, and let $C \subseteq S$ be a block of $\pi_{\rho\vee\sigma}$. Then $\rho|_C$ and $\sigma|_C$ are commuting equivalences on C. By Prop. 6.60(2) and Lemma 6.64 we have $\rho|_C \circ \sigma|_C = \rho|_C \vee \sigma|_C = U_C$, so $\rho|_C$ and $\sigma|_C$ are independent by Lemma 6.63.

Conversely, assume that, for each block C of the partition $\pi_{\rho\vee\sigma}$, the restrictions $\pi_\rho|_C$ and $\pi_\sigma|_C$ are independent. Let $a(\rho\vee\sigma)b$. Take $C \in \pi_{\rho\vee\sigma}$ such that $a, b \in C$, take $A \in \pi_\rho|_C$ and $B \in \pi_\sigma|_C$ such that $a \in A$ and $b \in B$. By the independence of $\pi_\rho|_C$ and $\pi_\sigma|_C$ we have $A \cap B \neq \emptyset$. Let $c \in A \cap B$; then $\langle a, c\rangle \in \rho|_C \subseteq \rho$ and $\langle c, b\rangle \in \sigma|_C \subseteq \sigma$, so $a\rho\sigma b$. Since always $\rho\sigma \subseteq \rho \vee \sigma$, we get $\rho\sigma = \rho \vee \sigma$, whence ρ and σ commute by Prop. 6.60(2). □

(b) Semantically independent relations:

Another notion of independence arises in computer science: Given two fragments of an imperative program, it is interesting to know if they are "independent", i.e. if their computations are mutually irrelevant. Program fragments recognized as independent may be executed in either order or in parallel. In the following we will call "fragments of an imperative program" simply statements, and in fact single assignments are the smallest units for which independence can eventually be assured. The kind of independence one is looking for in this context basically is of a semantic nature, whatever the underlying syntactical form of the program.

Consider the following simple example taken from [10]:

$$\textbf{var } x, y : \{0, 1\}; \quad \alpha_1 : x := x \oplus 1; \quad \alpha_2 : y := y \ominus 1; \tag{A}$$

where x and y can take as value 0 or 1, and the operators $\oplus$ and $\ominus$ denote cyclic addition and subtraction (i.e. $1 \oplus 1 = 0, 0 \ominus 1 = 1$). Apparently, the two assignments α_1 and α_2 do not interfere with each other as they are variable-disjoint. Therefore, they could be executed fully concurrently; and executing them in any order does not change the final result, whence they are commutative.

An appropriate semantic model of statements are binary relations on a state space X, relating input to output states: For $x, y \in X$, $x\rho y$ may be interpreted as "if started in the input state x, then the statement ρ may terminate in the output state y". See e.g. [,]. To keep things simple we will limit ourselves in presenting formal definitions as long as the principle ideas remain clearly visible.

The state space for example (A) is $S := Val(x) \times Val(y) = \{0, 1\} \times \{0, 1\} = \{00, 01, 10, 11\}$ writing e.g. 10 for $x = 1$ and $y = 0$. By $val(x, s) \in Val(x)$ we denote the value of x in state $s \in S$. Note that each variable x induces an equivalence on S, denoted by ρ_x in the following way: $s\rho_x s' :\leftrightarrow val(x, s) = val(x, s')$. Assignments have an associated *meaning relation* m (which we simplify to deterministic expressions e_i, i.e. ones that have one well defined value in a given state, whence m actually becomes a function):

Definition 6.66. ([10, Definition 2.2]) *Let* ***var*** $x_1 : Val(x_1), \ldots, x_n : Val(x_m)$ *be a declaration,* $m \geq 1$, *and* $S = Val(x_1) \times \ldots \times Val(x_m)$ *the associated state space. Let* $\alpha : (y_1, \ldots, y_k) := (e_1, \ldots, e_k)$; *denote a multiple assignment such that the* $y_1, \ldots, y_k$ *are mutually distinct variables and* α *contains only variables from* $x_1, \ldots, x_m$. *Then the* **meaning relation** *of* α *is given by* $(s', s) \in m(\alpha) \subseteq S \times S$ *iff*

(1) $\forall j, 1 \le j \le k\,[val(y_j, s) = val(e_j, s')]$*, and*
(2) $\forall x \in \{x_1, \ldots, x_m\} - \{y_1, \ldots, y_k\}\,[val(x, s) = val(x, s')]$*,*

where $val(e_j, s')$ *denotes the value of* e_j *in state* s'*.*

This definition mirrors the usual meaning of an assignment: In the final state s each y_j contains the value of e_j in the inital state s', and the values of all other variables remain unchanged.

So the meaning relations $m(\alpha_i) \subseteq S \times S$ for example (A), $\alpha_1 :\ x := x \oplus 1$ and $\alpha_2 :\ y := y \ominus 1$; are given as follows:

$$m(\alpha_1) = \{\langle 00, 10\rangle, \langle 10, 00\rangle, \langle 01, 11\rangle, \langle 11, 01\rangle\}$$
$$m(\alpha_2) = \{\langle 00, 01\rangle, \langle 01, 00\rangle, \langle 10, 11\rangle, \langle 11, 10\rangle\}.$$

We see that e.g. $val(y, 00) = val(y, 10) = 0$, so $\langle 00, 10\rangle \in \rho_y$. In Fig. 6.6 relation $m(\alpha_1)$ is depicted by solid, and $m(\alpha_2)$ by dashed arrows, and the equivalence classes of ρ_x are shown as solid rectangles, and those of ρ_y as dashed ones.

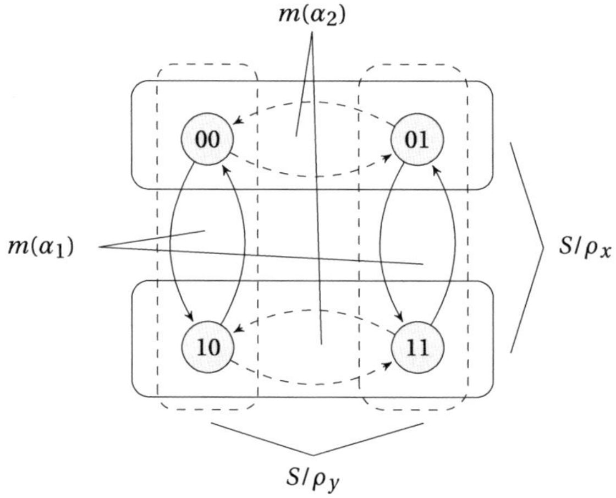

FIG. 6.6: Visualizing example (A).

What immediately catches one's eye is the perfect symmetry of the picture: α_1 and α_2 "*behave well*" ([10]) with respect to the two partitions S/ρ_x and S/ρ_y induced by the variables x and y, which means that $m(\alpha_1)$ connects only states within an equivalence class $[\rho_y]_s$, and $m(\alpha_2)$ connects only states within a $[\rho_x]_s$ for some $s \in S$. So we have $m(\alpha_1) \subseteq \rho_y$ and $m(\alpha_2) \subseteq \rho_x$. Note that the previous discussion has been based on the role of variables; but this is a syntax-driven viewpoint which we want to avoid.

So we will now invert the connection between variables and partitions on S, keeping their possible interpretation as variables in mind as a motivation. The intended definition of semantic independence can be phrased as follows ([10]): Let S be an arbitrary set (the state space), and let $m_1, m_2 \subseteq S \times S$. We will say m_1 and m_2 to be semantically independent on S if there exist two partitions S/ρ_1 and S/ρ_2 which are like partitions induced by variables and with respect to which m_1 and m_2 behave just as in the example given above. These considerations result in

Definition 6.67. ([10]) *Let $\tau_1, \tau_2 \in \mathcal{B}_S$; then τ_1 and τ_2 are are called* **semantically independent** *if there exist $\rho_1, \rho_2 \in \mathsf{Eq}_S$ such that*

(1) ρ_1 *and* ρ_2 *are orthogonal;*
(2) *(a)* $\tau_1 \subseteq \rho_2$
  (b) $\tau_2 \subseteq \rho_1$
(3) *(a)* $\rho_2 \cap (\rho_1 \tau_1 \rho_1) \subseteq \tau_1$
  (b) $\rho_1 \cap (\rho_2 \tau_2 \rho_2) \subseteq \tau_2$

As one can see from clause (a), (a special case of) independent equivalences are part of this definition; this requirement arises from the desire to regard ρ_1 and ρ_2 as distinct variables. Requirement (b)(1) means that τ_1 perturbes the state at most like ρ_2 (i.e. changes only the first state component), and analogous for τ_2. (c)(1) finally expresses that any perturbation of the state by ρ_1 (i.e. change of the second state component) before or after, does not affect τ_1. A more detailed analysis can be found below at the end of this section.

The next theorem shows that also semantical independence of two relations implies that they commute.

Theorem 6.68. ([10]) *Let S be a set and let $\tau_1, \tau_2 \subseteq S \times S$ be semantically independent; then τ_1 and τ_2 commute.*

Proof. Let $s\tau_1\tau_2 t'$; then there is an s' such that $s\tau_1 s'$ and $s'\tau_2 t'$. By Def. 6.67 there are equivalence relations $\rho_1, \rho_2 \in \mathsf{Eq}_S$ such that $s\rho_2 s'$ and $s'\rho_1 t'$. Let $s', t' \in X \in S/\rho_1$ and $s, s' \in Y \in S/\rho_2$. Further, let $s \in X' \in S/\rho_1$ and $t' \in Y' \in S/\rho_2$. The orthogonality of ρ_1 and ρ_2 implies $|X' \cap Y'| = 1$.

Let $t \in X' \cap Y'$; then $s\rho_1 t$, as well as $s\rho_2 s'$, $s'\tau_2 t'$ and $t'\rho_2\breve{}\, t$. Since equivalences are symmetric, we get $s\rho_2\tau_2\rho_2 t$, and by Def. 6.67(c)(2) we get $s\tau_2 t$. Similarly, Def. 6.67(c)(1) gives $t\tau_1 t'$, whence $s\tau_2\tau_1 t'$ and $\tau_1\tau_2 \subseteq \tau_2\tau_1$. The proof of the inequality $\tau_2\tau_1 \subseteq \tau_1\tau_2$ follows by symmetry. □

It would be very astonishing, if the extremely regular appearance of Fig. 6.6 (we called it "perfect symmetry") had not already been noticed elsewhere. In fact, orthogonal equivalences are a special case of factor congruences, viz. in algebras without operations (in [48] they are accordingly called *factor equivalences*). This is shown in

Proposition 6.69. ([48, Proposition 3]) *Two equivalences ρ and σ on a set X are factor equivalences iff the intersection of any two equivalence classes $[x]_\rho$ and $[y]_\sigma$, $x, y \in X$, always contains exactly one element.*

Proof. Let ρ and σ be a pair of factor equivalences on X, i.e. according to Def. 10.11 it holds that

$$\rho \cap \sigma = I_X, \quad \rho \circ \sigma = U_X, \quad \text{and} \quad \rho \circ \sigma = \sigma \circ \rho;$$

and let h be the natural map $h : X \longrightarrow X/\rho \times X/\sigma,\ x \mapsto \langle [x]_\rho, [x]_\sigma \rangle$.

For $x, y \in X$, $h(x) = h(y)$ implies $[x]_\rho = [y]_\rho$ and $[x]_\sigma = [y]_\sigma$, so $\langle x, y\rangle \in \rho \cap \sigma = I_X$; thus $x = y$, i.e. h is injective. Furthermore, $\langle [x]_\rho, [x]_\sigma\rangle = \langle [y]_\rho, [y]_\sigma\rangle$ iff $x\rho y \wedge x\sigma y$, i.e. x and y belong to the same equivalence classes under both ρ and σ. Therefore, $|[x]_\rho \cap [y]_\sigma| \le 1$ iff $x = y$, depending on whether this intersection is empty or not. Since $\rho \circ \sigma = U_X$ there is a $z \in X$ with $x\rho z\sigma y$; hence $h(z) = \langle [z]_\rho, [z]_\sigma\rangle = \langle [x]_\rho, [y]_\sigma\rangle$, i.e. h is surjective, and $z \in [x]_\rho \cap [y]_\sigma$. Thus $|[x]_\rho \cap [y]_\sigma| \ge 1$.

Conversely, assume $|[x]_\rho \cap [y]_\sigma| = 1$ for any $x, y \in X$. This implies $\forall x, y \in X \exists z \in X\, [x\rho z\sigma y]$, whence $\rho \circ \sigma = U_X$. Since $\forall x, y \in X\, [x\rho x \wedge y\sigma y]$ we get $\{x, y\} \subseteq [x]_\rho \cap [y]_\sigma$, and by the assumption we get $x = y$, i.e. $\rho \cap \sigma = I_X$. And since $\rho \circ \sigma = U_X$ is an equivalence, ρ and σ commute by Prop. 6.60(1). □

Proposition 6.70. ([48, Proposition 2]) *If $X = A \times B$, then the equivalences ρ, σ on X given by (for $a, a' \in A$ and $b, b' \in B$)*

$$\langle a, b\rangle \rho \langle a', b'\rangle :\Leftrightarrow a = a' \quad \textit{and} \quad \langle a, b\rangle \sigma \langle a', b'\rangle :\Leftrightarrow b = b' \tag{6.19}$$

are factor equivalences on X, and $A \cong X/\rho$ and $B \cong X/\sigma$.

Proof. ρ and σ are the same as $\ker \pi_1$ and $\ker \pi_2$, respectively, of Prop. 10.10. The equivalence classes of ρ are of the form $\{a\} \times B$, for each $a \in A$, so $A \cong X/\rho$. □

Besides factor equivalences, the second ingredient in Def. 6.67 are two pairs of inclusions with the following structure:

$$\tau \subseteq \sigma \text{ and } \sigma \cap \rho\tau\rho \subseteq \tau. \tag{*}$$

These inclusions are equivalent to the equation

$$\sigma \cap \rho\tau\rho = \tau. \tag{**}$$

(Obviously, $\tau \subseteq \rho\tau\rho$ for any reflexive relation ρ. So (*) implies $\tau \subseteq \sigma \cap \rho\tau\rho$, whence (**) follows. Conversely, assuming (**), we trivially get $\sigma \cap \rho\tau\rho \subseteq \tau$; moreover, from (**) and $\tau \subseteq \rho\tau\rho$ we get $\tau \subseteq \sigma \cap \tau \subseteq \sigma \cap \rho\tau\rho \subseteq \sigma$.)

Using the ρ and σ as defined in (6.19) and considering each program state as a pair $\langle a, b\rangle \in X = A \times B$, equation (**) says that for any two pairs $\langle a, b\rangle, \langle a', b'\rangle \in A \times B$, $\langle a, b\rangle \tau \langle a', b'\rangle$ iff

$$\textit{(a)}\ \langle a, b\rangle \sigma \langle a', b'\rangle \quad \textit{and} \quad \textit{(b)}\ \langle a, b\rangle\, \rho\tau\rho\, \langle a', b'\rangle. \tag{6.20}$$

The latter condition is equivalent to: $\langle a, b\rangle \rho x, x\tau y$ and $y\rho\langle a', b'\rangle$ for some $x, y \in A \times B$, i.e. to: there exist $x = \langle a, c\rangle$ and $y = \langle a', c'\rangle$ for some $c, c' \in B$. Thus

$$\langle a, b\rangle\, \rho\tau\rho\, \langle a', b'\rangle \leftrightarrow \exists c, c' \in B\, [\langle a, c\rangle \tau \langle a', c'\rangle]. \tag{6.21}$$

By 6.20(a) the statement τ started in $\langle a, b\rangle$ terminates in state $\langle a', b'\rangle$ if and only if the second state component remains unchanged, i.e. $b = b'$. And because of 6.20(b) and 6.21 the same computation results if the second component b of the input state is arbitrarily changed to some $c \in B$ (note that $c = c'$ since $\tau \subseteq \sigma$ by (*)). Thus τ neither changes nor depends on the second state component.

Putting together the foregoing considerations, we can rephrase Def. 6.67 as follows:

Definition 6.71. ([48, Definition 4]) *Let $\tau_1, \tau_2 \in \mathcal{B}_S$; then τ_1 and τ_2 are are called* **semantically independent** *if there exists a pair of factor equivalences $\rho, \sigma \in \mathsf{Eq}_S$ such that*

(1) $\tau_1 = \sigma \cap \rho\tau_1\rho$,
(2) $\tau_2 = \rho \cap \sigma\tau_2\sigma$.

We have just found that τ_1 neither changes nor depends on the second state component. An analogous results holds for τ_2 and the first state component. This does very well fit to our intuitive understanding: Two statements τ_1 and τ_2 are semantically independent if they do not interfere, i.e. each of them acts on a separate state component. For a still more general notion of semantic independence see [,].

6.2.3.2 *Regular Equivalences*

Note in advance: "*Regular equivalence*" with respect to a relation α is the name that has been given in the area of *Social Network Analysis* (SNA) to an equivalence relation ρ that commutes with α: $\rho\alpha = \alpha\rho$ (from which it follows that $\rho\alpha\rho = \alpha\rho^2 = \alpha\rho$, since equivalences are idempotent). Thus, a regular equivalence in the current context is not an equivalence that is a regular relation in the sense of Section 6.2.2, i.e. a relation ρ such that there is a relation α with $\rho\alpha\rho = \rho$.

Social sciences are another paradigmatic application area for graph theory (and thus indirectly also for relations). Both have been used as models for some decades, especially for the description and analysis of social networks. Intuitively, a *social network* is a set of individuals (actors, agents) together with one or more relationships between them. Take, for example, members of a family, then possible relationships are: husband, wife, married to, likes, mother, sister, etc. Or take the organizational structure of a company: Let there be an overall boss, say BOSS, of a company COMP, and the set of its employees. Eventually, like all other people, the BOSS is also an employee of COMP; but maybe (s)he is the owner of COMP. Other relationships are manager, project leader, worker, driver, etc. with various employees of COMP involved.

The model of graphs and relations is simple, but powerful, and does become mathematically interesting as soon as scientists ask about the structure of a social network, with the most natural question being that for the *structural position* or *social role* of individuals. Intuitively one can say that two actors have the same position to the extent that their patterns of relationships with other actors are similar. Generally spoken, SNA defines two nodes (or other more elaborate structures) as similar if they fall in the same "equivalence class"; thus, various kinds of equivalence relations defined on the population under consideration are used to define roles that members can have. Each approach specifies the graph theoretic properties actors must have in order to be considered equivalent in terms of the roles they play. For much more details than can be presented here, see e.g. [21, 32, 49]

Three particular definitions of equivalence have been particularly useful in applying graph theory to the understanding of social roles and structural positions: *structural, automorphic*, and *regular*. In this order, their definitions become less restrictive of what it means for two actors to be alike. And using less strict definitions we are able to understand social networks at increasing levels of abstraction.

Before going into more details, a word of caution is needed: What is discussed in this short section is the algebraic starting point only for further considerations on network roles and structures. There are, of course, also many non-algebraic aspects that have to be taken into account as e.g. sensitivity of definitions with respect to small changes, metrics for approximations of the types of "equivalence", various clustering approaches etc. For these topics interested readers are referred to the pertinent literature.

Definition 6.72. ([27])

(a) *Let V be a set of individuals (vertices). A* **role assignment** *for V is a surjective map $r : V \longrightarrow W$ onto some set of roles W. A role assignment defines a partition of V via the sets $r^{-1}(w)$, $w \in W$, as classes.*

(b) *Let $G = (V, E)$ be a graph, and $r : V \longrightarrow W$ a role assignment. The* **role graph** *$R = (W, F)$ is the graph with vertex set W and edges $F := \{\langle r(u), r(v)\rangle : \exists u, v \in V\,[\langle u, v\rangle \in E]\} \subseteq W \times W$.*

The ultimate goal of SNA is to find a meaningful role assignment for a given network. As its name indicates, a role graph illustrates the roles and their relations, which means that it is some form of network compression. Two goals of role analysis are to find out which individuals are similar, and to reduce complexity. The hope is that the role graph highlights essential and more persistent network structures: While individuals come and go, and behave rather irregularly, roles are expected to remain stable and to exhibit more regular patterns of interaction. (See [27, p. 218].)

Defining roles and positions in form of equivalence relations is a huge advance over the sometimes quite imprecise and contradictory verbal treatments often found in literature. It is, however, a kind of idealization that may not be fully realized in real world social networks which are often quite messy and/or may be badly measured. In many situations exact equivalence may not exist, and it can become necessary to use measures of approximate equivalence. Nonetheless, the precisely defined formal concepts can still serve as a guideline for generalizations and improvements.

Before giving the rigorous mathematical definitions we start with an informal presentation of the concepts and illustrate them using a standard example of SNA literature. (See e.g. [49, Fig. 12.1]; the same example can also be found on `https://en.wikipedia.org/wiki/Similarity_(network_science)`.

- The historically first approach has been defined in [29] under the name *structural equivalence*, and is the most simple, but also most restrictive one: Two individuals are role equivalent if they have identical neighborhoods, i.e. exactly the same links to and from other individuals in the network. Structural equivalence identifies actors that have the same position, or who are, disregarding any other properties, completely substitutable for each other.

- The idea of *automorphic equivalence* is that sets of actors can be equivalent by being embedded in local structures that have the same patterns of ties, i.e. this way we can find parallel structures, or a sort of structural replication.
- Two individuals are regularly equivalent if they are equally related to equivalent others: they have neighbors who are themselves similar, but do not necessarily have to share them. *Regular equivalence* classes describe the social roles that are the basic building blocks of all social institutions. The idea of a social role, which is "institutionalized" by normative and sanctioned relationships to other roles is at the very core of the entire sociological perspective.

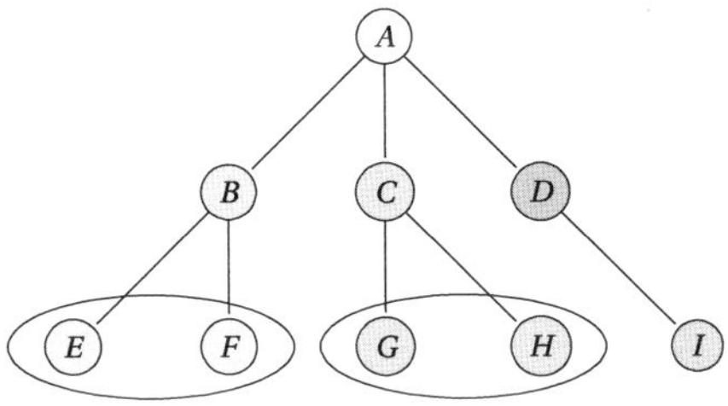

FIG. 6.7: A social network example: Structural equivalence,

Definition 6.73. *Let $G = (V, E)$ be a graph, and $r : V \longrightarrow W$ a role assignment. Then r defines* **structural equivalence** *of vertices if for all $a, b \in V$ we have (using notation for $E \subseteq V \times V$ as a relation) $r(a) = r(b) \rightarrow aE = bE \wedge Ea = Eb$, i.e. two individuals are structurally equivalent, denoted as $a \equiv_{se} b$, iff they have identical fore- and aftersets.*

Thus, structural equivalence is a wholly local concept: One does not need any knowledge of the rest of the network, and an actor's position can never be affected by changes in the network that happen more than two links away. So if there are identically organized departments in a company or a university, or subgroups of a big project team (i.e. uniform parallel substructures), corresponding members in these subnets (as e.g. head of department or group leader) will not qualify as having the same role. Moreover, structural equivalence does not allow comparison of positions and roles between different populations. Of course, as can easily be seen from Fig. 6.7, this is by far too restrictive to be of more than conceptual value for improvements and generalizations: There are seven equivalence classes of which only two have more than one member.

The crucial step, of course, is to abandon the requirement that equivalent individuals must have exactly the *same* neighbors. As alternative we can follow the idea of determining roles (or social positions) by meaningful definitions of a *similar enough* neighborhood configuration.

Definition 6.74. *(***Automorphic equivalence***) Let $G = (V, E)$ be a graph; two nodes $a, b \in V$ are* automorphically equivalent, *$a \equiv_{ae} b$, iff there is an automorphism $h : V \longrightarrow V$ such that $h(a) = b$.*

Automorphic equivalence depends on the notion of isomorphism between graphs. A *graph isomorphism* between two graphs $G = (V, E)$ and $H = (V', E')$ is a mapping $h : V \longrightarrow V'$ such that for all $a, b \in V$, $aEb \iff h(a)E'h(b)$. Since they differ only in labeling, isomorphic graphs are identical with respect to all graph-theoretic attributes. There always exists at least the trivial automorphism $id_V : V \longrightarrow V$.

Intuitively, actors are automorphically equivalent if we can permute the graph in such a way that exchanging the two actors has no effect on the distances among all actors in the graph. Unlike structural equivalence, automorphic equivalence is independent of proximity: automorphically equivalent actors may be neighbors, adjacent, or completely unreachable from each other. Moreover, whereas structural equivalence focuses on what individuals might be exchanged without modifying the social relations described by a network, automorphic equivalence asks what sets of actors are substitutable as sub-graphs, in relation to other sub-graphs. but retaining the relational structure as a whole.

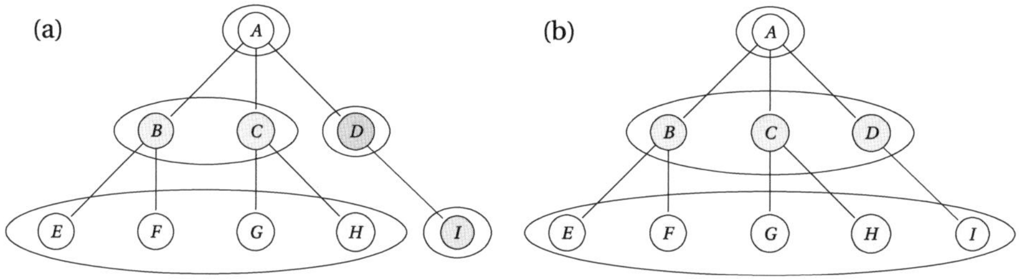

FIG. 6.8: A social network example: (a) automorphic and (b) regular equivalence.

Definition 6.75. (**Regular equivalence**) *Let $G = (V, E)$ be a graph, and $\equiv$ an equivalence relation on V; $\equiv$ is called* **regular** *iff, for all $a, b, c \in V$, $a \equiv b$ implies*

$$(aEc \rightarrow \exists d \in V\,[bEd \wedge c \equiv d]) \quad \wedge \quad (cEa \rightarrow \exists d \in V\,[dEb \wedge c \equiv d]).$$

That two nodes $a, b \in V$ are regular equivalent, *is denoted by $a \equiv_{re} b$.*

Regular equivalence is the least restrictive of the three most commonly used definitions of equivalence, but probably the most important one for sociologists. It does not require actors to have identical ties to identical other actors (as required by structural equivalence) or to be structurally indistinguishable (as required by automorphic equivalence). Instead, actors are regularly equivalent if they are equally related to equivalent others. Fig. 6.8 clearly illustrates this difference between focusing on substructures and roles, respectively: Even though actors B and C are not structurally equivalent (they do have the same boss A, but not the same workers), they do seem to be equivalent in a different sense: both have the same boss and two workers. And, in fact, $B \equiv_{ae} C$. But if you look at D you will recognize that like B and C it has a tie to the boss class $\{A\}$ as well as to the workers' class, though to only one member of it; but this doesn't matter, as there is a tie to some member of the third class. Consequently, $B \equiv_{re} C \equiv_{re} D$.

What is shown in Fig. 6.8(b) is the partition that in some sense could be viewed as the "best" one. However, there are more than only this one regular equivalence. To keep things small, consider the sub-tree including the C- and D-branches. Then π_I with only singleton classes, and π_U, which groups all actors in one class, are regular equivalences, but also {{A}, {C,D}, {G,H,I}} and {{A}, {C}, {D},{G,H}, {I}}. So, regular equivalences alone are not sufficient for finding adequate role assignments and supplementary methods must be used; see e.g. [15].

We still have to keep our promise concerning the commuting property of regular equivalences. This is done in the next proposition, with notation changed from SNA terminology back to our general formalism used for relations

Proposition 6.76. *Let $\rho, \alpha \in \mathcal{B}_S$, and ρ be a regular equivalence with respect to a relation α; then $\rho\alpha = \alpha\rho$.*

Proof. According to Def. 6.75 we have $a\rho b$, iff for all $a, b, c \in S$,

$$(\dagger)\ a\alpha c \to \exists d \in S[b\alpha d \wedge c\rho d] \quad \text{and} \quad (\ddagger)\ c\alpha a \to \exists d \in S[d\alpha b \wedge c\rho d].$$

Assume $a\alpha\rho b$; then $\exists z \in S[a\alpha z\rho b]$, and using ($\ddagger$), $a\alpha z$ implies $\exists d \in S[d\alpha b \wedge a\rho d]$, whence $a\rho\alpha b$. □

To be honest, this last result is only one of three reasons why regular equivalences have been mentioned here. In addition to the commuting property, it was (1) a fine opportunity to have a short look at an interesting and well established application area of graph theory and relations, and it is (2) interesting to arrive at the same concept from quite another point of view in Section 17.2 under the name of a *bisimulation* relation.

6.2.3.3 *Idempotent Relations*

Obviously, idempotent relations are a special case of commuting relations. Idempotent relations are also a special case of transitive relations. We did already meet an important class of idempotent relations, viz. equivalences. In fact it has been shown in Prop. 3.53(3), that all preorders, i.e. the reflexive and transitive relations, exhibit this property, and also coreflexive relations are idempotent by Prop. 3.53(5). It is known that preorders are a proper subset of idempotent relations which are a proper subset of the set of transitive relations (see [24, p. 83] or [11, p. 169]). Examples are given immediately after Prop. 6.81.

In the following we will use the terms "*relation*" and "*matrix*" interchangeably, depending on where the concepts and terminology under consideration arise from. In Def. 2.49 we introduced the notation E_A for the subset of idempotents of a semigroup A. In the following we will simply use Idm_X to denote the set of idempotent binary relations on an arbitrary finite, but fixed set X, $|X| \geq 2$, instead of the more cumbersome $E_{\mathcal{B}_X}$. In fact, thinking in terms of Boolean matrices, only their order $|X| = n$ is relevant.

Before we start investigating characteristic properties of idempotent relations themselves, we will have a look at the structure of Idm_X.

Corollary 6.77. *The set of idempotents* Idm_X *forms a complete lattice.*

Proof. $\{\emptyset, I_X, U_X\} \subseteq \mathsf{Idm}_X$, whence, for all idempotents χ, $\emptyset \subseteq \chi \subseteq U_X$; so Idm_X is a poset with a least and a greatest element. For $A = \{\chi_1, \ldots, \chi_K\} \subseteq \mathsf{Idm}_X$ we define $\bigvee A = \mathbf{t}(\chi_1 \cup \ldots \cup \chi_K)$, for which, by Prop. 3.64(3), we have

$$\mathbf{t}(\chi_1 \cup \ldots \cup \chi_K) = \bigcup_{1 \le i_k \le K,\ n \ge 1} \chi_{i_1} \chi_{i_2} \cdots \chi_{i_n}.$$

Next we have to ensure that $\mathbf{t}(\chi_1 \cup \ldots \cup \chi_K) \in \mathsf{Idm}_X$. Once again, for sake of simplicity we demonstrate the proof idea for $K = 2$:

$$\begin{aligned}
[\mathbf{t}(\rho \cup \sigma)]^2 &= [\textstyle\bigcup_{i \ge 1} (\rho \cup \sigma)^i]^2 \\
&= [(\rho \cup \sigma) \cup (\rho^2 \cup \sigma\rho \cup \rho\sigma \cup \sigma^2) \cup (\rho^3 \cup \sigma\rho^2 \cup \rho\sigma\rho \cup \ldots) \cup \ldots]^2 \\
&= (\rho \cup \sigma)^2 \cup (\rho^2 \cup \sigma\rho \cup \rho\sigma \cup \sigma^2)(\rho \cup \sigma) \cup (\rho^3 \ldots \\
&= (\rho^2 \cup \sigma\rho \cup \rho\sigma \cup \sigma^2) \cup (\rho^2 \cup \sigma\rho \cup \rho\sigma \cup \sigma^2)(\rho \cup \sigma) \cup (\rho^3 \ldots \\
&= (\rho \cup \sigma\rho \cup \rho\sigma \cup \sigma) \cup (\rho \cup \sigma\rho \cup \rho\sigma \cup \sigma)(\rho \cup \sigma) \cup (\rho^3 \ldots \\
&= (\rho \cup \sigma\rho \cup \rho\sigma \cup \sigma) \cup (\rho\sigma\rho \cup \sigma\rho\sigma) \cup (\rho^3 \ldots \\
&= \textstyle\bigcup_{1 \le i_k \le 2,\ n \ge 1} \chi_{i_1} \chi_{i_2} \cdots \chi_{i_n} \qquad \text{(with } \chi_1 := \rho, \chi_2 := \sigma) \\
&= \mathbf{t}(\rho \cup \sigma).
\end{aligned}$$

So we have proved that, for any $\emptyset \neq A \subseteq \mathsf{Idm}_X$ the join $\bigvee A$ exists, whence Idm_X is a complete join semilattice. From lattice theory it is known that if for a poset P $\bigvee A$ exists for all $A \subseteq P$, then P is a complete lattice (see [, p 24, Lemma 14] for the dual result): Let $B \subseteq P$ be the set of lower bounds of A; then, by hypothesis, $l := \bigvee B$ exists. If $a \in A$, then $a \ge b$ for all $b \in B$, whence $a \ge l$, and $l \in B$. Thus l is the greatest member of B, that is, $l = \bigwedge A$. □

Idm_X is however not a sublattice of Eq_X since the meet operations differ as we will see soon. In order to establish this fact, we need some preparation: In the following we write $I_\rho := \{a \in X : a\rho a\} \subseteq X$, $a \sim b := a(\rho \cap \sigma)b$, and $\hat{\sim} := a(\rho \wedge \sigma)b$.

Lemma 6.78. *Let* $\rho, \sigma \in \mathsf{Idm}_X$, *with* X *finite and* $|X| = k$.

(1) $a\rho b$ *iff* $\exists m \in I_\rho\, [a\rho m \wedge m\rho b]$*; and* $\rho \neq \emptyset$ *implies* $I_\rho \neq \emptyset$ ([45, Lemma 1]).

(2) $\rho \cap \sigma$ *is transitive.*

Proof. (1): By the idempotency of ρ, for any a, b such that $a\rho b$ and each $n \ge 1$, we have a sequence $a = x_0, x_1, \ldots, x_n = b$ with $x_i \rho x_{i+1}$ for $0 \le i \le n-1$. And since ρ is transitive by definition, $x_i \rho x_j$ whenever $i < j$. Taking $n > k$, $x_i = x_j$ must hold for some $i < j$. So $a\rho x_i \wedge x_i \rho b$, and $x_i \in I_\rho$.

(2): $(\rho \cap \sigma)^2 = \rho^2 \cap \rho\sigma \cap \sigma\rho \cap \sigma^2 = \rho \cap \sigma \cap \rho\sigma \cap \sigma\rho \subseteq \rho \cap \sigma$. □

Following [45, p. 298,(3)] we define

$$\rho \wedge \sigma := \{\langle a, b\rangle \in \rho \cap \sigma : \text{for some } m \in I_{\rho\cap\sigma} \text{ both } \langle a, m\rangle \text{ and } \langle m, b\rangle \in \rho \cap \sigma\},$$

and translated to our shorthand notation this gives

$$a \hat{\sim} b :\Longleftrightarrow a \sim b \wedge \exists m \in I_{\rho\cap\sigma}\, [a \sim m \wedge m \sim b].$$

Proposition 6.79. *Let* $\rho, \sigma \in \mathsf{Idm}_X$.

(1) $\rho \wedge \sigma$ *is idempotent.*
(2) $\rho \wedge \sigma$ *is the greatest lower bound of* ρ *and* σ *in* Idm_X.

Proof. (1): Let $a \mathrel{\hat{\sim}}^2 b$; then $\exists c \in X\,[a \mathrel{\hat{\sim}} c \mathrel{\hat{\sim}} b]$, whence by definition of $\hat{\sim}$ we get

$$\exists c \in X\left[\exists m \in I_{\rho\cap\sigma}\,[a \sim m \wedge m \sim c] \;\wedge\; \exists m' \in I_{\rho\cap\sigma}\,[c \sim m' \wedge m' \sim b]\right],$$

that is $\exists c \in X\, \exists m, m' \in I_{\rho\cap\sigma}$ such that

$$a\rho m \wedge a\sigma m \wedge m\rho c \wedge m\sigma c \;\wedge\; c\rho m' \wedge c\sigma m' \wedge m'\rho b \wedge m'\sigma b.$$

By transitivity, $m\rho c \wedge c\rho m' \;\rightarrow\; m\rho m'$, so that $a\rho m\rho m'\rho b$, i.e. $a\rho m \wedge m\rho^2 b$; and similarly we have $a\sigma m \wedge m\sigma^2 b$. Thus, $a(\rho\cap\sigma)m \wedge m(\rho^2 \cap \sigma^2)b$ which means $\exists m \in I_{\rho\cap\sigma}\,[a \sim m \wedge m \sim b]$. Furthermore, since $\rho \wedge \sigma \subseteq \rho\cap\sigma$ by definition, we get $(\rho \wedge \sigma)^2 \subseteq (\rho\cap\sigma)^2 \subseteq \rho\cap\sigma$ by monotonicity and Lemma 6.78(2); so $a \mathrel{\hat{\sim}}^2 b$ implies $a \sim b$. Altogether we thus have $a \mathrel{\hat{\sim}}^2 b \subseteq a \mathrel{\hat{\sim}} b$.

For the other inclusion, assume $a \mathrel{\hat{\sim}} b$; this gives $a \sim b \;\wedge\; \exists m \in I_{\rho\cap\sigma}\,[a \sim m \;\wedge\; m \sim b]$. Since also $m \sim m$ we get $a \mathrel{\hat{\sim}} m \wedge m \mathrel{\hat{\sim}} b$, whence $a \mathrel{\hat{\sim}}^2 b$.

(2): We have to show that, for any $\lambda \in \mathsf{Idm}_X$ with $\lambda \subseteq \rho\cap\sigma$, also $\lambda \subseteq \rho \wedge \sigma$. Let $a\lambda b$; then by Lemma 6.78(1) $\exists m \in I_\lambda\,[a\lambda m \wedge m\lambda b]$. Thus, since $\lambda \subseteq \rho\cap\sigma$ has been assumed, we get $a \sim b \;\wedge\; \exists m \in I_{\rho\cap\sigma}\,[a \sim m \wedge m \sim b]$. Therefore, $a \mathrel{\hat{\sim}} b$ and together with (1) the assertion follows. □

In the following we will present various characterizations of idempotent relations of finite order to be found in literature, as well as a few of their properties. We begin with listing obvious and easy to prove facts:

Proposition 6.80.

(1) *Idempotent relations are transitive.*
(2) *Converses of idempotents are idempotent.*
(3) *Idempotent relations are regular.*
(4) *An idempotent* $\chi \neq I$ *cannot posses an inverse.*
(5) *Let* π *be a permutation relation; then* ρ *is idempotent iff* $\pi\rho\pi^{\smallsmile}$ *is.*

Proof. (2): $\lambda = \lambda^2$ implies $\lambda^{\smallsmile} = (\lambda\lambda)^{\smallsmile} = \lambda^{\smallsmile}\lambda^{\smallsmile}$.
(4): Assuming $\chi\chi^{-1} = I$ implies $\chi = \chi I = \chi\chi\chi^{-1} = \chi\chi^{-1} = I$.
(5): $\rho^2 = \rho$ implies $\pi\rho\pi^{\smallsmile}\pi\rho\pi^{\smallsmile} = \pi\rho^2\pi^{\smallsmile} = \pi\rho\pi^{\smallsmile}$. Conversely, from $(\pi\rho\pi^{\smallsmile})^2 = \pi\rho\pi^{\smallsmile}$ we get $\pi\rho^2\pi^{\smallsmile} = \pi\rho\pi^{\smallsmile}$ as above. Multiplying the last equation with $\pi^{\smallsmile}$ from the left and with π from the right we get $\rho^2 = \rho$. □

We have already met examples of idempotent relations in former chapters, and for sake of completeness we repeat these results in

Proposition 6.81.

(1) *Preorders are idempotent.*

(2) *Equivalence relations are idempotent.*
(3) *Coreflexive relations are idempotent.*
(4) *Let ρ and σ be partial equivalence relations on a set S, then $\rho\sigma$ is idempotent.*
(5) *If $\rho\,\breve{}\rho = \rho$ then ρ is idempotent.*

Proof. (1): By Prop. 3.53(3).
(2): By (1).
(3) By Prop. 3.53(5).
(4) By Prop. 4.32(5).
(5) From the hypothesis we get $\rho\breve{} = (\rho\breve{}\rho)\breve{} = \rho\breve{}\rho = \rho$, whence ρ is symmetric. Therefore $\rho^2 = \rho\breve{}\rho = \rho$. □

The inclusions preorders ⊂ idempotents ⊂ transitive relations are proper: $\begin{pmatrix} 0 & 1 \\ 0 & 1 \end{pmatrix}$ is idempotent but not reflexive, and $\begin{pmatrix} 0 & 1 \\ 0 & 0 \end{pmatrix}$ is transitive but not idempotent.

Various characterizations of the idempotency of relations of finite order can be found in literature. Among the earliest ones are [37], based on graph-theoretic considerations, and [39], based on a viewpoint from order relations. The latter one is presented in Th. 6.82 below, but before we can do so we need some new notions.

Let ρ be a preorder, i.e. a reflexive and transitive relation on a set A. An element $a \in A$ is called *ρ-strict* if $a\rho a' \wedge a'\rho a$ imply $a = a'$. Generalizing the notion from orders to arbitrary relations we say that b **covers** a in ρ, denoted as $a \prec_\rho b$, if $a\rho b$ and $a\rho c\rho b \to c = a \vee c = b$. Two elements $a, b \in A$ are called *ρ-neighbors* if $a \prec_\rho b \vee b \prec_\rho a$. A subset $B \subseteq A$ is called *ρ-permissible*, if all $b \in B$ are ρ-strict and B does not contain ρ-neighbors.

A binary relation $\sigma \in \mathcal{B}_A$ is called a **pseudo-order** if $\sigma = \rho - I_B$, with $\rho \in \mathcal{B}_A$ a preorder and $B \subseteq A$ a ρ-permissible subset of A.

Theorem 6.82. *$\sigma \in \mathcal{B}_A$ is idempotent iff it is a pseudo-order relation.*

Proof. [⇐] Let $\sigma = \rho - I_B$ be a pseudo-order relation, with ρ a preorder and $B \subseteq A$ a ρ-permissible subset of A, and let $a\sigma b \wedge b\sigma c$. Then, since clearly $\sigma \subseteq \rho$, $a\rho b \wedge b\rho c$, whence $a\rho c$ by transitivity of ρ. Now assume aI_Bc; then $a \in B$ and $a = c$ by definition, whence we get $c\sigma b \wedge b\sigma c$. Since $a = c \in B$ is ρ-strict, we have to conclude that $a = b$ and aI_Bb, a contradiction to our hypothesis that $a\sigma b$. So $\neg aI_Bc$ and $a\sigma c$, i.e. $\sigma^2 \subseteq \sigma$.

Conversely, let $a\sigma b$. If $a \notin B \vee b \notin B$ then either $a\sigma a$ or $b\sigma b$. In both cases we get $a\sigma^2 b$. And assuming the contrary, i.e. $\{a, b\} \subseteq B$, we get $a\rho b$, since $\sigma \subseteq \rho$. As a preorder, ρ is idempotent by Prop. 3.53(3), so $a\rho c\rho b$ for some $c \in A$. Since a, b are not ρ-neighbors (due to $B \subseteq A$ being assumed to be ρ-permissible) it follows that $a \neq c$ and $b \neq c$, whence $a\sigma c\sigma b$. Thus $\sigma \subseteq \sigma^2$, and so σ is idempotent.

[⇒] Let σ be idempotent such that $\rho = \sigma \cup I_A$, and $B = A - \mathsf{fld}(\sigma \cap I_A)$. Then ρ is reflexive and $\rho^2 = \sigma^2 \cup \sigma I_A \cup I_A\sigma \cup I_A^2 = \sigma \cup I_A = \rho$, i.e. ρ is also transitive and thus a preorder. Clearly, $\sigma = \rho - I_B$.

We have to show that B is ρ-permissible: Let $a \in B$ and $a\rho b \wedge b\rho a$. $a \neq b$ implies $a\sigma b \wedge b\sigma a$, whence $a\sigma a$ since σ was assumed to be idempotent, a contradiction to $\sigma = \rho - I_B$. Therefore $a = b$, i.e. a is ρ-strict. Now we consider two distinct elements

$a, b \in B$ with $a\rho b$. We have $a\sigma b$, whence $a\sigma c\sigma b$ for some $c \in A$. $c = a \vee c = b$ implies $\neg a\sigma c \vee \neg c\sigma b$ so that we get $a \neq c \wedge c \neq b$, whence a and b are not ρ-neighbors. □

We conclude this section with some further characterizations of the idempotency of Boolean relations and matrices:

Proposition 6.83. *Let $B \in \mathbf{B}_n$ be reflexive with index k. Then B is idempotent iff $B = \mathrm{adj}(B)$.*

Proof. Assume $B = \mathrm{adj}(B)$. By Cor. 5.32(1) we have $k \leq n-1$, whence, by Cor. 5.32(3), B^{n-1} is idempotent. And in Prop. 5.37 it has been shown that $\mathrm{adj}(B) = B^{n-1}$. Conversely, if B is idempotent, then $B = B^{n-1} = \mathrm{adj}(B)$ by Prop. 5.37. □

Proposition 6.84. ([31, Lemmas 1.7, 1.8])

(1) Let $\alpha \in \mathcal{B}_X$; then $\alpha^2 = \alpha$ iff $A\alpha = A$ for every $A \in V(\alpha)$ iff $\alpha A = A$ for every $A \in W(\alpha)$.
(2) Let $\emptyset \neq A, B \subseteq X$, such that $A \cap B \neq \emptyset$, and let $\alpha = A \times B$. Then $V(\alpha) = \{\emptyset, B\}$ and $W(\alpha) = \{\emptyset, A\}$; hence $\alpha \in \mathrm{Idm}_X$.
(3) Partial order relations on subsets of X are reduced idempotents.

Proof. (See page 174 for the definitions of the row and column space of a relation.) (1): Assume $A \in V(\alpha)$; then there is a $B \subseteq X$ such that $A = B\alpha$. Thus, $A\alpha = B\alpha\alpha = B\alpha = A$. Conversely, $x\alpha \in V(\alpha)$ for every $x \in X$. By assumption, $x\alpha = (x\alpha)\alpha = x(\alpha\alpha)$, whence $\alpha^2 = \alpha$.

(2): By definition of α, $a\alpha = B$ for all $a \in A$ and $a\alpha = \emptyset$ for all $a \in \overline{A}$. Thus, $V(\alpha) = \{\emptyset, B\}$. Since $A \cap B \neq \emptyset$ we get

$$B\alpha = \textstyle\bigcup_{x\in B} x\alpha = \left(\bigcup_{x\in A\cap B} x\alpha\right) \cup \left(\bigcup_{x\in B-A} x\alpha\right) = B \cup \emptyset = B.$$

and α is idempotent by (1).

(3): See Prop. 9.40. □

References

[1] Bandelt, H.-J.: Regularity and complete distributivity, Semigroup Forum **19**, 123-126 (1980).

[2] Bandler, W., Kohout, L.J.: Fuzzy relational products as a tool for analysis and synthesis of the behaviour of complex natural and artificial systems. In Wang, P.P., Chang S.K. (Eds.): *Fuzzy Sets: Theory and Applications to Policy Analysis and Information Systems*, 341-367, Plenum Press (1980).

[3] Bandler, W., Kohout, L.J.: On the universality of the triangle superproduct and the square product of relations. Int. J. of General System **25**, 399-403 (1997).

[4] Berghammer R.: *Ordnungen, Verbände und Relationen mit Anwendungen*, 2nd ed., Springer Vieweg, Wiesbaden (2012).

[5] Berghammer R., Gritzner T.F., Schmidt G.: Prototyping relational specifications using higher-order objects. In Heering J., Meinke K., Möller B., Nipkow T. (Eds): *Higher-Order Algebra, Logic, and Term Rewriting. HOA 1993.* Lecture Notes in Computer Science, vol 816, 56-75, Springer, Berlin, Heidelberg (1994).

[6] Berghammer R., Hoffmann, T.: Deriving relational programs for computing kernels by reconstructing a proof of Richardson's theorem, Science of Computer Programming **38** 1-25 (2000).

[7] Berghammer R., Hoffmann, T., Leoniuk, B., Milanese, U.: Prototyping and Programming with Relations, Electronic Notes in Theoretical Computer Science **44**(3), 27-50 (2003).

[8] Berghammer, R., Winter, M.: Embedding mappings and splittings with applications, Acta Informatica **47**, 77-110 (2010).

[9] Berghammer, R., Winter, M.: Decomposition of relations and concept lattices. Fundamenta Informaticae, **126**, 37-82 (2013).

[10] Best, E., Lengauer, C.: Semantic independence, Science of Computer Programming **13**, 23-50 (1989).

[11] Boyd, J.P..: Structural Similarity, Semigroups and Idempotents, Social Networks **5**, 157-172 (1983).

[12] Costa, J.P.: On the lattice of equivalence relations (2006).

[13] De Baets, B., Kerre, E.: A revision of Bandler-Kohout compositions of relations, Math. Pannonica **4**, 59-78 (1993).

[14] De Baets, B., Kerre, E.: Fuzzy relations and applications, Advances in Electronics and Electron Physics **89**, 255-324 (1994).

[15] Doreian, P., Batagelj, V., Ferligoj, A.: *Generalized Blockmodeling*, Cambridge University Press (2005).

[16] Dubreil, P., Dubreil-Jacotin, M.-L.: Théorie algébrique des relations d'équivalence, J. Math. Pures Appl. **18**, 63-95 (1939).

[17] Finberg, D., Mainetti, M., Rota, G.-C.: The logic of commuting equivalence relations. In Ursini, A., Agliano, P. (Eds): *Logic and Algebra* (Pontignan, 1994), 69-96, Marcel Dekker, New York (1996).

[18] Fränzle, M., Lengauer, C.: Semantic Independence. In Padua, D. (Ed.): *Encyclopedia of Parallel Computing*, 1803-1810, Springer Science+Business Media (2011).

[19] Fränzle, M., von Stengel, B., Wittmüss, A.: A generalized notion of semantic independence, Information Processing Letters **53**, 5-9 (1995).

[20] Grätzer, G.: *General lattice theory*, Birkhäuser Verlag, Basel und Stuttgart (1978).

[21] Hanneman, R.A., Riddle, M.: *Introduction to social network methods*, University of California, Riverside (2005) (see also `http://faculty.ucr.edu/~hanneman/`).

[22] Kahl, W.: Collagories: Relation-algebraic reasoning for gluing constructions, The Journal of Logic and Algebraic Programming **80**, 297-338 (2011).

[23] Khchérif, R., Gammoudi, M., Jaoua, A.: Using Difunctional Relations in Information Organization, Information Sciences, **125**, 153-166 (2000).

[24] Kim, K.H.: *Boolean Matrix Theory and Applications*, Marcel Dekker, Inc., New York and Basel (1982).

[25] Kohout, L.J., Bandler, W.: Fuzzy relational products in knowledge engineering. In Novák, V., Ramík, J. et.al. (Eds.): *Fuzzy Approach to Reasoning and Decision-Making*, 51-66, Springer Science+Business Media Dordrecht (1992).

[26] Kushilevitz, E.: Nisan, N.: *Communication Complexity*, Cambridge University Press (1997).

[27] Lerner. J.: Role assignments. In Brandes, U., Erlebach, T. (Eds): *Network Analysis - Methodological Foundations*, 216-252, Springer-Verlag Berlin Heidelberg (2005).

[28] Longstaff : Strongly reflexive lattices, J. London Math. Soc. (2) **11**, 491-498 (1975).

[29] Lorrain, F., White, H.C.: . Structural equivalence of individuals in social networks, Journal of

Mathematical Sociology **1**, 49-80, (1971).

[30] Luce, R.D.: A note on Boolean matrix theory, Proc. American Math. Soc. **3**, 382-388, (1952).

[31] Namnak, C., Preechasilp, P.: Natural partial orders on the semigroup of binary relations, Thai J. Mathematics 4, 39-50 (2006).

[32] Pattison, P.: *Algebraic Models for Social Networks*, Cambridge University Press (1993).

[33] Posey, A.E.: On difunctional and circular relations, Pi Mu Epsilon J. **6**, 394-399 (1977).

[34] Raney, G.N.: A Subdirect-Union Representation for Completely Distributive Complete Lattices, Proc. American Math. Soc. **4**, 518-522 (1953).

[35] Raney, G.N.: Tight Galois connections and complete distributivity, Trans. American Math. Soc. **97**, 418-426 (1960).

[36] Riguet, J.: Riguet, J. Relations binaires, fermetures, correspondences de Galois, Bull. Soc. Math. France **76**, 114-155 (1948).

[37] Rosenblatt, D.: On the graphs of finite idempotent Boolean relation matrices, J. Res. Nat. Bureau of Standards **67B**, 249-256 (1963).

[38] Rutherford, D.: Inverses of Boolean matrices, Proc. of the Glasgow Math. Assoc. **6**, 49-53 (1963).

[39] Schein,B.M.: A construction for idempotent binary relations, Proc. Japan Acad. **46**, 246-247 (1970).

[40] Schein, B.M.: Regular elements of the semigroup of all binary relations, Semigroup Forum **13**, 95-102 (1976).

[41] Schmidt, G.: Rectangles, fringes, and inverses. In Berghammer, R., Möller, B., Struth, G. (Eds.) *Relations and Kleene Algebra in Computer Science*. Lecture Notes in Computer Science, Vol. 4988, 352-366, Springer-Verlag Berlin Heidelberg (2008).

[42] Schmidt, G.: *Relational Mathematics*, Cambridge University Press, Cambridge UK (2011).

[43] Schmidt, G., Ströhlein, T.: *Relations and Graphs*, Springer-Verlag, Berlin Heidelberg (1993).

[44] Schmidt, G., Ströhlein, T.: Relation algebras-concept of points and representability, Discrete Mathemathics **54**, 83-92 (1985).

[45] Shmuely, Z.: The lattice of idempotent binary relations, Algebra Universalis **9**, 297-303 (1979).

[46] Skornjakov, L.A.: *Elements of Lattice Theory*, Hindustan Publishing Corporation (India) (1977).

[47] Sobera, J.: Mixed pseudo-associativities of Bandler-Kohout compositions of relations, Kybernetika **43**, 143-156 (2007).

[48] von Stengel, B.: An algebraic characterization of semantic independence, Information Processing Letters **39**, 291-296 (1991).

[49] Wasserman, S., Faust, K.: *Social Network Analysis-Methods and Applications*, Cambridge University Press (1994).

[50] Wedderburn, J.H.M.: Boolean Linear Associative Algebra, Annals of Mathematics **35**, 185-194 (1934).

[51] Winter M.: Decomposing relations into orderings. In Berghammer R., Möller B., Struth G. (Eds.): *Relational and Kleene-algebraic Methods in Computer Science*. Lecture Notes in Computer Science, Vol. 3051, 265-277, Springer, Berlin (2004).

[52] Zaretskiĭ, K.A.: The semigroup of binary relations, Mat. Sbornik **61**, 291-305 (1963).

PART 3

Relations - Concrete, many-valued

Chapter 7

Many-valued Extensions: Preliminaries

Despite of their modeling power, binary relations are only a special two-valued (or crisp) case of a more general class of relations where the two-element Boolean algebra $\mathbf{2} := \{0 < 1\}$ is replaced by a certain type of lattice L as a valuation space. In order to be well prepared for many-valued extensions of classical two-valued concepts in general, an enhancement of the basic formal prerequisites becomes necessary.

It is easily seen that despite of the many theoretical investigations and despite of their modeling power, binary relations as introduced in the preceding chapters are only a special two-valued (or crisp) case of relations. Therefore, whatever else a compilation on binary relations claiming to be reasonably comprehensive may offer, not mentioning their many-valued variants would it make incomplete.

In order to be well prepared for many-valued extensions of classical two-valued concepts, an enhancement of the basic prerequisites from Chapter 2 becomes necessary. We have decided to include the additional material in a separate part in order to have all of our many-valued digressions packed into one place. Thus readers not interested in these generalizations can skip Part III. (There is one exception: categories of fuzzy relations are considered in Section 11.4.4.)

Dealing with many-valued concepts often leads to situations where definitions and properties result in formulas identical to their two-valued counterparts. Once again for sake of completeness I will repeat the most basic facts in the many-valued version. This is why this chapter maybe tends to get a little bit redundant compared to others. But no doubt it is interesting to see which of the classical results carry over to the many-valued case when the two-element Boolean algebra $\mathbf{2} := \{0 < 1\}$ is replaced by a lattice L of a certain type as the valuation space.

On the syntactical level we start from scratch. Let A be a non empty set. It is well-known that there is a bijective correspondence between the family of all subsets of A and the family $\mathbf{2}^A$ of all functions from A to $\mathbf{2} = \{0,1\}$, which assigns to each $B \subseteq A$ its **characteristic function** $\chi_B : A \longrightarrow \mathbf{2}$ defined as

$$\chi_B(a) = \begin{cases} 1\,, & \text{if } a \in B \\ 0\,, & \text{otherwise.} \end{cases} \tag{7.1}$$

The enhancement to many-valued sets can best be achieved by using a lattice L with $|L| > 2$ as valuation space: Using functions $\mu : A \to L$ yields generalized characteristic functions. According to [14], μ is then called an **L-fuzzy set**, and more specific also an L-fuzzy set over A denoted μ_A. The term *fuzzy* has first been used in a mathematical context for the concept of a **fuzzy set** introduced in Zadeh's 1965 paper [37], where L is taken to be **unit interval** $[0,1] \subseteq \mathbb{R}$. Henceforth we will use the term '*fuzzy*' as a common and well-known synonym for '*many-valued*'.

$\mu_A(x) = l$ is usually read as "*x is element of A with degree l*" or "*the proposition '$x \in A$' has truth degree l*". We will not discuss here possible interpretations of these statements; instead we are interested in what are desirable and reasonable properties for L as a set of generalized truth values. This is the topic of the next section.

7.1 Graded Truth Values and their Structures

In this section we enter the area of *many-valued logic* - a term that is used to cover a very broad spectrum of mathematical constructs, ranging from rigorous formal calculi to what is called approximate reasoning within the fuzzy community. We will stay somewhere in the middle and focus on the algebraic structures suitable for sets of generalized truth values. Our considerations follow the expositions in [2, Section 2.2] and [33, Section 3.1]; see also [10].

Let $\{\varphi_i : i \in I\}$ be a set of propositions; the truth value of a proposition φ is denoted by $\|\varphi\|$. For the example given above we simply write $\|x \in A\| = l$.

We will now try to explain how certain natural logical assumptions reflect themselves in corresponding properties of the structure of truth values. Obviously, the graded truth approach directly leads to the requirement that the set L of truth values be a poset with a minimum 0 and a maximum 1 as the values which have to serve for the special case of classical two-valued logic. Considering sentences like "*there is an $i \in I$ such that φ_i*", a nearby reformulation of the two-valued case is to take the supremum of the $\|\varphi_i\|, i \in I$, and dually for universal propositions. Together with the properties of disjunction and conjunction this requires L to be a complete lattice.

Of course, as it is the case for classical propositional logic, we also want the many-valued generalization to be *truth functional*, i.e. the truth value of a compound formula can be calculated from the truth values of its parts. Furthermore, it is known that both $\{\neg, \&\}$ and $\{\neg, \to\}$ are both *functionally complete* sets of connectives for propositional logic (i.e. all propositional formulas can be expressed using only connectives from such a set).

We first consider conjunction and denote it by & in order to have the symbol $\wedge$ available for lattice operations. Writing $\otimes$ for the binary operation on L corresponding to & we have $1 \otimes 1 = 1, 1 \otimes 0 = 0 \otimes 1 = 0 \otimes 0 = 0$ as compatibility requirements due to the two-valued case. Furthermore we want to have $\|\varphi \& \psi\| = \|\psi \& \varphi\|$; assuming truth functionality leads to the requirement of commutativity of $\otimes$: $\|\varphi\| \otimes \|\psi\| = \|\varphi \& \psi\| = \|\psi \& \varphi\| = \|\psi\| \otimes \|\varphi\|$.

Since we get associativity of $\otimes$ in an analogous way, $(L, \otimes, 1)$ should have the structure of a commutative monoid. Moreover, it is intuitively appealing to require that the bigger the truth degrees of two propositions the bigger the truth degree of their conjunction: $\otimes$ should be non-decreasing, i.e. for $a, b, c, d \in L$ if $a \leq b \;\&\; c \leq d \rightarrow a \otimes c \leq b \otimes d$.

We now turn to implication, not only because, like conjunction, it also forms (together with negation) a functionally complete set of connectives, but also because conjunction and implication together play an important role for inference via the rule called **modus ponens**. Modus ponens says that if φ and $\varphi \rightarrow \psi$ are valid propositions, then we are allowed to infer that also ψ is a valid proposition. An equivalent reformulation of classical modus ponens is : If φ is valid in degree at least a, and $\varphi \rightarrow \psi$ in degree at least b, then ψ is valid in (at least) degree $\min\{a, b\}$, $a, b \in \{0, 1\}$, where min corresponds to conjunction.

Clearly we set $\|\varphi\| = 1$ iff φ is valid, i.e. $\|\varphi\| = 0$ iff φ is invalid. Then two-valued modus ponens is sound in the following sense: If $a \leq \|\varphi\|$ and $b \leq \|\varphi \rightarrow \psi\|$ then $\min\{a, b\} \leq \|\psi\|$ (the inferred proposition is at least as true as the conjunction of the antecedents). This way, the role of $\otimes$ is (r1): to get the lower estimation of the validity of the inferred formula ψ from the lower estimations of the validities of φ and $\varphi \rightarrow \psi$. It is easy to see that (r2): min also gives the highest possible estimation under the condition that the inference rule still be sound.

Denoting by $\curvearrowright$ the binary operation on L corresponding to $\rightarrow$, requirement (r1) translates to:

$$\text{if } a \leq \|\varphi\| \text{ and } b \leq \|\varphi \rightarrow \psi\| \text{ then } a \otimes b \leq \|\psi\|;$$

using truth-functionality this gives

$$\text{if } a \leq \|\varphi\| \text{ and } b \leq \|\varphi\| \curvearrowright \|\psi\| \text{ then } a \otimes b \leq \|\psi\|.$$

Putting $a := \|\varphi\|$ and denoting $c := \|\psi\|$ results in the special case

$$b \leq a \curvearrowright c \text{ implies } a \otimes b \leq c.$$

By requirement (r2) we want to become the inference by modus ponens as strong as possible. From $a := \|\varphi\|$ and $\|\varphi \rightarrow \psi\|$ we get $a \otimes \|\varphi \rightarrow \psi\|$ as a lower estimation of $c := \|\psi\|$. Since a is given and $\otimes$ is non-decreasing, a greater value of $\|\varphi \rightarrow \psi\|$ leads to a bigger value of $a \otimes \|\varphi \rightarrow \psi\|$. So the requirement that modus ponens be as powerful as possible yields that $a \curvearrowright c$ should be the largest possible value that leads to a lower estimation of c. That is, we want that whenever $a \otimes b \leq c$ (i.e. b is a possible candidate for $\|\varphi\| \curvearrowright \|\psi\| = a \curvearrowright c$) then $b \leq a \curvearrowright c$. Putting together what has been derived from the requirements (r1) and (r2) we get

$$a \otimes b \leq c \quad \text{iff} \quad b \leq a \curvearrowright c. \tag{7.2}$$

Why we gave such a detailed discussion should immediately become clear, if you compare the following reformulation of (7.2) via two functions $\otimes_a, \curvearrowright_a : L \longrightarrow L$ with the first two statements of Prop. 7.10.

$$\otimes_a(b) \leq c \quad \text{iff} \quad b \leq \curvearrowright_a(c). \tag{7.3}$$

The above considerations amount to an intuitive justification for the fact that residuated lattices play a prominent role as truth value spaces generalizing the classical Boolean algebra with values $\{0, 1\}$.

7.2 Order and Lattice Theory: Supplements

In this section we present some more formal prerequisites that are necessary for developing many-valued concepts further in a reasonable way. The structure to be used as valuation space instead of **2** must be general enough to provide a certain plausible range for many-valued models and must also be rich enough to encompass a couple of significant examples.

Important Remark: Based on my own experience during the compilation of the stuff to be included in this chapter, I decided to bother its readers with some remarks on notation and naming. But I am rather sure that these hints will be helpful in understanding of, and finding their way through, related literature. These hints are given under the heading 'CAVEAT'.

7.2.1 *More on Orders and Residuation*

We now turn to one of the central notions that we need in the area of fuzzy relations (and also of various fuzzy logic systems): *residuation.* We have already taken a short look at residuals in Section 3.5.2 where they have been defined for two-valued binary relations. In the following we will consider residuation in a more abstract and algebraic context, viz. in the area of ordered sets and semigroups. Much more can be found in the books of Blyth and Janowitz [5, 6].

7.2.1.1 *Duality*

An important concept in the context or order is that of duality (see e.g. [15, p. 47], or [28, p. 10]): Let $(A, \leq)$ be a poset; the converse $\leq^{\smile}$ is called the **dual poset** of A, denoted by $(A^{\mathrm{op}}, \leq)$ or by the short form A^{op} only. It has the same underlying set but the order relation is the opposite of $\leq$, i.e. $x \leq y$ iff $x \geq y$. Defining $(A, \leq)$, $x \leq y$ iff $y \geq x$, as is usually done for partial orders, $\leq$ and $\geq$ become dual orders on A.

An *order-theoretical statement* S is a statement with $\leq$ or $\geq$ as the only signs occurring besides logical symbols and variables. The *dual statement D(S)* is obtained by replacing each occurrence of $\leq$ by $\geq$, or $\geq$ by $\leq$, respectively. If S is true for $(A, \leq)$, we say that A has property S. Do, however, note that the dual order of $\leq$ is *not* the dual relation $\leq^d$ in the sense of Def. 3.30.

Concepts defined or valid for a poset $(A, \leq)$ have always counterparts for the dual poset $(A, \geq)$: e.g. lower and upper bound, infimum and supremum, or maximum and minimum. Furthermore, if a statement S holds for a poset $(A, \leq)$, then $D(S)$ holds for the dual poset $(A, \geq)$. For instance: '*x is an upper bound for $B \subseteq A$ in* $(A, \leq)$' if and only if '*x is a lower bound for $B \subseteq A$ in* $(A, \geq)$'. This allows us to formulate the following

Duality Principle for posets: *If a statement is true for a poset A, then its dual is true for* A^{op}.

Since any poset can be regarded as the dual of some other poset we can conclude that *if a statement is true for all posets, then its dual is also true for all posets.* The duality principle allows us to simplify proofs of statements involving both a concept and its dual: we need to prove only one half of the statement since the other "follows by duality" as shown by the following consideration: If an order-theoretical statement S is a consequence of a set $\mathcal{P}$ of order-theoretical statements, and if we exchange each statement of $\mathcal{P}$ by its dual, thus obtaining a set $D(\mathcal{P})$, and if we do so also in each sentence of the proof, which derives S from $\mathcal{P}$, then we obtain a proof of $D(S)$ from $D(\mathcal{P})$.

7.2.1.2 *Residuated Mappings*

In the following we shall assume for posets $(A, \leq_A)$ and $(B, \leq_B)$ that $A, B \neq \emptyset$. And since it will always be clear from the context which of the two order symbols applies, no confusion will arise from simply using $\leq$ in both cases.

Definition 7.1. ([6, p. 5]) *Let $(A, \leq)$ be a poset.*

(a) A subset $\emptyset \neq D \subseteq A$ is called a **down-set** *(or* **ideal***) of A if for $y \in A$*

$$x \in D \,\&\, y \leq x \;\rightarrow\; y \in D.$$

The set $x^{\downarrow} = \{y \in A : y \leq x\}$ is called the **principal down-set** *(or* **principal ideal***) generated by x.*

(b) Dually, we define an **up-set** *(or* **filter***) of A as a subset $\emptyset \neq F \subseteq A$ such that for $y \in A$*

$$x \in F \,\&\, y \geq x \;\rightarrow\; y \in F.$$

The set $x^{\uparrow} = \{y \in A : y \geq x\}$ is called the **principal up-set** *(or* **principal filter***) generated by x.*

From now on only in rare cases both of the two existing dual forms of definitions or properties will be presented. We begin with a characterization of isotone mappings in

Proposition 7.2. ([6, THEOREM 2.1]) *Let $f : A \longrightarrow B$ be any mapping between posets $(A, \leq)$ and $(B, \leq)$; then the following conditions are equivalent:*

(1) f is isotone;
(2) the pre-image under f of every principal ideal of B is either empty or is an ideal of A;
(3) the pre-image under f of every principal filter of B is either empty or is a filter of A.

Proof. (1) $\rightarrow$ (2): Suppose f is isotone, and let $y \in B$ and $D = f^{-1}[y^{\downarrow}]$. If $x \in D$, then for each $z \in A$ with $z \leq x$ we have $f(z) \leq f(x) \leq y$. Thus $z \in D$, whence D is an ideal of A.
(2) $\rightarrow$ (1): Note that for each $y \in A$ trivially $f(y) \leq f(y)$, so $y \in f^{-1}[f(y)^{\downarrow}]$ which, by assumption is an ideal of A. Thus, $x \leq y \;\rightarrow\; x \in f^{-1}[f(y)^{\downarrow}] \;\rightarrow\; f(x) \leq f(y)$, i.e. f is isotone. □

Let A and B be ordered sets. Then the set B^A of all mappings $f : A \longrightarrow B$ can be ordered by defining $f \leq g :\leftrightarrow \forall x \in A[f(x) \leq g(x)]$. If $f, g, h : A \longrightarrow B$ are mappings with $g \leq h$ then $gf \leq hf$ and, for isotone f, $fg \leq fh$ hold.

In the next definition we introduce a kind of mappings that will soon turn out to strengthen the conditions equivalent to the isotonicity of f given in the previous lemma: the pre-images of principal ideals (filters) will again be principal ideals (filters). To show this property we follow the presentation in [6, p. 6ff] and [5, Section 1.3].

Definition 7.3. *Let $(A, \leq)$ and $(B, \leq)$ be posets, and f be an isotone mapping $f : A \longrightarrow B$.*

(a) f is said to be **residuated** *if there is an isotone mapping $g : B \longrightarrow A$ such that*

$$gf \geq \mathrm{id}_A \quad \text{and} \quad fg \leq \mathrm{id}_B. \tag{7.4}$$

(b) f is said to be **dually residuated** *if there is an isotone mapping $g : B \longrightarrow A$ such that*

$$\mathrm{id}_A \geq gf \quad \text{and} \quad \mathrm{id}_B \leq fg. \tag{7.5}$$

As shown below in Prop. 7.4(1), the mapping $g : B \longrightarrow A$ mentioned in (1) is unique and is called the **residual of** f, denoted as f^+ and given by $f^+(y) = \max\{x \in A : f(x) \leq y\}$. The pair (f, g) is also called a **residuated pair**.

Before we list various properties of residuated mappings, we give some examples (see Section 2.3.6 for examples of Galois connections):

(1) As already mentioned on page 33, a function $f : X \longrightarrow Y$ between sets X and Y gives rise to two associated functions between the powersets of X and Y: The image map $f : \mathbf{2}^X \longrightarrow \mathbf{2}^Y$ is defined by $f[S] = \{f(x) : x \in S\} \subseteq Y$ for all $S \subseteq X$, and the pre-image map $f^{-1} : \mathbf{2}^Y \longrightarrow \mathbf{2}^X$, where $f^{-1}[T] = \{x : f(x) \in T\} \subseteq X$, for all $T \subseteq Y$. The image map and its associated pre-image map form a residuated pair (f, f^{-1}).
(2) If X is any set and $S \subseteq X$ then $\lambda_S : \mathbf{2}^X \longrightarrow \mathbf{2}^X$ defined by $\lambda_S(X) = S \cap X$ is residuated with residual $\lambda_S^+ : \mathbf{2}^X \longrightarrow \mathbf{2}^X$ given by $\lambda_S^+(Y) = Y \cup \overline{S}$.
(3) If, for $m \in \mathbb{N}$, we define $f_m : \mathbb{N} \longrightarrow \mathbb{N}$ by $f_m(n) := mn$, then f_m is residuated with $f_m^+(p) = \lfloor \frac{p}{m} \rfloor$ with $\lfloor q \rfloor$ denoting the integer part of $q \in \mathbb{Q}$.

Proposition 7.4. *Let $f : A \longrightarrow B$ be a residuated mapping.*

(1) An isotone mapping $g : B \longrightarrow A$ such that $gf \geq \mathrm{id}_A$ and $fg \leq \mathrm{id}_B$ is unique.
(2) Residuated mappings preserve existing suprema.
(3) $f f^+ f = f$ and $f^+ f f^+ = f^+$.
(4) Let $g : B \longrightarrow C$ be another residuated mapping; then also $fg : A \longrightarrow C$ is residuated and $(fg)^+ = f^+ g^+$.

Proof. (1): Assume that there are two mappings g and g' satisfying (7.4). Then $g = \mathrm{id}_A\, g \leq (g'f)g = g'(fg) \leq g'\mathrm{id}_B = g'$. Similarly we also have $g' \leq g$, whence $g = g'$.

(2): Let $x = \bigvee_{i \in I} x_i$ exist in A, then, since f is isotone, $f(x) \geq f(x_i)$ for every $i \in I$. Using (7.4) a few times we get: If $y \geq f(x_i)$ for all $i \in I$ then $g(y) \geq g(f(x_i)) \geq x_i$ for each $i \in I$. So $g(y) \geq \bigvee_{i \in I} x_i = x$, and $y \geq f(g(y)) \geq f(x)$. Thus we have shown that $\bigvee_{i \in I} f(x_i)$ exists and that $f(x) = f(\bigvee_{i \in I} x_i)$.

(3): By Prop. 2.83.

(4): Both fg and gf are isotone, and by the monotonicity of f and g together with $f^+f \geq \mathrm{id}_A$, $ff^+ \leq \mathrm{id}_B$, $g^+g \geq \mathrm{id}_B$, $gg^+ \leq \mathrm{id}_C$ we get

$$(f^+ g^+)(gf) \geq f^+ \mathrm{id}_B f = f^+ f \geq \mathrm{id}_A,$$
$$(gf)(f^+ g^+) \leq g\,\mathrm{id}_B\, g^+ = gg^+ \geq \mathrm{id}_C.$$

Thus, using the uniqueness of residuals shown in (1), $(gf)^+$ exists and is f^+g^+. □

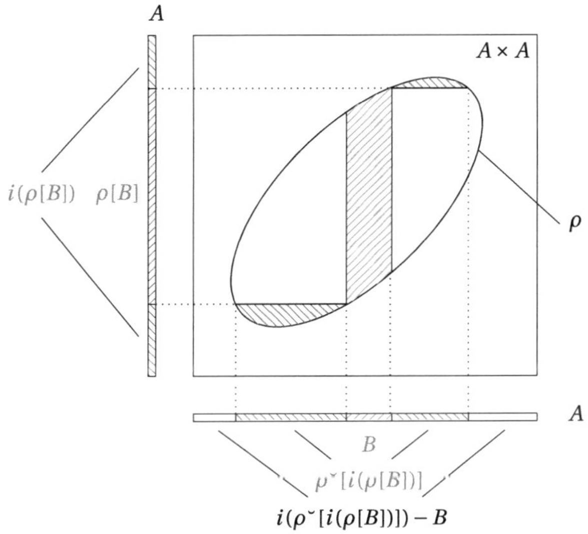

FIG. 7.1: Illustrating the mapping h_ρ^+.

The next theorem shows how all the residuated mappings on the power set of a given non-empty set A (equivalently, on a complete atomic Boolean algebra) can be described using relations:

Theorem 7.5. ([5, Theorem 1.4]) *Let $\emptyset \neq A$ and $\rho \in \mathcal{B}_A$. The map $h_\rho : \mathbf{2}^A \longrightarrow \mathbf{2}^A$ defined by $h_\rho(B) = B\rho = \rho[B]$, $B \subseteq A$, is residuated. Every residuated mapping $f : \mathbf{2}^A \longrightarrow \mathbf{2}^A$ is of this form for some $\rho \in \mathcal{B}_A$.*

Proof. Clearly, h_ρ is isotone. Let i be the antitone mapping $i : \mathbf{2}^A \longrightarrow \mathbf{2}^A, B \mapsto \overline{B}$ for each $B \subseteq A$. Then the mapping $h_\rho^+ := i\, h_{\rho^{\smile}}\, i$ is isotone and we have $h_\rho h_\rho^+ \leq \mathrm{id}_{\mathbf{2}^A}$ and $h_\rho^+ h_\rho \geq \mathrm{id}_{\mathbf{2}^A}$. Therefore, h_ρ^+ is the residual of h_ρ. It is extremely helpful to use illustrations like Fig. 7.1 to support the proof (based on [4, p. 190]): Note that $h_\rho^+ h_\rho(B) = i\, h_{\rho^{\smile}}\, i\, h_\rho(B) = i(\rho^{\smile}[i(\rho[B])])$. From the figure we can easily see that $h_\rho^+ h_\rho(B) \supseteq B$ or, equivalently, $h_\rho^+ h_\rho \geq \mathrm{id}_B$.

Now let $f : \mathbf{2}^A \longrightarrow \mathbf{2}^A$ be a residuated mapping. We define a binary relation ρ_f on A by $x \rho_f y :\Leftrightarrow y \in f(\{x\})$. Then $h_{\rho_f}(\{x\}) = x\rho_f = \{y \in A : x \rho_f y\} = f(\{x\})$, whence f and h_{ρ_f} agree on singletons. If $g : \mathbf{2}^A \longrightarrow \mathbf{2}^A$ is any residuated mapping then for every

$\emptyset \neq B \subseteq A$ we have $g(B) = g\big(\bigcup_{x\in B}\{x\}\big) = \bigcup_{x\in B} g(\{x\})$, since g is isotone. If $C = \bigcup_{x\in B} g(\{x\})$ then clearly $g(B) \supseteq C$. On the other hand we have $g(\{x\}) \subseteq C$ for every $x \in B$ and so $\{x\} \subseteq g^+(C)$; hence $B = \bigcup_{x\in B}\{x\} \subseteq g^+(C)$ and therefore $g(B) \subseteq C$, i.e. $g(B) = C$. Applying this equality to both f and h_{ρ_f} and using the fact that they agree on singletons gives $f(B) = \bigcup_{x\in B} f(\{x\}) = \bigcup_{x\in B} h_{\rho_f}(\{x\}) = h_{\rho_f}(B)$, so $f = h_{\rho_f}$. □

Lemma 7.6. ([6, THEOREM 2.2]) *Let A, B, C be posets, and $f : A \longrightarrow B$ and $g : A \longrightarrow C$ be mappings. If $h : B \longrightarrow C$ is isotone, the following conditions are equivalent:*

(1) $g \leq hf$;
(2) $\forall x \in B\,\big[f^{-1}[x^{\downarrow}] \subseteq g^{-1}[h(x)^{\downarrow}]\big]$.

Proof. (1) → (2): If $f^{-1}[x^{\downarrow}] = \emptyset$ there is nothing to prove. But for $y \in f^{-1}[x^{\downarrow}]$ we have $f(y) \leq x$, so that, h being isotone, $g(y) \leq h(f(y)) \leq h(x)$ whence $y \in g^{-1}[h(x)^{\downarrow}]$.
(2) → (1): For each $y \in A$ we have $y \in f^{-1}[f(y)^{\downarrow}]$, since $f(y) \leq f(y)$. And from the assumption we get $f^{-1}[f(y)^{\downarrow}] \subseteq g^{-1}[h(f(y))^{\downarrow}]$, whence $g(y) \leq h(f(y))$. □

Lemma 7.7. ([6, THEOREM 2.3]) *Let A, B, C be posets, and $f : B \longrightarrow A$ and $g : C \longrightarrow A$ be mappings. The following conditions are equivalent:*

(1) *there exists a $h : C \longrightarrow B$ such that $fh \leq g$ (note that we do not require h to be isotone);*
(2) $\forall x \in C\,\big[f^{-1}[g(x)^{\downarrow}] \neq \emptyset\big]$.

Proof. (1) → (2): From (1) we get $\forall x \in C\,\big[f(h(x)) \leq g(x)\big]$ which implies $\forall x \in C\,\big[h(x) \in f^{-1}[g(x)^{\downarrow}]\big]$, whence (2) holds.
(2) → (1): If (2) holds a mapping $h : C \longrightarrow B$ can be defined by associating with each $x \in C$ a chosen element $h(x) \in f^{-1}[g(x)^{\downarrow}]$. But then $\forall x \in C\,[f(h(x)) \leq g(x)]$, i.e. $fh \leq g$. □

Corollary 7.8. ([6, p. 8, COROLLARY]) *Let A, B be posets, and $f : A \longrightarrow B$ an isotone mapping. The following conditions are equivalent:*

(1) *there exists a $h : B \longrightarrow A$ such that $fh \leq \mathrm{id}_B$;*
(2) *the pre-image under f of every principal ideal of B is an ideal of A.*

Proof. (1) → (2): Since f is isotone, (2) follows by Prop. 7.2.
(2) → (1): That f is isotone, follows by Prop. 7.2. Applying Lemma 7.7 to f and id_B we get that there exists $h : B \longrightarrow A$ such that $fh \leq \mathrm{id}_B$. □

Lemma 7.9. *Let A, B be posets, and $f : A \longrightarrow B$ be a mapping. Then the following obviously holds for ideals*

(1) $x \in x^{\downarrow}$;
(2) $x \leq y \leftrightarrow x^{\downarrow} \leq y^{\downarrow}$;
(3) $x \in f^{-1}[y^{\downarrow}] \leftrightarrow f(x) \leq y$.

We are now prepared to give the previously announced characterization of residuated mappings analogous to that of Prop. 7.2:

Proposition 7.10. *Let* $(A, \leq)$ *and* $(B, \leq)$ *be posets and* $f : A \longrightarrow B$ *be any mapping; then the following conditions are equivalent:*

(1) f *is residuated;*
(2) there exists a function $g : B \longrightarrow A$ *such that* $f(x) \leq y \leftrightarrow x \leq g(y)$*;*
(3) the pre-image of every principal ideal of B is a principal ideal of A, i.e. $\forall y \in B \exists x \in A\left[f^{-1}[y^{\downarrow}] = x^{\downarrow}\right]$*;*
(4) f *is isotone and for every* $y \in B$ *the set* $\{x \in A : f(x) \leq y\}$ *has a maximal element.*

Proof. (1) → (2): By Def. 7.3 f is isotone and there is an isotone mapping $g : B \longrightarrow A$ such that $gf \geq \mathrm{id}_A$ and $fg \leq \mathrm{id}_B$. We then have $f(x) \leq y \rightarrow x = \mathrm{id}_A(x) \leq g(f(x)) \leq g(y)$ and $x \leq g(y) \rightarrow f(x) \leq f(g(y)) \leq \mathrm{id}_B(y) = y$. Thus, $x \leq g(y)$ iff $f(x) \leq y$.

(2) → (3): $x \leq g(y)$ iff $f(x) \leq y$ immediately implies $f^{-1}[y^{\downarrow}] = g(y)^{\downarrow}$ by Lemma 7.9(3).

(3) → (1): f is isotone by Prop. 7.2. By Prop. 3.9(4) f^{-1} is isotone; therefore, given $y \in B$ the $x \in A$ with $f^{-1}[y^{\downarrow}] = x^{\downarrow}$ is unique by the antisymmetry of $\leq_A$ and Lemma 7.9(2). So we can define a mapping $g : B \longrightarrow A$ which is isotone since f^{-1} is, by $g(y) = x$. By Lemma 7.9(1) $g(y) \in g(y)^{\downarrow}$, but $g(y)^{\downarrow} = x^{\downarrow} = f^{-1}[y^{\downarrow}]$, whence $f(g(y)) \leq y$ by Lemma 7.9(3), i.e. $fg \leq \mathrm{id}_B$.
On the other hand we have $x \in f^{-1}[f(x)^{\downarrow}]$ by Lemma 7.9(3); but by definition of g $f^{-1}[f(x)^{\downarrow}] = g(f(x))^{\downarrow}$ so that $\forall x \in A[x \leq g(f(x))]$, i.e. $gf \geq \mathrm{id}_A$.

(3) → (4): f is isotone by Prop. 7.2. Furthermore, $\emptyset \subset \{x\} \subseteq x^{\downarrow}$, and x is maximal in $x^{\downarrow}$. By Lemma 7.9(3) $x \in f^{-1}[y^{\downarrow}] \leftrightarrow f(x) \leq y$; but by assumption $\exists x_0 \in A\left[f^{-1}[y^{\downarrow}] = x_0^{\downarrow}\right]$ with x_0 maximal.

(4) → (3): f is isotone by assumption, i.e. $\tilde{A} := f^{-1}[y^{\downarrow}]$ is an ideal of A by Prop. 7.2. By Lemma 7.9(3) $x \in \tilde{A} \leftrightarrow f(x) \leq y$, and, again by assumption, the set $\{x \in A : f(x) \leq y\}$ has a maximal element. Thus $\exists x_0 \in \tilde{A}$ such that $\tilde{A} = x_0^{\downarrow}$. □

Without proof we mention the dual to Prop. 7.10:

Proposition 7.11. *Let* $(A, \leq)$ *and* $(B, \leq)$ *be posets and* $f : A \longrightarrow B$ *be any mapping; then the following conditions are equivalent:*

(1) f *is residual;*
(2) there exists a function $g : B \longrightarrow A$ *such that* $f(x) \geq y \leftrightarrow x \geq g(y)$*;*
(3) the pre-image of every principal filter of B is a principal filter of A, i.e. $\forall y \in B \exists x \in A\left[f^{-1}[y^{\downarrow}] = x^{\downarrow}\right]$*;*
(4) f *is isotone and for every* $y \in B$ *the set* $\{x \in A : f(x) \geq y\}$ *has a minimal element.*

The unique g that satisfies either of the equivalent conditions of Prop. 7.11 is called the **dual residual of** f, is denoted by $f^{\flat}$ and given by $f^{\flat}(y) = \min\{x \in A : f(x) \geq y\}$. The pair (f, g) is also called a **residual pair**.

CAVEAT 1: Remember that we have already introduced two kinds of pairs of functions in Section 2.3.6. And we have learned that the two different concepts of *Galois connections* and *monotone Galois connections* often appear under the first as their common name. Here we have to state that a single concept bears two different names. Moreover, all the concepts involved are very similar; so it is advisable to arrange them in an overview to get a clear picture.

TABLE 7.1: Galois connections and residuation: overview.

Galois Connections		f, g	**Residuation**	
(antitone)	$\mathrm{id}_A \leq gf$ & $\mathrm{id}_B \leq fg$ $a \leq g(b) \leftrightarrow b \leq f(a)$	antitone		
dual	$\mathrm{id}_A \geq gf$ & $\mathrm{id}_B \geq fg$ $a \geq g(b) \leftrightarrow b \geq f(a)$	antitone		
monotone	$\mathrm{id}_A \leq gf$ & $\mathrm{id}_B \geq fg$ $a \leq g(b) \leftrightarrow b \geq f(a)$	isotone	$\mathrm{id}_A \leq gf$ & $\mathrm{id}_B \geq fg$ $a \leq g(b) \leftrightarrow b \geq f(a)$	residuated map
		isotone	$\mathrm{id}_A \geq gf$ & $\mathrm{id}_B \leq fg$ $a \geq g(b) \leftrightarrow b \leq f(a)$	dually residuated map

Table 7.1 shows that monotone Galois connections and residuated maps define the same concept. Obviously, it is just the naming of the former that breaks the symmetry of the whole scheme. Note that the notion of a **dual Galois connection** has not been introduced earlier in this book and is rarely to be met in literature. In e.g. [12] it appears under this name, and in [21] as *dual Galois pair.*

To enable a concise description of the relationships between the concepts summarized in Table 7.1 we will use the following self-explanatory shorthands (leaving the respective partial orders implicit): "*(A, f, g, B) is a Galois connection*" for: "*the pair (f, g) of functions $f : A \longrightarrow B$ and $g : B \longrightarrow A$ is a Galois connection between the posets $(A, \leq)$ and $(B, \leq)$*" etc. Then the results as shown in Table 7.2 hold (see [21]).

7.2.2 *More on Lattices*

We have already become acquainted with the most basic concepts of lattice theory in Def. 2.54, and also have presented related properties subsequently. Here we will enlarge our knowledge with things needed in our analysis of fuzzy relations later on.

Definition 7.12. *Let L be a lattice, and $a, b \in L$.*

(a) If the set $T(a, b) := \{x \in L : a \wedge x \leq b\}$ has a greatest element, it is called the **relative pseudocomplement** *of a in b, denoted by $a \curvearrowright b$, and a is called relatively pseudocomplemented. L is called a* **relatively pseudocomplemented lattice** *if $a \curvearrowright b$ exists for all $a, b \in L$.*

TABLE 7.2: The relationships between Galois-type connections.

(A,f,g,B) is a			
(1) Galois connection	iff	(B,g,f,A)	is a Galois connection
(2) Galois connection	iff	$(A^{\mathrm{op}},f,g,B^{\mathrm{op}})$	is a dual Galois connection
(3) dual Galois connection	iff	(B,g,f,A)	is a dual Galois connection
(4) residuated pair	iff	$(B^{\mathrm{op}},g,f,A^{\mathrm{op}})$	is a residuated pair
(5) residuated pair	iff	$(A^{\mathrm{op}},f,g,B^{\mathrm{op}})$	is a residual pair
(6) residuated pair	iff	(A,f,g,B^{op})	is a Galois connection
(7) residuated pair	iff	(B,g,f,A)	is a residual pair
(8) residual pair	iff	(A,f,g,B^{op})	is a dual Galois connection
(9) residual pair	iff	$(B^{\mathrm{op}},g,f,A^{\mathrm{op}})$	is a residual pair

(b) *If L is a lattice with bottom element 0 then an element $x \in L$ is* **pseudocomplemented** *if $T(a,0) := \{x \in L : a \wedge x \leq 0\}$ has a greatest element $a \curvearrowright 0$, denoted by $\neg a$. L is called a* **pseudocomplemented lattice** *if $\neg a$ exists for all $a \in L$.*
(c) *A relatively pseudocomplemented lattice with a smallest element 0 is called* **Heyting algebra** *or* **Heyting lattice**.
(d) *A* **Brouwerian lattice** *is the dual of a Heyting algebra (we have $\vee$ instead of $\wedge$ and $\geq$ instead of $\leq$), i.e. a lattice L such that, for every pair $a,b \in L$, the set $\{x \in L : a \vee x \geq b\}$ has a smallest element $a \curvearrowleft b$.*

CAVEAT 2: Unfortunately you cannot necessarily rely on the names of the concepts involved. Whereas *relatively pseudocomplemented* is unambiguous, the name *Brouwerian* is not: What is usually called a *relatively pseudocomplemented* lattice got the name *Brouwerian* lattice in Birkhoff [3, p. 45], a very influential book by an authority in the field. This term has become very popular in the fuzzy community starting with [30], and is still in use (see [35]). The definition of *Heyting algebra* as given above appears in e.g. in [1, p. 174], or [5, p. 112].

Another synonym for *relatively pseudocomplemented* is **implicative** (in e.g. [19, 31] or [8]), and Heyting algebras are sometimes called **pseudo-Boolean algebras** (in e.g. [16, 27, 29]).

Heyting algebras serve as the algebraic models of propositional intuitionistic logic in the same way Boolean algebras model propositional classical logic. They generalize Boolean algebras in the sense that a Heyting algebra satisfying the excluded middle law ($a \vee \overline{a} = 1$) or, equivalently, the double negation law ($\overline{\overline{a}} = a$), is a Boolean algebra.

Examples for relatively pseudocomplemented and Heyting lattices are:

- Every chain becomes a relatively pseudocomplemented lattice by defining $a \curvearrowright b = 1$ whenever $a \leq b$, and $a \curvearrowright b = b$ otherwise.
- Every Boolean algebra is a Heyting lattice when putting $a \curvearrowright b = \overline{a} \vee b$.

- (For logicians only.) The Lindenbaum algebra of propositional intuitionistic logic is a Heyting algebra.

We will now have a look at some properties of relatively pseudocomplemented lattices:

Proposition 7.13.

(1) Any relatively pseudocomplemented lattice L is distributive.
(2) Any finite distributive lattice L is relatively pseudocomplemented.

Proof. (1): For given $a, b, c \in L$ let $d = (a \wedge b) \vee (a \wedge c)$. Since $a \wedge b \leq d$ and $a \wedge c \leq d$ we have $b \leq a \curvearrowright d$ and $c \leq a \curvearrowright d$, whence $b \vee c \leq a \curvearrowright d$. Thus $a \wedge (b \vee c) \leq a \wedge a \curvearrowright d \leq d = (a \wedge b) \vee (a \wedge c)$. Together with Prop. 2.60(L5) this implies $a \wedge (b \vee c) = (a \wedge b) \vee (a \wedge c)$ which is (L6) of the same proposition, and this is equivalent to the second distributive law (L6'), whence we have shown the distributivity of L.

(2): In a finite distributive lattice the join $u = \bigvee\{x_u : a \wedge x_u \leq b\}$ satisfies $a \wedge u = a \wedge \bigvee x_u = \bigvee(a \wedge x_u) \leq b$. □

Proposition 7.14. *Let L be a relatively pseudocomplemented lattice, and $a, b \in L$*

(1) L has a greatest element 1.
(2) $a \leq b$ iff $a \curvearrowright b = 1$;
(3) $a \curvearrowright a = 1, \quad a \curvearrowright 1 = 1, \quad 0 \curvearrowright a = 1$;
(4) $b \leq a \curvearrowright b$;
(5) $1 \curvearrowright a = a$;
(6) A Heyting algebra is a bounded relatively pseudocomplemented lattice.

Proof. (1): For any b we have $a \wedge b \leq a$, and thus $b \leq a \curvearrowright a$ by definition of $\curvearrowright$; so $a \curvearrowright a$ is the greatest element in L, for any a.
(2): $a \curvearrowright b = 1 \;\leftrightarrow\; 1 \leq a \curvearrowright b \;\leftrightarrow\; a \wedge 1 \leq b$.
(3): Direct consequences of (2).
(4): $a \wedge b \leq b \;\rightarrow\; b \leq a \curvearrowright b$.
(5): By (1) and Def. 7.12(c).
(6): By Def. 7.12(c), a Heyting algebra is a relatively pseudocomplemented lattice with a smallest element 0, and by (1) it has a greatest element 1. □

Definition 7.15. *Let Q be a complete lattice, $S \subseteq Q$ any subset of Q, and $a \in Q$. Then Q is called a*

(a) **quantale** *if it has an associative binary operation $*$ that distributes over arbitrary joins, i.e.* $a * \left(\bigvee_{s \in S} s\right) = \bigvee_{s \in S}(a * s) \quad \left(\bigvee_{s \in S} s\right) * a = \bigvee_{s \in S}(s * a)$;
(b) **frame** *(or* **locale***) if Q is a quantale with $* = \wedge$.*
(c) A quantale Q is called **unital** *if there exists $e \in Q$ such that $(Q, *, e)$ is a monoid;* **commutative** *if $*$ is commutative; and* **idempotent** *if $*$ is idempotent.*
(d) A unital quantale Q is called **integral** *if e is the greatest element in Q.*

Frames are exactly complete Heyting algebras with the Heyting implication given by $a \rightarrowtail b := \bigvee\{s \in L : a \wedge s \leq b\}$, and also a typical example of an integral commutative quantale. Quantales are sometimes referred to as complete residuated semigroups.

Definition 7.16. *A* **residuated lattice** *is an algebra* $\mathscr{L} = (L, \wedge, \vee, \otimes, \rightarrowtail, 0, 1)$ *of type* $\langle 2,2,2,2,0,0\rangle$ *such that*

(1) $(L, \wedge, \vee, 0, 1)$ *is a bounded lattice;*
(2) $(L, \otimes, 1)$ *is a commutative monoid;*
(3) *the operation* $\rightarrowtail$ *is a* **residuation** *operation with respect to* $\otimes$, *i.e*
$$a \otimes b \leq c \leftrightarrow a \leq b \rightarrowtail c \quad \textbf{(adjointess property)}.$$

In comparison with an ordinary lattice, this structure is enriched by the two operations $\otimes, \rightarrowtail : L \times L \longrightarrow L$ called **adjoint couple**. The operations $\otimes$ (called *multiplication*) and $\rightarrowtail$ (called *residuum*) are intended for modeling conjunction and implication of the corresponding logical calculus, and supremum ($\bigvee$) and infimum ($\bigwedge$) are intended for modeling the existential and universal quantifier, respectively.

If $(L, \wedge, \vee, 0, 1)$ is a complete lattice then $\mathscr{L}$ is called a complete residuated lattice. Another operation in residuated lattices is **negation** $\neg : L \longrightarrow L$ defined by $\neg a := a \rightarrowtail 0$.

CAVEAT 3: The definition given above is a very common one, at least in the area of many-valued logic (see e.g. [20, 36]) and lattice theory (e.g. [13], an example for the non-commutative variant). The only thing to be observed, however, is that the original definition in [34, p. 339] does comprise commutativity due to the condition M3. Sometimes commutativity is abandoned, so take care of what is really meant by *residuated lattice* in a text under consideration; see e.g. [7, 13, 18]. Other names appearing in the field are: *integral residuated lattice-ordered commutative monoid* or *bounded commutative integral residuated lattice* which both emphasize the presence of commutativity, and that the unit of the monoidal reduct is the maximal element of the lattice *(integral).*

That residuated lattices offer a structure that is both general and rich enough can be seen from the examples given below. They include Boolean and Heyting algebras as well as the most studied and applied residuated structures of membership values in many-valued approaches (see [20] and [24]):

- Heyting algebras can be interpreted as residuated lattices if we take $a \otimes b = a \wedge b$ and $a \rightarrowtail b = \bigvee\{c : a \wedge c \leq b\}$.
- Every Boolean algebra is a residuated lattice when putting $a \rightarrowtail b = \overline{a} \vee b$. Note that in this case $a \otimes b = a \wedge b$.
- The **Łukasiewicz algebra** $([0,1], \wedge, \vee, \otimes_L, \rightarrowtail_L, 0, 1)$ with $a \otimes_L b = 0 \vee (a + b - 1)$ and $a \rightarrowtail_L b = 1 \wedge (1 - a + b)$.
- The **Gödel algebra** $([0,1], \wedge, \vee, \otimes_G, \rightarrowtail_G, 0, 1)$ with $\otimes_G = \wedge$ and $a \rightarrowtail_G b = 1$ if $a \leq b$ and $a \rightarrowtail_G b = b$ otherwise.
- A **BL-algebra** is a residuated lattice which satisfies *divisibility*: $a \wedge b = a \otimes$

$(a \curvearrowright b)$, and *prelinearity*: $(a \curvearrowright b) \vee (b \curvearrowright a) = 1$.

- An **MV-algebra** is a BL-algebra satisfying the law of double negation: $a = \neg\neg a$.
- A Boolean algebra is a residuated lattice which is both a Heyting algebra and an MV-algebra.
- Boolean, Łukasiewicz and Gödel algebras are all BL-algebras (with negation operations defined as $\neg_L\, a = 1 - a$, and $\neg_G\, a = 1$, if $a = 0$ and $\neg_G\, a = 0$, if $a > 0$, respectively).

Next we give some important properties of residuated lattices, mainly following the presentation in [2], where also the proofs omitted below can be found.

Proposition 7.17. *Let* $(L, \wedge, \vee, \otimes, \curvearrowright, 0, 1)$ *be a complete residuated lattice.*

(1) *For a given multiplication* $\otimes$, *there is at most one operation* $\curvearrowright$ *satisfying the adjointness property (see Def. 7.16(c)).*
(2) $x \otimes (x \curvearrowright y) \leq y, \quad y \leq x \curvearrowright (x \otimes y), \quad x \leq (x \curvearrowright y) \curvearrowright y;$
(3) $x \leq y$ *iff* $x \curvearrowright y = 1;$
(4) $x \curvearrowright x = 1, \quad x \curvearrowright 1 = 1, \quad 0 \curvearrowright x = 1;$
(5) $1 \curvearrowright x = x;$
(6) $x \otimes y \leq x, \quad x \leq y \curvearrowright x;$
(7) $(x \otimes y) \curvearrowright z = x \curvearrowright (y \curvearrowright z);$
(8) $x \curvearrowright y$ *is the greatest element in the set* $\{z : x \otimes z \leq y\}$.
(9) $x \otimes y$ *is the least element in the set* $\{z : x \leq y \curvearrowright z\}$.

Proof. (1): Assume $\rightsquigarrow$ is another operation satisfying the adjointness property; then, for any x, $x \leq y \curvearrowright z$ iff $x \otimes y \leq z$ iff $x \leq y \rightsquigarrow z$, whence obviously $y \curvearrowright z = y \rightsquigarrow z$. Similarly, $\curvearrowright$ uniquely determines $\otimes$.

(2): Using commutativity of $\otimes$ and adjointness we have $x \curvearrowright y \leq x \curvearrowright y$ iff $(x \curvearrowright y) \otimes x = x \otimes (x \curvearrowright y) \leq y$, from which it follows by adjointess that $x \leq (x \curvearrowright y) \curvearrowright y$.

(3): $x \leq y$ iff $1 \otimes x \leq y$ iff $1 \leq x \curvearrowright y$ iff $1 = x \curvearrowright y$.

(4): Direct consequences of (3).

(5): We have $x \leq 1 \curvearrowright x$ iff $x = 1 \otimes x \leq x$. The other inequality we get as follows: $1 \curvearrowright x \leq 1 \curvearrowright x$ gives $1 \curvearrowright x = 1 \otimes (1 \curvearrowright x) \leq x$, using (2).

(6): By (4) we have $x \curvearrowright x = 1$, whence $y \leq x \curvearrowright x$, and $x \otimes y \leq x$ by adjointness. Thus $x \otimes y = y \otimes x \leq x$, whence $x \leq y \curvearrowright x$, again by adjointness.

(7): We prove both inequalities and begin with

$$(x \otimes y) \curvearrowright z \leq x \curvearrowright (y \curvearrowright z)$$
$$\text{iff} \quad x \otimes [(x \otimes y) \curvearrowright z] \leq y \curvearrowright z$$
$$\text{iff} \quad (x \otimes y) \otimes [(x \otimes y) \curvearrowright z] \leq z,$$

which holds by the first inequality of (2). Assuming that the other inequality holds, we get

$$x \curvearrowright (y \curvearrowright z) \leq (x \otimes y) \curvearrowright z$$
$$\text{iff} \quad (x \otimes y) \otimes [x \curvearrowright (y \curvearrowright z)] \leq z \qquad \text{(commutativity, adjointness)}$$
$$\text{iff} \quad y \otimes x \otimes [x \curvearrowright (y \curvearrowright z)] \leq z \qquad \text{(commutativity, associativity, adjointness)}$$

iff $x \otimes [x \curvearrowright (y \curvearrowright z)] \leq (y \curvearrowright z)$. (adjointness)
But this last inequality again holds by (2).

(8): By (2) $x \otimes (x \curvearrowright y) \leq y$, whence $x \curvearrowright y \in \{z : x \otimes z \leq y\}$, and, by adjointness, $x \otimes z \leq y$ implies $z \leq x \curvearrowright y$.
(9): By (2) $x \leq y \curvearrowright (x \otimes y)$, whence $x \otimes y \in \{z : x \leq y \curvearrowright z\}$, and, by adjointness, $x \leq y \curvearrowright z$ implies $x \otimes y \leq z$. □

Proposition 7.18. (Monotonicity properties of $\otimes$ and $\curvearrowright$) *In a complete residuated lattice* $(L, \wedge, \vee, \otimes, \curvearrowright, 0, 1)$ *the following holds:*

(1) $y \leq y' \rightarrow x \otimes y \leq x \otimes y'$;
(2) $y \leq y' \rightarrow x \curvearrowright y \leq x \curvearrowright y'$;
(3) $x \leq x' \rightarrow x' \curvearrowright z \leq x \curvearrowright z$.

Proof. (1): From Prop. 7.17(2) we get $y' \leq x \curvearrowright (x \otimes y')$, whence, by the assumption $y \leq y'$, also $y \leq x \curvearrowright (x \otimes y')$, and this is equivalent to $x \otimes y \leq x \otimes y'$.
(3) By adjointness we have $x' \curvearrowright z \leq x \curvearrowright z$ iff $x \otimes (x' \curvearrowright y) \leq y$. But $x \leq x'$ implies $x \otimes (x' \curvearrowright y) \leq x' \otimes (x' \curvearrowright y) \leq y$, by (1) and Prop. 7.17(2). □

So, by commutativity of $\otimes$, it is isotone in both arguments, and therefore has all the constituting properties of a t-norm on L (see Def. 8.6). $\curvearrowright$ is antitone in the first and isotone in the second argument.

Proposition 7.19. (Distributivity of $\otimes, \curvearrowright$ over $\wedge, \vee$) *Let* $(L, \wedge, \vee, \otimes, \curvearrowright, 0, 1)$ *be a complete residuated lattice, and I an index set. Then*

(1) $x \otimes \bigvee_{i \in I} y_i = \bigvee_{i \in I}(x \otimes y_i)$;
(2) $x \otimes \bigwedge_{i \in I} y_i \leq \bigwedge_{i \in I}(x \otimes y_i)$;
(3) $x \curvearrowright \bigwedge_{i \in I} y_i = \bigwedge_{i \in I}(x \curvearrowright y_i)$;
(4) $\bigvee_{i \in I} x_i \curvearrowright y = \bigwedge_{i \in I}(x_i \curvearrowright y)$.

Proof. (1): We have $x \otimes y_i \leq x \otimes \bigvee_{i \in I} y_i$, for each $i \in I$, whence $\bigvee_{i \in I}(x \otimes y_i) \leq x \otimes \bigvee_{i \in I} y_i$. On the other hand, $x \otimes y_i \leq \bigvee_{i \in I}(x \otimes y_i)$; so $y_i \leq x \curvearrowright \bigvee_{i \in I}(x \otimes y_i)$ for each $i \in I$. Therefore, $\bigvee_{i \in I} y_i \leq x \curvearrowright \bigvee_{i \in I}(x \otimes y_i)$, whence $x \otimes \bigvee_{i \in I} y_i \leq \bigvee_{i \in I}(x \otimes y_i)$.
(3): Since $\curvearrowright$ is monotone in the second argument, we get $x \curvearrowright \bigwedge_{i \in I} y_i \leq x \curvearrowright y_i$, for all $i \in I$; so $x \curvearrowright \bigwedge_{i \in I} y_i \leq \bigwedge_{i \in I}(x \curvearrowright y_i)$. Let $z := \bigwedge_{i \in I}(x \curvearrowright y_i)$; then, $z \leq x \curvearrowright y_i$, hence $z \otimes x \leq y_i$, for all $i \in I$. Thus, $z \otimes x \leq \bigwedge_{i \in I} y_i$. Therefore, $z \leq x \curvearrowright \bigwedge_{i \in I} y_i$, which gives the converse inequality. □

Another important operation in the context of residuated lattices is **biresiduation** defined as $x \rightleftharpoons y = x \curvearrowright y \wedge y \curvearrowright x$ with the following properties:

Corollary 7.20.

(1) $x \rightleftharpoons y = y \rightleftharpoons x$;
(2) $x \rightleftharpoons x = 1$;
(3) $x \rightleftharpoons 1 = x$;
(4) $x = y$ *iff* $x \rightleftharpoons y = 1$;
(5) $(x \rightleftharpoons y) \otimes (y \rightleftharpoons z) \leq x \rightleftharpoons z$;
(6) $x \otimes y \leq x \wedge y \leq x \rightleftharpoons y$.

Proof. (2): By Prop. 7.17(4).

(3): By Prop. 7.17(4) and (5).

(4): Assuming $x = y$, then $x \rightleftharpoons y = 1$ by (2). For the other direction assume $x \rightleftharpoons y = x \curvearrowright y \wedge y \curvearrowright x = 1$. Then $x \curvearrowright y = 1$ and $y \curvearrowright x = 1$, and by Prop. 7.17(3) $x \leq y \wedge y \leq x$.

(5): By adjointness, $(x \rightleftharpoons y) \otimes (y \rightleftharpoons z) \leq x \rightleftharpoons z$ is equivalent to $x \otimes (x \rightleftharpoons y) \otimes (y \rightleftharpoons z) \leq z$. Using Prop. 7.18(1) and Prop. 7.17(2) twice, the second inequality follows from $x \otimes (x \rightleftharpoons y) \otimes (y \rightleftharpoons z) \leq x \otimes (x \curvearrowright y) \otimes (y \curvearrowright z) \leq y \otimes (y \curvearrowright z) \leq z$.

(6). By Prop. 7.17(6) we get $x \otimes y \leq x \wedge y$ and $x \leq y \curvearrowright x, y \leq x \curvearrowright y$, whence $x \wedge y \leq x \rightleftharpoons y$. □

Proposition 7.21. *Let* $(L, \wedge, \vee, \otimes, \curvearrowright, 0, 1)$ *be a complete residuated lattice; for the negation operation* $\neg a := a \curvearrowright 0$ *the following holds:*

(1) $a \leq b \rightarrow \neg b \leq \neg a \rightarrow \neg\neg a \leq \neg\neg b$;
(2) $a \leq \neg\neg a, \quad a \leq b \rightarrow a \leq \neg\neg b$;
(3) $\neg 0 = 1, \quad \neg 1 = 0$;
(4) $a \otimes \neg a = 0, \quad a \wedge \neg a = 0$;
(5) $\neg(a \otimes b) = a \curvearrowright \neg b = b \curvearrowright \neg a$;
(6) $\neg\neg 0 = 0, \quad \neg\neg 1 = 1$.
(7) $a \wedge b = 0 \leftrightarrow a \leq \neg b$;
(8) $a \leq b \rightarrow a \wedge \neg b = 0$.
(9) $\neg a = \neg\neg\neg a$;

Proof. (1): By Prop. 7.18(3), $a \leq b$ implies $b \curvearrowright 0 \leq a \curvearrowright 0$.

(2): By Prop. 7.17(2), $a \leq (a \curvearrowright 0) \curvearrowright 0 = \neg a \curvearrowright 0 = \neg\neg a$; the implication follows by combining this result with (1).

(3): By Prop. 7.17(4) and (5), $\neg 0 = 0 \curvearrowright 0 = 1$, and $\neg 1 = 1 \curvearrowright 0 = 0$.

(4): By Prop. 7.17(2), $a \otimes (a \curvearrowright 0) \leq 0$. By definition of $\neg$ we have $a = 0$, or $\neg a = 0$ otherwise.

(5): By Prop. 7.17(7), $\neg(a \otimes b) = (a \otimes b) \curvearrowright 0 = a \curvearrowright (b \curvearrowright 0)$.

(6): By (3), $0 = \neg 1 = \neg\neg 0$.

(7): $a \leq \neg b$ follows by the definition of the latter as the greatest element such that $a \wedge b = 0$. Conversely, $a \leq \neg b$ implies $a \wedge b \leq \neg b \wedge b = 0$ by (4).

(8): $a \leq b$ implies $\neg b \leq \neg a$ by (1); thus $\neg b \wedge a = 0$ by (7).

(9): By (2) $\neg a \leq \neg\neg\neg a$. Conversely, by monotonicity of $\wedge$, (2), and (4), $\neg\neg\neg a \wedge a \leq \neg\neg\neg a \wedge \neg\neg a = 0$. So by (7) we get $\neg\neg\neg a \leq \neg a$. □

(As can be seen from (2) the law of double negation does not hold in general for residuated lattices, i.e. $\neg$ is not involutive. However, by (6) we have $\neg\neg 0 = 0$ and $\neg\neg 1 = 1$, i.e. the restriction to the two-valued case would work. In contrast, MV-algebras satisfy the double negation property, whence they are used for investigations of fuzzy relation algebras in [25, 26]; see also [22].)

Proposition 7.22. ([2, Theorem 2.28]) *Let $(L, \wedge, \vee, \otimes, 0, 1)$ be a complete bounded lattice, such that $(L, \otimes, 1)$ is a commutative monoid. Then the following conditions are equivalent:*

(1) There exists an operation $\curvearrowright : L \times L \longrightarrow L$ satisfying adjointness:
$x \otimes y \leq z$ iff $x \leq y \curvearrowright z$.
(2) For any $x, y \in L$ the set $\{z : x \otimes z \leq y\}$ has a greatest element.
(3) $x \otimes \bigvee_{i \in I} y_i = \bigvee_{i \in I}(x \otimes y_i)$ holds in L.
(4) With $x \curvearrowright y := \bigvee\{z : x \otimes z \leq y\}$, $\otimes$ and $\curvearrowright$ satisfy adjointness.

Proof. (4) → (1): Obvious, i.e. L is a residuated lattice.

(1) → (2): By Prop. 7.17(8).

(1) → (3): By Prop. 7.19(1).

(2) → (4): Assume (2) and define $\curvearrowright$ as done in (4). Then $x \curvearrowright y$ is the greatest element in $\{z : x \otimes z \leq y\}$, i.e. $x \otimes z \leq y$ implies $z \leq x \curvearrowright y$. And if $z \leq x \curvearrowright y$ then $x \otimes z \leq x \otimes (x \curvearrowright y) \leq y$. Hence $\otimes$, $\curvearrowright$ satisfy adjointness.

(3) → (4): Assume (3) and define $\curvearrowright$ as done in (4). If $x \otimes z \leq y$ then $z \leq x \curvearrowright y$ by definition of $x \curvearrowright y$. And if $z \leq x \curvearrowright y$ then $x \otimes z \leq x \otimes (x \curvearrowright y) = x \otimes \bigvee\{z' : a \otimes z' \leq y\} = \bigvee\{a \otimes z' : a \otimes z' \leq y\} \leq y$, whence (4) holds. □

In order to understand why residuated lattices get their prominent role in the area of fuzzy relations, note that Sanchez (following Birkhoff [3] in his seminal paper [30]) used the term *Brouwerian* for what usually is called a *relatively pseudocomplemented* lattice (see Def. 7.12). This terminology has been prevalent since then in literature on fuzzy relation equations and inequalities; see e.g. [9, 11, 23] for usage in books. The term is still used in many papers, e.g. in [32]; but in rare cases you can find the other name, viz. *Heyting* lattice as e.g. in [17]. Now recall that a **Heyting algebra**, in the terminology of [30], is nothing else than a Brouwerian, i.e. relatively pseudocomplemented, lattice with a zero element (standing for falsehood, or "*is not*"), and thus a residuated lattice with $a \otimes b := a \wedge b$.

References

[1] Balbes, R., Dwinger, P.: *Distributive Lattices*, University of Missouri Press (1974).

[2] Bělohlàvek, R.: *Fuzzy Relational Systems-Foundations and Principles*, Springer Science +Business Media, New York (2002).

[3] Birkhoff, G.: *Lattice Theory*, 3rd ed., American Mathematical Society, Providence, Rhode Island (1973).

[4] Blyth, T.S.: Residuated mappings, Order **1**, 187-204 (1984).

[5] Blyth, T.S.: *Lattices and Ordered Algebraic Structures*, Springer-Verlag, London (2005).

[6] Blyth, T.S., Janowitz, M.F.: *Residuation Theory*, Pergamon Press, Oxford (1972).

[7] Ciungu, L.C.: *Non-commutative multiple-valued logic algebras*, Springer International Publishing Switzerland (2014).

[8] Curry, H.B.: *Foundations of Mathematical Logic*, Dover Publications Inc. New York (1977).

[9] DiNola, A., Sessa, S., Pedrycz, W., Sanchez, E.: *Fuzzy Relation Equations and Their Applications to Knowledge Engineering*, Kluwer Academic Publishers, Dordrecht (1989).

[10] Dubois, D., Esteva, F., Godo, L., Prade, H.: Fuzzy-set based logics - an history-oriented presentation of their main developments. In Gabbay, D.M., Woods, J.: *Handbook of the History of Logic, Vol 8: The Many Valued and Nonmonotonic Turn in Logic*, 325-449 (2007).

[11] Dubois, D., Prade, H.: *Fuzzy Sets and Systems:Theory and Applications*, Academic Press, New York, (1980).

[12] Dunn, J.M.: Gaggle theory: An abstraction of Galois connections and residuation, with applications to negation, implication, and various logical operators. In van Eijck, J. (Ed.): *Logics in AI*, Springer Berlin Heidelberg, 31-51 (1991).

[13] Galatos, N., Jipsen, P., Kowalski, T., & Ono, H.: *Residuated Lattices: An Algebraic Glimpse at Substructural Logics*, Elsevier, Amsterdam (2007).

[14] Goguen, J.: L-fuzzy sets, J. Math. Anal. Appl. **18**, 145-174 (1967).

[15] Harzheim, E.: *Ordered Sets*, Springer Science+Business Media (2005).

[16] Hazewinkel, M. (ed): *Encyclopaedia of Mathematics* Vol 7, Kluwer Academic Publishers, Dordrecht (1991).

[17] Jiménez, J., Montes, S., Šešelja, B., & Tepavčević, A.: Fuzzy correspondence inequations and equations. Fuzzy Sets and Systems **239**, 81-90 (2014).

[18] Jipsen, P., Tsinakis, C.: A survey of residuated lattices. In Martinez , J. (Ed.): *Ordered algebraic structures* Springer Science+Business Media Dordrecht, 19-56 (2002).

[19] Nemitz, W.C.: Implicative semi-lattices, Transactions of the American Mathematical Society **117**, 128-142 (1965).

[20] Novák, V., Perfilieva, I., Močkoř, J.: *Mathematical Principles of Fuzzy Logic*, Kluwer Academic Publishers, Boston/Dordrecht (1999).

[21] Orłowska, E., Rewitzky, I.: Algebras for Galois-style connections and their discrete duality, Fuzzy Sets and Systems **161**, 1325-1342 (2010).

[22] Pan, X., Xu, Y.: On the algebraic structure of binary lattice-valued fuzzy relations. Soft Computing, **17**, 411-420 (2013).

[23] Peeva, K., Kyosev, Y.: *Fuzzy Relational Calculus. Theory, Applications and Software*, World Scientific, Singapore (2004).

[24] Perfilieva, I., Nosková, L.: System of fuzzy relation equations with inf- $\rightarrow$ composition: Complete set of solutions. Fuzzy Sets and Systems **159**, 2256-2271 (2008).

[25] Popescu, A.: Many-valued relation algebras, Algebra universalis **53**, 73-108 (2005).

[26] Popescu, A.: Some algebraic theory for many-valued relation algebras, Algebra universalis **56**, 211-235 (2007).

[27] Rasiowa, H.: *Algebraic Approach to Non-classical Logics*, North Holland, Amsterdam (1974).

[28] Roman, S.: *Lattices and Ordered Sets,* Springer Science+Business Media, New York (2008).

[29] Rybakow, V.V.: *Admissibility of Logical Inference Rules,* Elsevier Science B.V., Amsterdam(1997).

[30] Sanchez, E.: Resolution of composite fuzzy relation equations, Information and Control **30**, 38-48 (1976).

[31] Smith, D.P.: Meet-Irreducible Elements in Implicative Lattices, Proc. American Math. Soc. **34**, 57-62 (1972).

[32] Sun, F., Wang, X.P., Qu, X.B.: Minimal join decompositions and their applications to fuzzy relation equations over complete Brouwerian lattices, Information Sciences **224**, 143-151 (2013).

[33] Turunen,E.: *Mathematics Behind Fuzzy Logic,* Physica Verlag, Heidelberg(1999).

[34] Ward, M., Dilworth, R.P.: Residuated lattices, Transactions American Math. Soc. **45**, 335-354 (1939).

[35] Winter, M.: *Goguen Categories - A Categorical Approach to L-fuzzy Relations,* Springer, Dordrecht, The Netherlands (2007).

[36] Xu, Y., Ruan, D., Qin, K. & Liu, J.: *Lattice valued logic,* Springer-Verlag Berlin Heidelberg (2003).

[37] Zadeh, L.A.: Fuzzy sets, Inform. and Control **8**, 338-353 (1965).

Chapter 8

Fuzzy Sets and Fuzzy Relations

For the topic of *many-valued relations* (more convenient and well known as *fuzzy relations*) the following aspects are of relevance in the context of this book: On the syntactic level the concept of (binary) fuzzy relations is a generalization of the classical concept of (binary) crisp relations; so the general formalism must reduce to the two-valued one. On the semantic level fuzzy relations admit for a much more fine grained modeling of relationships between two objects under consideration. Instead of only asserting the existence or absence of a relationship, information about its strength can be given. And from the viewpoint of applications, the role of fuzzy relations in approximate reasoning and fuzzy control is the most interesting aspect; it has lead to a vivid research in fuzzy relation equations and inequalities.

The current chapter presents many-valued versions of some important crisp concepts around binary relations, and thus becomes something like a generalizing mirror of many of the other things mentioned in this book. Additionally the revolutionary approach of fuzzy models is illustrated by the paradigm of fuzzy control.

Technically spoken, the transition from a two-valued to a many-valued conception of a set or a relation can best be approached via a generalization of their characteristic function: Let $A = \{a_1, a_2, \ldots, a_m\}$ and $B = \{b_1, b_2, \ldots, b_n\}$ two sets. In Def. 3.19 a Boolean matrix $M_\rho = (m_{ij}) \in \mathfrak{M}^{\mathbf{2}}_{m\times n}$ has been defined to represent a relation $\rho \subseteq A \times B$, in case $m_{ij} = 1$ iff $a_i \rho b_j$. This means that $m_{ij} = 1$ iff $\langle a_i, b_j\rangle \in \rho$ and $m_{ij} = 0$ otherwise. $M_\rho : A \times B \to \mathbf{2}$ can thus be viewed as the (classical) characteristic function of ρ. The crucial steps towards many-valued concepts are (1): the replacement of $\mathbf{2}$ by another more general poset, usually a complete residuated lattice L, and (2): to choose suitable generalized set-theoretical and logical operations on L for intersection and union, and for conjunction, disjunction, negation and implication, respectively.

8.1 Fuzzy Sets: Basic Definitions and Properties

We start with introducing the most basic terminology and formal aspects of fuzzy set theory (henceforth often abbreviated to FST)[1]. This is necessary for appreciating the idea and motivation that lies behind the step from the two-valued yes-no or black-and-white description of phenomena to a more fine grained many-valued model. To first of all give you an impression of what is revolutionary in the fuzzy or graded approach, and which has therefore been felt to be a big challenge to the traditional view, we quote from [49]:

> Given the deeply entrenched tradition of scientific thinking which equates the understanding of a phenomenon with the ability to analyze it in quantitative terms, one is certain to strike a dissonant note by questioning the growing tendency to analyze the behavior of humanistic systems as if they were mechanistic systems governed by difference, differential, or integral equations.
>
> ...
>
> An alternative approach ... is based on the premise that the key elements in human thinking are not numbers, but labels of fuzzy sets, that is, classes of objects in which the transition from membership to non-membership is gradual rather than abrupt.

Thus we have a paradigm shift from a quantitative to a qualitative, but only seemingly imprecise, approach for the analysis and description of phenomena, because this approach is still being cast into a classical precise and quantitative language. Nevertheless, FST had to overcome a lot of opposition, especially during its early history, e.g. a heavy dispute if fuzzy sets have a right to exist on their own in addition to probability theory. Let us start by fixing the new concept in

Definition 8.1. *Let X be a crisp set, and L be a complete bounded lattice with $|L| \geq 2$. An L-**fuzzy set** (also **fuzzy set**) A over (in, of) X is (given by) a map $\mu_A : X \to L$. X is called the **universe of discourse**, comprising all objects under consideration. μ_A is also called the **membership function** of A, and $\mu_A(x) \in L$ the **membership degree** of x in A, often also denoted by the shorthand $A(x)$.*

Fuzzy sets are defined in a way such that both the classical two-valued and also a number of many-valued variants are subsumed. If we want to emphasize that a set X is to be understood in the classical sense, we will call it a **crisp set**: The universe of discourse of a fuzzy set is an example. The set of all L-fuzzy sets over a universe of discourse X is denoted by $\mathcal{F}^L(X)$.

Note from the definition above that the terminology for this basic notion is a little bit sloppy in literature (or pragmatic, if you want): Sometimes a fuzzy set is identified with, and sometimes is said to be described by, its membership function. But this is not really confusing: To remedy this terminological inconsistency, one may consider the notion of

[1] For more details start with a look into the early classic [14] from 1980 which still carries a lot of Zadeh's original motivation and intents but also does already hint to the formal development directions to emerge later on. You may continue with [23].

a fuzzy set an extra-mathematical notion that is not defined itself; only the notion of a membership function of a fuzzy set is defined ([3], p. 79).

Henceforth we will not always distinguish between Zadeh's original version of a fuzzy set (with $L = [0,1] \subseteq \mathbb{R}$) and the more general concept of L-fuzzy sets, and we will often simply use the term '*fuzzy*' to indicate that we are dealing with a many-valued variant. Please notice that Def. 8.1 is formulated in a way ($|L| \geq 2$) such that it comprises classical sets as a special case for which the usual characteristic function acts as membership function.

The use of lattices as valuation spaces imparts to the theory a special flavor which emphasizes order theoretic aspects: It may be helpful to think of fuzzy sets as sets that are ordered according to the membership degrees of their elements. The question now is: What is described with such a generalized characteristic function? A look into the early days of fuzzy set theory shows that many of the first examples are related to human perceptual equipment like TALL, COLD, LOUD, HIGH etc. A degree of membership in a fuzzy set HEAVY may be thought of as a degree of compatibility or typicality of a quantitative "objective" measurement like 3KG, 90KG or 3 TONS with a linguistic label "*heavy*" in a given context under consideration. Membership degrees can also be interpreted as degree of truth, e.g. of the statement "*She earns a* HIGH *salary.*". For some more epistemological considerations on fuzziness see Section 12.2 or [42]. Furthermore, think of e.g. a set of LARGE numbers, a CHEAP car, or of experiencing a GREAT LOSS during a stock exchange crash. All these examples clearly indicate that the choice of a 'fuzzy label' is highly subjective and context dependent. But it is just this flexibility of human language that allows for efficient communication: RED stands for a lot of different wavelengths of the spectrum, among them many pairs of values which we cannot discriminate when using our inborn perceptual equipment only; and most of the time in every-day life we also have no need to do so.

Just as classical relations are sets, fuzzy relations are fuzzy sets. It makes therefore sense to continue with some basic notions for fuzzy sets. Since (classical) sets are a special case of fuzzy sets, viz. their crisp two-valued variant, their operations must appropriately be subsumed under the formalism of fuzzy sets. If you translate the definitions for the basic set operations into operations on their characteristic functions, you will more or less automatically arrive at the corresponding definitions for fuzzy sets as can be seen from

Definition 8.2. *Let S, T be two L-fuzzy sets over a universe X with membership function μ_S and μ_T, respectively.*

(a) S is called a **fuzzy subset** *of T, denoted $S \subseteq T$, if $\forall x \in X[\mu_S(x) \leq \mu_T(x)]$.*
(b) S and T are said to be equal, $S = T$, if $S \subseteq T$ and $T \subseteq S$, i.e. if $\forall x \in X[\mu_S(x) = \mu_T(x)]$.
(c) We say that S is a **proper fuzzy subset** *of T, $S \subset T$, if $S \subseteq T$ but $S \neq T$.*
(d) The **intersection** *of two fuzzy sets is $S \cap T$ with $\mu_{S \cap T}(x) = \mu_S(x) \wedge \mu_T(x)$.*
(e) The **union** *of two fuzzy sets is $S \cup T$ with $\mu_{S \cup T}(x) = \mu_S(x) \vee \mu_T(x)$.*

(f) *The* **product** *of two fuzzy sets is* $S \otimes T$ *with* $\mu_{S\otimes T}(x) = \mu_S(x) \otimes \mu_T(x)$.
(g) *The* **complement** *of a fuzzy set* S *is* $\overline{S}(x) = 1 - S(x)$.

Apart from complement, the above definitions clearly are suitable for all lattices L usually involved (see Def. 2.54(c)), so you should read them in a generic way just as meet and join in a lattice which, however, must obey the involution and De Morgan Laws (see Prop. 2.66) to become compatible when restricted to the two-element Boolean algebra. $\overline{S}$ is also called the *standard complement* of S (in the unit interval), and is then the default candidate if nothing else is specified.

A few more concepts will be helpful for later considerations:

Definition 8.3. *Let A be an L-fuzzy set over a universe X, and let 0 be the minimal and 1 be the maximal element of L.*

(a) *The* **support of** *A is the set defined as* $\mathsf{supp}(A) := \{x \in X : A(x) > 0\} \subseteq X$.
(b) *The* **level set of** *A is the set defined as* $\Lambda(A) := \{A(x) \in L : x \in X\} \subseteq L$.
(c) *Let* $0 < \alpha \in L$*; the* α**-cut** *(or* α**-level set**) *of A is the crisp set* $A_\alpha := \{x \in X : A(x) \geq \alpha\}$.
(d) *Let* $0 < \alpha \in L$*; the* α**-restriction** *of A is the fuzzy set* $A_{|\alpha}$ *with* $A_{|\alpha}(x) = A(x)$ *if* $A(x) \geq \alpha$ *and* $A_{|\alpha}(x) = 0$ *otherwise, i.e. it equals* $A_{|A_\alpha}$, *the restriction of A to* A_α.
(e) *The* **height of** *A is* $\mathsf{hgt}(A) := \bigvee_{x\in X} A(x)$.
(f) *A fuzzy set A is called* **normal** *if* $\mathsf{hgt}(A) = 1$.
(g) *The* **core of** *A is the set* $\mathsf{core}(A) := \{x \in X : A(x) = 1\} \subseteq X$. *$A$ is called* **modal** *if* $\mathsf{core}(A) \neq \emptyset$.

This definition is a first illustration of the increase in possibilities that comes along with the many-valued approach: Having left behind the severe restriction of two-valued classification, one can now decide to take only those (pairs of) elements into consideration whose degree of compatibility (strength of relatedness) exceeds a certain threshold. And one can do this in two ways, both of which limit the number of candidates for further processing: Applying the concept of α-cut of a fuzzy set you will continue your considerations (or computations) using smaller crisp sets but you will lose information contained in the membership degrees; and applying α-restrictions you get fuzzy sets with smaller support.

α-cuts have some interesting properties, among them a relationship between fuzzy sets and families of subsets of the underlying universe; see e.g. [41]. In the following let A_L denote the set $\{A_\alpha : \alpha \in L\} \subseteq \mathbf{2}^X$.

Proposition 8.4. (Decomposition for L-fuzzy sets) *Any $A \in \mathcal{F}^L(X)$ can be decomposed using its α-cuts as follows:*

$$A(x) = \bigvee_{\alpha\in\Lambda(A)} \{\alpha : A_\alpha(x) = 1\}$$

(i.e. the supremum on the right hand side exists for every $x \in X$ and equals $A(x)$).

Proof. Assume $A(x) = \beta$; then $A_\beta(x) = 1$. If $A_\alpha(x) = 1$, then $A(x) \geq \alpha$ and $\beta \geq \alpha$. But $\beta \in \{\alpha \in L : A_\alpha(x) = 1\}$, and it is the greatest element of this set. Thus $A(x) = r = \bigvee_{\alpha\in\Lambda(A)}\{\alpha :$

$A_\alpha(x) = 1\}$. □

Proposition 8.5. *Let $A, B \in \mathcal{F}^L(X)$, let $(L, \wedge, \vee, \otimes, 0, 1)$ be a complete bounded lattice, $K \subseteq L$, and $\alpha, \beta \in L$. Then*

(1) $a \leq \beta$ *implies* $A_\beta \subseteq A_\alpha$;
(2) $\bigcap_{\alpha \in K} A_\alpha = A_{\bigvee_{\alpha \in K} \alpha}$;
(3) $\bigcup_{\alpha \in L} A_\alpha = X$;
(4) *for each* $x \in X, \bigcap\{A_\alpha : x \in A_\alpha\} \in A_L$;
(5) $(A \cap B)_\alpha = A_\alpha \cap B_\alpha$ *and* $(A \cup B)_\alpha = A_\alpha \cup B_\alpha$.

Proof. (1): If $a \leq \beta$, then $A_\beta(x) = 1$ implies $A_\alpha(x) = 1$; thus $A_\beta \subseteq A_\alpha$.
(2): $x \in A_{\bigvee_{\alpha \in K} \alpha} \leftrightarrow A(x) \geq \bigvee_{\alpha \in K} \alpha \leftrightarrow \forall \alpha \in K[A(x) \geq \alpha] \leftrightarrow x \in \bigcap_{\alpha \in K} A_\alpha$.
(3): $A_0 = X$.
(4): Let $x \in X$; then $x \in A_\alpha \leftrightarrow A(x) \geq \alpha \leftrightarrow A_\alpha(x) = 1$. By Prop. 8.4 $A(x) = \bigvee_{\alpha \in \Lambda(A)}\{\alpha : A_\alpha(x) = 1\}$, and by (2) $\bigcap\{A_\alpha : x \in A_\alpha\} = A_{\bigvee\{\alpha \in \Lambda(A) : A_\alpha(x) = 1\}} \in A_L$. □

8.2 Many-valued Connectives on Complete lattices

A lot has been written, and there have been many debates, about the possible meanings of membership degrees. But this is not our concern here, is also far beyond our goals, and probably would need another book, or at least a separate chapter, for proper treatment. Here we give a short overview on the syntactical handling of membership degrees by many-valued connectives.

Subsequently to Def. 2.65 we have already mentioned $\mathbf{2} = (\{0, 1\}, \vee, \wedge, \bar{\ }, 0, 1)$, with $\vee, \wedge$ denoting max, min, and $\overline{x} = 1 - x$, as the simplest Boolean algebra having more than one element. Replacing $\{0, 1\}$ with $[0, 1]$ one gets the basic operations of classical fuzzy set theory but $([0, 1], \vee, \wedge, \bar{\ }, 0, 1)$ is no longer a Boolean but only a De Morgan algebra since 0 and 1 are the only complemented elements.

If one views $\mathbf{2} = (\{0, 1\}, \vee, \wedge, \bar{\ }, 0, 1)$ as an imploded very special case of more general structures, the question is about what are consistent "explosions" of it. E.g. interpreting $\vee = max$ and $\wedge$ as normal multiplication, we also get operations compatible with the two-valued case. Indeed, there exist infinitely many such binary operations on the unit interval $[0, 1]$, both for intersection (called *t-norms*) and union (called *t-conorms*) of fuzzy sets. For details see [23, Chapter 3] or [22].

Besides using t-norms and t-conorms, it is furthermore common to interpret a fuzzy logic implication $\rightarrow$ in a way that conjunction and implication form an adjoint pair. This is reasonable: for two propositions P and Q, $P \rightarrow Q$ should be understood as the weakest proposition such that P together with $P \rightarrow Q$ implies Q.

We have already mentioned above that there are infinitely many t-norms and t-conorms on $[0, 1]$ by which min and max can be appropriately replaced. On the other hand, $[0, 1]$ is a chain, i.e. a totally ordered set (additionally bearing suitable arithmetical operations); this surely is a restriction too strong to be met when applying fuzzy sets to human classification abilities and behavior. Therefore lattices of a more general character become interesting which, however, must have "enough suitable structure" to allow for building a reasonable calculus. So we have to find a way to combine the flexibility of the valuation structure with a plurality of operators in lattices other than the unit interval.

Though our concern of interest in this topic is mainly focused on binary relations, be aware that we do in fact consider a whole family of logical operators so that we also have a look to many-valued logic in general. As announced at the very beginning of this chapter, as the second task we have "to choose suitable generalized set-theoretical and logical operations on L for intersection and union, and for conjunction, disjunction, negation and implication, respectively". And this has to be done in a way that yields reasonable properties for the many-valued and, furthermore, reduces to the two-valued case. The following considerations borrow from [11,30,45]

Definition 8.6. *Let* $(L, \wedge, \vee, 0, 1)$ *be a bounded complete lattice. A* **triangular norm** *(short* **t-norm***) on L is a commutative and associative mapping* $\sqcap : L \times L \longrightarrow L$ *such that for all* $a, b, c \in L$

(1) the following boundary condition holds: $1 \sqcap a = a$*;*
(2) $b \leq c$ *implies* $a \sqcap b \leq a \sqcap c$*.*

(For mnemonic ($\sqcap$ corresponds to conjunction) and optical reasons (to avoid mixed notation like e.g. $\bigvee_I T(a, b_i)$) I will use $\sqcap$ as a generic symbol.)

Note that the structure of the lattice L heavily influences which and how many t-norms on L can be defined. There exist at least two t-norms on L: the meet $a \wedge b$, and the drastic intersection

$$a \sqcap_D b = \begin{cases} a \wedge b & \text{if } 1 \in \{a, b\} \\ 0 & \text{otherwise.} \end{cases}$$

Algebraically speaking, $\sqcap$ is a t-norm iff $(L, \sqcap, 1, \leq)$ is a commutative partially ordered integral monoid. (A partially ordered monoid is called **integral** if 1 is its neutral element.) If L is also residuated then $\otimes$ is a t-norm on L since it is commutative, associative and $x \otimes 1 = x$ by Def. 7.16, and isotone in both arguments by Prop. 7.18.

Examples for important t-norms on the unit interval $[0, 1]$ are:

(1) the *Gödel t-norm:* $a \sqcap_G b := \min\{a, b\}$;
(2) the *algebraic product:* $a \sqcap_P b := ab$;
(3) the *bounded difference* or *Łukasiewicz t-norm:* $a \sqcap_L b := \max\{a + b - 1, 0\}$;

Definition 8.7. *Let* $(L, \wedge, \vee, 0, 1)$ *be a bounded complete lattice. A* **triangular conorm** *(short* **t-conorm***) on L is a commutative and associative mapping* $\sqcup : L \times L \longrightarrow L$ *such that for all* $a, b, c \in L$

(1) the following boundary condition holds: $0 \sqcup a = a$*;*
(2) $b \leq c$ *implies* $a \sqcup b \leq a \sqcup c$*.*

Examples for important t-conorms on the unit interval $[0, 1]$ are:

(1) the *Gödel t-conorm:* $a \sqcup_G b := \max\{a, b\}$;
(2) the *algebraic sum:* $a \sqcup_P b := a + b - ab$;
(3) the *bounded sum* or *Łukasiewicz t-conorm:* $a \sqcup_L b := \min\{a + b, 1\}$;

There exist at least two t-conorms on L: the join $a \sqcup b = a \vee b$, and the drastic union

$$a \sqcup_D b = \begin{cases} a \vee b & \text{if } 0 \in \{a, b\} \\ 1 & \text{otherwise.} \end{cases}$$

Proposition 8.8. *Let $(L, \wedge, \vee, 0, 1)$ be a bounded complete lattice.*

(1) For t-norms $\sqcap$ we have $a \sqcap_D b \le a \sqcap b \le a \wedge b$.
(2) For t-conorms $\sqcup$ we have $a \vee b \le a \sqcup b \le a \sqcup_D b$.

Proof. (1): We have $a \sqcap b \le 1 \sqcap b = b$ and $a \sqcap b = b \sqcap a \le 1 \sqcap a = a$, whence $a \sqcap b \le a \wedge b$. Furthermore, $a \sqcap_D 1 = a = a \sqcap 1$, $1 \sqcap_D b = b = 1 \sqcap b$, and $a \sqcap_D b = 0 \le a \sqcap b$ otherwise. Hence, for all $a, b \in L$, $a \sqcap_D b \le a \sqcap b$. □

The notion of a t-norm plays the role of intersection, or in a logical context, that of conjunction; its dual is that of union, or disjunction. In the case of crisp sets (two-valued logic), union and intersection (disjunction and conjunction) are related via complement (negation) by the well-known De Morgan formulas. So the notion of complement, or negation, is our next candidate for generalization.

Definition 8.9. *Let $(L, \wedge, \vee, 0, 1)$ be a bounded complete lattice.*

(a) A **negator** *on L is an mapping $\eta : L \longrightarrow L$ such that for all $a, b \in L$*

- *η is order reversing: $a \le b \rightarrow \eta(b) \le \eta(a)$;*
- *$\eta(0) = 1$ and $\eta(1) = 0$.*

(b) A negator such that $\eta(\eta(a)) = a$ is called **strong** *or* **involutive**.

From Prop. 7.21 we see that negation as defined for residuated lattices is a negator in the above sense, though not an involutive one. We say that $\sqcap$ and $\sqcup$ are *dual with respect to the negator η* and denote this by $(\sqcap, \sqcup, \eta)$, called a **De Morgan triple**, iff

$$\eta(a \sqcup b) = \eta(a) \sqcap \eta(b) \quad \text{and} \quad \eta(a \sqcap b) = \eta(a) \sqcup \eta(b).$$

If we let N denote the standard complementation $\mathsf{N}(x) = 1 - x$ on the unit interval, then both $(\sqcap_L, \sqcup_L, \mathsf{N})$ and $(\sqcap_G, \sqcup_G, \mathsf{N})$ are De Morgan triples on $[0, 1]$.

The following proposition gives some evidence that we are on a good way with the generalized concepts introduced so far.

Proposition 8.10. ([23], Theorem 3.20) *Let $(L, \wedge, \vee, 0, 1)$ be a bounded complete lattice. Given a t-norm $\sqcap$ and a strong negator η on L, the binary operation $\sqcup$ defined by*

$$a \sqcup b = \eta\big(\eta(a) \sqcap \eta(b)\big). \tag{8.1}$$

is a t-conorm on L such that $(\sqcap, \sqcup, \eta)$ is a De Morgan triple.

Proof. The proof proceeds by routine calculations in five steps.
(a) $a \sqcup 0 = \eta(\eta(a) \sqcap \eta(0)) = \eta(\eta(a) \sqcap 1) = \eta(\eta(a)) = a$.
(b) Assume $b \le d$; then we have $\eta(b) \ge \eta(d)$, whence, $\eta(a) \sqcap \eta(b) \ge \eta(a) \sqcap \eta(d)$ by monotonicity of $\sqcap$. Thus $a \sqcup b = \eta(\eta(a) \sqcap \eta(b)) \le \eta(\eta(a) \sqcap \eta(d)) = a \sqcup d$ which shows the monotonicity of $\sqcup$.
(c) $a \sqcup b = \eta(\eta(a) \sqcap \eta(b)) = \eta(\eta(b) \sqcap \eta(a)) = b \sqcup a$.
(d)
$$\begin{aligned}
a \sqcup (b \sqcup c) &= \eta(\eta(a) \sqcap \eta(b \sqcup c)) \\
&= \eta(\eta(a) \sqcap \eta[\eta(\eta(b) \sqcap \eta(c))]) \\
&= \eta(\eta(a) \sqcap (\eta(b) \sqcap \eta(c))), \quad \text{since } \eta \text{ is involutive} \\
&= \eta((\eta(a) \sqcap \eta(b)) \sqcap \eta(c)), \quad \text{since } \sqcap \text{ is associative} \\
&= \eta(\eta[\eta(\eta(a) \sqcap \eta(b))] \sqcap \eta(c)), \quad \text{since } \eta \text{ is involutive} \\
&= \eta(\eta[a \sqcup b] \sqcap \eta(c)) \\
&= (a \sqcup b) \sqcup c.
\end{aligned}$$
(e)
$$\begin{aligned}
\eta[a \sqcup b] &= \eta[\eta(\eta(a) \sqcap \eta(b))] = \eta(a) \sqcap \eta(b), \text{ and} \\
\eta[a \sqcap b] &= \eta[\eta(\eta(a)) \sqcap \eta(\eta(b))] = \eta(a) \sqcup \eta(b).
\end{aligned}$$
□

Proposition 8.11. *Let* $(L, \wedge, \vee, 0, 1)$ *be a bounded complete lattice, and let* $\sqcap$ *and* $\sqcup$ *be a t-norm and a t-conorm on L, respectively. Then the following holds:*

(1) $\sqcap$ *is idempotent, iff* $\sqcap = \wedge$.
(2) $\sqcup$ *is idempotent, iff* $\sqcup = \vee$.
(3) $\sqcap$ *and* $\sqcup$ *satisfy the absorption law* $(a \sqcup b) \sqcap a = a$ *iff* $\sqcap = \wedge$.
(4) $\sqcap$ *and* $\sqcup$ *satisfy the absorption law* $(a \sqcap b) \sqcup a = a$ *iff* $\sqcup = \vee$.

Proof. (1): If $\sqcap = \wedge$, then obviously, $x \wedge x = x$. Conversely, by Prop. 8.8(1), monotonicity and assuming idempotency we get $x \wedge y = (x \wedge y) \sqcap (x \wedge y) \le x \sqcap y \le x \wedge y$.
(3): Taking $b = 0$ yields $a = (a \sqcup 0) \sqcap a = a \sqcap a$, for all $a \in L$, whence $\sqcap = \wedge$ by (1). Now assume $\sqcap = \wedge$; then $(a \sqcup b) \wedge a = a$, since $a \le a \vee b \le a \sqcup b$ by Prop. 8.8(2). □

What is still missing are results about the distributivity properties of t-norms and t-conorms. Let $\sqcap$ and $\sqcup$ be a t-norm and a t-conorm on a lattice L. Following [22, Chapter 2.1], we say that $\sqcup$ *is distributive over* $\sqcap$, if

$$a \sqcup (b \sqcap c) = (a \sqcup b) \sqcap (a \sqcup c),$$

and that $\sqcap$ *is distributive over* $\sqcup$, if

$$a \sqcap (b \sqcup c) = (a \sqcap b) \sqcup (a \sqcap c).$$

Proposition 8.12. *Let* $(L, \wedge, \vee, 0, 1)$ *be a bounded complete lattice, and let* $\sqcap$ *and* $\sqcup$ *be a t-norm and a t-conorm on L, respectively. Then the following holds:*

(1) $\sqcup$ *is distributive over* $\sqcap$ *iff* $\sqcap = \wedge$.
(2) $\sqcap$ *is distributive over* $\sqcup$ *iff* $\sqcup = \vee$.
(3) $\sqcap$ *and* $\sqcup$ *are distributive over each other iff* $\sqcap = \wedge$ *and* $\sqcap = \vee$.

Proof. (1): Suppose that $\sqcup$ is distributive over $\sqcap$; then we have $x = x \sqcup (0 \sqcap 0) = (x \sqcup 0) \sqcap (x \sqcup 0) = x \sqcap x$, whence $\sqcap = \wedge$ by Prop. 8.11(1). Conversely, let $\sqcap = \wedge$; then $a \sqcup (b \wedge$

$c) \le a \sqcup b$ and $a \sqcup (b \wedge c) \le a \sqcup c$, thus $a \sqcup (b \wedge c) \le (a \sqcup b) \wedge (a \sqcup c)$. Now assume that $a \sqcup (b \wedge c) < (a \sqcup b) \wedge (a \sqcup c)$; then, if $b \le c$, we get $a \sqcup b < (a \sqcup b) \wedge (a \sqcup c)$, and, if $b > c$, $a \sqcup c < (a \sqcup b) \wedge (a \sqcup c)$, a contradiction. Therefore, any t-conorm on L distributes over the the meet operation $\wedge$. □

Proposition 8.13. *Let $(L, \wedge, \vee, 0, 1)$ be a bounded complete lattice, let A, B, C denote L-fuzzy sets over a set X, and let $(\sqcap, \sqcup, \eta)$ be a De Morgan triple (on L). Then the following holds:*

(1) *$\sqcap$ is commutative, associative, and idempotent:*

$$A \sqcap B = B \sqcap A, \quad A \sqcap (B \sqcap C) = (A \sqcap B) \sqcap C, \quad A \sqcap A = A;$$

(2) *$\sqcup$ is commutative, associative, and idempotent:*

$$A \sqcup B = B \sqcup A, \quad A \sqcup (B \sqcup C) = (A \sqcup B) \sqcup C, \quad A \sqcup A = A;$$

(3)

$$A \sqcap B \subseteq A \subseteq A \sqcup B,$$
$$A \sqcap \emptyset = \emptyset, \quad A \sqcup \emptyset = A,$$
$$A \sqcap X = A, \quad A \sqcup X = X;$$

(4) *Distributivity laws:*

$$A \sqcap (B \sqcup C) = (A \sqcap B) \sqcup (A \sqcap C), \quad \text{if } \sqcap \text{ is distributive over } \sqcup,$$
$$A \sqcup (B \sqcap C) = (A \sqcup B) \sqcap (A \sqcup C), \quad \text{if } \sqcup \text{ is distributive over } \sqcap;$$

(5) *De Morgan laws:*

$$\eta(A \sqcap B) = \eta(A) \sqcup \eta(B), \quad \eta(A \sqcup B) = \eta(A) \sqcap \eta(B).$$

8.3 Fuzzy Relations: Basic Operations and Properties

The concept of fuzzy relations is one of the most fundamental notions in fuzzy set theory, and a major mathematical tool in numerous application areas, where they usually appear in form of fuzzy relation equations or inequalities. It seems thus worthwhile to have a separate look at many-valued relations and their properties, and especially on the topic of fuzzy relation equations. I will try to present enough information to give a proper overview on both mathematical aspects and application possibilities of fuzzy relations as a solid base for further investigation.

In this section I not only will introduce the basic concepts in a technical sense, but will also take the opportunity to discuss some important semantical, by which I mean non-formal, aspects that automatically arise by going the crucial step from two-valued to many-valued conceptions: If we let $L^+ := \{a \in L : a > 0\}$ denote the *positive* elements of L, this step means a drastic change from $|L^+| = 1$ to $|L^+| > 1$, i.e. we now have more than only one way to say "*yes*", "*has*" or "*is*", allowing for finer grained judgements.

Of course, you may ask: Given the modeling power of (the usual) two-valued binary relations, why shall one bother with fuzzy relations? A confessing fuzzyist (as I am) would probably argue that, using fuzzy relations, you can build FAR BETTER models that describe various aspects of reality MUCH MORE ACCURATE because of abandoning the old tradition of black-or-white or all-or-none judgments and reasoning. There are many more examples of that kind: Simply take any adjective and combine it with a linguistic modifier in an appropriate way, as e.g. MUCH HIGHER, A LITTLE BIT LOUDER, SLIGHTLY COOLER, CLOSE TO, VERY CLOSE TO, or CONSIDERABLY LARGER THAN etc.

As an example consider the $[0,1]$-fuzzy relation MUCH GREATER THAN on $\mathbb{R}$, $\gg \subseteq \mathbb{R}\times\mathbb{R}$, given by

$$\gg (x,y) := \begin{cases} (1+\frac{100}{(x-y)^2})^{-1}, & x > y \\ 0 & \text{otherwise.} \end{cases}$$

You can see from this example that membership functions usually can be adapted to a special context in an easy way: change the numerator from 100 to e.g. 300 or 653, or you may decide to choose some suitable linear function.

Let X and Y be two sets. In the classical two-valued theory, a relation from X to Y is a subset of the Cartesian product of the two sets, $\rho \subseteq X \times Y$. In order to achieve a seamless inclusion of the special two-valued case in the new general framework, we try to follow the classical way as close as possible. So, as a first approach, we proceed as follows: We define the **Cartesian product** of two fuzzy sets A and B of X and Y, respectively, as the fuzzy set $A \times B$ of $X \times Y$ with

$$A \times B(x,y) = A(x) \wedge B(y). \tag{8.2}$$

Following [38], let S be a crisp set and $T \subseteq S$. Then a relation ρ on S is also a relation on T if $x\rho y$ implies $x \in T$ and $y \in T$, for all x, y. Restated in terms of characteristic functions this gives: $R(x,y) = 1 \rightarrow T(x) = T(y) = 1$, or equivalently $R(x,y) \leq T(x) \wedge T(y)$, which finally results in

Let A and B be two fuzzy sets over X and Y, respectively. A (binary) **fuzzy relation** *ρ from A to B is a fuzzy subset of $A \times B$.*

In other words, for ρ to be a fuzzy relation from A to B, it is required that the degree of membership of any pair of elements does not exceed the degree of membership of either of the elements in A and B, respectively. This definition is used in e.g. [29].

Though this definition would be a perfect twin of the bivalent variant (a fuzzy relation is a fuzzy subset of the Cartesian product of two fuzzy sets), it rules out some data from forming fuzzy relations. Consider the following example: Let $U = \{Adam, Eve\}$ be our universe of discourse, and suppose we are given two fuzzy sets over U as follows:

$$\text{BLOND} := \{\langle Adam, 0.2\rangle, \langle Eve, 0.1\rangle\}, \text{ and}$$

$$\text{LIGHT BROWN} := \{\langle Adam, 0.4\rangle, \langle Eve, 0.3\rangle\}.$$

Comparing the Cartesian product of BLOND and LIGHT BROWN (left matrix) with a possible description of the HAIR COLOR-SIMILARITY of Adam and Eve (right matrix) shows that the latter one fails to obey the restrictions for being a fuzzy relation from BLOND to LIGHT BROWN (using obvious abbreviations):

BL×LIBR	Adam	Eve
Adam	0.2	0.2
Eve	0.1	0.1

HC-SIM	Adam	Eve
Adam	1	0.8
Eve	0.8	1

Note that this can occur only in valuation spaces L with $|L^+| > 1$ and is, therefore, impossible in the bivalent case where all non-zero values are equal to 1. Be it out of deliberation or out of conceiving fuzzy relations as fuzzy sets over Cartesian products of crisp sets, the following definition is commonly used:

Definition 8.14. *Let X and Y denote two crisp sets, and $(L, \wedge, \vee, \otimes, \curvearrowright, 0, 1)$ be a complete residuated lattice. A (binary) L-**fuzzy relation** ρ over $X \times Y$ (from X to Y) is an L-fuzzy subset of $X \times Y$.*

The requirement from above, viz. that the degree of membership of any pair of elements does not exceed the degree of membership of either of the elements in X and Y, respectively, is now fulfilled automatically.

Since fuzzy relations are fuzzy sets, they obey the same set-theoretic laws as shown in Prop. 8.13. The most prominent kinds of fuzzy relations clearly arise for $L = \mathbf{2}$ as the classical variant of (Boolean, two-valued) *relations*; for $L = [0,1]$ as *fuzzy relations* as originally defined in [47], and for a more general L as *L-fuzzy relations* in [17].

Written in form of generalized characteristic functions we get relations as $\mathbf{2}^{X\times Y}$, fuzzy relations as $[0,1]^{X\times Y}$, and L-fuzzy relations as $L^{X\times Y}$, respectively. Analogous to $\mathcal{B}_{X\times Y}$ for classical two-valued relations, we will write $\mathcal{F}_{X\times Y}$ and $\mathcal{F}^L_{X\times Y}$ in the second and third case to denote the set of all fuzzy relations of the respective kind. If $X = Y$ we simply write $\mathcal{F}^L_X$.

The set L of membership values is required to be a complete residuated lattice; so we can consider fuzzy relations on arbitrary, i.e. not necessarily finite sets. However, whenever we work with fuzzy subsets and relations on a finite set, the completeness assumption becomes needless and can be omitted.

As in the case of crisp binary relations, every fuzzy relation ρ on a set X can be represented by a directed fuzzy graph G_ρ with X as its set of nodes such that there is a weighted arc $\langle x, y\rangle$ connecting each pair of nodes x, y, with $\rho(x, y)$ being its associated weight. Following the common habit, arcs with weight 0 are not shown when drawing G_ρ.

In most cases the translation of basic classical two-valued concepts into corresponding many-valued ones can be performed schematically on the formal or syntactical level. As an example let us look at the notion of the *domain* of a relation $\rho \subseteq X \times Y$ which has been defined in Def. 3.5 as

$$\mathsf{dom}(\rho) := \{x \in X : \exists y \in Y[x\rho y]\} = \{x \in X : \exists y \in Y[\rho(x, y) = 1]\} \subseteq X.$$

In terms of membership functions this is equivalent to $\mu_{\mathsf{dom}(\rho)}(x) = \max\{y \in Y : \rho(x, y)\}$. Analogous considerations lead to

Definition 8.15. *Let ρ be a fuzzy relation from X to Y, and A be a fuzzy set over X, and $(L, \wedge, \vee, \otimes, \curvearrowright, 0, 1)$ be a complete residuated lattice.*

*(a) The **afterset** of $x \in X$ is the fuzzy set $x\rho$ over Y defined as*

$$x\rho(y) = \rho(x, y).$$

(b) *The* **foreset** *of* $y \in Y$ *is the fuzzy set* ρy *over* X *defined as*

$$\rho y(x) = \rho(x, y).$$

(c) *The* **image** *of* A *under* ρ *is the fuzzy set* $\rho[A]$ *over* Y *defined as*

$$\rho[A](y) = \bigvee_{x \in X} \big(A(x) \otimes \rho(x, y)\big).$$

(d) *The* **domain** *of* ρ *is the fuzzy set* $\mathsf{dom}(\rho)$ *over* X *defined as*

$$\mathsf{dom}(\rho)(x) = \bigvee_{y \in Y} \rho(x, y) = \mathsf{hgt}(x\rho).$$

(e) *The* **range** *of* ρ *is the fuzzy set* $\mathsf{rng}(\rho)$ *over* Y *defined as*

$$\mathsf{rng}(\rho)(y) = \bigvee_{x \in X} \rho(x, y) = \mathsf{hgt}(\rho y).$$

Conversion and composition of fuzzy relations can be defined in a nearly self-evident way:

Definition 8.16. *Let* ρ *and* σ *be two* L*-fuzzy relations from* A *to* B *and from* B *to* C*, respectively, and* $(L, \wedge, \vee, \otimes, \frown, 0, 1)$ *a complete residuated lattice.*

(a) *The* L*-fuzzy relation* $\rho\,\breve{}$ *defined as* $\rho\,\breve{}(a, b) = \rho(b, a)$ *is called the* **converse** *of* ρ *and is an* L*-fuzzy relation from* B *to* A*.*

(b) *The* **product** *(or* **composition***) of* ρ *and* σ *is the* L*-fuzzy relation* $\rho \circ \sigma$ *from* A *to* C *defined by*

$$\rho \circ \sigma(a, c) = \bigvee_{b} \big(\rho(a, b) \otimes \sigma(b, c)\big). \tag{8.3}$$

We simplify notation to remain close to the one used in the two-valued context and will write

$$\mu_\rho(a, b) = \rho(a, b) = a\rho b, \tag{8.4}$$

and use the second or third variant at will; so (8.3) will then become

$$a\rho\sigma c = \bigvee_{b} a\rho b \otimes b\sigma c.$$

Taking $\vee = \max$ and $\otimes = \min$, (8.3) yields the so-called **standard composition**, since it is based on the standard t-conorm and t-norm. (Remember that max and min give the only pair of a t-norm and a t-conorm that distribute over each other; see Prop. 8.12).

Representation of fuzzy relations is analogous to the classical case as described in Def. 3.19.

Definition 8.17.

(a) *A* $m \times n$ *matrix with values from a lattice* L *is called an* L**-valued matrix***. The set of all* $m \times n$ L*-valued matrices is denoted by* $\mathfrak{M}^L_{m \times n}$*.*

(b) *Let* ρ *be an* L*-fuzzy relation from* $A = \{a_1, a_2, \ldots, a_m\}$ *to* $B = \{b_1, b_2, \ldots, b_n\}$*.* $M = (m_{ij}) \in \mathfrak{M}^L_{m \times n}$ *is a* **representation of** ρ *in case* $m_{ij} = \rho(a_i, b_j)$*, and is then denoted by* M_ρ*.*

(c) *Let* $M = (m_{ij}) \in \mathfrak{M}^L_{m \times n}$ *and* $P = (p_{ij}) \in \mathfrak{M}^L_{n \times p}$*. The* **product of** L**-valued matrices** *M and P, denoted by MP is the matrix* $Q = (q_{ij}) \in \mathfrak{M}^L_{m \times p}$ *with*

$$q_{ij} = \bigvee_{1 \le k \le n} (m_{ik} \otimes p_{kj}). \tag{8.5}$$

Proposition 8.18. *Let $\rho \subseteq A \times B$ and $\sigma \subseteq B \times C$ be fuzzy relations from A to B, and from B to C, respectively, with $A = \{a_1, a_2, \dots, a_m\}$, $B = \{b_1, b_2, \dots, b_n\}$, and $C = \{c_1, c_2, \dots, c_p\}$. Furthermore, let $R = (r_{ij}) \in \mathfrak{M}^L_{m\times n}$ and $S = (s_{ij}) \in \mathfrak{M}^L_{n\times p}$ be representations of ρ and σ. Then $P = RS \in \mathfrak{M}^L_{m\times p}$ is a representation of $\rho\sigma$.*

Having transferred the most basic operations for relations to the many-valued context, one may now ask about properties of fuzzy relations and their operations. A few answers are given below, together with some sample proofs.

Proposition 8.19. *Let ρ and σ be two L-fuzzy relations from S to T.*

(1) *If $\rho \subseteq \sigma$, then $\mathsf{dom}(\rho) \subseteq \mathsf{dom}(\sigma)$ and $\mathsf{rng}(\rho) \subseteq \mathsf{rng}(\sigma)$.*
(2) $\mathsf{dom}(\rho \cup \sigma) = \mathsf{dom}(\rho) \cup \mathsf{dom}(\sigma)$.
(3) $\mathsf{dom}(\rho \cap \sigma) \subseteq \mathsf{dom}(\rho) \cap \mathsf{dom}(\sigma)$.
(4) $\mathsf{rng}(\rho \cup \sigma) = \mathsf{rng}(\rho) \cup \mathsf{rng}(\sigma)$.
(5) $\mathsf{rng}(\rho \cap \sigma) \subseteq \mathsf{rng}(\rho) \cap \mathsf{rng}(\sigma)$.

Proof.

$$
\begin{aligned}
(1)\text{:}\quad \rho \subseteq \sigma &\to \forall x \in S, y \in T[x\rho y \le x\sigma y] \\
&\to \textstyle\bigvee_{y\in T} x\rho y \le \bigvee_{y\in T} x\sigma y \\
&\to \mathsf{dom}(\rho) \subseteq \mathsf{dom}(\sigma).
\end{aligned}
$$

$$
\begin{aligned}
(3)\text{: } \mathsf{dom}(\rho\cap\sigma)(x) &= \textstyle\bigvee_{y\in T} x\rho\cap\sigma y \\
&= \textstyle\bigvee_{y\in T} (x\rho y \wedge x\sigma y) \\
&\le \textstyle\bigvee_{y\in T} x\rho y \wedge \bigvee_{y\in T} x\sigma y && \text{by Prop. 2.59(L4)} \\
&= [\mathsf{dom}(\rho)](x) \wedge [\mathsf{dom}(\sigma)](x) \\
&= [\mathsf{dom}(\rho) \cap \mathsf{dom}(\sigma)](x).
\end{aligned}
$$

□

We finish this section with a well known and often cited example from the area of fuzzy relations that, finally, initiated most of the investigations concerning fuzzy relation equations and inequalities, viz. the already mentioned work of Sanchez [40]. Suppose we are interested in what patients have what diseases. The medical knowledge can be structured by using three sets appropriately chosen for our considerations: a set P of patients, a set D of diseases, and a set S of symptoms of diseases.

Furthermore, we set up three fuzzy relations as follows: $\pi \subseteq P \times S$ with $\pi(p, s)$ indicating the strength of symptom s for patient p, $\sigma \subseteq S \times D$ with $\sigma(s, d)$ informing about the explanatory power of symptom s for disease d, and $\delta \subseteq P \times D$ with $\delta(p, d)$ as the degree to which patient p has disease d. Usually the data observed in this setup are about symptoms of patients (from medical examination) and about what symptoms are characteristic for what diseases (from medical expert knowledge). What we want to infer from this is information about the disease of patient p.

8.3.1 *Basic Operations on Fuzzy Relations*

This section summarizes some facts from what could be called the *fuzzy relation calculus*. That means, we will list the most important relationships that hold for operations on fuzzy relations, and some more generalizations of theorems from the two-valued case. Though proofs are simple and do closely follow the ones for the two-valued case, we give some examples to get familiar with the notation we use in the many-valued context. We start with the many-valued version of propositions 3.13 and 3.17 and will see that most of the properties carry over.

If not stated otherwise, in the following we let L denote a complete residuated lattice $(L, \wedge, \vee, \otimes, \curvearrowright, 0, 1)$. If appropriate we will change our notation to differentiate between the lattice theoretical operators $\wedge, \vee$ and the logical operators $\&, \|$ in order to improve readability.

Proposition 8.20. *For any L-fuzzy relations ρ and σ we have*

(1) $\mathsf{dom}(\rho^{\smile}) = \mathsf{rng}(\rho)$.
(2) $\mathsf{rng}(\rho^{\smile}) = \mathsf{dom}(\rho)$.
(3) *If* $\rho \subseteq S \times T$, *then* $\rho^{\smile} \subseteq T \times S$.
(4) $\rho^{\smile\smile} = \rho$.
(5) $\overline{\rho}^{\smile} = \overline{\rho^{\smile}}$.
(6) *If* $\rho \subseteq \sigma$, *then* $\rho^{\smile} \subseteq \sigma^{\smile}$.
(7) $(\rho \cap \sigma)^{\smile} = \rho^{\smile} \cap \sigma^{\smile}$.
(8) $(\rho \cup \sigma)^{\smile} = \rho^{\smile} \cup \sigma^{\smile}$.
(9) $x\rho = \rho^{\smile}x, \quad \rho y = y\rho^{\smile}$.

Proof. (1): $\mathsf{dom}(\rho^{\smile})(x) = \bigvee_y \rho^{\smile}(x,y) = \bigvee_y \rho(y,x) = \mathsf{rng}(\rho)(x)$.
(5): $x\overline{\rho}^{\smile} y = y\overline{\rho}x = y\rho x \curvearrowright 0 = x\rho^{\smile} y \curvearrowright 0 = x\overline{\rho^{\smile}}y$.
(7): $x(\rho \cap \sigma)^{\smile} y = y(\rho \cap \sigma)x = y\rho x \wedge y\sigma x = x\rho^{\smile} y \wedge x\sigma^{\smile} y = x(\rho^{\smile} \cap \sigma^{\smile})y$.
(9): $x\rho(y) = \rho(x,y) = \rho^{\smile}(y,x) = \rho^{\smile}x(y)$. □

Definition 8.21. *Let A, B be two L-fuzzy sets over sets X and Y, respectively. The* **product** *$A \otimes B$ is a fuzzy set over $X \times Y$ defined by $A \otimes B(x,y) = A(x) \otimes B(y)$.*

Theorem 8.22. *Let ρ, σ, and τ be L-fuzzy relations such that the mentioned operations are possible. Then the following holds:*

(1) $I_S\rho = \rho = \rho I_T$.
(2) $\emptyset\rho = \emptyset = \rho\emptyset$.
(3) *Associativity of relation product:* $\rho(\sigma\tau) = (\rho\sigma)\tau$.
(4) *Subdistributivity of relation product over intersection:*
$$\rho(\sigma \cap \tau) \subseteq (\rho\sigma) \cap (\rho\tau) \quad \text{and} \quad (\rho \cap \sigma)\tau \subseteq (\rho\tau) \cap (\sigma\tau).$$
(5) *Monotonicity of relation product: If* $\sigma \subseteq \tau$, *then*
$$\rho\sigma \subseteq \rho\tau \quad \text{and} \quad \sigma\rho \subseteq \tau\rho.$$
(6) *Distributivity of relation product over union:*
$$\rho(\sigma \cup \tau) = (\rho\sigma) \cup (\rho\tau) \quad \text{and} \quad (\rho \cup \sigma)\tau = (\rho\tau) \cup (\sigma\tau).$$
(7) *If* $\rho \subseteq \sigma$ *and* $\tau \subseteq \omega$, *then* $\rho\tau \subseteq \sigma\omega$.

(8) $(\rho\sigma)\breve{} = \sigma\breve{}\rho\breve{}$.

(9) $\rho\sigma = \bigcup_{y\in Y}(\rho y \otimes y\sigma)$.

(10) If $\otimes = \wedge$ *(as e.g. in Boolean and Heyting algebras; see p. 213) then* $(\rho \circ \sigma)_\alpha = \rho_\alpha \circ \sigma_\alpha$.

Proof.

(The example proofs given below are routine applications of the laws that hold in residuated lattices; compare e.g. the proof of (3) with the corresponding proof in Th. 3.16. The proof of (5), in the compact notation introduced in (8.4) above, is actually an easy transfer of the proof for two-valued relations, and (7) is a mere copy; again see Th. 3.16.)

(1): $aI_S\rho c = \bigvee_b aI_S b \otimes b\rho c = aI_S a \otimes a\rho c = a\rho c$, since 1 is the unit for $\otimes$.

$$
\begin{aligned}
(3): [\rho(\sigma\tau)](x,w) &= \bigvee_y[\rho(x,y)\otimes(\sigma\tau)(y,w)]\\
&= \bigvee_y[\rho(x,y)\otimes\bigvee_z\big(\sigma(y,z)\otimes\tau(z,w)\big)]\\
&= \bigvee_y\bigvee_z[\rho(x,y)\otimes\big(\sigma(y,z)\otimes\tau(z,w)\big)]\\
&= \bigvee_z[\bigvee_y\big(\rho(x,y)\otimes\sigma(y,z)\big)\otimes\tau(z,w)\big)]\\
&= \bigvee_z[(\rho\sigma)(x,z)\otimes\tau(z,w)\big]\\
&= [(\rho\sigma)\tau](x,w)
\end{aligned}
$$

(5): Assume $x\rho\sigma z$ (i.e. > 0). By (8.3) $x\rho\sigma z = \bigvee_y(x\rho y \otimes y\sigma z)$. By hypothesis we have $\sigma \subseteq \tau$, whence $y\sigma z \le y\tau z$, for all y, z. Thus, by monotonicity of $\otimes$, $x\rho\sigma z \le x\rho\tau z$, i.e. $\rho\sigma \subseteq \rho\tau$.

(6): Using Prop. 7.19(1) we get $x\big[\rho\circ(\bigcup_i\sigma_i)\big]y = x\rho z \otimes z(\bigcup_i\sigma_i)y = x\rho z \otimes \bigvee_i z\sigma_i y = \bigvee_i(x\rho z \otimes z\sigma_i y) = x\bigcup_i \rho\sigma_i\, y$.

(7): By (5) $\rho\subseteq\sigma \wedge \tau\subseteq\omega \to \rho\tau\subseteq\rho\omega \wedge \rho\omega\subseteq\sigma\omega$, whence $\rho\tau\subseteq\sigma\omega$.

(8): $x(\rho\sigma)\breve{}z = z(\rho\sigma)x = \bigvee_y(z\rho y \otimes y\sigma x) = \bigvee_y(x\sigma\breve{}y \otimes y\rho\breve{}z) = x\sigma\breve{}\rho\breve{}z$.

(9) $x\rho\sigma z = \bigvee_{y\in Y}(x\rho y \otimes y\sigma z) = \bigvee_{y\in Y}\big(\rho y(x)\otimes y\sigma(z)\big) = \bigvee_{y\in Y}\big[(\rho y \otimes y\sigma)(x,z)\big] = \bigcup_{y\in Y}(\rho y \otimes y\sigma)$.

$$
\begin{aligned}
(10): \langle x,z\rangle \in (\rho\sigma)_\alpha &\leftrightarrow x\rho\sigma z \ge \alpha\\
&\leftrightarrow \bigvee_{y\in Y}(x\rho y\otimes y\sigma z)\ge\alpha\\
&\leftrightarrow \exists_{y\in Y}[(x\rho y\otimes y\sigma z)\ge\alpha]\\
&\leftrightarrow \exists_{y\in Y}[(x\rho y\wedge y\sigma z)\ge\alpha], \qquad \text{since } \otimes=\wedge\\
&\leftrightarrow \exists_{y\in Y}[x\rho y\ge\alpha \;\&\; y\sigma z\ge\alpha]\\
&\leftrightarrow \exists_{y\in Y}[\langle x,y\rangle\in\rho_\alpha \;\&\; \langle y,z\rangle\in\sigma_\alpha]\\
&\leftrightarrow \langle x,z\rangle\in\rho_\alpha\sigma_\alpha\,.
\end{aligned}
$$

□

In Section 3.4 the SCHRÖDER equivalences and related results have been called the "most important results in the calculus of relations". The above considerations have shown that the operations with fuzzy sets and fuzzy relations have a lot of the properties that we know from their two-valued classical variants. So a natural question is for the validity of the SCHRÖDER equivalences $\rho\sigma\subseteq\tau \;\leftrightarrow\; \rho\breve{}\,\overline{\tau}\subseteq\overline{\sigma} \;\leftrightarrow\; \overline{\tau}\sigma\breve{}\subseteq\overline{\rho}$, the modular laws, and the Dedekind rule and their interrelationships for L-fuzzy relations.

Theorem 8.23. *Let L denote a* HEYTING *algebra (interpreted as residuated lattice), and* ρ, σ, *and* τ *L-fuzzy relations; as usual, we assume that the constructs are well formed. Then the following holds:*

(1) **SCHRÖDER *rule*** *for L-fuzzy relations:*
  (a) $\rho\sigma \subseteq \tau \;\leftrightarrow\; \rho^{\smile}\overline{\tau} \subseteq \overline{\sigma}$,
  (b) $\rho\sigma \subseteq \tau \;\leftrightarrow\; \overline{\tau}\sigma^{\smile} \subseteq \overline{\rho}$.
  (c) $\rho^{\smile}\overline{\tau} \subseteq \overline{\sigma} \;\leftrightarrow\; \overline{\tau}\sigma^{\smile} \subseteq \overline{\rho}$.
(2) **DEDEKIND *rule***: $\rho\sigma \cap \tau \subseteq (\rho \cap \tau\sigma^{\smile})(\sigma \cap \rho^{\smile}\tau)$.
(3) **Modular laws**:
  (a) $\rho\sigma \cap \tau \subseteq (\rho \cap \tau\sigma^{\smile})\sigma$,
  (b) $\rho\sigma \cap \tau \subseteq \rho(\sigma \cap \rho^{\smile}\tau)$.

(To make a difference in notation between the logical negation in $\neg\exists[\ldots]$ and $\neg x\rho y = x\rho y \curvearrowright 0 \in L$, the latter one is simply written as $x\overline{\rho}y$. Remember that if in a HEYTING algebra $\otimes$ coincides with the meet $\wedge$ then it gets a residuated lattice (see p. 213); so we will use & and $\|$ for logical *and* and *or* to get the proof readable. Note that we need $\wedge$ instead of $\otimes$ for concluding backward from line $(*)$ to line $(\dagger)$; see Cor. 7.20(6).)

Proof. *(1)* SCHRÖDER *rule* (a):

$$
\begin{aligned}
\rho\sigma \subseteq \tau &\leftrightarrow \forall x,y[x\rho\sigma y \le x\tau y]\\
&\leftrightarrow \forall x,y[\textstyle\bigvee_z(x\rho z \wedge z\sigma y) \le x\tau y]\\
&\leftrightarrow \forall x,y[\forall z[x\rho z \wedge z\sigma y \le x\tau y]] && (\dagger)\\
&\leftrightarrow \forall x,y[\forall z[x\rho z \le x\overline{\tau}y \;\|\; z\sigma y \le x\tau y]] && (*)\\
&\leftrightarrow \forall x,y[\forall z[z\rho^{\smile}x \le x\overline{\tau}y \;\|\; x\overline{\tau}y \le z\overline{\sigma}y]]
\end{aligned}
$$

Using monotonicity and Prop. 7.21(4), from the first inequality we get: $z\rho^{\smile}x \wedge x\overline{\tau}y \le x\tau y \wedge x\overline{\tau}y = 0 \le z\overline{\sigma}y$. So $\bigvee_x(z\rho^{\smile}x \wedge x\overline{\tau}y) \le z\overline{\sigma}y$. In the second case $z\rho^{\smile}x \wedge x\overline{\tau}y \le x\overline{\tau}y \le z\overline{\sigma}y$; so again $\bigvee_x(z\rho^{\smile}x \wedge x\overline{\tau}y) \le z\overline{\sigma}y$. Together this yields, $\rho^{\smile}\overline{\tau} \subseteq \overline{\sigma}$.

Conversely assume $\rho^{\smile}\overline{\tau} \subseteq \overline{\sigma}$. Then

$$
\begin{aligned}
\rho^{\smile}\overline{\tau} \subseteq \overline{\sigma} &\leftrightarrow \forall y,z[\textstyle\bigvee_x(z\rho^{\smile}x \wedge x\overline{\tau}y) \le z\overline{\sigma}y]\\
&\leftrightarrow \forall y,z[\forall x[z\rho^{\smile}x \wedge x\overline{\tau}y \le z\overline{\sigma}y]]\\
&\leftrightarrow \forall x,y,z[x\rho z \le z\overline{\sigma}y \;\|\; x\overline{\tau}y \le z\overline{\sigma}y]\\
&\leftrightarrow \forall x,y,z[x\rho z \le z\overline{\sigma}y \;\|\; z\sigma y \le x\tau y]
\end{aligned}
$$

From the first inequality we get $x\rho z \wedge z\sigma y \le z\overline{\sigma}y \wedge z\sigma y = 0 \le x\tau y$, using monotonicity and Prop. 7.21(4), so $\bigvee_z(x\rho z \wedge z\sigma y) \le x\tau y$, i.e. $\rho\sigma \subseteq \tau$. The second one also implies $\bigvee_z(x\rho z \wedge z\sigma y) \le x\tau y$, i.e. we get $\rho\sigma \subseteq \tau$ again.

(1): SCHRÖDER *rule* (b): The proof is similar to the one above, so we present only parts of it:

$$
\begin{aligned}
\rho\sigma \subseteq \tau &\leftrightarrow \ldots \text{ (as above)}\\
&\leftrightarrow \forall x,y[\forall z[x\rho z \le x\overline{\tau}y \;\|\; z\sigma y \le x\tau y]](*)\\
&\leftrightarrow \forall x,y[\forall z[x\overline{\tau}y \le x\overline{\rho}z \;\|\; y\sigma^{\smile}z \le x\tau y]]
\end{aligned}
$$

Using monotonicity and Prop. 7.21(1), from the first inequality we get: $x\overline{\tau}y \wedge y\sigma\breve{}z \leq x\overline{\tau}y \leq x\overline{\rho}z$. So $\bigvee_y(x\overline{\tau}y \wedge y\sigma\breve{}z) \leq x\overline{\rho}z$. In the second case $x\overline{\tau}y \wedge y\sigma\breve{}z \leq x\overline{\tau}y \wedge x\tau y = 0 \leq x\overline{\rho}z$; so, again $\bigvee_y(x\overline{\tau}y \wedge y\sigma\breve{}z) \leq x\overline{\rho}z$, whence $\overline{\tau}\sigma\breve{} \subseteq \overline{\rho}$.

(2): DEDEKIND *rule*:

$$\begin{aligned} x(\rho\sigma \cap \tau)y &= \textstyle\bigvee_z(x\rho z \wedge z\sigma y) \wedge x\tau y \\ &= \textstyle\bigvee_z(x\rho z \wedge z\sigma y \wedge y\sigma\breve{}z \wedge z\rho\breve{}x) \wedge x\tau y \\ &= \textstyle\bigvee_z(x\rho z \wedge (x\tau y \wedge y\sigma\breve{}z) \wedge (z\rho\breve{}x \wedge x\tau y) \wedge z\sigma y) \\ &\leq \textstyle\bigvee_z(x\rho z \wedge \bigvee_y(x\tau y \wedge y\sigma\breve{}z) \wedge \bigvee_x(z\rho\breve{}x \wedge x\tau y) \wedge z\sigma y) \\ &= \textstyle\bigvee_z[x(\rho \cap \tau\sigma\breve{})z \wedge z(\sigma \cap \rho\breve{}\tau)y] \\ &= x(\rho \cap \tau\sigma\breve{})(\sigma \cap \rho\breve{}\tau)y \end{aligned}$$

(3): *Modular laws*: By the DEDEKIND rule (2) we have $\rho\sigma \cap \tau \subseteq (\rho \cap \tau\sigma\breve{})(\sigma \cap \rho\breve{}\tau)$. Applying the monotonicity property of relation composition to the right side yields the assertions: $(\rho \cap \tau\sigma\breve{})(\sigma \cap \rho\breve{}\tau) \subseteq (\rho \cap \tau\sigma\breve{})\sigma$ and $(\rho \cap \tau\sigma\breve{})(\sigma \cap \rho\breve{}\tau) \subseteq \rho(\sigma \cap \rho\breve{}\tau)$ □

Some trivial but important results concerning complements and used in the following theorem are given in the next lemma. Note that the converse direction does not hold in (1).

Lemma 8.24. *Let L denote a* HEYTING *algebra, and let A, B be two L-fuzzy sets over X. Then*

(1) $A \subseteq B \;\rightarrow\; A \cap \overline{B} = \emptyset$;
(2) $A \cap B = \emptyset \;\leftrightarrow\; A \subseteq \overline{B}$;
(3) $A \subseteq \overline{\overline{A}}$;
(4) $A \cap \overline{A} = \emptyset$.

Proof. (1): By Prop. 7.21(8) we get $A(x) \leq B(x) \;\rightarrow\; A(x) \wedge \neg B(x) = 0$.
(2): By Prop. 7.21(7) $A(x) \wedge B(x) = 0$ iff $A(x) \leq \neg B(x)$.
(3): By Prop. 7.21(2) $A(x) \leq \neg\neg A(x)$.
(4): By Prop. 7.21(4). □

Theorem 8.25. *Let L denote a* HEYTING *algebra, and ρ, σ, and τ L-fuzzy relations; then the following holds:*

(1) The two modular laws together are equivalent to the DEDEKIND *rule.*
(2) The DEDEKIND *rule implies the following parts of the* SCHRÖDER *equivalences Th. 8.23(1):*
(a) $\rho\sigma \subseteq \tau \;\rightarrow\; \rho\breve{}\,\overline{\tau} \subseteq \overline{\sigma}$,
(b) $\rho\sigma \subseteq \tau \;\rightarrow\; \overline{\tau}\sigma\breve{} \subseteq \overline{\rho}$.
(3) The DEDEKIND *rule implies the following weakened parts of the* SCHRÖDER *equivalences Th. 8.23(1):*
(a) $\rho\breve{}\,\overline{\tau} \subseteq \overline{\sigma} \;\rightarrow\; \rho\sigma \subseteq \overline{\overline{\tau}}$,
(b) $\overline{\tau}\sigma\breve{} \subseteq \overline{\rho} \;\rightarrow\; \rho\sigma \subseteq \overline{\overline{\tau}}$.

Proof. (1): The proof of Th. 3.27(1) can be used verbatim.

(2): Again following Th. 3.27 we show that the DEDEKIND rule implies $\rho\sigma \subseteq \tau \ \rightarrow\ \rho^{\smile}\overline{\tau} \subseteq \overline{\sigma}$. Now assume that the DEDEKIND rule holds; then by (1) also the modular laws are valid. From $\rho\sigma \subseteq \tau$ we get $\rho\sigma \cap \overline{\tau} = \emptyset$ by Lemma 8.24(1). This result together with Th. 8.23(3)(b) yields $\rho^{\smile}\overline{\tau} \cap \sigma \subseteq \rho^{\smile}(\overline{\tau} \cap \rho\sigma) = \rho^{\smile}\emptyset = \emptyset$, i.e. $\rho^{\smile}\overline{\tau} \cap \sigma = \emptyset$. Thus $\rho^{\smile}\overline{\tau} \subseteq \overline{\sigma}$ by Lemma 8.24(2). Similarly we have with Th. 8.23(3)(a) $\overline{\tau}\sigma^{\smile} \cap \rho \subseteq (\overline{\tau} \cap \rho\sigma)\sigma^{\smile} = \emptyset$, and thus $\overline{\tau}\sigma^{\smile} \subseteq \overline{\rho}$.

(3): Assume $\rho^{\smile}\overline{\tau} \subseteq \overline{\sigma}$. Then $\rho^{\smile}\overline{\tau} \cap \sigma = \emptyset$ by Lemma 8.24(2). Again using Th. 8.23(3)(b) we get $\rho\sigma \cap \overline{\tau} \subseteq \rho(\sigma \cap \rho^{\smile}\overline{\tau}) = \rho\emptyset = \emptyset$, whence $\rho\sigma \subseteq \overline{\overline{\tau}}$ by Lemma 8.24(2). Similarly, from $\overline{\tau}\sigma^{\smile} \subseteq \overline{\rho}$ we get $\overline{\tau}\sigma^{\smile} \cap \rho = \emptyset$, and using Th. 8.23(3)(a) this gives $\rho\sigma \cap \overline{\tau} \subseteq (\rho \cap \overline{\tau}\sigma^{\smile})\sigma = \emptyset\sigma = \emptyset$, whence $\rho\sigma \subseteq \overline{\overline{\tau}}$. □

So we see that as soon as $\tau = \overline{\overline{\tau}}$ we are back to the classical result as proven in Th. 3.27; this is the case for $L = \mathbf{2}$, and for MV-algebras in general.

8.3.2 *Basic Properties of Fuzzy Relations*

Roughly following the presentation for two-valued relations we now turn to properties a fuzzy relation may have, but we will have to confine ourselves to the most important ones, and also to a few aspects only that arise out of a many-valued approach. Why this is necessary you will shortly see from the few examples and their discussion given below. But these discussions are important, because they should (1) give you a feeling how the many-valued approach enriches the flexibility and modeling possibilities of your formal apparatus, but can also lead to new situations that have to be considered; and (2) draw your attention to the fact that formal generalizations have to be done carefully, i.e. they also have to account for semantical aspects, if they should not degenerate to a mere mathematical playground.

In Def. 3.21 and Prop. 3.50 two equivalent definitions for reflexivity have been given as follows: $\forall x \in X[x\rho x]$ and $I_X \subseteq \rho$. This has also been done for symmetry and transitivity. The question is how to generalize the two-valued concepts appropriately. Good examples for reflexivity can be found in [14, Chapter II.3] and [28] where the following variants are listed:

Definition 8.26. *Let ρ be an L-fuzzy relation on a set X. Then ρ is called*

(a) **reflexive** *iff for all $x \in X$, $\rho(x,x) = 1$;*
(b) **ϵ-reflexive** *iff for all $x \in X$, $\rho(x,x) \geq \epsilon > 0$;*
(c) **strictly ϵ-reflexive** *iff for all $x \in X$, $\rho(x,x) = \epsilon > 0$;*
(d) **strongly reflexive** *iff for all $x \in X$, $\rho(x,x) \geq \bigvee_{y\in X} \rho(x,y) = \mathsf{dom}(\rho)(x)$;*
(e) **weakly reflexive** *iff for all $x \in X$, $\rho(x,x) > 0$;*
(f) **symmetric** *iff for all $x, y \in X$, $\rho(x,y) = \rho(y,x)$.*
(g) **transitive** *iff for all $x, z \in X$, $\rho(x,z) \geq \bigvee_{y\in X} \big(\rho(x,y) \otimes \rho(y,z)\big)$.*

Let us have a short discussion of the variants for the concept of reflexivity given above: (a) is a direct translation of the two-valued definition and can easily be justified. It just expresses identity: a thing is indisputably identical to itself, whence a two-valued classification suffices to describe this, and thus $I_X \subseteq \rho$ results as definition of reflexivity also in the many-valued context.

(b) can be seen as an innocent arithmetical generalization (remember that $L = [0,1] \subset \mathbb{R}$ is the very first lattice used in FST). But unless we take $\epsilon = 1$, an ϵ-reflexive identity I_ϵ, i.e. $I_\epsilon(x,x) \geq \epsilon$ for all $x \in X$ and $I_\epsilon(x,y) = 0$ for $x \neq y$, will not be a unit for composition with an arbitrary fuzzy relation; in fact, $I_\epsilon\rho = \rho I_\epsilon = \rho$ does not hold as soon as $\mathsf{hgt}(\rho) > \epsilon$.

Using strong reflexivity (d) we get a ρ-specific diagonal I_ρ, where $I_\rho(x,x)$ becomes dependent also from the elements $y \neq x$ to which x is ρ-related, and can no longer be interpreted as a "*degree of identity*", but is an upper bound for the degree of relatedness of x to the other elements of X.

As the second and last example for the manifold of extension possibilities within a many-valued approach we give, without further discussion, the following plausible variants of a generalized notion of symmetry. An L-fuzzy relation ρ on a set X is called

- *weakly* ϵ*-symmetric* iff $\forall x, y \in X[x\rho y \geq \epsilon \rightarrow y\rho x > 0]$;
- (ϵ)*-symmetric* iff $\forall x, y \in X[x\rho y > 0\ (\geq \epsilon) \rightarrow y\rho x > 0\ (\geq \epsilon)]$;
- *strongly symmetric*, iff for all $x, y \in X[x\rho y = y\rho x]$;
- *strongly* ϵ*-symmetric*, iff for all $x, y \in X[x\rho y \wedge y\rho x \geq \epsilon \rightarrow x\rho y = y\rho x]$.

Note that in this terminology the notion of symmetry in the sense of Def. 8.26(f) appears as strongly symmetric; and in the Boolean two-valued context all of weakly 1-symmetric, 1-symmetric, symmetric, strongly symmetric, and strongly 1-symmetric coincide.

By Def. 8.26(f), symmetry is interpreted in the same way as in the two-valued case, viz. by *equality* of the two values for *all* possible pairs of elements, or if you want, as a kind of *perfect* symmetry. Despite the many-valued approach used, symmetry of a fuzzy relation, and also equality of fuzzy sets (see Def. 8.2(b)), are sharp all-or-nothing concepts - thus, in a way, they have remained "non-fuzzy". In both cases one might think of three more relaxed versions that emerge by a simple combinatorial process: On one side you have usual, sharp or APPROXIMATE EQUALITY of membership degrees, and on the other side you have *all* or NEARLY ALL, ALMOST ALL, or MOST of the elements possibly concerned. For whatever reason, a very weak form of symmetry may be sufficient for some purpose: It is okay for you, if for MOST pairs of elements (i.e. for e.g. 75%, 80%, ..., or 90%) the membership degrees in ρ and $\rho^{\smile}$ are VERY CLOSE TO each other. And you might eventually call a fuzzy relation ρ with this property *almost weak symmetric* and, moreover, quantify this with a degree of symmetry. Clearly, this kind of reasoning can also be applied to other properties.

Do *not* smile at what you have just read. If you take the original semantic intentions of fuzzy set theory seriously, this is a typical rather than abstruse aspect of applying its principal concepts; see e.g. [18, Section 18.6] or [3, p. 82ff, Section 4.1.2], where graded properties are formally represented by many-valued predicates in the language of a many-valued logical system. And if you want to generalize, on the syntactical level, from the collapsed two-valued to an appropriate many-valued definition of a concept, you will have to be careful in selecting a suitable way of 'explosion'. Furthermore, a detailed analysis of the above way of thinking shows, that even the introduction of a higher level kind of flexibility, as in e.g. combining a high percentage of elements with sharp equality of membership grades, does not lead beyond two-valued mathematics and logic: They still serve as the metalanguage in which various models offering additional possibilities for finer grained descriptions are formulated.

8.4 Fuzzy Equivalence Relations

It is hardly to be denied that much - if not most - of human knowledge is of qualitative nature. This kind of knowledge is cast in so-called *natural* languages, which are a highly economic tool for everyday communication since they include notions that allow for flexible interpretation. But, as it has been pointed out in [42, p. 488], this flexibility of natural language is not of an arbitrary nature: it rests on the capability for generalization and abstraction as basic presuppositions for the evolution of language. Only on basis of these two processes we get a natural explanation for the emergence of fuzziness as follows:

- One of the main assumptions of *hypothetical realism*[2] – as well as of common sense and natural science – is the existence of a real world that is structured and is at least partially knowable by perception and experience.
- Structure means a certain kind of order, i.e. a temporal or spatial coincidence or succession of states or events. So living beings, and humans in particular, have the possibility for, and also the ability of *re*-cognizing objects and events: Only this makes the emergence of predicators and language as a whole both plausible and reasonable at all.
- *Re*-cognition in turn requires for comparisons, the result of which can be either 'discriminate' or 'make equal' (i.e. regard as equivalent).
- Fuzziness, finally, arises by the possibility of regarding as equal some attribute values (that are discriminable in principle and, hence, if necessary) in certain situations. In general, this can be done without having to fear any bad consequences or serious problems for the individual and, consequently, for the species too. (If you want e.g. to cross a road and you see a car coming along, you will neither be interested in its exact distance nor its exact speed; the only thing which is relevant for you is whether this distance is large enough and the speed is sufficiently low.)

[2]Described in Vollmer[43, pp. 188, 191]as follows: "All knowledge is hypothetical, i.e., conjectural, fallible, preliminary. There exists a real world, independent of our consciousness; it is structured, coherent, and quasi-continuous; it is at least partially knowable and explainable by perception, experience, and intersubjective science. ... Our forms of intuition and our categories of experience, the qualities of our perceptions, and the basic elements of inference, of classification, of abstraction, are, according to evolutionary epistemology, innate, ontogenetically a priori and constitutive for mesocosmic knowledge."

Very similar points of view have been described already in 1772 long before fuzzy sets have been conceptualized, and also 2004, when fuzzy sets had nearly reached the age of 40:

> In reality, all arguments from experience are founded on the similarity which we discover among natural objects, and by which we are induced to expect effects similar to those which we have found to follow from such objects. ... none but a fool or madman will ever pretend to dispute the authority of experience, or to reject that great guide of human life, ... From causes which appear similar we expect similar effects. This is the sum of all our experimental conclusions ([20], chapter on *Cause and Effect*).

> Similarity is perhaps the most important phenomenon accompanying human perception and reasoning. On the one hand, similarity allows humans to think on an abstract level. Instead of storing individual elements (events, objects encountered in the past, etc.), people store classes of similar elements. Disregarding individual elements and the subsequent reasoning with similarity-based classes of these elements makes it possible to reason with complex evidence. On the other hand, similarity allows humans to act when they face a situation not encountered previously. Usually, people try to act as in a known situation which is most similar to the actual one. ([4], first paragraph)

Approximate equality, similarity, indistinguishability, or nearness thus being at the epistemological roots of fuzziness, it is, of course, interesting to have a look at how these concepts are handled within FST as a kind of generalized equality. In the following considerations we will prefer the term '*similarity*'.

There are two principle approaches to be found in literature to formally describe similarity: A qualitative one by a crisp relation $\sigma \subseteq S \times S$, with $x\sigma y$ describing that x and y are similar, and a quantitative one using a mapping $s : S \times S \longrightarrow \mathbb{R}^+$ that associates a positive similarity index to x and y, with $s(x, y)$ the higher the more similar the two elements are. Since fuzzy sets can be regarded as a means to model (and calculate with) what might be called '*quantified qualities*' they should prove to be a suitable tool to handle similarity in an appropriate way.

Fuzzy equivalence relations have originally been introduced in Zadeh [48] under the name **similarity relations** as min-transitive fuzzy relations that assign to each pair$\langle x, y\rangle \in S \times S$ a truth degree $x\theta y \in [0,1]$.

Definition 8.27. *Let* $(L, \wedge, \vee, \otimes, \frown, 0, 1)$ *be a complete residuated lattice, and S a non-empty set.*

(a) An **fuzzy tolerance** *is a reflexive and symmetric L-fuzzy relation.*
(b) θ *is called a* **fuzzy equivalence**, *or* **similarity relation**, *on S if* θ *is a fuzzy tolerance that is transitive, i.e.* $x\theta y \otimes y\theta z \le x\theta z$ *for all* $x, y, z \in S$. *The set of all L-fuzzy equivalence relations on S is denoted by* $\mathsf{Eq}^L{}_S$.
(c) A fuzzy equivalence is called a **fuzzy equality** *if* $x\theta y = 1 \rightarrow x = y$.

In case that L is a Heyting lattice then the following correspondence with crisp equivalence relations can be shown:

Proposition 8.28. *Let $(L, \wedge, \vee, \otimes, \curvearrowright, 0, 1)$ be a complete Heyting lattice, i.e. $\otimes = \wedge$. Then $\rho \in \mathsf{Eq}^L{}_S$ iff $\rho_\alpha \in \mathsf{Eq}_S$ for all $\alpha \in L^+$.*

Proof. We have to show that ρ is reflexive, symmetric, and transitive iff ρ_α is for all $\alpha \in L^+$. We prove the assertion for transitivity only. If ρ is $\wedge$-transitive and $\{\langle x, y\rangle, \langle y, z\rangle\} \subseteq \rho_\alpha$ for any fixed $\alpha \in L^+$, then $x\rho y \geq \alpha$ and $y\rho z \geq \alpha$, and by the assumption of $\wedge$-transitivity we have $x\rho z \geq x\rho y \wedge y\rho z = \alpha$. Thus $\langle x, z\rangle \in \rho_\alpha$. (It is here where we need $\otimes = \wedge$: Since $\wedge$ is the only idempotent t-norm by Prop. 8.11, and also the greatest one by Prop. 8.8, only $\wedge$ allows us to derive $x\rho z \geq \alpha$.)
Conversely assume that ρ_α is an equivalence on S for all $\alpha \in L^+$, and let $x\rho y \wedge y\rho z = \alpha$. Then $\{\langle x, y\rangle, \langle y, z\rangle\} \subseteq \rho_\alpha$, whence $\langle x, z\rangle \in \rho_\alpha$ since ρ_α is transitive. Thus $x\rho z \geq \alpha$. □

It has been shown in Section 4.3.1 that a crisp equivalence relation groups the elements of its universe into disjoint classes, and that there is a one-to-one correspondence between partitions and crisp equivalence relations on a set S. A fuzzy equivalence ρ on a set S induces a crisp partition π_{ρ_α} of S, for each $\alpha \in L^+$. This clearly means that a fuzzy equivalence groups elements into crisp sets whose members are similar to each other at least to degree $\alpha \in L^+$: two elements $x, y \in S$ belong to the same block of the partition π_{ρ_α} if $x\rho y \geq \alpha$. If $\alpha \geq \beta$, then $\pi_{\rho_\alpha} \leq \pi_{\rho_\beta}$, i.e. π_{ρ_α} is a refinement of π_{ρ_β}. Such a nested family of partitions generated by α-cuts of S can be represented in the form of partition tree.

Besides this nice connection to crisp concepts, fuzzy equivalences and their properties are of interest on their own. So we now turn to proper many-valued generalizations of some other crisp concepts connected to equivalence relations. As a first straightforward result we are led to

Definition 8.29. *Let θ be a fuzzy equivalence relation on a non-empty set S.*

(a) The aftersets $x\theta$, denoted as $[x]_\theta \in \mathcal{F}^L(S)$ or simply by $[x]$ if no confusion about the equivalence involved can occur, is called the **fuzzy equivalence class** *of x induced by θ, and is given by $[x]_\theta(y) = x\theta y$.*

(b) The set of all fuzzy equivalence classes is called the **quotient set** *of S modulo θ, denoted by S/θ.*

For crisp equivalence relations, by Prop. 4.14 we have $[x] = [y]$ iff $x\rho y$; the fuzzy counterpart to this is shown in

Proposition 8.30. ([31]) *Let θ be a fuzzy equivalence relation on a set S. Then the following holds:*

(1) $[x]_\theta = [y]_\theta$ iff $x\theta y = 1$;
(2) If $[x]_\theta \neq [y]_\theta \rightarrow \mathsf{hgt}([x]_\theta \cap [y]_\theta) < 1$;
(3) $x\theta y = \mathsf{hgt}([x]_\theta \otimes [y]_\theta)$.

Proof. (1): If $[x]_\theta = [y]_\theta$ then $x\theta y = [x]_\theta(y) = [y]_\theta(y) = y\theta y = 1$. Conversely, from $x\theta y = 1$ we get by transitivity of θ that, for each $s \in S$, $[x]_\theta(s) = x\theta s \geq x\theta y \otimes y\theta s = y\theta s = [y]_\theta(s)$.

Similarly, $[y]_\theta(s) \geq [x]_\theta(s)$, whence $[x]_\theta = [y]_\theta$.

(2): Assume $[x]_\theta \neq [y]_\theta$ and $\mathsf{hgt}([x]_\theta \cap [y]_\theta) = 1$. From the last equation we conclude that $\exists s \in S[[x]_\theta(s) = 1 = [y]_\theta(s)]$. Then, by transitivity and symmetry of θ,

$$x\theta y \geq \bigvee_{s\in S}(x\theta s \otimes s\theta y) = \bigvee_{s\in S}\big([x]_\theta(s) \otimes [y]_\theta(s)\big) = 1,$$

a contradiction to (1).

(3): $\mathsf{hgt}([x]_\theta \otimes [y]_\theta) = \bigvee_{s\in S}\big([x]_\theta(s) \otimes [y]_\theta(s)\big) = \bigvee_{s\in S}(x\theta s \otimes s\theta y) \ = x\theta y.$ □

Similarly to crisp equivalence relations, also their fuzzy counterpart is a very important concept. So, before proceeding on the formal level, some excursion into semantical aspects is advisable, providing another example for problems that eventually remain hidden at a first look on the generalization from the two-valued to a many-valued formalism. There have been some debates of whether the properties of reflexivity, symmetry, and transitivity are appropriate for a concept of similarity. The first two are commonly agreed upon: Reflexivity expresses the fact that every object is completely indistinguishable from, or similar to, itself. Symmetry says that the degree in which x is indistinguishable from y coincides with the degree in which y is indistinguishable from x; so it models "*similar to each other*" which is different from "*the son resembles his father*" which is sometimes used as a counterexample (and then implicitly incorporates a temporal aspect).

From an epistemological point of view, transitivity is the most interesting property and has been a point of disagreement long before fuzzy sets have appeared. Already at the beginning of the twentieth century Poincaré (in [36], Chapter II on *Mathematical Magnitude and Experiment*) discussed the problems that arise if one follows the common intuition that indistinguishability be transitive. The so-called **Poincaré paradox**, usually symbolized as $A = B \wedge B = C \wedge A \neq C$, which is the name given to Poincaré's observation that indistinguishability might behave non-transitive in real life, can be illustrated by the following example: A microscope or a telescope has a certain resolution so that we can only distinguish between two points, if their distance exceeds a certain value, say d. So it is possible that we are not able to distinguish v_1 from v_2 , v_2 from v_3 , $\ldots$, v_{m-1} from v_m, but can easily do so between v_1 and v_m.

Now, if we try model such a situation e.g. in the context of height in cm by a fuzzy equivalence, then we get $151 \approx 152 = \ldots 199 \approx 200 = 1$ and by transitivity $151 \approx 200 = 1$, which is of course absurd. But if we take a value $\alpha \neq 1$ but close to it, how could we justify this choice for perfectly indistinguishable pairs? Besides, assuming $v_i \approx v_{i+1} = \alpha$, we conclude from transitivity that $v_1 \approx v_m \geq \alpha^{m-1}$. If $\otimes = \wedge$ this does not diminish the value of $v_1 \approx v_m$, so L-fuzzy equivalences with L a Heyting lattice will not be suitable. Further discussions on this topic are beyond the current scope, whence we have to refer the reader to [16, 37] for details.

To become well prepared for the considerations to follow below, we need some more definitions. We follow mainly the presentations in [6, 21, 31, 37].

Definition 8.31. *Let $\mu \in \mathcal{F}^L(X)$ and $\theta \in \mathsf{Eq}^L{}_X$.*

(a) *μ is called* **extensional** *(or* **observable***) w.r.t. θ if*
$$\mu(x) \otimes \theta(x,y) \le \mu(y) \text{ for all } x, y \in X.$$
(b) *μ is called* **indistinguishable** *(or* **indiscernible***) w.r.t. θ if*
$$\mu(x) \otimes \mu(y) \le \theta(x,y) \text{ for all } x, y \in X.$$
(c) *If μ is normal, extensional and indistinguishable w.r.t. θ, then it is called a* **fuzzy point** *with respect to θ.*

Some comments on this definition are in order: Interpreting an ordinary equivalence $\approx$ on X in the sense that equivalent elements cannot be distinguished or may be identified, then only those subsets $A \subseteq X$ satisfying $x \in A$ and $x \approx y \rightarrow y \in A$ "behave well" with respect to $\approx$. Extensionality generalizes this condition, and thus is the most important property a fuzzy set can fulfill with respect to a fuzzy equivalence. Similarly, indistinguishability generalizes $x, y \in A \rightarrow x \approx y$.

Lemma 8.32. *Let $\mu \in \mathcal{F}^L(X)$ and $\theta \in \mathsf{Eq}^L{}_X$.*

(1) *μ is extensional w.r.t. θ iff $\mu(x) \rightleftharpoons \mu(y) \ge \theta(x,y)$ for all $x, y \in X$.*
(2) *If μ is normal, then it is indistinguishable w.r.t. θ iff it is contained in an equivalence class of θ.*
(3) *μ is a fuzzy point w.r.t. θ iff it is an equivalence class of θ.*

Proof. (1): Let μ be extensional w.r.t. θ; then the extensionality condition $\mu(x) \otimes \theta(x,y) \le \mu(y)$ can be rewritten as $\theta(x,y) \le \mu(x) \curvearrowright \mu(y)$. Therefore,
$$\theta(x,y) = \theta(x,y) \wedge \theta(y,x) \le (\mu(x) \curvearrowright \mu(y)) \wedge (\mu(y) \curvearrowright \mu(x)) = \mu(x) \rightleftharpoons \mu(y).$$
For the other direction assume $\theta(x,y) \le \mu(x) \rightleftharpoons \mu(y)$. Then $\theta(x,y) \le \mu(x) \curvearrowright \mu(y)$, and by adjointness $\mu(x) \otimes \theta(x,y) \le \mu(y)$.

(2): Let $a \in X$ be such that $\mu(a) = 1$. If μ is indistinguishable with respect to θ, then $\mu(a) \otimes \mu(x) = \mu(x) \le \theta(a,x) = [a]_\theta(x)$; so $\mu \subseteq [a]_\theta$, for every $x \in X$. Conversely, if $\mu \subseteq [a]_\theta$ for some $a \in X$, then $\mu(x) \le \theta(a,x)$ and $\mu(y) \le \theta(a,y)$ for every $x, y \in X$, whence $\mu(x) \otimes \mu(y) \le \theta(a,x) \otimes \theta(a,y) \le \theta(x,y)$.

(3): If μ is a fuzzy point w.r.t. θ, then by (2) $\mu \subseteq [a]_\theta$, with $a \in X$ be such that $\mu(a) = 1$. And the extensionality of μ w.r.t. θ yields $[a]_\theta(x) = \theta(a,x) = \mu(a) \otimes \theta(a,x) \le \mu(x)$. Therefore, $[a]_\theta \subseteq \mu$, whence $\mu = [a]_\theta$.
Conversely, let $\mu = [a]_\theta$ for some $a \in X$; then $\mu(a) = 1$ and μ is normal. From the $\otimes$-transitivity of θ we get $\mu(x) \otimes \theta(x,y) .= \theta(a,x) \otimes \theta(x,y) \le \theta(a,y) = \mu(y)$ and $\mu(x) \otimes \mu(y) = \theta(x,a) \otimes \theta(a,y) \le \theta(x,y)$. for any $x, y \in X$. Thus, μ is a fuzzy point. □

Definition 8.33. *Let $\mu \in \mathcal{F}^L(X)$, and $C \subseteq \mathcal{F}^L(X)$ for a set X. Following [6] we define fuzzy relations as follows:*

(a) $S_\mu(x,y) = 1$ *if* $x = y$, *and* $= \mu(x) \otimes \mu(y)$ *if* $x \neq y$;
(b) $T_\mu(x,y) = 1$ *if* $\mu(x) = \mu(y)$, *and* $= \mu(x) \otimes \mu(y)$ *if* $\mu(x) \neq \mu(y)$;

(c) $E_\mu(x,y) = \mu(x) \rightleftharpoons \mu(y)$;

(d) $E_\mathcal{C} = \bigcap_{\mu\in\mathcal{C}} E_\mu$, *i.e.* $E_\mathcal{C}(x,y) = \bigwedge_{\mu\in\mathcal{C}}(\mu(x) \rightleftharpoons \mu(y))$;

(e) $S_\mathcal{C}$ *denotes the least* $\theta \in \mathsf{Eq}^L{}_X$ *such that* $S_\mu \subseteq \theta$ *for all* $\mu \in \mathcal{C}$ *(such an equivalence exists since* $U_X \in \mathsf{Eq}^L{}_X$*).*

Lemma 8.34. *All of* S_μ, T_μ, E_μ*, and* $E_\mathcal{C}$ *are fuzzy equivalence relations.*

Proof. Obviously, S_μ and T_μ are both reflexive and symmetric, and E_μ is so by Cor. 7.20(1) and (2). As an example we prove the transitivity of S_μ: We have to show $S_\mu(x,y) \otimes S_\mu(y,z) \le S_\mu(x,z)$. If $x = z$ then $S_\mu(x,z) = 1$, and the inequality trivially holds. Now assume $x \ne z$. The product $S_\mu(x,y) \otimes S_\mu(y,z)$ equals (a) $S_\mu(y,z) = S_\mu(x,z)$ if $x = y$; (b) $S_\mu(x,y) = S_\mu(x,z)$ if $y = z$; and (c) $\mu(x) \otimes \mu(y) \otimes \mu(z) \le \mu(x) \otimes \mu(z) = S_\mu(x,z)$ if $x \ne y$ and $y \ne z$ (using Prop. 7.17(vi)). Similar to Prop. 4.18(1), the intersection of fuzzy equivalences is again a fuzzy equivalence. □

The next proposition characterizes all fuzzy equivalence relations having a given fuzzy set as an equivalence class.

Proposition 8.35. ([6, Theorem 3.1]) $\mu \in \mathcal{F}^L(X)$ *is an equivalence class of some* $\theta \in \mathsf{Eq}^L{}_X$ *iff it is normal. In this case, the closed interval* $[S_\mu, E_\mu]$ *is the set of all* $\theta \in \mathsf{Eq}^L{}_X$ *such that* $\mu \in X/\theta$.

Proof. Let $\mu = [x]_\theta$; then by reflexivity of θ we have $x\theta x = 1$ for all $x \in X$, whence $\mathsf{hgt}([x]_\theta) = 1$. Conversely, let $\mu \in \mathcal{F}^L(X)$ be normal; then it is an equivalence class of any of the fuzzy equivalence relations S_μ, T_μ, and E_μ defined above. Assume $\mu(a) = 1$; then for e.g. S_μ we get $[a]_{S_\mu}(x) = 1 = \mu(a)$ for $x = a$, and $[a]_{S_\mu}(x) = \mu(a) \otimes \mu(x) = \mu(x)$ otherwise.

By Lemma 8.32(1) we have that μ is extensional w.r.t. θ iff $\theta(x,y) \le \mu(x) \rightleftharpoons \mu(y) = E_\mu(x,y)$ for all $x, y \in X$. Thus, the set of all $\theta \in \mathsf{Eq}^L{}_X$ such that μ is extensional w.r.t. θ is the principal ideal $(E_\mu]$ of $\mathsf{Eq}^L{}_X$ generated by E_μ. Furthermore, for any $\theta \in \mathsf{Eq}^L{}_X$ and $x, y \in X, x \ne y$ we have that μ is indistinguishable with respect to θ iff $S_\mu \subseteq \theta$, whence the set of all $\theta \in \mathsf{Eq}^L{}_X$ such that μ is indistinguishable w.r.t. θ is the principal filter $[S_\mu)$ of $\mathsf{Eq}^L{}_X$ generated by E_μ. By Lemma 8.32(3), the set of all $\theta \in \mathsf{Eq}^L{}_X$ such that μ is an equivalence class of θ is the intersection $[S_\mu) \cap (E_\mu] = [S_\mu, E_\mu]$ (note that by Cor. 7.20(vi) $S_\mu \subseteq E_\mu$). □

Now we give a partial fuzzy counterpart of Th. 4.15, in which properties of the set of equivalence classes have been shown for the crisp case.

Proposition 8.36. *Let* $\theta \in \mathsf{Eq}^X{}_L$ *for a set* X*. Then the following holds:*

(1) $[x]_\theta \ne \emptyset$ *and* $\mathsf{hgt}([x]_\theta) = 1$;

(2) *For* $x, y \in X$*, either* $[x]_\theta = [y]_\theta$ *or* $\mathsf{hgt}([x]_\theta \cap [y]_\theta) < 1$;

(3) $\bigcup\{[x]_\theta : x \in X\} = X$;

Proof. (1): Since θ is reflexive, $[x]_\theta(x) = 1$ for each $x \in X$.
(2): By Prop. 8.30(1),(2).
(3): Since for each $x \in X$ we have $[x]_\theta(x) = 1$, $x \in \mathsf{core}([x]_\theta)$, and we get $X \subseteq \bigcup\{[x]_\theta : x \in X\}$. On the other hand, $[x]_\theta \subseteq X$ gives $\bigcup\{[x]_\theta : x \in X\} \subseteq X$. □

As you can see, from the three characteristic properties of crisp partitions, the one concerning disjointness of fuzzy equivalence classes is missing: the main distinction between fuzzy and crisp equivalence classes is that the former may have nonempty intersections. So Th. 4.15(4), i.e. X/θ is a partition of X, cannot be transferred. But one can in fact proceed in a "disjointness-free" way as e.g. in [6, 31]. It is well-known that $\mathcal{A} \subseteq \mathbf{2}^X$ is a partition of a crisp set X if and only if it is the set of all equivalence classes of some equivalence relation on X. Likewise one can define a **fuzzy partition** as a family $\mathcal{B}$ of normal fuzzy subsets of a set X if it is the set of all distinct equivalence classes of some fuzzy equivalence relation on X. A family $\mathcal{C} \subseteq \mathcal{B}$ is said to be a **fuzzy semi-partition** of X. Following [6], fuzzy semi-partitions can be characterized in several ways.

To this end two more notions are introduced in the following definition, which can often be encountered explicitly (e.g. as *natural T-indistinguishability operator* in [37]) or implicitly ([6, 21]) in connection with fuzzy equivalences and their discussion. The subsequent presentation of results follows the one of [6].

Definition 8.37. *Let $\mu, \nu \in \mathcal{F}^L(X)$.*

(a) *The* **degree of compatibility** *of μ and ν is* $\mathsf{cmp}(\mu,\nu) = \bigvee_{x\in X}(\mu(x) \otimes \nu(x))$.
(b) *The* **degree of equality** *of μ and ν is* $\mathsf{equ}(\mu,\nu) = \bigwedge_{x\in X}(\mu(x) \rightleftharpoons \nu(x))$.

Proposition 8.38. ([6, Theorem 3.2]) *Let $\mathcal{C} \subseteq \mathcal{F}^L(X)$ be a set of normal fuzzy subsets of X; then the following conditions are equivalent:*

(1) *$\mathcal{C}$ is a fuzzy semi-partition of X;*
(2) *for all $\mu, \nu \in \mathcal{C}$, $\mathsf{cmp}(\mu,\nu) \le \mathsf{equ}(\mu,\nu)$;*
(3) *for all $x, y \in X$, $\bigvee_{\mu\in\mathcal{C}}(\mu(x)\otimes\mu(y)) \le \bigwedge_{\nu\in\mathcal{C}}(\nu(x) \rightleftharpoons \nu(y))$;*
(4) *$S_\mathcal{C} \subseteq E_\mathcal{C}$.*

Moreover, if any of these conditions holds, then the set of all fuzzy equivalence relations on X having $\mathcal{C}$ as fuzzy semi-partition is given by the interval $[S_\mathcal{C}, E_\mathcal{C}] \subseteq \mathsf{Eq}^L{}_X$.

Proof. (1) → (2): Let $\mathcal{C} \subseteq X/\theta$ for some $\theta \in \mathsf{Eq}^L{}_X$, and $\mu, \nu \in \mathcal{C}$. Then $\mu = [a]_\theta$ and $\nu = [b]_\theta$, for some $a, b \in X$. Using Prop. 8.30(3) we have

$$\mathsf{cmp}(\mu,\nu) = \bigvee_{x\in X}(\mu(x)\otimes\nu(x)) = \mathsf{hgt}(\mu\otimes\nu) = \mathsf{hgt}([a]_\theta \otimes [b]_\theta) = \theta(a,b).$$

Now take an arbitrary $y \in X$; then by Prop. 8.35

$$\theta(a,b) \le E_{[y]_\theta}(a,b) = [y]_\theta(a) \rightleftharpoons [y]_\theta(b) = [a]_\theta(y) \rightleftharpoons [b]_\theta(y) = \mu(y) \rightleftharpoons \nu(y)$$

Therefore, $\theta(a,b) \le \bigwedge_{y\in X}(\mu(y) \rightleftharpoons \nu(y)) = \mathsf{equ}(\mu,\nu)$.

(2) → (3): Note that (2) can be stated as

$$\text{for all } \mu,\nu \in \mathcal{C} \text{ and all } x, y \in X,\ \mu(x)\otimes\nu(x) \le \mu(y) \rightleftharpoons \nu(y).$$

For arbitrary $\mu, \nu \in \mathcal{C}$ and $x, y \in X$ this yields

$$\mu(x) \otimes \nu(x) \le \mu(y) \curvearrowright \nu(y) \quad \text{and} \quad \mu(y) \otimes \nu(y) \le \mu(x) \curvearrowright \nu(x).$$

By the adjunction property we get from the first inequality $\mu(x) \otimes \nu(x) \le \mu(y) \curvearrowright \nu(y) \leftrightarrow \mu(x) \otimes \nu(x) \otimes \mu(y) \le \nu(y) \leftrightarrow \mu(x) \otimes \mu(y) \otimes \nu(x) \le \nu(y) \leftrightarrow \mu(x) \otimes \mu(y) \le \nu(x) \curvearrowright \nu(y)$; then, by a similar argumentation for the second inequality, in sum we have

$$\mu(x) \otimes \mu(y) \le \nu(x) \curvearrowright \nu(y) \quad \text{and} \quad \mu(x) \otimes \mu(y) \le \nu(y) \curvearrowright \nu(x).$$

Therefore, for all $\mu, \nu \in \mathcal{C}$ and all $x, y \in X$, $\mu(x) \otimes \mu(y) \le \nu(x) \rightleftharpoons \nu(y)$.

(3) $\rightarrow$ (4): For $\mu \in \mathcal{C}$, and $x, y \in X$ with $x \ne y$, (3) implies $\mu(x) \otimes \mu(y) \le \bigwedge_{\nu \in \mathcal{C}} (\nu(x) \rightleftharpoons \nu(y)) = E_{\mathcal{C}}(x, y)$; thus $S_\mu \subseteq E_{\mathcal{C}}$, for every $\mu \in \mathcal{C}$, whence $S_{\mathcal{C}} \subseteq E_{\mathcal{C}}$.

(4) $\rightarrow$ (1): For any $\theta \in [S_{\mathcal{C}}, E_{\mathcal{C}}]$ we have $\theta \in [S_\mu, E_\mu]$, for every $\mu \in \mathcal{C}$. By Prop. 8.35, any $\mu \in \mathcal{C}$ is an equivalence class of θ.

Finally, if $\mathcal{C}$ is a fuzzy semi-partition of X, i.e. $S_{\mathcal{C}} \subseteq E_{\mathcal{C}}$, then $[S_{\mathcal{C}}, E_{\mathcal{C}}]$ is the intersection $\bigcap_{\mu \in \mathcal{C}} [S_\mu, E_\mu]$, and by Prop. 8.35 this is the set of all fuzzy equivalence relations on X having every $\mu \in \mathcal{C}$ as one of its equivalence classes. □

Corollary 8.39. ([6, Corollaries 3.1 and 3.2]) *Let $\mathcal{C} \subseteq \mathcal{F}^L(X)$ be a fuzzy semi-partition of X. Then*

(1) *for all $\mu, \nu \in \mathcal{C}$, $\mathsf{cmp}(\mu, \nu) = \mathsf{equ}(\mu, \nu)$;*
(2) *for all $\mu, \nu \in \mathcal{C}$ with $\mu \ne \nu$, the following two assertions hold:*
  (A1) $\mathsf{hgt}(\mu \otimes \nu) < 1$, *and*
  (A2) $\nu(x) = \mathsf{hgt}(\mu \otimes \nu)$ *for every $x \in \mathsf{core}(\mu)$;*
(3) *$\mathcal{C}$ does not contain two different members with the same core.*

Proof. (1): Let $\mu, \nu \in \mathcal{C}$; so each of μ, ν is normal and thus has a non-empty core. For $a \in \mathsf{core}(\mu)$, by Prop. 7.14(3) and (5) we have

$$\bigwedge_{x \in X} (\mu(x) \rightleftharpoons \nu(x)) \le \mu(a) \rightleftharpoons \nu(a) = \nu(a) = \mu(a) \otimes \nu(a) \le \bigvee_{x \in X} (\mu(x) \otimes \nu(x)),$$

i.e. $\mathsf{equ}(\mu, \nu) \le \mathsf{cmp}(\mu, \nu)$. Together with Prop. 8.38(2) this proves the assertion.

(2): Noting that $\mathsf{hgt}(\mu \otimes \nu) = \mathsf{cmp}(\mu, \nu)$, (A1) follows from (1) and Cor. 7.20(4), and (A2) is an immediate consequence of the proof of (1).

(3): (2)(A1) is equivalent to $\bigvee_{x,y \in X} (\mu(x) \otimes \nu(y)) < 1$, i.e. there are no $x, y \in X$ such that $\mu(x) = \nu(y) = 1$. □

Proposition 8.40. ([6, Theorem 3.3]) *Let $\mathcal{C}$ be a fuzzy semi-partition of X, and $x, y \in X, x \ne y$. Then for $S_{\mathcal{C}}$ the following holds: $S_{\mathcal{C}}(x, x) = 1$ and $S_{\mathcal{C}}(x, y) = \bigvee_{\mu \in \mathcal{C}} (\mu(x) \otimes \mu(y))$ for any two different $x, y \in X$.*

Proof. Let $\sigma \in \mathcal{F}^L_X$ with $\sigma(x, x) = 1$ and $\sigma(x, y) = \bigvee_{\mu \in \mathcal{C}} (\mu(x) \otimes \mu(y))$ for $x \ne y$. We will show both $\sigma \subseteq S_{\mathcal{C}}$ and $S_{\mathcal{C}} \subseteq \sigma$ to hold.

By Def. 8.33 $S_\mu \subseteq S_{\mathcal{C}}$, for every $\mu \in \mathcal{C}$, and the definitions of S_μ and σ give $\sigma \subseteq S_\mu$ proving the first inequality. Furthermore, due to the properties assumed for σ, $S_\mu \subseteq \sigma$, for every $\mu \in \mathcal{C}$. We will show that σ is also transitive, i.e. a fuzzy equivalence relation.

But then $S_{\mathcal{C}} \subseteq \sigma$ follows since by Def. 8.33 it is the smallest $\theta \in \mathsf{Eq}^L{}_X$ such that $S_\mu \subseteq \theta$ for all $\mu \in \mathcal{C}$.

Since $\mathcal{C}$ is a fuzzy semi-partition of X, $\mathcal{C} = \{[a]_\theta\}_{a\in A}$ for some $\theta \in \mathsf{Eq}^L{}_X$ and $A \subseteq X$. Then, $\bigvee_{\mu\in\mathcal{C}}(\mu(x) \otimes \mu(y)) = \bigvee_{\mu\in\mathcal{C}}(\theta(x,a) \otimes \theta(a,y))$, and for arbitrary $x, y, z \in X$, we have

$$\begin{aligned}
\sigma(x,y) \otimes \sigma(y,z) &= \left[\bigvee_{a\in A}(\theta(x,a) \otimes \theta(a,y))\right] \otimes \left[\bigvee_{b\in A}(\theta(y,b) \otimes \theta(b,z))\right] \\
&\leq \left[\bigvee_{a\in A}(\theta(x,a) \otimes \theta(a,y))\right] \otimes \theta(y,z) \\
&= \bigvee_{a\in A}(\theta(x,a) \otimes \theta(a,y) \otimes \theta(y,z)) \\
&\quad \text{(by transitivity of } \theta \text{ and Prop. 7.19(1))} \\
&\leq \bigvee_{a\in A}(\theta(x,a) \otimes \theta(a,z)) \\
&= \sigma(x,z).
\end{aligned}$$

□

As has been shown in Prop. 8.38 a fuzzy equivalence relation corresponding to a fuzzy semi-partition need not necessarily be uniquely determined, but in some cases it is as shown in

Corollary 8.41. ([6, Corollary 3.3]) *Let $\mathcal{C} \subseteq \mathcal{F}^L(X)$ be a set of normal fuzzy subsets of X; then the following conditions are equivalent:*

(1) There exists a unique $\theta \in \mathsf{Eq}^L{}_X$ having $\mathcal{C}$ as fuzzy semi-partition;
(2) for any two different $x, y \in X$: $\bigvee_{\mu\in\mathcal{C}}(\mu(x) \otimes \mu(y)) = \bigwedge_{\nu\in\mathcal{C}}(\nu(x) \rightleftharpoons \nu(y))$;
(3) $S_{\mathcal{C}} = E_{\mathcal{C}}$.

Proof. (1) and (3) are equivalent by Prop. 8.38(1),(4). (2) and (3) are equivalent by Prop. 8.40, since by Def. 8.33(d) $\bigwedge_{\nu\in\mathcal{C}}(\nu(x) \rightleftharpoons \nu(y)) = E_{\mathcal{C}}$. □

Now we are ready to characterize fuzzy partitions as follows:

Theorem 8.42. ([6, Theorem 3.4]) *Let $\mathcal{C} \subseteq \mathcal{F}^L(X)$ be a set of normal fuzzy subsets of a set X; then the following conditions are equivalent:*

(1) $\mathcal{C}$ is a fuzzy partition of X;
(2) $\mathcal{C}$ is a fuzzy semi-partition of X, and satisfies the following condition:
  (A3) *for every $x \in X$ there is a $\mu \in \mathcal{C}$ with $\mu(x) = 1$;*
(3) $\mathcal{C}$ satisfies the following two conditions:
  (A4) *for every $x \in X$ there is a unique $\mu_x \in \mathcal{C}$ with $\mu_x(x) = 1$, and*
  (A5) *$\mu_x(y) = \mathsf{hgt}(\mu_x \otimes \mu_y)$ for all $x, y \in X$.*

Moreover, if any of the above five conditions is satisfied, then $S_{\mathcal{C}} = E_{\mathcal{C}}$ and it is the unique fuzzy equivalence relation on X such that $\mathcal{C}$ is its fuzzy partition.

Proof. (1) → (2): If $\mathcal{C}$ is a fuzzy partition of some $\theta \in \mathsf{Eq}^L{}_X$, then for every $x \in X$ $[x]_\theta \in \mathcal{C}$ and $[x]_\theta = 1$.

(2) → (3): By (A3) for every $x \in X$ there is an $\mu \in \mathcal{C}$ with $\mu(x) = 1$. If there is another $\nu \in \mathcal{C}$ with $\nu(x) = 1$, then $\mathsf{hgt}(\mu \otimes \nu) \geq \mu(x) \otimes \nu(x) = 1$, which contradicts Cor. 8.39(A1). Therefore, (A4) holds. (A5) is just a reformulation of Cor. 8.39(A2).

(3) $\to$ (1): Using (A4) and (A5) we define a fuzzy relation $\theta \in \mathcal{F}_X^L$ by $\theta(x,y) = \mathsf{hgt}(\mu_x \otimes \mu_y)$. Clearly, θ is reflexive and symmetric. To prove transitivity we use (A5) and get $\theta(x,y)\otimes\theta(y,z) = \mathsf{hgt}(\mu_x\otimes\mu_y)\otimes\mathsf{hgt}(\mu_y\otimes\mu_z) = \mu_x(y)\otimes\mu_z(y) \le \mathsf{hgt}(\mu_x\otimes\mu_z) = \theta(x,z)$. Hence $\theta \in \mathsf{Eq}^L{}_X$ and by (A5) $\theta(x,y) = [x]_\theta(y)$ for all $x, y \in X$, whence $\mathcal{C} = X/\theta$.

Now assume that any of the conditions (1)-(3) above holds. Then by (A4), (A5) and Cor. 8.39(1) we have

$$\begin{aligned}\bigvee_{\mu\in\mathcal{C}}(\mu(x)\otimes\mu(y)) &= \bigvee_{a\in A}(\mu_a(x)\otimes\mu_a(y)) = \bigvee_{a\in A}(\mu_x(a)\otimes\mu_y(a))\\ &= \bigwedge_{a\in A}(\mu_x(a)\rightleftharpoons\mu_y(a)) = \bigwedge_{a\in A}(\mu_a(x)\rightleftharpoons\mu_a(y))\\ &= \bigwedge_{\nu\in\mathcal{C}}(\nu(x)\rightleftharpoons\nu(y)).\end{aligned}$$

By Cor. 8.41 we conclude that $S_\mathcal{C} = E_\mathcal{C}$, and that it is the unique $\theta \in \mathsf{Eq}^L{}_X$ with $\mathcal{C} = X/\theta$. □

Observe that conditions (A3) and (A4) are, respectively, equivalent to
(A3′) $\{\mathsf{core}(\mu)\}_{\mu\in\mathcal{C}}$ is a covering of X; and
(A4′) $\{\mathsf{core}(\mu)\}_{\mu\in\mathcal{C}}$ is a partition of X.

8.5 Fuzzy Relation Equations and Inequalities

Fuzzy relation equations (FREQs) and inequalities have a long tradition in Fuzzy Set Theory, starting in the mid-seventies with two papers by Sanchez [39, 40] in the area of medical research. Since then, fuzzy relation equations have intensively been investigated from both a theoretical and application oriented point of view. As a result much more papers of both kinds have been published than a single person can work through with reasonable effort, if it is possible at all.

Aside from pushing forward research in many-valued logic, fuzzy relation equations are the second area where a vast progress in theory has been achieved under the roof of Fuzzy Set Theory during the last four decades. During this period, fuzzy relations have also found a wide field of applications and nowadays are used in e.g. fuzzy control, approximate reasoning, decision-making, fuzzy information retrieval, and fuzzy pattern recognition, to name a few. Therefore this topic must not be missing in a text on relations.

We restrict our considerations to the questions of solvability of sup-$\sqcap$ fuzzy relation equations and inequalities ($\sqcap$ being some t-norm on L) and also to the structure of their solution sets on complete BROUWERian lattices, i.e. what often is called "*relatively pseudocomplemented lattice*"; see Def. 7.12. FREQs of this kind are in line with what has been initiated in Sanchez [39, 40], have been thoroughly analyzed over the years, and are a solid basis for various applications. Algorithms and their complexity, or properties of FREQs on more general or more special algebraic structures, or other kinds of fuzzy relation equations and inequalities will not be our concern here. For these topics we have to refer interested readers to e.g. [1, 2, 10, 32].

8.5.1 *Types of Fuzzy Relation Equations*

The various types of FREQs naturally originate from the variants of relation composition. In Def. 6.1 we learned about the circle $\rho\circ\sigma$, the subdirect $\rho\lhd\sigma$, and the superdirect product $\rho\rhd\sigma$ for classical two-valued relations, defined as follows:

(a) $\rho\circ\sigma = \{\langle x,z\rangle : \exists y\in Y[x\rho y \wedge y\sigma z\}$,
(b) $\rho\lhd\sigma = \{\langle x,z\rangle : \forall y\in Y[x\rho y \rightarrow y\sigma z\}$,
(c) $\rho\rhd\sigma = \{\langle x,z\rangle : \forall y\in Y[y\sigma z \rightarrow x\rho y\}$.

For a complete bounded lattice $(L,\vee,\wedge,0,1)$, the classical logical operations of conjunction, disjunction, and implication can be extended in the following way:

Definition 8.43.

(a) *A* **conjunctor** $\mathcal{C}: L\longrightarrow L$ *is a binary operation on L that is isotone in both arguments such that* $\mathcal{C}(0,1)=\mathcal{C}(1,0)=0$ *and* $\mathcal{C}(1,1)=1$. *A* **border conjunctor** *additionally satisfies the neutrality principle, i.e.* $\mathcal{C}(1,b)=b$ *for all* $b\in L$.
(b) *A* **disjunctor** $\mathcal{D}: L\longrightarrow L$ *is a binary operation on L that is isotone in both arguments such that* $\mathcal{D}(0,1)=\mathcal{D}(1,0)=1$ *and* $\mathcal{D}(0,0)=0$. *A* **border disjunctor** *additionally satisfies the neutrality principle, i.e.* $\mathcal{D}(0,b)=b$ *for all* $b\in L$.
(c) *A* **implicator** $\mathcal{I}: L\longrightarrow L$ *is a binary operation on L that is antitone in the first, and isotone in the second argument such that* $\mathcal{I}(0,0)=\mathcal{I}(1,1)=1$ *and* $\mathcal{I}(1,0)=0$. *A* **border implicator** *additionally satisfies the neutrality principle, i.e.* $\mathcal{I}(1,b)=b$ *for all* $b\in L$.

Note that, for all $a,b\in L$, the following holds: For any conjunctor we have $\mathcal{C}(a,0)=\mathcal{C}(0,b)=0$, for any disjunctor $\mathcal{D}(a,1)=\mathcal{C}(1,b)=1$, and for any implicator $\mathcal{I}(a,1)=\mathcal{I}(0,b)=1$. These boundary properties are also referred to as *absorption principles.* It is evident that $a\wedge b$ is a conjunctor, $a\vee b$ is a disjunctor, and the $a\curvearrowright b$ is an implicator on a complete Brouwerian lattice.

If you, furthermore, remember that the supremum and infimum of L model the existential and universal quantifier, respectively, then we get the many-valued versions of the above compositions as follows:

(a) $\rho\circ\sigma(x,z) = \sup_{y\in Y}\mathcal{C}(\rho(x,y),\sigma(y,z))$,
(b) $\rho\lhd\sigma(x,z) = \inf_{y\in Y}\mathcal{I}(\rho(x,y),\sigma(y,z))$,
(c) $\rho\rhd\sigma(x,z) = \inf_{y\in Y}\mathcal{I}(\sigma(y,z),\rho(x,y))$.

Since very often a t-norm $\mathcal{T}$ is used as a conjunctor we get the two basic groups of sup-$\mathcal{C}$ or sup-$\mathcal{T}$, and inf-$\mathcal{I}$ equations. In the sections below we will focus on the first group, for which (8.6) is an example, and we will outline the principle strategies that have been applied to tackle the resolution problem for FREQs.

For more details on both logical operators and further types of FREQs you may best start with [25] which, though in principle based mainly on the older work [10] and [50], gives a comprehensive and well readable overview with a lot of references.

8.5.2 *The Lattice Based Approach*

For sake of simplicity, in what follows let X, Y and Z denote three non-empty finite sets with $|X| = m, |Y| = n$ and $|Z| = r$; remember that, for any $n \in \mathbb{N}$, we denote by $\underline{n}$ the set $\{1,2,\ldots,n\}$. Let L a complete BROUWERian lattice (in the terminology of Sanchez [39]), i.e. a complete residuated lattice $(L, \wedge, \vee, \otimes, \curvearrowright, 0, 1)$ where $\otimes = \wedge$, or HEYTING lattice, in our terminology.

Assume that $\rho \in \mathcal{F}^L_{X\times Y}$, $\sigma \in \mathcal{F}^L_{Y\times Z}$, and $\tau \in \mathcal{F}^L_{X\times Z}$. Let $R = (r_{ij}) \in \mathfrak{M}^L_{m\times n}$, $S = (s_{jk}) \in \mathfrak{M}^L_{n\times r}$ and $T = (t_{mr}) \in \mathfrak{M}^L_{m\times r}$ be representations of ρ, σ, and τ, respectively. Furthermore, let the three relations be constrained by

$$\rho \circ \sigma = \tau, \qquad i.e.\ t_{ik} = \bigvee\nolimits_j \left(r_{ij} \wedge s_{jk}\right) \tag{8.6}$$

If only two of the three relations are given, there is a "problem": The *direct resolution problem*, if we know both ρ and σ; we can get τ by direct computation. If however the result τ and only one of the relations on the left side are known, then we have an *inverse resolution problem*, viz. (1) $\chi \circ \sigma = \tau$ or (2) $\rho \circ \chi = \tau$ with χ unknown. Whereas the direct resolution problem can simply be solved by calculation and has a unique solution, inverse problems are far from being trivial and a solution is guaranteed neither to exist nor to be unique.

Before we are going to present results known about solutions for equation (8.6), note the following decomposition possibilities to reduce the complexity of the problem and its structure ([10, 23, 25, 50]):

(1) If we have a method for solving (1) $\chi \circ \sigma = \tau$, we can also use it for solving (2) $\rho \circ \chi = \tau$: Rewrite the last equation as $\chi^{\smile} \circ \rho^{\smile} = \tau^{\smile}$ and try to solve this to get $\chi^{\smile}$ by the solution method for (1); if there is such a solution then clearly $\chi^{\smile\smile} = \chi$ is a solution for (2).

(2) Let $\mathscr{S}(\sigma,\tau) := \{\rho : \rho\sigma = \tau\}$ or, equivalently, $\mathscr{S}(S,T) = \{R : R\circ S = T\}$ (correspondingly $\mathscr{S}(\rho,\tau) := \{\sigma : \rho\sigma = \tau\}$) denote the set of solutions for equation (8.6). Then it is easy to see that (8.6) can be partitioned into a set of simpler equations as follows: If $S \in \mathscr{S}(R,T)$, i.e. $R \circ S = T$ then we have

$$\begin{bmatrix} r_{11}\ r_{12}\ \cdots\ r_{1n} \\ r_{21}\ r_{22}\ \cdots\ r_{2n} \\ \vdots \\ r_{m1}\ r_{22}\ \cdots\ r_{mn} \end{bmatrix} \circ \begin{bmatrix} s_{1k} \\ s_{2k} \\ \vdots \\ s_{nk} \end{bmatrix} = \begin{bmatrix} t_{1k} \\ t_{2k} \\ \vdots \\ t_{mk} \end{bmatrix} \tag{8.7}$$

for all $k \in \underline{r}$. As an equivalent vector formulation we have, for all $k \in \underline{r}$, with $\boldsymbol{s}_k = (s_{ik})_{i\in\underline{n}}$, and $\boldsymbol{t}_k = (t_{ik})_{i\in\underline{m}}$

$$R \circ \boldsymbol{s}_k = \boldsymbol{t}_k. \tag{8.8}$$

If $\mathscr{S}_k(R, \boldsymbol{t}_k) = \{\boldsymbol{s}_k : R \circ \boldsymbol{s}_k = \boldsymbol{t}_k\}$ then we get the solutions S as one-row matrices

$$S = \left[\boldsymbol{s}_1\ \boldsymbol{s}_2\ \cdots\ \boldsymbol{s}_r\right],$$

where $\boldsymbol{s}_k \in \mathscr{S}_k(R, \boldsymbol{t}_k)$ for all $k \in \underline{r}$.

(3) Clearly, (8.7) is a system of m homogenous and independent equations in n variables s_j over L of the form

$$\begin{bmatrix} r_{i1}\, r_{i2} \cdots r_{in} \end{bmatrix} \circ \begin{bmatrix} s_1 \\ s_2 \\ \vdots \\ s_n \end{bmatrix} = t_i, \qquad \text{or } \bigvee_{j \in \underline{n}} (r_{ij} \wedge s_j) = t_i. \tag{8.9}$$

(4) And this, in turn, means that we finally deal with n elementary equations of the form $r_j \wedge s_j = t_{i_j}$ or, in a more simple formulation: with n equations $r \wedge s = t$, or, equivalently, with n pairs of inequalities $r \wedge s \leq t$ and $r \wedge s \geq t$.

8.5.2.1 *Starting with the Begin*

As already mentioned, investigation on fuzzy relational equations started in the mid-seventies with Sanchez's groundbreaking papers [39, 40] in which he has presented results that remained an important cornerstone throughout all the time since then. We begin with some definitions necessary for the subsequent presentation of these results, for which we do only partially adhere to the notation that originated from the papers just mentioned and which has commonly been used in early contributions as e.g. [12]. Therefore it makes sense to insert

CAVEAT 4: In [39, 40], and also in related papers, $S \circ R$ is written for $\rho\sigma$, R^{-1} for $\rho^{\smile}$, and $a\alpha b$ instead of $a \frown b$.

Let $\rho \in \mathcal{F}^L_{X \times Y}$, $\sigma \in \mathcal{F}^L_{Y \times Z}$, and $\tau \in \mathcal{F}^L_{X \times Z}$ denote L-fuzzy relations.

Definition 8.44. *The* **@-composition of relations** ρ *and* σ *is defined by*

$$\rho @ \sigma(x, z) = \bigwedge\nolimits_y (x\rho y \frown y\sigma z) \tag{8.10}$$

The following results are due to [39]; see also [32, Chapter 2.2] for an elaborate exposition (with a notation that avoids the change of factors in relation composition).

Proposition 8.45.

(1) $\sigma \subseteq \rho^{\smile} @ (\rho\sigma)$;
(2) $\rho \subseteq (\sigma @ (\rho\sigma)^{\smile})^{\smile}$;
(3) $\rho(\rho^{\smile} @ \tau) \subseteq \tau$;
(4) $(\sigma @ \tau^{\smile})^{\smile} \sigma \subseteq \tau$.

Proof. Noting that $\otimes = \wedge$, and using monotonicity properties of $\frown$ we get

(1):
$$\begin{aligned}
(\rho^{\smile} @ (\rho\sigma))(y, z) &= \bigwedge\nolimits_x \left[x\rho y \frown (x\rho\sigma z) \right] \\
&= \bigwedge\nolimits_x \left[x\rho y \frown \left(\bigvee\nolimits_t (x\rho t \wedge t\sigma z) \right) \right] \\
&= \bigwedge\nolimits_x \left[x\rho y \frown \left((x\rho y \wedge y\sigma z) \vee \bigvee\nolimits_{t \neq y} (x\rho t \wedge t\sigma z) \right) \right] \\
&\geq \bigwedge\nolimits_x \left[x\rho y \frown (x\rho y \wedge y\sigma z) \right] \qquad \text{by Prop. 7.18(2)} \\
&\geq \bigwedge\nolimits_x \left[y\sigma z \right] \qquad \text{by Prop. 7.17(2), second inequality}
\end{aligned}$$

$$
\begin{aligned}
&&&= y\sigma z \\
(3):\quad & \rho(\rho^{\smile}@\tau)(x,z) &&= \bigvee_y \left[x\rho y \wedge y(\rho^{\smile}@\tau)z\right] \\
&&&= \bigvee_y \left[x\rho y \wedge \bigwedge_t (t\rho y \curvearrowright t\tau z)\right] \\
&&&= \bigvee_y \left[x\rho y \wedge (x\rho y \curvearrowright x\tau z) \wedge \bigwedge_{t\neq x} (t\rho y \curvearrowright t\tau z)\right] \\
&&&\leq \bigvee_y \left[x\rho y \wedge (x\rho y \curvearrowright x\tau z)\right] \\
&&&\leq \bigvee_y \left[x\tau z\right] \qquad \text{by Prop. 7.17(2), first inequality} \\
&&&= x\tau z
\end{aligned}
$$

□

Theorem 8.46. *The fundamental results on the solution sets of fuzzy relation equations are:*

(1) *Let $S_\sigma := \mathscr{S}(\rho,\tau)$; then $S_\sigma \neq \emptyset$ iff $\hat{\sigma} := \rho^{\smile}@\tau \in S_\sigma$; then $\hat{\sigma}$ is the greatest element in S_σ.*

(2) *Let $S_\rho := \mathscr{S}(\sigma,\tau)$; then $S_\rho \neq \emptyset$ iff $\hat{\rho} := (\sigma@\tau^{\smile})^{\smile} \in S_\rho$; then $\hat{\rho}$ is the greatest element in S_ρ.*

Proof. (1): Assume $S_\sigma \neq \emptyset$; then there is a relation σ' with $\rho\sigma' = \tau$. We get

$$
\sigma' \subseteq \rho^{\smile}@(\rho\sigma') = \rho^{\smile}@\tau = \hat{\sigma} \tag{8.11}
$$

from Prop. 8.45(1). Thus $\tau = \rho\sigma' \subseteq \rho(\rho^{\smile}@\tau)$, which, together with Prop. 8.45(3), gives $\rho(\rho^{\smile}@\tau) = \tau$, i.e. $\hat{\sigma} \in S_\sigma$. And (8.11) shows that $\hat{\sigma}$ is the greatest solution.

(2): The proof is analogous to (1) using Prop. 8.45(2) and (4). □

Before we have a closer look at the solution sets of fuzzy relation equations in Section 8.5.2.3, a few words on fuzzy relation inequalities are in order.

8.5.2.2 *Fuzzy Relation Inequalities*

Lemma 8.47. *Let $\rho \in \mathcal{F}^L_{X\times Y}$ and $\sigma_1, \sigma_2 \in \mathcal{F}^L_{Y\times Z}$ denote L-fuzzy relations such that $\sigma_1 \subseteq \sigma_2$. Then $\rho@\sigma_1 \subseteq \rho@\sigma_2$.*

Proof. By Prop. 7.18(2) we have $x\rho y \curvearrowright y\sigma_1 z \leq x\rho y \curvearrowright y\sigma_2 z$ for each $y \in Y$, so $\rho@\sigma_1(x,z) \leq \rho@\sigma_2(x,z)$. □

Theorem 8.48. ([40, p. 225]) *Let $\rho \in \mathcal{F}^L_{X\times Y}$, $\sigma \in \mathcal{F}^L_{Y\times Z}$, and $\tau \in \mathcal{F}^L_{X\times Z}$ denote L-fuzzy relations. Then the following holds:*

(1) *Let $S_{\sigma/\subseteq} := \{\sigma : \rho\sigma \subseteq \tau\}$; then $\hat{\sigma} := \rho^{\smile}@\tau$ is the greatest element in $S_{\sigma/\subseteq}$.*

(2) *Let $S_{\rho/\subseteq} := \{\rho : \rho\sigma \subseteq \tau\}$; then $\hat{\rho} := (\sigma@\tau^{\smile})^{\smile}$ is the greatest element in $S_{\rho/\subseteq}$.*

Proof. (1): $S_{\sigma/\subseteq} \neq \emptyset$ since $\emptyset \in S_{\rho/\subseteq}$. Suppose $\sigma \in S_{\sigma/\subseteq}$, then $\rho\sigma \subseteq \tau$, whence $\rho^{\smile}@(\rho\sigma) \subseteq \rho^{\smile}@\tau$ by Lemma 8.47. Therefore, by Prop. 8.45(1), we get $\sigma \subseteq \rho^{\smile}@\tau$, and Prop. 8.45(3) gives $\rho(\rho^{\smile}@\tau) \subseteq \tau$. Thus, $\hat{\sigma} \in S_{\sigma/\subseteq}$, and $\sigma \subseteq \hat{\sigma}$ for all $\sigma \in S_{\sigma/\subseteq}$.

(2): $S_{\rho/\subseteq} \neq \emptyset$ since $\emptyset \in S_{\rho/\subseteq}$. Note that Prop. 8.20(6) implies $\rho \subseteq \sigma \leftrightarrow \rho^{\smile} \subseteq \sigma^{\smile}$. Suppose $\rho \in S_{\rho/\subseteq}$, then $\rho\sigma \subseteq \tau$. But $\rho\sigma \subseteq \tau \leftrightarrow (\rho\sigma)^{\smile} \subseteq \tau^{\smile}$, whence $\sigma@(\rho\sigma)^{\smile} \subseteq \sigma@\tau^{\smile}$ by Lemma 8.47. This implies $(\sigma@(\rho\sigma)^{\smile})^{\smile} \subseteq (\sigma@\tau^{\smile})^{\smile}$, so $\rho \subseteq (\sigma@\tau^{\smile})^{\smile}$ by Prop. 8.45(2). Finally, Prop. 8.45(4) gives $(\sigma@\tau^{\smile})^{\smile}\sigma \subseteq \tau$; so $\hat{\rho} \in S_{\rho/\subseteq}$ and $\rho \subseteq \hat{\rho}$ for all $\rho \in S_{\rho/\subseteq}$. □

8.5.2.3 *Minimal Solutions*

Now that we have given in Th. 8.46 necessary and sufficient conditions for the existence of solutions of equation (8.6), what can be said about the structure of the solution sets? Some trivial but important facts are listed in

Proposition 8.49. *Assume $S_\rho \neq \emptyset$, $|S_\rho| \geq 2$, and $\rho_1, \rho_2 \in S_\rho$. Then*

(1) $\rho \in [\rho_1, \rho_2]$ (i.e. $\rho_1 \subseteq \rho \subseteq \rho_2$) implies $\rho \in S_\rho$.
(2) $\rho_1 \cup \rho_2 \in S_\rho$.

Proof. (1): $\tau = \rho_1\sigma \subseteq \rho\sigma \subseteq \rho_2\sigma = \tau$.
(2): $(\rho_1 \cup \rho_2)\sigma = \rho_1\sigma \cup \rho_2\sigma = \tau \cup \tau = \tau$. □

Let $\check{S}_\rho$ denote the set of minimal elements in $S_\rho \neq \emptyset$. If we know the minimal solutions of (8.6) then, by what has just been shown, the set of all solutions is given by

$$\mathscr{S}(\rho,\tau) = \bigcup_{\check{\rho}\in\check{S}_\rho} [\check{\rho}, \hat{\rho}].$$

So minimal solutions play a crucial role in describing all solutions of relational equations, and we are therefore left with the task to find a general method for obtaining them. Unfortunately, the problem of finding minimal solutions is by far more complicated than that of finding the greatest one as will become clear from the considerations below: Neither the decision about their existence for a given class of fuzzy relations, nor the construction of efficient algorithms for their computation is a trivial task. Accordingly, throughout the last four decades, a great many of research papers have been devoted to this topic. The top-down problem decomposition from (8.9) suggests to search the problem solution in a bottom-up way, and this has really been done starting with Zhao [50], followed by De Baets [8–10] and others (see e.g. [25, 44] for an overview).

Before we can proceed we need some more concepts that play an essential role in the analysis of the minimal solution problem. Looking back to equation (8.9) it is clear that any solution $\boldsymbol{s}_k = (s_{ik})_{i\in\underline{n}}$ means a join decomposition $\bigvee_{j\in\underline{n}}(r_{ij} \wedge s_j)$ of t_i. Therefore many researchers discuss conditions for the existence of minimal solutions of (8.9) via considering decompositions of the form $\bigvee_{j\in\underline{n}}(r_j \wedge s_j) = t$.

Definition 8.50. *Let L be a complete lattice.*

(a) An element $0 \neq a \in L$ is called **join-irreducible** *if, for all $b, c \in L$,*

$$a = b \vee c \rightarrow a \in \{b, c\}.$$

(b) A set $A \subseteq L$ of join-irreducible elements of L, with $|A| \geq 2$, such that $b = \bigvee_{a\in A} a$, is called a **join decomposition** *of b, and $b \in L$ is called* **join-decomposable**.

(c) *A join decomposition A of $b \in L$ is called* **irredundant** *if there is no join decomposition A' of b such that $A' \subseteq A$.*

All nonzero elements of a chain are join-irreducible and all nonzero elements of a finite distributive lattice or a direct product of a finite number of chains are either join-irreducible or join-decomposable. On the other hand, it is known that every element in a finite lattice is a join of join-irreducible elements with the convention that $\bigvee \emptyset = 0$.

Definition 8.51. *Let L be a complete lattice, and f be a mapping $f : L \longrightarrow L$;*

(a) *L is called* **completely join-distributive** *if it is distributive and*
$$b \wedge (\bigvee_{a \in A} a) = \bigvee_{a \in A}(b \wedge a).$$
(b) *f is called a* **sup-morphism** *(an* **inf-morphism***) if for any non-empty $A \subseteq L$ we have*
$$f(\bigvee_{a \in A} a) = \bigvee_{a \in A} f(a) \qquad \big(f(\bigwedge_{a \in A} a) = \bigwedge_{a \in A} f(a) \big).$$
(c) *f is called a* **homomorphism** *if it is both a sup-morphism and inf-morphism.*

Lemma 8.52. *The following are well known results from lattice theory:*

(1) *Let L be a distributive lattice and $b \in L$ be a join-irreducible element. Then $b \leq \bigvee_{i \in \underline{n}} a_i$ iff $\exists i \in \underline{n}\, [b \leq a_i]$.*
(2) *In a distributive lattice L, the irredundant join-decomposition of an element $b \in L$ is unique.*
(3) *A Brouwerian lattice is distributive.*
(4) *A complete lattice is Brouwerian if and only if it is completely join-distributive.*

Proof. (1): By the assumption we have $b = b \wedge \bigvee_{i \in \underline{n}} a_i$, and by distributivity $b = \bigvee_{i \in \underline{n}}(b \wedge a_i)$. Since b is join-irreducible, $b = b \wedge a_i$ for some i, i.e. $b \leq a_i$.

(2): Let L be a distributive lattice; assume that an element $b \in L$ has two irredundant join-decompositions, i.e.
$$b \;=\; x_1 \vee \cdots \vee x_r \;=\; y_1 \vee \cdots \vee y_s.$$
Then, by (1) we get for a given x_i that for some $j \in \underline{s}$ $y_j \geq x_i$, and similarly $x_k \geq y_j$. The irredundancy of the join-decompositions implies that the x_i and y_i are equal in pairs, and $r = s$.

(3): By Prop. 7.13.

(4): See [5, p. 128, THEOREM 24]. □

Our representation follows De Baets [8–10] which generalizes the results of Sanchez reported above by replacing the meet operation by t-norms on L with certain properties. Note that these results are therefore valid in residuated lattices L since $\otimes$ is a t-norm on L.

Definition 8.53. ([8, 10]) *Let $(L, \leq)$ be a poset; a subset $R \subseteq L$ is called a* **root system** *of L if there is an element $m \in R$ and an antichain $O \subseteq m^{\downarrow}$, the principal ideal generated by*

m, such that $R = \bigcup_{o \in O}[o, m]$. m is called the **stem**, *and the elements $o \in O$ the* **offshoots** *of R.*

A root system is called *finitely generated* if its set of offshoots is finite. In a root system R with stem m and set of offshoots O, m is the greatest element, and the elements of O are the minimal elements of R. Hence, a root system clearly is an order convex subset that is completely determined by its greatest element and its minimal elements. Below we will see that the solution set of sup-$\sqcap$ FREQs, if not empty, is a root system under some rather general restrictions. In [8,9] the following results have been stated without proof:

Lemma 8.54.

(1) (Intersection of finitely generated root systems)
Let $\{R_i : i \in I\}$ be a family of finitely generated root systems of a complete lattice L with stem m_i and set of offshoots O_i. If $\bigcap_{i \in I} R_i \neq \emptyset$, then this intersection is a root system with stem $m = \inf_{i \in I} m_i$ and with offshoots the minimal elements of $\{\sup_{i \in I} o_i : \forall i \in I[o_i \in O_i \wedge o_i \leq m]\}$.

(2) (Finite intersection of finitely generated root systems)
If the intersection of a finite family of finitely generated root systems of a complete lattice is non-empty, then it is a finitely generated root system.

(a) Single Equations:

Let $\sqcap$ be a t-norm on a bounded poset; then for the map $\sqcap_a$ defined by $\sqcap_a(x) = a \sqcap x$ we have $\mathsf{rng}(\sqcap_a) \subseteq [0, a]$: for all $x \in L$, $a \sqcap x \leq a \wedge x$ by Prop. 8.8(1), and clearly $a \wedge x \in [0, a]$ for all $x \in L$. The partial mapping $\sqcap_a$ is called **maximally surjective** if $\mathsf{rng}(\sqcap_a) = [0, a]$. $\wedge_a$ is an example of a maximally surjective map: for $b \in [0, a]$ we clearly have $a \wedge b = b$, so $b \in \mathsf{rng}(\sqcap_a)$. The t-norm $\sqcap$ is called **maximally surjective** iff all its partial mappings $\sqcap_a$ are.

The most basic equation we consider is the equation $a \sqcap x = b$. In order to describe its solution set, two operators are associated to a t-norm $\sqcap$ as follows:

$$a \check{\sqcap} b = \bigvee\{c \in L : a \sqcap c \leq b\},$$
$$a \hat{\sqcap} b = \bigwedge\{c \in L : a \sqcap c \geq b\}.$$

Proposition 8.55. ([8, Propositions 12-14]) *Let L be a complete lattice, $\sqcap$ be a t-norm on L, and $a, b \in L$.*

(1) *If $\sqcap_a$ is a sup-morphism, then the solution set of the inequality $a \sqcap x \leq b$ is $[0, a \check{\sqcap} b]$.*
(2) *If $\sqcap_a$ is a inf-morphism, then the solution set of the inequality $a \sqcap x \geq b$ is $[a \hat{\sqcap} b, 1]$ if $a \geq b$, and $\emptyset$ otherwise.*
(3) *If $\sqcap_a$ is a homomorphism, then the solution set of the equation $a \sqcap x = b$ is $[a \hat{\sqcap} b, a \check{\sqcap} b]$ if $b \in \mathsf{rng}(\sqcap_a)$, and $\emptyset$ otherwise.*

Proof. (1): Due to the monotonicity of $\sqcap_a$ and the definition of $\check{\sqcap}$ it suffices to show that $a \check{\sqcap} b$ is a solution of $a \sqcap x \leq b$. The set $\{c \in L : a \sqcap c \leq b\}$ is not empty since 0 belongs to it. Thus, since $\sqcap_a$ is a sup-morphism,

$$a \sqcap (a \check{\sqcap} b) \;=\; a \sqcap \bigvee\{c \in L : a \sqcap c \leq b\} \;=\; \bigvee\{a \sqcap c : c \in L,\; a \sqcap c \leq b\} \;\leq b.$$

(2): Assume $a \geq b$; then due to the monotonicity of $\sqcap_a$ and the definition of $\hat{\sqcap}$ it suffices to show that $a \hat{\sqcap} b$ is a solution of $a \sqcap x \geq b$. The set $\{c \in L : a \sqcap c \geq b\}$ is not empty since 1 belongs to it. Thus, since $\sqcap_a$ is an inf-morphism,

$$a \sqcap (a \hat{\sqcap} b) \;=\; a \sqcap \bigwedge\{c \in L : a \sqcap c \geq b\} \;=\; \bigwedge\{a \sqcap c : c \in L,\; a \sqcap c \geq b\} \;\geq b.$$

(3): Immediate consequence of (1) and (2): If $b \notin \mathsf{rng}(\sqcap_a)$, then there is no $c \in L$ such that $a \sqcap c = b$. Otherwise, we have $a \sqcap c = b$ for at least one $c \in L$. So $c \in [0, a \check{\sqcap} b] \cap [a \hat{\sqcap} b, 1]$, and by construction $a \hat{\sqcap} b \leq c \leq a \check{\sqcap} b$, whence $[a \hat{\sqcap} b, a \check{\sqcap} b]$ is the set of solutions. □

Corollary 8.56. *Let L be a complete lattice, $\sqcap$ be a t-norm on L, and $a, b \in L$. If $\sqcap_a$ is a maximally surjective homomorphism, then the solution set of the equation $a \sqcap x = b$ is $[a \hat{\sqcap} b, a \check{\sqcap} b]$ if $a \geq b$, and $\emptyset$ otherwise.*

(b) Equations of n variables:

We now turn to equations of the form (8.9): Given $b \in L$ and $\boldsymbol{a} = (a_1, a_2, \ldots, a_n)$, determine $\boldsymbol{x}^T = (x_1, x_2, \ldots, x_n)^T$ such that $\boldsymbol{a} \circ \boldsymbol{x} = b$, i.e. $\bigvee_{i \in \underline{n}} (a_i \sqcap x_i) = b$.

Note that both $\boldsymbol{a}$ and $\boldsymbol{x}$ can be regarded as fuzzy sets over $\underline{n}$, i.e. $\boldsymbol{a}, \boldsymbol{x} \in \mathcal{F}^L(\underline{n})$. It is well known that $(\mathcal{F}^L(\underline{n}), \subseteq)$ is a complete lattice if L is; thus $\mathscr{S}(\boldsymbol{a}, \boldsymbol{b}) \subseteq \mathcal{F}^L(\underline{n})$.

Proposition 8.57. ([8, Propositions 16-18]) *Let L be a complete Brouwerian lattice, $\sqcap$ be a t-norm on L, $b \in L$, and $\boldsymbol{a} \in \mathcal{F}^L(\underline{n})$.*

(1) If the mappings $\sqcap_a$ are sup-morphisms then the solution set of the inequality $\bigvee_{i \in \underline{n}} (a_i \sqcap x_i) \leq b$ is given by $[\emptyset, \mathbf{g}]$, with $\mathbf{g} \in \mathcal{F}^L(\underline{n})$ defined by $g_i = a_i \check{\sqcap} b$.

(2) Let b be a join-irreducible element of L. If the mappings $\sqcap_a$ are inf-morphisms then the solution set of the inequality $\bigvee_{i \in \underline{n}} (a_i \sqcap x_i) \geq b$ is non-empty iff $b \leq \bigvee_{i \in \underline{n}} a_i$. If it is non-empty then it is a finitely generated root system with stem $\mathbf{1}_n = \chi_{\underline{n}}$ and with offshoots the elements of

$$O = \{\boldsymbol{o}_k \in \mathcal{F}^L(\underline{n}) : b \leq a_k\},$$

where $\boldsymbol{o}_k(i) = a_k \hat{\sqcap} b$ for $i = k$, and $\boldsymbol{o}_k(i) = 0$ otherwise.

(3) Let b be a join-decomposable element of L with $b = \sup_{j \in J} b_j$. If the mappings $\sqcap_a$ are inf-morphisms and the solution set of the inequality $\bigvee_{i \in \underline{n}} (a_i \sqcap x_i) \geq b$ is non-empty then it is a root system with stem $\mathbf{1}_n$ and with offshoots the elements of

$$O = \{\textstyle\bigcup_{j \in J} \boldsymbol{o}_k^j \in \mathcal{F}^L(\underline{n}) : b_j \leq a_k\},$$

where $\boldsymbol{o}_k^j(i) = a_k \hat{\sqcap} b_j$ for $i = k$, and $\boldsymbol{o}_k^j(i) = 0$ otherwise. If the join-decomposition is finite, then this root system is also finitely generated.

Proof. (1): By Prop. 8.55(1), the solution set of the inequality $a \sqcap x \le b$ is $[0, a \check{\sqcap} b]$. Observing the equivalences

$$\bigvee_{i\in\underline{n}}(a_i \sqcap x_i) \le b \;\leftrightarrow\; \forall i \in \underline{n}\,[a_i \sqcap x_i \le b] \;\leftrightarrow\; \forall i \in \underline{n}\,[0 \le x_i \le a_i \check{\sqcap} b],$$

the assertion follows immediately.

(2): By Lemma 8.52(1),

$$\bigvee_{i\in\underline{n}}(a_i \sqcap x_i) \ge b \leftrightarrow \exists k \in \underline{n}[a_k \sqcap b_k \ge b],$$

and from Prop. 8.55(2) we get

$$\bigvee_{i\in\underline{n}}(a_i \sqcap x_i) \ge b \leftrightarrow \exists k \in \underline{n}[b \le a_k \;\&\; a_k \hat{\sqcap} b \le x_k].$$

Again with Lemma 8.52(1) it follows that $\sup_{i\in\underline{n}}(a_i \sqcap x_i) \ge b$ is solvable iff $b \le \sup_{i\in\underline{n}} a_i$, whence the solution set is given by $\bigcup_{\boldsymbol{o}_k\in O}[\boldsymbol{o}_k, \mathbf{1}_n]$. For the solution set to be a root system, O must be an antichain: If $b = 0$ then obviously $O = \{\emptyset\}$ (since $a \hat{\sqcap} 0 = \bigwedge\{c \in L : a \sqcap c \ge b\} = 0$). For $b > 0$ it follows (since $\forall a \in L\, \forall b \in \mathrm{L}-\{0\}[a \hat{\sqcap} b > 0]$) that $\boldsymbol{o}_k(k) > 0$, and thus the elements of O form an antichain.

(3): Note that

$$\bigvee_{i\in\underline{n}}(a_i \sqcap x_i) \ge b = \bigvee_{j\in J} b_j \;\leftrightarrow\; \forall j \in J[\bigvee_{i\in\underline{n}}(a_i \sqcap x_i) \ge b_j].$$

We complete the proof by applying (2) (to the join-irreducible elements of b's join-decomposition) and Lemma 8.54. □

Combining the results obtained for the inequalities we get the solution sets for the corresponding equations $\bigvee_{i\in\underline{n}}(a_i \sqcap x_i) = b$ (which clearly is the intersection of the solution sets of the two corresponding inequalities) as stated in the following proposition.

Proposition 8.58. ([8, Theorems 2 and 3]) *Let L be a complete Brouwerian lattice, $\sqcap$ be a t-norm on L, $b \in L$, and $\boldsymbol{a} \in \mathcal{F}^L(\underline{n})$.*

(1) *Let b be a join-irreducible element of L. If the mappings $\sqcap_a$ are homomorphisms, and the solution set of the equation $\bigvee_{i\in\underline{n}}(a_i \sqcap x_i) = b$ is non-empty then it is a finitely generated root system with stem $\boldsymbol{g} \in \mathcal{F}^L(\underline{n})$ defined by $g_i = a_i \check{\sqcap} b$ and with offshoots the elements of*

$$O = \{\boldsymbol{o}_k \in \mathcal{F}^L(\underline{n}) : b \le a_k \wedge \boldsymbol{o}_k \subseteq \boldsymbol{g}\},$$

where $\boldsymbol{o}_k(i) = a_k \hat{\sqcap} b$ for $i = k$, and $\boldsymbol{o}_k(i) = 0$ otherwise.
If $\sqcap$ is also maximally surjective, then the set of offshoots can be written as

$$O = \{\boldsymbol{o}_k \in \mathcal{F}^L(\underline{n}) : b \le a_k\}.$$

A necessary and sufficient solvability condition therefore is that $\boldsymbol{g}$ is a solution.

(2) *Let b be a join-decomposable element of L with $b = \bigvee_{j\in J} b_j$. If the mappings $\sqcap_a$ are homomorphisms and the solution set of the inequality $\bigvee_{i\in\underline{n}}(a_i \sqcap x_i) \ge b$ is non-empty then it is a root system with stem $\boldsymbol{g} \in \mathcal{F}^L(\underline{n})$ defined by $g_i = a_i \check{\sqcap} b$ and with offshoots the elements of*

$$S = \{\bigcup_{j\in J} \boldsymbol{o}_k^j \in \mathcal{F}^L(\underline{n}) : b_j \le a_k \wedge \boldsymbol{o}_k^j \subseteq \boldsymbol{g}\},$$

where $\boldsymbol{o}_k^j(i) = a_k \hat{\sqcap} b_j$ for $i = k$, and $\boldsymbol{o}_k(i) = 0$ otherwise. If the join-decomposition *is finite, then this root system is also finitely generated, and if* $x \hat{\sqcap} y \le x \check{\sqcap} y$*, then the offshoots are the minimal elements of the*

$$T = \{\bigcup_{j\in J} \boldsymbol{o}_k^j \in \mathcal{F}^L(\underline{n}) : b_j \le a_k\}.$$

A necessary and sufficient solvability condition therefore is that $\mathbf{g}$ *is a solution.*

Proof. (1): Let $O = \{\bigcup_{j\in J} \boldsymbol{o}_k^j \in \mathcal{F}^L(\underline{n}) : b_j \le a_k\}$, and assume that the solution set is non-empty. Then, by Prop. 8.57(1) and (2), it is given by $[\emptyset, \mathbf{g}] \cap \bigcup_{\boldsymbol{o}_k \in O}[\boldsymbol{o}_k, \mathbf{1}_n]$, which can be written as

$$\bigcup_{\boldsymbol{o}_k \in O}[\boldsymbol{o}_k, \mathbf{g}] \text{ and also as } \bigcup_{\boldsymbol{o}_k \in O'}[\boldsymbol{o}_k, \mathbf{g}],$$

with $O' = \{\boldsymbol{o}_k : \boldsymbol{o}_k \in O \wedge \boldsymbol{o}_k \subseteq \mathbf{g}\}$. If $\sqcap$ is maximally surjective, then $O = O'$, since $a_k \hat{\sqcap} b \le a_k \check{\sqcap} b$ for $a_k \ge b$ by Cor. 8.56.

(2): Let O denote the minimal elements of T, and assume that the solution set is non-empty. Then, by Prop. 8.57(1) and (3), it is given by

$$[\emptyset, \mathbf{g}] \cap \bigcup_{\boldsymbol{o} \in O}[\boldsymbol{o}, \mathbf{1}_n] = \bigcup_{\boldsymbol{o} \in O'}[\boldsymbol{o}, \mathbf{g}]$$

with $O' = \{\boldsymbol{o} : \boldsymbol{o} \in O \wedge \boldsymbol{o} \subseteq \mathbf{g}\}$. Let $\boldsymbol{o} = \bigcup_{j\in J} \boldsymbol{o}_k^j \in O$, then $\boldsymbol{o} \subseteq \mathbf{g}$ iff $\forall j \in J[\boldsymbol{o}_k^j \subseteq \mathbf{g}]$. The elements of O' are the minimal elements of S. If $\hat{\sqcap} \le \check{\sqcap}$, then $\boldsymbol{o}_k^j \subseteq \mathbf{g}$, since $b_j \le b$ implies $a_k \hat{\sqcap} b_j \le a_k \hat{\sqcap} b \le a_k \check{\sqcap} b$ by the monotoncity of $\sqcap$. □

(c) System of equations:

Let a system $SE_m = \{\mathrm{EQ}_j : j \in \underline{m}\}$ of m equations be given as

$$\mathrm{EQ}_j : \quad \bigvee a_{ji} \sqcap x_i = b_j, \ j \in \underline{m}.$$

For the solution set of SE_m we have

Proposition 8.59. ([10, Section 6.4.4]) *Let L be a complete Brouwerian lattice,* $\sqcap$ *be a t-norm on L,* $\mathbf{b} \in \mathcal{F}^L(\underline{m})$*, and* $\mathbf{a}_j \in \mathcal{F}^L(\underline{n})$*, such that each* b_j *is either a join-irreducible element of L or has a finite join-decomposition. If the mappings* $\sqcap_a$ *are homomorphisms, and the solution set of* SE_m *is non-empty then it is a root system of* $(\mathcal{F}^L(\underline{n}), \subseteq)$*. Moreover, if the index set J is finite, then this root system is finitely generated.*

Suppose the solution set of SE_m *is non-empty, and that the* EQ_j *have solution sets with stem* G_j *and offshoots* O_j*; then the solution set of* SE_m *is the root system with stem and offshoots given by*

$$G = \bigcap_{j\in\underline{m}} G_j \text{ and the minimal elements of } \{\bigcup_{j\in\underline{m}} N_j \in O_j : N_j \subseteq G\}.$$

A necessary and sufficient solvability condition therefore is that G is the greatest solution.

8.6 Applications of Fuzzy Relations

We include a section on applications of fuzzy relations here instead of Part V. Thus we keep things on many-valued concepts together in one place, and readers not interested in these aspects can skip Part III as a whole.

At the begin of Section 8.4 above we have asserted that large parts of human knowledge are qualitative and cast in natural languages expressions and sentences. As a consequence, there can be no doubt that human reasoning proceeds in such qualitative terms using suitable mechanisms. Formal models within FST for describing that kind of information processing run under the heading of **approximate reasoning**. Following the historical development we will outline its main concepts and their usage by presenting two classical concepts of FST: the generalized modus ponens conceived in [49], and fuzzy rule based systems initiated by the idea of fuzzy control as presented in [26].

8.6.1 *Inference Rules*

An *argument* is a sequence of statements (or propositions), of which all but the final one are called *premises*, and the final proposition is called the *conclusion*. An argument is called *valid* if the truth of all its premises implies that also the conclusion is true.

A **rule of inference** is a template for building valid arguments and is usually given in one of the two forms shown below. The rule is called **valid** with respect to the semantics of (classical) logic if true premises yield a conclusion that is also true (under a certain interpretation). In other words such a rule states that whenever in the course of some logical derivation the given premises have been obtained, the specified conclusion can be taken for granted as well. That is what we are usually interested in when searching for a proof. For some more details about the intricate relationship between the truth of statements and the validity of arguments see the section on 'Validity and Truth' in [7, p. 27ff].

Premise p_1
Premise p_2
...
Premise p_n

Conclusion q $\quad\therefore$

In a more compact notation this can be written as

$$p_1, p_2, \ldots, p_n \vdash q.$$

So an inference rule is valid if and only if $p_1 \wedge p_2 \wedge \ldots \wedge p_n \to q$ is a tautology.

Remember that in Chapter 1 that we have already presented some detailed considerations about a certain class of inference rules, viz. the so-called categorical syllogisms. Now we will focus on **hypothetical syllogisms** which contain one hypothetical proposition (or conditional) as premise (mixed), or only conditionals (pure):

- **modus ponens**: $p, p \to q \vdash q$, (MP)
- **modus tollens**: $p \to q, \neg q \vdash \neg p$,
- Pure hypothetical syllogism: $p \to q, q \to r \vdash p \to r$.

Within the area of approximate reasoning and its application in fuzzy control, decision-making etc., generalized versions of modus ponens are of utmost significance. The original extension of inference from two-valued logic to fuzzy logic was given by Zadeh's generalized modus ponens (GMP). In [49], another seminal work of Zadeh, he introduced the idea of formulating control algorithms by logical rules of qualitative nature. Such rules have the structure "*If A then B*", where A and B denote fuzzy sets over corresponding universes, whence they are also called fuzzy conditionals.

As an example consider *linguistic variables* (i.e variables whose possible values are defined in linguistic terms) like TEMPERATURE or SPEED with possible values say low, medium, and high, for sake of simplicity. An elementary piece of information is e.g. "TEMPERATURE *is* high", and typical fuzzy conditionals are "*If* SPEED *is* low *then* TEMPERATURE *is* high'. These are the elementary building blocks out of which fuzzy rule bases and fuzzy inference rules are built. If we have suitable mechanisms that can process qualitative information of this kind, then we can use the knowledge and experience of human experts in solving certain problems, instead of working with e.g. complex systems of differential equations.

Let us now try to find a formulation of the two-valued form of MP that is suitable for generalization to the fuzzy case by following the proven standard strategy of identifying sets with their characteristic functions (see [23, Section 8.6]): Let x and y be two variables that take their values from sets U and V, respectively, and assume that for all values $u \in U, v \in V$ the variables are related by a function $y = f(x)$. If the value of x is in a given set $A \subseteq U$, then one can infer that $f(x) \in \{v \in V : v = f(x), x \in A\}$.

We generalize this by assuming that the variables are related by an arbitrary relation $\rho \subseteq U \times V$, and let furthermore $A \subseteq U$ and $B \subseteq V$. Then, given $x = u$ we can infer that $v \in u\rho$, and given $u \in A$ we have $v \in B = \bigcup_{u\in A} u\rho = \{v \in V : u \in A \wedge u\rho v\}$. Thus, in terms of characteristic functions we get, for all $v \in V$,

$$B(v) = \bigvee_{u\in A} \big(A(u) \wedge \rho_{\rightarrow}(u, v)\big), \quad i.e. \quad B = A \circ \rho_{\rightarrow}, \tag{8.12}$$

where $\rho_{\rightarrow}$ models the IF-THEN-rule, and A models the fact (obtained by a measurement or observation).

Generalizing the above further by letting A be a fuzzy set over U and ρ be a fuzzy relation over $U \times V$, (8.12) is called the **compositional rule of inference** (CRI), and can clearly be written as the relation equation $B = A \circ \rho_{\rightarrow}$. We have finally arrived at a formal model for a generalized version of MP which allows us to handle e.g. inferences like

Temperature is high.
If temperature is high then fan speed is medium.

Fan speed is medium. $\therefore$

The relation $\rho_{\rightarrow}$ which is implicit in fuzzy conditionals can be modeled by using a conjunctor $\mathcal{C}$ and an implicator $\mathcal{I}$ (see Def. 8.43) so that the general form of the CRI appears as

$$B(y) = \bigvee_{x\in U} \mathcal{C}\Big(A(x), \mathcal{I}\big(A(x), B(y)\big)\Big). \tag{8.13}$$

An entry $\rho_{\rightarrow}(u, v)$ is thought of the strength of the rule 'IF u THEN v', or the degree to which v is related to u.

8.6.2 *Fuzzy Rule Based Systems*

Among all the techniques developed using fuzzy sets, fuzzy IF-THEN rules are by far the most visible due to their wide range of successful applications in many disciplines such as control systems, decision making, pattern recognition, etc. Among these fuzzy control has been the most popular application of rule based approximate reasoning, since it has been used around the mid nineties as a marketing argument for e.g. washing machines, cameras, or cars. Fuzzy control[3] consists of a set of techniques for transforming such qualitative rules into precise control actions. To describe its main concepts and strategies and their properties is beyond the current scope, but we will give a short overview on how relations and relation equations contribute to the evaluation of rule bases in general. For a short introduction to fuzzy control see e.g [24],[15, Chapter 1], or [23, Section 12.2], and for more details [13,27] or [46].

Fuzzy IF-THEN statements can be regarded in two ways: as *fuzzy mapping rules* or as *fuzzy implication rules.* A fuzzy mapping rule describes a functional relationship between inputs and an output using linguistic terms. Each rule approximates a small segment of the function, and the entire function is approximated by a set of rules. Such a collection of fuzzy mapping rules is sometimes called a *fuzzy rule-based model,* and is basic to the so-called Mamdani-approach to fuzzy control to which we now turn.

From the example above it can be seen that GMP is very near to classical MP, since its formal structure is $A, A \rightarrow B \vdash B$, i.e. it still assumes an exact formal matching of the fact with the antecedent of the implication. In the spirit of fuzzy set theory (compare the quotations on page 241), more can be achieved than only this mere symbolic matching: Fuzzy rules allow to draw a conclusion for a given input that was possibly never encountered before. Using the following running example in approximate reasoning literature,

This tomato is red.
If a tomato is red then it is ripe.

This tomato is ripe. $\therefore$

what can be said of a tomato that is slightly red? Put more formally, we want to make inferences of the form

[3]Quite generally spoken, **control theory** is concerned with influencing systems in order to achieve that certain output quantities take a desired course. The nature of the system does not matter, only its dynamic behavior is of interest. In classical control theory, which is mostly concerned with technical systems, the system that will be influenced is called the *plant.* But systems could as well be social, chemical or biological, as, for instance, the system of a national economy with the rate of inflation as output quantity. ([27])

x is A'.
If x is A then z is C.

---

z is C'. $\therefore$

So the formal structure of GMP thus becomes $A', A \to C \vdash C'$, where A and A' are fuzzy sets over the same universe describing "*similar*" meanings.

Now we will take a short look on the role and structure of fuzzy rule bases in fuzzy control. In fuzzy control terminology, a rule is a fuzzy conditional in which the antecedent is a restriction in its application domain and the consequent is some control action for the system under control. In cases when rules need a more complex structure than the one we have considered until now, one can use

$$\text{If } x_1 \text{ is } A_1 \text{ and ... and } x_m \text{ is } A_m \text{ then } y \text{ is } C\,.$$

for the description of a multidimensional antecedent. If we take the Cartesian product $A = A_1 \times \cdots \times A_m$, which gives $A(x_1, \ldots, x_m) = \bigwedge_{i \in \underline{m}} A_i(x_i)$ by (8.2), then we have again the simple form of a fuzzy conditional that we have used above in (8.13).

It has been said above that the rule base of the most common approach to fuzzy controllers (Mamdani & Assilian [26]) is regarded as collection of fuzzy mapping rules, i.e. its IF-THEN statements do not represent a causal relationship nor an implication in the usual sense. They convey the idea that for the rule's antecedent, the consequent is related to it in a "reasonable" way such that the rule contributes to the piecewise definition of a function (see [27]):

$$f(x_1, x_2, \ldots, x_n) := \begin{cases} C_1 & \text{if } x_1 \text{ is } A_{11} \text{ and...and } x_m \text{ is } A_{1m}, \\ \vdots & \\ C_n & \text{if } x_1 \text{ is } A_{n1} \text{ and...and } x_m \text{ is } A_{nm}. \end{cases}$$

Each elementary piece of expert knowledge, i.e. each fuzzy conditional, contributes to the overall conclusion to the amount its antecedent matches with a given fact or observation.

Henceforth, in order to achieve a clear representation of the main concepts, we will restrict the subsequent considerations to simple rules. Thus, a rule base $\mathcal{R} = \{R_1, \ldots, R_n\}$ of a Mamdani controller with rules of the form R_k: 'If x is A_k then y is C_k' comprises information about n of possible value pairs for a functional relationship:

$$A_1 \text{ and } C_1 \text{ OR } \cdots \text{ OR } A_n \text{ and } C_n. \tag{8.14}$$

This means that we have to use a disjunctor for aggregating the single conclusions of all rules in $\mathcal{R}$, and how much the conclusion from each rule does contribute to the overall result must be influenced by how much its antecedent matches with the fact.

Given the observation "x is A'" and the rule R_i: "If x is A_i then y is C_i" we first calculate the *compatibility* (or *matching*) *degree* between antecedent A_i and A' as

$$\gamma_i^{A'} = \bigvee\nolimits_{x \in U} \left(A'(x) \wedge A_i(x) \right) \tag{8.15}$$

and subsequently apply it to the fuzzy set C_i describing the rule's consequent to get the output fuzzy set of the rule as

$$C_i'(y) = \gamma_i^{A'} \wedge C_i(y). \tag{8.16}$$

This means clipping or cutting off C_i. Note that using product instead of minimum would result in some kind of scaling C_i down. $\gamma_i^{A'}$ is also called *activation* or *firing degree* of a rule. (8.16) is called the *CM approach* (for **C**ompatibility or **M**atching). According to (8.14) the overall result is given by

$$C'(y) = \bigvee_{i \in \underline{n}} C'_i(y) \tag{8.17}$$

Recalling that in the Mamdani view a rule "If x is A then y is C" has been rephrased as "A and C", i.e. $\rho_\rightarrow(a,c) = a \wedge c$, the following result holds:

Proposition 8.60. ([23, Section 11.2]) *For every fuzzy input A', the CM approach (8.17) is equivalent to the CRI approach (8.12).*

Proof. Let 'If x is A then y is C' be a fuzzy conditional, and let A' represent an observed value. Using (8.16) and (8.17), with the CM approach the membership degree of an element y in C' is calculated as

$$\begin{aligned}
C'(y) &= \textstyle\bigvee_{i \in \underline{n}} C'_i(y) \\
&= \textstyle\bigvee_{i \in \underline{n}} \left(\bigvee_{x \in U} \left[A'(x) \wedge A_i(x) \right] \wedge C_i(y) \right) \\
&= \textstyle\bigvee_{i \in \underline{n}} \bigvee_{x \in U} \left[A'(x) \wedge A_i(x) \wedge C_i(y) \right] \\
&= \textstyle\bigvee_{x \in U} \bigvee_{i \in \underline{n}} \left[A'(x) \wedge \left(A_i(x) \wedge C_i(y) \right) \right] \\
&= \textstyle\bigvee_{x \in U} \left[A'(x) \wedge \bigvee_{i \in \underline{n}} \left(A_i(x) \wedge C_i(y) \right) \right] \\
&= \textstyle\bigvee_{x \in U} \left[A'(x) \wedge \rho_\rightarrow(x,y) \right] \\
&= (A' \circ \rho_\rightarrow)(y).
\end{aligned}$$

□

We have now demonstrated how, in the context of the Mamdani interpretation of fuzzy rules, fuzzy relations arise in a natural way. But given a rule base $\mathcal{R} = \{R_1, \ldots, R_n\}$, what we really like to know is a fuzzy relation ρ which is a solution for the system of fuzzy relation equations

$$A_i \circ \rho = C_i, \quad i \in \underline{n}. \tag{8.18}$$

In other words, we are interested in a function that is in agreement with the n data points (A_i, C_i) given by $\mathcal{R}$. A relation ρ with the property (8.18) is also called a *correct model* or *interpolation* of $\mathcal{R}$. In general, such a correct model may not exist; so it is interesting to know necessary and sufficient conditions for when it does.

Let $\rho \in \mathcal{F}^L_{X\times Y}$, $\sigma \in \mathcal{F}^L_{Y\times Z}$, and $\tau \in \mathcal{F}^L_{X\times Z}$ denote three L-fuzzy relations. From Th. 8.46(1) we know that $S_\sigma := \mathscr{S}(\rho,\tau) \neq \emptyset$ iff $\hat{\sigma} := \rho\breve{}@\tau \in S_\sigma$, and that $\hat{\sigma}$ then is the greatest solution of $\rho\sigma = \tau$. Now, by Def. 8.44 we have

$$\rho@\sigma(x,z) = \textstyle\bigwedge_y \left(x\rho y \mathbin{\curvearrowright} y\sigma z \right).$$

Adapting this to the rule base configuration we get: $A_i \in \mathcal{F}^L(U)$, $\rho \in \mathcal{F}^L_{U\times V}$, and $C_i \in \mathcal{F}^L(V)$. Furthermore, the greatest solution $\hat{\rho}_i \in \mathcal{F}^L_{U\times V}$ of $A_i \circ \rho = C_i$ is $A_i\breve{}@C_i = A_i@C_i$, i.e. $\hat{\rho}_i(u,v) = A_i(u) \mathbin{\curvearrowright} C_i(v)$.

Proposition 8.61. *A system of fuzzy equations (8.18) has a solution $\hat{\rho}$ iff*

$$A_i \circ \hat{\rho} = C_i,\ i \in \underline{n} \tag{8.19}$$

with $\hat{\rho} = \bigcap_{i\in\underline{n}}(A_i@C_i)$. In case a solution exists, $\hat{\rho}$ is the greatest one.

Proof. Obviously, if (8.19) holds true, then the system (8.18) has $\hat{\rho}$ as a solution. Conversely, assume that ρ' is a solution of (8.18); then $\rho' \subseteq A_i@C_i$ for each $i \in \underline{n}$, whence $\rho' \subseteq \hat{\rho}$. Using monotonicity and subdistributivity of relation composition we get for each $j \in \underline{n}$

$$\begin{aligned} C_j &= A_j \circ \rho' \\ &\subseteq A_j \circ \hat{\rho} \\ &= A_j \circ \textstyle\bigcap_{i\in\underline{n}}(A_i@C_i) \\ &\subseteq \textstyle\bigcap_{i\in\underline{n}}\big(A_j \circ (A_i@C_i)\big) \\ &= C_j \cap \textstyle\bigcap_{j\neq i\in\underline{n}}\big(A_j \circ (A_i@C_i)\big) \\ &\subseteq C_j. \end{aligned}$$

Hence, $A_j \circ \hat{\rho} = C_j$. Thus, given any solution ρ' of the system (8.18), (8.19) holds and $\rho' \subseteq \hat{\rho}$, and $\hat{\rho}(u,v) = \bigwedge_{i\in\underline{n}}\big(A_i(u) \curvearrowright C_i(v)\big)$. □

The proposition above gives us a result about the solvability of certain systems of fuzzy relation equations in connection to the Mamdani approach to fuzzy control in terms of sharp/exact equality; but, as we have noted already earlier in Section 8.3.2 on properties of fuzzy relations, this is not fully in the spirit of fuzzy set theory. Accordingly, there have also been investigations about approximate solutions in case exact ones do not exist; see e.g. [19, 33]. There are also other necessary and sufficient conditions that avoid the check via $\hat{\rho}$ ([34]), among them ones that allow to get a decision directly in terms of the input-output fuzzy sets only ([35]).

References

[1] Bartl, E., Bělohlàvek, R.: Hardness of Solving Relational Equations, IEEE Transactions on Fuzzy Systems **23**, 2435-2438 (2015).

[2] Bartl, E., Klir G.J.: Fuzzy relational equations in general framework, Int. J. of General Systems **43**, 1-18 (2014).

[3] Bělohlàvek, R.: *Fuzzy Relational Systems-Foundations and Principles,* Springer Science +Business Media, New York (2002).

[4] Bělohlàvek, R., Funioková, T.: Similarity and Fuzzy Tolerance Spaces, J. Logic Computat. **14**, 827-855 (2004).

[5] Birkhoff, G.: *Lattice Theory,* 3rd ed., American Mathematical Society, Providence, Rhode Island (1973).

[6] Ćirić, M., Ignjatović, J., & Bogdanović, S.: Fuzzy equivalence relations and their equivalence classes, Fuzzy Sets and Systems **158**, 1295-1313 (2007).

[7] Copi, I.M., Cohen, C., McMahon, K.: *Introduction to Logic,* 14th ed., Pearson Education Limited, Harlow, Essex (2014).

[8] De Baets, B.: An Order-Theoretic Approach to Solving Sup-T Equations. In Ruan,D. (Ed.), *Fuzzy Set Theory and Advanced Mathematical Applications*, Springer Science+Business Media, New York, 67-87 (1995).

[9] De Baets, B.: Sup-T Equations: State of the Art. In Kaynak,O., Zadeh, L. et al.(Eds), *Computational Intelligence: Soft Computing and Fuzzy-Neural Integration with Applications*, Springer Verlag, Berlin Heidelberg, 80-93 (1998).

[10] De Baets, B.: Analytical solution methods for fuzzy relational equations. In Dubois, D., Prade, H.(Eds.), *Fundamentals of fuzzy sets*, Springer Science+Business Media, New York, 291-340 (2000).

[11] D'eer, L., Verbiest, N., Cornelis, C., Godo, L.: A comprehensive study of implicator-conjunctor-based and noise-tolerant fuzzy rough sets: Definitions, properties and robustness analysis, Fuzzy Sets and Systems **275**, 1-38 (2015).

[12] Di Nola, A.: On solving relational equations in Brouwerian lattices, Fuzzy Sets and Systems **34**, 365-376 (1990).

[13] Driankov, D., Hellendoorn, H., Reinfrank, M.:*An Introduction to Fuzzy Control*, 2nd ed., Springer-Verlag Berlin Heidelberg (1996).

[14] Dubois, D., Prade, H.: *Fuzzy Sets and Systems:Theory and Applications*, Academic Press, New York, (1980).

[15] Feng, G.: *Analysis and Synthesis of Fuzzy Control Systems*, CRC Press, Boca Raton (2010).

[16] Gerla, G.: Approximate Similarities and Poincaré Paradox, Notre Dame J. of Formal Logic **49**, 203-226 (2008).

[17] Goguen, J.: L-fuzzy sets, J. Math. Anal. Appl. **18**, 145-174 (1967).

[18] Gottwald, S.: *A Treatise on Many-Valued Logics*, Research Studies Press Ltd., Baldock, England (2001).

[19] Gottwald, S.: Fuzzy control and fuzzy relation equations. A unified view as interpolation problem. In: *NAFIPS 2004*, Proc. Ann. Meeting North Amer. Fuzzy Inform. Proc. Soc.; IEEE Comp. Soc., Piscataway/NJ, 270-275 (2004).

[20] Hume, D.: *An Enquiry Concerning Human Understanding*, Hackett Publ Co., Cambridge, MA 1993 (1772).

[21] Klawonn, F.: Fuzzy points, fuzzy relations and fuzzy functions. In: V. Novák and I. Perfilieva (Eds.), *Discovering World with Fuzzy Logic*, Physica-Verlag, Heidelberg, 431-453 (2000).

[22] Klement, E.P., Mesiar, R., Pap, E.: *Triangular Norms*, Kluwer Academic Publishers, Dordrecht (2000).

[23] Klir, G., Yuan, B.: *Fuzzy Sets and fuzzy Logic: Theory and Applications*, Prentice Hall, New Jersey, (1995).

[24] Lee, C.C.: Fuzzy logic in control systems: fuzzy logic controller. I, IEEE Transactions on Systems, Man and Cybernetics **20**, 404-418 (1990).

[25] Li, P., Fang, S.C.: A survey on fuzzy relational equations, part I: classification and solvability, Fuzzy Optimization and Decision Making **8**, 179-229 (2009).

[26] Mamdani, E.H., Assilian; S.: An experiment in linguistic synthesis with a fuzzy logic controller, Int. J. of Man-Machine Studies **7**, 1-13 (1975).

[27] Michels, K., Klawonn, F., Kruse, R., Nürnberger, A.: *Fuzzy Control-Fundamentals, Stability and Design of Fuzzy Controllers*, Springer-Verlag Berlin Heidelberg (2006).

[28] Montes, S., Jiměnez, J., Gil, P.: ϵ-Partitions and α-Equivalences, Mathware & Soft Computing **5**, 201-211 (1998).

[29] Mordeson, J.N., Mair, P.S.: *Fuzzy Graphs and Fuzzy Hypergraphs*, Physica Verlag, Heidelberg

(2000).

[30] Nguyen, H.T., Walker, E.A.: *A First Course in Fuzzy Logic,* 3rd ed., Taylor & Francis, Boca Raton (2005).

[31] Ovchinnikov, S.V.: Structure of fuzzy binary relations, Fuzzy Sets and Systems **6**, 169-195 (1981).

[32] Peeva, K., Kyosev, Y.: *Fuzzy Relational Calculus. Theory, Applications and Software,* World Scientific, Singapore (2004).

[33] Perfilieva, I.: Fuzzy function as an approximate solution to a system of fuzzy relation equations, Fuzzy Sets and Systems **147**, 363-383 (2004).

[34] Perfilieva, I., Novák, V.: System of fuzzy relation equations as a continuous model of IF-THEN rules. Information Sciences **177**, 3218-3227 (2007).

[35] Perfilieva, I., Tonis, A.: Compatibility of systems of fuzzy relation equations. Int. J. of General Systems **29**, 511-528 (2000).

[36] Poincaré, H.: La Science et l'Hypothèse (1902).

[37] Recasens, J.: *Indistinguishability Operators-Modelling Fuzzy Equalities and Fuzzy Equivalence Relations,* Springer-Verlag Berlin Heidelberg (2010).

[38] Rosenfeld, A.: Fuzzy Graphs, in: Zadeh, L.A., Fu, K.S., Tanaka, K. & Shimura, M.(eds), *Fuzzy Sets and their Application to Cognitive and Decision Processes,* Academic Press, New York, 77-95 (1975).

[39] Sanchez, E.: Resolution of composite fuzzy relation equations, Information and Control **30**, 38-48 (1976).

[40] Sanchez, E.: Solutions in composite fuzzy relation equations: Applications to medical diagnosis in Brouwerian logic, in: Gupta, M.M., Saridis, M.M. & Gaines, B.R. (Eds.), *Fuzzy Automata and Decision Processes,* Elsevier/North-Holland, New York, 221-234 (1977).

[41] Šešelja, B., Tepavčević, A.: On a construction of codes by P-fuzzy sets, Rev. Res. Fac. Sci. Univ. Novi Sad **20**, 71-80 (1990).

[42] Toth, H.: Fuzziness: From epistemic considerations to terminological clarification, Int. J. of Uncertainty, Fuzziness and Knowledge-Based Systems **5**, 481-503 (1997).

[43] Vollmer, G.: On supposed circularities in an empirically oriented epistemology. In Radnitzky, G. and Bartley, W.W. III (Eds.), *Evolutionary Epistemology, Theory of Rationality, and the Sociology of Knowledge,* Open Court, La Salle, Illinois, 163-200 (1987).

[44] Wang, X.: Method of solution to fuzzy relation equations in a complete Brouwerian lattice, Fuzzy Sets and Systems **120**, 409-414 (2001).

[45] Wang, X., Xiong, Q.: The solution set of a fuzzy relational equation with sup-conjunctor composition in a complete lattice, Fuzzy Sets and Systems **153**, 249-260 (2005).

[46] Wang, L.-X.: *A Course in Fuzzy Systems and Control,* Prentice Hall, Upper Saddle River, New Jersey (1996).

[47] Zadeh, L.A.: Fuzzy sets, Information and Control **8**, 338-353 (1965).

[48] Zadeh, L.A.: Similarity relations and fuzzy orderings, Information Sciences **3**, 177-200 (1971).

[49] Zadeh, L.A.: Outline of a new approach to the analysis of complex systems and decision processes, IEEE Transactions on Systems, Man and Cybernetics **3**, 28-44 (1973).

[50] Zhao, C.K.: On matrix equations in a class of complete and completely distributive lattice, Fuzzy Sets and Systems **22**, 303-320 (1987).

PART 4

Relations - Abstract

Chapter 9

Semigroups and Binary Relations

Its title "*Semigroups and Binary Relations*" does already suggest that this chapter is the paradigmatic crossing point of the meta and object level in the context of binary relations: On the meta level, relations are used in the study of the structure and properties of semigroups, the Green's relations and the natural partial orders being important examples. And on the object level, semigroup concepts are used to investigate the structure and properties of $(\mathcal{B}_X, \cdot)$.

9.1 Some Basic Concepts of Semigroup Theory

In Chapter 3 we have already learned that $(\mathcal{B}_S, \cdot, I_S)$ is a monoid; see Prop. 3.18. So semigroups provide a suitable context for describing collections of binary relations closed with respect to composition, since not every such collection also has a unit. A nearby question then is how concepts of semigroup theory carry over to semigroups of binary relations. Therefore, as a first step, we have to enhance our knowledge about semigroups.

The concept of a semigroup is a very simple one which is determined by only two requirements: A set S becomes a groupoid $(S, \cdot)$ if $\cdot$ is a binary operation on S, usually called multiplication, and a semigroup if, in addition, $\cdot$ is associative. Despite, or just because of, their mathematical simplicity, the concepts of semigroups and relations yielded a vast amount of correlated concepts and results, some of which will be considered below.

9.1.1 *Various Kinds of Semigroups*

Before going on with semigroups have a look back on what has been said about commutative and idempotent semigroups in Def. 2.49.

Definition 9.1. *Let S be a semigroup.*

(a) An element $a \in S$ is called **regular** *if $\exists x \in S[axa = a]$. If every element of S is regular, then S is called a* **regular semigroup**.

(b) *Two elements a and b of S are said to be* **inverse elements** *of each other if* $aba = a$ *and* $bab = b$. *If, for every* $a \in S$, *there is a unique inverse* $b \in S$, *denoted by* a^{-1}, *then S is called an* **inverse semigroup** .

(c) *An algebra* $(A, \cdot, ^*)$ *of type* $\langle 2, 1\rangle$ *such that* $(A, \cdot)$ *is a semigroup and such that* $\forall a, b \in A[(a^*)^* = a \wedge (a \cdot b)^* = b^* \cdot a^*]$, *is called* **involution semigroup**.

(d) *An element* $z \in S$ *is called a* **left zero** *of S if* $\forall a \in S[za = z]$ *and a* **right zero** *of S if* $\forall a \in S[az = z]$. *An element of* $z \in S$ *is called a* **zero element** *of S, denoted by* 0, *if it is both a left and a right zero element, i.e.* $\forall a \in S[0a = a0 = 0]$. *A semigroup (with zero) in which the product of any two elements is zero is called* **zero semigroup**.

(Note that in the terminology of Def. 6.41, the x occurring in item (a) has been called a *generalized inverse*, and b of item (b) a *pseudoinverse* of a.)

Example: Unlike a group, a semigroup S can have a zero element which is then unique. Consider the following set $B \subseteq \boldsymbol{B}_2$:

$$B := \left\{ \begin{pmatrix} 0 & 0 \\ 0 & 0 \end{pmatrix}, \begin{pmatrix} 1 & 0 \\ 0 & 0 \end{pmatrix}, \begin{pmatrix} 0 & 0 \\ 0 & 1 \end{pmatrix} \right\}. \tag{9.1}$$

B forms a zero semigroup under Boolean matrix multiplication (see Def. 3.19).

By defining a product in a set S by $\forall x, y \in S[xy = x]$, S becomes a semigroup, in which each element $x \in S$ is a right identity and at the same time a left zero. Moreover, $E_S = S$. Note that, contrary to a semigroup, a group has exactly one idempotent: its identity.

Proposition 9.2. *Let S be a semigroup.*

(1) *S is regular, i.e.* $\exists x \in S[axa = a]$, *iff every element a has at least one inverse b, in the sense that* $aba = a$ *and* $bab = b$.

(2) *If* $aba = a$ *then* $e = ab$ *and* $f = ba$ *are idempotent, and* $ea = a = af$.

(3) *If* e, f, ef *and* fe *are idempotent elements of S, then ef and fe are inverses of each other.*

(4) *An idempotent e is an inverse of itself.*

Proof. (1): If S is regular then $a(xax)a = axa(xa) = axa = a$ and $(xax)a(xax) = x(axa)(xax) = xa(xax) = x(axa)x = xax$; so xax is an inverse for a. Conversely, for every $a \in S$, there is a b with $aba = a$; so S is regular.

(2): We have $e^2 = (ab)(ab) = (aba)b = ab = e$.

(3): $(ef)(fe)(ef) = ef^2e^2f = (ef)^2 = ef$, and by symmetry $(fe)(ef)(fe) = fe$.

(4): Take $a = b = e$ in Def. 9.1(b). □

Theorem 9.3. ([3, Theorem 1.17]), [13, Theorem 3]) *The idempotents of a regular semigroup S commute if, and only if, every element of S has a unique inverse.*

Proof. Let S be a regular semigroup in which the idempotents commute, and let u and v be inverses of x. Then we have $u = uxu = u(xvx)u = (ux)(vx)u$, where ux, vx are idempotents by Prop. 9.2(2). Since idempotents commute, from the last equation we get

$$u = (vx)(ux)u = v(xux)u = vxu = (vxv)xu = v(xv)(xu),$$

where xv and xu are again idempotents. Thus $u = v(xu)(xv) = v(xux)v = vxv = v$, whence the inverse of x is unique.

Conversely, we now assume that S be a regular semigroup in which every element has a unique inverse. Let $e, f \in E_S$, and let a denote the unique inverse of ef, i.e.

$$(ef)a(ef) = ef \quad \text{and} \quad a(ef)a = a.$$

Let $b = ae$. Then

$$(ef)b(ef) = efae^2f = efaef = ef, \quad \text{and} \tag{9.2}$$

$$b(ef)b = ae^2fae = aefae = ae = b, \tag{9.3}$$

so b is also an inverse of ef. But by uniqueness we have $a = b = ae$; and similarly we get $fa = a$. Hence $a^2 = (ae)(fa) = a(ef)a = a$. By Prop. 9.2(4), an idempotent is a inverse of itself, whence, again using uniqueness, if follows that $a = ef$. Thus we have shown that $e, f \in E_S$ implies $ef, fe \in E_S$, and by Prop. 9.2(3) they are inverses of each other. Therefore, ef and fe are both inverses of ef, and, by uniqueness, $ef = fe$. □

9.1.2 *Ideals and Green's Relations*

We start by noting that the definition of the groupoid product can easily be extended to subsets: Let $(S,\cdot)$ denote a groupoid, and let $A, B \subseteq S$. Then we define the product of A and B as $AB := \{ab \in S : a \in A, b \in B\} \subseteq S$. Singletons in products of subsets will be written without braces, e.g. aB instead of $\{a\}B$. Thus $AB = \bigcup\{aB : a \in A\} = \{\bigcup Ab : b \in B\}$.

Every semigroup can easily be extended to a monoid: Let 1 be a symbol not representing any element of S. Then we extend the binary operation from S to $S \cup \{1\}$ as follows: $11 = 1$ and $1a = a1 = a$ for every $a \in S$. This passage from S to $S \cup \{1\}$ is called the *adjunction of an identity element to S*. By S^1 we mean S if it is a semigroup with identity, and $S \cup \{1\}$ otherwise.

Definition 9.4. *Let $(S,\cdot)$ be a groupoid.*

(a) *A* **left ideal** *of S is a subset $\emptyset \subset L \subseteq S$ such that $SL \subseteq L$, and a* **right ideal** *of S is a subset $\emptyset \subset R \subseteq S$ such that $RS \subseteq R$. $\emptyset \subset I \subseteq S$ is a* **two-sided ideal** *(or simply* **ideal**) *if it is simultaneously a left and a right ideal.*

(b) *For a subset $\emptyset \subset A \subseteq S$ the intersection of all left ideals of S containing A is a left ideal of S containing A and contained in every other such left ideal of S. It is called the* **left ideal of** S **generated by A**. *A left ideal is called* **principal left ideal** *if it is generated by a single element. Analogous definitions describe the* **principal right** *and* **principal two-sided** *ideals.*

(c) *S is called a* **simple groupoid** *if S itself is the only ideal.*

Without proofs we list some simple facts about ideals:

- In commutative groupoids the concepts of left ideal, right ideal and two-sided ideal coincide.
- $R \subseteq S$ is a right ideal if for each $r \in R$ and $s \in S$, $rs \in R$.
- The left ideal of a semigroup S generated by A is $A \cup SA = S^1 A$, and the (two-sided) ideal of S generated by A is $A \cup SA \cup AS \cup SAS = S^1 AS^1$.
- Since $AA \subseteq AS \subseteq A$ and $AA \subseteq SA \subseteq A$, ideals are always subsemigroups.

Green's relations, introduced in 1951 in [6], are essential for analyzing the structure of semigroups. For a semigroup S four preorder relations are defined as follows:

$$\begin{aligned}
s \leq_{\mathcal{L}} t \quad &\text{iff} \quad s = ut \text{ for some } u \in S^1,\\
s \leq_{\mathcal{R}} t \quad &\text{iff} \quad s = tu \text{ for some } u \in S^1,\\
s \leq_{\mathcal{J}} t \quad &\text{iff} \quad s = utv \text{ for some } u, v \in S^1,\\
s \leq_{\mathcal{H}} t \quad &\text{iff} \quad s \leq_{\mathcal{R}} t \text{ and } s \leq_{\mathcal{L}} t.
\end{aligned}$$

These definitions can be reformulated in terms of ideals as follows (e.g. $s = ut$ for some $u \in S^1$ iff $s \in S^1 t$ iff $\{s\} \subseteq S^1 t$ iff $S^1 s \subseteq S^1 S^1 t \subseteq S^1 t$):

$$\begin{aligned}
s \leq_{\mathcal{L}} t \quad &\Longleftrightarrow \quad S^1 s \subseteq S^1 t\\
s \leq_{\mathcal{R}} t \quad &\Longleftrightarrow \quad sS^1 \subseteq tS^1\\
s \leq_{\mathcal{J}} t \quad &\Longleftrightarrow \quad S^1 sS^1 \subseteq S^1 tS^1\\
s \leq_{\mathcal{H}} t \quad &\Longleftrightarrow \quad sS^1 \subseteq tS^1 \text{ and } S^1 s \subseteq S^1 t.
\end{aligned} \tag{9.4}$$

The equivalence relations associated with the four preorders are called the **Green's relations**. They allow us to study the "large scale" structure of a semigroup via its principal ideals, and are denoted by $\mathcal{R}, \mathcal{L}, \mathcal{J}$, and $\mathcal{H}$, respectively. We thus have

$$\begin{aligned}
s\mathcal{L}t &\Longleftrightarrow S^1 s = S^1 t,\\
s\mathcal{R}t &\Longleftrightarrow sS^1 = tS^1,\\
s\mathcal{J}t &\Longleftrightarrow S^1 sS^1 = S^1 tS^1,\\
s\mathcal{H}t &\Longleftrightarrow s\mathcal{R}t \wedge s\mathcal{L}t,\\
s\mathcal{D}t &\Longleftrightarrow \exists r \in S^1\,[s\mathcal{L}r \wedge r\mathcal{R}t].
\end{aligned}$$

Hence, two elements $s, t \in S$ are e.g. $\mathcal{R}$-equivalent if they generate the same principal right ideal. Clearly, $\mathcal{H} = \mathcal{L} \cap \mathcal{R}$, and $\mathcal{D} = \mathcal{L} \circ \mathcal{R}$.

Proposition 9.5. *For the relations $\mathcal{L}$, $\mathcal{R}$, and $\mathcal{J}$ on a semigroup the following holds:*

(1) $s\mathcal{L}t$ *iff* $\exists x, y \in S^1[xs = t \wedge yt = s]$;
(2) $s\mathcal{R}t$ *iff* $\exists x, y \in S^1[sx = t \wedge ty = s]$;
(3) $s\mathcal{J}t$ *iff* $\exists x, y, u, v \in S^1[xsu = t \wedge ytv = s]$.

Proof. (1): $s\mathcal{L}t$ implies $S^1 s = S^1 t$. Since $t \in S^1 t$, we get $t \in S^1 s$ and so there exists $p \in S^1$ such that $ps = t$. Similarly, there is a $q \in S^1$ with $qt = s$. Conversely, assume that there exist $p, q \in S^1$ with $ps = t$ and $qt = s$. Then $S^1 s = S^1 qt \subseteq S^1 t$ and similarly $S^1 t = S^1 ps \subseteq S^1 s$. Hence $S^1 s = S^1 t$. □

The following definition is a specialization of the more general Def. 10.4 to semigroups:

Definition 9.6. *Let ρ be a binary relation on a semigroup S. Then ρ is said to be*

(a) **left compatible** *if $\forall a,b,c \in S[a\rho b \rightarrow ca\,\rho\, cb]$;*
(b) **right compatible** *if $\forall a,b,c \in S[a\rho b \rightarrow ac\,\rho\, bc]$;*
(c) **compatible** *if $\forall a,b,c,d \in S[a\rho b \wedge c\rho d \rightarrow ac\,\rho\, bd]$;*
(d) A left (right) compatible equivalence relation on S is called **left congruence** *(***right congruence***), and a compatible equivalence relation a* **congruence**.

Proposition 9.7. *A relation ρ on S is a congruence if and only if it is both a left and a right congruence.*

Proof. Assume that ρ is both a left and a right congruence. Let $a,b,c,d \in S$ such that $a\rho b$ and $c\rho d$. Since ρ is a right congruence, we get $ac\rho bc$; and since it is also a left congruence, $bc\rho bd$. Thus, from the transitivity of ρ, $ac\,\rho\, bd$.

Conversely, let ρ be a congruence, and $a,b \in S$ such that $a\rho b$. For each $c \in S$ we have $c\rho c$, and ρ being a congruence implies $ca\,\rho\, cb$ and $ac\,\rho\, bc$. □

Corollary 9.8. *$\mathcal{L}$ is a right, and $\mathcal{R}$ a left congruence.*

Proof. $a\mathcal{L}b \rightarrow S^1a = S^1b \rightarrow S^1ac = S^1bc \rightarrow ac\mathcal{L}bc$. Similarly for $\mathcal{R}$. □

Proposition 9.9. *The relations $\mathcal{L}$ and $\mathcal{R}$ commute: $\mathcal{L}\circ\mathcal{R} = \mathcal{R}\circ\mathcal{L}$.*

Proof. Assume $x\,\mathcal{L}\circ\mathcal{R}\,y$. Then $\exists z \in S[x\mathcal{L}z \wedge z\mathcal{R}y]$; therefore we have $l,l',r,r' \in S$ such that $x = lz$, $z = l'x$, $z = yr$, and $y = zr'$. With $t := lzr'$ we get $t = lz\cdot r' = xr'$ and $x = lz = lyr = lzr'r = tr$, i.e. $x\mathcal{R}t$. On the other hand, $t = l\cdot zr' = ly$ and $y = zr' = l'xr' = l'lzr' = l't$, i.e. $y\mathcal{L}t$. Thus we have shown $x\,\mathcal{R}\circ\mathcal{L}\,y$, whence $\mathcal{L}\circ\mathcal{R} \subseteq \mathcal{R}\circ\mathcal{L}$. The inclusion in the other direction can be proved in a similar way. □

Proposition 9.10. *Let S be a semigroup. $\mathcal{D}$ is the smallest equivalence relation on S that contains both $\mathcal{L}$ and $\mathcal{R}$.*

Proof. $\mathcal{D}$ is an equivalence relation by Prop. 9.9 and Prop. 4.18(3). Moreover, by Prop. 3.53(2) we have $\mathcal{L}\cup\mathcal{R} \subseteq \mathcal{D}$. Let γ be another equivalence with $\mathcal{L}\cup\mathcal{R} \subseteq \gamma$; then $\mathcal{L} \subseteq \gamma$ and $\mathcal{R} \subseteq \gamma$, whence $\mathcal{D} = \mathcal{L}\circ\mathcal{R} \subseteq \gamma\circ\gamma = \gamma$ by Prop. 4.18(2). □

By the definition of the Green's relations and Prop. 9.10 we thus get the following relationships:

$$\mathcal{H} \subseteq \mathcal{L}\cap\mathcal{R} \subseteq \mathcal{L}\cup\mathcal{R} \subseteq \mathcal{D} \subseteq \mathcal{J}. \tag{9.5}$$

For the last inclusion note that $s\mathcal{D}t$ implies that there exists $u \in S$ with $s\mathcal{R}u\mathcal{L}t$. Then $sS^1 = uS^1$ and $S^1u = S^1t$, whence $S^1sS^1 = S^1tS^1$. And by Prop. 4.19(iv), $\mathcal{D} = \mathcal{L}\vee\mathcal{R}$.

In general, the inclusion $\mathcal{D} \subseteq \mathcal{J}$ may be strict. In certain classes of semigroups, however, equality holds: For groups G we have

$$\mathcal{H} = \mathcal{L} = \mathcal{R} = \mathcal{D} = \mathcal{J} = G \times G,$$

and in an Abelian semigroup

$$\mathcal{H} = \mathcal{L} = \mathcal{R} = \mathcal{D} = \mathcal{J}.$$

The next proposition implies that $\mathcal{D} = \mathcal{J}$ for finite semigroups.

Proposition 9.11. *If S is a periodic semigroup then $\mathcal{D} = \mathcal{J}$.*

Proof. Since $\mathcal{D} \subseteq \mathcal{J}$ is valid for all semigroups, it suffices to show that $\mathcal{J} \subseteq \mathcal{D}$. Assume $a\mathcal{J}b$; then there exist $u, v, r, s \in S^1$ such that

$$ras = b \quad \text{and} \quad ubv = a. \tag{9.6}$$

To prove the desired result we need to find an element $c \in S$ such that $a\mathcal{L}c\mathcal{R}b$. From the equations (9.6) we get, for all $i \geq 0$, by mutual repeated replacement of a and b that

$$a = (ur)^i a(sv)^i \quad \text{and} \quad b = (ru)^i b(vs)^i. \tag{9.7}$$

Since S is periodic, by Prop. 5.13 there is a $m \geq 0$ such that $(ur)^m \in E_S$. Taking $c = ra$ we get

$$a = (ur)^m a(sv)^m = (ur)^m (ur)^m a(sv)^m = (ur)^m a = (ur)^{m-1} uc,$$

whence $a\mathcal{L}c$.

In addition, $cs = ras = b$; then, choosing k such that $(vs)^k \in E_S$, by (9.7) we get

$$\begin{aligned} c &= ra = r(ur)^{k+1} a(sv)^{k+1} = r(ur)^k urasv(sv)^k = (ru)^{k+1} ras(vs)^k v \\ &= (ru)^{k+1} b(vs)^k (vs)^k v = (ru)^{k+1} b(vs)^{k+1} (vs)^{k-1} v \\ &= b(vs)^{k-1} v. \end{aligned}$$

Thus $c\mathcal{R}b$ as required, whence $a\mathcal{D}b$. This proves $\mathcal{J} \subseteq \mathcal{D}$. □

It is common practice to denote the equivalence classes of an element $a \in S$ with respect to the relations $\mathcal{L}$, $\mathcal{R}$, $\mathcal{J}$, $\mathcal{H}$, and $\mathcal{D}$ by L_a, R_a, I_a, H_a, and D_a, respectively. So, in the notation of this book, $R_a = a\mathcal{R} = [a]_{\mathcal{R}}$ etc. From these, the $\mathcal{D}$-classes play a very prominent role in semigroup theory; so we take a closer look.

9.1.2.1 *The Structure of $\mathcal{D}$-Classes*

From their definitions we have $\mathcal{L} \subseteq \mathcal{D}$ and $\mathcal{R} \subseteq \mathcal{D}$. Thus each $\mathcal{D}$-class of a semigroup S is a union of $\mathcal{L}$-classes and also a union of $\mathcal{R}$-classes; see (9.5), Def. 2.9 and Prop. 4.20(1). Moreover, as is shown in Prop. 9.12 below, every $\mathcal{L}$-class and every $\mathcal{R}$-class belonging to the same $\mathcal{D}$-class have a nonempty intersection.

Proposition 9.12. *An $\mathcal{L}$-class and an $\mathcal{R}$-class intersect if and only if they are contained in the same $\mathcal{D}$-class: $a\mathcal{D}b \iff R_a \cap L_b \neq \emptyset \iff L_a \cap R_b \neq \emptyset$.*

Proof. Let R and L be an $\mathcal{R}$- and $\mathcal{L}$-class respectively, such that $R \cap L \neq \emptyset$. Then, for any $a \in R \cap L$, $R = R_a$ and $L = L_a$, whence R and L are both contained in D_a. Conversely, if R and L are in the same $\mathcal{D}$-class of S, there is a $c \in S$ such that $a\mathcal{R}c$ and $c\mathcal{L}b$, i.e. $c \in R \cap L$. □

$\mathcal{D}$-classes can thus be conveniently visualized by what is called an *egg-box picture* in Clifford and Preston [3, p. 48]. Each row of an egg-box represents an $\mathcal{R}$-class, each column an $\mathcal{L}$-class, and each cell represents an $\mathcal{H}$-class.

Proposition 9.13. (Green's Lemma)

(1) *[$\mathcal{R}$-version] Let S be a semigroup, and $a, b \in S$ such that $a\mathcal{R}b$, and let $s, t \in S^1$ such that $as = b$ and $bt = a$. Then the right translations $\rho_s|_{L_a}$ and $\rho_t|_{L_b}$ on S determined by $s, t \in S^1$ are mutually inverse $\mathcal{R}$-class preserving bijections from L_a onto L_b and from L_b onto L_a, respectively.*

(2) *[$\mathcal{L}$-version] Let S be a semigroup, and $a, b \in S$ such that $a\mathcal{L}b$, and let $s, t \in S^1$ such that $sa = b$ and $tb = a$. Then the left translations $\lambda_s|_{R_a}$ and $\lambda_t|_{R_b}$ on S determined by $s, t \in S^1$ are mutually inverse $\mathcal{L}$-class preserving bijections from R_a onto R_b and from R_b onto R_a, respectively.*

Proof. (1): The right translation $\rho_s : S \longrightarrow S$ thus maps a to b. Moreover, it maps L_a into L_b, for $x \in L_a$ implies $xs\mathcal{L}as$, since $\mathcal{L}$ is a right congruence by Cor. 9.8. Hence, $xs \in L_{as} = L_b$. Similarly we get that ρ_t maps L_b into L_a, and considering the composition $\rho_s\rho_t : L_a \longrightarrow L_a$ one has $x\rho_s\rho_t = uast = ubt = ua = x$ for any $x = ua \in L_a$. Thus $\rho_s\rho_t$ is the identity map of L_a, and similarly $\rho_t\rho_s$ that of L_b. So $\rho_s|_{L_a}$ and $\rho_t|_{L_b}$ are mutually inverse bijections from R_a onto R_b and from R_b onto R_a, respectively.

In addition, if $x \in L_a$ then for $y = x\rho_s \in L_b$ we have $y = xs$ and $x = yt$, i.e. $y\mathcal{R}x$. Therefore, ρ_s is $\mathcal{R}$-class preserving. It bijectively maps each $\mathcal{H}$-class in L_a onto the corresponding $\mathcal{R}$-equivalent $\mathcal{H}$-class in L_b. Analogous results hold for ρ_t. □

$\mathcal{D}$-classes have a number of remarkable properties that help in understanding the structure and the resulting properties of a semigroup. Our presentation follows the notes of Cain [2, Chapter 3] and Pin [20, Chapter 5]. But first we have to mention

Proposition 9.14. *Idempotents of a semigroup S are left (right) identities for their $\mathcal{R}$ ($\mathcal{L}$)-classes.*

Proof. Assuming $e \in E_S$ and $a \in R_e$, there exists a $b \in S^1$ such that $eb = a$. Therefore, $ea = eeb = eb = a$; so e is a left identity for R_e. □

Proposition 9.15. *Let $a, b \in S$ with $a\mathcal{D}b$. Then $ab \in L_b \cap R_a$ if and only if $L_a \cap R_b$ contains an idempotent.*

Proof. From $ab \in L_b \cap R_a$ we get $ab\mathcal{R}x$. Hence there is a $r \in S^1$ such that $abr = a$. By Green's Lemma Prop. 9.13, $\rho_b|_{L_a} : L_a \longrightarrow L_{ab}$ and $\rho_r|_{L_{ab}} : L_{ab} \longrightarrow L_a$ are mutually inverse $\mathcal{R}$-class preserving bijections between L_a and L_{ab}. And since $ab\mathcal{L}b$ these maps are in fact mutually inverse $\mathcal{R}$-class preserving bijections between L_a and L_b. Hence $(br)^2 = brbr = b\rho_r|_{L_b}\rho_b|_{L_a}\rho_r|_{L_b} = b\rho_r|_{L_b} = br$; thus $br \in E_S$. Furthermore, $br = b\rho_r|_{L_b} \in L_a \cap R_b$.

Conversely, suppose that $L_a \cap R_b$ contains an $e \in E_S$. Then $eb = b$ by Prop. 9.14. And $e\mathcal{R}b$ implies that $\rho_b|_{L_a} : L_e \longrightarrow L_b$ is an $\mathcal{R}$-class preserving bijection by Prop. 9.13. So $ab \in R_a \cap L_b$. □

To give an example, we illustrate this proposition with an eggbox:

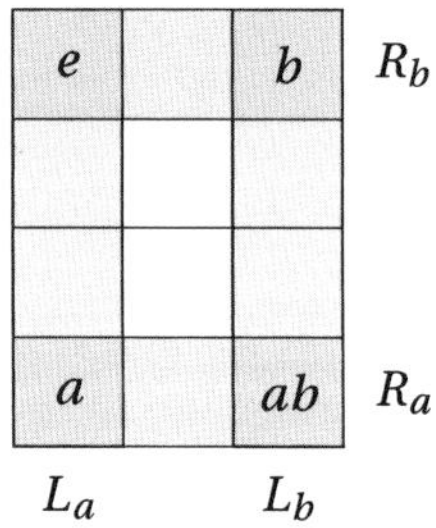

Proposition 9.16. *If $a \in S$ is regular, then each element of D_a is regular. (A $\mathcal{D}$-class is called* **regular** *if all its elements are regular.)*

Proof. For a regular element $a \in S$ there exists $b \in S$ such that $aba = a$. Assume $c\mathcal{L}a$; then there are $s, t \in S^1$ with $sc = a$ and $ta = c$. Hence $c = ta = taba = cbsc$, so c is regular. Thus, every element of L_a is regular. A dual argument shows the analogous property for R_b, which proves the assertion. □

Proposition 9.17.

(1) A regular $\mathcal{D}$-class contains an idempotent.
(2) If a $\mathcal{D}$-class D contains an idempotent, then also each of its $\mathcal{L}$- and $\mathcal{R}$-classes.
(3) In a regular $\mathcal{D}$-class, each $\mathcal{L}$-class and each $\mathcal{R}$-class contains an idempotent.

Proof. (1): Let $a \in S$ be regular, i.e. $aba = a$ for some $b \in S$. So we have $a(ba) = a \wedge ba = (ba)$, i.e. $(ba)\mathcal{L}a$ by Prop. 9.5(1), taking $x = a,\ s = ba,\ t = a$, and $y = b$. Since $a\mathcal{R}a$, we get $(ba)\mathcal{D}a$, i.e. $ba \in D_a$. But $(ba)^2 = b(aba) = ba$, so ba is an idempotent in D_a.

(2): Let $e, s \in D$ and e be idempotent. Then $e\mathcal{R}r \wedge r\mathcal{L}s$ for some $r \in D$. By Prop. 9.5(2) we have $rv = e$ and $eu = r$ for some $u, v \in S^1$, so we get $er = e(eu) = e^2u = eu = r$ from the latter. But $vrvr = v(rv)r = ver = vr$, whence vr is idempotent. Furthermore, $r(ur) = er = r$. Therefore, $r\mathcal{L}ur$ and $L_s = L_r = L_{ur}$, whence L_s contains an idempotent. The assertion for R_s can be proved by taking into account that, by definition of $\mathcal{D}$, $\mathcal{D} = \mathcal{L}\circ\mathcal{R} = \mathcal{R}\circ\mathcal{L}$ and Prop. 9.9.

(3): Obvious consequence from (1) and 2. □

Proposition 9.18. *Let S be a semigroup and $s \in S$. Then the following conditions are equivalent:*

(1) s is regular.
(2) R_s contains an idempotent.
(3) L_s contains an idempotent.
(4) D_s contains an idempotent.

Proof. If s is regular then D_s is a regular $\mathcal{D}$-class by Prop. 9.16. So we have (1) $\to$ (2) $\wedge$ (3) $\wedge$ (4), by Prop. 9.17. Next we show that (2) $\to$ (1): Let e be an idempotent such that $s\mathcal{R}e$. By Prop. 9.5(2) we have $ev = s$ and $su = e$ for some $u, v \in S^1$, so we get $es = e(ev) = e^2 v = ev = s$. We let $t = usu$; then $sts = susus = ees = es = s$, whence s is regular. Therefore, (1) $\leftrightarrow$ (2), and in an analogous way one shows (1) $\leftrightarrow$ (3).

To complete the proof we have to show that (4) implies (1). By Prop. 9.10 we have $\mathcal{L} \cup \mathcal{R} \subseteq \mathcal{D}$; so, if $s\mathcal{R}e$ or $s\mathcal{L}e$, we are done by (2) or (3). Assuming $e \in D_s - (L_s \cup R_s)$ we need a general proof: Let $s\mathcal{D}e$; then $\exists t \in S^1\, [s\mathcal{L}t \wedge t\mathcal{R}e]$. So by Prop. 9.5 we have $(us = t \wedge vt = s) \wedge (tx = e \wedge ey = t)$ for some $u, c, x, y \in S^1$. Moreover, $t \in R_e$, so by Prop. 9.14 we have $et = t$, so $et = ey$. We can thus conclude as follows: $s = vt = vey = vet = vtxt = sxt = sxus$, whence s is regular. □

As a standard result on inverse semigroups we note

Proposition 9.19. ([19, II.1.2 Theorem]) *For a semigroup S the following statements are equivalent:*

(1) S is an inverse semigroup.
(2) S is regular, and its idempotents commute.
(3) Each $\mathcal{L}$-class and each $\mathcal{R}$-class contains exactly one idempotent.

Proof. (1) $\Rightarrow$ (2): By Def. 9.1(b), for every $a \in S$, there is a unique inverse $b \in S$, such that $aba = a$ and $bab = b$; so S is regular. And by Th. 9.3 the idempotents of S commute.

(2) $\Rightarrow$ (3): From Prop. 9.18 we know that there is an idempotent in each $\mathcal{L}$-class and each $\mathcal{R}$-class. Let $e, f \in E_S$. If $e\mathcal{L}f$ then by Prop. 9.14 e, f are right identities, so we have $e = ef$ and $f = fe$, and since, by hypothesis, idempotents commute, $e = f$ follows. A similar argument holds for $\mathcal{R}$-classes.

(3) $\Rightarrow$ (1): Let x, y be inverses of some $a \in S$; then $xa, ya \in E_S$. Since $axa = aya = a$, we get $xa\mathcal{L}a\mathcal{L}ya$. So under the assumption about L_a we have $xa = ya$, and by a symmetric argument $ax = ay$. Therefore, $x = xax = yax = yay = y$. □

9.1.3 *Natural Partial Order*

The concept of a **natural partial order** on a semigroup has been developed in several steps, starting at approximately the same time as the development of the Green's relations in the mid-fifties ([23]). A partial order on a semigroup $(S, \cdot)$ is called *natural* if it is defined by means of the multiplication of S ([8, 16]).

For a semigroup S and $a \in S$, we set $V_{S,a} = V_a = \{b \in S : a$ and b form a pair of inverse elements$\}$. Remember that it is customary to use E_S to denote the set of idempotents of a semigroup S. The usual ordering of E_S is defined as follows ([3, p. 23]):

$$e \le f \quad \text{if and only if} \quad e = ef = fe. \tag{9.8}$$

(To convince ourselves that $\le$ is a partial order on E_S, let $e, f, g \in E_S$. Since $ee = e$ we have $e \le e$. $e \le f$ and $f \le e$ imply $ef = fe = e$ and $fe = ef = f$; so $e = f$. Finally, if $e \le f$

and $f \leq g$ then $ef = fe = e$ and $fg = gf = f$. Hence, $eg = (ef)g = e(fg) = ef = e$ and $ge = g(fe) = (gf)e = fe = e$, i.e. $e \leq g$.)

In 1952 VAGNER [24] introduced a natural partial order on inverse semigroups by

$$a \leq b \quad \text{iff} \quad a = eb, \quad \text{for some } e \in E_S. \tag{9.9}$$

In 1980 this concept was extended to the much larger class of regular semigroups (note that inverse semigroups are regular by Prop. 9.19) in HARTWIG [7] and NAMBOORIPAD [17]:

$$a \leq_N b \quad \text{iff} \quad a = eb = bf, \quad \text{for some } e, f \in E_S, \tag{9.10}$$

is a partial order on S which extends the usual ordering of E_S, but is not compatible with multiplication in general.

In [16] the natural order is extended to arbitrary semigroups by defining

$$a \leq_M b \quad \text{iff} \quad a = b \text{ or } a = xa = xb = by, \quad \text{for some } x, y \in S^1. \tag{9.11}$$

See [4, (1)], which adds the $a = b$ part to gain reflexivity of $\leq_M$ in the general case. For the orders defined in (9.8) and (9.9) (since $a = (aa^{-1})a$ and $aa^{-1} \in E_S$) reflexivity holds. Concerning (9.10) from Prop. 9.2(1) and (2) we get that a has at least one inverse b, i.e. $aba = a$ and $bab = b$. Then $f = ba$ and $e = ab$ are idempotent and $a = ea = af$, whence $a \leq_N a$.

Note that from (9.11) we get $ay = x \cdot by = x \cdot by = xa = a$. Clearly, $\leq_M$ coincides with the usual ordering (9.8) when restricted to E_S. That it equals $\leq_N$ on regular semigroups is shown in

Lemma 9.20. ([16, Lemma 1]) *For a regular semigroup S the following statements are equivalent:*

(1) $a \leq_N b$;
(2) $a \leq_M b$;
(3) $a = eb = bx$ *for some* $e \in E_S$, $x \in S$.

Proof. (1) $\Rightarrow$ (2): $a = eb = bf$ for some $e, f \in E_S$ clearly implies $a = xb = by$ for some $x, y \in S^1$, viz. $x = e$ and $y = f$. Furthermore, $ea = e(eb) = eb = a$.

(2) $\Rightarrow$ (3): For $a' \in V_a$ we have $a = aa'a = aa'(xb) = eb$, where $e = aa'x \in E_S$, since $(aa'x)^2 = (aa')(xa)(a'x) = (aa')a(a'x) = a(a'x)$. And by assumption, $a = by$ for some $y \in S$.

(3) $\Rightarrow$ (1): For $a' \in V_a$ we have $a = aa'a = (bx)a'a = bf$, for some $x \in S^1$, where $f = xa'a \in E_S$: From $a = eb$ we get $ea = e^2b = eb = a$, whence $f^2 = (xa'a)^2 = (xa')(eb)(xa'a) = (xa'e)(bx)a'a = (xa'e)a = (xa')a = f$. □

We still have to show that $\leq_M$ as defined in (9.11) is indeed a partial order:

Proposition 9.21. $a \leq_M b :\leftrightarrow a = b$ *or* $a = xa = xb = by$ *for some* $x, y \in S^1$ *is a partial order.*

Proof. Reflexivity: Obvious since $a = a$.
Antisymmetry: Assume $a \leq_M b$ and $b \leq_M a$; then we have $a = xa = xb$ and $b = xb = xa$, for some $x \in S^1$ whence $a = b$.
Transitivity: Assume $a \leq_M b$ and $b \leq_M c$; then, for some $x, y, w, z \in S^1$, we have $a = xa = xb = by$ and $b = wb = wc = cz$. So $a = x \cdot wc = cz \cdot y$ and $xw \cdot a = xw \cdot by = x \cdot wb \cdot y = x \cdot by = xa = a$. □

Note that defining some natural partial order $\leq$ on a semigroup S does not automatically mean that S is a partially ordered semigroup (see Def. 2.85), which requires that $\forall a, b, c \in S[a \leq b \rightarrow ac \leq bc \wedge ca \leq cb]$.

9.2 Algebraic Properties of Relations on Finite Sets

9.2.1 *The Semigroup of Binary Relations $\mathcal{B}_X$*

From the definitions and results given earlier in this book (e.g. Prop. 3.18) we do already know the following:

- $(\mathcal{B}_X, \cdot, \breve{}, I_X)$ is an involution monoid,
- $(\mathcal{B}_X, \cdot)$ is a semigroup with zero element $\emptyset$,
- $(\mathcal{B}_X, \cdot; \subseteq)$ is a partially ordered semigroup.

We are, of course, interested in how these and other concepts of semigroup theory manifest themselves in semigroups of binary relations. Before we take a closer look at algebraic properties of $\mathcal{B}_X$, we turn to the perhaps not so popular fact that $\mathcal{B}_X$ forms a Baer semigroup. In Chapter 2, page 62 we did already meet the left and right annihilators of a subset $A \subseteq S$ of a semigroup S with zero defined as

$$L(A) = \{x \in S : \forall a \in A[xa = 0]\} \text{ and } R(A) = \{x \in S : \forall a \in A[ax = 0]\}.$$

Definition 9.22. *A semigroup S with zero element 0 is called a* **Baer semigroup** *if for each $x \in S$ there exist idempotents $e, f \in S$ such that $R(x) = \{y \in S : xy = 0\} = eS$ and $L(x) = \{y \in S : yx = 0\} = Sf$.*

A semigroup with zero is thus a Baer semigroup if and only if the right annihilator of every $x \in S$ is an idempotent-generated principal right ideal, and analogous for the left annihilator. In particular we have $S = R(0) = L(0)$, whence there exist idempotents $e, f \in S$ with $S = eS = Sf$. Then e is a left, and f is a right identity; therefore $e = f$ by Prop. 2.43 and so S has an identity.

Proposition 9.23. ([1, p. 35, Ex. 2.26]) *$\mathcal{B}_X$ is a Baer semigroup with zero element $\emptyset$.*

Proof. Let $\sigma, \tau \in \mathcal{B}_X$ with $\sigma\tau = \emptyset$. $z \in \mathsf{rng}(\sigma)$ implies $z \notin \mathsf{dom}(\tau)$, i.e. $\mathsf{rng}(\sigma) \subseteq \overline{\mathsf{dom}(\tau)}$, or equivalently $\mathsf{rng}(\sigma) \cap \mathsf{dom}(\tau) = \emptyset$. Conversely, if $\mathsf{rng}(\sigma) \cap \mathsf{dom}(\tau) = \emptyset$ then clearly there are no $x, y \in X$ with $x\sigma\tau y$, whence

$$\sigma\tau = \emptyset \quad \text{iff} \quad \mathsf{rng}(\sigma) \cap \mathsf{dom}(\tau) = \emptyset \quad \text{iff} \quad \mathsf{rng}(\sigma) \subseteq \overline{\mathsf{dom}(\tau)}.$$

Let $S \subseteq X$ and $\tau \in \mathcal{B}_X$ such that $\tau = I_S\tau$; then, using Th. 3.17(9), $\mathsf{rng}(\tau) = \mathsf{rng}(I_S\tau) = \tau[\mathsf{rng}(I_S)] = \tau[S] = I_S\tau[S] \subseteq I_S[X] = S$, since $\tau[S] \subseteq X$. Otherwise $\mathsf{rng}(\tau) \subseteq S$ implies $x\tau y \to yI_Sy$, whence $\tau = \tau I_S$. Thus $\tau \in \mathcal{B}_X I_S$ iff $\mathsf{rng}(\tau) \subseteq S$.

Therefore, for every $\sigma \in \mathcal{B}_X$, for the left annihilator we have $L(\sigma) = \mathcal{B}_X I_S$, with $S = \overline{\mathsf{dom}(\sigma)}$. Since $\tau\sigma = \emptyset \leftrightarrow \sigma^\smile\tau^\smile = \emptyset$ a dual argument gives $R(\sigma) = I_T\mathcal{B}_X$, with $T = \overline{\mathsf{dom}(\sigma)^\smile} = \overline{\mathsf{rng}(\sigma)}$. Note that for each $S \subseteq X$, $I_S \in \mathcal{B}_X$ is coreflexive and thus idempotent by Prop. 3.53(5). Hence $\mathcal{B}_X$ is a Baer semigroup. □

9.2.1.1 *Subsemigroups of* $\mathcal{B}_X$

Remember that due to Def. 4.1 an element $\rho \in \mathcal{B}_X$ is called a **partial map** on X if for all $x \in \mathsf{dom}(\rho)$ we have $|x\rho| = 1$, that is, if

$$\forall x, y_1, y_2 \in X[(\langle x, y_1\rangle \in \rho \wedge \langle x, y_2\rangle \in \rho) \to y_1 = y_2] \tag{9.12}$$

Note that this condition vacuously holds for the empty relation $\emptyset$, which is therefore included into the set of the partial maps on X. The set of all partial maps of X is denoted by $\mathcal{P}_X$. For $\alpha \in \mathcal{P}_X$ it is a convenient and common habit to let $x\alpha$ denote the unique y such that $x\alpha y$. In semigroup theory, this concept runs under the name of a **partial transformation** on a set X, defined as a function $\alpha : A \longrightarrow B$, with $A, B \subseteq X$.

A partial transformation is called **transformation**, if $\mathsf{dom}(\alpha) = X$, and a **permutation** if it is a bijective transformation. The set of all transformations of X is called the **full transformation semigroup** on X, denoted by $\mathcal{T}_X$; and the permutations on a set X form the **symmetric group** $\mathcal{S}_X$. Much more on basic results on transformations than is included below can be found in [5]. The set $\mathcal{I}_X := \{\alpha \in \mathcal{P}_X : \alpha \text{ is injective}\}$ is called the **symmetric inverse semigroup**.

Since the nature of elements of X is not important for the current considerations, instead of X the set $\underline{n}$ can be used, where $n = |X|$. In a similar way as in Chapter 5 for Boolean matrices $\boldsymbol{B}_n$ we shall also write $\mathcal{P}_n$, $\mathcal{T}_n$, and $\mathcal{S}_n$. A common notation for partial transformations is in form of a two-line array: As an example, let $\epsilon_n : \underline{n} \longrightarrow \underline{n}$ be defined by $i\epsilon_n = i$, or $\epsilon_n(i) = i$, for all $i \in \underline{n}$. This is often written as

$$\epsilon_n = \begin{pmatrix} 1\,2 \cdots n \\ 1\,2 \cdots n \end{pmatrix}, \text{ generally: } \alpha = \begin{pmatrix} 1 & 2 & \cdots & n \\ 1\alpha & 2\alpha & & n\alpha \end{pmatrix}.$$

For denoting the composition of transformations one can apply either the relational or the functional notation, and we will use them at liberty. For composition we get $\beta(\alpha(x)) = x\alpha\beta$, whence right properties of functions will become left properties of relations, and vice versa.

Obviously, ϵ_n is the identity element in both $\mathcal{P}_n$ and $\mathcal{T}_n$ and corresponds to the relation $I_{\underline{n}}$. The partial transformation $\mathbf{0}_n : \emptyset \longrightarrow \underline{n}$ is the zero element of $\mathcal{P}_n$, since $\alpha\mathbf{0}_n = \mathbf{0}_n\alpha = \mathbf{0}_n$.

The importance of $\mathcal{T}_X$ in semigroup theory stems from the following embedding property (see e.g. [9, Theorem 1.1.2] or [14, Chapter I.3.9]):

Proposition 9.24. *If S is a semigroup and $X = S^1$ then there is an isomophism $\phi : S \longrightarrow \mathcal{T}_X$.*

Proof. For each $a \in S$, we define a map $\rho_a : S^1 \longrightarrow S^1$ by $x\rho_a = xa,\ x \in S^1$. Thus $\rho_a \in \mathcal{T}_X$, whence there is a map $\alpha : S \longrightarrow \mathcal{T}_X$ given by $a\alpha = \rho_a,\ a \in S$. This map is bijective since

$$\forall a, b \in S[a\alpha = b\alpha \rightarrow \rho_a = \rho_b \rightarrow \forall x \in S^1[xa = xb]] \rightarrow 1a = 1b \rightarrow a = b.$$

α is a homomorphism, since

$$\forall x \in S^1[x(\rho_a\rho_b) = (x\rho_a)\rho_b = (xa)b = x(ab) = x\rho_{ab},$$

hence $(a\alpha)(b\alpha) = (ab)\alpha$. □

We will now study some properties of $\mathcal{P}_n$, starting with

Proposition 9.25. ([9, Proposition 1.4.3]) *For $\alpha, \beta \in \mathcal{P}_n$ we have*

$$\forall x \in \mathsf{dom}(\alpha\beta)[x(\alpha\beta) = (x\alpha)\beta].$$

Proof. Note that $x\rho\sigma y$ if and only if there is a $z \in X$ such that $x\rho z$ and $z\sigma y$. Since ρ, σ and $\rho\sigma$ are all partial maps, we get $z = x\rho, y = z\sigma$, and $y = x(\rho\sigma)$. Thus $x(\rho\sigma) = y = z\sigma = (x\rho)\sigma$. □

$|\mathsf{rng}(\alpha)|$ is called the **rank** of the partial transformation $\alpha \in \mathcal{P}_n$, denoted as rank(α). Thus rank(α) is the number of different elements in the second line of its two-line array representation.

Proposition 9.26. ([5, Proposition 2.3.3]) *For $n > 1$, $\mathcal{T}_n$ does not have right zeros. $\alpha \in \mathcal{T}_n$ is a left zero if it is a constant transformation. $\mathcal{T}_n$ contains exactly n left zeros.*

Proof. Obviously, $0_j(\alpha(k)) = 0_j(k) = j$ for any $\alpha \in \mathcal{T}_n$ and all $k \in \underline{n}$. So $0_j\alpha = 0_j$, i.e. any constant transformation is a left zero in $\mathcal{T}_n$, whence $\mathcal{T}_n$ contains at least n different left zeros.

Note that constant transformations are the only ones with rank 1. Let $\alpha \in \mathcal{T}_n$ be of rank at least 2. Then, for any $j \in \underline{n}$ and all $k \in \underline{n}$, $\alpha(0_j(k)) = \alpha(j)$, i.e. the rank of $\alpha 0_j$ is 1. Hence the equality $\alpha 0_j = \alpha$ cannot hold, whence α is not a left zero of $\mathcal{T}_n$. Thus $\mathcal{T}_n$ contains exactly n left zeros. □

Theorem 9.27. ([5, Theorem 2.6.3]) *The semigroups $\mathcal{P}_n$, $\mathcal{T}_n$, and $\mathcal{I}_n$ are regular.*

(1) *$\mathcal{P}_n$ is a regular submonoid of $\mathcal{B}_X$, $|X| = n$.*
(2) *$\mathcal{T}_n$ is a regular submonoid of $\mathcal{P}_n$.*
(3) *$\mathcal{I}_n$ is a regular submonoid of $\mathcal{P}_n$.*
(4) *$\mathcal{S}_n$ is a subsemigroup of $\mathcal{T}_n$.*

Proof. (1): Let $\alpha, \beta \in \mathcal{P}_n$, and assume $x\alpha\beta y_1$ and $x\alpha\beta y_2$. Then

$$\exists z_1, z_2 \in X[x\alpha z_1 \wedge z_1\beta y_1 \wedge x\alpha z_2 \wedge z_2\beta y_2].$$

Condition (9.12) applied to α implies that $z_1 = z_2$, and then applied to β that $y_1 = y_2$. Thus, $\alpha\beta \in \mathcal{P}_n$. (Compare this with the relational proof in Cor. 4.3(1).) Furthermore, $I_n \in \mathcal{P}_n$, and composition is associative.

To see that $\mathcal{P}_n$ is regular, let $\alpha \in \mathcal{P}_n$. We then construct a $\beta \in \mathcal{P}_n$ as follows: for each $x \in \mathsf{rng}(\alpha)$ we take some $y \in X$ with $\alpha(y) = x$ and set $\beta(x) = y$, and for $x \notin \mathsf{rng}(\alpha)$ we define $\beta(x) = x_1$. Then, with $x = \alpha(z)$, i.e. $x \in \mathsf{rng}(\alpha)$, we get $\alpha(\beta(\alpha(z))) = \alpha(\beta(x)) = \alpha(z)$. Thus $\alpha\beta\alpha = \alpha$.

(2): By (1).

(3): Obviously, $\emptyset, I_n \in \mathcal{I}_n$. For $\alpha \in \mathcal{I}_n$ we define $\beta \in \mathcal{I}_n$ as follows: $\mathsf{dom}(\beta) := \mathsf{rng}(\alpha)$, and for each $x \in \mathsf{rng}(\alpha)$ we define β as in (1). Since α is a partial bijection, viz. from $\mathsf{dom}(\alpha)$ to $\mathsf{rng}(\alpha)$, β is the inverse bijection from $\mathsf{rng}(\alpha)$ to $\mathsf{dom}(\alpha)$, so $\beta \in \mathcal{I}_n$. As above in (1), we have $\alpha\beta\alpha = \alpha$.

(4): It is well known that the permutations of a set form a group. □

We note the following elementary result needed below and also later:

Lemma 9.28. *If $\alpha \in \mathcal{P}_X$ then $1_{\mathsf{dom}(\alpha)} \subseteq \alpha\alpha^{-1}$ and $\alpha^{-1}\alpha = 1_{\mathsf{rng}(\alpha)}$. Moreover, if $\alpha \in \mathcal{I}_X$ then $\alpha\alpha^{-1} = 1_{\mathsf{dom}(\alpha)}$*

Proof. If $x \in \mathsf{dom}(\alpha)$ and $x\alpha = y$ then $\langle y, x\rangle \in \alpha^{-1}$, whence $\langle x, x\rangle \in \alpha\alpha^{-1}$. In case $\alpha \in \mathcal{I}_X$, $y\alpha^{-1} = x$, so equality holds.

Let $\langle x, y\rangle \in \alpha^{-1}\alpha$; then $x\alpha^{-1} = z$ and $z\alpha y$ for some $z \in \mathsf{dom}(\alpha)$. So $z\alpha = x$ and therefore $x = y \in \mathsf{rng}(\alpha)$. Conversely, if $x = z\alpha \in \mathsf{rng}(\alpha)$ then $z\alpha x$ and $x\alpha^{-1}z$, hence $x\alpha^{-1}\alpha x$, i.e. $1_{\mathsf{rng}(\alpha)} \subseteq \alpha^{-1}\alpha$. □

Next we will show that $\mathcal{I}_n$ is indeed an inverse semigroup, which explains its name. This demonstrates that partial transformations (and thus binary relations) provide us with examples of various kinds of semigroups.

Proposition 9.29. [9, Theorem 5.1.5]) *$\mathcal{I}_n$ is an inverse semigroup.*

Proof. $\mathcal{I}_n$ is closed under composition by Prop. 2.24(2). Each $\alpha \in \mathcal{I}_n$ is a bijection from $\mathsf{dom}(\alpha)$ onto $\mathsf{rng}(\alpha)$, whence there is an inverse map $\alpha^{-1} \in \mathcal{I}_n$: simply let $y'\alpha^{-1} = y$ iff $y\alpha = y'$. Then $\alpha^{-1} = \alpha^{\smile}$, and

$$\mathsf{dom}(\alpha^{\smile}) = \mathsf{rng}(\alpha) \text{ and } \mathsf{rng}(\alpha^{\smile}) = \mathsf{dom}(\alpha).$$

Moreover, $\alpha\alpha^{\smile} = 1_{\mathsf{dom}(\alpha)}$ and $\alpha^{\smile}\alpha = 1_{\mathsf{rng}(\alpha)}$ by Lemma 9.28. So $\alpha\alpha^{\smile}\alpha = \alpha$ and $\alpha^{\smile}\alpha\alpha^{\smile} = \alpha^{\smile}$, and $\alpha^{\smile}$ is inverse to α in $\mathcal{I}_n$, whence $\mathcal{I}_n$ is regular.

We now show that idempotents in $\mathcal{I}_n$ commute and thus get that it is an inverse semigroup by Prop. 9.19. If α is idempotent then by Th. 3.17(10) and (11) we have

$$\mathsf{dom}(\alpha^2) = [\mathsf{rng}(\alpha) \cap \mathsf{dom}(\alpha)]\alpha^{\smile} = \mathsf{dom}(\alpha) = \mathsf{rng}(\alpha)\alpha^{\smile}.$$

Since α^{-1} is injective we have $\mathsf{rng}(\alpha) \cap \mathsf{dom}(\alpha) = \mathsf{rng}(\alpha)$, i.e. $\mathsf{rng}(\alpha) \subseteq \mathsf{dom}(\alpha)$. And similarly

$$\mathsf{rng}(\alpha^2) = [\mathsf{rng}(\alpha) \cap \mathsf{dom}(\alpha)]\alpha = \mathsf{rng}(\alpha) = \mathsf{dom}(\alpha)\alpha,$$

implying $\mathsf{dom}(\alpha) \subseteq \mathsf{rng}(\alpha)$: So $\mathsf{dom}(\alpha) = \mathsf{rng}(\alpha) =: A \subseteq X$, and $x\alpha^2 = x\alpha = x\,id_A\alpha$ for $x \in A$. Since α is injective, this implies $\alpha = id_A$ by Prop. 2.23(4). So α is a subidentity on X, and by Prop. 3.53(7) subidentities commute. □

9.2.2 The Green's Relations on $\mathcal{B}_X$

Let $\emptyset \neq S$ be a set and $\mathcal{A} \subseteq \mathbf{2}^S$ not containing $\emptyset$. A subcollection $\mathcal{C} \subseteq \mathcal{A}$ is called **independent** if no member of $\mathcal{C}$ is a union of other members of $\mathcal{C}$. $\mathcal{C}$ is said to *generate* $\mathcal{A}$ if each member of $\mathcal{A}$ is a union of members of $\mathcal{C}$. If $\mathcal{C} \subseteq \mathcal{A}$ is independent and generates $\mathcal{A}$ then it will be called a **basis** of $\mathcal{A}$.

In connection with matrices it is common to speak of rows and columns: Rows and columns of any $\beta \in \boldsymbol{B}_n$ are elements of the set $V_n := \{0,1\}^n$, i.e. *Boolean vectors* of length n, naturally corresponding to the aftersets $x\beta$ and foresets βy.

For $\alpha \in \mathcal{B}_X$ let $V(\alpha) := \{A\alpha : A \subseteq X\}$, called the **row space** of α. Then $V(\alpha) \subseteq V_n$ is the collection consisting of the zero vector $\mathbf{0}$ and all possible unions of (one or more) non-zero rows of α. Then the **row basis** of α, denoted by $rb(\alpha)$, is the set of all non-zero vectors in $V(\alpha)$ that are not union of other vectors in $V(\alpha)$; and $|rb(\alpha)|$ is called the **row rank** of α, denoted by $r(\alpha)$. Note that each vector in $rb(\alpha)$ must be a row of α, and that $V^*(\alpha) := \{x\alpha : x \in \alpha X\}$ is the basis of $V(\alpha)$. There are, of course, the dual notions of the **column space** $W(\alpha) := \{\alpha A : A \subseteq X\}$, the **column basis** $cb(\alpha)$, and the **column rank** of α, denoted by $c(\alpha)$. For every $\alpha \in \boldsymbol{B}_n$, $V(\alpha)$ and $W(\alpha)$ are complete lattices under inclusion of Boolean vectors.

Definition 9.30. *A binary relation $\alpha \in \mathcal{B}_X$ is said to be*

(a) **row reduced** *if its nonempty rows form an independent set;*
(b) **column reduced** *if its nonempty columns form an independent set;*
(c) a **reduced relation** *if it is both row and column reduced.*

So ρ is row reduced if none of its rows $x\rho$ can be written as a union of other rows in ρ, i.e. for all $x \in X$ and $A \subseteq X$, $\emptyset \neq x\alpha = A\alpha = \bigcup_{a\in A} a\rho$ implies $x \in A$. As an example take the relation $\rho := \begin{pmatrix} 1&0&1&0\\ 0&1&0&1\\ 1&1&0&0\\ 1&1&1&0 \end{pmatrix}$ on $X = \{x_1, x_2, x_3, x_4\}$. It is column reduced but not row reduced $(x_4\rho = x_1\rho \cup x_3\rho)$, and has column rank 4 and row rank 3.

An early classical result is the characterization of the Green's relations in terms of Boolean matrices, which can be used to investigate the ideal structure of $\mathcal{B}_X$ or to obtain tests for regularity of a binary relation; see e.g. [10, 11, 22]. Remember that according to Prop. 9.5 for $\mathcal{B}_X$ the following holds:

$$\begin{array}{lll} \alpha\mathcal{L}\beta & \text{iff} & \exists\sigma,\tau \in \mathcal{B}_X\,[\sigma\alpha = \beta \wedge \tau\beta = \alpha];\\ \alpha\mathcal{R}\beta & \text{iff} & \exists\sigma,\tau \in \mathcal{B}_X\,[\alpha\sigma = \beta \wedge \beta\tau = \alpha];\\ \alpha\mathcal{J}\beta & \text{iff} & \exists\sigma,\tau,\chi,\nu \in \mathcal{B}_X\,[\sigma\alpha\chi = \beta \wedge \tau\beta\nu = \alpha]. \end{array}$$

So we have $R_\alpha = [\alpha]_{\mathcal{R}} = \{\beta \in \mathcal{B}_X : \alpha\mathcal{R}\beta\} = \alpha\mathcal{R}$.

Lemma 9.31. *Let $\alpha, \beta \in \mathcal{B}_X$. Then*

(1) $\alpha\mathcal{L}\beta \leftrightarrow V(\alpha) = V(\beta) \leftrightarrow rb(\alpha) = rb(\beta)$;
(2) $\alpha\mathcal{R}\beta \leftrightarrow W(\alpha) = W(\beta) \leftrightarrow cb(\alpha) = cb(\beta)$;
(3) $\alpha\mathcal{D}\beta \leftrightarrow V(\alpha)$ and $V(\beta)$ are isomorphic lattices.

Obviously, $V(\alpha) = V(\beta)$ iff $rb(\alpha) = rb(\beta)$; thus, row and column spaces can be replaced by the row and column bases, respectively. It is, however, much more convenient to obtain a characterization of the relations $\mathcal{L}$ and $\mathcal{R}$ directly in terms of rows and columns as follows:

Lemma 9.32. ([22, LEMMA 1.3]) *Let* $\alpha, \beta \in \mathcal{B}_X$.

(1) $\alpha\mathcal{L}\beta \leftrightarrow \forall x \in X \exists H, K \subseteq X[x\alpha = H\beta \wedge x\beta = K\alpha]$;
(2) $\alpha\mathcal{R}\beta \leftrightarrow \forall x \in X \exists H, K \subseteq X[\alpha x = \beta H \wedge \beta x = \alpha K]$;
(3) $\alpha\mathcal{L}\beta \rightarrow X\alpha = X\beta$ *and* $\alpha\mathcal{R}\beta \rightarrow \alpha X = \beta X$.

Proof. (1) $\alpha\mathcal{L}\beta$ implies $\exists\gamma,\delta \in \mathcal{B}_X\ [\gamma\alpha = \beta \wedge \delta\beta = \alpha]$ by definition of $\mathcal{L}$. For $x \in X$ we have $x\alpha = x(\delta\beta) = (x\delta)\beta$ and $x\beta = x(\gamma\alpha) = (x\gamma)\alpha$, whence $x\alpha = H\beta$ and $x\beta = K\alpha$ for $H = x\delta$ and $K = x\gamma$, respectively.

Conversely, by assumption for $x \in X$ there are $H, K \subseteq X$ such that $x\alpha = H\beta \wedge x\beta = K\alpha$. Let $x\alpha = H_x\beta \in V(\alpha)$; then $H_x\beta = x\alpha \in V(\beta)$. For $A\alpha \in V(\alpha)$ we get $A\alpha = \bigcup_{x\in A} x\alpha$ by Prop. 3.9(5), whence $A\alpha = \bigcup_{x\in A} H_x\beta$. Again using Prop. 3.9(5) yields $\bigcup_{x\in A} H_x\beta = (\bigcup_{x\in A} H_x)\beta$, so $A\alpha \in V\beta$, i.e. $V(\alpha) \subseteq V(\beta)$. An analogous argument shows $V(\beta) \subseteq V(\alpha)$. Now apply Lemma 9.31(1).

(3) If $\alpha\mathcal{L}\beta$ and $x \in X$, then there exist $y \in X$ and $H \subseteq X$ such that $x \in y\alpha = H\beta \subseteq X\beta$. Thus $X\alpha \subseteq X\beta$. Similarly for $X\beta \subseteq X\alpha$. □

Recall that $V^*(\alpha) = \{x\alpha : x \in \alpha X\}$. Following [22] we are now going to characterize the Green's relations on reduced elements of $\mathcal{B}_X$.

Proposition 9.33. ([22, THEOREM 1.5]) *Let* $\alpha, \beta \in \mathcal{B}_X$ *be row reduced relations. Then the following statements are equivalent:*

(1) $\alpha\mathcal{L}\beta$;
(2) $V^*(\alpha) = V^*(\beta)$;
(3) $\alpha = \rho\beta$, *where* ρ *is a one-to-one mapping of* αX *onto* βX.

Proof.
(1) ⇒ (2): This follows from Lemma 9.31(1), since $V^*(\alpha)$ is the basis for $V(\alpha)$.

(2) ⇒ (3): Let $\rho : \alpha X \longrightarrow \beta X$ be defined by $x\rho = y$ iff $x\alpha = y\beta$. Then ρ is one-to-one from αX onto βX, and $\alpha = \rho\beta$.

(3) ⇒ (1): From (3) it follows that $\rho^{\smile}\alpha = \beta$. But then α and β generate the same principal left ideal, whence $\alpha\mathcal{L}\beta$. □

Since $\alpha\mathcal{R}\beta$ iff $\alpha^{\smile}\mathcal{L}\beta^{\smile}$, there is a dual assertion formed from the above proposition by replacing "row" by "column", "$\mathcal{L}$" by "$\mathcal{R}$", and "$\rho\beta$" by "$\beta\rho$", where ρ is from $X\beta$ onto $X\alpha$. As a corollary we then get

Corollary 9.34. ([22, COROLLARY 1.6]) *Let* $\alpha, \beta \in \mathcal{B}_X$ *be reduced relations. Then the following statements are equivalent:*

(1) $\alpha\mathcal{H}\beta$;
(2) $V^*(\alpha) = V^*(\beta)$ *and* $V^*(\alpha^{\smile}) = V^*(\beta^{\smile})$;
(3) $\alpha = \beta\rho = \sigma\beta$, *where* ρ *is an one-to-one mapping of* $X\beta$ *onto* $X\alpha$ *and* σ *is a one-to-one mapping of* αX *onto* βX.

Lemma 9.35. ([22, LEMMA 1.7]) *Let* $\alpha \in \mathcal{B}_X$ *be row (column) reduced; then every* $\beta \in R_\alpha$ *(*L_α*) is also row (column) reduced. In particular, if* $\alpha, \beta \in \mathcal{B}_X$ *are reduced and* $\alpha\mathcal{D}\beta$*, then every* $\gamma \in L_\alpha \cap R_\beta$ *is reduced.*

Proof. Let $\alpha \in \mathcal{B}_X$ be row reduced, i.e. for all $x \in X$ and $A \subseteq X$, $\emptyset \neq x\alpha = A\alpha$ implies $x \in A$. We prove $\beta \in R_\alpha$ to be row reduced by showing that $V^*(\beta) = \{x\beta : x \in \beta X\}$ is independent. Let $y \in \beta X$ and $\emptyset \neq B \subseteq \beta X$ such that $y\beta = B\beta$. By Lemma 9.32(3) we have $y \in \beta X = \alpha X$ and $B \subseteq \beta X = \alpha X$. Since $\beta \in R_\alpha$, $\alpha = \beta\gamma$ for some $\gamma \in \mathcal{B}_X$. Therefore, $y\alpha = y\beta\gamma = B\beta\gamma = B\alpha$. Since α has been assumed to be row reduced, $V^*(\alpha) = \{x\alpha : x \in \alpha X\}$ is independent, and hence $y \in B$. Therefore also $V^*(\beta)$ is independent, whence β is row reduced. □

Lemma 9.36. ([21, LEMMA 1.5]) *Let* $\alpha \in \mathcal{B}_X$*, and assume that* L_α *(*R_α*) contains a row (column) reduced relation. If* L_α *(*R_α*) contains an idempotent, then also a row (column) reduced idempotent.*

Proof. Assume that L_α contains an idempotent; then α is regular by Prop. 9.18. So D_α is a regular $\mathcal{D}$-class by Prop. 9.16, and thus every element in L_α is regular. Let β be a row reduced member of L_α. Since β is regular, we have $\beta\rho\beta = \beta$ for some $\rho \in \mathcal{B}_X$, whence $\rho\beta$ is idempotent. Moreover, $\beta\rho\beta = \beta$ and $(\rho\beta\rho)\beta = \rho\beta$, so $\rho\beta\mathcal{L}\beta$, and since $\beta\mathcal{L}\alpha$, we get $\rho\beta\mathcal{L}\alpha$, i.e. $\rho\beta \in L_\alpha$. β has been assumed row reduced, so there is a permutation π on X with $\beta\pi\beta = \beta$; then $\pi\beta \in L_\alpha$ is idempotent and row reduced. A similar argument holds for columns. □

Theorem 9.37. ([22, THEOREM 1.8]) *Let* $\alpha, \beta \in \mathcal{B}_X$ *be reduced relations; then* $\alpha\mathcal{D}\beta$ *iff* $\alpha = \sigma\beta\rho$*, with* σ, ρ *being one-to-one mappings of* αX *onto* βX *and* $X\beta$ *onto* $X\alpha$*, respectively.*

Proof. Assume $\alpha\mathcal{D}\beta$; then there exists a $\gamma \in \mathcal{B}_X$ such that $a\mathcal{L}\gamma$ and $\gamma\mathcal{R}\beta$, which is reduced by Lemma 9.35. By Lemma 9.32(3), $X\alpha = X\gamma$ and $\gamma X = \beta X$. From Prop. 9.33 and its dual we get $\alpha = \sigma\gamma$, with σ a one-to-one map of αX onto γX, and $\gamma = \beta\rho$, with ρ a one-to-one map of $X\beta$ onto $X\gamma$. But then $\alpha = \sigma\gamma = \sigma\beta\rho$.

Conversely, if $\alpha = \sigma\beta\rho$, with ρ as specified, we can set $\gamma = \beta\rho$. Then, by Prop. 9.33 and its dual, $\alpha\mathcal{L}\gamma$ and $\gamma\mathcal{R}\beta$, i.e. $\alpha\mathcal{D}\beta$. □

If X is finite the partial one-to-one mappings in Prop. 9.33 and Th. 9.37 can be taken to be restrictions of permuations of X to the respective domains. So our results and their duals can be reformulated as follows (with $\mathcal{S}_X$ denoting the symmetric group on X):

Corollary 9.38. ([22, THEOREM 1.9]) *Let* $\alpha, \beta \in \mathcal{B}_X$ *be reduced relations on a finite set* X*. Then*

(1) $\alpha\mathcal{L}\beta$ *iff* $\alpha = \rho\beta,\ \rho \in \mathcal{S}_X$
(2) $\alpha\mathcal{R}\beta$ *iff* $\alpha = \beta\sigma,\ \sigma \in \mathcal{S}_X$,
(3) $\alpha\mathcal{H}\beta$ *iff* $\alpha = \rho\beta = \beta\sigma,\ \rho,\sigma \in \mathcal{S}_X$,
(4) $\alpha\mathcal{D}\beta$ *iff* $\alpha = \rho\beta\sigma,\ \rho,\sigma \in \mathcal{S}_X$.

In Prop. 9.17 it has been shown that a regular $\mathcal{D}$-class of a semigroup contains an idempotent. For the semigroup $\mathcal{B}_X$ this can be strengthened to:

Lemma 9.39. ([22, LEMMA 2.3]) *Let D be a regular $\mathcal{D}$-class of $\mathcal{B}_X$ which contains reduced elements. Then D contains a reduced idempotent.*

Proof. Let $\alpha \in D$ be a reduced relation, then there is the row reduced $\alpha \in L_\alpha$, whence, by Lemma 9.35, every $\beta \in L_\alpha$ is column reduced. And by Lemma 9.36, if there is a row reduced relation ρ in L_α, then also a row reduced idempotent α_r. But, as just shown, α_r is also column reduced. So D contains a reduced idempotent. □

Remember that a partial order ρ on a set X is characterized by the following properties: $I_X \subseteq \rho, \rho^2 \subseteq \rho$, and $\rho \cap \rho^{\smile} \subseteq I_X$. The set of partial orders on X is non-empty, since $\emptyset$ and I_X are examples which have two properties in common: they are reduced and idempotent. That this is not accidental is shown in

Proposition 9.40. ([22, THEOREM 2.4]) *The class of reduced idempotents in $\mathcal{B}_X$ coincides with the class of partial order relations on subsets of X.*

Proof. Let α be a partial order on $Y \subseteq X$; then α is a preorder that is additionally antisymmetric, and therefore idempotent by Prop. 3.53(3). If $x\alpha = H\alpha$ for some $x \in Y$ and $H \subseteq Y$, then, by reflexivity, $x \in x\alpha = H\alpha$, whence $x \in h\alpha$ for some $h \in H$. But, again by reflexivity, $h \in H\alpha = x\alpha$. Thus $x\alpha h$ and $h\alpha x$, so $x\,\alpha \cap \alpha^{\smile}\,h$ and thus $x = h$ by the antisymmetry of α. Therefore $x \in H$, whence α is row reduced. Similarly we get that α is also column reduced.

Conversely, let α be a reduced idempotent. Then $x\alpha^2 = (x\alpha)\alpha = x\alpha$, and since α is reduced we get $x \in x\alpha$ for all $x \in \alpha X = \mathsf{dom}(\alpha)$. Hence $x \in \bigcup_{x\in X} x\alpha = X\alpha = \mathsf{rng}(\alpha)$. So $\alpha X \subseteq X\alpha$, and similarly $X\alpha \subseteq \alpha X$ follows from $\alpha(\alpha x) = \alpha x$. We let $Y = X\alpha = \alpha X$. From $x \in x\alpha$ for all $x \in Y$ it follows that $I_Y \subseteq \alpha$. And since idempotent relations clearly are transitive, it remains to show that α is antisymmetric. Assume $x\alpha y$ with $x \neq y$; then $x\alpha \neq y\alpha$ since α is assumed reduced. Thus there is a $z \in X$ such that either $x\alpha z \wedge \neg\, y\alpha z$ or $\neg\, x\alpha z \wedge y\alpha z$. If we also had $y\alpha x$, the first case would give $y\alpha x\alpha z$, i.e. $y\alpha z$, a contradiction to $\neg\, y\alpha z$. So we conclude that the second alternative must hold; but for it we get $x\alpha y\alpha z$, i.e. $x\alpha z$, a contradiction to $\neg\, x\alpha z$. Altogether we cannot keep the assumption $y\alpha x$, whence α must be antisymmetric. □

9.2.2.1 *The Green's Relations on Subsemigroups of $\mathcal{B}_X$*

Let S be a semigroup; we do already know that an element $a \in S$ is called regular if $a \in aSa$, and have already considered various properties of regular elements and semigroups in Section 6.2.2 and Section 9.1.1. Now we will look at regularity under the viewpoint of Green's relations.

We first have a look at the principal ideals in the submonoids $\mathcal{T}_X$, $\mathcal{P}_X$, and $\mathcal{I}_X$ of $\mathcal{B}_X$, for which we follow [3, Chapter 2.2] for $\mathcal{T}_X$, and [5] for the general case. We will use the notation of the former, i.e. we write $x\alpha$ instead of the functional notaion $\alpha(x)$ used in the latter, since this finally gives the common terminology for $\mathcal{L}$ and $\mathcal{R}$ in Th. 9.44. So, as already remarked above on page 282, left and right properties change from the results as stated in [5] to here.

Theorem 9.41. ([5, Theorem 4.2.1]) *Let S denote any of the semigroups $\mathcal{T}_X$, $\mathcal{P}_X$, and $\mathcal{I}_X$. Then for each $\alpha \in S$ the principal left ideal generated by α has the form $S\alpha = \{\beta \in S : \mathsf{rng}(\beta) \subseteq \mathsf{rng}(\alpha)\}$.*

Proof. Let $A := \{\beta \in S : \mathsf{rng}(\beta) \subseteq \mathsf{rng}(\alpha)\}$, and $\gamma \in S$. Then $x\gamma\alpha = (x\gamma)\alpha$ by definition, so $\mathsf{rng}(\gamma\alpha) \subseteq \mathsf{rng}(\alpha)$. Thus $S\alpha \subseteq A$.

Conversely, assume $\beta \in A$, thus $\mathsf{rng}(\beta) \subseteq \mathsf{rng}(\alpha)$. Then define γ as follows: $\mathsf{dom}(\gamma) = \mathsf{dom}(\beta)$ and for each $b \in \mathsf{rng}(\beta)$ let γ map all elements of $\beta b = \{a \in X : a\beta = b\}$ to some $c_b \in \alpha b$, then $a\beta = b$ iff $a\gamma c_b \alpha b$, i.e.$\gamma\alpha = \beta$. Thus $\beta \in S\alpha$, whence $A \subseteq S\alpha$.

(Note that in the case $S = \mathcal{I}_X$, from the injectivity of α and β we get

$$x_1 \neq x_2 \rightarrow y_1 = \beta(x_1) \neq \beta(x_2) = y_2 \rightarrow z_{y_1} \neq z_{y_2} \rightarrow \gamma(x_1) \neq \gamma(x_2),$$

for all $x_1, x_2 \in \mathsf{dom}(\gamma)$. Therefore $\gamma \in \mathcal{I}_X$.) □

Before describing the principal right ideals in the semigroups under consideration, we first associate a binary relation κ_α with each $\alpha \in \mathcal{P}_X$ as follows:

$$x\kappa_\alpha y :\Longleftrightarrow x\alpha = y\alpha \quad \text{or} \quad x, y \notin \mathsf{dom}(\alpha) = \alpha X.$$

Clearly, the equivalence classes of κ_α are $\overline{\mathsf{dom}(\alpha)}$ and all pre-images of the elements in $\mathsf{rng}(\alpha)$. Note that for $\alpha \in \mathcal{T}_X$ the right condition always evaluates to false, so κ_α becomes equal to $\alpha\alpha^{\smile}$.

Theorem 9.42. ([5, Theorem 4.2.4]) *Let S denote any of the semigroups $\mathcal{T}_X$, $\mathcal{P}_X$, and $\mathcal{I}_X$. Then for each $\alpha \in S$ the principal right ideal generated by α has the form*

$$\alpha S = \{\beta \in S : \mathsf{dom}(\beta) \subseteq \mathsf{dom}(\alpha) \wedge \kappa_\alpha \subseteq \kappa_\beta\}. \tag{9.13}$$

Proof. Let A denote the right hand side of (9.13), and let $\beta \in \alpha S$. Then, for $x \in \mathsf{dom}(\beta)$ and some $\gamma \in S$, we get $x\beta = x\alpha\gamma = (x\alpha)\gamma$, whence $x \in \mathsf{dom}(\alpha)$ and thus $\mathsf{dom}(\beta) \subseteq \mathsf{dom}(\alpha)$. Moreover, $x, y \in \mathsf{dom}(\beta)$ and $x\alpha = y\alpha$ imply $x\beta = (x\alpha)\gamma = (y\alpha)\gamma = y\beta$. Thus $\kappa_\alpha \subseteq \kappa_\beta$, so $\alpha S \subseteq A$.

Conversely, assume $\beta \in A$. We set $M = \mathsf{dom}(\beta)\,\alpha$ and define $\gamma : M \longrightarrow X$ by

$$(y\alpha)\gamma = y\beta. \tag{9.14}$$

(Note that this mapping is well defined: If $x = y_1\alpha = y_2\alpha$ then $y_1\kappa_\alpha y_2$ which implies $y_1\kappa_\beta y_2$. Thus $y_1\beta = y_2\beta$, whence $x\gamma$ is uniquely defined.)

If $S = \mathcal{T}_X$, we have $\mathsf{dom}(\beta) = X$ and $M = X\alpha = \mathsf{rng}(\alpha)$, and we can extend γ to a total transformation in an arbitrary way, also be denoted by γ. If $S \neq \mathcal{T}_X$ we consider γ as a partial transformation on X with domain M. In case $S = \mathcal{I}_X$ and $y_1\alpha = x_1 \neq x_2 = y_2\alpha$ are in M, we have $y_1 \neq y_2$, and the injectivity of β implies that $x_1\gamma = y_1\beta$ and $x_2\gamma = y_2\beta$ are also different, whence $\gamma \in \mathcal{I}_X$.

Now we first consider the case $y \notin \mathsf{dom}(\beta)$: Then, from the definition of γ, it follows that either $y \notin \mathsf{dom}(\alpha)$, or $y\alpha \notin M$: Assume that $y \in \mathsf{dom}(\alpha)$ and $y\alpha \in M$. Then there is a $y_1 \in \mathsf{dom}(\beta)$ such that $y\alpha = y_1\alpha$, i.e. $y\kappa_\alpha y_1$. Therefore $y\kappa_\beta y_1$, so $y \in \mathsf{dom}(\beta)$, a contradiction. This means that for $y \notin \mathsf{dom}(\beta)$ the equality $y\alpha\gamma = (y\alpha)\gamma$ implies $y \notin \mathsf{dom}(\gamma\alpha)$. Otherwise, assume $y \in \mathsf{dom}(\beta)$: Then $y\alpha \in M$ and (9.14) implies $y\alpha\gamma = (y\alpha)\gamma = y\beta$. So $\beta = \alpha\gamma \in \alpha S$, whence $A \subseteq \alpha S$. □

Theorem 9.43. ([5, Theorem 4.2.8]) *Let S denote any of the semigroups $\mathcal{T}_X$, $\mathcal{P}_X$, and $\mathcal{I}_X$. Then for each $\alpha \in S$ the principal ideal generated by α has the form*

$$S\alpha S = \{\beta \in S : \mathrm{rank}(\beta) \leq \mathrm{rank}(\alpha)\}. \tag{9.15}$$

Proof. Let again A denote the right-hand side of (9.15). Recall that $\mathrm{rank}(\alpha) := |\mathsf{rng}(\alpha)|$, and $\mathsf{rng}(\rho\sigma) \subseteq \mathsf{rng}(\sigma)$ for arbitrary relations ρ and σ by Th. 3.17(7). So $\mathrm{rank}(\gamma\alpha\beta) \leq \mathrm{rank}(\beta)$, and at best β does a permutation of $\mathsf{rng}(\gamma\alpha)$, whence $\mathrm{rank}(\beta) \leq \mathrm{rank}(\gamma\alpha) \leq \mathrm{rank}(\alpha)$. Thus $\gamma\alpha\beta \in A$, i.e. $S\alpha S \subseteq A$.

To prove the converse inclusion assume $\mathsf{rng}(\alpha) = \{a_1, a_2, \dots, a_k\}$ and $\beta \in A$ such that $\mathsf{rng}(\beta) = \{b_1, b_2, \dots, b_m\}$; then $m \leq k$ by definition of A. For each $i = 1, \dots, k$ we choose some $c_i \in A_i := \{x \in X : \alpha(x) = a_i\}$ and define $\gamma \in \mathcal{P}_X$ as follows: $\mathsf{dom}(\gamma) = \mathsf{dom}(\beta)$, and for all $y \in \{z \in X : \beta(z) = b_j\}$, $j = 1, \dots, m$, we set $\gamma(y) = c_j \in A_j$. Note that if β is in $\mathcal{T}_X$ or $\mathcal{I}_X$, then by construction also γ. Now let δ be any permutation satisfying $\delta(a_i) = b_i$, $i = 1, \dots, m$, then $\gamma\alpha\delta = \beta$: For $y \in \{z \in X : \beta(z) = b_j\}$ we get $\beta(y) = b_j$, and $\gamma(y) = c_j, \alpha(c_j) = a_j, \delta(a_j) = b_j$. Therefore we have $\beta \in S\alpha S$ and $A \subseteq S\alpha S$. □

Theorem 9.44. ([5, Theorem 4.5.1]) *Let S denote any of the semigroups $\mathcal{T}_n$, $\mathcal{P}_n$, and $\mathcal{I}_n$, and $\alpha, \beta \in S$. Then the following holds*

(1) *$\alpha\mathcal{L}\beta$ if and only if $\mathsf{rng}(\alpha) = \mathsf{rng}(\beta)$;*
(2) *$\alpha\mathcal{R}\beta$ if and only if $\mathsf{dom}(\alpha) = \mathsf{dom}(\beta)$ and $\kappa_\alpha = \kappa_\beta$;*
(3) *$\alpha\mathcal{H}\beta$ if and only if $\mathsf{dom}(\alpha) = \mathsf{dom}(\beta)$ and $\mathsf{rng}(\alpha) = \mathsf{rng}(\beta)$ and $\kappa_\alpha = \kappa_\beta$;*
(4) *$\alpha\mathcal{J}\beta$ if and only if $\mathrm{rank}(\alpha) = \mathrm{rank}(\beta)$.*
(5) *$\alpha\mathcal{D}\beta$ if and only if $\mathrm{rank}(\alpha) = \mathrm{rank}(\beta)$;*

Proof. (1): By (9.4), $\alpha \leq_\mathcal{L} \beta \iff S\alpha \subseteq S\beta$ and $\beta \leq_\mathcal{L} \alpha \iff S\beta \subseteq S\alpha$, whence $\alpha\mathcal{L}\beta$ iff $S\alpha = S\beta$. But using Th. 9.41, we get $S\alpha = \{\gamma \in S : \mathsf{rng}(\gamma) \subseteq \mathsf{rng}(\alpha)\} = \{\gamma \in S : \mathsf{rng}(\gamma) \subseteq \mathsf{rng}(\beta)\} = S\beta$ iff $\mathsf{rng}(\alpha) = \mathsf{rng}(\beta)$.

(2): Similar to (1) using Th. 9.42.

(3): Follows from (1) and (2) since $\mathcal{H} = \mathcal{R} \cap \mathcal{L}$.

(4): Follows from Th. 9.43.

(5): From $\mathcal{D} \subseteq \mathcal{J}$ and (4) we get that $\alpha\mathcal{D}\beta$ implies $\mathrm{rank}(\alpha) = \mathrm{rank}(\beta)$. Conversely, assume $\mathrm{rank}(\alpha) = \mathrm{rank}(\beta)$. Set $\mathsf{rng}(\alpha) = \{a_1, a_2, \dots, a_k\}$ and $\mathsf{rng}(\beta) = \{b_1, b_2, \dots, b_k\}$, and let $\gamma(x) =$

b_j if and only if $\alpha(x) = a_j$, $j = 1,\ldots,k$. Then $\kappa_\alpha = \kappa_\beta$ and thus $\alpha\mathcal{R}\gamma$ by (2). Moreover, $\mathsf{rng}(\gamma) = \mathsf{rng}(\beta)$, so $\gamma\mathcal{L}\beta$ by (1). Thus $\langle\alpha,\beta\rangle \in \mathcal{R}\circ\mathcal{L} = \mathcal{D}$. □

Note that if $S = \mathcal{T}_X$, then $\mathsf{dom}(\alpha) = \mathsf{dom}(\beta) = X$, so the first condition in (9.13) can be omitted, and we get the familiar form for assertion (2) of the above theorem. If $S = \mathcal{I}_X$, then the restriction of κ_α to $\mathsf{dom}(\alpha)$ becomes the equality relation. Hence $\mathsf{dom}(\beta) \subseteq \mathsf{dom}(\alpha)$ automatically gives $\kappa_\alpha \subseteq \kappa_\beta$, whence the second condition in (9.13) can be omitted in this case.

9.2.3 *The Natural Partial Order on $\mathcal{B}_X$*

We recall from (9.11) the definition of the natural partial order $\leq_M$ for the context of $\mathcal{B}_X$ (using a simplified notation):

$$\alpha \leq \beta \quad := \quad \alpha = \beta \text{ or } \alpha = \rho\alpha = \rho\beta = \beta\sigma \quad \text{for some } \rho,\sigma \in \mathcal{B}_X.$$

Our first result concerns the comparability of converses:

Proposition 9.45. ([18, part of Theorem 2.3]) *For $\alpha,\beta \in \mathcal{B}_X$ we have $\alpha \leq \beta$ iff $\alpha^{\smile} \leq \beta^{\smile}$*

Proof. Assume $\alpha \leq \beta$: then for some $\rho,\sigma \in \mathcal{B}_X$ we have $\alpha = \rho\alpha = \rho\beta = \beta\sigma$ which implies $\alpha\sigma = \rho\beta\sigma = \rho\alpha = \alpha$. Therefore, $\alpha^{\smile} = \alpha^{\smile}\rho^{\smile} = \beta^{\smile}\rho^{\smile} = \sigma^{\smile}\beta^{\smile}$ and $\alpha^{\smile} = \sigma^{\smile}\alpha^{\smile}$. In sum, $\alpha^{\smile} = \sigma^{\smile}\alpha^{\smile} = \sigma^{\smile}\beta^{\smile} = \beta^{\smile}\rho^{\smile}$, whence we have shown $\alpha^{\smile} \leq \beta^{\smile}$. Conversely, given that $\alpha^{\smile} \leq \beta^{\smile}$ we have $\alpha^{\smile} = \sigma^{\smile}\alpha^{\smile} = \sigma^{\smile}\beta^{\smile} = \beta^{\smile}\rho^{\smile}$. So $\alpha^{\smile} = \sigma^{\smile}\alpha^{\smile} = \sigma^{\smile}\beta^{\smile}\rho^{\smile} = \alpha^{\smile}\rho^{\smile}$. Therefore, $\alpha = \alpha\sigma = \rho\beta = \rho\alpha = \beta\sigma$, i.e. $\alpha \leq \beta$. □

It can be shown that there is a purely equational equivalent to the definition of $\leq$ in $\mathcal{B}_X$. Before we note an easy consequence of Prop. 3.34 in

Lemma 9.46. *Let $\alpha,\beta \in \mathcal{B}_X$ be given.*

(1) If $\{\chi : \beta\chi = \alpha\} \neq \emptyset$ then $\alpha/\beta = \overline{\beta^{\smile}\overline{\alpha}}$ is the greatest solution with respect to $\subseteq$.
(2) If $\{\chi : \chi\beta = \alpha\} \neq \emptyset$ then $\alpha\backslash\beta = \overline{\overline{\alpha}\beta^{\smile}}$ is the greatest solution with respect to $\subseteq$.

Proof. (1): By Prop. 3.34(5) we have $\chi \subseteq (\alpha/\beta)$. So $\alpha = \beta\chi \subseteq \beta(\alpha/\beta) \subseteq \alpha$, the last inequality again following from Prop. 3.34(5).

(2): Similar using Prop. 3.34(6). □

Proposition 9.47. ([4, Theorem 3.1]) *For $\alpha,\beta \in \mathcal{B}_X$ the following holds:*

$$\alpha \leq \beta \quad \textit{iff} \quad \alpha = \alpha(\alpha/\beta) = \beta(\alpha/\beta) = (\alpha\backslash\beta)\beta.$$

Proof. Suppose $\alpha \leq \beta$; so there are $\rho,\sigma \in \mathcal{B}_X$ such that $\alpha = \rho\alpha = \rho\beta = \beta\sigma$. This implies $\alpha\sigma = \rho\beta\sigma = \rho\alpha = \alpha$. So by Lemma 9.46 we get $\alpha = \beta(\alpha/\beta) = (\alpha\backslash\beta)\beta$. Moreover $\alpha = \rho\alpha = \rho\beta(\alpha/\beta) = \alpha(\alpha/\beta)$. The converse direction is obvious: Taking $\sigma = (\alpha/\beta)$ and $\rho = (\alpha\backslash\beta)$ we get $\alpha = \rho\beta = \beta\sigma$. and $\alpha = \alpha\sigma$. From the latter we get $\alpha = (\rho\beta)\sigma = \rho(\beta\sigma) = \rho\alpha$. □

The next proposition gives another characterization of when $\alpha, \beta \in \mathcal{B}_X$ are comparable under $\leq$ which uses the concepts of row and column space as introduced at the begin of Section 9.2.2.

Lemma 9.48. *Let $\alpha, \beta \in \mathcal{B}_X$ and $T \subseteq X$. Then*

(1) $T(\alpha\beta) = (T\alpha)\beta$ *and* $(\alpha\beta)T = \alpha(\beta T)$.
(2) $V(\alpha\beta) \subseteq V(\beta)$ *and* $W(\alpha\beta) \subseteq W(\alpha)$.

Proof. (1): Use Prop. 3.12(4).

(2): The column space of α has been defined above as $W(\alpha) := \{\alpha A : A \subseteq X\}$, where $\alpha A := \bigcup_{x \in A} \alpha x$. Using (1) we then have $W(\alpha\beta) = \{(\alpha\beta)B : B \subseteq X\} = \{\alpha(\beta B) : B \subseteq X\} \subseteq \{\alpha C : C \subseteq X\} = W(\alpha)$. □

Proposition 9.49. ([18, Theorem 2.3]) *For $\alpha, \beta \in \mathcal{B}_X$ the following statements are equivalent:*

(1) $\alpha \leq \beta$.
(2) $\alpha^{\smile} \leq \beta^{\smile}$.
(3) $V(\alpha) \subseteq V(\beta)$, $W(\alpha) \subseteq W(\beta)$, *and for all* $A, B \subseteq X$ *we have:* $A\alpha = B\beta \rightarrow A\alpha = B\alpha$ *and* $\alpha A = \beta B \rightarrow \alpha A = \alpha B$.

Proof. In the light of Prop. 9.45 we have to show the following two implications:
(2) $\Rightarrow$ (3): Assume $\alpha^{\smile} \leq \beta^{\smile}$. Then there are $\rho, \sigma \in \mathcal{B}_X$ such that $\alpha^{\smile} = \rho\alpha^{\smile} = \rho\beta^{\smile} = \beta^{\smile}\sigma$, so $\alpha = \alpha\rho^{\smile} = \beta\rho^{\smile} = \sigma^{\smile}\beta$. By Lemma 9.48 we get $V(\alpha) = V(\sigma^{\smile}\beta) \subseteq V(\beta)$ and $W(\alpha) = W(\beta\rho^{\smile}) \subseteq W(\beta)$. And for $A, B \subseteq X$, $A\alpha = B\beta$ implies $A\alpha = A\alpha\rho^{\smile} = B\beta\rho^{\smile} = B\alpha$, and similarly $\alpha A = \beta B$ implies $\alpha A = \alpha B$.

(3) $\Rightarrow$ (1): For each $x \in X$ we have $x\alpha \in V(\alpha) \subseteq V(\beta)$ and $\alpha x \in W(\alpha) \subseteq W(\beta)$. Then there exist sets $A_x, B_x \subseteq X$ such that $x\alpha = A_x\beta$ and $\alpha x = \beta B_x$, and by assumption $x\alpha = A_x\alpha$ and $\alpha x = \alpha B_x$. Now we define $\rho, \sigma \in \mathcal{B}_X$ by $x\rho = A_x$ and $\sigma x = B_x$ for $x \in X$ and thus get

$$x\rho\beta = A_x\beta = x\alpha = A_x\alpha = x\rho\alpha \quad \text{and} \quad \alpha x = \beta B_x = \beta\sigma x.$$

So $\alpha = \rho\alpha = \rho\beta = \beta\sigma$, whence $\alpha \leq \beta$. □

Using this result together with a few more lemmas the authors of [18] obtain characterizations of when $\alpha \in \mathcal{B}_X$ is maximal, and when $\alpha \in \mathcal{B}_X \backslash \emptyset$ is minimal in $\mathcal{B}_X \backslash \emptyset$ with respect to $\leq$.

9.2.3.1 *The Natural Partial Order on Subsemigroups of $\mathcal{B}_X$*

It has been mentioned above in Section 9.1.3 that the concept of a natural partial order has first been studied in inverse, and later in regular semigroups. We will now take a short glance at what this means for subsemigroups of $\mathcal{B}_X$, and begin by considering the partial order on $\mathcal{I}_X$ defined by set inclusion: $\alpha \leq \beta := \alpha \subseteq \beta$, i.e. $\alpha \leq \beta$ iff α is a restriction of β:

$$\mathsf{dom}(\alpha) \subseteq \mathsf{dom}(\beta) \quad \text{and} \quad x\alpha = x\beta \text{ for all } x \in \mathsf{dom}(\alpha). \tag{9.16}$$

The underlying idea is to characterize $\leq$ using relationships of mappings in $\mathcal{I}_X$ in terms of equality of products, which can then be used as the definition of the relationship for abstract inverse semigroups, and as basis for generalizations to wider classes of semigroups.

Lemma 9.50. ([8, p. 44]) *Let $\alpha, \beta \in \mathcal{I}_X$, then the following conditions are equivalent:*

(1) $\alpha \leq \beta$
(2) $\alpha = \alpha\alpha^{-1}\beta$
(3) $\alpha = \epsilon\beta$, *for some* $\epsilon \in E_{\mathcal{I}_X}$
(4) $\alpha\alpha^{-1} = \beta\alpha^{-1}$
(5) $\alpha = \alpha\beta^{-1}\alpha$
(6) $\alpha = \beta\alpha^{-1}\alpha$
(7) $\alpha = \beta\epsilon$, *for some* $\epsilon \in E_{\mathcal{I}_X}$

Proof. (1) $\Rightarrow$ (2): Since $\alpha\alpha^{-1} = 1_{\mathsf{dom}(\alpha)}$ by Lemma 9.28, and $x\alpha = x\beta$ for all $x \in \mathsf{dom}(\alpha)$ by (9.16).

(2) $\Rightarrow$ (3): $\alpha\alpha^{-1} \in E_{\mathcal{I}_X}$.

(3) $\Rightarrow$ (4): Assume $\alpha = \epsilon\beta$; then $\alpha\alpha^{-1} = \epsilon(\beta\beta^{-1})\epsilon$. But $\beta\beta^{-1} \in E_{\mathcal{I}_X}$, and by Th. 9.3 idempotents commute, so $\epsilon(\beta\beta^{-1})\epsilon = \beta(\beta^{-1}\epsilon^2) = \beta(\epsilon\beta)^{-1} = \beta\alpha^{-1}$.

(4) $\Rightarrow$ (5): Assume $\alpha\alpha^{-1} = \beta\alpha^{-1}$. Then $\alpha^{-1} = \alpha^{-1}\alpha\alpha^{-1} = \alpha^{-1}\beta\alpha^{-1}$, and taking the inverses of both sides gives the result.

(5) $\Rightarrow$ (1): $\alpha\beta^{-1} \in E_{\mathcal{I}_X}$.

(1) $\Rightarrow$ (6): Since $\alpha^{-1}\alpha = 1_{\mathsf{rng}(\alpha)}$ by Lemma 9.28, and $x\alpha = x\beta$ for all $x \in \mathsf{dom}(\alpha)$ by (9.16).

(6) $\Rightarrow$ (7): $\alpha^{-1}\alpha \in E_{\mathcal{I}_X}$.

(7) $\Rightarrow$ (3): It is not hard to show that $\beta\epsilon\beta^{-1} \in E_{\mathcal{I}_X}$. But then $(\beta\epsilon\beta^{-1})\beta = \beta\epsilon(\beta^{-1}\beta) = (\beta\beta^{-1}\beta)\epsilon = \beta\epsilon = \alpha$. So there is an $\epsilon' \in E_{\mathcal{I}_X}$ with $\epsilon'\beta = \alpha$. □

In (9.10) $\leq$ has been defined for regular semigroups as $a \leq_N b :\leftrightarrow\ a = eb = bf$, for some $e, f \in E_S$. Properties of this order for $\mathcal{T}_X$ have been investigated in [12], including a characterization of all the minimal and maximal elements, and for $\mathcal{P}_X$ in [15].

References

[1] Blyth, T.S.: *Lattices and Ordered Algebraic Structures*, Springer-Verlag, London (2005).

[2] Cain, A.J.: Nine Chapters on the Semigroup Art, Lecture Notes, version 0.65.18 (2017-03-21), `www-groups.mcs.st-and.ac.uk/~alanc/pub/c_semigroups/` (2017).

[3] Clifford, A.H., Preston, G.B.: *The Algebraic Theory of Semigroups, Vol. 1*, 2nd ed., reprinted with corrections, American Mathematical Society, Providence, Rhode Island (1977).

[4] FitzGerald, D.G.: Mitsch's order and inclusion for binary relations and partitions, Semigroup Forum **87**, 161-170 (2013).

[5] Ganyushkin, O., Mazorchuk, V.: *Classical Finite Transformation Semigroups - An Introduction*, Springer-Verlag London Limited (2009).

[6] Green, J.A.: On the structure of semigroups, Ann. of Math. **54**, 163-172 (1951).

[7] Hartwig, R.: How to partially order regular elements. Math. Japon. **35**, 1-13 (1980).

[8] Higgins, P.M.: *Techniques of Semigroup Theory*, Oxford University Press, New York (1992).

[9] Howie, J.M: *Fundamentals of Semigroup Theory*, Clarendon Press, Oxford (1995).

[10] Konieczny, J.: Green's equivalences in finite semigroups of binary relations, Semigroup Forum **48**, 235-252 (1994).

[11] Konieczny, J.: The semigroup generated by regular Boolean matrices, Southeast Asian Bulletin of Mathematics **25**, 627-641 (2002).

[12] Kowol, G., Mitsch, H.: Naturally ordered transformation semigroups, Monatshefte für Mathematik **102**, 115-138 (1986).

[13] Lawson, M.V.: *Inverse semigroups-The theory of partial symmetries*, World Scientific Publishing, Singapore (1998).

[14] Ljapin, E.S.: *Semigroups*, American Mathematical Society, Providence, Rhode Island (1963).

[15] Marques-Smith, M.P.O., Sullivan, R.P.: Partial orders on transformation semigroups, Monatshefte für Mathematik, **140**, 103-118 (2003).

[16] Mitsch, H.: A natural partial order for semigroups. Proc. American Math. Soc. **97**(3), 384-388 (1986).

[17] Nambooripad, K.S.: The natural partial order on a regular semigroup, Proc. Royal Soc. Edinburgh **23**, 249-260 (1980).

[18] Namnak, C., Preechasilp, P.: Natural partial orders on the semigroup of binary relations, Thai J. of Mathematics **4**, 39-50 (2006).

[19] Petrich, M.: *Inverse Semigroup*, John Wiley & Sons, (1984).

[20] Pin, J.-E.: *Mathematical foundations of automata theory*, available at `https://www.irif.fr/~jep/PDF/MPRI/MPRI.pdf`, June 2019.

[21] Plemmons, R.: Generalized Inverses of Boolean Relation Matrices, SIAM J. Appl. Math. **20**, 426-433 (1971).

[22] Plemmons, R., West, M.T.: On the semigroup of binary relations, Pacific J. of Mathematics **35**, 743-753 (1970).

[23] Sangkhanan, K., Sanwong, J.: Partial orders on semigroups of partial transformations with restricted range, Bull. Aust. Math. Soc. **86**, 100-118 (2012).

[24] Vagner, V.: Generalized groups. Dokl. Akad. Nauk. SSSR **84**, 1119-1122 (1952), (in Russian).

Chapter 10

Relation Algebras

In this chapter we turn to the abstract algebraic structures known as *relation algebras*. (Be careful: You must not mistake a *relation* for a *relational* algebra, a concept from database theory.) Of course, the stuff presented is far from being a comprehensive account; much more on concepts and results on relation algebras can be found in the monographs by Givant [9, 10], Hirsch and Hodkinson [13], and Maddux [23]. Nonetheless, this chapter should provide a suitable starting point for continuing with these books or other advanced material. After their definition, important and useful properties of relation algebras, as well as examples will be presented.

Although this chapter will include not more than what may be called the general basic knowledge on relation algebras, you will probably recognize a big step towards a considerably higher level of abstraction, identifiable through the number of new concepts to be introduced. So we will often find ourselves in a situation where a detailed exposition of concepts, and proofs of their properties, reaches beyond the introductory level that is aimed for in this book. Therefore, some of the proofs of what follows will have to be omitted; instead references will be given where to find them and also to other additional material.

This way we will try to give a sketchy overview on some central algebraic aspects, and hope to provide a rough guideline on what is going on in the higher realms of universal algebra in general, and of relation algebras in particular. But first of all, we start with a short look towards familiar areas.

10.1 Algebras of relations

As we have learned in Chapter 1 about some milestones of the history of relations, developing a mathematical theory of binary relations began in 1864 with [28] by AUGUSTUS DE MORGAN, followed approximately ten years later by CHARLES SANDERS PEIRCE ([29]). Their work has been highly extended by ERNST SCHRÖDER in 1895 with the 649 pages of [30]. The basic objects of these investigations are binary relations between elements of some given base set S, using as fundamental operations the Boolean operations of set theory, where the complement $\overline{\rho}$ of a relation ρ is taken with respect to the universal (greatest) relation $U_S := S \times S$. Also the operations of composition and conversion, specific to binary relations, have been considered.

This is also the way we have been following in our presentation of the topic: In Def. 3.3 (binary) relations ρ have been introduced as sets of ordered pairs, usually forming a subset of a Cartesian product $S \times T$. Thus the set of all relations from S to T, denoted by $\mathcal{B}_{S\times T}$, has been identified to be the power set $\mathcal{P}(S \times T)$. So, on pure set-theoretical grounds, $\mathcal{B}_{S\times T}$ bears the structure of a Boolean algebra $(\mathcal{B}_{S\times T}, \cup, \cap, ^{-})$, with $\emptyset$ as least and $U_{S\times T}$ as greatest element.

Later on the operations of product and conversion have been introduced on $\mathcal{B}_{S\times T}$, and we have seen in Prop. 3.18 that $(\mathcal{B}_S, \cdot, I_S)$ is a monoid. Also some basic properties of the relational operations have been mentioned, as well as their interplay with the set theoretical operations. Putting all this together on a more general level, we arrive at the following definition ([23], p. 24):

Definition 10.1. *Let $\mathcal{R}$ be a set of binary relations on a set S, and $1 \in \mathcal{R}$, satisfying the following conditions:*

(1) $\rho \subseteq 1$ for each $\rho \in \mathcal{R}$, i.e. $\bigcup \mathcal{R} = 1$;
(2) $\mathbb{1} \in \mathcal{R}$ is the identity relation on the field of 1;
(3) $\mathcal{R}$ is closed under $\cap$, $\cup$ and complementation $^{-}$ relative to 1;
(4) $\mathcal{R}$ is closed under composition and conversion;

The structure $\mathfrak{Re} = (\mathcal{R}, \cup, \cap, ^{-}, \emptyset, 1, \cdot, ^{\smile}, \mathbb{1})$ is called an **algebra of binary relations**[1] *with base S. The class of algebras of binary relations will be denoted by* BRA.

The following example is borrowed from [13, p. 101]: Let $A = \{0,1,2\}$ and let $\mathcal{R} \subseteq A \times A$ be given as follows:

$$\begin{aligned}\mathcal{R} = \{\ &\emptyset, \quad \{\langle 2,2\rangle\}, \quad \{\langle 0,0\rangle,\langle 1,1\rangle\}, \quad \{\langle 0,1\rangle,\langle 1,0\rangle\},\\ &\{\langle 0,0\rangle,\langle 1,1\rangle,\langle 2,2\rangle\}, \quad \{\langle 2,2\rangle,\langle 0,1\rangle,\langle 1,0\rangle\},\\ &\{\langle 0,0\rangle,\langle 1,1\rangle,\langle 0,1\rangle,\langle 1,0\rangle\}, \quad \{\langle 0,0\rangle,\langle 1,1\rangle,\langle 2,2\rangle,\langle 0,1\rangle,\langle 1,0\rangle\}\\ \}\end{aligned}$$

The identity and unit of $\mathcal{R}$ are, respectively,

- $\mathbb{1} = \{\langle 0,0\rangle,\langle 1,1\rangle,\langle 2,2\rangle\}$ and
- $1 = \{\langle 0,0\rangle,\langle 1,1\rangle,\langle 2,2\rangle,\langle 0,1\rangle,\langle 1,0\rangle\}$.

For all elements $\alpha \in \mathcal{R}$ we have $\alpha = \alpha^{\smile}$. Since $\mathcal{R}$ is closed under composition, $(\mathcal{R}, \cup, \cap, ^{-}, \emptyset, 1, \cdot, ^{\smile}, \mathbb{1})$ is an algebra of binary relations with base A.

Note that according to Def. 10.1, each algebra of binary relations $\mathfrak{Re}$ contains a set S, called the **base of** $\mathfrak{Re}$. An algebra of binary relations is called **full** if it is of the form $(\mathcal{B}_S, \cup, \cap, ^{-}, \emptyset, S \times S, \cdot, ^{\smile}, I_S)$, and is then denoted by $\mathfrak{Re}(S)$. Every subalgebra of $\mathfrak{Re}(S)$ is

[1] Here I follow the naming of [3] or [7], because it clearly emphasizes that the universes considered are sets of binary relations. In e.g. [13], Def. 3.1 and in contributions by Maddux the term **proper relation algebra** is used instead. Later we will see that elements of *relation algebras* need not be binary relations.

referred to as an algebra of binary relations on S. An algebra of binary relation is called **square** if its unit is of the form $U \times U$. (The example from above shows that this need not always be the case.)

Proposition 10.2. *Let* $\mathfrak{Re}(S) = (\mathcal{R}, \cup, \cap, \bar{\ }, \emptyset, 1, \cdot, \breve{\ }, \mathbb{1})$ *be an algebra of binary relation with base* S. *Then the unit* 1 *is an equivalence relation over* S.

Proof. $\mathbb{1} \subseteq 1$, since the latter is the biggest element of $\mathfrak{Re}(S)$, whence 1 is reflexive. Since $\mathcal{R}$ is closed under conversion, $1^{\smile} \in \mathcal{R}$ and $1^{\smile} \subseteq 1$, whence 1 is symmetric; and since $\mathcal{R}$ is closed under composition, $1^2 \in \mathcal{R}$ and $1^2 \subseteq 1$, i.e. 1 is also transitive. □

10.2 Some Basic Concepts from Universal Algebra

In Section 2.3 we have introduced some basic notions and results concerning algebraic structures, and have illustrated them giving some popular and important examples. Before we will climb the next abstraction level concerning relations, we have to master some algebraic preparations. In doing so, this section becomes a kind of balancing between providing not too much beyond what is needed later on and, notwithstanding, offering enough information (including some sample proofs) to raise the reader's appetite and to provide a basis for carrying on with some more advanced aspects of *universal algebra.*

Next we will consider the main methods used to assemble new algebras from those already at hand: these are the formation of subalgebras, of homomorphic images, and of direct products.

10.2.1 *Subalgebras and Homomorphic Images*

Let f be an n-ary operation on a set A, and $B \subseteq A$. Then also $B^n \subseteq A^n$, so f defines a mapping of B^n into A, namely the restriction $f_{|B^n}$. If $f[B^n] \subseteq B$, B is called **closed with respect to** f.

Definition 10.3. *Let* $\mathfrak{A} = (A, \Omega^A)$ *be an algebra.*

(a) A subset of the universe A *of* $\mathfrak{A}$*, which is closed with respect to each operation of* $\mathfrak{A}$*, is called a* **subuniverse** *of* $\mathfrak{A}$.
(b) The algebra $\mathfrak{B} = (B, \Omega^B)$ *is called a* **subalgebra** *of* $\mathfrak{A} = (A, \Omega^A)$*, if* $\mathfrak{A}$ *and* $\mathfrak{B}$ *are similar,* B *is a subuniverse of* A*, and* B *is closed with respect to every operation* $f^A \in \Omega^A$.

Note that if $\mathfrak{B}$ is a subalgebra of $\mathfrak{A}$, then B is a nonempty subuniverse of $\mathfrak{A}$. However, the empty set may be a subuniverse of $\mathfrak{A}$, but it is not the underlying set of any subalgebra. In fact, the empty set is a subuniverse of $\mathfrak{A}$ if and only if $\mathfrak{A}$ has no nullary basic operations, since if $\mathfrak{A}$ has nullary operations then every subuniverse contains them as well.

Well known examples are the ring of integers, which is a subalgebra of the ring of all rational numbers, which in turn is a subalgebra of the ring of all real numbers. On the other hand, if we take the algebra $(\mathbb{Z},-)$, and narrow it down to the set $\mathbb{N}$ of natural numbers, then we get a subrelational structure $(\mathbb{N},-)$ but clearly not a subalgebra. But e.g. $\{-2,-1,0,1,2\}$ is closed under subtraction and thus forms a subalgebra of $(\mathbb{Z},-)$.

Sometimes it is useful to deal with only a part of the operations of an algebra, e.g. one may want to discuss the additive structure of a ring, or view a group as being merely a monoid. Let $\mathfrak{A}=(A,\Omega^A)$ be an algebra and $\mathfrak{B}=(B,\Omega^B)$ be another algebra such that Ω^B is a subsequence of Ω^A, then $\mathfrak{B}$ is called a **reduct** of $\mathfrak{A}$, and $\mathfrak{A}$ an **expansion** of $\mathfrak{B}$.

10.2.2 *Congruences and Quotient Algebras*

Definition 10.4. *Let $\mathfrak{A}=(A,\Omega^A)$ be an algebra, and ρ be a relation on A.*

(a) Then ρ is said to be **compatible** *with the n-ary operation $f^A\in\Omega^A$ if for $a_i,b_i\in A$ we have*

$$\forall 1\le i\le n\,[a_i\rho b_i] \rightarrow f^A(a_1,a_2,\dots,a_n)\,\rho\, f^A(b_1,b_2,\dots,b_n).$$

(b) ρ is said to be **compatible on** *$\mathfrak{A}$ (also: to have the* **substitution property***) if ρ is compatible with each operation $f^A\in\Omega^A$.*

(c) A compatible equivalence relation on $\mathfrak{A}$ is called a **congruence***, and its equivalence classes are referred to as* congruence classes. *The set of all congruences on an algebra $\mathfrak{A}$ is denoted by* $\mathsf{Con}(\mathfrak{A})$. .

Examples of congruences on $\mathfrak{A}$ are the relations I_A and U_A. An algebra $\mathfrak{A}$ is called **simple** if it is non-trivial, and $\mathsf{Con}(\mathfrak{A})=\{I_A,U_A\}$. Compatibility of an equivalence relation can be illustrated as shown in Fig. 10.1 (after [5, Fig. 10]).

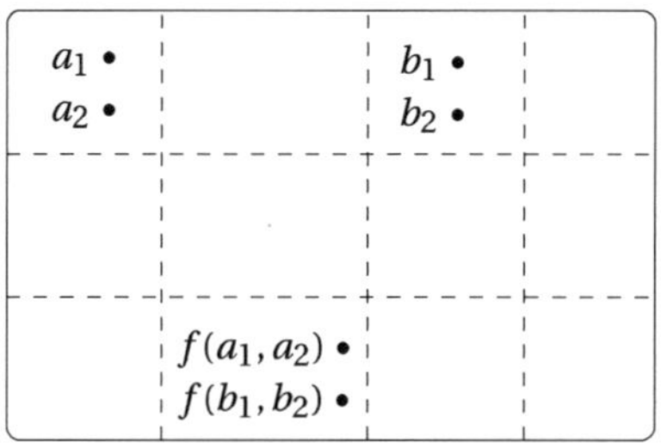

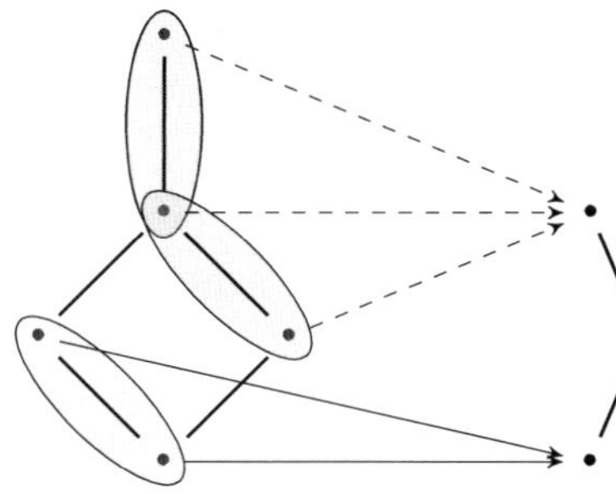

FIG. 10.1: Compatibility of a congruence. FIG. 10.2: Homomorphism and its kernel.

It is a well-known fact from universal algebra that the kernel of every homomorphism is a congruence relation (shown in the following theorem), and that every congruence relation is the kernel of a homomorphism (see Th. 10.8 below).

Proposition 10.5. *Let $\mathfrak{A}=(A,\Omega^A)$ and $\mathfrak{B}=(B,\Omega^B)$ be two similar algebras and $g:\mathfrak{A}\longrightarrow\mathfrak{B}$ be a homomorphism. Then* $\mathsf{ker}(g)$ *is a congruence on $\mathfrak{A}$.*

Proof. If $\langle a_i, b_i\rangle \in \mathsf{ker}(g)$, $i \in \underline{n}$, and $f^A \in \Omega^A$ is an n-ary operation, then

$$\begin{aligned} g(f_i^A(a_1,a_2,\ldots,a_n)) &= f_i^B(g(a_1),\ldots,g(a_n)) \\ &= f_i^B(g(b_1),\ldots,g(b_n)) \\ &= g(f_i^A(b_1,b_2,\ldots,b_n)), \end{aligned}$$

hence $\langle f_i^A(a_1,a_2,\ldots,a_n), f_i^A(b_1,b_2,\ldots,b_n)\rangle \in \mathsf{ker}(g)$. Obviously, $\mathsf{ker}(g)$ is an equivalence relation, and thus a congruence on $\mathfrak{A}$. □

As an example, Fig. 10.2 shows a 5-element lattice together with a homomorphism as given by the arrows. Its kernel is indicated in the diagram by ellipses covering the two congruence classes.

From now on we will follow the common habit to denote congruences on algebras by θ as it is done in e.g. [4,5,] and others; do not confuse this with the type θ of an algebra $\mathfrak{A}$ as introduced in Def. 2.45 above.

Definition 10.6. *Let $\mathfrak{A}$ be an algebra and $\theta \in \mathsf{Con}(\mathfrak{A})$.*

(a) The **quotient algebra** *of $\mathfrak{A}$ by θ, denoted $\mathfrak{A}/\theta$, is the algebra with carrier A/θ and whose operations satisfy*

$$f^{A/\theta}(a_1/\theta,\ldots,a_n/\theta) = f^A(a_1,\ldots,a_n)/\theta,$$

where $a_1,\ldots,a_n \in A$ and f is an n-ary operation.

(b) The **natural mapping** *$m_\theta : \mathfrak{A} \longrightarrow \mathfrak{A}/\theta$ is defined by $m_\theta(a) = a/\theta$. m_θ is also called* **canonical homomorphism**.

Note that quotient algebras of $\mathfrak{A}$ have the same type as $\mathfrak{A}$.

Proposition 10.7. *Let $\mathfrak{A}$ be an algebra and $\theta \in \mathsf{Con}(\mathfrak{A})$; then the natural mapping m_θ is an epimorphism.*

Proof. Clearly, m_θ is onto. For f an n-ary function symbol and $a_1,a_2,\ldots,a_n \in A$ we have

$$\begin{aligned} m_\theta(f_i^A(a_1,a_2,\ldots,a_n)) &= f_i^A(a_1,a_2,\ldots,a_n)/\theta \\ &= f^{A/\theta}(a_1/\theta,\ldots,a_n/\theta) \\ &= f^{A/\theta}(m_\theta(a_1),\ldots,m_\theta(a_n)), \end{aligned}$$

so m_θ is a homomorphism. □

By the results obtained so far we are in the position to demonstrate that every congruence relation is determined by a homomorphism.

Theorem 10.8. (Homomorphism Theorem) *Let $\mathfrak{A}$ and $\mathfrak{B}$ be algebras, and $g : \mathfrak{A} \longrightarrow \mathfrak{B}$ be an epimorphism. Then there is an isomorphism $h : \mathfrak{A}/\mathsf{ker}(g) \longrightarrow \mathfrak{B}$ defined by $g = hm_g$, where m_g is the natural homomorphism $m_g : \mathfrak{A} \longrightarrow \mathfrak{A}/\mathsf{ker}(g)$.*

Proof. For sake of notational convenience we set $\theta := \ker(g)$. By Prop. 10.5 $\theta \in \mathrm{Con}(\mathfrak{A})$. Note that due to the definition $g = hm_g$ we must have $g(a) = h(a/\theta)$, so by Prop. 2.27 h is bijective. It remains to show that h is a homomorphism: Suppose f is an n-ary function symbol and $a_1, a_2, \ldots, a_n \in A$. Then

$$\begin{aligned} h(f^{A/\theta}(a_1/\theta,\ldots,a_n/\theta)) &= h(f^A(a_1,a_2,\ldots,a_n)/\theta) \\ &= g(f^A(a_1,a_2,\ldots,a_n)) \\ &= f^B(g(a_1),\ldots,g(a_n)) \\ &= f^B(h(a_1/\theta),\ldots,h(a_n/\theta)), \end{aligned}$$

so h is a homomorphism. □

The Homomorphism Theorem, together with Lemma 2.48, thus says that every homomorphic image of an algebra $\mathfrak{A}$ is isomorphic to a quotient algebra of $\mathfrak{A}$. And this means that the "external" problem of finding all homomorphic images of $\mathfrak{A}$ can be transformed to the "internal" problem of finding all congruences on $\mathfrak{A}$.

10.2.3 *Direct Products and Factor Congruences*

The two techniques presented above in sections (10.2.1) and (10.2.2), formation of subalgebras and homomorphic images, do not give means to create algebras of a larger cardinality or complexity than the one with which the construction started.

We now turn to the third of the three basic ways of constructing new algebras from old ones, the *direct product.* It is the only one that increases complexity, or at least the size of the algebras, and differs from the other two in that it offers the possibility of combining a system of many algebras into a single one.

Definition 10.9. *Let $\mathfrak{A}_i = (A_i, \Omega^{A_i})$, $i \in \underline{k}$, be a family of similar algebras. Their* **direct product** *$\mathfrak{A}_1 \times \cdots \times \mathfrak{A}_k$ is the algebra $\mathfrak{A} = (A, \Omega^A)$ with carrier $A = \prod_{i \in \underline{k}} A_i$ and n-ary operations $f^A \in \Omega^A$, defined by*

$$\pi_i\big(f^A(a_1,\ldots,a_n)\big) = f^{A_i}\big(\pi_i(a_1),\ldots,\pi_i(a_n)\big).$$

for each $a_k := \langle a_{k_1},\ldots,a_{k_n}\rangle \in A$.

To simplify matters, we will usually restrict our consideration to two factors since the general case is evident. The direct product then looks as follows: Let $\mathfrak{A}_1 = (A_1, \Omega^{A_1})$ and $\mathfrak{A}_2 = (A_2, \Omega^{A_2})$ be two algebras of the same type. Then $\mathfrak{A}_1 \times \mathfrak{A}_2$ is the algebra with carrier $A_1 \times A_2$ such that for each n-ary operation f and $a_i \in A_1, a_i' \in A_2, 1 \le i \le n$,

$$f^{A_1 \times A_2}(\langle a_1, a_1'\rangle,\ldots,\langle a_n, a_n'\rangle) = \langle f^{A_1}(\langle a_1,\ldots,a_n\rangle), f^{A_2}(\langle a_1',\ldots,a_n'\rangle)\rangle.$$

That both $\mathfrak{A}_1$ and $\mathfrak{A}_2$ are homomorphic images of $\mathfrak{A}_1 \times \mathfrak{A}_2$ is shown in

Proposition 10.10. *The mapping $\pi_i : A_1 \times A_2 \longrightarrow A_i$ is a surjective homomorphism from $\mathfrak{A}_1 \times \mathfrak{A}_2$ to $\mathfrak{A}_i$, $i = 1,2$. Furthermore, in* $\mathrm{Con}\,(\mathfrak{A}_1 \times \mathfrak{A}_2)$ *we have*

(1) $\ker \pi_1 \cap \ker \pi_2 = I_{(A_1 \times A_2)^2}$,
(2) $\ker \pi_1 \vee \ker \pi_2 = U_{(A_1 \times A_2)^2}$.

(3) $\ker\pi_1 \circ \ker\pi_2 = \ker\pi_2 \circ \ker\pi_1$,

Proof. Obviously, π_i is surjective, $i = 1,2$. Let $a_i \in A_1, a_i' \in A_2, 1 \le i \le n$, and let f be an n-ary operation; then

$$\begin{aligned}\pi_1\big(f^{A_1\times A_2}(\langle a_1,a_1'\rangle,\ldots,\langle a_n,a_n'\rangle)\big) &= \pi_1\big(\langle f^{A_1}(a_1,\ldots,a_n), f^{A_2}(a_1',\ldots,a_n')\rangle\big)\\ &= f^{A_1}(a_1,\ldots,a_n)\\ &= f^{A_1}\big(\pi_1(\langle a_1,a_1'\rangle),\ldots,\pi_1(\langle a_n,a_n'\rangle)\big),\end{aligned}$$

whence π_1 is a homomorphism. This also holds for π_2 by an analogous argument.

(1): For $i = 1,2$ we have

$$\big\langle\langle a_1,a_2\rangle,\langle b_1,b_2\rangle\big\rangle \in \ker\pi_i \text{ iff } \pi_i(\langle a_1,a_2\rangle) = \pi_i(\langle b_1,b_2\rangle) \text{ iff } a_i = b_i,$$

whence

$$\begin{aligned}\ker\pi_1 &= \{\big\langle\langle a,a_2\rangle,\langle a,b_2\rangle\big\rangle : a\in A_1, a_2,b_2\in A_2\};\\ \ker\pi_2 &= \{\big\langle\langle a_1,b\rangle,\langle b_1,b\rangle\big\rangle : a_1,b_1\in A_1, b\in A_2\}.\end{aligned}$$

Thus $\ker\pi_1 \cap \ker\pi_2 = \{\big\langle\langle a,b\rangle,\langle a,b\rangle\big\rangle : a\in A_1, b\in A_2\} = I_{(A_1\times A_2)^2}$.

(2): For any $\langle a_1,a_2\rangle,\langle b_1,b_2\rangle \in A_1\times A_2$ we have $\langle a_1,a_2\rangle \ker\pi_1 \langle a_1,b_2\rangle \ker\pi_2 \langle b_1,b_2\rangle$. Thus

$$\ker\pi_1\circ\ker\pi_2 = \{\big\langle\langle a,b\rangle,\langle c,d\rangle\big\rangle : a,c\in A_1, b,d\in A_2\} = U_{(A_1\times A_2)^2}.$$

So $\ker\pi_1\circ\ker\pi_2$ is an equivalence on $(A_1\times A_2)^2$, and by Prop. 6.60(2), $\ker\pi_1\circ\ker\pi_2 = \ker\pi_1 \vee \ker\pi_2$.

(3): By (2) $\ker\pi_1\circ\ker\pi_2$ is an equivalence, whence by Prop. 6.60(1) $\ker\pi_1$ and $\ker\pi_2$ commute. □

Definition 10.11. *A congruence θ on $\mathfrak{A}$ is a* **factor congruence** *if there is a congruence θ' on $\mathfrak{A}$ such that*

$$\theta\cap\theta' = I_A,\quad \theta\circ\theta' = U_A,\quad \textit{and}\quad \theta\circ\theta' = \theta'\circ\theta.$$

θ and θ' are called a pair of factor congruences *on $\mathfrak{A}$.*

Two sets X and Y are called **isomorphic**, written as $X \cong Y$, if there is a bijection from X to Y. We will now have a look, in terms of equivalences, at when a set X is isomorphic to the Cartesian product $A\times B$ of two sets A and B.

Theorem 10.12. *If θ and θ' is a pair of factor congruences on $\mathfrak{A}$, then $\mathfrak{A} \cong \mathfrak{A}/\theta \times \mathfrak{A}/\theta'$ under the map $h : A \longrightarrow A/\theta\times A/\theta'$, $a \mapsto \langle a/\theta, a/\theta'\rangle$.*

Proof. Let $a,b\in A$. Then from $h(a) = h(b)$ it follows that $a/\theta = b/\theta$ and $a/\theta' = b/\theta'$, so $\langle a,b\rangle \in \theta\cap\theta' = I_A$: thus $a = b$, i.e. h is injective. Since $\theta\circ\theta' = U$ there is a $c\in A$ with $a\theta c\theta' b$; hence $h(c) = \langle c/\theta, c/\theta'\rangle = \langle a/\theta, b/\theta'\rangle$, i.e. h is onto. Finally, for $a_1,a_2,\ldots,a_n$ and any n-ary operation f

$$\begin{aligned}h\big(f^A(a_1,a_2,\ldots,a_n)\big) &= \langle f^A(a_1,a_2,\ldots,a_n)/\theta, f^A(a_1,a_2,\ldots,a_n)/\theta'\rangle\\ &= \langle f^{A/\theta}(a_1/\theta,\ldots,a_n/\theta), f^{A/\theta'}(a_1/\theta',\ldots,a_n/\theta')\rangle\\ &= f^{A/\theta\times A/\theta'}\big(\langle a_1/\theta,a_1/\theta'\rangle,\ldots,\langle a_n/\theta,a_n/\theta'\rangle\big)\end{aligned}$$

$$= f^{A/\theta \times A/\theta'}\big(h(a_1),\dots,h(a_n)\big).$$

Therefore, h is an isomorphism. □

Definition 10.13. *An algebra* $\mathfrak{A}$ *is called* **directly indecomposable** *if* $\mathfrak{A}$ *is not isomorphic to a direct product of two nontrivial algebras.*

Corollary 10.14. *An algebra* $\mathfrak{A}$ *is directly indecomposable iff the only factor congruences on it are* I_A *and* U_A.

Theorem 10.15. *Every finite algebra* $\mathfrak{A} = (A, \Omega^A)$ *is isomorphic to a direct product of directly indecomposable algebras.*

Proof. If $\mathfrak{A}$ is trivial then it is indecomposable. Proceeding by induction on $|A|$, suppose that $\mathfrak{A}$ is nontrivial such that every $\mathfrak{B} = (B, \Omega^B)$ with $|B| < |A|$ is isomorphic to a product of indecomposable algebras. If $\mathfrak{A}$ is indecomposable we are finished. If not, then $\mathfrak{A} \cong \mathfrak{A}_1 \times \mathfrak{A}_2$, with $1 < |A_1|, |A_2|$. So $|A_1|, |A_2| < |A|$, and by the induction hypothesis we have

$$\mathfrak{A}_1 \cong \mathfrak{B}_1 \times \cdots \times \mathfrak{B}_m \quad \text{and} \quad \mathfrak{A}_2 \cong \mathfrak{C}_1 \times \cdots \times \mathfrak{C}_n,$$

where all $\mathfrak{B}_i$ and $\mathfrak{C}_j$ are indecomposable. Therefore,

$$\mathfrak{A} \cong \mathfrak{B}_1 \times \cdots \times \mathfrak{B}_m \times \mathfrak{C}_1 \times \cdots \times \mathfrak{C}_n.$$

□

Definition 10.16.

(a) *Let* $\mathfrak{A}_i = (A_i, \Omega^{A_i})$, $i \in I$, *be a family of similar algebras, and* $\mathfrak{B}$ *be a subalgebra of their direct product* $\mathfrak{A} := \prod_{i \in I} \mathfrak{A}_i$. *Then* $\mathfrak{B}$ *is said to be* **subdirect product** *if, for each* $i \in I$, *the* i*-th projection restricted to* B *maps* $\mathfrak{B}$ *onto* $\mathfrak{A}_i$, *i.e.* $\pi_i|_B$ *is surjective.*

(b) *A* **subdirect representation** *of the algebra* $\mathfrak{A}$ *is an embedding* $f : \mathfrak{A} \longrightarrow \prod_{i \in I} \mathfrak{A}_i$ *(for some index set* I *and some algebras* $\mathfrak{A}_i$*) such that* $\pi_i f$ *maps* $\mathfrak{A}$ *onto* $\mathfrak{A}_i$, *for each* $i \in I$.

(c) *An algebra* $\mathfrak{A}$ *is said to be* **subdirectly irreducible** *if for every subdirect representation* $f : \mathfrak{A} \longrightarrow \prod_{i \in I} \mathfrak{A}_i$ *at least one of the homomorphisms* $\pi_i f : \mathfrak{A} \longrightarrow \mathfrak{A}_i$ *is an isomorphism.*

10.2.4 *Varieties and Equational Definability*

The study of classes of algebras of the same type closed under one or more constructions is a major topic in universal algebra. Among them *varieties* are the most prominent, and they have been the primary objects of study in universal algebra for the last few decades. We start our considerations with

Definition 10.17. *Let* $\mathcal{K}$ *be a class of similar algebras, and let* $\mathfrak{A}, \mathfrak{B}$ *denote algebras. Then the following notation is common:*

$\mathbf{H}(\mathcal{K}) := \{\mathfrak{A} : \mathfrak{A}$ *is a homomorphic image of some* $\mathfrak{B} \in \mathcal{K}\}$;
$\mathbf{S}(\mathcal{K}) := \{\mathfrak{A} : \mathfrak{A}$ *is isomorphic to a subalgebra of some* $\mathfrak{B} \in \mathcal{K}\}$;
$\mathbf{P}(\mathcal{K}) := \{\mathfrak{A} : \mathfrak{A}$ *is isomorphic to a direct product of some* $\mathfrak{B}_1, \ldots, \mathfrak{B}_k \in \mathcal{K}\}$;
$\mathbf{I}(\mathcal{K}) := \{\mathfrak{A} : \mathfrak{A}$ *is an isomorphic image of some* $\mathfrak{B} \in \mathcal{K}\}$.

A function mapping classes of algebras (of the same type) to classes of algebras (of the same type) are called **class operators**. If O and P are two such operators on classes of algebras, OP denotes their composition, defined by $OP(\mathcal{K}) = O(P(\mathcal{K}))$. Furthermore, let $\leq$ be the usual partial ordering: $O \leq P \iff O(\mathcal{K}) \subseteq P(\mathcal{K})$ for all classes of algebras $\mathcal{K}$ A class of similar algebras $\mathcal{K}$ is called **closed** under an operator O if $O(\mathcal{K}) \subseteq \mathcal{K}$.

Lemma 10.18. ([32, p. 168, Lemma 13]) *For the operators* **H**, **S**, *and* **P** *the following holds.*

(1) **H**, **S**, *and* **P** *are closure operators;*
(2) $\mathbf{SH} \leq \mathbf{HS}$;
(3) $\mathbf{PS} \leq \mathbf{SP}$;
(4) $\mathbf{PH} \leq \mathbf{HP}$;

Proof. (1): Immediate from Def. 10.17.

(2): Assume $\mathfrak{A} \in \mathbf{SH}(\mathcal{K})$. Then there is an algebra $\mathfrak{C} \in \mathcal{K}$ and an image $\mathfrak{B}$ of $\mathfrak{C}$ such that $\mathfrak{A}$ is a subalgebra of $\mathfrak{B}$. Let h be a surjective homomorphism from $\mathfrak{C}$ to $\mathfrak{B}$. Since $h^{-1}(\mathfrak{A})$ is a subalgebra of $\mathfrak{C}$, we have $h^{-1}(\mathfrak{A}) \in \mathbf{S}(\mathcal{K})$, and because of $\mathfrak{A} = h(h^{-1}(\mathfrak{A}))$ we get $\mathfrak{A} \in \mathbf{HS}(\mathcal{K})$.

(3): Assume that $\mathfrak{A} \in \mathbf{PS}(\mathcal{K})$. Then $\mathfrak{A}$ is either trivial, and then $\mathfrak{A} \in \mathbf{SP}(\mathcal{K})$; or there are algebras $\mathfrak{A}_i \in \mathbf{S}(\mathcal{K})$ such that $\mathfrak{A} = \prod_i \mathfrak{A}_i$. Then for each $\mathfrak{A}_i \in \mathbf{S}(\mathcal{K})$ there is an algebra $\mathfrak{B}_i \in \mathcal{K}$ such that $\mathfrak{A}_i$ is a subalgebra of $\mathfrak{B}_i$. But then $\mathfrak{A}$ obviously is a subalgebra of $\prod_i \mathfrak{B}_i$, whence $\mathfrak{A} \in \mathbf{SP}(\mathcal{K})$.

(4): If $\mathfrak{A} \in \mathbf{PH}(\mathcal{K})$, then there is a familiy $\{\mathfrak{A}_i : i \in \underline{k}\}$ of algebras $\mathfrak{A}_i \in \mathbf{H}(\mathcal{K})$ such that $\mathfrak{A} = \prod_{i \in I} \mathfrak{A}_i$. By definition, there are algebras $\mathfrak{B}_i$ such that $\mathfrak{A}_i$ is a homomorphic image of $\mathfrak{B}_i$ for all $i \in \underline{k}$. Let $h_i : \mathfrak{B}_i \longrightarrow \mathfrak{A}_i$ be a surjective homomorphism for each $i \in \underline{k}$, and put $\mathfrak{B} = \prod_{i \in I} \mathfrak{B}_i$. Then there is exactly one homomorphism $h : \mathfrak{B} \longrightarrow \mathfrak{A}$ such that, for all $i \in \underline{k}$, $h\pi_i^A = \pi_i^B h$. But h is surjective, whence $\mathfrak{A} \in \mathbf{HP}(\mathcal{K})$. □

A class $\mathcal{K}$ is called a **variety** if it is closed under each of **H**, **S**, and **P**. And since each of **H**, **S**, and **P** is a closure operator, $\mathcal{K}$ is a variety if and only if $\mathbf{H}(\mathcal{K}) = \mathbf{S}(\mathcal{K}) = \mathbf{P}(\mathcal{K}) = \mathcal{K}$. This means that the basic algebraic constructions of forming homomorphic images, subalgebras, and products can always be carried out within $\mathcal{K}$.

The intersection of a class of similar varieties of type τ is again a variety; and since also all algebras of type τ form a variety, it follows that for every class $\mathcal{K}$ of similar algebras there is a smallest variety containing $\mathcal{K}$, denoted by $\mathbf{V}(\mathcal{K})$, and called the **variety generated by** $\mathcal{K}$.

Theorem 10.19. (Tarski's theorem) *If* $\mathcal{K}$ *is any abstract class of algebras, then* $\mathbf{HSP}(\mathcal{K})$ *is the variety generated by* $\mathcal{K}$, *i.e.* $\mathbf{V} = \mathbf{HSP}$.

Proof. Using the idempotency of each operator and Lemma 10.18(2)-(4) we get

$$\mathbf{H}(\mathbf{HSP}(\mathcal{K})) = \mathbf{HSP}(\mathcal{K});$$
$$\mathbf{S}(\mathbf{HSP}(\mathcal{K})) \subseteq \mathbf{HSSP}(\mathcal{K}) = \mathbf{HSP}(\mathcal{K});$$
$$\mathbf{P}(\mathbf{HSP}(\mathcal{K})) \subseteq \mathbf{HPSP}(\mathcal{K}) \subseteq \mathbf{HSPP}(\mathcal{K}) = \mathbf{HSP}(\mathcal{K}).$$

Therefore $\mathbf{HSP}(\mathcal{K})$ is a variety. Being a closure operator each of $\mathbf{H}, \mathbf{S}$ and $\mathbf{P}$ is extensive, so clearly $\mathcal{K} \subseteq \mathbf{HSP}(\mathcal{K})$. If $\mathcal{V}$ is a variety with $\mathcal{K} \subseteq \mathcal{V}$ then, by the isotonicity of closure operators, $\mathbf{HSP}(\mathcal{K}) \subseteq \mathbf{HSP}(\mathcal{V}) = \mathcal{V}$. Thus $\mathbf{HSP}(\mathcal{K})$ is the smallest variety containing $\mathcal{K}$. □

Without getting more formal than necessary (the interested reader may consult e.g. [1, Chapter 1], [25, Section 4.11] or [32, Section 3.2]) we now turn to what is the fascinating thing about varieties in a more colloquial manner: If you have a look back to the definitions of the algebraic structures introduced beginning in Section 2.3.3 you will notice that they are all defined via a set of equations (note that you may substitute e.g. $A \cup B = B$ for $A \subseteq B$ etc. if necessary). Such equations have the form $t_1 \doteq t_2$, where t_1 and t_2 are so-called terms, and $\doteq$ is the symbol for *equality* in first-order logic.

Put in a more abstract framework, recall from Def. 2.45 that a type is a pair $\tau = (\Omega, \theta)$, with $\Omega = \{f_i | i \in I\}$ a set of operation symbols and a mapping $\theta : I \longrightarrow \mathbb{N}$ indicating their arity. Denoting by $\Omega_{(k)}$ the set of operation symbols of arity k, let X be a set of objects called *variables* such that $X \cup \Omega_{(0)} \neq \emptyset$, where $\Omega_{(0)}$ is the set of constants of Ω. The set $T(X)$ of *terms of type* τ *over* X is the smallest set such that

- $X \cup \Omega_{(0)} \subseteq T(X)$, and
- if $t_1, t_2, \ldots, t_n \in T(X)$ and $f_i \in \Omega_{(n)}$ (i.e. $\theta(i) = n$) then also $f_i(t_1, t_2, \ldots, t_n) \in T(X)$.

An equation of type τ is a string of the form $s \doteq t$, where s, t are two terms of type τ.

So we get a relatively simple vocabulary for the definition of algebraic structures: we have constants, n-ary function symbols f_i, and a binary relation symbol $\doteq$. In order to specify what is meant by the notion of an *equational* (or *equationally definable*) class, we need to have a short look towards some basic model theoretical concepts. (For an introductory text see [14].)

Definition 10.20.

(a) A (first-order) **language** *L is given by specifying*

- *a set of constant symbols $\mathcal{C}$;*
- *a set of function symbols $\mathcal{F}$ and a positive integer n_f for each $f \in \mathcal{F}$;*
- *a set of relation symbols $\mathcal{R}$ and a positive integer n_R for each $R \in \mathcal{R}$,*

where the numbers n_f and n_R indicate the arity of the respective function and relation symbols.

*(b) An L-***structure** *is a pair $\mathcal{A} = \langle A, v\rangle$ consisting of a non-empty set A, called the* universe, *or* underlying set, *of $\mathcal{A}$, and a mapping $v : \sigma \mapsto \sigma^{\mathcal{A}}$ defined for every non-logical symbol of L, that assigns*

- *an element $c^{\mathcal{A}} \in A$ to each constant $c \in \mathcal{C}$;*

- *a function* $f^{\mathcal{A}} : A^{n_f} \longrightarrow A$ *for each* $f \in \mathcal{F}$;
- *a relation* $R^{\mathcal{A}} \subseteq A^{n_R}$ *for each* $R \in \mathcal{R}$.

(c) *Given an L-structure* $\mathcal{A} = \langle A, v\rangle$, *an* $\mathcal{A}$-**assignment** *is a function* $\beta : \{x_i : i \in \mathbb{N}\} \longrightarrow A$ *from the set of all variables in L into the universe A of* $\mathcal{A}$.

(d) *An L*-**interpretation** *is a pair* $\mathcal{I} = \langle \mathcal{A}, \beta\rangle$ *formed by an L-structure* $\mathcal{A}$ *and an* $\mathcal{A}$-*assignment* β.

This modest model theoretic background suffices to answer the only interesting question that can be asked about an equation (of type τ as introduced above): is it valid under a certain interpretation or not?

Given an algebra $\mathfrak{A} = (A, \Omega^A)$ of type τ, then we can form an L-structure $\mathcal{A}^{\mathfrak{A}} = \langle A, v\rangle$ where each $c \in \mathcal{C}$ is mapped to an element $a_i \in A$, and to each function symbol $f \in \mathcal{F}$ is assigned an n_f-ary function $f^{\mathfrak{A}} \in \Omega^A$. Let $\mathcal{I}^{\mathfrak{A}} = \langle \mathcal{A}^{\mathfrak{A}}, \beta\rangle$ be an L-interpretation such that $\beta(x_i) = a_i$. We write $t(x_1, x_2, \ldots, x_n)$ to indicate that the variables occuring in t, if any, are among $x_1, x_2, \ldots, x_n$.

Now we define an n-ary operation $t^{\mathfrak{A}} : A^n \longrightarrow A$ on A by induction on the term structure as follows:

- if $t = c \in \mathcal{C}$, then $t^{\mathfrak{A}}(a_1, a_2, \ldots, a_n) = c^{\mathfrak{A}} \in A$;
- if $t = x_i$, then $t^{\mathfrak{A}}(a_1, a_2, \ldots, a_n) = a_i \in A$, i.e. $t^{\mathfrak{A}}$ is the i-th projection map;
- if $t = f\big(t_1(x_1, x_2, \ldots, x_n), \ldots, t_k(x_1, x_2, \ldots, x_n)\big)$, then
$$t^{\mathfrak{A}}(a_1, a_2, \ldots, a_n) = f^{\mathfrak{A}}\big(t_1^{\mathfrak{A}}(a_1, a_2, \ldots, a_n),\ \ldots,\ t_k^{\mathfrak{A}}(a_1, a_2, \ldots, a_n)\big).$$

As we have seen above, an equation of type τ is a string of the form $s \doteq t$, where s, t are two terms of type τ. Let $\mathfrak{A}$ be an algebra of type τ, and let s, t be two terms of type τ in the variables $x_1, x_2, \ldots, x_n$. Then an n-tuple $(a_1, a_2, \ldots, a_n)$ satisfies the equation $s \doteq t$ if $s^{\mathfrak{A}}(a_1, a_2, \ldots, a_n) = t^{\mathfrak{A}}(a_1, a_2, \ldots, a_n)$, and we say that $s \doteq t$ is **true in** $\mathfrak{A}$ if it is satisfied by all n-tuples $(a_{i_1}, \ldots, a_{i_n}) \in A^n$, denoted by $\mathfrak{A} \models s \doteq t$.

The equation $s \doteq t$ is said to be true in a class $\mathcal{K}$ of algebras of type τ, iff it is true in every member of $\mathcal{K}$, denoted $\mathcal{K} \models s \doteq t$. If E is a set of equations of type τ, we say E is true in $\mathfrak{A}$, denoted $\mathfrak{A} \models E$, iff every equation in E is true in $\mathfrak{A}$, and that E is true in $\mathcal{K}$, denoted $\mathcal{K} \models E$, iff every equation in E is true in $\mathcal{K}$.

Definition 10.21. *Let* $\mathcal{K}$ *be a class of algebras, and E be a set of equations of type* τ. *We put* $\Theta(\mathcal{K}) := \{s \doteq t : \mathcal{K} \models s \doteq t\}$ *and* $\mathbf{Mod}(E) = \{\mathfrak{A} : \mathfrak{A} \models E\}$.

(a) $\mathcal{K}$ *is called an* **equational class** *iff* $\mathcal{K} = \mathbf{Mod}(\Gamma)$ *for some set* Γ *of equations of type* τ. $\mathcal{K}$ *is then said to be* **defined**, *or* **axiomatized by**, Γ.

(b) *E is called an* **equational theory** *of* $\mathcal{L}$ *iff* $E = \Theta(\mathcal{L})$ *for some class* $\mathcal{L}$ *of algebras of type* τ.

Mod(E) is called the **model class** of E, and $\mathfrak{A} \in \mathbf{Mod}(E)$ is called a **model** of E. Clearly, for two sets E and E' of equations we have $\mathbf{Mod}(E) \subseteq \mathbf{Mod}(E')$ whenever $E' \subseteq E$. A class $\mathcal{K}$ of similar algebras is called **abstract** if it is closed under the formation of isomorphic images, i.e. $\mathfrak{A} \in \mathcal{K}$ whenever $\mathfrak{A}$ is isomorphic to an $\mathfrak{A}' \in \mathcal{K}$. Note that **Mod**$(E)$ is always abstract for any set E of equations.

Theorem 10.22. (Birkhoff's Variety Theorem, **HSP** theorem) *Suppose $\mathcal{K}$ is a class of similar algebras. Then the following statements are equivalent:*

(1) $\mathcal{K}$ is an equational class..
(2) $\mathcal{K}$ is a variety.

Proof. (This is beyond the level of this book, therefore I give a rough sketch and some hints only.)
It can be shown that equations are preserved under the formation of subalgebras, homomorphic images, and direct products (for detailed proofs see [32, pp. 161/162]). For an algebra $\mathfrak{A}$ we then get that $\mathfrak{A} \in \mathcal{K}$ whenever it is a subalgebra, or the image, of an algebra $\mathfrak{B} \in \mathcal{K}$, or if $\mathfrak{A}$ is the direct product of algebras $\mathfrak{B}_i \in \mathcal{K}$. Thus, every equational class is a variety.
The proof of the converse direction is harder. In addition to what we have learned up to here, the concept of free algebras and their properties are needed. For details see e.g. [23, p. 137, Theorem 95]. □

10.2.5 *Boolean Algebras with Operators*

In Section 10.3 we will see that (nonassociative) relation algebras are Boolean algebras enriched with certain operations and constants in order to abstract the properties characteristic of the set of all relations $\mathcal{B}_S$ over some set S. The resulting structure is called a *Boolean algebra with operators* and defined as follows (see e.g. [16, Definition 2.13] or [13, Chapter 2.5]):

Definition 10.23. *Let $\mathfrak{B} = (B, +, \cdot, \bar{\ }, 0, 1)$ be a Boolean algebra.*

(a) An n-ary operation f on B is called **additive** *if it distributes over + in each of its arguments:*
$$f(x_1, \ldots, y+z, \ldots, x_n) = f(x_1, \ldots, y, \ldots, x_n) + f(x_1, \ldots, z, \ldots, x_n).$$
(b) An n-ary operation f on B is called **completely additive** *if, for every indexed set $\{y_i : i \in I\} \subseteq B$, if $\sum_{i\in I} y_i$ exists and $I \neq \emptyset$ then $\sum_{i\in I} f(y_i)$ also exists and*
$$f(x_1, \ldots, \textstyle\sum_{i\in I} y_i, \ldots, x_n) = \sum_{i\in I} f(x_1, \ldots, y_i, \ldots, x_n).$$
(c) An n-ary operation f on B is called **normal** *if for each of its arguments:*
$$f(x_1, \ldots, 0, \ldots, x_n) = 0.$$
(d) An n-ary **operator** *on $\mathfrak{B}$ is an n-ary operation that is additive; an operator on $\mathfrak{B}$ is called* **normal** *if it is a normal operation on B.*
(e) A (normal) **Boolean algebra with operators** *(BAO) is an algebra $(B, +, \cdot, \bar{\ }, 0, 1, \{f_i : i \in I\})$, where $(B, +, \cdot, \bar{\ }, 0, 1)$ is a Boolean algebra and the $f_i, i \in I$, are (normal) operators on $\mathfrak{B}$.*

Thus, when f is an unary operator it is required that

$$f(x+y) = f(x) + f(y),$$

and for binary operators both

$$f(x, y+z) = f(x, y) + f(x, z) \quad \text{and} \quad f(x+y, z) = f(x, z) + f(y, z)$$

must hold.

Examples which we do already know from Chapter 3 are: $(\rho \cup \sigma)^{\smile} = \rho^{\smile} \cup \sigma^{\smile}$ (see Prop. 3.13(8)), and $\rho \cdot (\sigma \cup \tau) = (\rho \cdot \sigma) \cup (\rho \cdot \tau)$ and $(\sigma \cup \tau) \cdot \rho = (\sigma \cdot \rho) \cup (\tau \cdot \rho)$ (see Th. 3.16(6)). Moreover, relational composition is a normal operator: $\emptyset\rho = \emptyset = \rho\emptyset$ (see Th. 3.16(2)). Another example of a binary operator is the Cartesian product on sets (see Th. 2.18), and for an unary operator the closure on sets of points in a topological space.

The notion of a BAO evolved out of the study of relation algebras and has been proposed by Jónsson and Tarski in their 1951 paper [16]. There are other types of algebras that consist of a Boolean algebra with additional operations, particularly arising from the algebraic treatments of various kinds of logics. The Boolean structure obviously serves to handle the propositional connectives.

For later purposes we have to prepare for a more general level of argumentation about Boolean algebras (see [22] or [23, p. 253]), starting with

Definition 10.24. *Let f and g be two functions on a Boolean algebra $\mathfrak{B}$. g is called a* **conjugate of** *f if, for all $x, y \in B$, $f(x) \cdot y = 0 \leftrightarrow x \cdot g(y) = 0$.*

Proposition 10.25. (Using [22, Theorem 14]) *Let f, g, and h be functions on a Boolean algebra $\mathfrak{B}$.*

(1) If g and h are conjugates of f, then $g = h$.
(2) The following statements are equivalent:

(a) g is conjugate of f,
(b) f is conjugate of g.

Proof. (1): Assume g and h are conjugates of f. Then, for all elements $x, y \in \mathfrak{B}$, $x \cdot g(y) = 0 \leftrightarrow f(x) \cdot y = 0 \leftrightarrow x \cdot h(y) = 0$. Thus,

$$x \cdot g(y) = 0 \leftrightarrow x \cdot h(y) = 0. \tag{10.1}$$

For some fixed $y \in \mathfrak{B}$ we set $x = \overline{g(y)}$. Then $x \cdot g(y) = \overline{g(y)} \cdot g(y) = 0$. By (10.1) we get $\overline{g(y)} \cdot h(y) = h(y) \cdot \overline{g(y)} = 0$, whence $h(y) \leq g(y)$ by Prop. 2.67(1). Similarly, set $x = \overline{h(y)}$ to get $g(y) \leq h(y)$.

(2): Immediate consequence from Def. 10.24, using the commutativity of $\cdot$ and $\leftrightarrow$. □

Lemma 10.26. ([22, Theorem 16]) *In every* ***Boolean*** *algebra $\mathfrak{B}$ the left and right translations $\lambda_x(y) = x \cdot y$ and $\rho_x(y) = y \cdot x$ are normal and completely additive (see Def. 2.86(a)).*

Proof. Clearly, $\lambda_x(0) = x \cdot 0 = 0 = 0 \cdot x = \rho_x(0)$, so normality is proved. To show that λ_x is completely additive, assume $\sum_{i \in I} y_i$ exists and $I \neq \emptyset$. For every $j \in I$ we have $y_j \leq \sum_{i \in I} y_i$, whence $x \cdot y_j \leq x \cdot \sum_{i \in I} y_i$. Therefore, $x \cdot \sum_{i \in I} y_i$ is an upper bound of $\{x \cdot y_i :$

$i \in I\}$, i.e. $\sum_{i\in I} x \cdot y_i \le x \cdot \sum_{i\in I} y_i$. We show that $x \cdot \sum_{i\in I} y_i$ is the least upper bound of $\{x \cdot y_i : i \in I\}$ to get the inequality $x \cdot \sum_{i\in I} y_i \le \sum_{i\in I} x \cdot y_i$. Suppose that z is another upper bound of $\{x \cdot y_i : i \in I\}$, i.e. $x \cdot y_i \le z$ for each $i \in I$; then, for every $i \in I$, $y_i \le \overline{x} + z$ by Prop. 2.67(2). Thus, $\sum_{i\in I} y_i \le \overline{x} + z$, and so $x \cdot \sum_{i\in I} y_i \le z$ again using Prop. 2.67(2). Now take $z = \sum_{i\in I} x \cdot y_i$. □

Lemma 10.27. (Part of [22, Theorem 18]) *Let $f : B \longrightarrow B$ be a function on a Boolean algebra $\mathfrak{B}$. If f has a conjugate then it is normal and completely additive.*

Proof. Assume that g is a conjugate to f. Then, since $0 \cdot g(1) = 0$ we get $f(0) \cdot 1 = f(0) = 0$ by definition, whence $f(0) = 0$. To show that f is completely additive, assume that $I \neq \emptyset$, $\{x_i : i \in I\} \subseteq B$, and that $\sum_{i\in I} x_i$ exists. Then the following equivalences hold:

	$f(\sum_{i\in I} x_i) \le y$	
iff	$f(\sum_{i\in I} x_i) \cdot \overline{y} = 0$	by Prop. 2.67(1)
iff	$\sum_{i\in I} x_i \cdot g(\overline{y}) = 0$	since f and g are conjugate
iff	$\sum_{i\in I} x_i \le \overline{g(\overline{y})}$	by Prop. 2.67(1)
iff	$x_i \le \overline{g(\overline{y})}$ for every $i \in I$	by Def. 2.52(a), of upper bound
iff	$x_i \cdot g(\overline{y}) = 0$ for every $i \in I$	by Prop. 2.67(1)
iff	$f(x_i) \cdot \overline{y} = 0$ for every $i \in I$	since f and g are conjugate
iff	$f(x_i) \le y$ for every $i \in I$	by Prop. 2.67(1)
iff	y is an upper bound of $\{f(x_i) : i \in I\}$	

If $y = f(\sum_{i\in I} x_i)$ then the first statement becomes true, so $f(\sum_{i\in I} x_i)$ is an upper bound of $F := \{f(x_i) : i \in I\}$. Reading bottom up gives: y is an upper bound of F iff $f(\sum_{i\in I} x_i) \le y$, i.e. $f(\sum_{i\in I} x_i)$ is the least upper bound of F. Thus $f(\sum_{i\in I} x_i) = \sum_{i\in I} f(x_i)$, hence f is completely additive. □

Theorem 10.28. ([22, Theorem 19]) *Let f and g be functions on a Boolean algebra $\mathfrak{B}$. The following statements are equivalent:*

(1) *f and g are conjugates of each other,*
(2) *for all $x, y \in B$*
    (a) $f(x \cdot \overline{g(y)}) \le f(x) \cdot \overline{y}$,
    (b) $g(y \cdot \overline{f(x)}) \le g(y) \cdot \overline{x}$.
(3) *$f(0) = 0, g(0) = 0$ and, for all $y, z \in B$,*
    (a) $f(y) \cdot z \le f(y \cdot g(z))$,
    (b) $g(z) \cdot y \le g(z \cdot f(y))$.

Proof.
(1) ⇔ (2): Suppose f and g are conjugate. Then f is completely additive by Lemma 10.27. Obviously, a completely additive function is additive, and additive functions are monotone, so $f(x \cdot \overline{g(y)}) \le f(x)$. From $x \cdot \overline{g(y)} \cdot g(y) = 0$, by assumption we get

$f(x \cdot \overline{g(y)}) \cdot y = 0$. Hence $f(x \cdot \overline{g(y)}) \leq \overline{y}$ by Prop. 2.67(1). Therefore, $f(x \cdot \overline{g(y)}) \leq f(x) \cdot \overline{y}$. By a symmetric argument we get $g(x \cdot \overline{f(y)}) \leq g(x) \cdot \overline{y}$, and we are done with (1) ⇒ (2).

Conversely, assume (2). By Prop. 2.67(1) $x \cdot g(y) = 0$ implies $x \cdot \overline{g(y)} = x$, so $f(x) = f(x \cdot \overline{g(y)}) \leq f(x) \cdot \overline{y} \leq \overline{y}$ by assumption (2)(a). Thus, $f(x) \cdot y = 0$. If $f(x) \cdot y = 0$, then $y = y \cdot \overline{f(x)}$, whence $g(y) = g(y \cdot \overline{f(x)}) \leq g(y) \cdot \overline{x} \leq \overline{x}$ by assumption (2)(b). Thus $x \cdot g(y) = 0$, and (2) ⇒ (1) is shown.

(1) ⇔ (3): If f and g are conjugate, then they are normal and additive by Lemma 10.27, and (2(a)) holds, as just shown. Then

$$\begin{aligned}
f(y) \cdot z &= f\big(y \cdot (g(z) + \overline{g(z)})\big) \cdot z && \text{since } x \cdot 1 = 1\\
&= f\big(y \cdot g(z) + y \cdot \overline{g(z)}\big) \cdot z\\
&= \big(f(y \cdot g(z)) + f(y \cdot \overline{g(z)})\big) \cdot z && \text{since } f \text{ is additive}\\
&\leq \big(f(y \cdot g(z)) + f(y) \cdot \overline{z}\big) \cdot z && \text{by (2)(a), monotonicity}\\
&= f(y \cdot g(z)) \cdot z + f(y) \cdot \overline{z} \cdot z && \text{by distributivity}\\
&= f(y \cdot g(z)) \cdot z\\
&\leq f(y \cdot g(z)).
\end{aligned}$$

Thus we have proved (3)(a); the argument for (3)(b) is similar. Thus (1) ⇒ (3) has been shown.

For the other direction assume (3). From $f(y) \cdot z = 0$ we get $g(z) \cdot y \leq g(z \cdot f(y)) = g(0) = 0$ by (3)(b) and the normality of g. Conversely, if $g(z) \cdot y = 0$ then $f(y) \cdot z \leq f(y \cdot g(z)) = f(0) = 0$ by (3)(a) and the normality of f. So f and g are conjugate, and we have shown (3) ⇒ (1). □

10.3 Relation Algebras

From a mathematical point of view, this chapter continues on a more abstract level where Section 1.3 has ended: with TARSKI's axiomatic algebraic approach to the calculus of relations. This approach has two important features that distinguish it from the older viewpoint ([20, p. 443]):

- First, PEIRCE and SCHRÖDER worked with the set of *all* binary relations on a universe of discourse U, i.e. with $\mathcal{B}_U$, whereas TARSKI's axioms may hold in models consisting of only *some* of them: any set of relations containing the identity relation on U and closed under union, complementation with respect to $U \times U$, relative multiplication, and conversion will comply with.
- The second distinguishing feature is that there are relation algebras of binary relations in which the universal relation need not relate every element of the universe of discourse to every other.

As it is customary in the context of relation algebras, and to emphasize their character as an abstract algebraic structure, we will use ';' to denote composition of $x, y \in A$ explicitly, and we will sometimes take the freedom to still call elements of the carrier A of a relation algebra "*relations*". But we must keep in mind that they can, but need not, represent relations in the common mathematical sense as we will see in examples given below. When concepts from the theory of Boolean algebras are applied to a relation algebra $\mathfrak{A}$, it is understood that they refer to the Boolean part of $\mathfrak{A}$.

The following definition is due to Tarski and Givant [31, §8.2]. Axioms (R1)-(R10) are equations, so relation algebras form a variety, that is, the class of relation algebras is closed under the formation of subalgebras, homomorphic images, and direct products. Therefore, one can construct other examples of relation algebras by applying these operations to given ones.

Definition 10.29. *A* **relation algebra** *is a structure* $\mathfrak{A} = (A, +, ^{-}, ;, ^{\smile}, \mathbb{1})$ *of type* $\langle 2, 1, 2, 1, 0\rangle$ *such that for any* $x, y, z \in A$*:*

(R1)	$x + y = y + x,$	*(+-commutativity)*
(R2)	$x + (y + z) = (x + y) + z,$	*(+-associativity)*
(R3)	$\overline{\overline{x} + y} + \overline{\overline{x} + \overline{y}} = x,$	*(Huntington's axiom)*
(R4)	$x; (y; z) = (x; y); z,$	*(;-associativity)*
(R5)	$(x + y); z = x; z + y; z\,,$	*(right ;-distributivity)*
(R6)	$x; \mathbb{1} = x,$	*(right identity law)*
(R7)	$x^{\smile\smile} = x,$	*($^{\smile}$-involution)*
(R8)	$(x + y)^{\smile} = x^{\smile} + y^{\smile},$	*($^{\smile}$-distributivity)*
(R9)	$(x; y)^{\smile} = y^{\smile}; x^{\smile},$	*($^{\smile}$-involution product)*
(R10)	$x^{\smile}; \overline{x; y} + \overline{y} = \overline{y}.$	*(*TARSKI/DE MORGAN *axiom)*

The class of all relation algebras will be denoted by RA.

The operation ; is referred to as *relative multiplication*, the operation $^{\smile}$ as *conversion*, and the element $\mathbb{1}$ as the *identity element.* := $\overline{\mathbb{1}}$ is called the **diversity** element. Equations (R1) - (R3), due to Huntington [15], assert that $(A, +, ^{-})$ is a Boolean algebra, so that $\mathfrak{A}$ is a Boolean algebra with operators. (R10) is the link between the Boolean and relative parts of $\mathfrak{A}$, and is equivalent to $x^{\smile}; \overline{x; y} \leq \overline{y}$, since $\leq$ is defined as $x \leq y :\iff x + y = y$. It can also be proven equivalent to $x^{\smile}; \overline{x; y} \cdot y = 0$ using (R1) - (R3).

Next we give some examples of relation algebras, including also ones where the base set does not consist of binary relations. Still more examples with detailed descriptions can be found in [9, Chapter 3].

(1) The full algebra of relations $(\mathcal{B}_S, \cup, ^{-}, \circ, ^{\smile}, I_S)$ (recall Th. 3.4 and definitions 3.10 and 3.11) is the standard example for relation algebras, where

$$\rho \cap \sigma = \overline{\overline{\rho} \cup \overline{\sigma}},$$
$$\rho \subseteq \sigma \text{ iff } \rho \cup \sigma = \sigma,$$
$$U_S = I_S \cup \overline{I_S},$$
$$\emptyset = \overline{U_S}.$$

(2) Let $(G, \cdot, ^{-1}, e)$ be a group, and let $X, Y \subseteq G$. Define

$$X; Y = \{x \cdot y : x \in X \wedge y \in Y\},$$
$$X^{\smile} = \{x^{-1} : x \in X\},$$
$$1' = \{e\},$$

and let $^{-G}$ denote the complementation relative to G, i.e. $\overline{X}^G = G - X$. Then $(\mathbf{2}^G, \cup, ^{-G}, ;, \smile, 1')$ is a relation algebra.

(3): The point algebra, the interval algebra: see Section 15.3.

Given a set of axioms, it is always interesting to know if it is **independent**, i.e. if each of the axioms is independent of the remaining axioms in the set, which means that it cannot be derived from them. The standard method for establishing the independence of an axiom is to construct a model in which the axiom fails and the remaining ones hold. If an axiom has been shown to be dependent, then it is usually removed from the set of axioms and derived as a theorem in the subsequent axiomatic development of the theory. It is known that the set of axioms given in Def. 10.29 is independent (see [2]).

Looking through the relevant literature one soon encounters various logically equivalent axiomatizations of relation algebras that differ mainly in which axioms are used to describe the Boolean algebra reduct (above done via (R1) - (R3)), and in how they relate the Boolean with the relational part. Maddux [20] gives an overview on some of them, including remarks on their historical development which show that Tarski himself contributed to the topic over more than three decades. Def. 10.29 is the last one Tarski can be said to have been involved in (he died in 1983), and it is parsimonious in that it uses none of the definable constants and operations.

Since equations (R1) - (R3) assert that $(A, +, ^{-})$ is a Boolean algebra, instead of proving assertions anew from these axioms we will freely reuse the results obtained in Section 2.3.4 or well known to hold for Boolean algebras. Similarly for (R4) and (R6), which state that $(A, ;, \mathbb{1})$ is a monoid. These two requirements are common to all axiomatizations of the concept of a relation algebra.

In order to prepare those readers who want to proceed to the primary literature for the diversity of definitions and notations they will have to expect, I will also use different, but equivalent, ways of describing one and the same notion. I think this is better than to strictly adhere to only one kind of presentation: It saves you some of the translation labor later on, and offers various angles already now.

A definition that is often used and which exhibits all the relevant operations and constants explicitly and also clearly keeps apart the set-theoretical and relational component is the following (see e.g. [19]; its equivalence to Def. 10.29 is shown in Th. 10.35 below):

Definition 10.30. *A structure* $\mathfrak{A} = (A, +, \cdot, ^{-}, 0, 1, ;, \breve{}, \mathbb{1})$ *of type* $\langle 2,2,1,0,0,2,1,0\rangle$ *is called a* **relation algebra** *iff the following holds:*

(1) $(A, +, \cdot, ^{-}, 0, 1)$ *is a Boolean algebra,*
(2) $(A, ;, \breve{}, \mathbb{1})$ *is an involution monoid, and*
(3) $\forall x, y, z \in A[(x; y) \cdot z = 0 \leftrightarrow (x^{\smile}; z) \cdot y = 0 \leftrightarrow (z; y^{\smile}) \cdot x = 0]$.

In addition to relation algebras, some more general classes of structures have been proposed and investigated in literature. The next definition follows [21], presenting another definition of relation algebras related to the context of the more general notions.

Definition 10.31. *Let* $\mathfrak{A} = (A, +, \cdot, \bar{\ }, 0, 1, ;, \breve{\ }, \mathbb{1})$ *be an algebra of type* $\langle 2, 2, 1, 0, 0, 2, 1, 0\rangle$ *and let* $\mathfrak{A} = (A, +, \cdot, \bar{\ }, 0, 1)$ *be a Boolean algebra. Then* $\mathfrak{A}$ *is called a*

(a) **nonassociative relation algebra** *if* $\mathfrak{A}$ *satisfies the* identity *and the* cycle law

(IL) $x = x; \mathbb{1} = \mathbb{1}; x,$
(CL) $x \cdot (y; z) = 0$ *iff* $y \cdot (x; z^{\smile}) = 0$ *iff* $z \cdot (y^{\smile}; x) = 0.$

NA *denotes the class of all nonassociative relation algebras.*

(b) **weakly associative relation algebra** *if* $\mathfrak{A} \in$ NA *and if it satisfies the* weak associative law

(WL) $((x \cdot \mathbb{1}); 1); 1 = (x \cdot \mathbb{1}); (1; 1) = (x \cdot \mathbb{1}); 1.$

WA *denotes the class of all weakly associative relation algebras.*

(c) **semiassociative relation algebra** *if* $\mathfrak{A} \in$ NA *and if it satisfies the* semiassociative law

(SL) $(x; 1); 1 = x; 1.$

SA *denotes the class of all semiassociative relation algebras.*

(d) **relation algebra** *if* $\mathfrak{A} \in$ NA *and if it satisfies the* associative law

(AL) $(x; y); z = x; (y; z).$

RA *denotes the class of all relation algebras.*

Note that a non-associative relation algebra is an algebra which satisfies all of the axioms listed in Def. 10.29 except eventually (R4). Thus, the term 'non-associative' does not entail that the algebra is not associative, merely that it need not be (i.e., a relation algebra is nothing else than an associative non-associative relation algebra).

Proposition 10.32. ([23, Theorem 234]) *Concerning the various classes of relation algebras the following holds:*

(1) For $\mathcal{K} \in \{\mathsf{NA}, \mathsf{WA}, \mathsf{SA}, \mathsf{RA}\}$, $\mathcal{K} = \boldsymbol{H}(\mathcal{K}) = \boldsymbol{S}(\mathcal{K}) = \boldsymbol{P}(\mathcal{K})$.
(2) $\mathsf{RA} \subseteq \mathsf{SA} \subseteq \mathsf{WA} \subseteq \mathsf{NA}$.

Proof. (2): From Def. 10.31 we can see that (WL) is a special case of (SL) which itself is a special case of the associativity axiom Def. 10.29(R4). □

We mention without proof that all inclusions in assertion (2) are proper.

Looking at Def. 10.31 above we see that algebras in NA are Boolean algebras enriched with certain operations and constants in order to abstract the properties characteristic for $\mathcal{B}_S$ for some set S. So each $\mathfrak{A} \in$ NA is an example of a BAO.

Next we will have a look at basic properties of the classes introduced in Def. 10.31; a much more detailed exposition can be found in e.g. [13, Chapter 5.1] or Maddux's papers [19, 21] for their algebraic, and [22, 23].for their arithmetic aspects.

10.3.1 *Arithmetic in Relation Algebras*

In order to convince ourselves that the concept of relation algebras is a suitable abstraction, we have to "check" it somehow. It has already been mentioned in example (1) above that full algebras of relations are indeed models of (R1) - (R10). So we are on a right way, but this is only a conditio sine qua non. Additionally we must check further results and important theorems that hold in algebras of relations. This means that we *must not* reuse results from the relational calculus presented in Chapter 3, but have to prove them anew from the axioms postulated above, and will do this selectively as it seems appropriate.

Sometimes it may be interesting in its own, which of the axioms of a relation algebra are needed to guarantee certain properties for its elements. For this, Maddux's seminal book [23] is an invaluable source, which is much more detailed than what will be presented below.

10.3.1.1 *Arithmetic in* NA

Proposition 10.33. *In a relation algebra* $\mathfrak{A} \in \mathsf{NA}$ *the following holds:*

(1) $x \le y \to x^{\smile} \le y^{\smile}$.
(2) $x \le y \leftrightarrow x^{\smile} \le y^{\smile}$.
(3) $x \le y^{\smile} \leftrightarrow x^{\smile} \le y$.
(4) $\mathbb{1}^{\smile} = \mathbb{1}$.
(5) $x + y^{\smile} = \mathbb{1} \leftrightarrow x^{\smile} + y = \mathbb{1}$.
(6) $\overline{x}^{\smile} = \overline{x^{\smile}}$.
(7) $0^{\smile} = 0,\ 1^{\smile} = 1$.
(8) $(x \cdot y)^{\smile} = x^{\smile} \cdot y^{\smile}$.
(9) $x^{\smile} = y^{\smile} \leftrightarrow x = y$.
(10) $x^{\smile} \cdot y = 0 \leftrightarrow x \cdot y^{\smile} = 0$.
(11) $x = \mathbb{1}; x$.

Proof. (1): $x \le y \leftrightarrow x + y = y \to ((x+y)^{\smile} = y^{\smile} \leftrightarrow x^{\smile} + y^{\smile} = y^{\smile} \leftrightarrow x^{\smile} \le y^{\smile}$, using the definition of $\le$ and (R8).

(2): By applying (1) twice we get $x \le y \to x^{\smile} \le y^{\smile} \to x^{\smile\smile} \le y^{\smile\smile} \leftrightarrow x \le y$.

(3): Substitute $x^{\smile}$ for x in (2) and apply (R7).

(4): Applying (R7), (R8) and laws of Boolean algebras we get $\mathbb{1} = \mathbb{1} + \mathbb{1}^{\smile} = \mathbb{1}^{\smile\smile} + \mathbb{1}^{\smile} = (\mathbb{1}^{\smile} + \mathbb{1})^{\smile} = \mathbb{1}^{\smile}$.

(5): $x^{\smile} + y = x^{\smile} + y^{\smile\smile} = (x + y^{\smile})^{\smile} = \mathbb{1}^{\smile} = \mathbb{1}$.

(6): We have $\overline{x}^{\smile} \le y \leftrightarrow \overline{x} \le y^{\smile}$ by (3), and by 2.64(4) we get $\overline{x} \le y^{\smile} \leftrightarrow \overline{\overline{x}} + y^{\smile} = x + y^{\smile} = \mathbb{1}$. Using (R8) and (4) this gives $(x + y^{\smile})^{\smile} = \mathbb{1}^{\smile} = \mathbb{1}$, i.e. $x^{\smile} + y = \mathbb{1}$. But $x^{\smile} + y = \overline{\overline{x^{\smile}}} + y = \mathbb{1} \leftrightarrow \overline{x^{\smile}} \le y$ using by Prop. 2.64(4) again. So finally

$$\overline{x}^{\smile} \le y \leftrightarrow \overline{x} \le y^{\smile} \leftrightarrow \overline{x^{\smile}} \le y.$$

Choosing first $y = \overline{x}^{\smile}$ and then $y = \overline{x^{\smile}}$, substitution in the equivalence between the first and the third expression then yields:

$$\overline{x}^{\smile} \le \overline{x}^{\smile} \leftrightarrow \overline{x^{\smile}} \le \overline{x}^{\smile} \quad \text{and} \quad \overline{x}^{\smile} \le \overline{x^{\smile}} \leftrightarrow \overline{x^{\smile}} \le \overline{x^{\smile}}.$$

(7) By definition $0 \le 0^{\smile}$. By (1) we get $0^{\smile} \le 0^{\smile\smile} = 0$.

(8) We have $xy \le x$ and $xy \le y$. So, by (1), $(xy)^{\smile} \le x^{\smile}$ and $(xy)^{\smile} \le y^{\smile}$, whence $(xy)^{\smile} \le x^{\smile}y^{\smile}$. Replacing x, y by their converses in the last formula gives $(x^{\smile}y^{\smile})^{\smile} \le xy$, which implies $x^{\smile}y^{\smile} \le (xy)^{\smile}$ using (1) and the involution property of $^{\smile}$.

(9): If $x = y$ then clearly $x^{\smile} = y^{\smile}$. Conversely, $x^{\smile} = y^{\smile}$ implies $x^{\smile\smile} = y^{\smile\smile}$, so $x = y$ using (R7).

(10): $x^{\smile} \cdot y = 0 \leftrightarrow (x^{\smile} \cdot y)^{\smile} = 0^{\smile} \leftrightarrow x^{\smile\smile} \cdot y^{\smile} = 0 \leftrightarrow x \cdot y^{\smile} = 0$, by applying, in that order (9), (7) and (8), and (R7).

(11): Using (4), (R7) and (R9) we get $\mathbb{1}; x = \mathbb{1}^{\smile}; x^{\smile\smile} = (x^{\smile}; \mathbb{1})^{\smile} = x^{\smile\smile} = x$. □

Proposition 10.34. *Let $\mathfrak{A} \in NA$ be a relation algebra; then the following holds:*

(1) $x; (y+z) = (x; y) + (x; z)$	*(distributivity of ; over +)*
(2) $x \le y \rightarrow z; x \le z; y \wedge x; z \le y; z$	*(monotonicity of ;)*
(3) $\overline{y; x}; x^{\smile} \le \overline{y}$	
(4) $x; y \le z \leftrightarrow x^{\smile}; \overline{z} \le \overline{y} \leftrightarrow \overline{z}; y^{\smile} \le \overline{x}$	*(Schröder rule)*
(5) $1; 1 = 1$	
(6) $a \le 1; a$ *and* $a \le a; 1$	

Proof.
(1) Due to (R5) it suffices to prove distributivity from the left, i.e. $x; (y+z) = x; y + x; z$.

$$\begin{aligned} x; (y+z) &= x^{\smile\smile}; (y^{\smile\smile} + z^{\smile\smile}) && \text{(R7)} \\ &= x^{\smile\smile}; (y^{\smile} + z^{\smile})^{\smile} && \text{(R8)} \\ &= ((y^{\smile} + z^{\smile}); x^{\smile})^{\smile} && \text{(R9)} \\ &= (y^{\smile}; x^{\smile} + z^{\smile}; x^{\smile})^{\smile} && \text{(R5)} \\ &= (y^{\smile}; x^{\smile})^{\smile} + (z^{\smile}; x^{\smile})^{\smile} && \text{(R8)} \\ &= x^{\smile\smile}; y^{\smile\smile} + x^{\smile\smile}; z^{\smile\smile} && \text{(R9)} \\ &= x; y + x; z && \text{(R7)} \end{aligned}$$

(2): Assume $x \le y$, i.e. $x + y = y$. Then, applying (1) we get $z; x + z; y = z; (x+y) = z; y$, thus $z; x \le z; y$. An analogous argument gives $x; z + y; z = (x+y); z = y; z$, thus $x; z \le y; z$.

$$\begin{aligned} \text{(3): } \overline{y; x}; x^{\smile} &= \overline{y^{\smile\smile}; x^{\smile\smile}}; x^{\smile} && \text{(R7)} \\ &= \overline{(x^{\smile}; y^{\smile})^{\smile}}; x^{\smile} && \text{(R9)} \\ &= (\overline{x^{\smile}; y^{\smile}})^{\smile}; x^{\smile} && \text{Prop. 10.33(6)} \\ &= (x^{\smile\smile}; \overline{x^{\smile}; y^{\smile}})^{\smile} && \text{(R9), (R7)} \\ &\le (\overline{y^{\smile}})^{\smile} && \text{(R10), Prop. 10.33(1)} \\ &= \overline{y^{\smile\smile}} = \overline{y} && \text{Prop. 10.33(6), (R7)} \end{aligned}$$

(4): We first show that $x; y \le z \rightarrow x^{\smile}; \overline{z} \le \overline{y} \wedge \overline{z}; y^{\smile} \le \overline{x}$. Assume $x; y \le z$; then by Prop. 2.67(1) $\overline{z} \le \overline{x; y}$, and by (2) and (R10) $x^{\smile}; \overline{z} \le x^{\smile}; \overline{x; y} \le \overline{y}$. From $\overline{z} \le \overline{x; y}$ we also get $\overline{z}; y^{\smile} \le \overline{x; y}; y^{\smile} \le \overline{x}$ by using (2) and (3).

Substituting $x^{\smile}$ for x in $x;y \le z \ \rightarrow\ x^{\smile};\overline{z} \le \overline{y}$ and applying (R7) gives $x^{\smile};y \le z \ \rightarrow\ x;\overline{z} \le \overline{y}$. Then substitute $\overline{z}$ for y and $\overline{y}$ for z to get $x^{\smile};\overline{z} \le \overline{y} \ \rightarrow\ x;\overline{\overline{y}} \le \overline{\overline{z}}$, whence $x^{\smile};\overline{z} \le \overline{y} \ \rightarrow\ x;y \le z$, completing the proof of the first equivalence.

(5): Since $\mathbb{1} \le 1$, using (2) and (R6) we get $1 = 1;\mathbb{1} \le 1;1 \le 1$.

(6): By (2), $a = \mathbb{1};a \le 1;a$. □

We are now ready to show the equivalence of definitions Def. 10.29 and Def. 10.30 of a relation algebra given above, using [17, Theorem 2.2].

Theorem 10.35. *Def. 10.29 and Def. 10.30 are equivalent.*

Proof. [⇒] A relation algebra given by Def. 10.29 is a structure $\mathfrak{A} = (A, +, ^{-}, ;, ^{\smile}, \mathbb{1})$ of type $\langle 2,1,2,1,0\rangle$, such that $(A, +, ^{-})$ is a Boolean algebra. For any $x, y \in A$ we set

$x \cdot y = \overline{\overline{x} + \overline{y}}$,	(Boolean multiplication)
$x \le y$ iff $x + y = y$,	(relation inclusion)
$1 = \mathbb{1} + \overline{\mathbb{1}}$,	(Boolean unit)
$0 = \overline{1}$.	(Boolean zero)

Straightforward calculations then show that $(A, +, \cdot, ^{-}, 0, 1)$ is a Boolean algebra, and by (R6), (R7), and (R9) $(A, ;, ^{\smile}, \mathbb{1})$ is an involution monoid.

If $(x;y) \cdot z = 0$, then by Prop. 2.67(1) $z \le \overline{x;y}$, whence $x^{\smile};z \le x^{\smile};\overline{x;y} \le \overline{y}$, using Prop. 10.34(2) and (R10). Thus $(x^{\smile};z) \cdot y = 0$ by Prop. 2.67(2). Conversely, by what we have just shown, $(x^{\smile};z) \cdot y = 0$ implies $(x^{\smile\smile};y) \cdot z = 0$, i.e. $(x;y) \cdot z = 0$. So we have shown $(x;y) \cdot z = 0 \leftrightarrow (x^{\smile};z) \cdot y = 0$, the first equivalence in Def. 10.30(3).

To prove the second equivalence we assume $z;y^{\smile} \cdot x = 0$. By applying the equivalence just shown, $z;y^{\smile} \cdot x = 0 \leftrightarrow z^{\smile};x \cdot y^{\smile} = 0$. Now

$$z^{\smile};x \cdot y^{\smile} = 0 \leftrightarrow (z^{\smile};x \cdot y^{\smile})^{\smile} = 0^{\smile} \leftrightarrow (z^{\smile};x)^{\smile} \cdot y^{\smile\smile} = 0^{\smile} \leftrightarrow x^{\smile};z \cdot y = 0^{\smile} = 0,$$

using Prop. 10.33(8) and (7).

[⇐] Assume $\mathfrak{A}$ to be relation algebra according to Def. 10.30; then its condition (3),

$$\forall x, y, z \in A[(x;y) \cdot z = 0 \leftrightarrow (x^{\smile};z) \cdot y = 0 \leftrightarrow (z;y^{\smile}) \cdot x = 0],$$

asserts that, for a fixed element a, the maps $x \mapsto a;x$ and $y \mapsto a^{\smile};y$ are conjugate, and also the maps $x \mapsto x;a$ and $y \mapsto y;a^{\smile}$. (If $f(x) = x;a$ and $g(x) = x;a^{\smile}$, then $0 = x;a \cdot z = f(x) \cdot z \ \leftrightarrow\ g(z) \cdot x = z;a^{\smile} \cdot x = 0$.) By Lemma 10.27, f preserves existing joins, so $f(x + y) = (x + y);a = f(x) + f(y) = x;a + y;a$, so (R5) holds.

Applying Def. 10.30(3) with $z = \mathbb{1}$, the second equivalence gives $x^{\smile};y = 0 \leftrightarrow y^{\smile};x = 0$, so the map $x \mapsto x^{\smile}$ is self-conjugate, hence (R8) holds. It remains to show (R10): Let $f(x) = a;x$ and $g(x) = a^{\smile};x$, then $0 = a;x \cdot z = f(x) \cdot z \ \leftrightarrow\ g(z) \cdot x = a^{\smile};z \cdot x = 0$, so f and g are conjugate. Taking $z = \overline{f(y)}$, we clearly have $f(y) \cdot z = 0$, and thus $g(z) \cdot y = 0$, i.e. $a^{\smile};z \cdot y = a^{\smile};\overline{f(y)} \cdot y = a^{\smile};\overline{a;y} \cdot y = 0 \ \leftrightarrow\ a^{\smile};\overline{a;y} \le \overline{y}$ by Prop. 2.67(1). □

Corollary 10.36. *In a relation algebra the following holds:*

(1) *For a fixed element a, the maps $f : x \mapsto a;x$ and $g : x \mapsto a^{\smile};x$ are conjugate.*

(2) For a fixed element a, the maps $f : x \mapsto x;a$ and $g : x \mapsto x;a^{\smile}$ are conjugate.

Proof. (1), (2): Follows from the [⇐] -part in the proof of Th. 10.35. □

Corollary 10.37. [23, p. 311, Theorem 295]) *In a relation algebra $\mathfrak{A}$ the following holds:*

(1) For every $x \in A$, the following statements are equivalent:

(a) The functions $f(\cdot) := x;(\cdot)$ and $g(\cdot) := x^{\smile};(\cdot)$ are conjugates of each other.
(b) $x;y \cdot z = 0 \leftrightarrow y \cdot x^{\smile};z = x^{\smile};z \cdot y = 0$, for all $y, z \in A$.
(c) $x;(\overline{y \cdot x^{\smile};z}) \leq x;\overline{y \cdot z}$ and $x^{\smile};(\overline{z \cdot x;y}) \leq x^{\smile};z \cdot \overline{y}$, for all $y, z \in A$.
(d) $x;0 = 0 = x^{\smile};0$, and
$x;y \cdot z \leq x;(y \cdot x^{\smile};z)$ and $x^{\smile};y \cdot z \leq x^{\smile};(y \cdot x;z)$, *for all $y, z \in A$.*

(2) For every $x \in A$, the following statements are equivalent:

(a) The functions $f(\cdot) := (\cdot);x$ and $g(\cdot) := (\cdot);x^{\smile}$ are conjugates of each other.
(b) $y;x \cdot z = 0 \leftrightarrow z;x^{\smile} \cdot y = 0$, for all $y, z \in A$.
(c) $(\overline{y \cdot z;x^{\smile}});x \leq y;x \cdot \overline{z}$ and $(\overline{z \cdot y;x});x^{\smile} \leq z;x^{\smile} \cdot \overline{y}$, for all $y, z \in A$.
(d) $0;x = 0 = 0;x^{\smile}$, and
$y;x \cdot z \leq (y \cdot z;x^{\smile});x$ and $z;x^{\smile} \cdot y \leq (z \cdot y;x);x^{\smile}$, *for all $y, z \in A$.*

Proof. Direct consequences from Def. 10.24, Th. 10.28, and Cor. 10.36. □

The fact that the Schröder rule holds in relation algebras by Prop. 10.34(4) gives a considerable amount of confidence that - with axioms (R1)-(R10) as basis - one should be able to derive all laws of the calculus of relations, the primary algebraic goal of the axiomatization. So, to enhance our confidence, we prove some more of the important laws valid in the calculus of relations also for relation algebras.

Proposition 10.38. *Let $\mathfrak{A}$ be a relation algebra; then the following holds:*

(1) $x;0 = 0;x = 0$	
(2) $x;(y \cdot z) \leq (x;y) \cdot (x;z)$	*subdistributivity of ; over ·*
$(x \cdot y);z \leq (x;z) \cdot (y;z)$	
(3) $x;y \cdot z \leq (x \cdot z;y^{\smile});y$	*modular law, Th. 3.26(2)(a)*
(4) $x;y \cdot z \leq x;(y \cdot x^{\smile};z)$	*modular law, Th. 3.26(2)(b)*
(5) $x;y \cdot z \leq (x \cdot z;y^{\smile});(y \cdot x^{\smile};z)$	*Dedekind rule, Th. 3.26(3)*

Proof.
(1): By Cor. 10.36 and Lemma 10.27 the maps $x \mapsto a;x$ and $x \mapsto x;a$ are normal.

(2): By Prop. 10.34(2).

(3): By Cor. 10.36(2) the functions $f(y) := y;x$ and $g(z) := z;x^{\smile}$ are conjugate. Then, by Th. 10.28(3)(a), we get $f(x) \cdot z = x;y \cdot z \leq f(x \cdot g(z)) = (x \cdot z;y^{\smile});y$.

(4): By Cor. 10.36(1) the functions $f(y) := x;y$ and $g(z) := x^{\smile};z$ are conjugate. Then, by Th. 10.28(3)(a), we get $f(y) \cdot z = x;y \cdot z \leq f(y \cdot g(z)) = x;(y \cdot x^{\smile};z)$.

$$
\begin{aligned}
(5)\colon\ x;y\cdot z &= [x;y\cdot z]\cdot z \\
&\le [(x\cdot z;y^{\smile});y]\cdot z && \text{by (3), monotonicity}\\
&= [(x\cdot z;y^{\smile});y\cdot z]\cdot z \\
&\le [(x\cdot z;y^{\smile});((x\cdot z;y^{\smile})^{\smile};z\cdot y)]\cdot z && \text{by (4) with } x := x\cdot z;y^{\smile}\\
&\le [(x\cdot z;y^{\smile});((x)^{\smile};z\cdot y)]\cdot z && \text{by monotonicity}\\
&\le (x\cdot z;y^{\smile});(y\cdot x^{\smile};z)
\end{aligned}
$$

□

The proof of the Dedekind rule is an example where we are allowed to simply transcribe the corresponding proof of Section 3.4, since all the relational arguments needed have already been shown and the rest are laws of Boolean algebra. Another well known theorem in NA is

Theorem 10.39. ([23, p. 311, Theorem 214]) *(The cycle laws)*
The statements given in Fig. 10.3, illustrated by oriented triangles with the standard position (2), are equivalent.

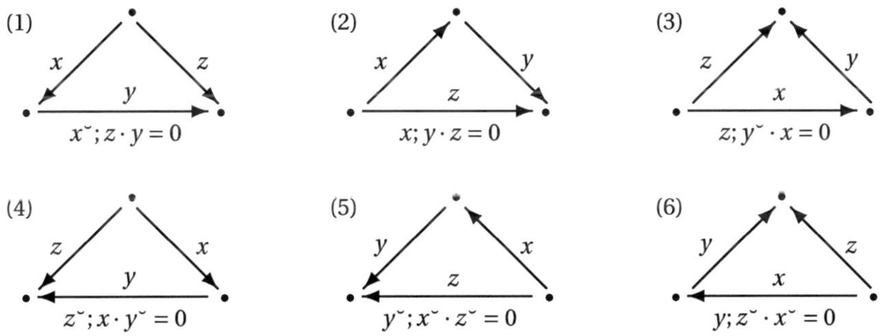

FIG. 10.3: The cycle laws.

Proof. The equivalences (1) $\leftrightarrow$ (2) $\leftrightarrow$ (3) hold by Th. 10.35. Applying the equivalence (1) $\leftrightarrow$ (2) to (4), gives (4) $\leftrightarrow$ (3), and applied to (5) it yields (5) $\leftrightarrow$ (6). Finally, from (3) $\leftrightarrow$ (2) applied to (6) we get (6) $\leftrightarrow$ (1). □

Corollary 10.40. ([13, Definition 3.10]) *(Peircean law)*

$$(x;y)\cdot z^{\smile}=0 \leftrightarrow (y;z)\cdot x^{\smile}=0.$$

Proof. Using the equivalence (1) $\leftrightarrow$ (3) of Th. 10.39 with a mutual renaming of y and z we get

$$
\begin{array}{llcll}
x^{\smile};y\cdot z & =0 & \leftrightarrow & y;z^{\smile}\cdot x & =0, \quad \text{thus, by setting } x\leftarrow x^{\smile},\ z\leftarrow z^{\smile},\\
x^{\smile\smile};y\cdot z^{\smile} & =0 & \leftrightarrow & y;z^{\smile\smile}\cdot x^{\smile} & =0, \quad \text{and, using involution of } {}^{\smile}\text{, this gives}\\
x;y\cdot z^{\smile} & =0 & \leftrightarrow & y;z\cdot x^{\smile} & =0.
\end{array}
$$

□

Proposition 10.41. *Let $\mathfrak{A} \in \mathsf{NA}$; then the following holds:*

(1) $r \leq s \wedge x \leq y \rightarrow r;x \leq s;y$.
(2) $x;y \cdot z \leq x;(y \cdot x^{\smile};z)$.
(3) $y;x \cdot z \leq (y \cdot z;x^{\smile});x$.
(4) $x \leq (x;x^{\smile});x$ *and* $x \leq x;(x^{\smile};x)$.

Proof.
(1): Using Prop. 10.34(2) we have $r \leq s \wedge x \leq y \rightarrow r;x \leq r;y \ \wedge \ r;y \leq s;y$, so $r;x \leq s;y$ by transitivity of $\leq$.

(2), (3): By Cor. 10.37(1)(d) and (2)(d).

(4):

x	$= x;\mathbb{1} \cdot x$	(R6)
	$\leq x;(\mathbb{1} \cdot x^{\smile};x)$	(2)
	$\leq x;(x^{\smile};x)$	(1)
x	$= \mathbb{1};x \cdot x$	Prop. 10.33(11)
	$\leq (\mathbb{1} \cdot x;x^{\smile});x$	(3)
	$\leq (x;x^{\smile});x$	(1)

□

10.3.1.2 *The Need for Associativity*

In view of the many results we could show above without using associativity axiom (R4) one might well be tempted to ask for properties and results for which associativity is needed. Some of them are provided in this section, mostly concerning the operation of relative sum or rather special properties of relative composition. Recall that the *relative sum* of two relations x and y has been defined in Section 1.3.1 as

$$\rho \dagger \sigma = \{\langle a,b\rangle : \text{ for every } c \in U, \text{ either } \langle a,c\rangle \in \rho \text{ or } \langle b,c\rangle \in \sigma\},$$

and the properties (B20) $\overline{x \dagger y} = \overline{x};\overline{y}$ and (B21) $\overline{x;y} = \overline{x} \dagger \overline{y}$ have been mentioned. In relation algebras we define

$$x \dagger y := \overline{\overline{x};\overline{y}} \tag{10.2}$$

according to [6, DEFINITION 1.13].

Lemma 10.42. ([6, THEOREM 1.23], [26, Theorem 2.21]) $\overline{(y;x)};x^{\smile} \subseteq \overline{y}$.

Proposition 10.43. ([23, Section 47] and [6, THEOREM 2.17])

(1) $(x \dagger y) \dagger z = x \dagger (y \dagger z)$.
(2) $x;(y \dagger z) \subseteq (x;y) \dagger z$ *and* $(x \dagger y);z \subseteq x \dagger (y;z)$.

Proof.
(1): $(x \dagger y) \dagger z = \overline{(\overline{x};\overline{y})} \dagger z = \overline{\overline{\overline{(\overline{x};\overline{y})}};\overline{z}} = \overline{(\overline{x};\overline{y});\overline{z}} = \overline{\overline{x};(\overline{y};\overline{z})} = \overline{\overline{x};\overline{\overline{(\overline{y};\overline{z})}}} = x \dagger \overline{(\overline{y};\overline{z})}$
$= x \dagger (y \dagger z)$

(2): From (10.2) we get

$$(\overline{x};\overline{y}) \cap (x \dagger y) = 0, \tag{10.3}$$

and by Lemma 10.42 we have $(\overline{y;z});z^{\smile} \subseteq \overline{y}$. So $\overline{x};(\overline{y;z});z^{\smile} \subseteq \overline{x};\overline{y}$ by monotonicity of ;, and thus

$$(\overline{x};(\overline{y;z});z^{\smile}) \cap (x \dagger y) \subseteq (\overline{x};\overline{y}) \cap (x \dagger y) \tag{10.4}$$

Combining (10.3) and (10.4) gives $(\overline{x};(\overline{y;z});z^{\smile}) \cap (x \dagger y) = 0$, from which we get $(x \dagger y);z \cap (\overline{x};(\overline{y;z})) = 0$ by using the equivalence of Th. 10.39(2),(3). So $(x \dagger y);z \subseteq \overline{(\overline{x};(\overline{y;z}))}$, and by (10.2) $(x \dagger y);z \subseteq x \dagger (y;z)$. □

Associativity is also needed to prove Prop. 10.50(3) and Prop. 10.52(6) below.

10.3.2 *Special Kinds of Elements*

Many of the notions applying to binary relations that we have learned about can be expressed abstractly by means of equations in the language of relation algebras as shown in the definition below. The contents of this section are based on results presented [9, Chapter 5] and [23, Chapters 6.4 and 6.15].

Definition 10.44. *An element a in a relation algebra $\mathfrak{A}$ is called*

(a) **reflexive** *if* $\mathbb{1} \leq a$;
(b) **symmetric** *if* $a^{\smile} \leq a$;
(c) **transitive** *if* $a;a \leq a$;
(d) **functional** *if* $a^{\smile};a \leq \mathbb{1}$;
(e) *a* **subidentity element** *if* $a \leq \mathbb{1}$;
(f) *an* **equivalence element** *if it is symmetric and transitive;*
(g) *a* **domain** *(or* **right ideal***)* **element** *if* $a = a;1$;
(h) *a* **range** *(or* **left ideal***)* **element** *if* $a = 1;a$;
(i) *an* **ideal element** *if* $1;a;1 = a$;
(j) *a* **square** *if* $a = a;1;a^{\smile}$;
(k) *a* **rectangle** *if* $a = a;1;a$.

In the following we collect some properties of some of the special elements just mentioned. For the considerations below we let $\mathsf{Idl}(\mathfrak{A})$ denote the set of ideals, and $\mathsf{Sy}(\mathfrak{A}), \mathsf{Tr}(\mathfrak{A})$ and $\mathsf{Eq}(\mathfrak{A})$ the sets of symmetric, transitive, and equivalence elements of $\mathfrak{A}$, respectively. Detailed information on equivalence and ideal elements will be given in separate sections below.

10.3.2.1 *Symmetric Elements*

Proposition 10.45. *The following conditions on an element a are equivalent in a relation algebra* $\mathfrak{A} \in \mathsf{NA}$.

(1) *a is symmetric.*
(2) $a^{\smile} = a$.
(3) $a = b + b^{\smile}$ *for some element b.*

Proof. From (1) by definition we get $a^{\smile} \leq a$, so $a = a^{\smile\smile} \leq a^{\smile}$ by Prop. 10.33(1), hence (2) follows. The other direction is trivial. Assuming (2) gives $a = a + a = a + a^{\smile}$, i.e. (3) holds. Conversely, if (3) is valid then $a^{\smile} = (b + b^{\smile})^{\smile} = b^{\smile} + b^{\smile\smile} = b^{\smile} + b = a$; thus (2) holds. □

Proposition 10.46. ([23, p. 300, Theorem 252]) *Let* $a, b \in \mathsf{Sy}(\mathfrak{A})$ *for a relation algebra* $\mathfrak{A} \in \mathsf{NA}$. *Then*

(1) $0, 1 \in \mathsf{Sy}(\mathfrak{A})$.
(2) $a + a^{\smile} \in \mathsf{Sy}(\mathfrak{A})$.
(3) $\overline{a} \in \mathsf{Sy}(\mathfrak{A})$.
(4) $a + b \in \mathsf{Sy}(\mathfrak{A})$.
(5) $a \cdot b \in \mathsf{Sy}(\mathfrak{A})$.

Proof. (1): Using (R7) and (R8) we get $0^{\smile} = 0 + 0^{\smile} = 0^{\smile\smile} + 0^{\smile} = (0^{\smile} + 0)^{\smile} = 0^{\smile\smile} = 0$. For 1 use $1 = 1 + 1^{\smile}$ to get an analogous proof.

(2): $a + a^{\smile} = a^{\smile\smile} + a^{\smile} = (a^{\smile} + a)^{\smile} = (a + a^{\smile})^{\smile}$.

(3): Using Prop. 10.33(6) we get $\overline{a} = \overline{a^{\smile}} = \overline{a}^{\smile}$. □

Proposition 10.47. ([23, p. 302, Theorems 258]) *Let* $a, b \in \mathsf{Sy}(\mathfrak{A})$ *for a relation algebra* $\mathfrak{A} \in \mathsf{NA}$. *Then*

(1) $a; b \in \mathsf{Sy}(\mathfrak{A})$ *iff* $a; b = b; a$
(2) $a; a \in \mathsf{Sy}(\mathfrak{A})$.

Proof. (1): First assume $a; b = b; a$; then using (R9) and Prop. 10.45 we get $(a; b)^{\smile} = b^{\smile}; a^{\smile} = b; a = a; b$, i.e. $a; b$ is symmetric. Conversely, if $a; b \in \mathsf{Sy}(\mathfrak{A})$; then $(a; b)^{\smile} = a; b = a^{\smile}; b^{\smile} = (b; a)^{\smile}$. So $a; b = b; a$ by Prop. 10.33(9). □

10.3.2.2 *Transitive Elements*

Proposition 10.48. *Let* $a, b \in \mathsf{Tr}(\mathfrak{A})$ *for a relation algebra* $\mathfrak{A} \in \mathsf{NA}$. *Then*

(1) $a; a \in \mathsf{Tr}(\mathfrak{A})$.
(2) $a \cdot b \in \mathsf{Tr}(\mathfrak{A})$.
(3) $\mathbb{1} \in \mathsf{Tr}(\mathfrak{A})$.

Proof. (1): We have $a; a \leq a$, so by monotonicity $(a; a); (a; a) \leq a; a$.

(2): By definition $a; a \leq a$ and $b; b \leq b$. Therefore, $(a \cdot b); (a \cdot b) \leq a; a \leq a$ and $(a \cdot b); (a \cdot b) \leq b; b \leq b$ by monotonicity of ;. Thus $(a \cdot b); (a \cdot b) \leq a \cdot b$.

(3): By (R6). □

Proposition 10.49. ([9, Lemma 5.5]) *The following conditions on an element* a *are equivalent for a relation algebra* $\mathfrak{A} \in \mathsf{NA}$.

(1) $a \in \mathsf{Tr}(\mathfrak{A})$.
(2) $a^{\smile}; \overline{a} \leq \overline{a}$.

(3) $\overline{a}; a^{\smile} \leq \overline{a}$.

Proof. Using Def. 10.30(3) we show the equivalence of (1) and (2) as follows: $a; a \leq a \leftrightarrow (a; a) \cdot \overline{a} = 0 \leftrightarrow (a^{\smile}; \overline{a}) \cdot a = 0 \leftrightarrow a^{\smile}; \overline{a} \leq \overline{a}$. An analogous argument yields the equivalence of (1) and (3). □

Some basic closure properties of the set of transitive elements are listed in

Proposition 10.50. ([9, Lemma 5.6]) *Let $a, b \in \mathsf{Tr}(\mathfrak{A})$ for a relation algebra $\mathfrak{A} \in \mathsf{RA}$. Then*

(1) $0, 1$ *and* $\mathbb{1}$ *are transitive..*
(2) $a^{\smile} \in \mathsf{Tr}(\mathfrak{A})$.
(3) *If* $a; b = b; a$ *then* $a; b \in \mathsf{Tr}(\mathfrak{A})$.

Proof. (1): The assertion follows for 0 by Prop. 10.38(1), for 1 Prop. 10.34(5), and for $\mathbb{1}$ by Prop. 10.33(11).

(2): By Prop. 10.33(1), $a; a \leq a$ implies $(a; a)^{\smile} = a^{\smile}; a^{\smile} \leq a^{\smile}$.

(3): $(a; b); (a; b) = a; b; a; b = (a; a); (b; b) \leq a; b$ (here associativity is used). □

10.3.2.3 *Equivalence Elements*

Note that the definition of equivalence elements given above does not require reflexiv ity. Thus, in relation algebra literature (beginning with [6, Definition 3.1]) equivalence elements differ slightly from what we are used to understand by equivalences: they correspond to *partial* equivalence relations (see Def. 4.12(b)). This allows also subidentity elements of $\mathfrak{A}$ to be in $\mathsf{Eq}(\mathfrak{A})$, which is reasonable since, speaking in terms of binary relations, coreflexive relations are reflexive on their field. (Only a total relation ρ on a set X that is symmetric and transitive is also reflexive, since $x\rho y$ implies $y\rho x$, whence $x\rho x$ for each $x \in X$).

Proposition 10.51. ([6, Theorems 3.2, 3.3]) *The following conditions are equivalent in* $\mathfrak{A} \in \mathsf{NA}$*:*

(1) $e \in \mathsf{Eq}(\mathfrak{A})$.
(2) $e; e = e = e^{\smile}$.
(3) $e = e; e^{\smile}$.
(4) $e; e^{\smile} \leq e$ *and* $e^{\smile}; e \leq e$.
(5) $e; \overline{e} \leq e$ *and* $\overline{e}; e \leq e$.

Proof. We first prove the equivalence of (1) - (3): Obviously, (2) implies (1) and (3) according to the definition of equivalence elements. Assume (1), then $e^{\smile} \leq e$ and $e; e \leq e$. But $e^{\smile} \leq e \rightarrow e^{\smile\smile} = e \leq e^{\smile}$ by Prop. 10.33(2) and (R7), so $e = e^{\smile}$. Furthermore, $e \leq (e; e^{\smile}); e = (e; e); e \leq e; e$ using Prop. 10.41(4), transitivity and monotonicity, whence, together with transitivity, we get (2). Finally, (3) implies $e^{\smile} = e^{\smile\smile}; e^{\smile} = e; e^{\smile} = e$ by (R7), (R9); thus (2) holds.

Now we turn to (4): Clearly, (2) implies (4). Conversely, suppose that (4) holds, then using the first inequality and Prop. 10.41(4) we get $e \le (e;e^{\smile});e \le e;e^{\smile} \le e$ by monotonicity of ;. So $e = e;e^{\smile}$, whence e is an equivalence element by (3), and therefore (2) holds.

Concerning the equivalence of (4) and (5) note that by the cycle laws Th. 10.39 we have $e;e^{\smile} \le e \leftrightarrow e;e^{\smile} \cdot \overline{e} = 0 \leftrightarrow \overline{e};e \cdot e = 0 \leftrightarrow \overline{e};e \cdot e = 0 \leftrightarrow \overline{e};e \le \overline{e}$. Similarly we get $e^{\smile};e \le e \leftrightarrow e;\overline{e} \le \overline{e}$. □

Proposition 10.52. ([8, p. 10, Lemma 1.14]) *For equivalence elements e, f in a relation algebra $\mathfrak{A}$ the following holds:*

(1) $0, 1, \mathbb{1} \in \mathsf{Eq}(\mathfrak{A})$.
(2) $e \cdot f \in \mathsf{Eq}(\mathfrak{A})$.
(3) $e \cdot f \le e;f$.
(4) $e;f = 0$ *iff* $e \cdot f = 0$
(5) *If* $e \le f$ *then* $e;f \le f$
(6) $e;f = f;e$ *iff* $e;f \in \mathsf{Eq}(\mathfrak{A})$, *with* $\mathfrak{A} \in \mathsf{RA}$.
(7) $e;1;e \in \mathsf{Eq}(\mathfrak{A})$, *with* $\mathfrak{A} \in \mathsf{RA}$..

Proof. (1): By Prop. 10.33(4),(7) we have $\mathbb{1} = \mathbb{1}^{\smile}$ and $0 = 0^{\smile}, 1 = 1^{\smile}$, and Prop. 10.38(1) implies $0;0 = 0$. By (R6) $\mathbb{1};\mathbb{1} = \mathbb{1}$, and by Prop. 10.34(5) $1;1 = 1$. So by Prop. 10.51 the assertion follows.

(2): Prop. 10.33(8) implies $(e \cdot f)^{\smile} = e^{\smile} \cdot f^{\smile} \le e \cdot f$, i.e. $e \cdot f$ is symmetric. From Prop. 10.41(1) and the transitivity of x and y we get $(x \cdot y);(x \cdot y) \le x;x \le x$ and $(x \cdot y);(x \cdot y) \le y;y \le y$, so $(x \cdot y);(x \cdot y) \le x \cdot y$ by the laws of Boolean algebra.

(3): By (2) $e \cdot f \in \mathsf{Eq}(\mathfrak{A})$, so $(e \cdot f) = (e \cdot f);(e \cdot f) \le e;f$ by Prop. 10.51(2) and monotonicity.

(4): This is [6, p. 359, THEOREM 3.11(ii)]. One directions is trivial: $e;f = 0$ implies $e \cdot f = 0$ by (3). The other direction is a little bit lengthy, but does not reveal new techniques, whence we omit it. Interested readers can find it in [26, Theorem 3.15(ii)] or [9, Lemma 5.22].

(5) $e \le f$ implies $e;f \le f;f \le f$.

(6): Analogous to the proof of Prop. 10.50(3), where (R4) is needed for transitivity: From $e;f \in \mathsf{Eq}(\mathfrak{A})$ we get $e;f = (e;f)^{\smile} = f^{\smile};e^{\smile} = f;e$. For the other direction note that $e;f = f;e$ implies $(e;f)^{\smile} = f^{\smile};e^{\smile} \le f;e = e;f$, and $(e;f);(e;f) = e;e;f;f = e;f$ by Prop. 10.51(2).

(7): We have $(e;1;e)^{\smile} = e^{\smile};1^{\smile};e^{\smile} \le e;1;e$ using monotonicity, and $e;1;e;e;1;e \le e;1;1;1;1;e = e;1;e$ using (1) and Prop. 10.51(2). □

Proposition 10.53. ([8, Lemma 1.9]) *Let a, b subidentity elements in $\mathfrak{A} \in \mathsf{NA}$. Then*

(1) $a^{\smile} = a$.
(2) $a;b = a \cdot b$.
(3) $a;a = a$.
(4) $a \in \mathsf{Eq}(\mathfrak{A})$.

Proof.

(1):

$$\begin{aligned} a &\le a;a^{\smile};a && \text{Prop. 10.41(4)}\\ &\le \mathbb{1};a^{\smile};\mathbb{1} && \text{subidentity, Prop. 10.34(2)}\\ &= a^{\smile} && \text{Prop. 10.33(11), (R6)} \end{aligned}$$

$$\begin{aligned} a^{\smile} &\le a^{\smile};a^{\smile\smile};a^{\smile} && \text{Prop. 10.41(4)}\\ &= a^{\smile};a;a^{\smile} && \text{(R7)}\\ &\le \mathbb{1}^{\smile};a;\mathbb{1}^{\smile} && \text{subidentity, Prop. 10.33(2), Prop. 10.34(2)}\\ &= a && \text{Prop. 10.33(4),(11), (R6)} \end{aligned}$$

(2): We have $a\cdot b \le (a\cdot b);(a\cdot b)^{\smile};(a\cdot b) \le a;(\mathbb{1}\cdot\mathbb{1})^{\smile};b = a;b$ using monotonicity properties and Prop. 10.41(4). For the other direction we have $a;b \le a;\mathbb{1}\cdot\mathbb{1};b = a\cdot b$.
(3): Follows from (2).
(4): Follows from (1) and (3). □

Compare the above with Prop. 3.53(5) and (6).

10.3.2.4 *Ideal Elements and Ideals*

Proposition 10.54. ([8, Lemma 1.7]) *Let $\mathfrak{A} \in \mathsf{RA}$ be a relation algebra. Then the following holds:*

(1) $a \in \mathsf{Idl}(\mathfrak{A})$ *iff* $a = a;1 = 1;a$ *(i.e. a is both a right and left ideal element).*
(2) $0, 1 \in \mathsf{Idl}(\mathfrak{A})$.
(3) If $a \in \mathsf{Idl}(\mathfrak{A})$ *then* $a = a^{\smile}$.
(4) If $a \in \mathsf{Idl}(\mathfrak{A})$ *then* $a \in \mathsf{Eq}(\mathfrak{A})$.
(5) $1;a;1 = 0$ *iff* $a = 0$*, for all* $a \in A$.
(6) $\mathsf{Idl}(\mathfrak{A})$ *is closed under* $+, \cdot$*, and* $^{-}$*, and therefore forms a subalgebra of* $\mathfrak{A}$.
(7) If a is an atom of $\mathfrak{A}$ *then* $1;a;1$ *is an atom in* $\mathsf{Idl}(\mathfrak{A})$.

Proof. (1): Assume $a \in \mathsf{Idl}(\mathfrak{A})$; then $a = 1;a;1$. Using Prop. 10.34(6) and monotonicity we get $a = 1;a;1 \ge a;1$, so $a = a;1$; similar for $a = 1;a$. Conversely, $a = 1;a$ implies $a;1 = (1;a);1$, but $a;1 = a$ by assumption.

(2): $0 \in \mathsf{Idl}(\mathfrak{A})$ by Prop. 10.38(1), and $1 \in \mathsf{Idl}(\mathfrak{A})$ by Prop. 10.34(5).

(3): By Prop. 10.41(4) we have $a^{\smile} \le a^{\smile};a^{\smile\smile};a^{\smile} = a^{\smile};a;a^{\smile} \le 1;a;1 = a$ (note that $\mathfrak{A} \in \mathsf{RA}$ is assumed, so associativity holds). Thus a is symmetric, and by Prop. 10.45(2) we get $a^{\smile} = a$.

(4): By (3) a is symmetric. Moreover, $a;a = (1;a;1);(1;a;1) = 1;a;1;a;1 = 1;a;a \le 1;a;1 = a$.

(5): Assume $a = 0$; then $1;0;1 = 1;0 = 0$ by Prop. 10.38(1). Conversely, $0 \le a$ and, by Prop. 10.34(2), $a = \mathbb{1};a;\mathbb{1} \le 1;a;1 = 0$. So $a = 0$.

(6): Let $a, b \in \mathsf{Idl}(\mathfrak{A})$. Then $a + b = a;1 + b;1 = (a+b);1$ by (R5), and $a + b = 1;a + 1;b = 1;(a+b)$ by Prop. 10.34(1). Using monotonicity of ; and Prop. 10.38(2) we get $a\cdot b = (1;a)\cdot(1;b) \ge 1;(a\cdot b) \ge \mathbb{1};(a\cdot b) = a\cdot b$, i.e. $a\cdot b = 1;(a\cdot b)$. Similar for $a\cdot b = (a\cdot b);1$.

Finally, for complementation, from $a = 1;a$ we get $\overline{a} = \overline{1;a}$; thus by Prop. 10.33(7) and (R10) $1^{\smile};(\overline{1;a}) = 1;(\overline{a}) \leq \overline{a}$. By Prop. 10.34(6) $\overline{a} \leq 1;\overline{a}$, whence $\overline{a} = 1;\overline{a}$. Similarly, $\overline{a} = \overline{a;1}$ implies $\overline{a;1};1^{\smile} = \overline{a};1 \leq \overline{a}$ using Prop. 10.34(3), whence $\overline{a} = \overline{a};1$.

(7): Using Th. 10.39(2) $\leftrightarrow$ (3), for $a, b \in A$ we get the first equivalence in

$$(1;a);1 \cdot b \neq 0 \quad \text{iff} \quad (b;1^{\smile}) \cdot (1;a) \neq 0 \quad \text{iff} \quad (1;b;1) \cdot a \neq 0. \tag{10.5}$$

The second equivalence can be shown as follows:

$$\begin{aligned}
(b;1^{\smile}) \cdot (1;a) &= (b;1) \cdot (1;a) && \text{by Prop. 10.33(7)} \\
&= (1;a) \cdot (b;1) && \text{by commutativity of } \cdot \\
&= (1^{\smile};a) \cdot (b;1) && \text{by Prop. 10.33(7)} \\
&= 1;(b;1) \cdot a && \text{by Th. 10.39(1)} \leftrightarrow (2)
\end{aligned}$$

Now suppose that a is an atom in $\mathfrak{A}$ and that $(1;a;1) \cdot (1;b;1) \neq 0$. Then replacing b by $1;b;1$ we get $(1;1;b;1;1) \cdot a = (1;b;1) \cdot a \neq 0$ from the last inequality in (10.5), i.e. $a \leq 1;b;1$ since $a \in \mathsf{At}(A)$. Hence, by monotonicity, $1;a;1 \leq 1;b;1$. By (4), since a is an atom, we get $1;a;1 \neq 0$, so it is an atom in $\mathsf{Idl}(\mathfrak{A})$. □

Lemma 10.55. ([26, Theorem 3.23]) *a is a left ideal element iff $a^{\smile}$ is a right ideal element.*

Proof. $a^{\smile}$ is a right ideal element iff $a^{\smile} = a^{\smile};1$ iff $a^{\smile\smile} = (a^{\smile};1)^{\smile}$, so $a = (a^{\smile};1)^{\smile}$. Therefore, $a = 1^{\smile};a^{\smile\smile} = 1;a$. □

Proposition 10.56. ([26, Theorem 3.35, Corollary 3.34]) *Let $\mathfrak{A} \in \mathsf{RA}$ be a relation algebra; then the following statements are equivalent:*

(1) *a is an ideal element.*
(2) *$a = 1;x;1$ for some $x \in A$.*
(3) *$a = a;1 = 1;a$.*
(4) *a and $a^{\smile}$ are both right ideal elements.*
(5) *$1;a;1 \leq a$.*

Proof. The (1) implies (2) is immediate from Def. 10.44 (i). Suppose (2) holds. Then $a;1 = 1;x;1;1 = 1;x;1 = a$ by Prop. 10.34(5). Similarly, $1;a = 1;1;x;1 = 1;x;1 = a$, whence we get (3). And assuming (3) we have $p = p;1 = (1;p);1$ (note that associativity comes into play), whence (1) is shown. The equivalence of (3) and (4) can be seen from the foregoing lemma. Finally we show (5) $\leftrightarrow$ (1): Obviously, if a is an ideal element, then $1;a;1 \leq a$. Conversely assume $1;a;1 \leq a$; then by Prop. 10.34(6) and monotonicity of ; we get $a \leq a;1 \leq 1;a;1$, so $a = 1;a;1$ which has been shown above to be equivalent to (1). □

In a relation algebra $\mathfrak{A}$ there is a close connection between ideal elements and ideals, so we continue with a general definition of the latter:

Definition 10.57.

(a) An **ideal** *in a Boolean algebra* $\mathfrak{B} = (B, \vee, \wedge, \overline{}, 0, 1)$ *is a non-empty subset* $I \subseteq B$ *such that*

(i) $0 \in I$,

(ii) $b \in I \rightarrow \forall a \in A[a \wedge b \in I]$.

(iii) $a, b \in I$ *then* $a \vee b \in I$.

(b) An **ideal** *in a relation algebra* $\mathfrak{A} = (A, +, \cdot, \overline{}, 0, 1, ;, \breve{}, \mathbb{1})$ *is a subset* $I \subseteq A$ *such that*

(i) I *is an ideal in the Boolean reduct* $(A, +, \cdot, \overline{}, 0, 1)$ *of* $\mathfrak{A}$,

(ii) $s \in I \rightarrow \forall r \in A[r; s \in I \wedge s; r \in I]$

(c) An ideal is called **principal** *if it has a greatest element.*

Condition (a)(i) may be replaced by the requirement that $\emptyset \subset I$, and (a)(ii) by the requirement that I be *downward closed*: if $s \in I$ and $r \leq s$, then $r \in I$. Recall that for $p \in A$ with $p^{\downarrow} := \{a \in A : a \leq p\}$ we denote the principal ideal generated by p; in the context of relation algebras usually the notation $A(p)$ is used for this, so we adhere to the latter.

Proposition 10.58. *In a relation algebra,* $A(p)$ *is an ideal if and only if* p *is an ideal element.*

Proof. Suppose $A(p)$ to be an ideal. Then $p \in A(p)$, and so $p; 1 \in I$ by Def. 10.57(b). Thus, $p; 1 \leq p$ by definition of $A(p)$, whence $p = p; 1$ using Prop. 10.34(6). Similarly, $p = 1; p$, whence p is an ideal element by Prop. 10.54(1).

Conversely, suppose p is an ideal element. Then $A(p)$ is closed downward by its definition. Furthermore, for $a, b \in A(p)$ we have $a \leq p$ and $b \leq p$, so $a + b \leq p$ by laws of Boolean algebra, whence $a + b \in A(p)$. Let $r \in A(p)$ and $s \in A$. Then $r \leq p$, whence $r; s \leq p; s \leq p; 1 = p$. So $r; s \in A(p)$, and similar for $s; r$. □

Ideals in relation algebras arise also from the perspective of homomorphisms. To gain a closer look we spell out some concepts and results obtained in Section 10.2.2 for Boolean algebras and consider some of their consequences in the current context.

Definition 10.59. *Let* $\mathfrak{A}$ *and* $\mathfrak{B}$ *be two Boolean algebras.*

(a) A **Boolean homomorphism** *is a mapping* $h : \mathfrak{A} \longrightarrow \mathfrak{B}$ *such that for all* $a, b \in A$

(1) $h(a \wedge b) = h(a) \wedge h(b)$,

(2) $h(a \vee b) = h(a) \vee h(b)$,

(3) $h(\overline{a}) = \overline{h(a)}$.

(b) A **Boolean congruence** *is an equivalence relation* θ *on a Boolean algebra* $\mathfrak{A}$ *that preserves the operations of meet, join, and complement: For all* $a, b, c, d \in A$, *whenever* $a\theta c$ *and* $b\theta d$, *then also*

(1) $(a \wedge c)\,\theta\,(b \wedge d)$,

(2) $(a \vee c)\,\theta\,(b \vee d)$,

(3) $\overline{a}\,\theta\,\overline{c}$.

The equivalence classes of a congruence θ are the sets $a/\theta := \{b \in A : a\theta b\}$, $a \in A$. Thanks to the preservation requirements for Boolean congruences it is possible to define analogous operations in $\mathfrak{A}/\theta$ in a natural way as follows:

$$(a/\theta) \wedge (b/\theta) = (a \wedge b)/\theta, \tag{10.6}$$

$$(a/\theta) \vee (b/\theta) = (a \vee b)/\theta, \tag{10.7}$$

$$\overline{(a/\theta)} = (\overline{a})/\theta. \tag{10.8}$$

It is easy to check that these operations are well defined, i.e. that they do not depend on the particular choice of elements in the equivalence classes used to define the operations. The algebra $\mathfrak{A}/\theta = (A/\theta, \wedge, \vee, \overline{\ })$, with the operations as given in (10.6) – (10.8), is again a Boolean algebra. As an example, for $a, b \in A$ we have $(a/\theta) \wedge (b/\theta) = (a \wedge b)/\theta = (b \wedge a)/\theta = (b/\theta) \wedge (a/\theta)$ etc.

Note the situation for the elements $0, 1 \in A$: We have $h(0) = h(a \wedge \overline{a}) = h(a) \wedge \overline{h(a)} = 0$; similarly $h(1) = 1$. This means that $0, 1 \in \mathsf{rng}(h)$ for every Boolean homomorphism h. Do also note that every Boolean operation can be defined in terms of $\wedge, \vee, \overline{\ }$, so it is also preserved by Boolean homomorphisms. Obviously, h is also order preserving: if $a \leq b$, i.e. $a \vee b = b$, then $h(b) = h(a \vee b) = h(a) \vee h(b)$, i.e. $h(a) \leq h(b)$.

Now we will turn to the analysis of the relationship between ideals and the canonical homomorphism. For this it might be useful to recall some basic facts on Boolean rings from Section 2.3.5.1, particularly the fact stated at the end of this section: that we can equivalently argue either in terms of Boolean algebras or in terms of Boolean rings. We also will need the notion of an ideal in a ring:

Definition 10.60. *An* **ideal** *in a ring* $\mathfrak{R} = (R, +, \cdot, -, 0, 1)$ *is a non-empty subset* $I \subseteq R$ *such that*

(1) $0 \in I$,
(2) if $a, b \in I$ *then* $a + b \in I$,
(3) if $a \in I, b \in A$ *then* $a \cdot b, b \cdot a \in I$.

The next lemma allows us to argue on ideals I either in terms of Boolean algebras or Boolean rings:

Lemma 10.61. [5, p. 127, Theorem 3.2] *Let* $\mathfrak{B} = (B, \vee, \wedge, \overline{\ }, 0, 1)$ *be a Boolean algebra. Then* $I \subseteq B$ *is an ideal in* $\mathfrak{B}$ *iff it is an ideal in* $\mathfrak{B}^*$.

Proof. All conclusions below follow directly Def. 10.57(a): Assume I is an ideal in $\mathfrak{B}$. Then $0 \in I$, and if $a, b \in I$ then also $a \wedge \overline{b}$ and $\overline{a} \wedge b$, whence $a \wedge \overline{b} \vee \overline{a} \wedge b = a + b \in I$. For $a \in I, b \in A$ we have $a \wedge b = a \cdot b \in I$.
All conclusions below follow directly Def. 10.60: Conversely, assume I is an ideal in $\mathfrak{B}^*$. Then $0 \in I$, and if $a, b \in I$ then also $a \cdot b$, whence $a + b + a \cdot b = a \vee b \in I$. For $a \in I, b \in A$ we have $a \cdot b = a \wedge b \in I$. □

Theorem 10.62. [5, p. 128, Theorem 3.5] *Let* $\mathfrak{B} = (B, \vee, \wedge, \overline{\ }, 0, 1)$ *be a Boolean algebra. A binary relation* θ *on* B *is a congruence on* $\mathfrak{B}$ *iff* $0/\theta$ *is an ideal, and* $a \theta b$ *iff* $a + b \in 0/\theta$ *for* $a, b \in B$.

Proof. (Note: We will use the properties of ideals in Boolean algebra (Def. 10.57(a)), Boolean rings (Def. 10.60) and Boolean congruences (Def. 10.59(b)) without explicitly mentioning them each time.)
[⇒] Assume θ to be a congruence on $\mathfrak{B}$. Then $0 \in 0/\theta$, and for $a, b \in 0/\theta$, we have $a\theta 0$ and $b\theta 0$, so $(a \vee b)\theta 0$. Let $a \in 0/\theta$ and $b \leq a$, then $a\theta 0$; thus $(a \wedge b)\theta(0 \wedge b)$, i.e. $b\theta 0$, whence $b \in 0/\theta$. Therefore $0/\theta$ is an ideal.

Moreover, $a\theta b$ implies $a+b\theta b+b$, i.e. $a+b\theta 0$, so $a+b \in 0/\theta$. Conversely, $a+b \in 0/\theta$ implies $a+b\theta 0$, so $((a+b)+b)\theta(0+b)$, i.e. $(a+(b+b)) = a\theta b$.

[⇐] First observe that θ is an equivalence relation: Since $a+a=0$, we have $a\theta a$ and, since $a+b=b+a$, if $a\theta b$ then $b\theta a$, for all $a, b \in B$. Assume $a\theta b$ and $b\theta c$; then $(a+b)+(b+c) = a+c \in 0/\theta$.

Next we have to show that θ is compatible with the operations of $\mathfrak{B}$. Let $a, b, c, d \in B$ such that $a\theta b$ and $c\theta d$. Then

$$\begin{aligned}(a \wedge c)+(b \wedge d) &= (a \cdot c)+(b \cdot d)\\ &= (a \cdot c)+(a \cdot d)+(a \cdot d)+(b \cdot d)\\ &= a \cdot (c+d)+(a+b) \cdot d\\ &\in 0/\theta,\end{aligned}$$

since all of $a, c+d, a+b$, and d are in $0/\theta$. Furthermore,

$$\begin{aligned}(a \vee c)+(b \vee d) &= (a+c+a \cdot c)+(b+d+b \cdot d)\\ &= (a+b)+(c+d)+(a \wedge c+b \wedge d)\\ &\in 0/\theta.\end{aligned}$$

Finally, $a+b \in 0/\theta$ implies $(1+a)+(1+b) \in 0/\theta$, whence $\overline{a}\theta\overline{b}$. So θ has shown to be a congruence. □

10.3.3 *Representability and other Algebraic Properties of* RA

This section deals with various properties of relation algebras (among them the topic of representation) without, however, going into details which are beyond the introductory scope of this book. Instead, references will be provided for the interested reader. By collocating some already mentioned concepts into one place, the following definition gives an overview on some prominent properties of relation algebras:

Definition 10.63. *A relation algebra* $\mathfrak{A} = (A, +, \cdot, \overline{\ }, 0, 1, ;, \breve{\ }, \mathbb{1})$ *is said to be*

- *(a)* **trivial** *if* $|A| = 1$;
- *(b)* **Boolean** *if* $x;y = x \cdot y$ *for all* $x, y \in A$;
- *(c)* **symmetric** *if* $x\breve{\ } = x$ *for all* $x \in A$;
- *(d)* **commutative** *if* $x;y = y;x$ *for all* $x, y \in A$;
- *(e)* **integral** *(or a relation algebra without zero divisors) if it is nontrivial, and if* $x;y \neq 0$ *for all* $x, y \in A$ *with* $x, y \neq 0$ *(equivalently:* $x;y = 0 \rightarrow x = 0 \vee y = 0$*);*
- *(f)* **atomic** *if the Boolean reduct of* $\mathfrak{A}$ *is atomic;*

(g) **simple** *if it is non-trivial and* $\mathsf{Con}(\mathfrak{A}) = \{I_A, U_A\}$.

Clearly, every symmetric relation algebra is commutative: $x;y = x^{\smile};y^{\smile} = (y;x)^{\smile} = y;x$. The next proposition shows how each Boolean algebra $\mathfrak{B}$ can be extended to a relation algebra. This allows for testing if an equation is valid in a relation algebra: an equation E cannot be derived from the axioms defining a relation algebra, unless the equation obtained from E by deleting $^{\smile}$ and replacing ; by $\cdot$ and $\mathbb{1}$ bay 1 is valid in all Boolean algebras.

Proposition 10.64. ([23, p. 291, Theorem 230]) *Let* $\mathfrak{B} = (B, +, \cdot, \overline{}, 0, 1)$ *be a Boolean algebra. Define the operations* $x;y = x \cdot y$ *and* $x^{\smile} = x$ *for all* $x, y \in B$. *Then* $(B, +, \cdot, \overline{}, 0, 1, ;, ^{\smile}, 1)$ *is a relation algebra, called the* **Boolean relation algebra of** $\mathfrak{B}$.

10.3.3.1 *Representation of Relation Algebras*

Let us start with some general considerations: In algebra, a *representation theorem* is a theorem that states that every abstract structure with certain properties is isomorphic to another, already familiar (abstract or concrete) structure. Such theorems thus characterize the class of models of a given axiomatic theory and are usually sought for to find an axiomatisation for a given class of structures, and/or to show that a given axiom system is complete for an intended class of models ([3, p. 127]).

One of the most famous examples is the **Stone representation theorem** for Boolean algebras which states that every Boolean algebra is isomorphic to a field of sets, i.e. to some subalgebra of a suitably chosen power set Boolean algebra. Studying abstract algebraic structures, isomorphic ones are generally identified. Under this viewpoint, the Stone representation theorem says that the study of all power set Boolean algebras does exhaust the world of Boolean algebras: Whatever holds for the former does so for *all* Boolean algebras. Another famous representation result is Cayley's theorem that every group is isomorphic to a group of permutations.

Recall from Section 10.1 that the concrete, set-theoretic objects of our study are algebras of relations (also called *proper relation algebras*), and that relation algebras are their abstract algebraic counterpart. So there is not much surprise in

Definition 10.65. *Let* $\mathfrak{A}$ *be a relation algebra.*

(a) *A* **representation of** $\mathfrak{A}$ *is an isomorphism from* $\mathfrak{A}$ *to an algebra of binary relations.* $\mathfrak{A}$ *is called* **representable** *if it has a representation.*
(b) *The class of all representable relation algebras is denoted by* RRA; *it is the closure of* BRA *under isomorphism.*

Thus, a *(square) representation* of a relation algebra $\mathfrak{A} = (A, +, \cdot, \overline{}, 0, 1, ;, ^{\smile}, \mathbb{1})$ is a one-to-one mapping $h : \mathfrak{A} \longrightarrow \mathcal{P}(U \times U)$ for some base set U, such that for all $a, b \in A$

(1) $h(a+b) = h(a) \cup h(b)$
(2) $h(\overline{a}) = U \times U - h(a)$

(3) $h(0) = \emptyset$
(4) $h(1) = U \times U$
(5) $h(\mathbb{1}) = id_U$
(6) $h(a^{\smile}) = \{\langle x, y\rangle \in U \times U : \langle y, x\rangle \in h(a)\}$
(7) $h(a;b) = \{\langle x, y\rangle \in U \times U : \exists z\,[\langle x, z\rangle \in h(a) \wedge \langle z, y\rangle \in h(b)]\}$.

So, a representation maps each $a \in A$ to a binary relation on a set U, and the algebraic operations on $\mathfrak{A}$ to operations on binary relations. Relation algebras are thus related to binary relations in (almost) the same manner as Boolean algebras are to sets, i.e. unary relations.

In connection with an axiom system two fundamental questions arise which, in the context of relation algebras, are as follows:

- *The Completeness Problem:* Is the set of axioms (R1) - (R10) complete in the sense that every law true of all algebras of binary relations is derivable from it?
- *The Representation Problem:* Is every relation algebra representable as an algebra of binary relations?

In 1950 Lyndon [18] proved that both questions must be answered with a "*no*", i.e. there are models of (R1) - (R10) that are not (isomorphic to) proper relation algebras, by constructing a large relation algebra containing 56 atoms. Despite this negative there are also some important positive results from which we mention the following two (see [11, p. 41]):

(1) Every relation algebra can be embedded into a complete, atomic relation algebra. (Note that for Boolean, but not for relation algebras this is equivalent to the well-known assertion that every Boolean algebra is representable.)
(2) Every atomic relation algebra in which the atoms are functional is representable.

Lyndon's negative answer immediately led to the question for additional axioms that might allow us to exactly define the isomorphism types of proper relation algebras. In particular, it was asked if RRA could possibly be axiomatized with only finitely many axioms. That this is not the case has been shown in Monk [27] in 1964. It is, however, known that RRA is a variety (see [13, Section 3.3.4] for the involved details), i.e. it can be axiomatized by (an infinite set of) equations. So, cast in an official form, the following holds:

Proposition 10.66.

(1) RA$\neq$ RRA.
(2) RRA *is a variety.*

A small non-representable relation algebra (for details see [23, p. 361] or [13, p. 140]), let us call it $\mathfrak{C}$, has been presented by McKenzie in his 1970 paper [24]. $\mathfrak{C}$ looks as follows:

it has four atoms $\mathbb{1}, a, a^{\smile}, b$, where $b^{\smile} = b$, and the composition table (omitting the row and column for $\mathbb{1}$) as shown in Table 10.1.

TABLE 10.1: McKenzie's non-representable relation algebra.

;	a	$a^{\smile}$	b
a	a	1	$a+b$
$a^{\smile}$	1	$a^{\smile}$	$a^{\smile}+b$
b	$a+b$	$a^{\smile}+b$	$\overline{b} = \mathbb{1}+a+a^{\smile}$

(It is easy to check the equation $\overline{b} = \mathbb{1}+a+a^{\smile}$: By Cor. 2.73 for $a \in \mathrm{At}\,\mathfrak{C}$, $c \in C$ we have $a \leq c \vee a \cdot c = 0$. So if both a, c are atoms we get $a \cdot c = 0$, whence $b \cdot (\mathbb{1}+a+a^{\smile}) = 0$. And by Lemma 2.75(3) $1 = (\mathbb{1}+a+a^{\smile}) + b$.)

Remark: Let $\mathfrak{B}$ and $\mathfrak{B}'$ denote two complete, atomic Boolean algebras with the same number of atoms. Recall that in Th. 2.80 it has been shown that $\mathfrak{B}$ and $\mathfrak{B}'$ are isomorphic, and in Th. 2.77 that $\mathfrak{B}$ is isomorphic to the powerset of its atoms: $\mathfrak{B} \simeq \mathcal{P}(\mathrm{At}\,B) \simeq \mathcal{P}(\mathrm{At}\,B') \simeq \mathfrak{B}'$. Thus the structure of a finite Boolean algebra is completely determined by the number of its atoms, say n: It has 2^n elements, since for a set X with $|X| = n$ we have $|\mathcal{P}(X)| = 2^n$. Of course, the same holds true for any finite relation algebra $\mathfrak{A}$. So the structure of $\mathfrak{A}$ depends entirely on the choice of $\mathbb{1}$ and on the way the operations ; and $^{\smile}$ are defined. Since every element in a finite Boolean algebra is the join of the atoms below it by Th. 2.76(1), and since ; and $^{\smile}$ are distributive over +, the action of ; and $^{\smile}$ in $\mathfrak{A}$ is completely determined by their restrictions to the atoms in its Boolean reduct.

We are now going to show that McKenzie's relation algebra cannot be represented by a proper relation algebra. Assume that h is a homomorphism of $\mathfrak{C}$ into a proper relation algebra $\mathfrak{B}$, where $\alpha = h(a), \alpha^{\smile} = h(a^{\smile})$, and $\beta = h(b)$. Fig. 10.4 will be very helpful in the argumentation.

(1) Since all the atoms of $\mathfrak{C}$ are non-zero, so is $h(b)$, whence there are $x, y \in B$ such that $x\beta y$.
(2) We have $b \leq a; a^{\smile} = a^{\smile}; a = 1$. Thus, $x\beta y \rightarrow \exists v, w \in B\,[x\alpha v \wedge v\alpha^{\smile} y \wedge x\alpha^{\smile} w \wedge w\alpha y]$. So we get $x\alpha v, y\alpha v, w\alpha x, w\alpha y$, whence $w\alpha; \alpha v$.
(3) But from $a; a = a$ we have $\alpha; \alpha = \alpha$, whence $w\alpha v$.
(4) Furthermore, $a \leq \mathbb{1}+a+a^{\smile} = b; b$, so there is a $u \in B$ with $w\beta u, u\beta v$, whence $u\beta w, v\beta u$ since $b^{\smile} = b$.
(5) $x\alpha v \wedge v\beta u$ gives $x\alpha; \beta u$, and from $y\alpha v \wedge v\beta u$ we get $y\alpha; \beta u$. So $\alpha; \beta \neq \emptyset$ is the join of the atoms below it by Th. 2.76(1). It follows that there must be two atoms $m, o \in \mathfrak{C}$ such that $\langle x, u\rangle \in \rho := h(m)$ and $\langle y, u\rangle \in \sigma := h(o)$.

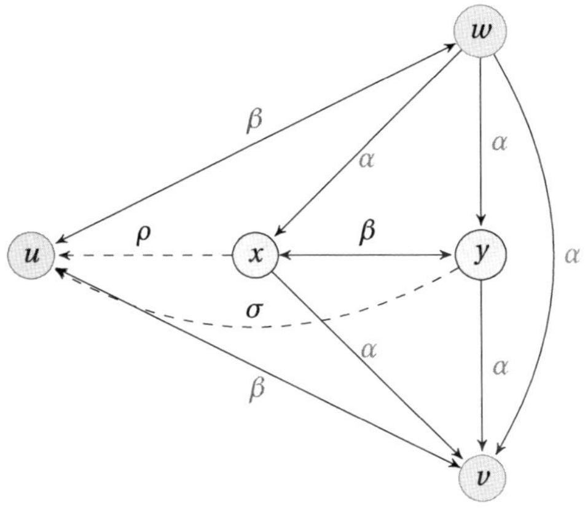

FIG. 10.4: The McKenzie relation algebra is not representable.

We have $x\rho u$, $x\alpha;\beta u$, and $x\alpha^{\smile};\beta u$. Since ρ is an atom this implies $\rho \subseteq \alpha;\beta = \alpha \cup \beta$ and $\rho \subseteq \alpha^{\smile};\beta = \alpha^{\smile} \cup \beta$ by the assumption that h is a homomorphism of $\mathfrak{C}$ into a proper relation algebra $\mathfrak{B}$. From the three atoms $\alpha, \alpha^{\smile}$ and β only the latter one does fulfill the two restrictions, so we conclude that $\rho = \beta$. Similarly we get $\sigma = \beta$. But then we have $x\beta u \wedge x\beta y\beta u$, whence $\beta \cap \beta;\beta \neq \emptyset$, on the other hand in $\mathfrak{C}$ we have $b;b = \overline{b}$, whence $b;b \cdot b = 0$, a contradiction. Hence $\mathfrak{C}$ is not representable.

We close this section with links to [10, Chapters 16 and 17] for much more detailed information related to the topic of representability of relation algebras, and to [10, Section 18.5] for the fact that RRA is a variety.

10.3.3.2 *Simple and Integral Relation Algebras*

Of the various types of relation algebras mentioned in Def. 10.63 simple relation algebras have some remarkable properties. We give a short account in form of a list of relevant results, usually omitting the more involved proofs (for which we refer the interested reader to mainly [9, Chapter 9]).

Recall that $\mathsf{Idl}(\mathfrak{A})$, the set of ideal elements in a relation algebra $\mathfrak{A}$, is closed under $+, \cdot$, and $^{-}$, and therefore forms a subalgebra of $\mathfrak{A}$ by Prop. 10.54(6).

Theorem 10.67. ([9, Theorem 8.26]) *Let $\mathfrak{A}$ be relation algebra and B the Boolean algebra of ideal elements in $\mathfrak{A}$. The function mapping each ideal M in $\mathfrak{A}$ to the Boolean ideal $B \cap M$ is an isomorphism from the lattice of ideals in $\mathfrak{A}$ to the lattice of Boolean ideals in B. The inverse of this function maps each Boolean ideal K in B to the ideal $\{r \in \mathfrak{A} : 1;r;1 \in K\}$ in $\mathfrak{A}$.*

From this theorem it follows that the lattice of ideals in $\mathfrak{A}$ is isomorphic to the lattice of Boolean ideals in the Boolean algebra of ideal elements in $\mathfrak{A}$, say B, so the simplicity of $\mathfrak{A}$ is equivalent to the assertion that B has exactly two ideals, i.e. is simple as a Boolean algebra.

Lemma 10.68. ([9, Lemma 9.1]) *A relation algebra is simple if and only if it has exactly two (distinct) ideal elements, 0 and 1, i.e. if and only if the unit is an ideal element atom.*

Proof. The set of elements below any given element r in a Boolean algebra obviously is the principal Boolean ideal generated by r. Therefore, a Boolean algebra is simple if and only if it consists of exactly two elements, 0 and 1. □

A consequence of this lemma is the fact that the notion of simplicity for relation algebras can be characterized by universal sentences in the (first-order) language of relation algebras:

Theorem 10.69. ([9, Theorem 9.2]) *The following conditions on a relation algebra $\mathfrak{A}$ are equivalent:*

(1) *$\mathfrak{A}$ is simple.*
(2) $0 \neq 1$*, and for all* $r \in \mathfrak{A}$*, if* $r \neq 0$ *then* $1;r;1 = 1$.
(3) $0 \neq 1$*, and for all* $r \in \mathfrak{A}$*, either* $r;1 = 1$ *or* $1;\overline{r} = 1$.
(4) $0 \neq 1$*, and for all* $r, s \in \mathfrak{A}$*, if* $r;1;s = 0$ *then* $r = 0$ *or* $s = 0$.

A consequence of the preceding theorem is the following

Corollary 10.70. ([9, Corollary 9.3]) *Every subalgebra of a simple relation algebra is simple.*

Theorem 10.71. ([9, Theorem 12.10]) *Every relation algebra is isomorphic to a subdirect product of simple relation algebras*

This theorem has important implications for the validity of equations and, more generally, conditional equations in relation algebras:

Theorem 10.72. ([9, Theorem 12.11]) *An equation or conditional equation is true in all relation algebras if and only if it is true in all simple relation algebras.*

Integral relation algebras turn out to be special case of simple relation algebras. The following holds:

Theorem 10.73. ([9, Theorem 9.7]) *The following conditions on a relation algebra $\mathfrak{A}$ are equivalent:*

(1) *$\mathfrak{A}$ is integral.*
(2) $0 \neq 1$*, and for all* $r \in \mathfrak{A}$*, if* $r \neq 0$ *then* $r;1 = 1$.
(3) $0 \neq 1$*, and for all* $r \in \mathfrak{A}$*, if* $r \neq 0$ *then* $1;r = 1$.
(4) *$\mathbb{1}$ is an atom.*

Corollary 10.74. ([9, Corollary 9.8]) *Every subalgebra of an integral relation algebra is integral.*

Corollary 10.75. ([9, Corollary 9.10]) *Every integral relation algebra is simple.*

Proof. Using Prop. 10.34(5) and Th. 10.73(2) we get $1;r;1 = 1;1 = 1$. So $\mathfrak{A}$ is simple by Th. 10.69(2) □

References

[1] Almeida, J.: *Finite Semigroups and Universal Algebra*, World Scientific, Singapore (1994).

[2] Andréka, H., Givant, S., Jipsen, P., Németi, I.: On Tarski's axiomatic foundations of the calculus of relations, Journal Symbolic Logic **82**, 966-994 (2017).

[3] Bennett, B., Düntsch, I.: Axioms, Algebras and Topology. In Aiello, M., Pratt-Hartmann, I., VanBenthem, J. (Eds), *Handbook of Spatial Logics*, Springer Dordrecht, 99-159 (2007).

[4] Bergman, C.: *Universal Algebra- Fundamentals and Selected Topics*, CRC Press, Boca Raton (2012).

[5] Burris, S., Sankappanavar, H.P.: *A Course in Universal Algebra*, The Millenium Edition, Update 2012; available online at `https://www.math.uwaterloo.ca/~snburris/htdocs/UALG/univ-algebra2012.pdf` (2012).

[6] Chin, L.H., Tarski, A.: Distributive and modular laws in the arithmetic of relation algebras, University of California Publications in Mathematics, new series, **1**, no. 9, 341-384 (1951).

[7] Frias, M.F.: *Fork Algebras in Algebra, Logic and Computer Science*, World Scientific Publishing, Singapore (2002).

[8] Givant, S.: *The Structure of Relation Algebras Generated by Relativizations*, American Mathematical Society, Providence, Rhode Island (1994).

[9] Givant, S.: *Introduction to Relation Algebras-Relation Algebras, Volume 1*, Springer International Publishing AG, (2017).

[10] Givant, S.: *Advanced Topics in Relation Algebras-Relation Algebras, Volume 2*, Springer International Publishing AG, (2017).

[11] Givant, S., Andréka, H.: Groups and Algebras of Binary Relations, The Bulletin of Symbolic Logic **8**, 38-64 (2002).

[12] Grätzer, G.: *Universal Algebra*, 2nd ed., Springer Science+Business Media (2008).

[13] Hirsch, R., Hodkinson, I.: *Relation Algebras by Games*, Elsevier Science B.V., Amsterdam (2002).

[14] Hodges, W.: *A shorter model theory*, Cambridge University Press, Cambridge, U.K. (1997).

[15] Huntington, E.V.: New sets of independent postulates for the algebra of logic, with special reference to Whitehead and Russell's Principia Mathematica, Trans. American Math. Soc. **35**, 274-304 (1933).

[16] Jónsson, B., Tarski, A.: Boolean algebras with operators. Part I, Amer. J. Math. **73**, 891-939 (1951).

[17] Jónsson, B.: Relation algebras and Schröder categories, Discrete Math. **70**, 27-45 (1988).

[18] Lyndon, R. C.: The representation of relational algebras, Annals of Mathematics (series 2) **51**, 707-729, (1950).

[19] Maddux, R.D.: Some varieties containing relation algebras, Trans. American Math. Soc. **272**, 501-526 (1982).

[20] Maddux, R.D.: The Origin of Relation Algebras in the Development and Axiomatization of the Calculus of Relations, Studia Logica **50**, 421-455 (1991).

[21] Maddux, R.D.: Pair-dense relation algebras, Trans. American Math. Soc. **328**, 83-131 (1991).

[22] Maddux, R.D.: Relation-algebraic semantics, Theoretical Computer Science **160**, 1-85 (1996).

[23] Maddux, R.D.: *Relation Algebras*, Elsevier Science B.V., Amsterdam (2006).

[24] McKenzie, R.N.: Representations of integral relation algebras, Michigan Mathematics Journal **17**, 279-287, (1970).

[25] McKenzie, R.N., McNulty, G.F., Taylor, W.F.: *Algebras, Lattices, Varieties*, Vol. 1, Wadsworth & Brooks/Cole (1987).

[26] Mirin, A.C.: *Relation Algebras*, Master Thesis, Mills College (2013).

[27] Monk, J.D.: On representable relation algebras, Michigan Mathematics Journal **11**, 207-210 (1964).

[28] De Morgan, A.: On the syllogism no. IV, and on the logic of relations, Trans. Cambridge Philos. Soc. **10**, 331-358 (1864).

[29] Peirce, C.S.: Description of a notation for the logic of relatives, resulting from an amplification of the conceptions of Boole's calculus of logic, Memoirs of the American Academy of Arts and Sciences, New Series **9**, 317-378 (1873). Reprinted in *Writings of Charles S. Peirce, A Chronological Edition, Volume 2, 1867-1871*, Indiana University Press, Bloomington and Indianapolis, 359-429 (1984).

[30] Schröder, E.: *Vorlesungen über die Algebra der Logik (Exakte Logik). Dritter Band: Algebra und Logik der Relative.* B.G. Teubner, Leipzig (1895), (reprinted by Chelsea Publishing Company, New York, 1966).

[31] Tarski, A., Givant, S.: *A Formalization of Set Theory without Variables*, American Mathematical Society (1987).

[32] Wechler, W.: *Universal Algebra for Computer Scientists*, Springer-Verlag, Berlin Heidelberg (1992).

Chapter 11

Relations and Categories

This chapter presents information about categories of relations and relations in categories. In its first section it includes (deliberately!) another introduction into the most basic concepts of category theory needed in subsequent parts. We then continue with a look at regular categories and their properties, and how they are linked to relations, followed by a short section on related concepts. In the last section we introduce the categorical concept of an allegory and investigate its properties.

When I had to decide about this chapter I envisioned the following three possibilities:

- to simply omit it because categories are not really an introductory topic - a very unsatisfactory solution because there should be a way to present some introductory information on relations in categories;
- to give a rather cursory presentation, including some keywords and some basic results only, without any proofs, but with detailed references for further self study - also unsatisfactory, and moreover unlikely to work well as a knowledge transfer;
- or, in line with my general aim, to provide a more self-contained exposition by adding most of the necessary information, supplemented with some links to related concepts.

Finally, based on the experience with Chapter 10, I preferred the last alternative since I always find it instructive to risk a look beyond one's own nose. So the resulting text presents many definitions, a considerable number of comprehensible sample proofs, and discussions which often relate the new concepts to the familiar context of sets and functions. But in return readers get offered a short introduction to the elegance and beauty of category theory, and also a better understanding not only of the category theoretical view on relations, but also of how things in category theory run in general.

And although this chapter finally grew much larger than I have expected when starting to elaborate on it, I hope to have found the right balance between "*just enough*" and "*not too much*" of category theory. And although many will certainly find the presentation below much too long, too formal and too theoretical, it is, in fact, only scratching on the surface of category theory. The chapter may well bear the subtitle: *A Beginner's Guide Only*, but, nonetheless, it is probably the hardest one to work through for the average reader, which means for people who are not specialists in category theory, a group to which I myself do belong.

I conclude this lengthy preamble with the first few sentences from the *Introduction* of Leinster [15] which give a wonderful characterization of category theory (and at the same time also a kind of warning for the innocent reader, indicating where he will have to climb to): "Category theory takes a bird's eye view of mathematics. From high in the sky, details become invisible, but we can spot patterns that were impossible to detect from ground level. How is the lowest common multiple of two numbers like the direct sum of two vector spaces? What do discrete topological spaces, free groups, and fields of fractions have in common? We will discover answers to these and many similar questions, seeing patterns in mathematics that you may never have seen before."

11.1 Categories: Basic Concepts

The first section is devoted to the presentation of those basic definitions and properties of categories, together with some examples, which appear in practically each introduction to the topic because they allow you to do the first steps towards categories by moving on familiar ground. More details on categories in general you can find in any of the books listed in the references, with [20] or the introductory chapters of [2, 10] probably being the most suitable for beginners.

Definition 11.1. *A* **category** $\mathcal{C}$ *consists of*

(a) *a collection* $\mathsf{obj}_{\mathcal{C}}$ *of* **objects** *of* $\mathcal{C}$*;*
(b) *a collection* $\mathsf{mor}_{\mathcal{C}}$ *of* **morphisms** *(or* **arrows***);*
(c) *operations assigning to each arrow* f *an object* $\mathsf{dom}\, f$*, called the* **domain** *(or* **source***) of* f*, and an object* $\mathsf{cod}\, f$*, called the* **codomain** *(or* **target***) of* f*. The collection of all arrows with domain* A *and codomain* B *is denoted by* $\mathsf{mor}_{\mathcal{C}}[A,B]$ *and called the* **hom-set** *between* A *and* B*;*
(d) *a* **composition operator** *assigning to each pair of arrows* f *and* g *with* $\mathsf{cod}\, f = \mathsf{dom}\, g$*, an arrow* $g \cdot f \in \mathsf{mor}_{\mathcal{C}}[\mathsf{dom}\, f, \mathsf{cod}\, g]$ *that satisfies the following associative law: for all arrows* $f : A \to B$*,* $g : B \to C$*, and* $h : C \to D$
$$h \cdot (g \cdot f) = (h \cdot g) \cdot f.$$
(e) *for each* $B \in \mathsf{obj}_{\mathcal{C}}$ *a morphism* $id_B \in \mathsf{mor}_{\mathcal{C}}[B,B]$*, called the* **identity on** B*, such that for* $f \in \mathsf{mor}_{\mathcal{C}}[A,B]$ *and* $g \in \mathsf{mor}_{\mathcal{C}}[B,C]$ *we have*
$$id_B \cdot f = f \text{ and } g \cdot id_B = g.$$

Common alternatives notations are: $A \in \mathcal{A}$ for $A \in \mathsf{obj}_{\mathcal{A}}$, and $f : A \longrightarrow B$ or $A \xrightarrow{f} B$ for $f \in \mathsf{mor}_{\mathcal{C}}[A,B]$; and simple concatenation gf for $g \cdot f$.

Note that the metalanguage used in our definition of a category is ordinary set theory: For us, the neutral term *collection* denotes just sets (we do not deal with proper classes), *operations* are functions, and *equality* is set-theoretical identity. The name *hom-set* is something of a misnomer as the collection of morphisms is, in general, not required to be a set. To differentiate, a category $\mathcal{C}$ is called **locally small** if all its hom-sets are sets, **small** if $\mathsf{obj}_{\mathcal{C}}$ also is a set, and **large** otherwise.

Given objects and morphisms, one often uses diagrams to depict the relationships between them: Roughly speaking, a **diagram** in a category is a collection of objects together with a collection of arrows between these objects. The following diagram has four objects and five arrows f, g, h, k, l:

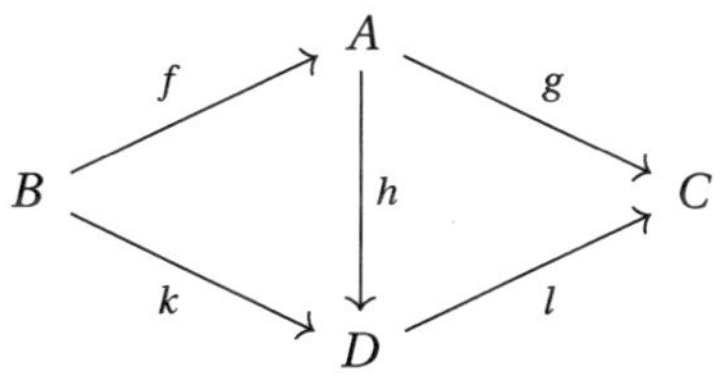

In this diagram there are also five composite morphisms, viz. gf, hf, lhf, lh, and lk, and three cells: the left and right hand triangle, and the outer rectangle.

A diagram as a whole is called a **commuting diagram** if whenever there are two paths *containing at least two arrows* from an object A to an object B, the morphism from A to B obtained by composing along one path is equal to the morphism obtained by composing along the other. In the diagram above, if the two triangles commute, then also the outer rectangle, since from $hf = k$ and $lh = g$ we conclude $gf = (lh)f = l(hf) = lk$.

From Def. 11.1 we immediately get the following three examples for commuting diagrams: the left one commutes by definition of composition, and the right diagram represents the identity laws. The associative law then asserts that diagrams like the middle one always commute.

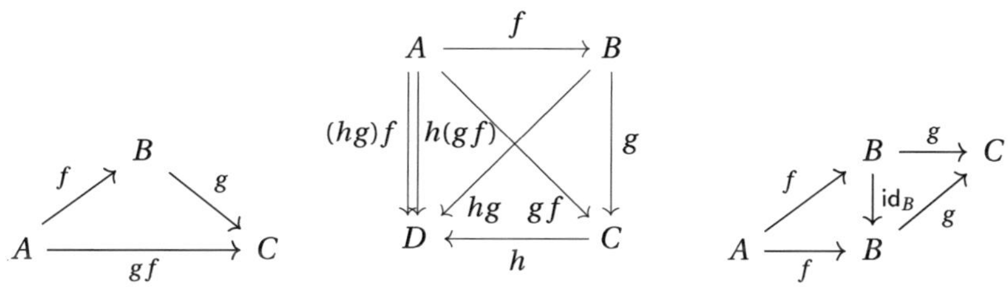

If you are a little bit familiar with algebra, then you have certainly got an idea of why categories are defined as they are. There exist rather familiar as well as exotic categories as the following examples will show.

(1) The prototype and motivating example for categories is **Set** which has sets as its objects and total functions between them as arrows. The composition of arrows is set-theoretic function composition, and the identity arrows are the identity functions. From Cor. 4.3(2) we know that totality is preserved by composition.

Obviously, for each set A, id_A is total and the associativity and the identity laws hold.

(2) Let P and Q be posets. A function $f : A \longrightarrow B$ is called *order-preserving* (or *monotone*) if $a \leq_P b$ implies $f(a) \leq_Q f(b)$. The category **Poset** has as objects all partially ordered sets and as morphisms all order-preserving total functions. What remains to be shown in addition to the results for **Set** is that composition of order-preserving functions, as well as the identity functions also have this property. But this follows trivially from the definitions: Let e.g. $a \leq_P b$, whence $f(a) \leq_Q f(b)$ and $g(f(a)) \leq_R g(f(b))$, and also $id_P(a) \leq_P id_P(b)$.

The two categories presented above are examples for categories with sets as objects and functions as arrows, where **Poset** is one of the most elementary cases where the objects bear some structure that has to be preserved by the morphisms. Such categories are called **concrete**.

There exists a lot of concrete categories ; see e.g. [21, Table 1.1] from which we list just a few in Table 11.1 below in order to demonstrate their broad spectrum.

TABLE 11.1: Examples of Concrete Categories.

Category	Objects	Morphisms
Set	sets	total functions
Pfn	sets	partial functions
Rel	sets	binary relations
Sgp	semigroups	homomorphisms
Mon	monoids	monoid homomorphisms
Grp	groups	group homomorphisms
Pre	pre-ordered sets	monotone maps
Pos	posets	monotone maps
Top	topological spaces	continuous maps
Vect	vector spaces	linear transformations

Concrete categories form an important group, but there are also many other interesting examples, as e.g. some finite categories:

(3) The category **0** has no objects and no arrows. So the identity and associativity laws are vacuously satisfied

(4) Given the category **1** with only one object, say a, and one arrow called f, then we find that its structure is completely determined: We clearly have $\mathsf{dom}\, f = \mathsf{cod}\, f = a$, $id_a = f$, and $f \cdot f = f$. This gives the identity law $id_a \cdot f = f \cdot id_a = f \cdot f = f$, and the associative law $(f \cdot f) \cdot f = f \cdot f = f \cdot (f \cdot f)$. Note that what a and f are does not have any influence on this, they can be anything you like. Making the definitions just given, one gets a structure that satisfies the axioms for a category.

(5) This is also true for the category **2** with two objects, two identity arrows and

the arrow from one object to the other, and also for the category **3** consisting of three identity arrows and three arrows forming a triangle between the objects:

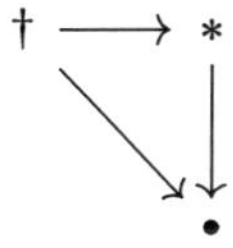

In each of the last three examples there is only one way that composites can be defined. This is because between any two objects a and b there is at most one arrow $a \longrightarrow b$, so once the dom and cod are known, the arrows are fixed. A category with this property is called a *preorder* category for obvious reasons: Let P be the set of objects of a preorder category and define a relation ρ on P by $a\rho b$ iff $a \longrightarrow b$, then ρ is a preorder relation. This brings us to another sort of category which is obtained by considering an individual algebraic structure as a category.

(6) A poset $(P, \leq)$ gives rise to a category whose objects are the elements of P and one arrow between those pairs of objects a and b with $a \leq b$. Composition of arrows is associative since there is at most one arrow between any given pair of objects. The identity law corresponds to reflexivity, and the existence of composite arrows reflects the transitivity of $\leq$. Note that regarding posets as categories makes **Poset** a category of categories.

(7) Any monoid $(M, \cdot, e)$ gives rise to a category with one object, viz M itself. The elements of $m \in M$ are represented as arrows $m : M \longrightarrow M$, putting $e := id_M$. Conversely, if $\mathcal{C}$ is a category with only one object a, and M is its collection of arrows, then $(M, \cdot, id_a)$ is a monoid.

In addition to categories of mathematical objects from various domains, there are also many ways to form new categories from given ones. Some examples of so-called *derived* categories are given in

Definition 11.2.

(a) A category $\mathcal{D}$ is a **subcategory** *of a category $\mathcal{C}$, if $\mathsf{obj}_{\mathcal{D}} \subseteq \mathsf{obj}_{\mathcal{C}}$ and for all $A, B \in \mathsf{obj}_{\mathcal{D}}$ it holds that $\mathsf{mor}_{\mathcal{D}}[A,B] \subseteq \mathsf{mor}_{\mathcal{C}}[A,B]$. $\mathcal{D}$ is called a* **full subcategory** *if for any $\mathcal{D}$-objects A, B $\mathsf{mor}_{\mathcal{D}}[A,B] = \mathsf{mor}_{\mathcal{C}}[A,B]$, i.e. $\mathcal{C}$ has no other arrows than the ones already in $\mathcal{D}$.*

(b) The **opposite** *or* **dual** *category $\mathcal{C}^{\mathrm{op}}$ of a category $\mathcal{C}$ has the same objects as $\mathcal{C}$ with all the arrows reversed: $\mathsf{obj}_{\mathcal{C}^{\mathrm{op}}} = \mathsf{obj}_{\mathcal{C}}$, and $f^{\mathrm{op}} \in \mathsf{mor}_{\mathcal{C}^{\mathrm{op}}}[A,B]$ iff $f \in \mathsf{mor}_{\mathcal{C}}[B,A]$. Identities are as in $\mathcal{C}$, and composition $\cdot^{\mathrm{op}}$ in $\mathcal{C}^{\mathrm{op}}$ is defined by $f \cdot^{\mathrm{op}} g = g^{\mathrm{op}} \cdot f^{\mathrm{op}}$.*

(c) Let $\mathcal{C}$ be a category, and $X \in \mathsf{obj}_{\mathcal{C}}$. Then the **slice category** *$\mathcal{C}/X$ over X is defined as follows:*

1. *The objects of $\mathcal{C}/X$ are all pairs (C, f) where $C \in \mathsf{obj}_{\mathcal{C}}$ and $f \in \mathsf{mor}_{\mathcal{C}}[C,X]$;*
2. *an arrow $g : (C, f) \longrightarrow (C', f')$ in $\mathcal{C}/X$ is a $\mathcal{C}$-arrow $g : C \longrightarrow C'$ such that $f = f' \cdot g$:*

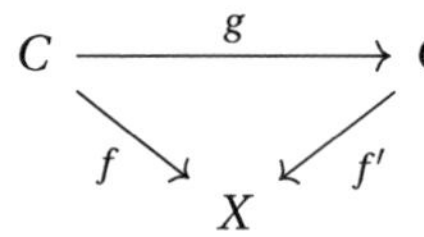

3. *the identity arrow on* (C, f) *is* $id_C : C \longrightarrow C \in \mathsf{mor}_{\mathcal{C}}[C, C]$;
4. *given two* $\mathcal{C}/X$*-arrows* $j:(A, f) \longrightarrow (B, g)$ *and* $k:(B, g) \longrightarrow (C, h)$, *their composition* $k \cdot j : (A, f) \longrightarrow (C, h)$ *is the arrow* $k \cdot j : A \longrightarrow C \in \mathsf{mor}_{\mathcal{C}}[A, C]$.

For a formal description of the concept of duality see any of the textbooks mentioned in the references, e.g. [1, p. 25f]. Here it must be mentioned that this notion is one of the cornerstones in category theory since it cuts the work in half: We get the dual N^{op} of any construct N in a category $\mathcal{C}$ by defining N in $\mathcal{C}^{\mathrm{op}}$ and reversing all the arrows. If T is a theorem for all categories $\mathcal{C}$, then its dual (obtained by reversing all the arrows of T) is again true for all categories, since $(\mathcal{C}^{\mathrm{op}})^{\mathrm{op}} = \mathcal{C}$. Two examples of dual concepts are monomorphisms and epimorphisms, and equalizers and coequalizers, to be defined among others in Section 11.1.1 below.

Clearly, the category **Fin** of finite sets and all functions between them is a full subcategory of **Set**, and **Set** is a subcategory of **Pfn**, but not a full subcategory. Let $\mathcal{C}$ be a poset $(P, \leq)$-category $\mathcal{P}$ as described in example 6 above. Then, for $p \in P$, the slice category $\mathcal{P}/p$ is just the principal ideal $x^{\downarrow} = \{q \in P : q \leq p\}$.

In Table 11.1 we have listed only some of the many mathematical domains that can be structured as categories. So categories themselves constitute a mathematical domain for which it makes sense to ask whether there is a category of categories. The answer is: yes, there is. Its objects are categories and its arrows are certain structure-preserving maps called *functors*. Since a category has two kinds of data, objects and arrows, a functor $F : \mathcal{C} \longrightarrow \mathcal{D}$ from category $\mathcal{C}$ to category $\mathcal{D}$ will need two components: one operating on objects, the other operating on arrows.

Definition 11.3. *Given categories* $\mathcal{C}$ *and* $\mathcal{D}$, *a* **functor** $F : \mathcal{C} \longrightarrow \mathcal{D}$ *is a map taking each* $\mathcal{C}$*-object* A *to a* $\mathcal{D}$*-object* $F(A)$ *and each* $\mathcal{C}$*-morphism* f *to a* $\mathcal{D}$*-morphism* $F(f)$ *such that for all* $\mathcal{C}$*-objects* A, B *and all composable* $\mathcal{C}$*-morphisms* f, g

(1) $F(f : A \longrightarrow B) = F(f) : F(A) \longrightarrow F(B)$,
(2) $F(id_A) = id_{F(A)}$,
(3) $F(g \cdot f) = F(g) \cdot F(f)$.

For each category $\mathcal{C}$ there is the obvious identity functor $id_{\mathcal{C}} : \mathcal{C} \longrightarrow \mathcal{C}$ taking every $\mathcal{C}$-object and every $\mathcal{C}$-arrow to itself. The composition of two functors $F : \mathcal{A} \longrightarrow \mathcal{B}$ and $G : \mathcal{B} \longrightarrow \mathcal{C}$ is given by separately composing their effects on objects and arrows: So $G \cdot F(A) = G(F(A))$ and each $\mathcal{A}$-arrow $f : A \longrightarrow B$ is mapped to $G(F(f)) : G(F(A)) \longrightarrow G(F(B))$. This composition operation is associative (since both the mapping of objects and that of morphisms is associative) and the identity functors are identities for the composition of functors. So functors have all the properties that morphisms in any category have.

The question arises how this underlying category should look like. It surely must have categories as objects and functors as morphisms: it is a category of categories. But a category of *all* categories would have to include itself, and we would run into the same kind of problems as identified by Russell's Paradox in set theory. So we define a category **Cat** as comprising as objects all small categories and as arrows the functors between them. Be aware that **Cat** itself is a large category, whence it cannot include itself.

Functors define mappings from one category to another. Going one step further, we have to look for structure-preserving mappings on the level of functors, called *natural transformations.* Note that this means to regard functors themselves as objects.

Definition 11.4. *Given categories $\mathcal{C}$ and $\mathcal{D}$ and* **functors** $F, G : \mathcal{C} \longrightarrow \mathcal{D}$, *a* **natural transformation** $\eta : F \Rightarrow G$ *is a function that*

(1) *assigns to every $\mathcal{C}$-object A a $\mathcal{D}$-arrow $\eta_A : F(A) \longrightarrow G(A)$ such that,*
(2) *for any $\mathcal{C}$-arrow $f : A \longrightarrow B$, $\eta_B \cdot F(f) = G(f) \cdot \eta_A$, i.e. the following diagram commutes in $\mathcal{D}$:*

$$\begin{array}{ccc} F(A) & \xrightarrow{\eta_A} & G(A) \\ \Big\downarrow{\scriptstyle F(f)} & & \Big\downarrow{\scriptstyle G(f)} \\ F(B) & \xrightarrow{\eta_B} & G(B) \end{array}$$

For any functor $F : \mathcal{B} \longrightarrow \mathcal{C}$, the identity natural transformation $id_F : F \Rightarrow F$ assigns to each $\mathcal{B}$-object B the $\mathcal{C}$-arrow $id_{F(B)} : F(B) \longrightarrow F(B)$.

For analyzing composition of natural transformations, let $F, G, H : \mathcal{C} \longrightarrow \mathcal{D}$ be functors , and $\tau : F \Rightarrow G$ and $\sigma : G \Rightarrow H$ be natural transformations. Then for any $\mathcal{C}$-arrow $f : A \longrightarrow B$ we get a composite diagram

$$\begin{array}{c} A \\ \Big\downarrow{\scriptstyle f} \\ B \end{array} \qquad \begin{array}{ccccc} F(A) & \xrightarrow{\sigma_A} & G(A) & \xrightarrow{\tau_A} & H(A) \\ \Big\downarrow{\scriptstyle F(f)} & & \Big\downarrow{\scriptstyle G(f)} & & \Big\downarrow{\scriptstyle H(f)} \\ F(B) & \xrightarrow{\sigma_B} & G(B) & \xrightarrow{\tau_B} & H(B) \end{array}$$

In this diagram both squares commute; so, by a well known result, the outer rectangle also does. Therefore, $(\tau \cdot \sigma) : F \Rightarrow H$, defined as $(\tau \cdot \sigma)_A := \tau_A \cdot \sigma_A$, is a natural transformation. Composition of natural transformations is associative, because the composition of the arrows which are components of a transformation is associative. Moreover, for each functor F, the natural transformation $id_F : F \Rightarrow F$ defined above is the identity. Thus, for every two categories $\mathcal{C}$ and $\mathcal{D}$, we can form a functor category $\mathsf{Func}(\mathcal{C}, \mathcal{D})$, whose objects are the functors from $\mathcal{C}$ to $\mathcal{D}$ and whose arrows are natural transformations between such functors.

We have now learned about the sine qua non of category theory: the concepts of categories themselves, of functors, and of natural transformations. In addition we have given a few examples of derived categories in Def. 11.2. All of the mentioned concepts pertain to external properties of categories. In the subsequent section we will now turn to some basic important concepts related to the internal structure of categories.

11.1.1 *Special Arrows and Structures*

In order to give you some more feeling about how the abstraction process in category theory works, the concepts to be introduced below will be motivated by considering analogies for sets and functions, and will be spelled out in some detail. Together with what you have mastered until here these concepts should provide a basis for proceeding to more advanced material.

11.1.1.1 *Mono-, Epi-, and Isomorphisms*

Beginning with the most simple case, recall that a function $f : X \longrightarrow Y$ is said to be

- *injective* (or *one-to-one*) when no two distinct inputs give the same output, i.e. for all $x, y \in X$, if $f(x) = f(y)$ then $x = y$; and
- *surjective* (or *onto*) if the codomain B is the range of f, i.e. for each $y \in Y$ there is some $x \in X$ such that $f(x) = y$.

Let $x \in A$ and $f \cdot g(x) = f \cdot h(x)$, i.e. $f(g(x)) = f(h(x))$. If f is injective, we get $g(x) = h(x)$, whence g and h, yielding the same output for every argument, are the same function. So we have shown that an injective function f is left cancellable:

$$\text{whenever } f \cdot g(x) = f \cdot h(x), \text{ then } g = h.$$

Generalizing this to the categorical setting results in

Definition 11.5.

(a) *An arrow $f : B \longrightarrow C$ in a category $\mathcal{C}$ is a* **monomorphism** *(or "is* **monic***") if, for any pair of arrows $g, h : A \longrightarrow B$, the equality $f \cdot g = f \cdot h$ implies that $g = h$.*

$$A \underset{h}{\overset{g}{\rightrightarrows}} B \xrightarrow{\quad f \quad} C$$

(b) *An arrow $f : C \longrightarrow B$ in a category $\mathcal{C}$ is an* **epimorphism** *(or "is* **epic***") if, for any pair of arrows $g, h : B \longrightarrow A$, the equality $g \cdot f = h \cdot f$ implies that $g = h$.*

$$A \underset{h}{\overset{g}{\leftleftarrows}} B \xleftarrow{\quad f \quad} C$$

Taking a look at the diagrams for epic and monic arrows, you immediately will recognize that the definitions of monomorphisms and epimorphisms in terms of arrows arise from each other by simply reversing the arrows involved. It is easy to check that f is a monomorphism in a category $\mathcal{C}$ if f^{op} is an epimorphism in $\mathcal{C}^{\mathrm{op}}$. We have $id_{\mathcal{C}^{\mathrm{op}}} = (id_{\mathcal{C}})^{\mathrm{op}}$ and $f^{\mathrm{op}} \cdot g^{\mathrm{op}} = (g \cdot f)^{\mathrm{op}}$. This is our first example of the notion of **duality** in category theory.

Proposition 11.6. *In* **Set**, *the monomorphisms are just the injective functions.*

Proof. Let $f : B \longrightarrow C$ be an injective function, and let $g, h : A \longrightarrow B$ be such that $f \cdot g = f \cdot h$ but $g \neq h$. Then $g(x) \neq h(x)$ for some $x \in A$. But since f is injective, $f(g(x)) \neq f(h(x))$, a contradiction, whence $g = h$ and f is a monomorphism. Conversely, let $f : B \longrightarrow C$ be a monomorphism. If f is not injective, then there are distinct elements $y, y' \in B$ with $f(y) = f(y')$. Let $A = \{x\}$, and let $g : A \longrightarrow B$, $x \mapsto y$, and $h : A \longrightarrow B$, $x \mapsto y'$. Then $f(g(x)) = f(h(x))$ and $g \neq h$, a contradiction to the assumption that f is a monomorphism. □

Since monomorphisms are left cancellable, one can prove two arrows g, h to be equal by finding a known monic f and proving that $f \cdot g = f \cdot h$. Similar arguments as the above ones show that the epimorphisms in **Set** are the surjective functions. Some other basic properties of morphisms are given in

Proposition 11.7. *Let C be a category and $f : A \longrightarrow B$ and $g : B \longrightarrow C$.*

(1) If f and g are monomorphisms, then also $g \cdot f$.
(2) If $g \cdot f$ is a monomorphism, then also f.
(3) If f and g are epimorphisms, then also $g \cdot f$.
(4) If $g \cdot f$ is a epimorphism, then also g.

Proof. (1): Consider a pair of arrows $p, q : C \longrightarrow D$. Assuming both f and g to be monic we get $(g \cdot f) \cdot p = (g \cdot f) \cdot q \rightarrow g \cdot (f \cdot p) = g \cdot (f \cdot q) \rightarrow f \cdot p = f \cdot q \rightarrow p = q$.
(2): Suppose $g \cdot f$ is a monomorphism, and moreover assume that $f \cdot p = f \cdot q$ for some arrows $p, q : C \longrightarrow A$. Then $(g \cdot f) \cdot p = g \cdot (f \cdot p) = g \cdot (f \cdot q) = (g \cdot f) \cdot q$, whence $p = q$ since $g \cdot f$ is a monomorphism. Thus, f is a monomorphism too. □

Obviously, an identity arrow is both a monomorphism and an epimorphism. Another concept that must be mentioned in connection with monics and epics are iso arrows:

Definition 11.8. *Let C be a category, and $f : A \longrightarrow B$, and $g : B \longrightarrow A$ be in mor_C.*

(a) f is called an **isomorphism**, *or* **iso arrow**, *if there is an arrow $g : B \longrightarrow A$ such that $g \cdot f = id_A$ and $f \cdot g = id_B$.*
(b) A category C is called **balanced** *iff every arrow which is both monic and epic is an isomorphism.*
(c) If $g \cdot f = id_B$ then f is called a **split monomorphism** *or* **section** *(of g), and g a* **split epimorphism** *or* **retraction** *(of f)*

In order to make diagrams more informative in a concise way, special kinds of arrows are used for depicting the various kinds of morphisms. It is common practice to use $f : A \rightarrowtail B$ for monic, and $f : A \twoheadrightarrow B$ for epic arrows $f : A \longrightarrow B$. Consequently, isomorphisms are shown as $f : A \rightarrowtail\!\!\!\twoheadrightarrow B$. Inclusions are written as $A \hookrightarrow B$. And for compact textual presentation we will use $\mathsf{mor}_{\mathcal{C}}$, $\mathsf{epi}_{\mathcal{C}}$, $\mathsf{mono}_{\mathcal{C}}$ and $\mathsf{iso}_{\mathcal{C}}$ for all morphisms, the epimorphisms, the monomorphisms, and the isomorphisms of a category $\mathcal{C}$.

Proposition 11.9. *Let $\mathcal{C}$ be a category and f and g be $\mathcal{C}$-arrows.*

(1) f has at most one iso-arrow g.
(2) Isomorphisms are always monic and epic.
(3) If f and g are both monic arrows with the same target, and each factors through the other, i.e. there are i, j such that $f = g \cdot i$ and $g = f \cdot j$, then the factors i and j are isomorphisms and inverse to each other.
(4) The composite of two isomorphisms is an isomorphism.
(5) Sections are monic and retractions are epic.

Proof. (1): Assuming $g \cdot f = g' \cdot f = id_A$ and $f \cdot g = f \cdot g' = id_B$ we get $g' = id_A \cdot g' = (g \cdot f) \cdot g' = g \cdot (f \cdot g') = g \cdot id_B = g$. So, when it exists, this g is called the *inverse of f*, and is denoted as $f^{-1} : B \longrightarrow A$. We write $f : A \simeq B$ if f is an isomorphism between the two objects.

(2): Let $f : A \longrightarrow B$ be an isomorphism and $g.h : C \longrightarrow A$ such that $f \cdot g = f \cdot h$; then $g = id_A \cdot g = (f^{-1} \cdot f) \cdot g = f^{-1} \cdot (f \cdot g) = f^{-1} \cdot (f \cdot h) = id_A \cdot h = h$, whence f is left cancellable. And with an analogous argument one can prove that iso's are always epic.

(3): Let $f : A \rightarrowtail C$ and $g : B \rightarrowtail C$ be two monic arrows, such that $f = g \cdot i$ and $g = f \cdot j$ for two arrows $i : A \longrightarrow B$ and $j : B \longrightarrow A$.

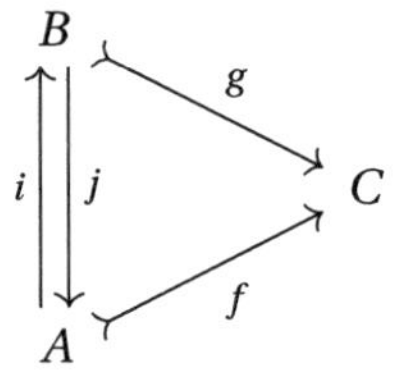

Then $f \cdot id_A = f = g \cdot i = f \cdot j \cdot i$, and, since f is monic, $j \cdot i = id_A$. Similarly, one gets $i \cdot j = id_B$. Therefore, i and j are each other's two-sided inverse, and both are isomorphisms.

(4): Assume f, g to be isomorphisms as shown below. Then $(g \cdot f) \cdot (f^{-1} \cdot g^{-1}) = g \cdot id_B \cdot g^{-1} = g \cdot g^{-1} = id_C$. Similarly $(f^{-1} \cdot g^{-1}) \cdot (g \cdot f) = id_A$, so $f^{-1} \cdot g^{-1}$ is the inverse $(g \cdot f)^{-1}$.

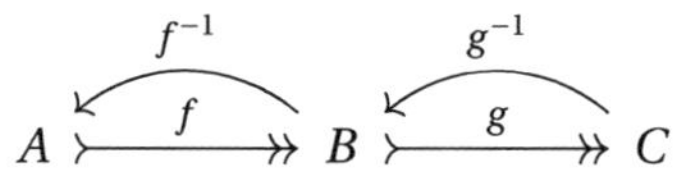

(5): Obviously, identities are isomorphisms, hence monic by (2). By Prop. 11.7(2) $g \cdot f = id_A$ implies that f is a monomorphism. A similar proof holds for retractions. $\square$

11.1.1.2 *Equalizers and Coequalizers*

Equalizers and coequalizers are our second example of dual notions in categories, with their full formal description given in

Definition 11.10.

(a) *An arrow* $e: E \longrightarrow A$ *in a category* $\mathcal{C}$ *is an* **equalizer** *of a pair of arrows* $f, g: A \longrightarrow B$, *if*

1. $f \cdot e = g \cdot e$;
2. *whenever* $e': E' \longrightarrow A$ *satisfies* $f \cdot e' = g \cdot e'$ *there is an arrow* $u: E' \longrightarrow E$ *such that* $e \cdot u = e'$ *which is unique:*

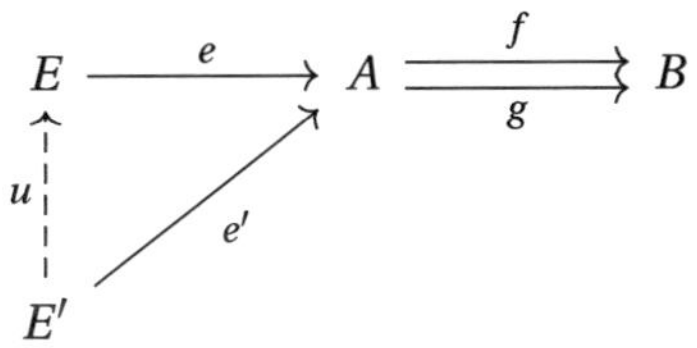

(b) *An arrow* $c: A \longrightarrow B$ *in a category* $\mathcal{C}$ *is a* **coequalizer** *of a pair of arrows* $f, g: C \longrightarrow A$ *if*

1. $c \cdot f = c \cdot g$;
2. *whenever* $c': A \longrightarrow B'$ *satisfies* $c' \cdot f = c' \cdot g$ *there is an arrow* $u: B \longrightarrow B'$ *such that* $u \cdot c = c'$ *which is unique:*

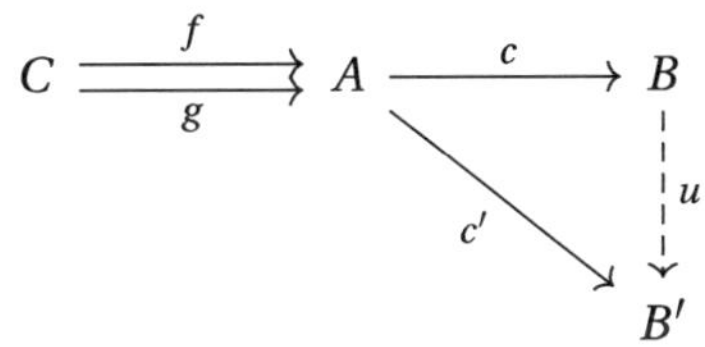

Proposition 11.11. *For any category* $\mathcal{C}$ *the following holds:*

(1) *If* $e: E \longrightarrow A$ *is an equalizer of some pair of arrows* $f, g: A \longrightarrow B$, *then* e *is monic.*
(2) DUALLY: *If* $c: A \longrightarrow B$ *is a coequalizer of some pair of arrows* $f, g: C \longrightarrow A$, *then* c *is epic.*
(3) *If both* $e_1: E_1 \longrightarrow A$ *and* $e_2: E_2 \longrightarrow A$ *are equalizers of some pair of arrows* $f, g: A \longrightarrow B$, *then there is a unique isomorphism* j *such that* $e_1 = e_2 \cdot j$.
(4) DUALLY: *Coequalizers of some pair of arrows* $f, g: C \longrightarrow A$ *are unique up to a unique isomorphism.*

Proof. (1): Consider a pair of arrows $h, i: E' \longrightarrow E$ such that $e \cdot h = e \cdot i$. Put $e' = e \cdot h = e \cdot i$. Then we have $f \cdot e' = f \cdot e \cdot h = g \cdot e \cdot h = g \cdot e'$, whence there is a unique $u: E' \longrightarrow E$ such that $e \cdot u = e'$. Therefore, the assumption $e \cdot h = e \cdot i$ implies $h = u = i$.

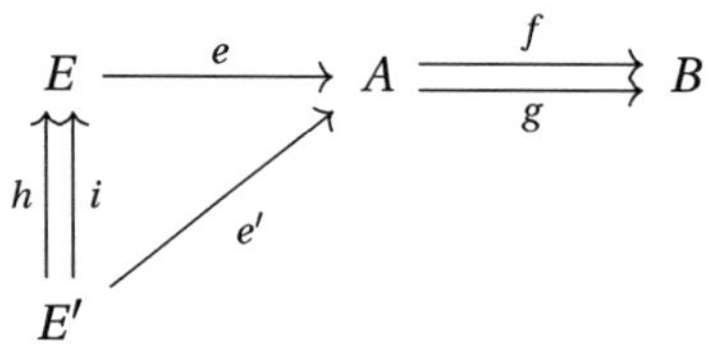

□

We can thus talk about "the" equalizer of two morphisms in the sense that an equalizer e can stand for any one of a class of isomorphic subobjects of A; see Section 11.1.1.4. If an equalizer exists for all pairs of morphisms in $\mathcal{C}$ with common domains and codomains, one says that $\mathcal{C}$ *has equalizers.*

As usual we take a look at **Set**: Assume we have a pair of functions $f, g : A \longrightarrow B$. We define $E \subseteq A$ as $E := \{x \in A : f(x) = g(x)\}$ and let $e : E \longrightarrow A$ be the inclusion map $i : E \hookrightarrow A$. Then $E \overset{i}{\hookrightarrow} A \overset{f}{\underset{g}{\rightrightarrows}} B$ is an equalizer. By construction, $f \cdot i = g \cdot i$. Let $h : C \longrightarrow A$ be such that $f \cdot h = g \cdot h$, i.e.

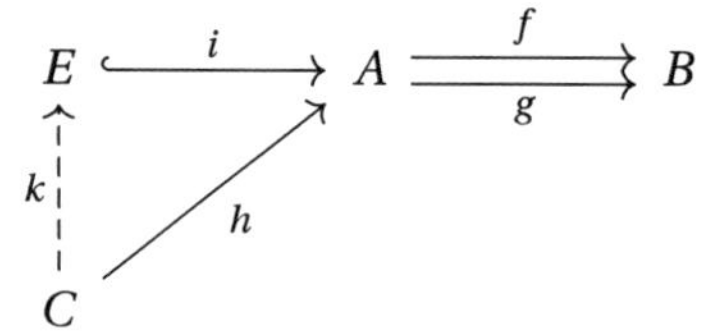

then there can be at most one way to fill in the broken arrow to make this diagram commute: If $h = i \cdot k$ for $c \in C$ we must have $i(k(c)) = h(c)$, which means $k(c) = h(c)$, i being the inclusion. But this does fit in, for $f(h(c)) = g(h(c))$, whence $h(c) \in E$, so k is unique.

Concerning coequalizers, let $\rho \subseteq A \times A$ be an equivalence relation.

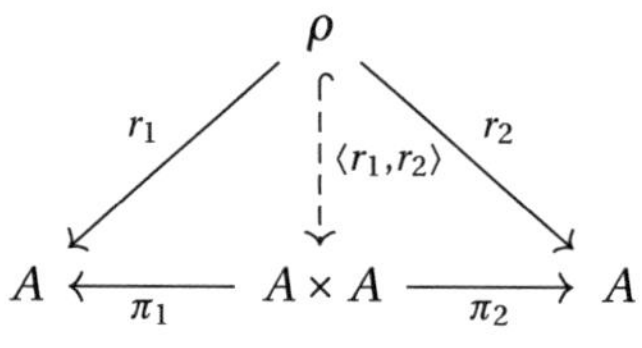

Then $p : A \longrightarrow A/\rho$, $a \mapsto [a]$, mapping each element a to its equivalence class, is a coequalizer of r_1 and r_2. For given an $f : A \longrightarrow B$ with

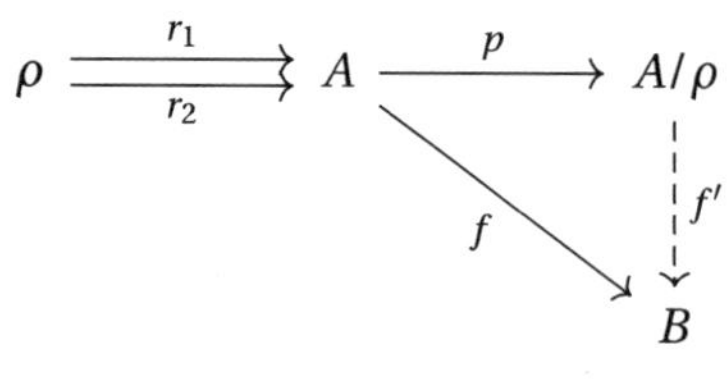

there exists a function f' such that $f'(p(x)) = f(x)$ whenever f respects ρ in the sense that $x\rho y$ implies $f(x) = f(y)$. Since, for all $\langle x, y \rangle \in \rho$, $f \cdot r_1(x, y) = f(x)$ and $f \cdot r_2(x, y) =$

$f(y)$ this means that $f \cdot r_1 = f \cdot r_2$. Moreover, if it exists, such a function f' is then necessarily unique, since p is an epimorphism.

11.1.1.3 *Initial and Terminal Objects*

Definition 11.12. *Let $\mathcal{C}$ be a category, and $I, T \in \mathsf{obj}_{\mathcal{C}}$.*

(a) I is called **initial** *if for every $A \in \mathsf{obj}_{\mathcal{C}}$ there is exactly one $f : I \longrightarrow A$.*
(b) T is called **terminal** *if for every $A \in \mathsf{obj}_{\mathcal{C}}$ there is exactly one $f : A \longrightarrow T$.*

A category need not have an initial object, but if it does have one, it is unique up to isomorphism. Indeed, it is unique up to unique isomorphism, as follows.

Proposition 11.13. *Let I and I' be initial objects of a category $\mathcal{C}$. Then there is a unique isomorphism $u : I \simeq I'$*

Proof. By assumption there are unique maps $f : I \longrightarrow I'$ and $f' : I' \longrightarrow I$. But both $f' \cdot f$ and id_I are in $\mathsf{mor}_{\mathcal{C}}[I, I]$, so $f' \cdot f = id_I$ since I is initial. Similarly, $f \cdot f' = id_{I'}$, whence f is an isomorphism. □

In **Set**, the empty set is the only initial object, and any singleton set $\{\bullet\}$ is a terminal object. In fact, for every set S, the empty function, say $f_\emptyset$, is the unique function from $\emptyset$ to S: since $\emptyset \times S = \emptyset$, we must have $f_\emptyset \subseteq \emptyset$, whence $f_\emptyset = \emptyset$ is the only subset satisfying this condition. Thus the category of sets has a unique initial but many terminal objects: for every set S, the rule $f(x) = \bullet$ gives a function $f : S \longrightarrow \{\bullet\}$ mapping every element of S to $\bullet$. Since $\bullet$ is the only possible output, this is the only possible such function.

11.1.1.4 *Elements, Subobjects and Images*

For all the concepts introduced in this section we not only give formal definitions but also some comments which are mainly due to [10, Sections 4.1 and 5.2].

(a) Subobjects and Elements:

Definition 11.14. *Let $\mathcal{C}$ be a category, and $X \in \mathsf{obj}_{\mathcal{C}}$.*

(a) A **subobject** *of X is a monic $\mathcal{C}$-arrow $m : A \rightarrowtail X$ with codomain X.*
(b) If $\mathcal{C}$ has a terminal object T, then an **element** *of X is defined as an arrow $e : T \longrightarrow X$.*

If A is a subset of B, then the inclusion function $i : A \hookrightarrow B$ is injective, and on the other hand any injective function $f : C \rightarrowtail B$ determines a subset of B, viz $f[C] = \{f(x) : x \in C\}$. It is clear that f induces a bijection between C and $f[C]$, so C is isomorphic to $f[C]$. Thus, in **Set** any monic arrow $f : A \rightarrowtail B$ generates an isomorphism between A and a subset of B. It is typical for category theory to care about identifications up to isomorphism only. In **Set** an object A together with a monic arrow $f : A \rightarrowtail B$ can therefore be treated as a subobject of B.

Having described subsets in terms of category theory, we can turn to elements: $x \in A$ can obviously be identified with the singleton subset $\{x\} \subseteq A$, and hence with an arrow $i : \{x\} \hookrightarrow A$, i.e. with an arrow from a terminal object to A. In the other direction, if $T = \{t\}$ is a terminal object in **Set**, a function $f : T \longrightarrow A$ determines an element of A, viz. the f-image of t. Thus, if a category $\mathcal{C}$ has a terminal object T, an element $x \in A$ is defined to be an arrow $x : T \longrightarrow A$.

Though things seem to run smoothly so far, it is worth to have a closer look on the categorial subobject concept: it will reveal that nearby generalizations can be deceptive, and that one has to proceed very carefully. The relation of set inclusion is a partial order on power sets, i.e. for a set A, $(\mathbf{2}^A, \subseteq)$ is a poset, and as we have seen above, it can be made into a category in which there is an arrow $A \longrightarrow B$ iff $A \subseteq B$.

So, if there are the two inclusions $A \hookrightarrow B \hookrightarrow C$ then also $A \hookrightarrow C$, as shown in the left diagram below. This suggests to define an "inclusion" relation between subobjects of an object C as follows: Given two monics $f : A \rightarrowtail C$ and $g : B \rightarrowtail C$, we put $f \subseteq g$ iff there is an arrow $h : A \longrightarrow B$ such that $f = g \cdot h$, i.e. $f \subseteq g$ iff f factors through g. Note that h is also monic by Prop. 11.7(2), so it will be a subobject of B.

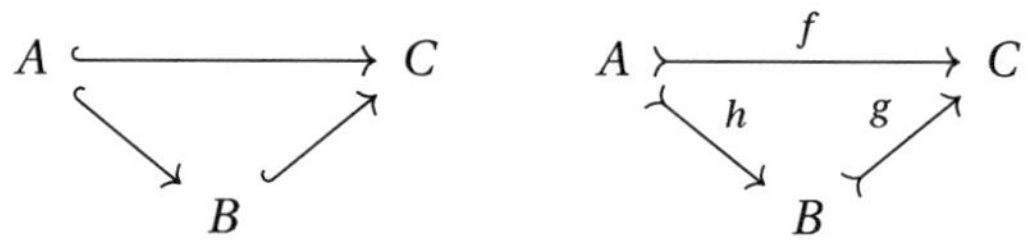

This inclusion concept for subobjects is

(i) *reflexive*: $f = f \cdot id_A$, so $f \subseteq f$;
(ii) *transitive*: Assume $f \subseteq g \subseteq h$; then we have $f = g \cdot k$ and $g = h \cdot i$. So $f = h \cdot (i \cdot k)$, whence $f \subseteq h$;
(iii) but *not antisymmetric*: $\subseteq$ as defined above will in general be a preordering on the subobjects, but not a partial ordering as it is in **Set**. Assume $f \subseteq g$ and $g \subseteq f$; then f and g factor through each other, say $f = g \cdot h$ and $g = f \cdot i$. By Prop. 11.9(3), $h : A \longrightarrow B$ is an iso, with i as inverse, so that $A \simeq B$. Thus, we needn't have $A = B$ which would be required for the arrows f, g to be identical: f and g are *isomorphic subobjects* only, written $f \simeq g$;
(iv) and singleton sets have *infinitely many subobjects*: In **Set**, $\{1\}$ would not only have two subobjects as expected (the empty set and itself), but there are as many monic arrows $S \longrightarrow \{1\}$ as there are singleton sets S.

So we are near, but not just on the right track: An obvious suggestion to get things right in the spirit of category theory is to regard two monic arrows $f : A \rightarrowtail X$ and $g : B \rightarrowtail X$ as representing the same subobject of X iff $A \simeq B$. By routine checks one can verify that $\simeq$ as defined above ($f \simeq g$ if $f = g \cdot h$ and $g = f \cdot i$ for some $\mathcal{C}$-morphisms i, h) is an equivalence relation. Thus we arrive at the following concept for subobjects:

Definition 11.15. *Let $\mathcal{C}$ be a category, and $X \in \mathsf{obj}_{\mathcal{C}}$. A* []**-subobject** *of X is a class of subobjects of X which factor through each other: $[f] := \{g : f \simeq g\}$. We then form the* *collection* $\mathbf{Sub}(X) = \{[f] : f$ *is a monic with* $\mathrm{cod}\, f = X\}$.

Next we define inclusion on $\mathbf{Sub}(X)$ by $[f] \subseteq [g]$ iff $f \subseteq g$. Assuming $[f] \subseteq [g]$ and $[g] \subseteq [f]$, we then have $f \subseteq g$ and $g \subseteq f$, hence $f \simeq g$. Thus $[f] = [g]$ as desired, and $(\mathbf{Sub}(X), \subseteq)$ is a poset. That this is what we really have aimed for is shown by the fact that in **Set** one has $\mathbf{Sub}(X) \simeq \mathbf{2}^X$. A proof of this can be found as [22, Theorem 65].

Our considerations on subobjects are not yet done with: It is common habit to blur the distinction between equivalence class and representative. That is, in order to simplify both notation and wording, one usually says "subobject f" for the "[]-subobject $[f]$" and $f \subseteq g$ when strictly speaking $[f] \subseteq [g]$ is meant. However, concerning identity, we must be precise: "$f \simeq g$" is used if f and g are the same subobject, i.e. $[f] = [g]$. and "$f = g$" if they are the same arrow.

(b) Images:

We begin our comments on images of an arrow by considering how this concept looks like in **Set**. A well known result in set theory is that any function $f : A \longrightarrow B$ can be decomposed into a surjection and an injection (see Cor. 2.29). So the following diagram commutes

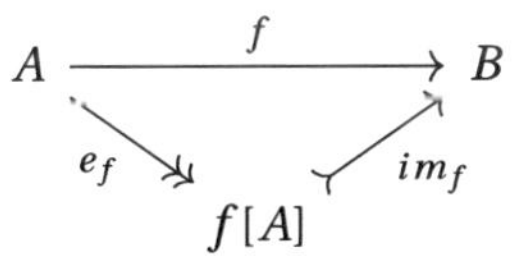

where $e_f(x) = f(x)$, for all $x \in A$, and im_f is called the image of f. It should be noted that sometimes the domain $f[A]$ of im_f is called the image of f. In **Set** the following holds:

Proposition 11.16. ([10, p. 112, THEOREM 1]) *im_f is the smallest subobject of B through which f factors, i.e. if, for any u and monic v, the diagram*

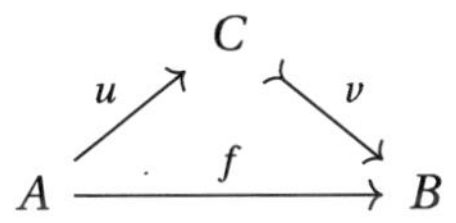

commutes, then there is a (unique) $k : f[A] \longrightarrow C$ such that

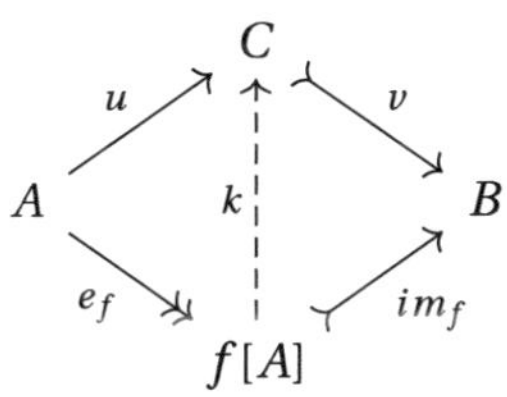

also commutes. Thus $im_f \subseteq v$.

Proposition 11.17. ([10, p. 114, THEOREM 2]) *$A \overset{e_f}{\twoheadrightarrow} f[A] \overset{im_f}{\rightarrowtail} B$ is an epi-monic factorization of $f : A \longrightarrow B$ which is unique up to unique isomorphism. That is, if $A \overset{u}{\twoheadrightarrow} C \overset{v}{\rightarrowtail} B$ such that $f = v \cdot u$, then there is exactly one arrow $k : f[A] \longrightarrow C$ that makes the following diagram commute. Moreover, k is an isomorphism.*

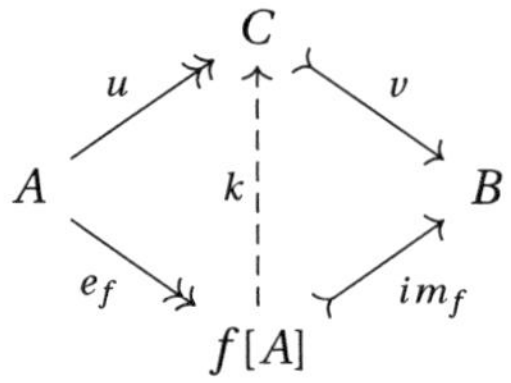

Generalizing the above from **Set** to an arbitrary category gives

Definition 11.18. *Let $\mathcal{C}$ be a category, and $A, B \in \mathsf{obj}_{\mathcal{C}}$. The* **image** *of a morphism $f : A \longrightarrow B$, denoted as $(\mathrm{Im} f, im_f)$, is the smallest subobject of B which f factors through. A category* **has images** *if every morphism $f : A \longrightarrow B$, has an image factorization as shown in*

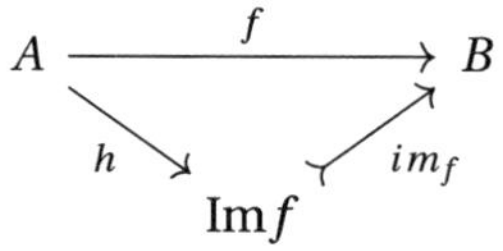

Proposition 11.19. *Let $\mathcal{C}$ be a category which has images, and $A, B, C \in \mathsf{obj}_{\mathcal{C}}$. For an arbitrary factorization $A \overset{u}{\longrightarrow} C \overset{v}{\rightarrowtail} B$. there is a (unique) $k : \mathrm{Im} f \longrightarrow C$ such that*

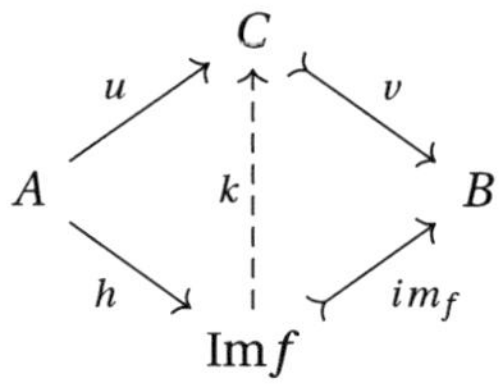

commutes.

11.1.1.5 *Products*

We continue with introducing the notion of a product of two objects in a category. If you think of A and B as sets in the definition below, it gives a characterization of the Cartesian product $A \times B$ of two sets in terms of arrows only, without any reference to ordered pairs.

Definition 11.20.

(a) *A* **product** *in a category $\mathcal{C}$ of two objects $A, B \in \mathsf{obj}_{\mathcal{C}}$ is a $\mathcal{C}$-object $A \times B$ together with a pair of arrows $\pi_A : A \times B \longrightarrow A$, $\pi_B : A \times B \longrightarrow B$, called* **projections**, *such* *that for* any *pair of arrows* $f : C \longrightarrow A$ *and* $g : C \longrightarrow B$ *there is a unique arrow* $u : C \longrightarrow A \times B$ *making the following diagram commute, i.e. such that* $f = \pi_A \cdot u$ *and* $g = \pi_B \cdot u$. *This unique* u *is denoted by* $\langle f, g \rangle$ *and is called the* **product arrow** *of* f *and* g *with respect to the projections* π_A *and* π_B.

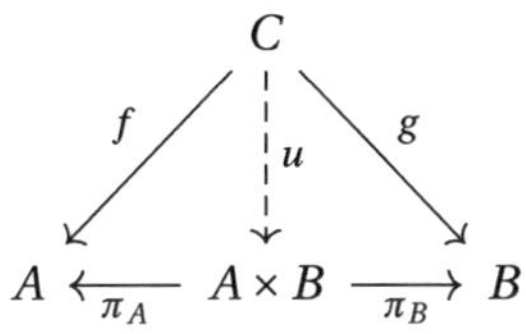

(b) DUALLY: *The* **coproduct** *of two objects* $A, B \in \mathsf{obj}_{\mathcal{C}}$ *is a* $\mathcal{C}$*-object denoted as* $A + B$.

The following proposition shows that the characterization of a product $A \times B$ in terms of arrows is unique up to isomorphism:

Proposition 11.21. *When they exist, products of objects are unique up to a unique projection-preserving isomorphism.*

Proof. Suppose that $A \xleftarrow{p_1} C \xrightarrow{p_2} B$ is a product of A and B. Then, there is a unique $u : C \longrightarrow A \times B$ such that such that $\pi_A \cdot u = p_1$ and $\pi_B \cdot u = p_2$, as shown in the left diagram below.

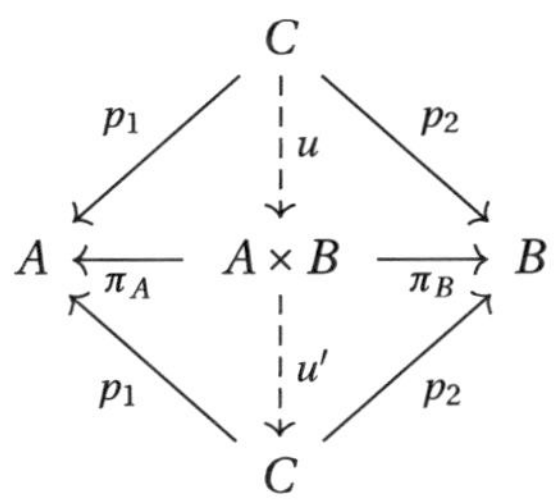

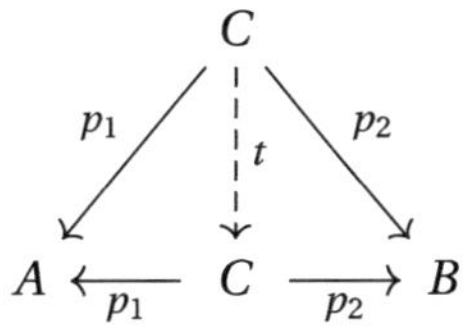

Since C is a product of A and B, there can be only one arrow $t : C \longrightarrow C$ such that the right diagram above commutes, and taking $t = id_C$ does the job. But the commutativity of the left diagram implies that $p_1 \cdot u' \cdot u = \pi_A \cdot u = p_1$, and by the uniqueness of t we must conclude that $u' \cdot u = id_C$. Interchanging the roles of C and $A \times B$ leads to $u \cdot u' = id_{A \times B}$, so $u : C \simeq A \times B$. □

In **Set**, the coproduct of sets A and B is their disjoint union, i.e. the union of two sets that are isomorphic to A and B, respectively, but disjoint. We put

$$A' := \{\langle a, 1\rangle : a \in A\} = A \times \{1\} \quad \text{and} \quad B' := \{\langle b, 2\rangle : b \in B\} = B \times \{2\}$$

and define $A + B = A' \cup B'$ with obvious coproduct injections

$$i_A = \langle a, 1\rangle \quad \text{and} \quad i_B = \langle b, 2\rangle.$$

Given any functions f and g as in

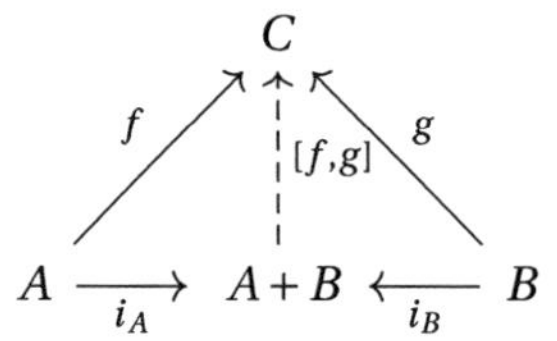

we define $[f,g](x,1) := f(x)$ and $[f,g](x,2) := g(x)$. If there is an h such that $h \cdot i_A = f$ and $h \cdot i_B = g$, then for any $\langle x, \mathsf{n}\rangle \in A+B$ we have $h(x,\mathsf{n}) = [f,g](x,\mathsf{n})$, whence $[f,g]$ is unique.

We have to distinguish the product arrow $\langle f, g\rangle$ from the arrow product $f \times g$ of arrows f and g in a category. As you will immediately see, we use the former concept to define the latter.

Definition 11.22. *Let $f : A \longrightarrow B$ and $g : C \longrightarrow D$ be two arrows. The* **arrow product** *$f \times g$ of f and g is the unique morphism $\langle f \cdot p_A, g \cdot p_C\rangle$ that makes the following diagram commute.*

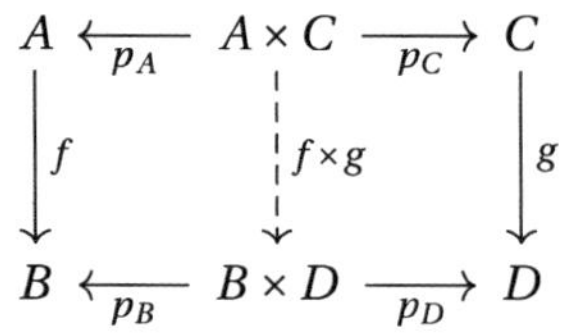

11.1.1.6 *Pullbacks*

Definition 11.23.

(a) *A* **pullback** *(or* **fiber product** *of A and B over C) in a category $\mathcal{C}$ of a pair of arrows with common target, say $f : A \longrightarrow C$ and $g : B \longrightarrow C$, consists of an object D (also denoted by $A \times_C B$) together with two $\mathcal{C}$-arrows $f' : D \longrightarrow B$ and $g' : D \longrightarrow A$ such that*

(1) *$f \cdot g' = g \cdot f'$;*

(2) *whenever two arrows $h : E \longrightarrow A$ and $j : E \longrightarrow B$ satisfy $f \cdot h = g \cdot j$ then there is a unique arrow $u : E \longrightarrow D$ such that $h = g' \cdot u$ and $j = f' \cdot u$:*

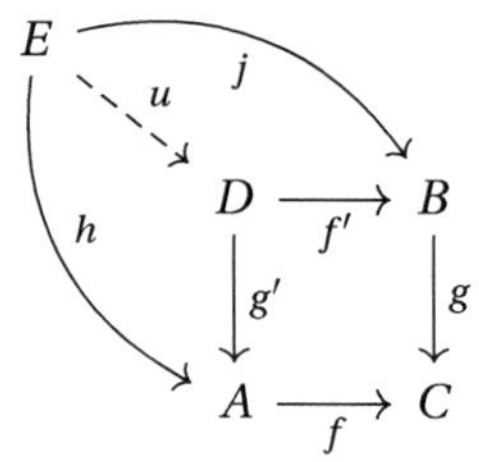

(b) DUALLY: *A commutative square*

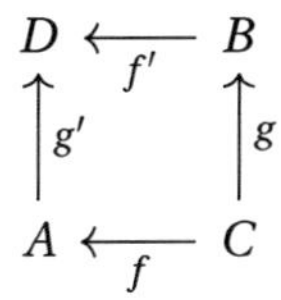

is called a **pushout** *if for any object E and any pair of arrows* $h\colon A \longrightarrow E$ *and* $j\colon B \longrightarrow E$ *with* $h\cdot f = j\cdot g$ *there is a unique arrow* $u\colon D \longrightarrow E$ *such that* $u\cdot g' = h$ *and* $u\cdot f' = j$.

In other words, for pullbacks the following holds: If h and j make the outer square in the diagram commute, then there is only one way to fill in the arrow u to make the whole diagram commute. The inner square of the pullback diagram, i.e. the objects A, B, C, D with arrows f, g, f', g', is called a *pullback square.* One also says that f' arises by "*pulling back f along g*", and g' by "*pulling back g along f*"

The notion of a pullback is very important and fundamental, and incorporates a number of well known constructions as is demonstrated with a few examples:

(1) In **Set** let a function $f\colon A \longrightarrow B$ be given, and $C \subseteq B$. Then $f^{-1}[C] \subseteq A$, and the diagram below is a pullback, with $f' := f|_{f^{-1}[C]}$, and $i\colon f^{-1}[C] \hookrightarrow A$ and $j\colon C \hookrightarrow B$ being inclusion functions.

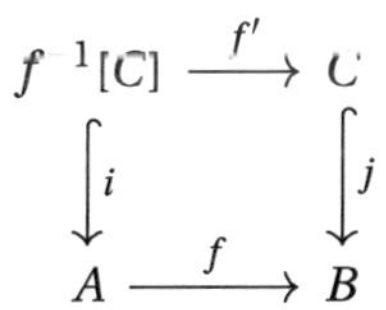

To verify the assertion, following [15, p. 116], we take any commutative square as shown on left side below; we must then show that there is a unique map $u\colon A \longrightarrow f^{-1}[C]$ such that the diagram on the right commutes.

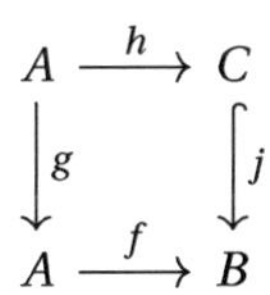

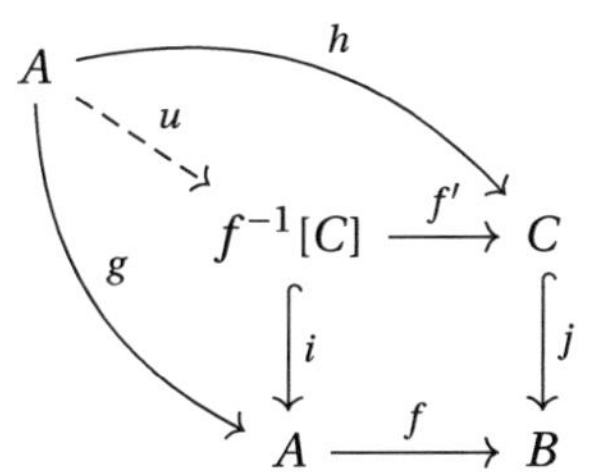

Concerning uniqueness, let u be a map making the diagram commute. Then we have $i(u(a)) = g(a)$, i.e. $u(a) = g(a)$, for all $a \in A$, and this determines u uniquely. And for existence, note that, for all $a \in A$, $f(g(a)) = j(h(a)) \in C$, whence $g(a) \in f^{-1}[C]$. So we may define $u\colon A \longrightarrow f^{-1}[C]$ by $u(a) = g(a)$, whence we get $i(u(a)) = u(a) = g(a)$ and, since f' is defined as $f' := f|_{f^{-1}[C]}$, $f'(u(a)) = f(u(a)) = f(g(a)) = j(h(a)) = h(a)$, for all $a \in A$. Therefore, $i\cdot u = g$ and $f'\cdot u = h$, as required

(2) Our second example is also from **Set**: Let $f: A \longrightarrow C$ and $g: B \longrightarrow C$ be a pair of arrows with common target. Using the pullback square as shown in Def. 11.23, and putting $D = \{\langle x, y\rangle : x \in A, y \in B, \text{ and } f(x) = g(y)\}$, f' and g' as the first and second projections of D, then D, f', g' is a pullback of f and g.

(3) For the last example, recall from Def. 2.21 the definition of the kernel relation of a function $f: A \longrightarrow B$, $\ker(f) = \{\langle x, x'\rangle \in A \times A : f(x) = f(x')\}$. In the light of our last example we can easily find that

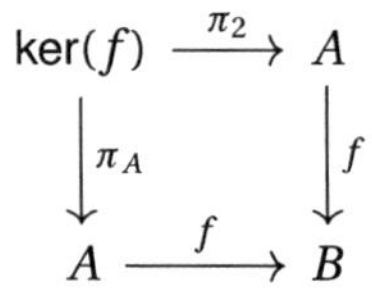

is a pullback square, with $\ker(f)$ arising by pulling back f along itself. The relation $\ker(f) \underset{\pi_1}{\overset{\pi_1}{\rightrightarrows}} A$ is called the *kernel pair* of f.

This last example leads to a general definition of the concept for an arbitrary category, which, together with some of their properties, we will need later on in the special case of regular categories. See [5, Section 2.5].

Definition 11.24. *In a category $\mathcal{C}$, the* **kernel pair** *of an arrow $f: A \longrightarrow B$ is (when it exists) the pullback (P, f_1, f_2) of f with itself, denoted as $(f_1, f_2): P \rightrightarrows A$.*

Proposition 11.25. *In a category $\mathcal{C}$, if the* **kernel pair** *$(f_1, f_2): P \rightrightarrows A$ of an arrow $f: A \longrightarrow B$ exists, then f_1 and f_2 are epimorphisms.*

Proof. By the definition of a pullback, from $f \cdot id_A = f \cdot id_A$ one gets a unique factorization u such that $f_1 \cdot u = id_A$ and $f_2 \cdot u = id_A$. Recall that an identity arrow is both a monomorphism and an epimorphism. So from the first equation we get that u is a monomorphism using Prop. 11.7(2), and from the second that f_1 and f_2 are epimorphisms by Prop. 11.7(4).

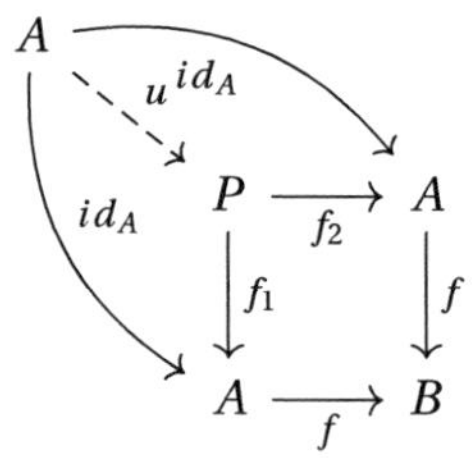

□

Proposition 11.26. ([5, Propositions 2.5.7 and 2.5.8]) *Let $\mathcal{C}$ be a category; then the following holds*

(1) *If a coequalizer f has a kernel pair, it is the coequalizer of its kernel pair.*
(2) *If a kernel pair has a coequalizer, it is the kernel pair of its coequalizer.*

Definition 11.27. *A set of morphisms* $\mathsf{M} \subseteq \mathsf{mor}_{\mathcal{C}}$ *in a category* $\mathcal{C}$ *is called* **pullback stable** *provided that for each pullback square in* $\mathcal{C}$

$$\begin{array}{ccc} P & \xrightarrow{f'} & B \\ \Big\downarrow{\scriptstyle g'} & & \Big\downarrow{\scriptstyle g} \\ A & \xrightarrow{f} & B \end{array}$$

$f \in \mathsf{M}$ *implies* $f' \in \mathsf{M}$.

Lemma 11.28. *Monomorphism are pullback stable.*

Proof. Let f be a monomorphism. If, for some $Q \in \mathsf{obj}_{\mathcal{C}}$, $h, k : Q \longrightarrow P$ are morphisms such that $f' \cdot h = f' \cdot k$, then we get

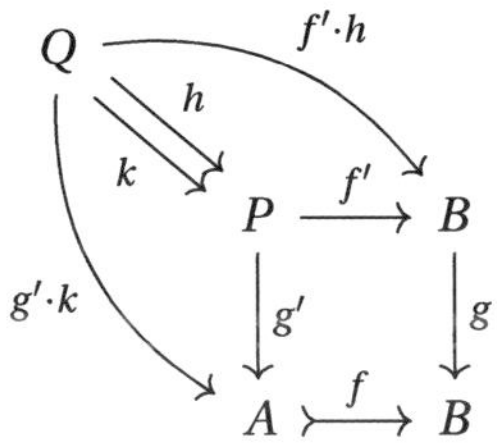

It follows that

$$f \cdot (g' \cdot h) = g \cdot (f' \cdot h) = g \cdot (f' \cdot k) = f \cdot (g' \cdot k).$$

Thus, $g' \cdot h = g' \cdot k$. But, by definition, there is a unique arrow $u : Q \longrightarrow P$ such that $g \cdot (f' \cdot u) = f \cdot (g' \cdot u)$, whence $h = k$, so f' is a monomorphism. □

Another important and often used property of pullbacks is the following

Lemma 11.29. (Pullback Lemma) *Let*

$$\begin{array}{ccccc} \bullet & \longrightarrow & \bullet & \longrightarrow & \bullet \\ \downarrow & & \downarrow & & \downarrow \\ \bullet & \longrightarrow & \bullet & \longrightarrow & \bullet \end{array}$$

be a commuting diagram. Then the following holds:

(1) *If the squares are pullbacks, then also the outer rectangle.*
(2) *If the outer rectangle and the right hand square are pullbacks, then also the left hand square.*

11.1.1.7 *Limits*

Turning to the concept of a *limit* is a further step along the steep climb of categories. As you will see shortly, limits are a kind of second-order abstraction in the abstracting world of categories, closely related to what is called a *universal property*. (The latter one will not be formally defined here, since that would require still more concepts and definitions. In case you are interested, you may have a look at [16, Chapter III].) Again resorting to the introduction of [15], we give an intuitive characterization enriched with some examples.

If you look back at the definitions of initial and terminal objects, products, equalizers and coequalizers, and pullbacks, you will easily detect as common pattern phrases of the form "*there exists a unique such-and-such satisfying some condition*". Properties such as this are called *universal* because they state how the object being described relates to the entire universe in which it lives. The crucial point is that, in principle, there can be only one object satisfying a given universal property. The words "*in principle*" mean that two objects satisfying the same universal property need not literally be equal, but they are always isomorphic. Note that we have shown the isomorphism of initial objects and products in Prop. 11.13 and Prop. 11.21, respectively.

As an example we take the category **Set**, and let **1** denote a set with one element, say $\{\bullet\}$. Then **1** has the following universal property:

for all sets X, there exists a unique mapping from X to **1**.

Such a map *exists*, because we can define $u : X \longrightarrow \mathbf{1}$ with $u(x) = \bullet$ for all $x \in X$, and this is the *unique* mapping: since there is no choice, any $m : X \longrightarrow \mathbf{1}$ must send each element of X to the single element $\bullet$ of **1**. The condition describing the universal property begins with the words "for all sets X", and therefore says something about the relationship between **1** and every set X, namely that there is a unique mapping from X to **1**. From Section 11.1.1.3 we know that one-element sets are terminal objects in **Set**, and similarly to Prop. 11.13 it can be shown that two terminal objects are isomorphic.

We continue with the definition of the notions of a *diagram* and a *cone*, both needed for the definition of a *limit*:

Definition 11.30. ([20, 1.2.1, 1.9.1])

(a) *A* **diagram** D *in a category* $\mathcal{C}$ *is a collection of vertices and directed edges labeled consistently with* $\mathcal{C}$*, i.e. if* $f : A \longrightarrow B \in \mathsf{mor}_{\mathcal{C}}[A, B]$ *then an edge with label* f *must be directed from node* A *to node* B *in* D.

(b) *Let* D *be a diagram in a category* $\mathcal{C}$*. A* **cone** *for* D *is a* $\mathcal{C}$*-object* C *and arrows* $c_i : C \longrightarrow D_i$ *(one for each* D_i *in* D*) such that for each* g *in* D $g \cdot c_i = c_j$*, i.e. the following diagram commutes. We use* $\{c_i : C \longrightarrow D_i\}$ *to denote a cone.*

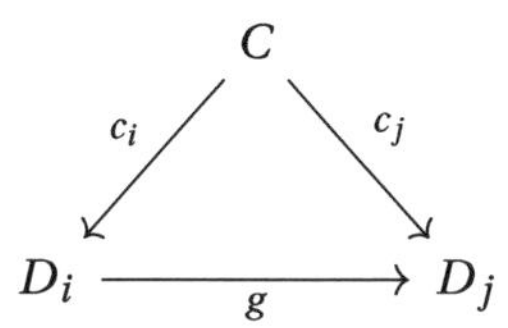

(c) *A* **limit** *for a diagram* D *is a cone* $\{l_i : L \longrightarrow D_i\}$ *such that any other cone* C *for* D *uniquely factors through it, i.e. if* $\{c_i : C \longrightarrow D_i\}$ *is another cone for* D *then there is a unique arrow* $u : C \longrightarrow L$ *such that* $l_i \cdot u = c_i$ *for all* i*, i.e.the following diagram commutes.*

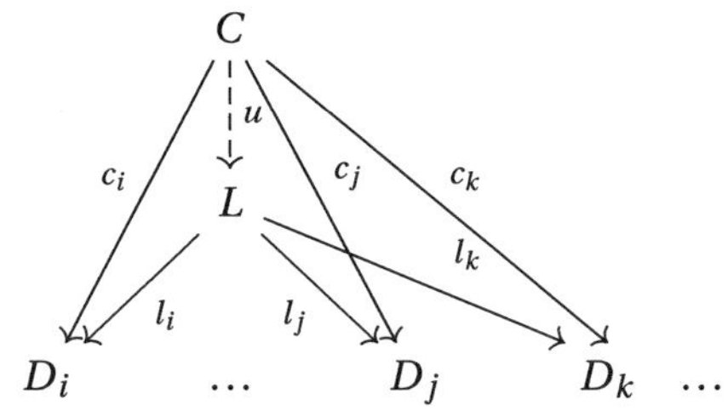

(d) DUALLY: *Let* D *be a diagram in a category* $\mathcal{C}$. *A* **cocone** *for* D *is a* $\mathcal{C}$*-object* C *and arrows* $c_i : D_i \longrightarrow C$ *(one for each* D_i *in* D*) such that for each* g *in* D $c_j \cdot g = c_i$. *A* **colimit** *for a diagram* D *is a cocone* $\{l_i : D_i \longrightarrow L\}$ *such that any other cocone* C *for* D *uniquely factors through it, i.e. there is a unique* $u : L \longrightarrow C$ *such that* $u \cdot l_i = c_i$ *for all* i.

We will now give some detailed examples of limits, namely terminal objects, products, equalizers, and pullbacks:

(1) Our first example consists of nothing: We denote the diagram having no objects and so, necessarily, no arrows by $\mathsf{D}_\emptyset$. A cone for it simply is any $\mathcal{C}$-object, since $\mathsf{D}_\emptyset$ has no nodes, whence the cone has no arrows. A limiting cone in this case is an object C with the additional requirement that for any $\mathcal{C}$-object C' (i.e. for any cone for $\mathsf{D}_\emptyset$) there is exactly one arrow from C' to C, i.e. C is a terminal object.

(2) Given two $\mathcal{C}$-objects A, B, let D be the diagram

$$A \qquad\qquad B$$

with two nodes, labeled A and B and no arrows. Then a cone for D is an object C with two arrows f_A and f_B of the form

$$A \xleftarrow{f_A} C \xrightarrow{f_B} B\,.$$

The cone $\{f_A : C \longrightarrow A, f_B : C \longrightarrow B\}$ is a limit for D if for any other cone $\{f'_A : C' \longrightarrow A, f'_B : C' \longrightarrow B\}$ for D there is a unique arrow $u : C' \longrightarrow C$ such that the left triangle below commutes for every D in D. Combining the two triangles for our two nodes A, B in D we get the right diagram below, which is nothing else than the definition of a product of A and B.

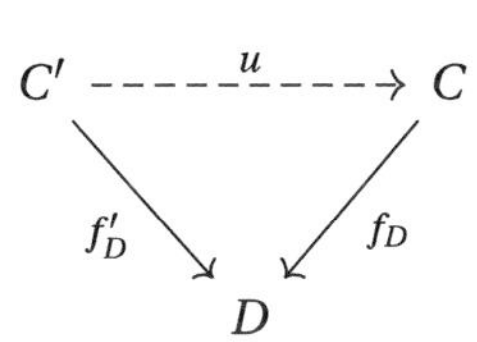

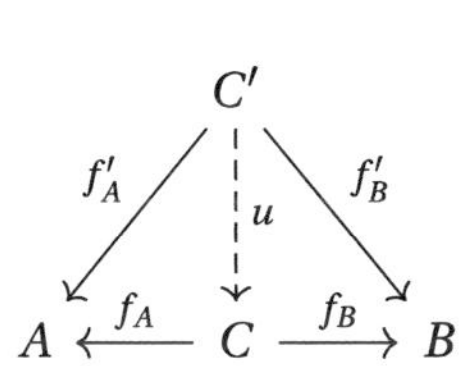

(3) Let D be the diagram $A \overset{f}{\underset{g}{\rightrightarrows}} B$. A D-cone is a pair $\{e : E \longrightarrow A, b : E \longrightarrow B\}$

such that the diagram

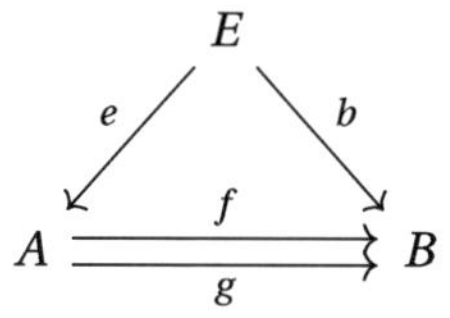

commutes. But this requires that $b = f \cdot e = g \cdot e$, whence b is determined by e. So it is equivalent to say that a D-cone in this case consists of an object E and an arrow $e : E \longrightarrow A$ such that $E \xrightarrow{e} A \overset{f}{\underset{g}{\rightrightarrows}} B$ commutes, i.e. $f \cdot e = g \cdot e$. It is then easy to see from the definitions that if $\{e : E \longrightarrow A\}$ is a limit of D then it is an equalizer of f and g.

(4) Similarly, pullbacks arise as limits starting with the diagram

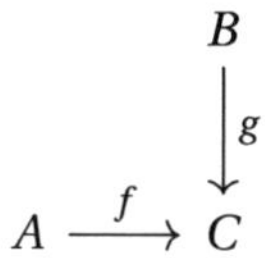

Up to now we have defined limits and have illustrated them by specific instances that we had already introduced before. But we have not yet said anything about properties of limits themselves. Without further comments we mention the following results:

(1) Given a diagram D in category $\mathcal{C}$, by $\mathcal{C}_{C(\mathsf{D})}$ we denote the *category of cones over* D, with objects the cones $\{f_i : C \longrightarrow D_i\}$ over D and with arrows $\{f_i : C \longrightarrow D_i\} \longrightarrow \{f'_i : C' \longrightarrow D_i\}$ any $\mathcal{C}$-arrow $f : C \longrightarrow C'$ such that $f'_i \cdot f = f_i$ for all i.
(2) A limit for D in $\mathcal{C}$ is a cone which is a terminal object in $\mathcal{C}_{C(\mathsf{D})}$.
(3) Limits over a given diagram D are unique up to a unique isomorphism commuting with the cone arrows.

Before we can proceed to the next section we need a last concept and some of its properties:

Definition 11.31. *A category $\mathcal{C}$* **has all finite limits** *if for any finite diagram* D *(i.e. containing finitely many objects $D_j \in \mathsf{obj}_{\mathcal{C}}$) $\mathcal{C}$ has a limit over* D. *A category with all finite limits is called* **finitely complete** *(a* **cartesian category** *in* [12]*).*

Proposition 11.32. ([5, Proposition 2.8.2]) *For a category $\mathcal{C}$ the following conditions are equivalent:*

(1) $\mathcal{C}$ is finitely complete.
(2) $\mathcal{C}$ has a terminal object and has all binary products and equalizers.
(3) $\mathcal{C}$ has a terminal object and pullbacks.

11.2 Relations and Regular Categories

Important note: A careful reading of the subsequent paragraph reveals that the common habit for denoting composition of morphism in category theory reverses the operands compared to the composition of relations. *In order to alleviate the access to category theory, we will adhere to its usual notation of composition (which is the notation of function composition).* Thus reading about categories should become much easier since it avoids the need of mirroring composition formulas each time they occur. So for a relation $\rho \subseteq A \times B$, instead of the familiar notation for relational composition $\rho I_B = \rho$, we get e.g. the categorical version $id_B \cdot \rho = \rho$, which is also inverse to the usual representation of morphisms in diagrams which are read from left to right.

In Table 11.1 we have already mentioned the category **Rel** which has sets as objects and binary relations as arrows, i.e. an arrow $\rho : A \longrightarrow B$ is a subset $\rho \subseteq A \times B$. The identity arrow on a set A is I_A, and given $\rho \subseteq A \times B$, and $\sigma \subseteq B \times C$, their composition as arrows is $\sigma \cdot \rho := \rho \circ \sigma$. Since composition of relations is associative, and $I_B \cdot \rho = \rho$ and $\sigma \cdot I_B = \sigma$, **Rel** is indeed a category.

In this section we will start out from a nearby view on relations and categories by abstracting the former as subobjects of some product of objects $A \times B$ in a category $\mathcal{C}$. As we already did in Section 11.1, in order not to get lost in the cosmos of categories, we will relate our considerations to the category **Set**, and thus motivate the generalizations to be introduced below.

11.2.1 *Regular Categories and Factorizations*

A standard context for "*relations in categories*" are the so-called *regular categories*, since they support a decently behaved calculus of relations. There are various equivalent definitions of this notion to be found in literature; see e.g. [6, Definition 2.1.1, Propositions 2.2.1 and 2.2.2] or [17, Chapter 25]. Below we use the one given in [12, Section A1.3, p. 20]. But first we have to introduce the concepts needed later on as e.g. orthogonality of arrows and various kinds of epimorphisms.

Definition 11.33. *Let $\mathcal{C}$ be a category, and $h : A \longrightarrow B$ and $j : C \longrightarrow D$ be two arrows in $\mathcal{C}$.*

(a) h is called **orthogonal** *to j, written $h \bot j$, if for any commutative square*

$$\begin{array}{ccc} A & \xrightarrow{h} & B \\ {\scriptstyle f}\downarrow & \swarrow {\scriptstyle d} & \downarrow{\scriptstyle g} \\ C & \xrightarrow[j]{} & D \end{array} \tag{11.1}$$

there exist a unique diagonal $d : B \longrightarrow C$ such that the two triangles commute, i.e. $d \cdot h = f$ and $j \cdot d = g$.

(b) Given $\mathsf{E}, \mathsf{M} \subseteq \mathsf{mor}_{\mathcal{C}}$, E is called **orthogonal** *to M, written $\mathsf{E} \bot \mathsf{M}$, if for any $e \in \mathsf{E}$ and $m \in \mathsf{M}$ we have $e \bot m$. For $\{f\} \bot \mathsf{M}$ we write $f \bot \mathsf{M}$.*

Definition 11.34. *An epimorphism $e : A \longrightarrow B$ in a category $\mathcal{C}$ is called*

(a) **regular** *if it is the coequalizer of a pair of arrows;*
(b) **extremal**, *or a* **cover**, *if $e = m \cdot f$ with m a monomorphism implies that m is an isomorphism, i.e. if e does not factor through any proper subobject of B.*
(c) **strong** *if $e \perp \mathsf{mono}_{\mathcal{C}}$.*

Covers will be depicted as $c : A \longrightarrow\!\!\triangleright B$. The covers in **Set** are precisely the surjective functions, and we have already shown in Cor. 2.29 that in **Set** any function can be decomposed into a surjection and an injection.

For a regular epimorphism e we thus have the following pattern: There exists a pair of arrows $f, g : C \longrightarrow A$ such that $e \cdot f = e \cdot g$, and whenever $e' : A \longrightarrow B'$ satisfies $e' \cdot f = e' \cdot g$ there is a unique arrow $u : B \longrightarrow B'$ such that $u \cdot e = e'$. And, for any pair of arrows $f', g' : B \longrightarrow D$, if $f' \cdot e = g' \cdot e$ then $f' = g'$.

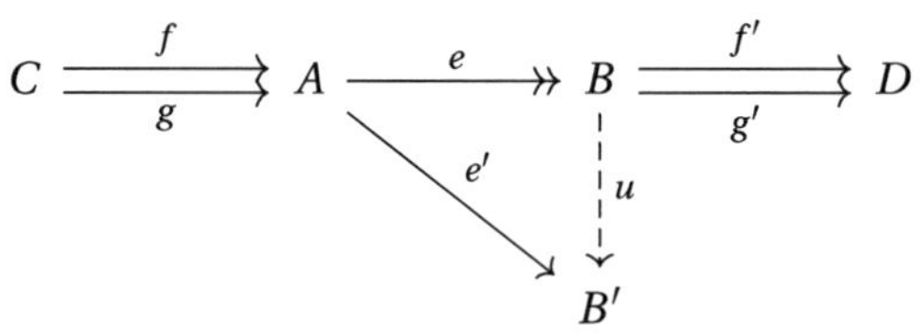

(11.2)

We continue with observing some properties of and relationships between the concepts just introduced:

Proposition 11.35. ([5, Section 4.3]) *In a category $\mathcal{C}$ the following holds:*

(1) Every regular epimorphism is extremal.
(2) If a composite $e \cdot f$ is an extremal (strong) epimorphism, e itself is an extremal (strong) epimorphism.
(3) A morphism which is both a monomorphism and an extremal (strong) epimorphism is an isomorphism.
(4) Every regular epimorphism is strong.
(5) Every strong epimorphism is extremal.
(6) The composite of two strong epimorphisms is a strong epimorphism.
(7) Any split epimorphism (retraction) is a regular epimorphism.

Proof. In the following let m denote any monomorphism.
(1): A consequence of (4) and (5), but to get familiar with category theoretical argumentation we include a direct proof. Let e be the coequalizer of the arrows f, g, and $e = m \cdot h$. From $m \cdot h \cdot f = e \cdot f = e \cdot g = m \cdot h \cdot g$ we get $h \cdot f = h \cdot g$. Thus there exists a unique u with $h = u \cdot e$.

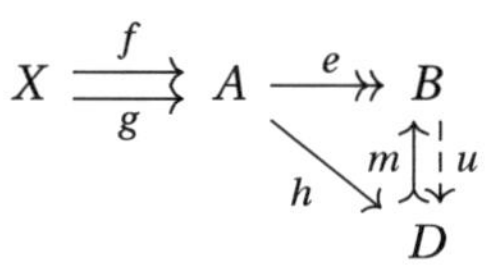

From $m \cdot u \cdot e = m \cdot h = e$ we conclude $m \cdot u = id_B$ since e is an epimorphism. $m \cdot u \cdot m = id_B \cdot m = m$ implies $u \cdot m = id_D$ since m is a monomorphism.

(3): If f is both a monomorphism and a strong epimorphism, then there is a unique d such that $d \cdot f = id_A$ and $f \cdot d = id_B$. So f is an isomorphism.

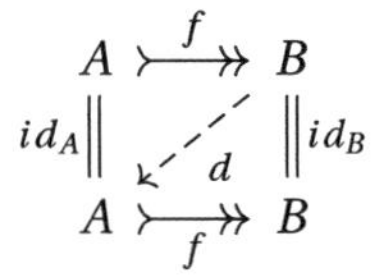

(4): Let e be the coequalizer of the arrows f, g, and assume $m \cdot h = j \cdot e$, i.e. the rectangle commutes. Then $m \cdot h \cdot f = j \cdot e \cdot f = j \cdot e \cdot g = m \cdot h \cdot g$ implies $h \cdot f = h \cdot g$.

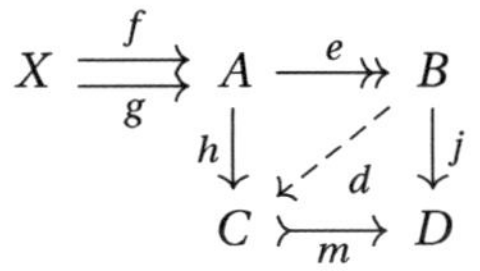

Therefore there is a unique d with $h = e \cdot d$, since e is coequalizer of f, g. But $h = d \cdot e$ implies $m \cdot d \cdot e = m \cdot h = j \cdot e$, so $m \cdot d = j$

(5): If f is a strong epimorphism with $f = m \cdot g$, then there is a unique d such that $d \cdot f = g$ and $m \cdot d = id_B$. Thus $m \cdot d \cdot m = id_B \cdot m = m \cdot id_B$, and since m is monic, we get $d \cdot m = id_B$, whence m is an isomorphism.

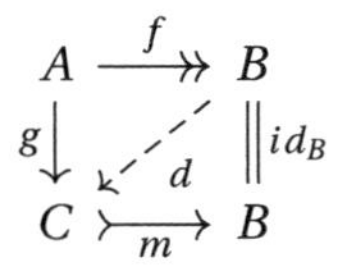

(6): Suppose $f, g \in \mathsf{epi}_C$ to be strong. Then, in the diagram below choose $z \cdot u = v \cdot f \cdot g$ with $z \in \mathsf{mono}_C$, i.e. the square $ACDE$ commutes. Then also $ABDE$ does, and since g is strong, there is w such that $w \cdot g = u$ and $z \cdot w = v \cdot f$. Therefore, the square $BCDE$ also commutes, and since f is strong there is a t with $t \cdot f = w$ and $z \cdot t = v$.

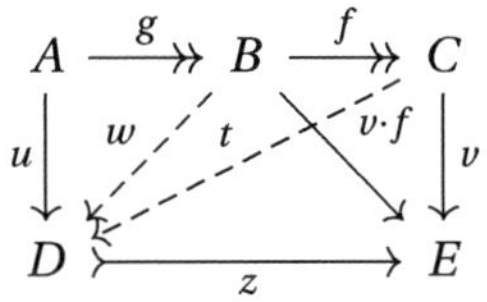

Moreover, $w \cdot g = t \cdot f \cdot g = u$, so it remains to show that t is the unique arrow that makes the two triangles ACD and CDE commute. Assume there is a $t' : C \longrightarrow D$ such that $t' \cdot f \cdot g = u = w \cdot g$; then $t' \cdot f = w$ since $g \in \mathsf{epi}_C$. But, as shown above, $t \cdot f = w$, so $t = t'$ since $f \in \mathsf{epi}_C$. By Prop. 11.7(3), $f \cdot g \in \mathsf{epi}_C$ with, as has just been shown, $f \cdot g \perp z$; and since z was an arbitrary monomorphism, $f \cdot g$ is strong.

(7): Let $r : B \longrightarrow A$ be a retraction such that $s : A \longrightarrow B$ and $r \cdot s = id_A$. Then $r \cdot (s \cdot r) =$

$id_A \cdot r = r = r \cdot id_B$. Furthermore, if $g : B \longrightarrow A'$ is such that $g \cdot (s \cdot r) = g \cdot id_B = g$, then $u = g \cdot s$ is the only arrow with $u \cdot r = g$. So s is a coequalizer of $s \cdot r$ and id_B.

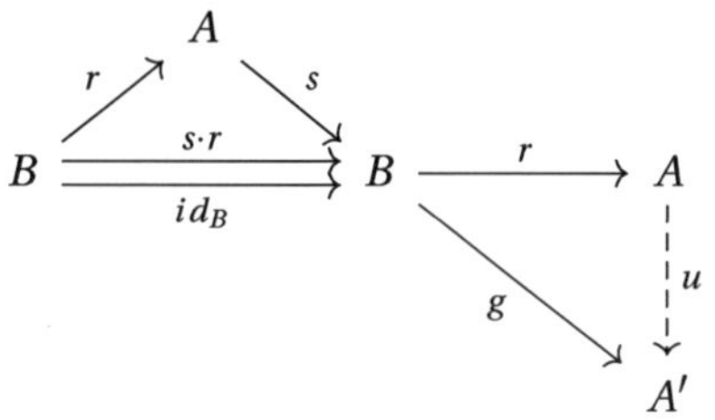

□

Lemma 11.36. ([5, Proposition 4.3.7(3)]) *In a finitely complete category $\mathcal{C}$ extremal and strong epimorphisms coincide.*

Proof. By Prop. 11.35(5) every strong epimorphism is extremal. Conversely suppose $f : A \longrightarrow B$ is an extremal epimorphism. Assume the square in the diagram to be a pullback (which exists due to Prop. 11.32), and $z \cdot u = v \cdot f$ with z a monomorphism.

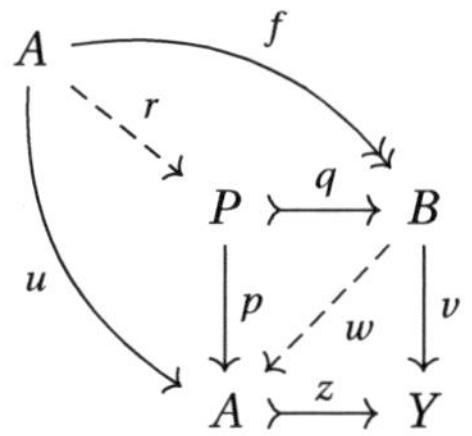

Then there is a unique r with $q \cdot r = f$ and $p \cdot r = u$. Since z is a monomorphism, q is monic as well by Lemma 11.28, and since f is extremal, q is an isomorphism. Then $w := p \cdot q^{-1} : B \longrightarrow A$ is the required unique arrow (since f is epimorphism). □

Definition 11.37. *A category is called a* **regular category** *if*

(1) *it is finitely complete,*
(2) *it has images, and*
(3) *the arrow opposite a cover in a pullback diagram is again a cover (often described as "covers are stable under pullback").*

Set (and more generally, any *topos*) is a regular category. The next proposition will allow us to formulate further results on regular categories in terms of either covers or regular epimorphisms. For sake of comprehensibility its proof is spelled out in full detail.

Proposition 11.38. ([12, Proposition 1.3.4]) *In a regular category $\mathcal{C}$ covers are exactly the regular epimorphisms.*

Proof. By Prop. 11.35(1) every regular epimorphism is a cover. Conversely, let $c : A \longrightarrow\!\!\triangleright\, B$ be a cover in a regular category. The pullback of c with itself (which exists due to

Prop. 11.32) is the kernel pair $(g,h): C \rightrightarrows A$; see Def. 11.24. So we have $c \cdot g = c \cdot h$. We will show that c is a coequalizer of g and h which makes it a regular epimorphism.

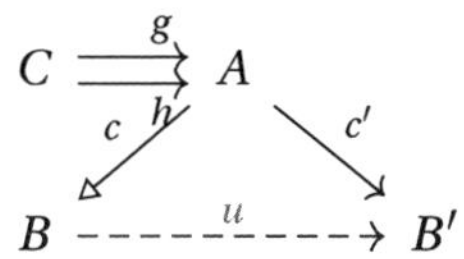

So if $c' : A \longrightarrow B'$ is a morphism such that $c' \cdot g = c' \cdot h$, then we have to show that there is an arrow u such that $u \cdot c = c'$. This u is unique since c is an epimorphism.

Since products exist in $\mathcal{C}$ we get

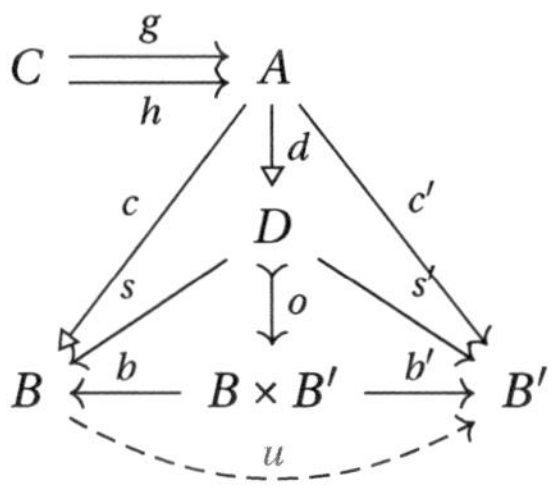

(11.3)

where $\langle c, c' \rangle$ has an image factorization $A \xrightarrow{d} D \rightarrowtail^{o} B \times B'$ since $\mathcal{C}$ has images. (Note that d is a cover since the triangle ADB can easily be extended to a pullback square $AADB$, and covers are stable under pullback.) If we can show s to be monic, then it is an isomorphism since c is a cover. But then $s' \cdot s^{-1} : B \longrightarrow B'$ gives a factorization of c' through c which is unique since c is epimorphism, and taking $u := s' \cdot s^{-1}$ we finally get $u \cdot c = c'$ as desired.

We expand (11.3) and finally restructure the resulting diagram as a whole as shown below: First suppose that we have $k, l : E \longrightarrow D$ such that $s \cdot k = s \cdot l$; then, to show s to be a monomorphism, we have to prove $k = l$. Moreover we redraw the kernel pair $(g,h): C \rightrightarrows A$ as a pullback. Further note that the left of the two diagrams contains the pullback square on the right. So we get

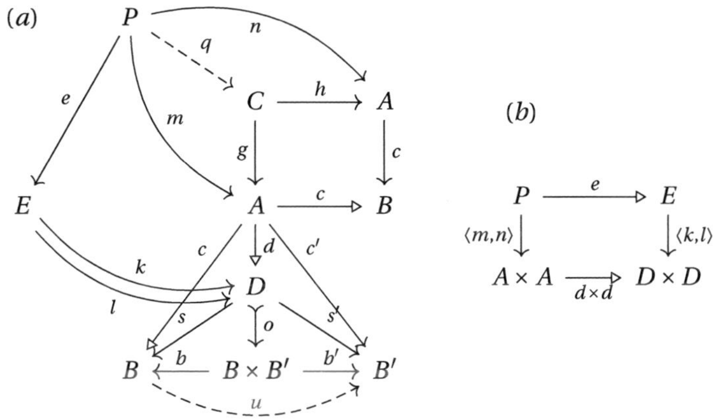

(11.4)

We can thus proceed in our proof with

$$c \cdot m = s \cdot d \cdot m = s \cdot k \cdot e = s \cdot l \cdot e = s \cdot d \cdot n = c \cdot n.$$

So, by Def. 11.23, there exists a unique q such that $g \cdot q = h \cdot q$.

Using (11.4)(a) we then get:

$$s \cdot k \cdot e = s \cdot d \cdot m = c \cdot m = c \cdot g \cdot q = c \cdot h \cdot q = c \cdot n = s \cdot d \cdot n = s \cdot l \cdot e. \tag{11.5}$$

From diagram (11.6) below we can see that $(d \times id_D) \cdot (id_A \times d) = d \times d$, and that both $id_A \times d : A \times A \longrightarrow A \times D$ and $d \times id_D : A \times D \longrightarrow D \times D$ are pullbacks of d, hence covers. And since covers are closed with respect to composition (by Prop. 11.35(6) and Lemma 11.36), $d \times d$ is a cover, and thus also e due to the pullback square (11.4)(b). So e in particular is an epimorphism whence from (11.5) we get $s \cdot k = s \cdot l$. Thus, since o is monic, $k = l$, i.e. s is monic too as needed to complete the proof. □

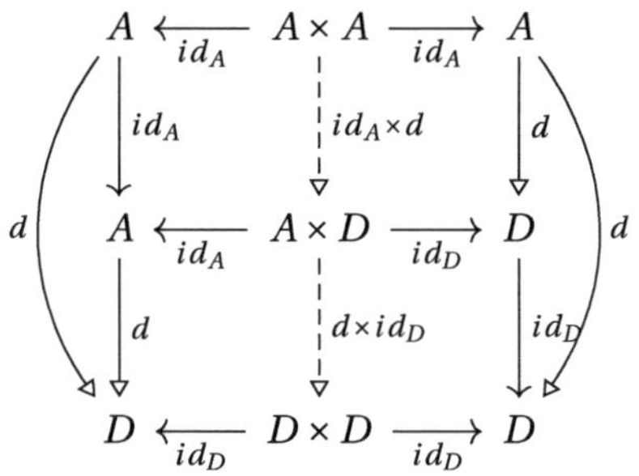

(11.6)

Definition 11.39. *Let $\mathcal{C}$ be a category, and let* $\mathsf{E}, \mathsf{M} \subseteq \mathsf{mor}_{\mathcal{C}}$.

(a) E/M *is called a* **factorization structure** *for $\mathcal{C}$, and $\mathcal{C}$ is called* E/M**-structured**, *whenever*

  (1) $\mathsf{mor}_{\mathcal{C}} = \mathsf{M} \cdot \mathsf{E}$, *i.e. every arrow f can be written as $f = m \cdot e$ with $m \in \mathsf{M}$ and $e \in \mathsf{E}$;*

  (2) $\mathsf{E} \perp \mathsf{M}$;

  (3) *For $i, i' \in \mathsf{iso}_{\mathcal{C}}$, $e \in \mathsf{E}$ and $m \in \mathsf{M}$ we have $i \cdot e \in \mathsf{E}$ and $m \cdot i' \in \mathsf{M}$.*

(b) E/M *is an* **epi-mono factorization structure** *if it is a factorization structure for $\mathcal{C}$ such that* $\mathsf{E} \subseteq \mathsf{epi}_{\mathcal{C}}$ *and* $\mathsf{M} \subseteq \mathsf{mono}_{\mathcal{C}}$.

It is easy to see that any category $\mathcal{C}$ has the (trivial) factorization structures $\mathsf{iso}_{\mathcal{C}}/\mathsf{mor}_{\mathcal{C}}$ and $\mathsf{mor}_{\mathcal{C}}/\mathsf{iso}_{\mathcal{C}}$. In **Set** the regular epimorphisms and monomorphisms form an epi-mono factorization structure, and they do so in any regular category as shown in Prop. 11.44(2) below: A regular category has a stable regular epi-mono factorization structure.

The notion of factorization system is self-dual in the sense that M/E is a factorization system in $\mathcal{C}^{op}$ if E/M is one in $\mathcal{C}$. Moreover, an immediate consequence of the orthogonality $\mathsf{E} \perp \mathsf{M}$ (sometimes also called *unique diagonalization property*) is that E/M-factorizations are unique up to a unique isomorphism: if $m \cdot e = m' \cdot e'$ with $e, e' \in \mathsf{E}$ and $m, m' \in \mathsf{M}$, then $i \cdot e = e'$ and $m' \cdot i = m$ for a unique isomorphism i; see Prop. 11.41 below.

Proposition 11.40. ([1, Chapter 14]) *Let $\mathcal{C}$ be an E/M-structured category, and let $e \in \mathsf{E}$ and $m \in \mathsf{M}$. Then the following holds:*

(1) If the diagram

$$\begin{array}{ccc} A & \xrightarrow{e} & B \\ {\scriptstyle id_A}\downarrow & \overset{d}{\swarrow} & \downarrow{\scriptstyle m} \\ A & \xrightarrow[f]{} & C \end{array}$$

commutes, then $e \in \mathsf{iso}_{\mathcal{C}}$ and $f \in \mathsf{M}$.

(2) $\mathsf{E} \cap \mathsf{M} = \mathsf{iso}_{\mathcal{C}}$.

(3) Each of E and M is closed under composition.

Proof. (1): The diagram

$$\begin{array}{ccc} A & \xrightarrow{e} & B \\ {\scriptstyle e}\downarrow & \overset{x}{\swarrow} & \downarrow{\scriptstyle m} \\ B & \xrightarrow[m]{} & C \end{array}$$

commutes for $x = id_B$. It also does if we insert $B \xrightarrow{d} A \xrightarrow{e} B$ as diagonal, i.e. we take $x = e \cdot d$. Since by Def. 11.39(a),2 the diagonal is unique, we get $e \cdot d = id_B$, whence e is an isomorphism. From the assumption diagram we see that $f = m \cdot e$, so from Def. 11.39(a),3 it follows that $f \in \mathsf{M}$.

(2): For any $f \in \mathsf{E} \cap \mathsf{M}$ there exists a diagonal d such that

$$\begin{array}{ccc} A & \xrightarrow{f} & B \\ {\scriptstyle id_A}\downarrow & \overset{d}{\swarrow} & \downarrow{\scriptstyle id_B} \\ A & \xrightarrow[f]{} & B \end{array}$$

commutes. So, by (1), $f \in \mathsf{iso}_{\mathcal{C}}$. Conversely, if $f \in \mathsf{iso}_{\mathcal{C}}$ and an E/M-factorization of f is $f = m \cdot e$, then for $d = f^{-1} \cdot m$ the diagram

$$\begin{array}{ccc} A & \xrightarrow{e} & \bullet \\ {\scriptstyle id_A}\downarrow & \overset{d}{\swarrow} & \downarrow{\scriptstyle m} \\ A & \xrightarrow[f]{} & B \end{array}$$

commutes. Thus by (1), $f \in \mathsf{M}$. By similar arguments we get $f \in \mathsf{E}$.

(3): Let $m_1 : A \longrightarrow B$ and $m_2 : B \longrightarrow C$ be morphism in M. If $m_2 \cdot m_1 = m \cdot e$ is an E/M-factorization, then one can construct diagonals successively such that the following diagrams commute:

$$\begin{array}{ccc} A & \xrightarrow{e} & \bullet \\ {\scriptstyle m_1}\big\downarrow & \overset{d_1}{\swarrow} & \big\downarrow{\scriptstyle m} \\ B & \xrightarrow[m_2]{} & C \end{array} \quad \text{and} \quad \begin{array}{ccc} A & \xrightarrow{e} & \bullet \\ {\scriptstyle id_A}\big\downarrow & \overset{d_2}{\swarrow} & \big\downarrow{\scriptstyle d_1} \\ A & \xrightarrow[m_1]{} & B \end{array} \quad \text{and thus} \quad \begin{array}{ccc} A & \xrightarrow{e} & \bullet \\ {\scriptstyle id_A}\big\downarrow & \overset{d}{\swarrow} & \big\downarrow{\scriptstyle m} \\ A & \xrightarrow[m_2\cdot m_1]{} & C \end{array}$$

Once again applying (1) we get $m_2 \cdot m_1 \in \mathsf{M}$. Similar for E. □

Proposition 11.41. ([1, 14.4 PROPOSITION]) *Let $\mathcal{C}$ be an E/M-structured category, then E/M-factorizations are unique up to isomorphism, i.e.*

(1) If $A \xrightarrow{e} C \xrightarrow{m} B$ and $A \xrightarrow{e'} D \xrightarrow{m'} B$ are E/M-factorizations of $f : A \longrightarrow B$, then there exists a (unique) isomorphism $h : C \longrightarrow D$, such that the diagram

$$\begin{array}{ccc} A & \xrightarrow{e} & C \\ {\scriptstyle e'}\big\downarrow & \overset{h}{\swarrow} & \big\downarrow{\scriptstyle m} \\ D & \xrightarrow[m']{} & B \end{array}$$

commutes.

(2) If $A \xrightarrow{e} C \xrightarrow{m} B$ is an E/M-factorization of $f : A \longrightarrow B$, and $h \in \mathsf{iso}_{\mathcal{C}}$, then $A \xrightarrow{h\cdot e} D \xrightarrow{m\cdot h^{-1}} B$ is also an E/M-factorization of $f : A \longrightarrow B$.

Proof. (1): Due to orthogonality there exist morphisms h and k such that the diagrams

$$\begin{array}{ccc} A & \xrightarrow{e} & C \\ {\scriptstyle e'}\big\downarrow & \overset{h}{\swarrow} & \big\downarrow{\scriptstyle m} \\ D & \xrightarrow[m']{} & B \end{array} \quad \text{and} \quad \begin{array}{ccc} A & \xrightarrow{e'} & D \\ {\scriptstyle e}\big\downarrow & \overset{k}{\swarrow} & \big\downarrow{\scriptstyle m'} \\ C & \xrightarrow[m]{} & B \end{array}$$

commute. Thus, also

$$\begin{array}{ccc} A & \xrightarrow{e} & C \\ {\scriptstyle e}\big\downarrow & \overset{k\cdot h}{\swarrow} & \big\downarrow{\scriptstyle m} \\ C & \xrightarrow[m]{} & B \end{array} \quad \text{and} \quad \begin{array}{ccc} A & \xrightarrow{e'} & D \\ {\scriptstyle e'}\big\downarrow & \overset{h\cdot k}{\swarrow} & \big\downarrow{\scriptstyle m'} \\ D & \xrightarrow[m']{} & B \end{array}$$

commute, and by uniqueness we get $k \cdot h = id_C$ and $h \cdot k = id_D$, whence h is an isomorphism.
(2): By Prop. 11.40 E and M are closed under composition with isomorphisms. □

In order to get an idea why regular categories are the favored context for the development of a category theoretical calculus of relations, it is necessary to have a closer look at some of their key properties. Before we can do this, we need to learn about some important properties of and relationships between the various kinds of epimorphisms.

Lemma 11.42. *Every split epimorphism is regular.*

Proof. If $e: A \twoheadrightarrow B$ has a left inverse $s: B \longrightarrow A$ such that $s \cdot e = id_A$, then $e \cdot id_A = e = e \cdot (s \cdot e)$. We show that e is the coequalizer of $s \cdot e, id_A : A \longrightarrow A$, and hence regular. Assume that $f: A \longrightarrow C$ with $f \cdot id_A = f = f \cdot (s \cdot e) = (f \cdot s) \cdot e$.

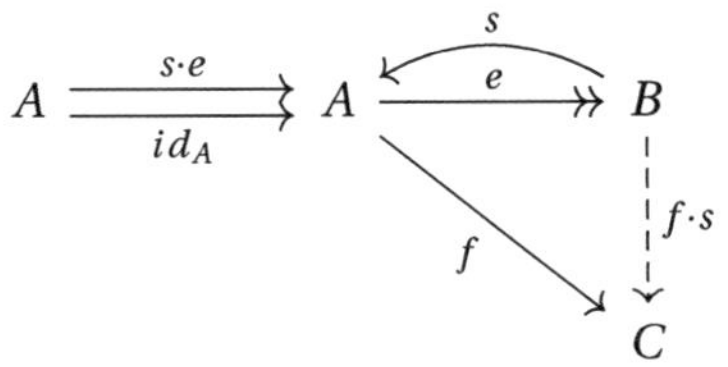

Now suppose that for some $g: B \longrightarrow C$ we have $f = g \cdot e$; then $g = g \cdot e \cdot s = f \cdot s$. Thus, $f \cdot s$ is the only possibility for f to factor through e. □

Corollary 11.43. *Let C be a regular category; Then the following holds:*

(1) An epimorphism is regular iff it is strong.
(2) If a composite $e \cdot f$ is a regular epimorphism, e itself is a regular epimorphism.
(3) The composite of two regular epimorphisms is a regular epimorphism.

Proof. (1): Because of Prop. 11.35(4) it remains to show that any strong epimorphism is regular. Let $e: A \twoheadrightarrow B$ be a strong epimorphism with factorization $e = i \cdot e'$,

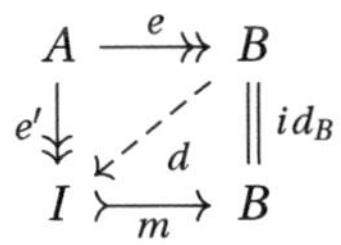

where e' is a regular epimorphism. Because e is strong, there is a unique arrow $d: B \longrightarrow I$ such that $e' = d \cdot e$ and $m \cdot d = id_B$. So m is a split epimorphism, whence regular by Lemma 11.42. C is finitely complete, thus there exists a pullback for each pair of arrows with common target by Prop. 11.32(2), and since regular epimorphisms are pullback stable, e is regular too. (Note that m is also strong by Prop. 11.35(4), and an isomorphism by Prop. 11.35(3).)

(2): Follows from (1) and Prop. 11.35(2).

(3): Follows from (1) and Prop. 11.35(6). □

Proposition 11.44. *Let C be a regular category, $m \in \mathsf{mono}_C$ and let $e \in \mathsf{epi}_C$ be any regular epimorphism in C. Then the following holds:*

(1) Every arrow f factors as a regular epimorphism followed by a monomorphism: $f = m \cdot e$. This factorization is unique up to isomorphism and is called the **image factorization** *of f.*
(2) f is a regular epimorphism iff f is a cover iff $f \perp \mathsf{mono}_C$.
(3) For $i, j \in \mathsf{iso}_C$, $i \cdot e$ is an epimorphism and $m \cdot j$ a monomorphism.

Proof. (1): See e.g. [3, 1.8.3. PROPOSITION], or [7, Theorem A.5.3].

(2): By Prop. 11.38 and Cor. 11.43.

(3): By Prop. 11.9(2), isomorphisms are always monic and epic. And by Prop. 11.7, if f, g are monics (or epics) then also their composition. □

As an immediate consequence we get that in a regular category every map factors as a cover followed by a mono.

11.2.2 *Relations in Regular Categories*

Definition 11.45. *Let $\mathcal{C}$ be a category with finite limits.*

(a) A **span** *in $\mathcal{C}$ is a pair of $\mathcal{C}$-arrows $(f : C \longrightarrow A, g : C \longrightarrow B)$, alternatively*

$$A \xleftarrow{\;f\;} C \xrightarrow{\;g\;} B \quad \text{or} \quad (f, g) : A \longrightarrow B$$

in case the common domain is not relevant.

(b) The morphisms f and g in a span are said to be **jointly monic** *if, for any D and parallel arrows $h, k : D \longrightarrow C$, $f \cdot h = f \cdot k$ and $g \cdot h = g \cdot k$ implies $h = k$.*

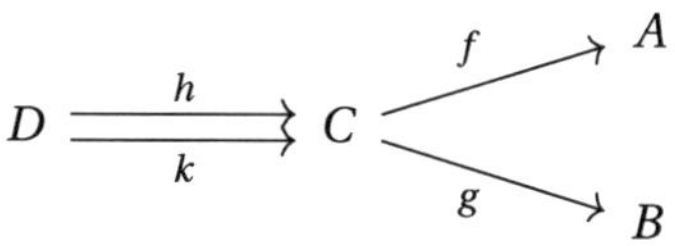

(c) Two spans $A_1 \xleftarrow{\;s_1\;} S \xrightarrow{\;s_2\;} A_2$ and $A_1 \xleftarrow{\;r_1\;} R \xrightarrow{\;r_2\;} A_2$ are said to be **isomorphic** *if there exists an isomorphism $i : S \rightarrowtail\!\!\!\!\twoheadrightarrow R$ such that the following diagram commutes.*

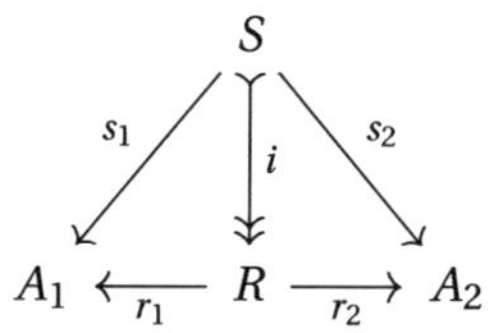

Obviously, f and g are jointly monic if f is monic or if g is (by Th. 2.33(17)). The projections π_A, π_B of a product $A \times B$ are jointly monic: Let $f_A : C \longrightarrow A$, and $h, k : C \longrightarrow A \times B$ such that $\pi_A \cdot h = \pi_A \cdot k$; then by the uniqueness condition in the definition of product, $h = \langle f, g \rangle = k$.

Proposition 11.46. *Let $A \xleftarrow{\;f\;} C \xrightarrow{\;g\;} B$ be a span, and $A \times B$ the product of A and B. Then f and g are jointly monic iff the induced morphism $\langle f, g \rangle : C \longrightarrow A \times B$ is monic.*

Proof. According to Def. 11.20 there is a unique arrow $u : C \longrightarrow A \times B$, denoted as $\langle f, g \rangle$, such that $f = \pi_A \cdot u$ and $g = \pi_B \cdot u$. Assume that one of f, g is monic; then by Prop. 11.7(2) $\langle f, g \rangle$ is monic too, so it can be considered a subobject of $A \times B$.

Consider the diagram below and assume that f, g are jointly monic, i.e. for any D and parallel arrows $h, k : D \longrightarrow C$, $f \cdot h = f \cdot k$ and $g \cdot h = g \cdot k$ implies $h = k$, but that none of them is monic. Since $A \times B$ is a product, for any pair $f' : D \longrightarrow A$, $g' : D \longrightarrow B$ there is a unique arrow $u' : D \longrightarrow A \times B$ such that $f' = \pi_A \cdot u'$ anf $g' = \pi_B \cdot u'$. This implies $u' = u \cdot h = u \cdot k$. Taking $f' := f \cdot h$ and $g' := g \cdot k$ we get $f' = \pi_A \cdot u \cdot h = \pi_A \cdot u \cdot k$ and $g' = \pi_B \cdot u \cdot h = \pi_B \cdot u \cdot k$, i.e. $f \cdot h = f \cdot k$ and $g \cdot h = g \cdot k$. By the assumption that f, g are jointly monic we get $h = k$, whence u is monic.

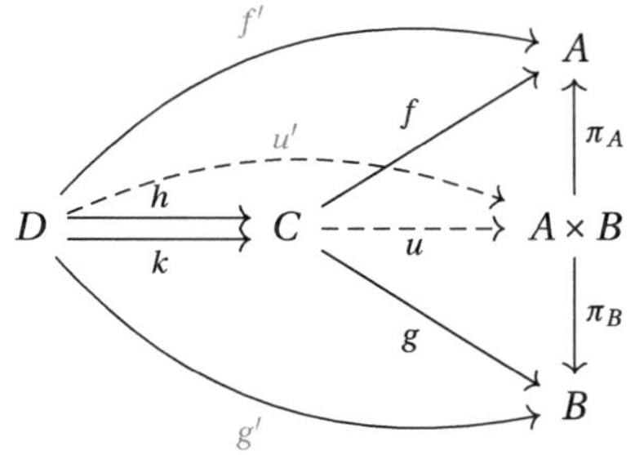

Conversely, assume u to be monic. We do already know that $u' = u \cdot h = u \cdot k$. Since u is monic, $h = k$ follows. So, by Th. 2.33(16) $f \cdot h = f \cdot k \wedge g \cdot h = g \cdot k \rightarrow h = k$, whence f and g are jointly monic. □

Definition 11.47. *Let $\mathcal{C}$ be a category with finite limits. For objects $A, B \in \mathsf{obj}_{\mathcal{C}}$ we form the category* **Span**$(A, B) \simeq \mathcal{C} / A \times B$ *consisting of equivalence classes of isomorphic objects of the slice category $\mathcal{C} / A \times B$. Hence an object of* **Span**(A, B) *can be represented by a span $(f, g) : A \longrightarrow B$.*

(1) A morphism $h : (f, g) \longrightarrow (f', g')$ in **Span**(A, B) *is a $\mathcal{C}$-morphism $h : C \longrightarrow D$ such that $f = f' \cdot h$ and $g = g' \cdot h$. The identity arrow of object $(f, g) : A \longrightarrow B$, $id_{(f,g)} : (f, g) \longrightarrow (f, g)$, is $id_{(f,g)} = id_C$.*

(2) Composition in **Span**(A, B) *is composition of the underlying morphisms of $\mathcal{C}$: $h' \cdot h : (f, g) \longrightarrow (f'', g'')$ with $f = f'' \cdot h' \cdot h$ and $g = g'' \cdot h' \cdot h$, and thus associative.*

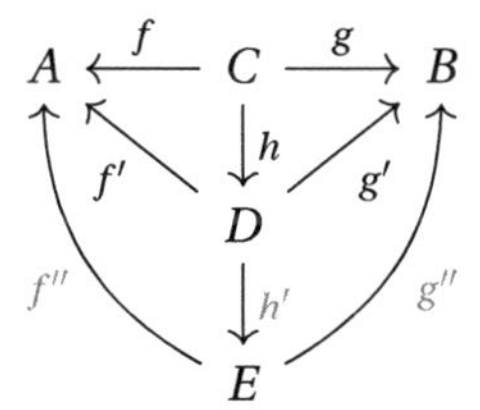

Definition 11.48. *Let $\mathcal{C}$ be a finitely complete category. A* **relation from** *A* **to** *B in $\mathcal{C}$ is a span $A \xleftarrow{f} C \xrightarrow{g} B$ which is jointly monic. In case $A = B$ the span is called a* **relation on** *A.*

From Prop. 11.46 we can see that a relation from A to B can equivalently be presented either as a monic to $A \times B$, i.e. as a subobject of $A \times B$, or as a jointly monic pair of maps to A and B, respectively. Recall from the discussion in Section 11.1.1.4 that a subobject of an object B in a category $\mathcal{C}$ is an isomorphism class of monomorphisms $m : A \rightarrowtail B$ to B. So a relation is in fact an equivalence class of monomorphisms to $A \times B$ which we denote by $\rho : A \nrightarrow B$, following [12, p. 130].

Let $\mathcal{C}$ be a regular category. If $A \xleftarrow{f} C \xrightarrow{g} B$ is a jointly monic span in $\mathcal{C}$ such that the induced morphism $\langle f, g\rangle : C \rightarrowtail A \times B$ belongs to $\rho : A \nrightarrow B$, then $\langle f, g\rangle$ is called a **tabulation of** ρ.

(Remark: Notation now becomes a little bit versatile since, dependent on the kind of information that is to be emphasized in text or diagrams, a few variants are useful: For spans we use

$$s : A \xleftarrow{f} C \xrightarrow{g} B \quad \text{or} \quad (f, g) : A \longrightarrow B,$$

and for a relation from A to B

$$\rho \rightarrowtail A \times B \quad \text{or} \quad [A f \rho g B] \quad \text{or} \quad A \xleftarrow{f} \rho \xrightarrow{g} B;$$

and if we also need to mention the source object we write it as a span

$$\rho : A \xleftarrow{f} C \xrightarrow{g} B.)$$

Since relations are defined as a special kind of spans, we obviously need another kind for composing spans if we want a suitable counterpart for the composition of relations as we know it from **Set**: Two relations

$$r : A \xleftarrow{r_0} R \xrightarrow{r_1} B \in \textbf{Span}(A, B) \text{ and } s : B \xleftarrow{s_0} S \xrightarrow{s_1} C \in \textbf{Span}(B, C)$$

are composed by forming the pullback of r_1 and s_0 as shown in

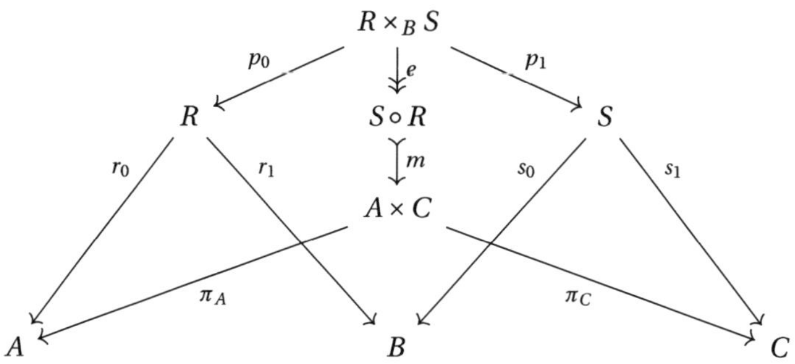

(11.7)

giving the **composition of** r **and** s as

$$s \odot r := (r_0 \cdot p_0, s_1 \cdot p_1) : A \longrightarrow C. \tag{11.8}$$

Taking the pullback itself is not enough, as $\langle r_0 \cdot p_0, s_1 \cdot p_1\rangle : R \times_B S \longrightarrow A \times C$ might not be monic. Thus one factorizes this arrow as

$$R \times_B S \overset{e}{\twoheadrightarrow} S \circ R \overset{m}{\rightarrowtail} A \times C$$

and considers the image $S \circ R$; see [6, Definition 2.8.3].

The set of relations from A to B is preordered by defining $\rho \leq \sigma$ as follows: Let $[A f \rho g B]$ and $[A r \sigma s B]$; then

$$\rho \leq \sigma \quad \text{if} \quad f = r \cdot t \wedge g = s \cdot t \tag{11.9}$$

for some (unique) $\mathcal{C}$-morphism $t : \rho \longrightarrow \sigma$ as shown in the diagram below. ρ is then called a **subrelation** of σ.

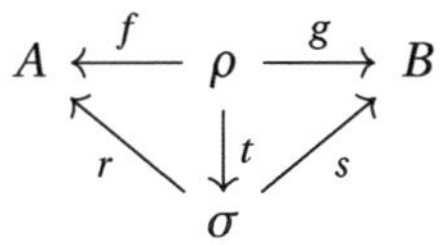

(11.10)

Every relation $[Af\rho gB]$ has a **converse relation** $\rho^{\smile}$ given by $[Bg\rho^{\smile}fA]$, and the relation $[A\,id_A\Delta_A id_A\,A]$ is called the *diagonal*, also *equality*, on A.

Now we are ready to look at the categorical version of some elementary properties, namely those by which equivalence relations are defined:

- A relation $[Af\rho gA]$ is said to be **reflexive** if $\Delta_A \leq \rho$, i.e. if $id_A = f \cdot t = g \cdot t$:

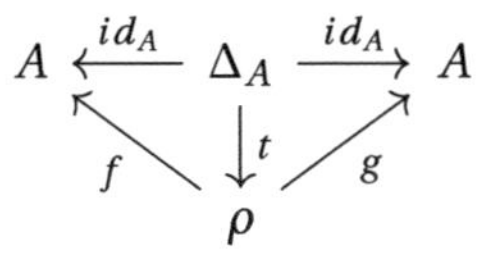

- A relation $[Af\rho gA]$ is said to be **symmetric** if $\rho^{\smile} \leq \rho$, i.e. $g = f \cdot t$ and $f = g \cdot t$:

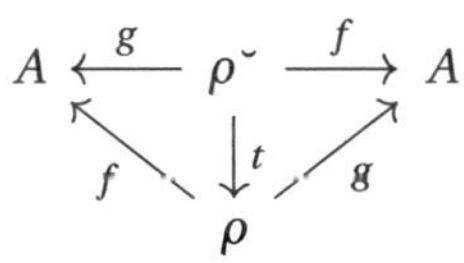

- A relation $[Af\rho gA]$ is said to be **transitive**, if $\rho \odot \rho \leq \rho$. We have

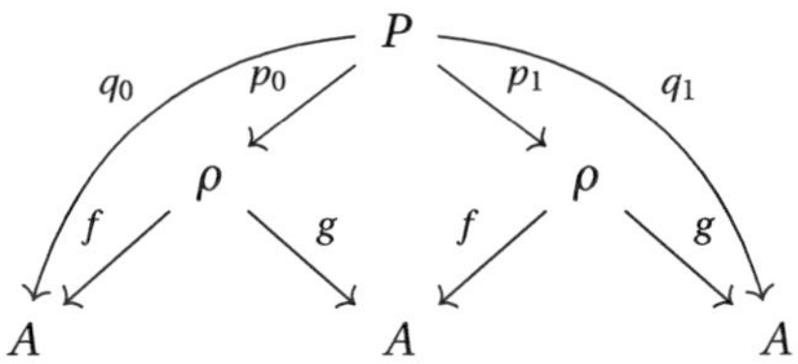

where $A \xleftarrow{q_0} P \xrightarrow{q_1} A$ with $(q_0, q_1) := (f \cdot p_0, g \cdot p_1) = \rho \odot \rho : A \longrightarrow A$, using (11.8). Thus

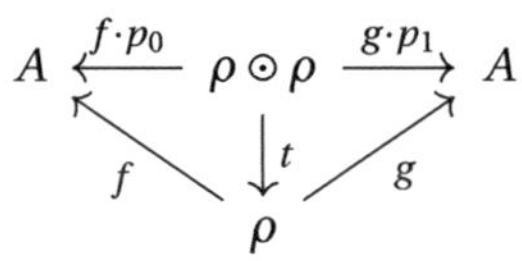

So we get, by the definition of $\leq$, that $[Af\rho gA]$ is transitive if $f \cdot p_0 = f \cdot t$ and $g \cdot p_1 = g \cdot t$ for some $t : \rho \odot \rho \longrightarrow \rho$.

We first consider the question if the diagonals Δ_A function as identities for composition, i.e. if $\rho \odot \Delta_A = \rho$ (or more exactly: if $\rho \odot \Delta_A \in \rho : A \nrightarrow B$) with the help of the following diagram:

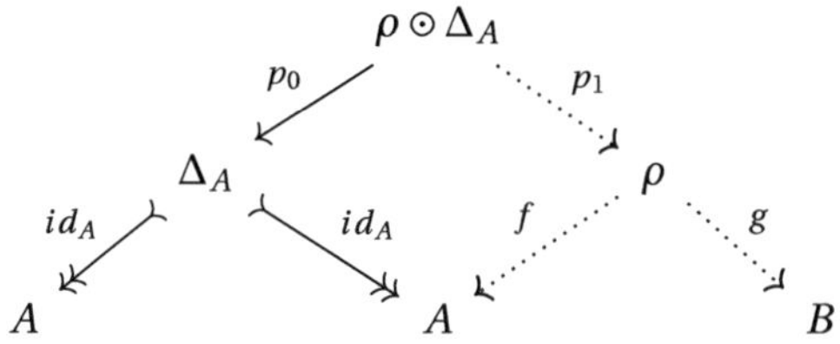

We will show that in the diagram above we have $p_1 : \rho \odot \Delta_A \rightarrowtail\!\!\!\twoheadrightarrow \rho$ which makes the two spans isomorphic by Def. 11.45(c). By Lemma 11.28 monomorphisms are pullback stable, so p_1 is monic since id_A is an isomorphism and hence both monic and epic by Prop. 11.9(2). id_A is in fact a regular epimorphism:

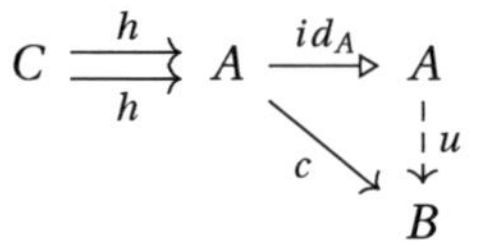

Clearly $id_A \cdot h = id_A \cdot h$, and also $c \cdot h = c \cdot h$; then there is indeed a u such that $c = u \cdot id_A = u$, viz. $u := c$ which is unique. So id_A is a cover by Prop. 11.38 and thus stable under pullback in $\mathcal{C}$. Therefore, p_1 is also a cover and hence an isomorphism by Prop. 11.35(3). So we have shown $\rho \odot \Delta_A \simeq \rho$ as desired.

A necessary property for forming a category out of the relations in a regular category $\mathcal{C}$ is the associativity of composition to which we now turn in

Lemma 11.49. ([12, Lemma 3.1.1]) *Let $\mathcal{C}$ be a regular category. If $\rho \rightarrowtail A \times B$, $\sigma \rightarrowtail B \times C$ and $\tau \rightarrowtail C \times D$ are relations in $\mathcal{C}$, then $\tau \odot (\sigma \odot \rho) = (\tau \odot \sigma) \odot \rho$.*

Proof. Let $\rho : A \xleftarrow{r} R \xrightarrow{r'} B$ and $\sigma : B \xleftarrow{s} S \xrightarrow{s'} C$ be two relations; their composition is given by forming the pullback of r' and s as shown above in diagram (11.7), yielding the span

$$\sigma \odot \rho := (r \cdot p, s' \cdot p') : R \times_B S \longrightarrow A \times C.$$

Let $\tau : C \xleftarrow{t} T \xrightarrow{t'} D$ be a third relation; then we similarly get

$$\tau \odot \sigma := (s \cdot q, t' \cdot q') : S \times_C T \longrightarrow B \times D.$$

So we get the following diagram including three pullback squares.

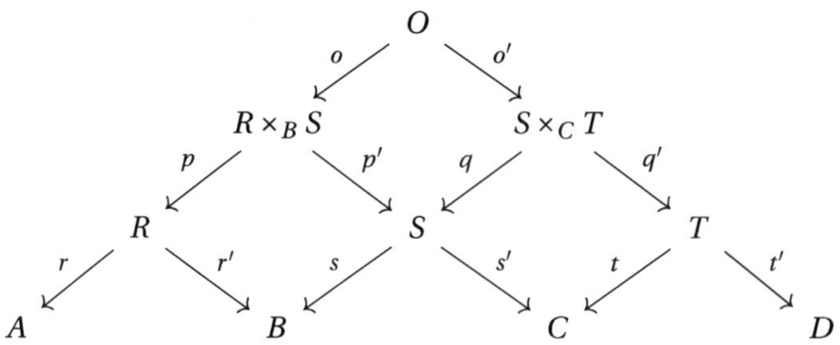

(11.11)

Recall that, by Prop. 11.44, each $\mathcal{C}$-arrow $f : A \longrightarrow B$ has a factorization into a regular epimorphism e followed by a monomorphism m via an object I, $A \xrightarrow{e} I \xrightarrow{m} B$, i.e. $f = m \cdot e$ with m called the *image of f*. What we have to show is that both relations $\tau \odot (\sigma \odot \rho)$ and $(\tau \odot \sigma) \odot \rho$ are given by the image $m : I \longrightarrow A \times D$ in the factorization of $\langle r \cdot p \cdot o, t' \cdot q' \cdot o' \rangle : O \longrightarrow A \times D$.

Let $R \times_B S \xrightarrow{e} I \xrightarrow{m} A \times C$ be the factorization of $\langle r \cdot p, s' \cdot p' \rangle : R \times_B S \longrightarrow A \times C$. To define $\tau \odot (\sigma \odot \rho)$ we need to form the pullback

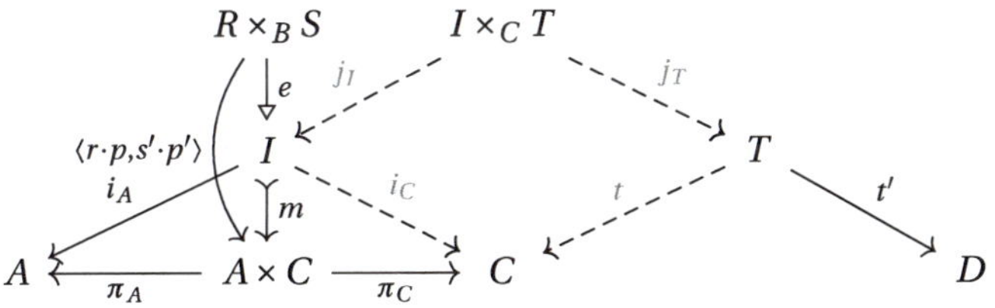

(11.12)

There exists also the pullback square containing the epimorphism e:

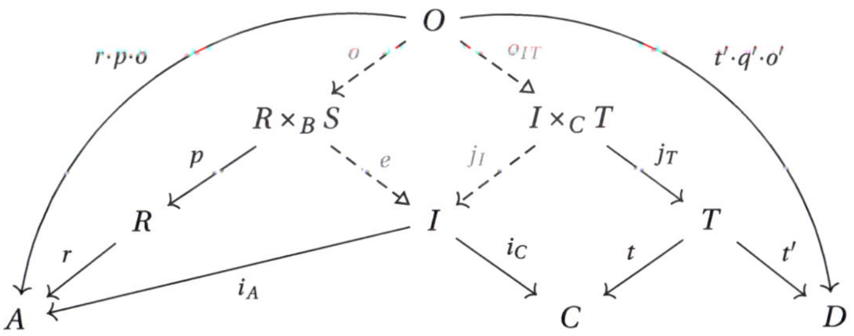

(11.13)

It is clear from diagram (11.13) that o_{IT} is a cover since covers are stable under pullbacks in $\mathcal{C}$ and that the image of $\langle i_A \cdot j_I, t' \cdot j_T \rangle : I \times_C T \longrightarrow A \times D$ is by definition a tabulation of $\tau \odot (\sigma \odot \rho)$.

Since the right square in diagram (11.13) is a pullback, o_{IT} is the unique arrow such that $j_I \cdot o_{IT} = e \cdot o$ and $j_T \cdot o_{IT} = q' \cdot o'$. Moreover, o_{IT} is a cover; so its epi-mono-factorization is in fact an epi-iso-factorization, i.e. the image of o_{IT} is an isomorphism $m_{IT} : \bullet_{IT} \rightarrowtail\!\!\!\twoheadrightarrow I \times_C T$, whence it is isomorphic to the image of $\langle i_A \cdot j_I, t' \cdot j_T \rangle : I \times_C T \longrightarrow A \times D$. So we can consider $(i_A \cdot j_I, t' \cdot j_T) \cdot o_{IT} : O \longrightarrow A \times D$.

Using $j_T \cdot o_{IT} = q' \cdot o'$ demonstrated just before, we get $(i_A \cdot j_I, t' \cdot j_T) \cdot o_{IT} = (i_A \cdot j_I \cdot o_{IT}, t' \cdot j_T \cdot o_{IT}) = (i_A \cdot e \cdot o, t' \cdot q' \cdot o') = (r \cdot p \cdot o, t' \cdot q' \cdot o')$. Similar calculations show that $(\tau \odot \sigma) \odot \rho$. is also given by the image of the same map. □

The next proposition shows the reverse of the former, demonstrating that associativity of relation composition in a category and regularity of this category are intimately connected.

Lemma 11.50. ([12, Lemma 3.1.1]) *Let $\mathcal{C}$ be a finitely complete category which has images. If relation composition as defined above is associative, then $\mathcal{C}$ is regular.*

Proof. Let $g : A \twoheadrightarrow B$ be a cover in $\mathcal{C}$, and $f : C \longrightarrow B$ be an arbitrary morphism. In the following we write $t_X : X \longrightarrow T$ for the unique morphism from X to the terminal object T which exists in $\mathcal{C}$ by Prop. 11.32. Consider the following relations:

$$\varphi : C \nrightarrow B, \text{ tabulated by } \langle id_C, f\rangle : C \rightarrowtail C \times B, \quad \text{i.e. } \varphi : C \xleftarrow{id_C} C \xrightarrow{f} B,$$

$$\psi : B \nrightarrow A, \text{ tabulated by } \langle g, id_A\rangle : A \rightarrowtail B \times A, \quad \text{i.e. } \psi : B \xleftarrow{g} A \xrightarrow{id_A} A,$$

$$\chi : A \nrightarrow T, \text{ tabulated by } \langle id_A, t_A\rangle : A \rightarrowtail A \times T, \quad \text{i.e. } \chi : A \xleftarrow{id_A} A \xrightarrow{t_A} T.$$

The proof in [12] is rather short, and its first part is given as: Since $m := \langle g, t_A\rangle : A \twoheadrightarrow B \times T \simeq B$ is a cover $\chi \odot \psi$ is tabulated by $\langle id_B, t_B\rangle : B \rightarrowtail B \times T$, and $(\chi \odot \psi) \odot \varphi$ is tabulated by $\langle id_C, t_C\rangle : C \longrightarrow C \times T$. For a layman in category-theory as I am, this is not quite self-explanatory; so I spent some effort to try to spell this out. The argumentation proceeds with the help of the diagram (11.14) below.

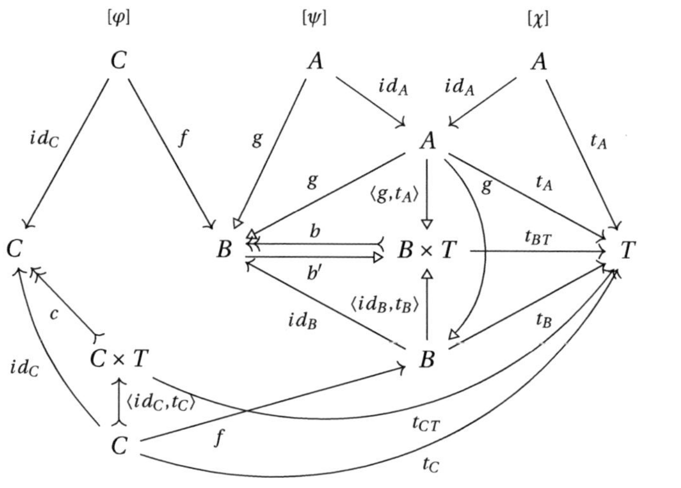

(11.14)

(a) By Prop. 11.32 $\mathcal{C}$ has all binary products, pullbacks, and a terminal object.
(b) For a terminal object T the projections $b : B \times T \longrightarrow B$ and $t_{BT} : B \times T \longrightarrow T$ are known to be isomorphisms.
(c) So there is an arrow $b' : B \longrightarrow B \times T$ with $b' \cdot b = id_{B\times T}$ and $b \cdot b' = id_B$ which is the retraction of b and hence a regular epimorphism and a cover by Prop. 11.35(7),(1).
(d) $\mathcal{C}$ is assumed finitely complete; thus covers and strong epimorphisms coincide by Lemma 11.36. And from Prop. 11.35(6) we get that the composition of strong epimorphisms is closed, whence $b' \cdot g$ is a cover. Since T is terminal we have $t_{BT} \cdot \langle g, t_A\rangle = t_A = t_{BT} \cdot b' \cdot g$, and since t_{BT} is a monomorphism it follows that $\langle g, t_A\rangle = b' \cdot g$, thus also a cover.

(e) Similarly we have $t_{BT} \cdot \langle id_B, t_B \rangle = t_B = t_{BT} \cdot b'$, so $\langle id_B, t_B \rangle = b'$, and thus a cover by Prop. 11.35(2).

So $\chi \odot \psi$ is tabulated by the image of $\langle id_B, t_B \rangle : B \rightarrowtail B \times T$, and in an analogous manner $(\chi \odot \psi) \odot \varphi$ is tabulated by the image of $\langle id_C, t_C \rangle : C \longrightarrow C \times T$. Note that $t_C = t_B \cdot f$ since T is a terminal object. The less detailed structure then is

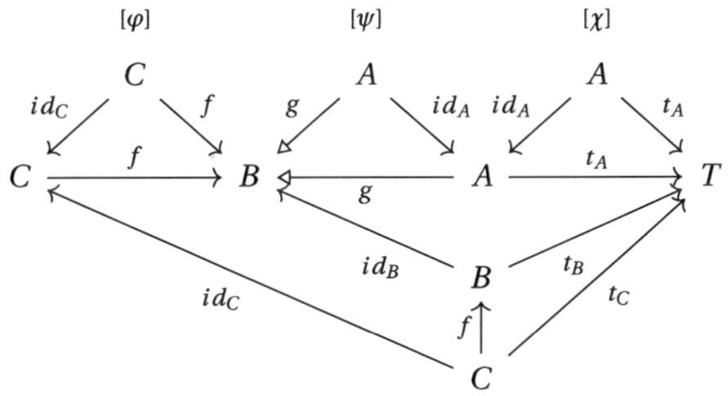

Similarly, $\chi \odot (\psi \odot \varphi)$ is tabulated by the image of $\langle h, t_P \rangle : P \longrightarrow C \times T$, containing the pullback square drawn with dashed arrows:

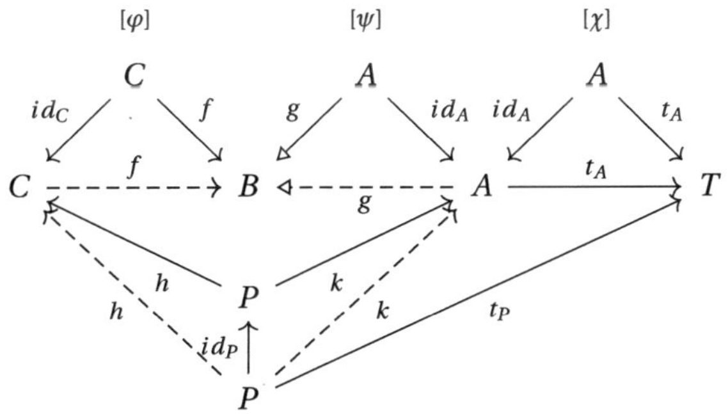

By assumption we have $(\chi \odot \psi) \odot \varphi = \chi \odot (\psi \odot \varphi)$, so h is an isomorphism, hence a split epimorphism, and thus a regular epimorphism and a cover by Prop. 11.35(7),(1). □

With the two foregoing lemmas we have thus established the following relationship between associativity of relation composition and regularity of categories:

Proposition 11.51. *Let $\mathcal{C}$ be a finitely complete category with images. The composition of relations in $\mathcal{C}$ as defined by* (11.7) *is associative iff $\mathcal{C}$ is regular.*

As demonstrated above, the composition of two spans r and s,

$$s \odot r := (r_0 \cdot p_0, s_1 \cdot p_1) : A \longrightarrow C$$

yields a relation by factorizing the arrow

$$\langle r_0 \cdot p_0, s_1 \cdot p_1 \rangle : R \times_B S \longrightarrow A \times C \quad \text{as} \quad R \times_B S \xrightarrow{e} I \xrightarrow{m} A \times C.$$

So the composition of two relations gives again a relation, and, by Lemma 11.49, it is associative, and its identities are the images of the arrows $\langle 1_A, 1_A \rangle : A \longrightarrow A \times A$. Thus, the relations in a regular category $\mathcal{C}$, denoted **Rel**($\mathcal{C}$), do indeed form a category.

But, of course, **Rel**($\mathcal{C}$) is more than just a category: The relations in $A \nrightarrow B$ are subobjects of a fixed object of $\mathcal{C}$. Therefore, they are partially ordered, and this partial ordering is preserved under composition. (Recall the definition of the preorder $\leq$ for relations in (11.10), which then induces a partial order on the equivalence classes.) Applying Th. 2.55 and Prop. 2.56 we can see that $A \nrightarrow B$ has the structure of a meet-semilattice.

As the last result in this section we state without proof the so-called *modular law*. Note that the order of factors within the products is the one of morphisms in categories; so the following corresponds to Th. 3.26(2)(b).

Proposition 11.52. ([12, Proposition 3.1.5]) *Let $\rho : A \nrightarrow B$, $\sigma : B \nrightarrow C$, and $\tau : A \nrightarrow C$ be three relations in a regular category. Then $\sigma \cdot \rho \cap \tau \leq (\sigma \cap \tau \cdot \rho^{\smile}) \cdot \rho$.*

11.3 Some Related Concepts

Before focusing on allegories in the final Section 11.4 of this chapter, we will take a slightly different view at **Rel**, and will mention some related concepts that appear in the context of *relations in categories*. For doing so we need to introduce some more notions of category theory and to present some of their properties, but we cannot formally define or prove all of them if we want to avoid the Sisyphean task of writing a still longer sub-book on categories. We therefore invite interested readers to follow the given links.

11.3.1 *Using Elements in the Categorical Setting*

In this section we analyze when and how it is possible to work within a convenient framework that allows us to treat relations almost as if they were relations between sets by referring to their elements. We closely follow the exposition of [18, Section 3.7], but specialized to proper epi-mono factorizations. Therefore, throughout this section, E/M-structured will always mean that $\mathsf{M} \subseteq \mathsf{mono}_{\mathcal{C}}$ and $\mathsf{E} \subseteq \mathsf{epi}_{\mathcal{C}}$.

Let $\mathcal{C}$ be a E/M-structured category, and $X \in \mathsf{obj}_{\mathcal{C}}$. By M/X we denote the full subcategory of $\mathcal{C}/X$ formed by the arrows of M (see Def. 11.2(c)). Instead of thinking in equivalence classes of isomorphic elements of M/X we shall refer to elements of M/X representing these classes as M-subobjects and write $m \simeq n$ for $m, n \in \mathsf{M}/X$ if there is an isomorphism i with $m \cdot i = n$. Recall the discussion at the end of Section 11.1.1.4(a).

We continue by clarifying what we mean by the term '*(generalized) element*':

Definition 11.53. *Let $\mathcal{C}$ be a category with epi-mono factorization structure E/M, $m \in \mathsf{M}$, and $B \in \mathsf{obj}_{\mathcal{C}}$.*

(a) *An arrow $b : X \longrightarrow B$ is called an X-**element** of B.*
(b) *If $m : M \rightarrowtail B$ is an M-subobject of B, we say that an X-element b **belongs to** m if there exists an arrow $c : X \longrightarrow M$ such that $b = m \cdot c$.*

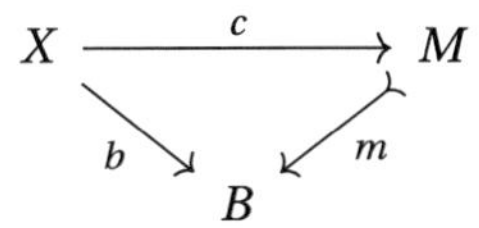

The following proposition contains the key result that allows us to develop a calculus of relations using elements.

Proposition 11.54. *Let $\mathcal{C}$ be a category with epi-mono factorization structure* E/M. *Any* M*-subobject is determined, for each object X, by the X-elements that belong to it.*

Proof. We first note that the elements of an M-subobject do not depend on the representation of that subobject: Let $m : M \rightarrowtail B$ and $m' : M' \rightarrowtail B$ represent the same subobject, i.e. $m' \cdot i = m$ for some isomorphism $i : M \longrightarrow M'$. Assuming that $b : X \longrightarrow B$ belongs to m, we have $m \cdot c = b$ for some $c : X \longrightarrow M$, so $m' \cdot i \cdot c = b$, whence b belongs to m'.

Now suppose that, for every $X \in \mathsf{obj}_{\mathcal{C}}$, m and m' have the same X-elements. Since $m \cdot id_M = m$, every M-subobject belongs to itself. So, in particular, m' belongs to m, and vice versa. Therefore, there exist arrows $i : M \longrightarrow M'$ and $j : M' \longrightarrow M$ such that $m' \cdot i = m$ and $m \cdot j = m'$. From Prop. 11.9(3) we then know that i, j are isomorphisms and inverse to each other. So we have $m \simeq m'$, i.e. m and m' represent the same M-subobject. □

Proposition 11.55. *Let $\mathcal{C}$ be a category with epi-mono factorization structure, such that* E *is stable under pullback. Then an X-element b of B belongs to the image of $f : A \longrightarrow B$ iff there is an epimorphism $p : Y \twoheadrightarrow X$ and an Y-element a of A such that the following square commutes.*

$$\begin{array}{ccc} Y & \overset{p}{\twoheadrightarrow} & X \\ {\scriptstyle a}\downarrow & & \downarrow{\scriptstyle b} \\ A & \underset{f}{\longrightarrow} & B \end{array}$$

Proof. Let $f = m \cdot e$ be an E/M-factorization of f. First assume that b belongs to m, i.e. $m \cdot i = b$ for some arrow i; then pulling back e along i yields the desired p and a, as shown in the left diagram below.

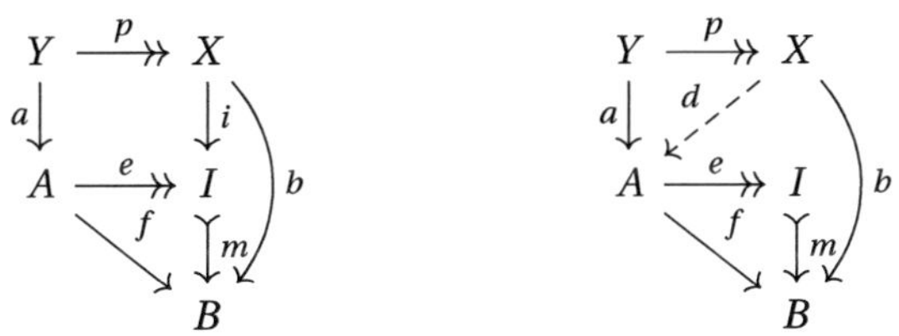

Conversely, if we have a commutative square $f \cdot a = b \cdot p$ with $p \in \mathsf{E}$, then, since $\mathcal{C}$ is E/M-structured, by the orthogonality of p and f (recall Def. 11.39) we get the right diagram above with $f \cdot d = m \cdot e \cdot d = b$, whence b belongs to m. □

11.3.2 *2-categories*

Writing $\mathsf{rel}[A,B]$ for the hom-set $\mathsf{mor}_{\mathbf{Rel}}[A,B]$ we already know that each $\mathsf{rel}[A,B]$ is a poset due to the subset relation $\subseteq$ between relations $\rho,\sigma \in \mathsf{rel}[A,B]$. So $(\mathbf{Rel},\subseteq)$ bears the structure of an *order enriched category*, that is a category in which hom-sets are partially ordered and composition is monotonic in both arguments. (Of course, we do already know that the set of all relations from A to B has a still richer structure since it equals the powerset $\mathbf{2}^{A\times B}$, and we will see the categorical counterparts of it in Section 11.4 below.) Taking inclusion of relations $\rho \subseteq \sigma$ as arrow between the morphisms ρ and σ leads us to the concepts so-called *2-morphisms* and *2-categories*.

2-categories and *bicategories* are two prominent kinds of structures that appear in connection with **Rel**; see e.g. [8, 19]. The notion of a *(strict) 2-category* generalizes that of category: a 2-category is a *higher category* where, on top of the objects and morphisms there are also the so-called *2-morphisms* between ordinary morphisms. Thus, a 2-category comprises three kinds of entities: objects (or *0-cells*), morphisms between objects (or *1-cells*), and 2-morphisms between morphisms(or *2-cells*). Morphisms can be composed in the way we have learned above, i.e. along the objects, while the 2-morphisms can be composed in two different directions: along objects, called *horizontal composition*, and along morphisms, called *vertical composition*.

In 2-categories each hom-set carries the structure of a category, whence they are **Cat**-*enriched* categories. A more general concept is that of a **bicategory** (or *weak 2-category*), where composition of morphisms is associative only up to a 2-isomorphism. In a more formal fashion we have

Definition 11.56.

(a) *A category C is called a* **2-category** *if between any two morphisms $f,g : A \longrightarrow B$ with common domain and codomain there exists at most one* **2-morphism** *written as*

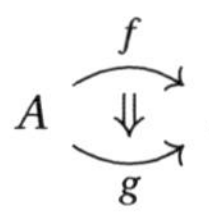

or more compact as $f \Rightarrow g$,

together with operations of **vertical** *composition,*

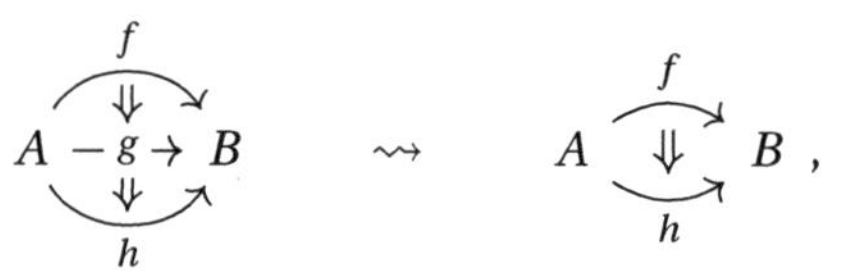

and **horizontal** *composition,*

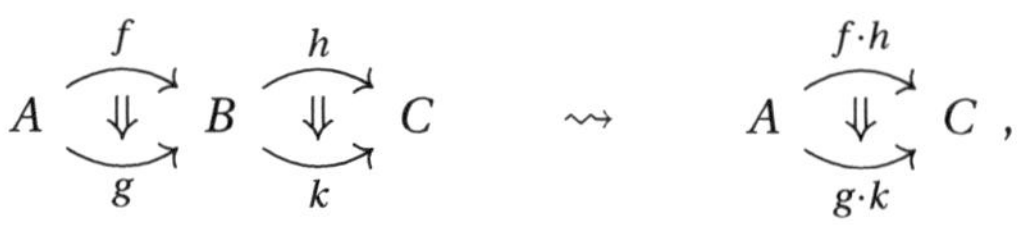

(b) *A 2-category* $\mathcal{C}$ *is called* **locally posetal** *if each hom-set* $\mathsf{mor}_{\mathcal{C}}[A,B]$ *is a poset under the relation* $\Rightarrow$.

Besides **Rel** also the spans in a category $\mathcal{C}$, **Span**($\mathcal{C}$), constitute an example of a 2-category: Its 0-cells are the objects of $\mathcal{C}$, its 1-cells are the spans, and its 2-cells are the arrows of the hom-categories **Span**(A,B).

In Section 11.2.2 above we have learned that the relations in a regular category $\mathcal{C}$ form a category **Rel**($\mathcal{C}$). Since the relations $\rho : A \nrightarrow B$ in **Rel**($\mathcal{C}$) are subobjects of $A \times B$ they have a natural partial order given by (11.9) which is preserved under composition. (Note that (11.9) gives in fact a meet-semilattice ordering since $\mathcal{C}$ is finitely complete.) Thus, **Rel**($\mathcal{C}$) is a locally posetal 2-category.

11.4 Relations and Allegories

Like a *category* can be viewed as the first abstraction step away from sets and functions between them, with **Set** as the prototypical example, so does the concept of *allegory* with respect to sets and relations. And not really surprising, **Set** turns out to be an important subcategory of **Rel**, the motivating standard example for allegories.

Our exposition in this section mainly rests on the book of Winter [24] since it comprises also the general case of L-fuzzy relations. The classical reference is Freyd and Scedrov [9]; there are also chapters on relations and allegories in [4] and [12].

11.4.1 *Allegories: The Basics*

An allegory is a category enriched with three operators which are inspired by the category of relations **Rel**: One can compare relations using a partial order $\sqsubseteq$, take the intersection of two relations with $\sqcap$, and transform a relation to its converse using an unary operator $\breve{\ }$. This results in the following definition.

(*Important remark:* Note that the order of factors in composition now is again the one of usual relational composition as in [24]. This makes sense since allegories are much more directly tailored towards relations from the outset than the approach via regular categories. Moreover, using this kind of notation avoids the cumbersome and error-prone mirroring between categorial and relational formulations. But be careful: In e.g. [9, 2.14] notation for morphism composition is used.)

Definition 11.57. ([24, Definition 4.4]) *A category* $\mathcal{C}$ *is called an* **allegory** *if the following holds:*

(1) *For all* $A, B \in \mathsf{obj}_{\mathcal{C}}$ *the hom-set* $\mathsf{mor}_{\mathcal{C}}[A,B]$ *is a meet semilattice, where meet and the induced ordering are denoted by* $\sqcap$ *and* $\sqsubseteq$, *respectively.*
(2) *There is a monotone operation* $\breve{\ }$ *assigning to each arrow* $r : A \longrightarrow B$ *a* **converse arrow** $r\breve{\ } : B \longrightarrow A$ *such that for all* $s : B \longrightarrow C$ *and* $t : A \longrightarrow C$

$$r\breve{\ }\breve{\ } = r,$$
$$(r \cdot s)\breve{\ } = s\breve{\ } \cdot r\breve{\ },$$

$$r \cdot s \cap t \subseteq r \cdot (s \cap r^{\smile} \cdot t) \qquad \textit{(modular law)}$$

(3) *For all arrows $r : A \longrightarrow B$ and $s, t : B \longrightarrow C$ we have subdistributivity of composition over meets:*

$$r \cdot (s \cap t) \subseteq r \cdot s \cap r \cdot t.$$

The prototypical example of an allegory is **Rel**, with subset inclusion of relations as its partial ordering on arrows, intersection as the meet of two arrows, and relational conversion as conversion of arrows. And as it has been the case with the abstract concept of *relation algebras* $\mathfrak{A}$, where elements of the carrier A are called "*relations*" without necessarily being ones in the usual mathematical sense, also arrows of any allegory might be referred to as "*relations*".

The structure imposed on allegories mirrors the property that the collection of binary relations on a fixed set is a meet semilattice, and that composition is subdistributive over meets. So, allegories certainly are a suitable kind of category to start with when considering binary relations. Again, a kind of additional check from the reverse point of view is to consider which properties known to hold for binary relations are derivable for allegories from their definition. Some of them are given in

Proposition 11.58. ([24, Lemma 4.5]) *Let $\mathcal{C}$ be an allegory, $A, B, C \in \mathsf{obj}_{\mathcal{C}}$, and $r, s \in \mathsf{mor}_{\mathcal{C}}[A, B]$, $t \in \mathsf{mor}_{\mathcal{C}}[B, C]$ and $u \in \mathsf{mor}_{\mathcal{C}}[A, C]$. Then*

(1) $id_A^{\smile} = id_A$.
(2) $(r \cap s)^{\smile} = r^{\smile} \cap s^{\smile}$
(3) $(r \cap s) \cdot t \subseteq r \cdot t \cap s \cdot t$
(4) $r \subseteq s \rightarrow r \cdot t \subseteq s \cdot t$ *and* $\qquad$ *(monotonicity of $\cdot$)*
$\quad t \subseteq u \rightarrow r \cdot t \subseteq r \cdot u$
(5) $r \subseteq r \cdot r^{\smile} \cdot r$
(6) $r \cdot s \cap t \subseteq (r \cap t \cdot s^{\smile}) \cdot s$ $\qquad$ *(modular law, dual variant)*
(7) $r \cdot s \cap t \subseteq (r \cap t \cdot s^{\smile}) \cdot (s \cap r^{\smile} \cdot t)$ $\qquad$ *(Dedekind rule)*
(8) *The two modular rules together are equivalent to the Dedekind rule.*

Proof. (1): $id_A^{\smile} = id_A^{\smile} \cdot id_A = id_A^{\smile} \cdot (id_A^{\smile})^{\smile} = (id_A^{\smile} \cdot id_A)^{\smile} = (id_A^{\smile})^{\smile} = id_A$.
(2): Using monotonicity of $^{\smile}$ and $r^{\smile\smile} = r$ we get

$$\begin{aligned} x \subseteq r^{\smile} \cap s^{\smile} &\leftrightarrow x \subseteq r^{\smile} \wedge x \subseteq s^{\smile} \\ &\leftrightarrow x^{\smile} \subseteq r \wedge x^{\smile} \subseteq s \\ &\leftrightarrow x^{\smile} \subseteq r \cap s \\ &\leftrightarrow x \subseteq (r \cap s)^{\smile} \end{aligned}$$

(3): Using (2) and Def. 11.57(2) we get

$$\begin{aligned} (r \cap s) \cdot t &= [(r \cap s) \cdot t]^{\smile\smile} \\ &= [t^{\smile} \cdot (r^{\smile} \cap s^{\smile})]^{\smile} \\ &\subseteq [t^{\smile} \cdot r^{\smile} \cap t^{\smile} \cdot s^{\smile}]^{\smile} \\ &= [(r \cdot t)^{\smile} \cap (s \cdot t)^{\smile}]^{\smile} \end{aligned}$$

$$= [(r \cdot t \cap s \cdot t)^{\smile}]^{\smile}$$
$$= r \cdot t \cap s \cdot t$$

(4): From (3) we get $(r \cap s) \cdot t \subseteq r \cdot t \cap s \cdot t$; assume $r \subseteq s$, then $r \cdot t = (r \cap s) \cdot t \subseteq r \cdot t \cap s \cdot t \subseteq r \cdot t$. Thus $r \cdot t \cap s \cdot t = r \cdot t$, whence $r \cdot t \subseteq s \cdot t$.

(5): Using the modular law we have

$$r = r \cdot id_B \cap r \subseteq r \cdot (id_B \cap r^{\smile} \cdot r) \subseteq r \cap r \cdot r^{\smile} \cdot r \subseteq r \cdot r^{\smile} \cdot r.$$

(6): By applying conversion to the modular law of Def. 11.57(b) we get $s^{\smile} \cdot r^{\smile} \cap t^{\smile} \subseteq (s \cap r^{\smile} \cdot t)^{\smile} \cdot r^{\smile} = (s^{\smile} \cap t^{\smile} \cdot r) \cdot r^{\smile}$. Now the assertion follows by appropriate renaming: $s^{\smile} := r, r^{\smile} := s, t^{\smile} := t$.

(7): $r \cdot s \cap t = (r \cdot s \cap t) \cap t$

$\subseteq (r \cap t \cdot s^{\smile}) \cdot s \cap t$	by (6)
$\subseteq (r \cap t \cdot s^{\smile}) \cdot (s \cap (r \cap t \cdot s^{\smile})^{\smile} \cdot t)$	by the modular law
$\subseteq (r \cap t \cdot s^{\smile}) \cdot (s \cap r^{\smile} \cdot t)$	by monotonicity of $^{\smile}$ and $\cdot$

(8): From the monotonicity of composition it is straightforward that the Dedekind rule implies the modular laws. Conversely, assume that the modular laws Def. 11.57(b) and (6) hold. Then

$r \cdot s \cap t = (r \cdot s \cap t) \cap t$

$\subseteq u \cdot s \cap t$	using (6), with $u := r \cap t \cdot s^{\smile}$
$\subseteq u \cdot (s \cap u^{\smile} \cdot t)$	using Def. 11.57(b)
$\subseteq u \cdot (s \cap r^{\smile} \cdot t)$	$u \subseteq r$, so $u^{\smile} \subseteq r^{\smile}$; monotonicity

□

11.4.2 *Special Properties of Arrows*

Recalling some basic facts that we have learned on relations in Prop. 3.50 the following should not come as a surprise: In **Rel** these definitions correspond to the set-theoretic ones.

Definition 11.59. ([24, Definition 4.6]) *Let $\mathcal{C}$ be an allegory, $A, B \in \mathsf{obj}_{\mathcal{C}}$, and $r \in \mathsf{mor}_{\mathcal{C}}[A, B]$. Then r is called*

(a) **univalent** *iff $r^{\smile} \cdot r \subseteq id_B$;*
(b) **total** *iff $id_A \subseteq r \cdot r^{\smile}$;*
(c) a **map** *iff it is univalent and total;*

Proposition 11.60. *Let $\mathcal{C}$ be an allegory, $A, B, C \in \mathsf{obj}_{\mathcal{C}}$, $r \in \mathsf{mor}_{\mathcal{C}}[A, B]$, and $s \in \mathsf{mor}_{\mathcal{C}}[B, C]$. Then the following holds*

(1) The class of univalent relations is closed under composition.
(2) The class of total relations is closed under composition.
(3) The class of maps is closed under composition.
(4) If $r \cdot s$ is total, so is r.

Proof. (1) $(r\cdot s)^{\smile}\cdot r\cdot s = s^{\smile}\cdot r^{\smile}\cdot r\cdot s \subseteq s^{\smile}\cdot s \subseteq id_C$.

(2) $r\cdot s\cdot (r\cdot s)^{\smile} = r\cdot s\cdot s^{\smile}\cdot r^{\smile} \supseteq r\cdot r^{\smile} \supseteq id_A$.

(3) Follows from (1) and (2).

(4): $id_A = id_A \cap r\cdot s\cdot (r\cdot s)^{\smile}$ $r\cdot s$ total

$= r\cdot s\cdot s^{\smile}\cdot r^{\smile} \cap id_A$

$\subseteq (r\cdot s\cdot s^{\smile} \cap id_A\cdot r)\cdot r^{\smile}$ Prop. 11.58(6), monotonicity

$\subseteq r\cdot r^{\smile}$.

□

For univalent and total arrows the modular laws and subdistributivity of composition over meets can be strengthened as follows(see Def. 11.57):

Proposition 11.61. *Let $\mathcal{C}$ be an allegory, and $A, B, C \in \mathsf{obj}_{\mathcal{C}}$. Then the following holds*

(1) For $r : A \longrightarrow B$ univalent and $s, t : B \longrightarrow C$: $r\cdot(s\cap t) = r\cdot s \cap r\cdot t$;
(2) For $r, s : A \longrightarrow B$ and $t : B \longrightarrow C$ total: $(r\cap s)\cdot t = r\cdot t \cap s\cdot t$.
(3) For $r : A \longrightarrow B$, $s : B \longrightarrow C$ univalent, and $t : A \longrightarrow C$: $r\cdot s \cap t = (r\cap t\cdot s^{\smile})\cdot s$.
(4) For $r : A \longrightarrow B$ total, $s : B \longrightarrow C$ and $t : A \longrightarrow C$: $r\cdot s\cap t = r\cdot(s\cap r^{\smile}\cdot t)$.

Proof. (1): Due to Def. 11.57,3 we only have to show one inequality only and get $r\cdot s\cap r\cdot t \subseteq r\cdot(s\cap r^{\smile}\cdot r\cdot t) \subseteq r\cdot(s\cap t)$ using the modular law first, and then that r is univalent, i.e. $r^{\smile}\cdot r\subseteq id_B$.

(2): Due to Prop. 11.58(3) we only have to show one inequality only and get $r\cdot t\cap s\cdot t \subseteq (r\cap s\cdot t\cdot t^{\smile})\cdot t \subseteq (r\cap s)\cdot t$ using Prop. 11.58(6) first, and then that t is total.

(3): We only have to show one inequality and get $(r\cap t\cdot s^{\smile})\cdot s \subseteq r\cdot s\cap t\cdot s^{\smile}\cdot s \subseteq r\cdot s\cap t$ using Prop. 11.58(4) first, and then that s is univalent.The other direction is Prop. 11.58(6). □

Lemma 11.62. ([24, Lemma 4.8]) *Let $\mathcal{C}$ be an allegory, $r : A \longrightarrow B$, $s : A \longrightarrow C$, and $t : D \longrightarrow B$ be relations; and let $f : B \longrightarrow C$ and $g : A \longrightarrow D$ be maps. Then*

(1) $r\cdot f \subseteq s$ iff $r \subseteq s\cdot f^{\smile}$;
(2) $g^{\smile}\cdot r\subseteq t$ iff $r\subseteq g\cdot t$.

Proof. (1): From $r\cdot f\subseteq s$ we get $r\subseteq r\cdot f\cdot f^{\smile} \subseteq s\cdot f^{\smile}$ since f is total. Conversely, $r\subseteq s\cdot f^{\smile}$ implies $r\cdot f\subseteq s\cdot f^{\smile}\cdot f \subseteq s$ since f is univalent. □

For later use we insert some results about *sub-* or *partial identities,* i.e. relations $r : A \longrightarrow A$ with $r\subseteq id_A$

Proposition 11.63. *Let $\mathcal{C}$ be an allegory, and $r, s : A \longrightarrow A$. be partial identities. Then we have*

(1) $r^{\smile} = r$,
(2) $r\cdot r = r$,
(3) $r\cdot s = r\cap s$.

Proof. (1): Using Prop. 11.58(5) we get $r \subseteq r \cdot r^{\smallsmile} \cdot r \subseteq id_A \cdot r^{\smallsmile} \cdot id_A \subseteq r^{\smallsmile}$, and similar for the other inclusion.

(2): Using Prop. 11.58(5) we get $r \subseteq r \cdot r^{\smallsmile} \cdot r \subseteq r \cdot id_A \cdot r = r \cdot r \subseteq r \cdot id_A = r$.

(3): We have $r \cdot s \subseteq r \cdot id_A = r$ and $r \cdot s \subseteq id_A \cdot s = s$, whence $r \cdot s \subseteq r \cap s$. By (2) we get $r \cap s = (r \cap s) \cdot (r \cap s) \subseteq r \cdot s$. using the fact that $r \cap s$ is a partial identity. □

11.4.3 *Regular Categories and Allegories*

Recall that in Section 11.2.1 our considerations have been based on the concepts of regular categories and epi-mono factorization structures (used in e.g. [11, 14] or [19]), whereas here the underlying concept is that of an allegory (introduced by Freyd and Scedrov [9]). Before changing from one framework to the other it is, of course, interesting to have a short look on the relationship between the two concepts. That there should be some definite connections must be expected by the fact that they both claim to provide a categorial framework for relations. We will sketch these relationships only, usually giving links to the more involved proofs in order to not overload the text.

We again emphasize that in our discussion of allegories we use the relational order for composition. Keep this in mind when studying the pertinent literature, where the usual way of defining tabulation is $g \cdot f^{\smallsmile} = \rho$ and $f^{\smallsmile} \cdot f \cap g^{\smallsmile} \cdot g = id_C$; compare this to Def. 11.64(a) below.

Definition 11.64. *Let $\mathcal{C}$ be an allegory, and $r : A \longrightarrow B$ be a relation in $\mathcal{C}$.*

(a) A pair of maps $f : C \longrightarrow A$ and $g : C \longrightarrow B$ is called a **tabulation of** *r iff $f^{\smallsmile} \cdot g = r$ and $f \cdot f^{\smallsmile} \cap g \cdot g^{\smallsmile} = id_C$.*

(b) A **unit** *in $\mathcal{C}$ is an object U such that 1_U is maximal in $\mathsf{mor}_{\mathcal{C}}[U,U]$, and for any $X \in \mathsf{obj}_{\mathcal{C}}$ there is an arrow $p : X \longrightarrow U$ with $1_U \le p^{\smallsmile}p$, i.e. p is total.*

(c) An allegory is called **unitary** *if it has a unit.*

(d) An allegory is called **tabular** *if every morphism has a tabulation.*

(e) An allegory is called **pretabular** *if every morphism is contained in one that has a tabulation.*

In the following let **Map**($\mathcal{C}$) denote the subcategory of maps in $\mathcal{C}$. If not indicated otherwise, $\mathcal{C}$ will denote an allegory throughout this section.

Lemma 11.65. **Map**($\mathcal{C}$) *is discretely ordered, i.e. if $f \subseteq g$ then $f = g$.*

Proof. Let f and g be maps such that $f \subseteq g$. Then by monotonicity of $^{\smallsmile}$ $f^{\smallsmile} \subseteq g^{\smallsmile}$; so $g \subseteq g f^{\smallsmile} f \subseteq g g^{\smallsmile} f \subseteq f$. □

Proposition 11.66. [12, Lemma 3.2.4] *If f, g tabulate r then $h^{\smallsmile}k \subseteq r$ iff there is a unique map j with $h = jf$ and $k = jg$.*

Corollary 11.67. [12, Corollary 3.2.5] *Tabulations are unique up to unique isomorphism: If f, g and f', g' both tabulate r, then there exists a unique isomorphism $u \in$* **Map**($\mathcal{C}$) *such that $f' = uf$ and $g' = ug$.*

Proposition 11.68. ([4, p 92]) *In any tabular allegory for a pair of maps $f : C \longrightarrow A$ and $g : C \longrightarrow B$ we have $ff^{\smallsmile} \cap gg^{\smallsmile} = id_C$ iff (f, g) is jointly monic in* **Map**$(\mathcal{C})$, *i.e. $hf = kf$ and $hg = kg$ implies $h = k$.*

Proof. Suppose $h \cdot f = k \cdot f$ and $h \cdot g = k \cdot g$ for maps $h, k : X \longrightarrow C$. If $ff^{\smallsmile} \cap gg^{\smallsmile} = id_C$ then, using Prop. 11.61(1),

$$h = h \cdot (ff^{\smallsmile} \cap gg^{\smallsmile}) = hff^{\smallsmile} \cap hgg^{\smallsmile} = kff^{\smallsmile} \cap kgg^{\smallsmile} = k \cdot (ff^{\smallsmile} \cap gg^{\smallsmile}) = k,$$

i.e. (f, g) is a monic pair.

Conversely, assume that (h, k) is a tabulation of $ff^{\smallsmile} \cap gg^{\smallsmile} : C \longrightarrow C$ which exists since $\mathcal{C}$ is tabular, and that (f, g) are jointly monic. We have to show that $ff^{\smallsmile} \cap gg^{\smallsmile} = id_C$. By Def. 11.64(a)

$$h^{\smallsmile} k = ff^{\smallsmile} \cap gg^{\smallsmile}. \tag{11.15}$$

So $h^{\smallsmile} k \subseteq ff^{\smallsmile}$ and $h^{\smallsmile} k \subseteq gg^{\smallsmile}$ (since $A \cap B \subseteq A$ and $A \cap B \subseteq B$). Using Lemma 11.62(2) and then (1) we get $h^{\smallsmile} k \subseteq gg^{\smallsmile}$ iff $k \subseteq hgg^{\smallsmile}$ iff $kg \subseteq hg$. Both kg and hg are maps by Prop. 11.60(3), whence we get $kg = hg$ by Lemma 11.65. Similarly, $kf = hf$ follows. But (f, g) is assumed to be jointly monic, so $h = k$. Thus by (11.15) $ff^{\smallsmile} \cap gg^{\smallsmile} = h^{\smallsmile} h \subseteq id_C$, where the inequality holds since h, as a map, is univalent. But since f and g are total, we have $id_C \subseteq ff^{\smallsmile}$ and $id_C \subseteq gg^{\smallsmile}$, whence $id_C \subseteq ff^{\smallsmile} \cap gg^{\smallsmile}$, and we have proved $ff^{\smallsmile} \cap gg^{\smallsmile} = id_C$. □

Proposition 11.69. ([12, Proposition 3.2.7], [18, Theorem A.20]) *Let $\mathcal{C}$ be a tabular allegory. Then* **Map**$(\mathcal{C})$ *has pullbacks, equalizers, images and covers are stable under pullbacks.*

The last proposition shows that **Map**$(\mathcal{C})$ is almost a regular category, with terminal objects being the only missing component. Recall that by Prop. 11.32(2) a category $\mathcal{C}$ is finitely complete iff it has a terminal object and all binary pullbacks. What we need is provided in

Lemma 11.70. ([12, Lemma 3.2.8]) *If U is a unit in an allegory $\mathcal{C}$, then it is a terminal object in* **Map**$(\mathcal{C})$. *The converse holds if $\mathcal{C}$ is tabular.*

Putting everything together we finally get as our main result

Theorem 11.71. ([12, Theorem 3.2.10])

(1) If $\mathcal{C}$ is a regular category, then **Rel**$(\mathcal{C})$ *is a unitary tabular allegory , and $\mathcal{C} \cong$* **Map**(**Rel**$(\mathcal{C})$).

(2) If $\mathcal{C}$ is a unitary tabular allegory, then **Map**$(\mathcal{C})$ *is a regular category, and $\mathcal{C} \cong$* **Rel**(**Map**$(\mathcal{C})$).

11.4.4 *Special Types of Allegories*

We have already learned that the set of all binary relations among others exhibits the structure of e.g. a distributive lattice with a least element. Following [24], this gives rise to a whole hierarchy of allegory specializations which will be described below. We will also have a short look on some characteristic properties of these variants together with some example proofs .

11.4.4.1 *Distributive Allegories*

Definition 11.72. *A* **distributive allegory** *is an allegory $\mathcal{C}$ which additionally satisfies the following conditions:*

(a) *The classes* $\mathsf{mor}_{\mathcal{C}}[A,B]$ *are distributive lattices with union* $\cup$ *and least element* $\perp_{AB}$.

(b) *For all relations* $r : A \longrightarrow B$ *we have* $r \cdot \perp_{BC} = \perp_{AC}$.

(c) *For all relations* $r : A \longrightarrow B$ *and* $s,t : B \longrightarrow C$ *we have* $r \cdot (s \cup t) = r \cdot s \cup r \cdot t$.

As already mentioned on page 231, $L = \mathbf{2}$ and $L = [0,1]$ are special cases of the general concepts of *L-fuzzy relations*, $(L, \wedge, \vee, \otimes, \curvearrowright, 0, 1)$ being a complete residuated lattice. Accordingly, **Rel** and $[0,1]$-**Rel** are special cases of L-**Rel** which has sets as objects and L-fuzzy relations as arrows. So it will suffice to state assertions for the latter one.

Corollary 11.73. *L-**Rel** is a distributive allegory.*

Proof. See Th. 8.22, with $0_{X\times Y}$, defined by $0_{X\times Y}(x,y) = 0$ for all $x \in X, y \in Y$, as least element. □

Proposition 11.74. ([24, Lemma 4.12]) *Let $\mathcal{C}$ be a distributive allegory, $A,B,C \in \mathsf{obj}_{\mathcal{C}}$, and let $r,s \in \mathsf{mor}_{\mathcal{C}}[A,B]$, $t \in \mathsf{mor}_{\mathcal{C}}[B,C]$. Then*

(1) $\perp_{AB}^{\smallsmile} = \perp_{BA}$,

(2) $\perp_{CA} \cdot r = \perp_{CB}$,

(3) $(r \cup s)^{\smallsmile} = r^{\smallsmile} \cup s^{\smallsmile}$,

(4) $(r \cup s) \cdot t = r \cdot t \cup s \cdot t$.

Proof. (1): For any $x \in \mathsf{mor}_{\mathcal{C}}[B,A]$, $\perp_{AB} \subseteq x^{\smallsmile}$, whence $\perp_{AB}^{\smallsmile} \subseteq x$.

(2): Using (1) and Def. 11.72(b) we get $\perp_{CA} \cdot r = (r^{\smallsmile} \cdot \perp_{CA}^{\smallsmile})^{\smallsmile} = (r^{\smallsmile} \cdot \perp_{AC})^{\smallsmile} = \perp_{BC}^{\smallsmile} = \perp_{CB}$.

(3): $r^{\smallsmile} \cup s^{\smallsmile} \subseteq (r \cup s)^{\smallsmile}$ is straightforward since $^{\smallsmile}$ is monotone by definition of an allegory. The converse inclusion follows from

$$
\begin{aligned}
(r \cup s)^{\smallsmile} \subseteq r^{\smallsmile} \cup s^{\smallsmile} &\leftrightarrow r \cup s \subseteq (r^{\smallsmile} \cup s^{\smallsmile})^{\smallsmile} \\
&\leftrightarrow r \subseteq (r^{\smallsmile} \cup s^{\smallsmile})^{\smallsmile} \text{ and } s \subseteq (r^{\smallsmile} \cup s^{\smallsmile})^{\smallsmile} \\
&\leftrightarrow r^{\smallsmile} \subseteq r^{\smallsmile} \cup s^{\smallsmile} \text{ and } s^{\smallsmile} \subseteq r^{\smallsmile} \cup s^{\smallsmile} .
\end{aligned}
$$

(4):
$$
\begin{aligned}
(r \cup s) \cdot t &= (t^{\smallsmile} \cdot (r \cup s)^{\smallsmile})^{\smallsmile} \\
&= (t^{\smallsmile} \cdot (r^{\smallsmile} \cup s^{\smallsmile}))^{\smallsmile} && \text{by (3)} \\
&= (t^{\smallsmile} \cdot r^{\smallsmile} \cup t^{\smallsmile} \cdot s^{\smallsmile})^{\smallsmile} && \text{by distributivity Def. 11.72(c)} \\
&= ((r \cdot t)^{\smallsmile} \cup (s \cdot t)^{\smallsmile})^{\smallsmile} \\
&= r \cdot t \cup s \cdot t && \text{by (3).}
\end{aligned}
$$

□

11.4.4.2 *Division Allegories*

The next step in the hierarchy of allegories are (variants of) division allegories which are characterized by the fact that $\cdot$ is a lower adjoint.

Definition 11.75.

(a) *A* **division allegory** $\mathcal{R}$ *is a distributive allegory such that for all relations* $r: B \longrightarrow C$ *and* $s: A \longrightarrow C$ *there is a relation* $s\backslash r: A \longrightarrow B$, *called the* **left residual** *of* s *and* r, *with* $t \cdot r \subseteq s \leftrightarrow t \subseteq s\backslash r$ *for all* $t: A \longrightarrow B$.

(b) *A* **Dedekind category** $\mathcal{R}$ *is a division allegory such that the classes* $\mathsf{mor}_{\mathcal{R}}[A, B]$ *are complete Brouwerian lattices (i.e. complete Heyting algebras) with a greatest element* $\top_{AB}$. *(They are also called* locally complete division allegories *in* [9].*)*

(c) *A* **Schröder category** $\mathcal{R}$ *is a Dedekind category in which the* $\mathsf{mor}_{\mathcal{R}}[A, B]$ *are (complete) Boolean algebras.*

The corollary below shows that L-**Rel** is also an example for a division and Dedekind category:

Corollary 11.76.

(1) L-**Rel** *is a division allegory.*
(2) L-**Rel** *is a Dedekind category.*

Proof. (1): By Th. 8.48(2) we have $\rho\sigma \subseteq \tau$ iff $\rho \subseteq (\sigma @ \tau^{\smile})^{\smile}$.

(2): By (1) L-**Rel** is a division allegory. The greatest element is the L-fuzzy relation 1_{XY}, with $1_{XY}(x, y) = 1$ for all $x \in X, y \in Y$. So every L-**Rel**$[A, B]$ is a complete Brouwerian lattice by Prop. 7.19(1). $\square$

The Dedekind category L-**Rel** is a Schröder category iff L is a Boolean algebra. But the substructure of crisp relations constitutes a Schröder category since $\overline{0} = 1$ and $\overline{1} = 0$ in any complete residuated lattice; see Prop. 7.21(6).

We continue with some properties of division allegories which we have shown to hold for binary relations in Section 3.5.2:

Corollary 11.77. ([24, Section 4.4]) *Let* $\mathcal{C}$ *be a division allegory,* $A, B, C \in \mathsf{obj}_{\mathcal{C}}$, *and let* $r, r' \in \mathsf{mor}_{\mathcal{C}}[A, B]$, $s, s' \in \mathsf{mor}_{\mathcal{C}}[B, C]$, *and* $t, t' \in \mathsf{mor}_{\mathcal{C}}[A, C]$. *Then we have*

(1) $\mathcal{C}$ *has right residuals.*
(2) $r \subseteq (r \cdot s)\backslash s$ *and* $s \subseteq r/(r \cdot s)$.
(3) $(s\backslash r) \cdot r \subseteq s$ *and* $r \cdot (t/r) \subseteq t$.
(4) *If* $r' \subseteq r$ *and* $s \subseteq s'$ *then* $s\backslash r \subseteq s'\backslash r'$.
(5) *If* $r' \subseteq r$ *and* $t \subseteq t'$ *then* $r/t \subseteq r'/t'$.
(6) $r\backslash id_B = r$ *and* $id_A/r = r$.

Proof. (1): Using Def. 11.57, 2. we have

$$r \cdot s \subseteq t \leftrightarrow s^{\smile} \cdot r^{\smile} \subseteq t^{\smile} \leftrightarrow s^{\smile} \subseteq t^{\smile}\backslash r^{\smile} \leftrightarrow s \subseteq (t^{\smile}\backslash r^{\smile})^{\smile}.$$

So $(t^\smile \backslash r^\smile)^\smile$ is the **right residual** of t and r, denoted by t/r.

(2): Obvious from $r \cdot s \subseteq r \cdot s$.

(3): By definition we have $t \cdot r \subseteq s \leftrightarrow t \subseteq s \backslash r$, so $(s \backslash r) \cdot r \subseteq s$.

(4): Using (3) we get $(s \backslash r) \cdot r' \subseteq (s \backslash r) \cdot r \subseteq s \subseteq s'$, so $s \backslash r \subseteq s' \backslash r'$.

(6): By (2) we have $r \subseteq (r \cdot id_B) \backslash id_B = r \backslash id_B$; and from (3) $r \backslash id_B = (r \backslash id_B) \cdot id_B \subseteq r$. □

By moving on to Dedekind categories we enter the area of complete lattices as underlying structure.Cor. 11.76(2) indicates that Dedekind categories are a suitable structure that embraces the general concept of L-fuzzy relations.

Proposition 11.78. *Let $\mathcal{D}$ be a Dedekind category. Then the following holds*

(1) $\top^\smile_{AB} = \top_{BA}$.
(2) $\top_{AA} \cdot \top_{AB} = \top_{AB} \cdot \top_{BB} = \top_{AB}$.
(3) $\top_{AB} = \top_{AB} \cdot \top_{BA} \cdot \top_{AB}$.

Proof. (1): For $t \in \mathsf{mor}_{\mathcal{D}}[B, A]$ we have $t^\smile \subseteq \top_{AB}$, whence $t \subseteq \top^\smile_{AB}$.

(2): $\top_{AB} = \top_{AB} \cdot id_B \subseteq \top_{AB} \cdot \top_{BB}$, and analogously $\top_{AB} = id_A \cdot \top_{AB} \subseteq \top_{AA} \cdot \top_{AB}$.

(3): Follows directly from (1) and Prop. 11.58(5). □

Proposition 11.79. ([24, Corollary 4.22]) *Let $\mathcal{D}$ be a Dedekind category, and let $r, r_i \in \mathsf{mor}_{\mathcal{D}}[A, B]$ and $s, s_i \in \mathsf{mor}_{\mathcal{D}}[B, C]$. Then*

(1) $\left(\bigcup_i r_i\right) \cdot s = \bigcup_i r_i \cdot s$ *and* $r \cdot \left(\bigcup_i s_i\right) = \bigcup_i r \cdot s_i$
(2) $\left(\bigcap_i r_i\right) \cdot s \subseteq \bigcap_i r_i \cdot s$ *and* $r \cdot \left(\bigcap_i s_i\right) \subseteq \bigcap_i r \cdot s_i$

Now we can conclude that L-fuzzy relations and Dedekind categories do indeed fit together quite well. However, it has been shown in [23] that the latter are too weak to express some standard properties of L-fuzzy relations, e.g. that a given L-fuzzy relation $\rho \in \mathcal{F}^L_{X \times Y}$ is a two-valued Boolean relation, also called 0-1-*crisp*, i.e. $\rho(x, y) \in \{0, 1\}$, for all $x \in X, y \in Y$.

This problem is studied in detail in [24, Section 4.10 and chapter 5], using two more kinds of categories, called *arrow* and *Goguen categories*. We close this chapter with a look at the main ideas of the former which start with enriching Dedekind categories with two operations that map every relation to its support and core, respectively (see Def. 8.3). In other terms, these operations map a relation to the greatest 0-1 crisp relation it contains and to the least 0-1 crisp relation it is included in.

For the following we need the notion of a *scalar* in a Dedekind category, the definition of which can be motivated as follows: Let us call an L-fuzzy relation a **scalar** on A (induced by $l \in L$) iff $\delta_l(x, y) = \begin{cases} l & \text{iff } x = y \\ 0 & \text{iff } x \neq y \end{cases}$. This obviously constitutes an isomorphism between L and the lattice of scalars, whence one can identify the elements of L and the scalar relations. Moreover, the set of scalars is closed under arbitrary intersections and unions. Clearly, both $\emptyset$ and id_A are scalar relations. Non-zero scalars d_l are strictly

l-reflexive in the sense of Def. 8.26(c). Scalar L-fuzzy relations can be characterized algebraically as follows:

Proposition 11.80. ([13, Proposition 3.2]) *ρ is a scalar L-fuzzy relation on X iff $\rho \subseteq I$ and $\rho U = U\rho$.*

Proof. For $\rho \subseteq I$ we have $\rho U(x,y) = \bigvee_{z\in X}(\rho(x,z) \wedge U(z,y)) = \rho(x,x) \wedge 1 = \rho(x,x)$ for all $x, y \in X$, and similarly $U\rho(x,y) = \rho(y,y)$. So, assuming ρ to be a scalar, we trivially get $\rho \subseteq I$, and, by what we have just shown, $\rho U(x,y) = \rho(x,x) = \rho(y,y) = U\rho(x,y)$. Conversely assume that $\rho \subseteq I$ and $\rho U = U\rho$. Thus $\rho(x,y) = 0$ for $x \neq y$. Moreover, $\rho(x,x) = \rho U(x,y) = U\rho(x,y) = \rho(y,y)$, whence ρ is a scalar L-fuzzy relation. □

Definition 11.81. *Let $\mathcal{D}$ be a Dedekind category; a relation $s : A \longrightarrow A$ is called a* **scalar** *on A iff $s \subseteq id_A$ and $\top_{AA} \cdot s = s \cdot \top_{AA}$.*

Lemma 11.82. *Let $\mathcal{D}$ be a Dedekind category, s be a scalar on A, and $r : A \longrightarrow X$ and $t : A \longrightarrow A$ be relations. Then the following holds:*

(1) $s \cdot r = r \cap s \cdot \top$ and $r \cdot s = r \cap \top \cdot s$; in particular, $s = id_A \cap s \cdot \top$.
(2) $s \cdot t = t \cdot s$.

Proof. (1): From $s \subseteq id_A$ we get $s \cdot r \subseteq r$, and $r \subseteq \top$ implies $s \cdot r \subseteq s \cdot \top$, so $s \cdot r \subseteq r \cap s \cdot \top$. By the modular law and Prop. 11.63 we get $r \cap s \cdot \top \subseteq s \cdot (s^{\smile} \cdot r \cap \top) = s \cdot (s^{\smile} \cdot r) = s \cdot r$.
(2): By (1) we have $s \cdot t = t \cap s \cdot \top_{AA} = t \cap \top_{AA} \cdot s = t \cdot s$. □

Definition 11.83. ([24, Definition 5.2]) *An* **arrow category** *$\mathcal{C}$ is a Dedekind category with $\top_{AB} \neq \bot_{AB}$ for all objects A and B together with two operations $^{\uparrow}$ and $^{\downarrow}$ such that the following holds:*

(1) $r^{\uparrow}, r^{\downarrow} : A \longrightarrow B$ for all $r : A \longrightarrow B$.
(2) $(^{\uparrow}, ^{\downarrow})$ is a dually residuated pair of maps.
(3) $(r^{\smile} \cdot s^{\downarrow})^{\uparrow} = r^{\uparrow\smile} \cdot s^{\downarrow}$ for all $r : B \longrightarrow A$ and $s : B \longrightarrow C$.
(4) $(q \cap r^{\downarrow})^{\uparrow} = q^{\uparrow} \cap r^{\downarrow}$ for all $q, r : A \longrightarrow B$.
(5) If $d \neq \bot_{AA}$ is a nonzero scalar, then $d^{\uparrow} = id_A$.

Recall from Table 7.1 that *dually residuated* here means $s \supseteq r^{\uparrow} \leftrightarrow r \subseteq s^{\downarrow}$ (called a *Galois correspondence* in [24, Definition 2.12]). As indicated above, the interpretation of $^{\uparrow}$ and $^{\downarrow}$ as support and core of an L-fuzzy relation (see Def. 8.3) gives the standard model as shown in

Theorem 11.84. ([24, Theorem 5.3]) *Let L be a complete Brouwerian lattice with $0 \neq 1$. Then L-**Rel** together with $^{\uparrow}$ and $^{\downarrow}$ is an arrow category.*

Proof. We show each of the five constituting properties separately:
(1): Obvious
(2): Suppose $\rho^{\uparrow} \subseteq \sigma$ and $\rho^{\uparrow}(x,y) \neq 0$. Then $1 = \rho^{\uparrow}(x,y) \leq \sigma(x,y)$, and thus $\rho(x,y) \leq 1 = \sigma^{\downarrow}(x,y)$. An analogous proof shows the reverse implication.

(3): We define two operations ${}^{\uparrow},{}^{\downarrow} : L \longrightarrow L$ by $x^{\uparrow} = 1$ if $x > 0$ and $x^{\uparrow} = 0$ otherwise; and by $x^{\downarrow} = 1$ if $x = 1$ and $x^{\downarrow} = 0$ otherwise. Then the following holds:

(a) $\rho^{\uparrow}(x, y) = \rho(x, y)^{\uparrow}$,
(b) $\rho^{\downarrow}(x, y) = \rho(x, y)^{\downarrow}$,
(c) $(x \wedge y^{\downarrow})^{\uparrow} = x^{\uparrow} \wedge y^{\downarrow}$, for all $x, y \in L$,
(d) $(\bigvee M)^{\uparrow} = \bigvee_{m \in M} x^{\uparrow}$, forall $M \subseteq L$.

Then the assertion is shown as follows:

$$\begin{aligned}
(\rho^{\smile}\sigma^{\downarrow})^{\uparrow}(x, y) &= (\rho^{\smile}\sigma^{\downarrow})(x, y)^{\uparrow} && \text{by (a)}\\
&= \big(\textstyle\bigvee_z(\rho^{\smile}(x, z) \wedge \sigma^{\downarrow}(x, y))\big)^{\uparrow} \\
&= \big(\textstyle\bigvee_z(\rho(z, x) \wedge \sigma(x, y)^{\downarrow})\big)^{\uparrow} && \text{by (b)}\\
&= \textstyle\bigvee_z\big((\rho(z, x) \wedge \sigma(x, y)^{\downarrow})\big)^{\uparrow} && \text{by (d)}\\
&= \textstyle\bigvee_z\big(\rho(z, x)^{\uparrow} \wedge \sigma(x, y)^{\downarrow}\big) && \text{by (c)}\\
&= \textstyle\bigvee_z\big(\rho^{\uparrow}(z, x) \wedge \sigma^{\downarrow}(x, y)\big) && \text{by (a), (b)}\\
&= \textstyle\bigvee_z\big(\rho^{\uparrow\smile}(x, z) \wedge \sigma^{\downarrow}(z, y)\big) \\
&= (\rho^{\uparrow\smile}\sigma^{\downarrow})(x, y).
\end{aligned}$$

(4): Similar, using the operations ${}^{\uparrow},{}^{\downarrow}$ defined above in (3) and their properties.

(5): Let δ_l be a scalar L-fuzzy relation, with $\delta_l(x, y) = l > 0$. Then

$$\delta_l^{\uparrow}(x, y) = \delta_l(x, y)^{\uparrow} = \left\{\begin{matrix} 1 \text{ iff } x = y \\ 0 \text{ iff } x \neq y \end{matrix}\right\} = I(x, y).$$

□

Moreover, we obviously have ρ is 0-1 crisp iff $\rho^{\uparrow} = \rho$ iff $\rho^{\downarrow} = \rho$.

Definition 11.85. *A relation $r : A \longrightarrow B$ in a Dedekind category is called a s-***crisp** *iff $s \cdot q \subseteq r \rightarrow q \subseteq r$ for all nonzero scalars s and all relations $q : A \longrightarrow B$.*

Concerning L-fuzzy relations the following connection between 0-1 crispness and s-crispness holds:

Proposition 11.86. ([24, Lemma 4.44]) *In L-***Rel** *all s-crisp relations are 0-1 crisp.*

Proof. Suppose $\rho : A \longrightarrow B$ is s-crisp with $\rho(x', y') = l > 0$. Consider the scalar $d_l : A \longrightarrow A$ induced by l and $\sigma : A \longrightarrow B$, with $\sigma(x, y) := \rho(x, y) \frown l$, the relative pseudocomplement of $\rho(x, y)$ in l, i.e. the greatest element in $\{x \in L : l \wedge x \leq \rho(x, y)\}$. Then, using Lemma 11.82(1), we get

$$(d_l\sigma)(x, y) = (d_l 1_{AB} \cap \sigma)(x, y) = l \wedge (\rho(x, y) \frown l) \leq \rho(x, y).$$

Since ρ is s-crisp we have $\sigma \subseteq \rho$, whence $1 = l \frown l = \rho(x', y') \frown l = \sigma(x', y') \leq \rho(x', y')$. Thus ρ is 0-1 crisp. □

The converse is, in general, not true, i.e. the notion of s-crispness coincides with 0-1 crispness only if an additional assumption holds in L as shown in the next proposition. For this we need the following notion: An element x of a lower semilattice L with least element 0 is called **linear** if $x \wedge y = 0 \to y = 0$, for all $y \in L$. (Note that in a linear poset every nonzero element has this property. The converse implication does not hold; consider the poset $\{0 \leq a \leq b, a \leq c\}$.)

Proposition 11.87. ([24, Lemma 4.45]) *In L-***Rel** *the following properties are equivalent:*

(1) *All 0-1 crisp relations are s-crisp.*
(2) *All nonzero elements of L are linear.*

Proof. Assume (1), and let $0 < k, l \in L$ such that $k \wedge l = 0$. Then d_k is nonzero and by Prop. 11.63(3) we have $d_k \cdot d_l = d_k \cap d_l = 0_{AA}$, where $0_{AA}(x, y) = 0$ for all $x, y \in A$ clearly is the least L-fuzzy relation on A. Since 0_{AA} is 0-1 crisp it is also s-crisp by assumption, whence $d_l \subseteq 0_{AA}$ and $l = 0$..

Conversely, assume (2) and suppose that $d_k\sigma \subseteq \rho$ for $0 < k \in L$, $\sigma : A \longrightarrow B$ and that $\rho : A \longrightarrow B$ is 0-1 crisp. To prove that ρ is s-crisp we have to show $\sigma \subseteq \rho$. Since $\rho(x, y) \in \{0, 1\}$ it suffices to prove $\rho(x, y) = 0 \to \sigma(x, y) = 0$. If $\rho(x, y) = 0$ we get $0 = \rho(x, y) \leq d_k\sigma(x, y) = k \wedge \sigma(x, y)$. Bus since k is linear by assumption we must have $\sigma(x, y) = 0$. □

So, under the assumption that all nonzero elements of L are linear, the notion of s-crispness coincides with 0-1 crispness for L-fuzzy relations. In particular, this holds for the classical case of $[0, 1]$-fuzzy relations.

References

[1] Adamek, J., Herrlich, H., Strecker, G.E.: *Abstract and Concrete Categories-The Joy of Cats* (2004), available at `http://katmat.math.uni-bremen.de/acc`.
[2] Awodey, S.: *Category Theory*, 2nd ed., Oxford University Press, New York (2010).
[3] Barr, M., Wells, C.: *Toposes, Triples and Theories*, (1984). Reprinted in Theory and Applications of Categories, No. 12 pp. 1-287. (2005), available at `http://www.tac.mta.ca/tac/reprints/articles/12/tr12abs.html`.
[4] Bird, R., de Moor, O.: *Algebra of Programming*, Prentice Hall Europe (1997).
[5] Borceux, F.: *Handbook of Categorical Algebra 1: Basic Category Theory*, Cambridge University Press (1994).
[6] Borceux, F.: *Handbook of Categorical Algebra 2: Categories and Structures*, Cambridge University Press (1994).
[7] Borceux, F., Bourn, D.: *Mal'cev, Protomodular, Homological and Semi-Abelian Categories*, Kluwer Academic Publishers (2004).
[8] Carboni, A., Kasangian S., Street, R.: Bicategories of spans and relations, Journal of Pure and Applied Algebra **33**, 259-267 (1984).
[9] Freyd, P.J., Scedrov, A.: *Categories, Allegories*, North-Holland, Amsterdam (1990).
[10] Goldblatt, R.: *Topoi-The Categorial Analysis of Logic*, Revised Edition, North-Holland, Amsterdam (1984).
[11] Jayewardene, R. Wyler, O.: Categories of Relations and Functional Relations, Applied Categorical Structures **8**, 279-305 (2000).

[12] Johnstone, P.T.: *Sketches of an Elephant-A Topos Theory Compendium*, Oxford University Press, (2003).

[13] Kawahara, Y., Furusawa, H.: Crispness in Dedekind categories, Bull. of Informatics and Cybernetics **33**, 1-18 (2001).

[14] Klein, A.: Relations in categories, Illinois J. Math. **14**, 536-550 (1970).

[15] Leinster, T.: *Basic Category Theory*, Cambridge University Press, Cambridge, UK (2014), available at `https://arxiv.org/abs/1612.09375`, and released under the Creative Commons Attribution-NonCommercial-ShareAlike 4.0 International licence (CC BY-NC-SA 4.0).

[16] Mac Lane, S.: *Categories for the Working Mathematician*, 2nd ed., Springer-Verlag New York (1998).

[17] McLarty, C.: *Elementary Categories, Elementary Toposes*, Clarendon Press, Oxford (1992).

[18] Milius, S.: Relations In Categories, Master thesis, York University, Toronto, Ontario (2000), available at `http://www8.cs.fau.de/staff/milius/thesis/thesis_a4.pdf`.

[19] Pavlovic, D.: Maps I: relative to a factorization system, Journal of Pure and Applied Algebra **99**, 9-34 (1995).

[20] Pierce, B.: *Basic Category Theory for Computer Scientists*, The MIT Press, Cambridge, Massachusetts (1991).

[21] Simmons, H.: *An Introduction to Category Theory*, Cambridge University Press, New York (2011).

[22] Smith, P.: *Category Theory-A Gentle Introduction*, version of January 29, 2018 (2018).

[23] Winter, M.: A new algebraic approach to L-fuzzy relations convenient to study crispness, Information Sciences **139**, 233-252 (2001).

[24] Winter, M.: *Goguen Categories: a Categorical Approach to L-fuzzy Relations*, Springer Science & Business Media (2007).

PART 5

Relations - Applied

Chapter 12

Semiotics and Semantics

Contrary to a probably widespread opinion in the so-called exact and engineering sciences, much - if not most - of human knowledge is of qualitative nature. This kind of knowledge usually is modeled in natural language, including notions that allow for flexible interpretation. One kind of such semantic indeterminism runs under the names *vagueness* or *fuzziness*. Using a simple ontological viewpoint enriched with few relations we will clarify some of the essential properties of these notions.

The contents of this chapter are largely based on the authors work [6] and [7] from the late nineties. Besides casting basic principles of knowledge acquisition and representation into a formal framework, it is shown in these papers that using an exact formal language can clarify the meaning of notions which are often discussed in some areas of mathematics and would otherwise themselves deserve one of these attributes. [6] focuses on the notions of '*vagueness*' viewed to be the same as '*fuzziness*', and '*ambiguity*'; and this will also make up the stuff of this chapter. The other paper also deals with the notions '*uncertainty*' and '*indeterminism*', and also with '*inexactness*' , '*imprecision*' and some more.

For developing further our first application example of relations we have to make commitments concerning the ontological structure we choose to work with. In this we follow the "Keep it small and simple" variant of the KISS-principle, leaving sophisticated philosophical and psychological discussions about this extremely important and interesting aspects of life to the experts in the field: Platonism, Nominalism and the like do not belong to the topics of this book. Instead we use as our starting point a simple but powerful viewpoint as e.g. presented in Bunge [2], and will shortly see that it offers a lot of interesting possibilities.

12.1 The Meaning Triangle

We take the word "*object*" in a general philosophical sense to stand for a primitive and undefined notion used in the following considerations, "for it is far too basic and important to be definable" ([2, p. 26]). Every *object* is either *factual* (as e.g. a tangible thing, an event, or a linguistic utterance) or *conceptual*, also called a *construct* (as e.g. concepts like "number", "set" etc., or propositional schemes like "x is a number" or propositions like "26 is a number"). The examples given already reveal what ingredients are needed for man to impose a kind of division and structure on the surrounding world. From the last few sentences the following ontological and epistemological principles can be inferred:

(1) One of the main assumptions of hypothetical realism – as well as of common sense and natural science – is the existence of a real world that is structured and is at least partially knowable by perception and experience. This is described by G. Vollmer as follows [8, p. 188, p. 191]

> All knowledge is hypothetical, i.e., conjectural, fallible, preliminary. There exists a real world, independent of our consciousness; it is structured, coherent, and quasi-continuous; it is at least partially knowable and explainable by perception, experience, and intersubjective science. ... Our forms of intuition and our categories of experience, the qualities of our perceptions, and the basic elements of inference, of classification, of abstraction, are, according to evolutionary epistemology, innate, ontogenetically *a priori* and constitutive for mesocosmic knowledge.

(2) Structure means a certain kind of order, i.e. a temporal or spatial coincidence or succession of states or events.
(3) So there exists the possibility for, and also the capability of *re*-cognizing objects and events: Only this makes the emergence of linguistic labels both plausible and reasonable at all.
(4) *Re*-cognition in turn requires for comparisons, the result of which can be either "*discriminate*" or "*make equal*" (i.e. regard as equivalent).
(5) Finally, we have to account for the fact that languages of one kind or the other are the only means by which living beings organize and communicate their knowledge and experience.

So we have arrived at a view that can roughly be cast in three layers as a first approach:

$$\textit{linguistic level} \xrightarrow{\mathcal{D}} \textit{conceptual level} \xrightarrow{\mathcal{R}} \textit{physical level}$$

$\mathcal{D}$ is called the **designation** (or *naming*) relation; $\mathcal{R}$ is the relation of **reference**, which gives information about what things a given construct is talking about at all. The most obvious interrelationship in everyday situations of communication and information is the one from the linguistic to the physical level, called **denotation**, arising from the composition of designation and reference: $\Delta := \mathcal{D} \circ \mathcal{R}$.

For the inner triangle in Fig. 12.1 the term **meaning triangle** has become popular through Ogden and Richards [3], but according to Sowa [5, p. 192] ARISTOTLE (once again!) was the first to make this distinction:

> Spoken words are symbols of experiences [pathēmata] in the psyche; written words are symbols of the spoken. As writing, so is speech not the same for all people. But the experiences themselves, of which these words are primarily

signs, are the same for everyone, and so are the objects of which those experiences are likenesses. (On Interpretation I6a4)

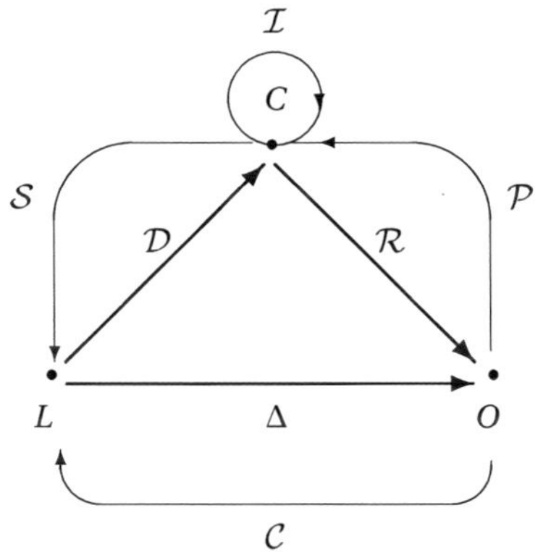

FIG. 12.1: The basic semantic and epistemic relations.

The components of this triangle are the following:

(1) the linguistic level, which usually appears as a set L of signs, words, symbols, linguistic labels, marks etc.;
(2) the conceptual level C: concepts, ideas;
(3) the object level O: anything we can think about (thus, especially, $C \subset O$ and $L \subset O$).

and the already mentioned relations

$\mathcal{D} \subset L \times C$ of *designation*,
$\mathcal{R} \subset C \times O$ of *reference*, and
$\Delta \subset L \times O$ of *denotation*.

The original triangle has been enriched in [6] with the following relations:

$\mathcal{P} \subset O \times C$ of *perception*, relating objects to concepts,
$\mathcal{S} \subset C \times L$ of *speech*, mapping concepts to words,
$\mathcal{C} \subset O \times L$ of *classification*, associating words to objects, and
$\mathcal{I} \subset C \times C$ of *intension* or *connotation*, relating concepts to other concepts.

Note that the requirements $C \subset O$ and $L \subset O$ are not a symptom of some kind of circularity in distinguishing the three levels, but are due to the necessity of enabling a distinction between meta-language and object-language, varying with the context of consideration.

The diagram in Fig. 12.1 induces several relationships (symbolized by $\widehat{=}$) between the relations included:

$$\mathcal{D} \circ \mathcal{R} \widehat{=} \Delta \widehat{=} (\mathcal{P} \circ \mathcal{S})^{\smile} \widehat{=} \mathcal{C}^{\smile} \tag{12.1}$$

The left of Δ corresponds to the viewpoint of logical analysis which will be the main topic in this chapter. However, we cannot really understand the importance and *nature* of the phenomena we will consider if we do not take into account the right hand side of (12.1) which can give us an explanation from the psychological and evolutionary viewpoint.

We thus have a set of tools that should allow us to analyze and describe a number of interesting and important properties of (everyday) language human beings use to communicate with each other. To begin with, applying the concepts of foreset and afterset of a relation, we get the following examples:

- $l\Delta$ is the set of objects denoted by l;
- Δo is the set of labels denoting o;
- $l\mathcal{D}$ is the set of concepts designated by l;
- $c\mathcal{R}$ is the set of objects referred to by c;
- $c\mathcal{S}$ is the set of labels naming c;
- $\mathcal{S}l$ is the set of concepts labeled as l;
- if $|\mathcal{D}c| > 1$, then $\mathcal{D}c$ is the set of synonyms for c.

Above we have defined **intension** (*compréhension* in French, and *Inhalt* or *Sinn* in German) as a relation from concepts to concepts, but have not yet said anything about its twin **extension** $\mathcal{E}$. One thing we have to take into account is the distinction between reference and extension ([2, p. 119f]). While both statements and predicates can be assigned referents, only the latter are also ascribed an extension. Some examples show that even in case of predicates we have to make a distinction between the extension and the reference class. We have e.g.

- PRIME NUMBER $\mathcal{R} = \mathbb{N}$, but
  PRIME NUMBER $\mathcal{E} = \{2, 3, 5, 7, ...\} \subset \mathbb{N}$;
- LOVES $\mathcal{R}$ = {Beatrice, Dante, Julia, Romeo, ...}, but
  LOVES $\mathcal{E}$ = {⟨Dante, Beatrice⟩, ⟨Romeo, Julia⟩, ...};
- CREATOR$\mathcal{R}$ = {God}, but
  CREATOR$\mathcal{E} = \emptyset$ for an atheist, and CREATOR$\mathcal{E}$ = {God} for a theist.

The reference class of an n-ary predicate is the set of objects concerned, whereas its extension is a set of ordered n-tuples of elements from this set. Furthermore, the notion of reference does not presuppose the concept of truth, but the concept of extension does. From a formal point of view we have $\mathcal{E} \subseteq C \times O^*$. Every extension is a set, but not vice versa: for e.g. {Paris, Aristotle} there does not exist a reasonable predicate.

12.2 Ambiguity, Vagueness, Uncertainty and the like

Language has a lot of properties that make it a highly economic, but sometimes also a puzzling vehicle. One of its most fascinating phenomena is the vagueness of predicates which has already been a topic of interest for a long time: see e.g. the papers of Russel [4], Black [1], and Zadeh [9]. (Note: In the considerations below we take VAGUENESS and FUZZINESS to be the same concept.) To proceed chronologically, we start with a contribution of CHARLES SANDERS PEIRCE[1] to the *Dictionary of Philosophy and Psychology, Vol. II* edited by J.M. Baldwin (New York: Macmillan, 1902):

> A proposition is vague when there are possible states of things concerning which it is intrinsically uncertain whether, had they been contemplated by the speaker, he would have regarded them as excluded or allowed by the proposition. By intrinsically uncertain we mean not uncertain in consequence of any ignorance of the interpreter, but because the speaker's habits of language were indeterminate; so that one day he would regard the proposition as excluding, another as admitting, those states of things. Yet this must be understood to have reference to what might be deduced from a perfect knowledge of his state of mind; for it is precisely because these questions never did, or did not frequently, present themselves that his habit remained indeterminate.

We continue with BERTRAND RUSSEL (1872-1970) [4] from 1923:

> Let us consider the various ways in which common words are vague, and let us begin with such a word as "red." It is perfectly obvious, since colours form a continuum, that there are shades of colour concerning which we shall be in doubt whether to call them red or not, not because we are ignorant of the meaning of the word "red," but because it is a word the extent of whose application is essentially doubtful. (p. 84)
>
> Vagueness and precision alike are characteristics which can only belong to a representation, of which language is an example. They have to do with the relation between a representation and that which it represents. Apart from representation, whether cognitive or mechanical, there can be no such thing as vagueness or precision; things are what they are, and there is an end of it. (p. 84/85)
>
> It would be a great mistake to suppose that vague knowledge must be false. On the contrary, a vague belief has a much better chance of being true than a precise one, because there are more possible facts that would verify it. (p. 91)

Black [1, p. 429/430] also focuses on vagueness, but extends his analysis to *generality* and *ambiguity*:

> This paper is written to show that while the vague symbol has a part to play in language which cannot be equally well performed by more accurate symbols from another level (wave lengths as substitute for names of colors) the transition to levels of higher accuracy can always in principle be made. ...
> A symbol's vagueness is held to consist in the existence of objects concerning which it is intrinsically impossible to say either that the symbol in question does, or does not, apply. ...
> It is claimed that all symbols whose application involves the recognition of sensible qualities are vague, and a typical case is constructed for convenience of reference. Vagueness is distinguished from generality and from ambiguity. The former is constituted by the application of a symbol to a multiplicity of objects in the fields of reference, the latter by the association of a finite number of

[1] CHARLES SANDERS PEIRCE not only did groundwork on relations but also laid foundations for semiotics, the science of signs, and is considered by many the greatest American philosopher.

> alternative meanings having the same phonetic form; but it is characteristic of the vague symbol that there are no alternative symbols in the language, and its vagueness is a feature of the boundary of its extension, and is not constituted by the extension itself.

We continue our vagueness journey through the twentieth century with Zadeh's seminal paper [9, p. 338/339] on fuzzy sets from 1965:

> More often than not, the classes of objects encountered in the real physical world do not have precisely defined criteria of membership. ...
>
> Clearly, the "class of all real numbers which are much greater than 1," or "the class of beautiful women", or "the class of tall men", do not constitute classes or sets in the usual mathematical sense of these terms. Yet, the fact remains that such imprecisely defined "classes" play an important role in human thinking, particularly in the domains of pattern recognition, communication of information, and abstraction. ...
>
> The concept in question is that of a *fuzzy set*, that is, a "class" with a continuum of grades of membership. ...
>
> A *fuzzy set (class)* A in X is characterized by a *membership (characteristic) function* $f_A(x)$ which associates with each point in X a real number in the interval $[0,1]$ with the value of $f_A(x)$ at x representing the "grade of membership" of x in A.

We finish with a short passage from Zeleny [10, p. 302] from 1984, thus having passed through nearly a century on investigation on vagueness:

> Any notion of fuzziness can be meaningful only in dependency on human systems, human observers, and human tools and strategies for describing reality: That is, human beings are the only and exclusive source of fuzziness in linguistic labeling - there are no other sources.

The inclusion of this completely informal section is intended to serve the following two purposes:

(1) Even for really great minds, though they had to say important things on vagueness, it is hard to do this in a concise way. None of them did attempt or succeed in building a formal framework to model this indispensable property of certain language components.

(2) And even "*if we want to theorize, we must start by building more or less sketchy models*". Following this advice from Bunge [2, p. 13] we are *naturally* led to a model formulated in *natural language.*

12.3 Some relational Clarifications

Now we will use the natural-language models presented in the last section to try some modest steps in direction of a formal model. Let us first summarize the essential ingredients from the given quotations in Table 12.1.

If at all, each of the authors ascribes VAGUENESS to a different kind of object, so we are left with the problem to find out what their real purpose could be. Only Black compares VAGUENESS to other properties:

- GENERALITY: $|l\Delta| > 1$, i.e. $l\Delta = \{o_1, o_2, \ldots, o_n\},\ n > 1$, i.e. denotation Δ is not a function.
- AMBIGUITY: $|l\mathcal{D}| > 1$, i.e. $l\mathcal{D} = \{c_1, c_2, \ldots, c_n\}, n > 1$, i.e. designation $\mathcal{D}$ is not a function.

TABLE 12.1: Characteristics of VAGUENESS

Author	**credited to**	**caused by**	**related to**	**example**	**others**
PEIRCE	proposition	speaker's indetermi-nate habit	perfect knowledge		is not igno-rance
RUSSEL	common words		belongs to represen-tation	"red"	is not igno-rance
BLACK	symbol			names of colors	no alter-native sym-bols
ZADEH	imprecisely defined "classes"			beautiful women, tall men	
ZELENY		linguistic labeling	describing reality		

(Below I will use different fonts to distinguish between CONCEPTS and linguistic labels designating them, if appropriate.) Black attributes all of the properties of GENERALITY, AMBIGUITY and VAGUENESS to 'symbols', which we take to denotate objects like linguistic labels, predicates etc., i.e. some $c \in L$. And while, in case of GENERALITY and AMBIGUITY, symbols are related to objects and meanings, respectively, no analogous association of that kind is made for the notion of VAGUENESS. Ambiguous labels are by definition – $\Delta := \mathcal{D} \circ \mathcal{R}$ – part of the general labels. Thus, it seems that GENERALITY in Black's sense is a concept far too unspecific to be of great value in the current analysis. And *ambiguity* is just another word for *homonymy*; thus, due to the context dependent constraints imposed by the current communication, one will usually not be uncertain about the designated concept.

Some typical properties of AMBIGUOUS labels l are the following: Let $l\mathcal{D} = \{c_1, c_2, \ldots, c_n\}, n > 1$; then

- $c_i\mathcal{I} \cap c_j\mathcal{I} = \emptyset$, for all $i \neq j$, and
- $c_i\mathcal{R} \cap c_j\mathcal{R} = \emptyset$, for all $i \neq j$.

That is to say that AMBIGUOUS labels point to different and disjoint parts of the world.

So we are still left with the most complex and interesting aspect of uncertainty, viz. VAGUENESS for which only Zadeh has made a proposal for a formal quantitative model. (Besides, the nineties have seen a number of successful implementations of this model under the name of '*fuzzy control*' in various devices as photo cameras, washing machines, cars etc.; recall Section 8.6.) Here we are interested in another aspect and its embedding in a formal framework that allows for a more concise and concrete discussion.

As we learned from Russel's quotation "*Vagueness and precision alike ... have to do with the relation between a representation and that which it represents*", so it is not a symbol s that can be vague or fuzzy, but (one of) the concept(s) $c \in s\mathcal{D}$ designated by s. Now remember Black claiming "*that all symbols whose application involves the recognition of sensible qualities are vague*", which fully corresponds to the examples that have nearly always been mentioned in fuzzy literature to be typical: TALL, COLD, LOUD, HIGH etc. What makes them VAGUE is the fact that implicitly we compare the linguistic with a presumably "more precise" numerical representation, given for the just mentioned examples in e.g. tallness in cm, temperature measured in degree Celsius, loudness in phone, and height in meters (for mountain, trees), km (for atmosphere), or some currency (for income).

So we have to provide some suitable formalism for capturing these ideas. When we say that a concept $c \in C$ is VAGUE or FUZZY, then we usually want to express that there is something unclear or left unfixed about what commonly is called its meaning or sense. Therefore, we may be uncertain whether c applies to an object $o \in O$ or not; but if we have to choose or to decide in a concrete situation, we must perform a classificational act that results in assigning a corresponding label $l \in c\mathcal{S}$ or not. To record the outcome of this classificational act we have to use some suitable sign system. We start formalization with

Definition 12.1. *Let A and B be two disjoint non-empty sets.*

(a) *B is called* refinement *of A if there is a mapping* $\hat{\Pi}: A \longrightarrow \mathbf{2}^B$, *the* **refinement mapping**, *such that*

   (1) $\hat{\Pi}(a) \neq \emptyset$ *for all* $a \in A$,
   (2) $\bigcup\{\hat{\Pi}(a) | a \in A\} = B$, *and*
   (3) $\hat{\Pi}(C) = \bigcup\{\hat{\Pi}(c) | c \in C\}$, *for* $C \subset A$.

(b) *A refinement mapping* $\hat{\Pi}$ *is called* **class preserving** *if* $a \neq a' \rightarrow \hat{\Pi}(a) \cap \hat{\Pi}(a') = \emptyset$.

(c) *A refinement mapping* $\hat{\Pi}$ *is called* **proper** *if* $\exists a \in A : |\hat{\Pi}(a)| \geq 2$.

Definition 12.2.

(a) *A* **classificational system** *is an ordered pair* $\mathbf{C} := \langle L, \perp \rangle$, *where*

   (1) *L denotes a set of linguistic labels, classificators, symbolic items etc. suitable for expressing the result of a classificational act;*

(2) $\perp$ *is a relation on L, where* $l_1 \perp l_2$ *intuitively means that* l_1 *and* l_2 classificationally reject *each other in the sense that they are mutually exclusive outcomes of a classificational act.*

(b) Let $\mathbf{C} = \langle L, \perp\rangle$ *and* $\mathbf{C}' = \langle L', \perp'\rangle$ *be two classificational systems.* $\mathbf{C}'$ *is called a* **refinement** *of* $\mathbf{C}$ *if* L' *is a refinement of* L.

According to our foregoing considerations, fuzziness is to be analyzed in connection with at least the two mappings $\mathcal{P}$ and $\mathcal{R}$. Using the example of colors, fuzzy concepts have the following properties (where the $\lambda_i, 1 \le i \le n$, are assumed to represent wavelengths perceived of the attribute color of objects $o_j, 1 \le j \le m$):

- $|\mathcal{P}c| > 1$: We have e.g. $\{\lambda_1, \lambda_2, \ldots, \lambda_n\} \mapsto$ RED, $n > 1$, i.e. $\mathcal{P}$ is not injective;
- $|c\mathcal{R}| > 1$: We have e.g. RED $\mapsto \{\lambda_1, \lambda_2, \ldots, \lambda_m\}$, $m > 1$, i.e. $\mathcal{R}$ is not a function.

That the above are only necessary but not sufficient conditions for the presence of VAGUENESS can be seen by substituting CAR for RED: We get $|\mathcal{P}\text{CAR}| > 1$ and $|\text{CAR}\mathcal{R}| > 1$, but CAR has never been claimed to be a fuzzy predicate. So there must be something else which is specific for vagueness. So we go on with

Definition 12.3. *A concept* $c \in C$ *is said to be* **coarse** *if there is a set* $B \subset C$ *which is a proper refinement of* $\{c\}$*, such that the following compatibility condition is satisfied*

$$c\mathcal{R} \cap \bigcup_{b \in B} b\mathcal{R} \neq \emptyset.$$

This condition clearly assures that all the involved concepts pertain to the same domain (of attributes, interest etc.), i.e. that they "*are speaking about the same things*". But the crucial point is that the concepts in the refining classification space can be ordered in a natural way. And this the case for all the typical examples of VAGUE, i.e. FUZZY, concepts like TALL, LOUD, HIGH, LONG, WARM, COLD, HEAVY, YOUNG, ..., since the refinement uses some suitable subset of the real numbers $\mathbb{R}$. What fuzzy sets do using the so-called *membership function* (recall the quotation from [9] on page 400) is to characterize the prototypicality of the refining concepts with respect to the coarse refined concept under consideration.

As an example from fuzzy control suppose you want to characterize the deviation of a measured from an expected value. A commonly used pattern looks as shown in Fig. 12.2.

Without going too much into the formal details, we can easily recognize the two classificational systems and a refinement mapping as indicated by the triangular membership functions:

- $\mathbf{C1} := \langle L_1, \perp_1\rangle = \langle\{NB, NM, NS, ZO, PS, PM, PB\}, \{\langle l_i, l_j\rangle | i \neq j\}\rangle$, and
- $\mathbf{C2} := \langle L_2, \perp_2\rangle = \langle\{r_0, r_1, \ldots, r_k\} \subset \mathbb{R}, \{\langle r_i, r_j\rangle | i \neq j\}\rangle$

So each concept in L_1 is coarse, and the compatibility condition of Def. 12.3 can reasonably assumed to be fulfilled, otherwise the modelling approach would be senseless.

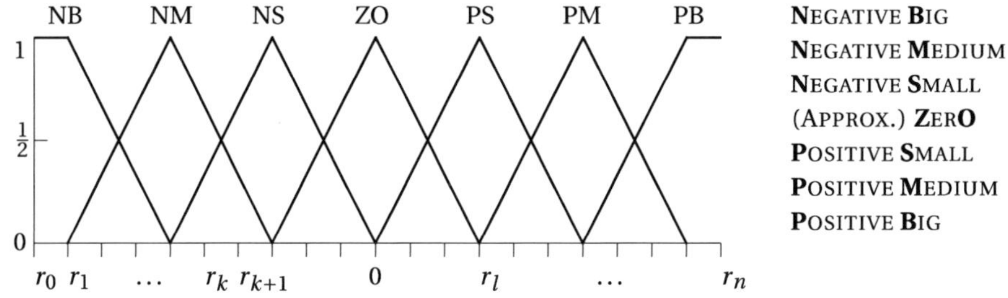

FIG. 12.2: Examples of fuzzy sets.

In contrast to AMBIGUITY one immediately notices that in case of vague concepts it is typical that

- $c_i\mathcal{I} \cap c_j\mathcal{I} \neq \emptyset$, for some $i \neq j$, with c_i, c_j being "semantical neighbors";

e.g. $\{r_k, r_{k+1}\} \subset$ NM$\mathcal{I} \cap$ NS$\mathcal{I}$, and by definition of coarseness

- $c_i\mathcal{R} \cap c_j\mathcal{R} \neq \emptyset$, for all $i \neq j$.

And the most typical value (i.e. refining concept) for POSITIVE SMALL is r_l.

Definition 12.4. *A concept $c \in C$ is said to be* **fuzzy** *if it is coarse and the corresponding refinement mapping goes from an unordered qualitative to an ordered classification space and is not class preserving.*

Let $c\mathcal{I} = \{c_1, c_2, \ldots, c_n\}$, $n \geq 2$; then some typical properties of fuzzy concepts c are the following:

- If $c_i, c_j \in c\mathcal{I}$ then $c_i\mathcal{I} \cap c_j\mathcal{I} \neq \emptyset, i \neq j$, for semantically nearby c_i, c_j.
- Only labels l designating vague concepts c, i.e. $l \in \mathcal{D}c$ can be combined with modifiers like very, more or less, rather ... .
- Usually, fuzzy concepts have a context dependent meaning: whereas 1 M always means the same, concepts like TALL, HIGH or LONG do not.
- Usually, fuzzy concepts can be applied to different universes of discourse, even if they are measured along the same physical scale as LONG and HIGH: we speak e.g. about high mountains, trees, or buildings. But both concepts do not only refer to object attributes measured in units of MM, CM or M etc. LONG e.g. also applies to durations, and HIGH also to incomes and tones.
- Thus, fuzzy concepts are flexible, universally applicable, and highly economic tools for human communication, due to the above mentioned properties.

Note the important consequence that FUZZINESS is not an absolute but a relative phenomenon that is inseparably connected to the usage of two different classificational systems, such that one of them is a refinement of the other. Remember that this is in total accordance with Russel [4] : VAGUENESS has to do "*with the relation between a representation and that which it represents*". Obviously, the recognition of the phenomenon of fuzziness arose out of the confrontation of the two most often used classificational systems, viz. *natural language* which we use in everyday discourse, and the *artificial man-made formal systems* used in the various scientific and engineering disciplines.

The crucial point is the change of the scale properties of the measurement device, which in practice nearly always goes from our natural equipment (eyes, ears, skin) to artificial measurement devices (centimeter sticks, thermometers, watches etc.) in order to get "exact" instead of "inexact" data. Neither would (choosing e.g. the context of room temperature) a refinement like

$$22° \text{C} \longmapsto \{21.5° \text{C}, 21.6° \text{C}, \ldots, 22.4° \text{C}\}$$

nor one like

$$\text{ADMISSIBLE} \longmapsto \{\text{COOL, PLEASANT, WARM}\}$$

induce VAGUENESS: due to the missing scale change we have to do with coarse concepts in both cases.

References

[1] Black, M.: Vagueness: An exercise in logical analysis, Philosophy of Science **4**, 427-455 (1937). (Reprinted in: Int. J. General Systems **17**, 107-128 (1990)).

[2] Bunge, M.: *Semantics I: Sense and Reference.* Vol. 1 of *Treatise on Basic Philosophy*, Reidel, Dordrecht/Boston (1974).

[3] Ogden, C.K. and Richards, I.A.: *The Meaning of Meaning*, 8th ed., Harcourt, Brace & World, Inc., New York (1946).

[4] Russell B.: Vagueness, Australasian Journal of Psychology and Philosophy **1**, 84-92 (1923).

[5] Sowa, J.F.: *Knowledge representation-logical, philosophical, and computational foundations*, Brooks/Cole, Thomson Learning (2000).

[6] Toth, H.: Fuzziness: From epistemic considerations to terminological clarification, Int. J. of Uncertainty, Fuzziness and Knowledge-Based Systems **5**, 481-503 (1997).

[7] Toth, H.: Towards fixing some 'fuzzy' catchwords: A terminological primer. In: Zadeh, L.A. and Kacprzyk, J. (Eds.), *Computing with Words in Information Intelligent Systems 1*, 154-179, Springer-Verlag Berlin Heidelberg (1999).

[8] Vollmer, G.: On supposed circularities in an empirically oriented epistemology. In: Radnitzky, G. and Bartley, W.W. III (Eds.), *Evolutionary Epistemology, Theory of Rationality, and the Sociology of Knowledge*, 163-200, Open Court, La Salle, Illinois (1987).

[9] Zadeh, L.A.: Fuzzy sets, Inform. and Control **8**, 338-353 (1965).

[10] Zeleny, M.: On the (ir)relevancy of fuzzy set theory, Human Systems Management **4**, 301-306 (1984).

Chapter 13

Information Systems and Rough Sets

Today's digital equipment allows for the collection of data from a wide variety of fields and their accumulation. The volume and complexity of these data is growing at a dramatic pace, but a huge amount of data does not mean even a small amount of knowledge. It has soon been recognized that new methods and tools are necessary to master the problems of organizing and evaluating the data. In order to describe and extract useful information, researchers have proposed many methods besides classical logic and probability theory, such as Fuzzy Set Theory [18], Rough Set Theory [12–14], Computing With Words [19], Formal Concept Analysis [6] and some more. And as you will see later on, or by a look on current research, many connections and interactions have emerged in the course of time.

The names chosen for the new approaches just mentioned do already hint towards the fact that various kinds of indeterminacy are inseparably connected with our usage of everyday language, often called incomplete, vague, or granular information. Remember the considerations of Chapter 12 where, based on some simple ontological commitments induced by the meaning triangle, some elementary building blocks of cognition, language, and knowledge have been considered. Even in this cursory discussion one important fact has immediately popped up: usually there is a one-to-many relation between a linguistic label and designated concepts, or between concepts and referenced objects etc. This leads to what is sometimes called lack of precision, exactness etc. of language; but on the other hand, indeterminacy makes our language an efficient and flexible tool in coping with everyday challenges and situations: If one has to cross a street, it is enough to know that, eventually, a car is approaching VERY FAST, instead of its exact speed measured in mph or km/h; the two different values "*25 mph*" and "*26 mph*" would not usually result in different behavioral patterns.

13.1 Information Systems

The first step if one aims for storing and evaluating data is to provide a scheme that supports both purposes sufficiently well. One obvious possibility has been proposed by Pawlak [11] under the name *information system*. It uses the notions of "object" (in the sense of Chapter 12) and "property" as the most elementary pieces of information, splitting the notion of property into two components: an attribute and its value. It is a common view that a concept is determined by its extension (or denotation) and intension (or connotation): The extension of a concept consists of the objects that are instances of this concept, and its intension is the set of the properties that are characteristic for the objects to which this concept applies.

Definition 13.1. *An* **information system**, *also* **information table**, *is a quadruple* $I = \langle O, A, V, f \rangle$ *such that*

(1) $O \neq \emptyset$ *is a finite set of* objects;
(2) $A \neq \emptyset$ *is a finite set of* attributes;
(3) $V = \bigcup_{a \in A} V_a$, *where* V_a *is the set of possible* values *that can be observed for attribute* a, $|V_a| > 1$;
(4) $f : O \times A \longrightarrow \mathcal{P}(V)$, *the* information map, *is a function which associates with any ordered pair* $\langle o, a \rangle$ *the value* $f(o,a) \subseteq \mathcal{P}(V_a)$ *attained by the attribute* a *for the object* o.

As a minimal requirement for an information system I we assume that the following coherence conditions hold:

- The information map f must be surjective: Values $v \in V$ such that there is no pair $\langle o, a \rangle \in O \times A$ with $v \in f(o,a)$ are of no interest with respect to the knowledge stored in I.
- For any attribute $a \in A$ there exist at least two objects o_1 and o_2 such that $f(o_1, a) \neq f(o_2, a)$, otherwise this attribute does not supply any information for the classification of the objects under consideration and can be suppressed.

If f is a total function, that is, $f(o,a)$ is defined for all pairs $\langle o, a \rangle \in O \times A$, then the information system I is called **complete**; otherwise, it is called **incomplete**.

According to Def. 13.1 we can consider each attribute $a \in A$ as a map $a : O \longrightarrow \mathcal{P}(V_a)$. The set $a(o)$ is called the *a-property* of o; if $|a(o)| > 1$ then this property is called a **nondeterministic property**, and a **deterministic property** otherwise. a is called a **nondeterministic attribute**, if there is an object o with a nondeterministic a-property, and a **deterministic attribute** otherwise. Let $B \subseteq A$; an object o is called *B-deterministic*, if all its b-properties for attributes in B are deterministic, i.e. $|b(o)| <= 1$ for all $b \in B$. Finally, an information system I is called **nondeterministic** if it contains nondeterministic properties, and a **deterministic information system** otherwise.

Any information system can be represented by a data table with rows labeled by objects and columns labeled by attributes, as shown in the example information system I_P. Any pair $\langle o, a \rangle$, where $o \in O$ and $a \in A$ defines the table entry consisting of the value $a(o)$.

Table 13.1 shows a simple information system I_P, collecting data about five persons P1, ..., P5 concerning their sex, age, and language capabilities with respect to e.g. the set of official languages of a conference {English, French, German}. We have

(a) $O = \{$P1, P2, P2, P4, P5$\}$,

TABLE 13.1: Information system I_P

Objects	***Sex***	***Age***	***Langs***
P1	Male	{28, 29, 30, 31, 32}	{English, German}
P2	Female	27	{English, French, German}
P3	Male	60	French
P4	Female	{55,56,57,58}	German
P5	Male	40	English

(b) $A = \{Sex, Age, Langs\}$,

(c) $V = V_{Sex} \cup V_{Age} \cup V_{Langs}$, where
V_{Sex} = {Male, Female}, $V_{Age} = \{27, 28, 29, 30, 31, 32, 55, 56, 57, 58, 60\}$, and
V_{Langs} = {English, French, German}.

Note that according to our definitions the *Langs*-property $Langs(o) = \emptyset$ is deterministic, meaning that o does not speak any of the official languages, which is a very definite information. Thus $\emptyset$ does not symbolize "unknown".

Clearly, I_P is a nondeterministic information system, with one deterministic attribute, viz. *Sex*, and two deterministic objects, viz. P3 and P5. The *Age*-property of P1 is $Age(P1) = \{28, 29, 30, 31, 32\}$, and its *Langs*-property is $Langs(P1)$ = {English, German}. Whereas both properties are nondeterministic, it is clear that for the age of P1 only one number can be the right one, but concerning P1's language capabilities we have no hint how to interpret the corresponding information: (i) English and German and no other language, (ii) English and German and possibly other languages, (iii) English or German but not both, (iv) English or German or both. The problem we have encountered is, whether a nondeterministic property is to be read in the disjunctive (*Age*), or in the conjunctive sense (*Langs*).

Information systems as introduced above are only a scheme how to organize data collected about a domain of interest. A natural question to ask is, if there is some information or knowledge that is not explicitly represented in an information system but can be derived from its content. Such information mirrors certain relationships among the objects in the form of *information relations*, determined only by the attributes of the system and without usage of external information. There are two groups of information relations: the ones that reflect various forms of indistinguishability or similarity of objects in terms of the properties considered, and the relations that indicate distinguishability of the objects.

Definition 13.2. *Let $I = \langle O, A, V, f\rangle$ be an information system, and $B \subseteq A$.*

(a) the **indiscernibility** *relation $\iota_B \subseteq O \times O$ is defined as $\langle x, y\rangle \in \iota_B$ iff $a(x) = a(y)$ for all $a \in B$;*

(b) the **similarity** *relation $\sigma_B \subseteq O \times O$ is defined as $\langle x, y\rangle \in \sigma_B$ iff $a(x) \cap a(y) \neq \emptyset$ for all $a \in B$.*

Intuitively, two objects are B-indiscernible whenever their a-properties determined by the attributes $a \in B$ are the same. In other words, based on the properties determined by B, these objects are all alike. Two objects are B-similar whenever these objects have a-properties in common, for all $a \in B$. In our example from Table 13.1 P2 and P4 are *Sex*-indiscernible, P1,P2 and P4 are *Langs*-similar, as well as P2 and P3; there are no *Age*-indiscernible persons.

Obviously, ι_B is an equivalence relation, and σ_B is a tolerance relation on O. Any subset $B \subseteq A$ determines an associated indiscernibility and similarity relation. Using some simple ideas concerning the organization and analysis of data collections, we have thus arrived at the starting point of rough set theory, the topic of the following sections. Of course, information systems do have other interesting application relevant or algebraic properties which, however, are not directly connected to the concept of rough sets. Readers interested in more details may consult e.g. [5, 8, 10, 11].

13.2 Rough Set Theory (RST)

13.2.1 *Basic Ideas of Rough Set Theory*

Consider the deterministic information system I_{P2} given in Table 13.2: Its set of objects is $O = \{P1, \ldots, P6\}$, and its set of attributes $A = \{Height, Hair, Eyes\}$. It can be seen that some objects have the same description, i.e. they are A-indiscernible, e.g. P2 and P3.

TABLE 13.2: Information system I_{P2}.

Objects	***Height***	***Hair***	***Eyes***
P1	short	blond	blue
P2	short	blond	brown
P3	short	blond	brown
P4	tall	dark	blue
P5	tall	dark	blue
P6	tall	red	blue

RST is based on two basic notions: the definability of concepts and the approximation of concepts. These two notions are conceived with respect to (i) an information system which is regarded to describe all the information available about a set of objects, and (2) the indiscernibility relations derivable from it. The idea behind is to regard a set of objects $P \subseteq O$ as the extension of some concept.

Definition 13.3.

(a) *Let U be a finite set, called the universe of discourse, and ρ be an equivalence relation on U. The pair $\langle U, \rho \rangle$ is called a* **Pawlak approximation space**. *(PA-space).*

(b) *$V \subseteq U$ is called a ρ-**definable set** if V is the union of equivalence classes $[u]_\rho$, and a ρ-**rough set** otherwise. The set of ρ-definable sets will be denoted by def_ρ^U.*

The blocks of the partition U/ρ are called *elementary sets* or *elementary granules.* In the rough set approach, the elementary sets are regarded as the basic building blocks (concepts) of our knowledge about reality. A ρ-rough set can be characterized by a pair of definable sets as follows:

Definition 13.4. *Let $\langle U, \rho\rangle$ be a PA-space, and $V \subseteq U$.*

(a) *The **lower ρ-approximation** of V is the set obtained by taking the union of all equivalence classes of U/ρ included in the set V:*

$$V_\rho^\perp = \bigcup\{[x]_\rho \in U/\rho : [x]_\rho \subseteq V\}.$$

(b) *The **upper ρ-approximation** of V is the set obtained by taking the union of all equivalence classes of U/ρ that have a nonempty intersection with the set V:*

$$V_\rho^\top = \bigcup\{[x]_\rho \in U/\rho : [x]_\rho \cap V \neq \emptyset\}.$$

With respect to any subset $V \subseteq U$, the universe can be divided into three disjoint regions using the lower and upper approximations:

- The **positive region** comprises all $x \in U$ that definitely belong to V:

$$\mathit{pos}_\rho^V = V_\rho^\perp;$$

- The **negative region** comprises all $x \in U$ that definitely do not belong to V:

$$\mathit{neg}_\rho^V = \mathit{pos}_\rho^{\overline{V}} = U - V_\rho^\top = \overline{V_\rho^\top};$$

- The ρ-**boundary** of V is the set of elements that possibly belong to V:

$$V_\rho^\sharp = V_\rho^\top - V_\rho^\perp.$$

Note that $V_\rho^\perp$ is the greatest definable set contained by V, and that $V_\rho^\top$ is the least definable set containing V. If $V \in \mathit{def}_\rho^U$ then $V_\rho^\perp = V = V_\rho^\top$ and $V_\rho^\sharp = \emptyset$.

The definition given above induces the following interpretation: Due to the granularity of the knowledge available, rough sets cannot be accurately described by using it. Therefore with every rough set two definable sets are associated, called lower and upper approximation: The lower (upper) approximation consists of all elements that surely (possibly) belong to the set, and the boundary region of the set comprises all elements that cannot be identified uniquely to belong to the set considered or its complement.

In the example of Table 13.2 the set {P1, P2} is not ρ-definable for any ρ, because any equivalence relation that describes P2 also describes P3, so P2 and P3 cannot be separated. Thus, {P1, P2} is a ρ-rough set for any ρ. Taking ι_A as indiscernibility relation, then we get

$$O/\iota_A = \{\{P1\},\{P2,P3\},\{P4,P5\},\{P6\}\};$$

$$\{P1,P2\}_{\iota_A}^\perp = \{P1\} \text{ and } \{P1,P2\}_{\iota_A}^\top = \{P1,P2,P3\}.$$

Let us have a further look on ρ-**rough sets**: Let us define the relation $\equiv_\rho$ on $\mathcal{P}(U)$ by $X \equiv_\rho Y \;:\Leftrightarrow\; X^{\perp}_{\rho} = Y^{\perp}_{\rho}$ and $X^{\top}_{\rho} = Y^{\top}_{\rho}$. Clearly, $X \equiv_\rho Y$ means that exactly the same elements belong certainly to X and to Y, and exactly the same elements belong possibly to X and to Y. It is easy to see that $\equiv_\rho$ is an equivalence relation on $\mathcal{P}(U)$.

Definition 13.5. *Let $\rho \in \mathsf{Eq}_U$; the equivalence classes of $\equiv_\rho$ are called ρ-**rough sets**.*

Since equivalence classes are uniquely determined by an arbitrary of its members, ρ-rough sets can be equivalently viewed as pairs $(X^{\perp}_{\rho}, X^{\top}_{\rho})$, $X \subseteq U$; such a pair is sometimes called an *ρ-approximation of X*, or simply a *rough set.*

Remark: Naming is a little bit sloppy in that the term "*rough set*" is used for an equivalence class $[X]_{\equiv_\rho}$, a pair of sets $(X^{\perp}_{\rho}, X^{\top}_{\rho})$, or X itself in case $X^{\sharp}_{\rho} \neq \emptyset$, $X \subseteq U$; see [2] for a still longer list.

Theorem 13.6. ([22, 24]) *Let $\langle U, \rho\rangle$ be a PA-space, and $X, Y \subseteq U$; then the following statements hold:*

(1L)	$U^{\perp}_{\rho} = U$	(Co-normality)
(1U)	$U^{\top}_{\rho} = U$	(Co-normality)
(2L)	$\emptyset^{\perp}_{\rho} = \emptyset$	(Normality)
(2U)	$\emptyset^{\top}_{\rho} = \emptyset$	(Normality)
(3L)	$X^{\perp}_{\rho} \subseteq X$	(Contraction)
(3U)	$X \subseteq X^{\top}_{\rho}$	(Extension)
(4L)	$(X \cap Y)^{\perp}_{\rho} = X^{\perp}_{\rho} \cap Y^{\perp}_{\rho}$	(Multiplication)
(4U)	$(X \cup Y)^{\top}_{\rho} = X^{\top}_{\rho} \cup Y^{\top}_{\rho}$	(Addition)
(5L)	$(X^{\perp}_{\rho})^{\perp}_{\rho} = X^{\perp}_{\rho}$	(Idempotency)
(5U)	$(X^{\top}_{\rho})^{\top}_{\rho} = X^{\top}_{\rho}$	(Idempotency)
(5La)	$X^{\perp}_{\rho} \subseteq (X^{\perp}_{\rho})^{\perp}_{\rho}$	
(5Ua)	$(X^{\top}_{\rho})^{\top}_{\rho} \subseteq X^{\top}_{\rho}$	
(6L)	$X \subseteq Y \rightarrow X^{\perp}_{\rho} \subseteq Y^{\perp}_{\rho}$	(Monotone)
(6U)	$X \subseteq Y \rightarrow X^{\top}_{\rho} \subseteq Y^{\top}_{\rho}$	(Monotone)
(7L)	$(\overline{X^{\perp}_{\rho}})^{\perp}_{\rho} = \overline{X^{\perp}_{\rho}}$	(Lower-complement relation)
(7U)	$(\overline{X^{\top}_{\rho}})^{\top}_{\rho} = \overline{X^{\top}_{\rho}}$	(Upper-complement relation)
(8L)	$\overline{X}^{\perp}_{\rho} = \overline{\overline{X}^{\top}_{\rho}}$	(Duality)
(8U)	$\overline{X}^{\top}_{\rho} = \overline{\overline{X}^{\perp}_{\rho}}$	(Duality)
(9L)	$(X \cup Y)^{\perp}_{\rho} \supseteq X^{\perp}_{\rho} \cup Y^{\perp}_{\rho}$	
(9U)	$(X \cap Y)^{\top}_{\rho} \subseteq X^{\top}_{\rho} \cap Y^{\top}_{\rho}$	
(10LU)	$X^{\perp}_{\rho} \subseteq X^{\top}_{\rho}$	(Appropriateness)
(11L)	$X \subseteq (X^{\top}_{\rho})^{\perp}_{\rho}$	
(11U)	$(X^{\perp}_{\rho})^{\top}_{\rho} \subseteq X$	

Proof.
(3L): If $x \in X_\rho^\perp$, then $[x]_\rho \subseteq X$; since $x \in [x]_\rho$ we get $x \in X$.
(3U): If $x \in X$ then $[x]_\rho \cap X \neq \emptyset$, because $x \in [x]_\rho \cap X$; so $x \in X_\rho^\top$.
(4U): We have
$$\begin{aligned} x \in (X \cup Y)_\rho^\top &\leftrightarrow [x]_\rho \cap (X \cup Y) \neq \emptyset \leftrightarrow [x]_\rho \cap X \cup [x]_\rho \cap Y \neq \emptyset \\ &\leftrightarrow [x]_\rho \cap X \neq \emptyset \vee [x]_\rho \cap Y \neq \emptyset \\ &\leftrightarrow x \in X_\rho^\top \vee x \in Y_\rho^\top \\ &\leftrightarrow x \in X_\rho^\top \cup Y_\rho^\top. \end{aligned}$$
(6U): $X \subseteq Y \leftrightarrow X \cup Y = Y$, so $(X \cup Y)_\rho^\top = Y_\rho^\top$, and $(X \cup Y)_\rho^\top = X_\rho^\top \cup Y_\rho^\top$ by (4U). Therefore, $X_\rho^\top \subseteq Y_\rho^\top$.
(8L): $x \in \overline{X}_\rho^\perp \leftrightarrow [x]_\rho \subseteq X \leftrightarrow [x]_\rho \cap \overline{X} = \emptyset \leftrightarrow x \notin \overline{X}_\rho^\top \leftrightarrow x \in \overline{\overline{X}_\rho^\top}$.
(9U): From $X \cap Y \subseteq X$ and $X \cap Y \subseteq Y$, using (6U) we get $(X \cap Y)_\rho^\top \subseteq X_\rho^\top$ and $(X \cap Y)_\rho^\top \subseteq Y_\rho^\top$. So $(X \cap Y)_\rho^\top \subseteq X_\rho^\top \cap Y_\rho^\top$.
(11L): Let $x \in X$ and $y \in x\rho$; then, by the symmetry of ρ, $x \in y\rho$, i.e. $x \in y\rho \cap X$. So $y \in X_\rho^\top$, whence $x\rho \subseteq X_\rho^\top$. Therefore, $x \in (X_\rho^\top)_\rho^\perp$. □

Definition 13.7. *Let $\langle U, \rho \rangle$ be an approximation space. The* **accuracy of the ρ-approximation** *of $V \subseteq U$ is the number*

$$acc_\rho^U = \frac{|V_\rho^\perp|}{|V_\rho^\top|}$$

Obviously, $0 \leq acc_\rho^U \leq 1$, and $acc_\rho^U = 1$ for ρ-definable sets.

13.2.2 *Generalized Relation Based Rough Sets*

In Section 13.1 we gave some hints about the motivation on why classical rough set theory as originally defined by Pawlak [12] is based on equivalence relations. This gives nice mathematical properties but may impose requirements too strong for some situations in practice. If we have a closer look at the resulting Def. 13.4 for the lower and upper approximation then it is easy to see that we can use mathematically equivalent characterizations as follows:

$$\begin{aligned} V_\rho^\perp &= \bigcup\{[x]_\rho \in U/\rho : [x]_\rho \subseteq V\} \\ &= \{x \in U : [x]_\rho \subseteq V\} \\ &= \{x \in U : \forall y \in U[x\rho y \rightarrow y \in V]\}; \end{aligned} \tag{13.1}$$

$$\begin{aligned} V_\rho^\top &= \bigcup\{[x]_\rho \in U/\rho : [x]_\rho \cap V \neq \emptyset\} \\ &= \{x \in U : [x]_\rho \cap V \neq \emptyset\} \\ &= \{x \in U : \exists y \in U[x\rho y \wedge y \in V]\}. \end{aligned} \tag{13.2}$$

Each of the equivalent definitions in equations (13.1) and (13.2) captures different aspects: the first two are based on a special kind of covering, viz. the partition of an equivalence relation, the third explicitly uses the equivalence relation itself, instead of the induced equivalence classes. So we get two straightforward ways to generalize the Pawlak definition rough sets: We can either ease restrictions on the relation or on the covering by e.g. using tolerance or arbitrary binary relations, or by introducing other models of covering-based rough sets. If we replace the partition U/ρ with a covering $\mathscr{C}$ of U, then $\langle U, \mathscr{C}\rangle$ is called a **covering approximation space** (CA-space).)

Both possibilities have attracted many researchers during the last two decades, and RST is still a very active area of investigation, resulting in a really very great number of contributions in journals and edited volumes. As consequence thereof we can only present the main basic ideas and results on the generalizations of Pawlak RST in this chapter and have to refer interested readers to the original literature as e.g. [3, 17] for covering-based rough sets, [16, 21] for relation-based generalizations, and [20, 23] for relationships between them.

13.2.2.1 *Using Arbitrary Relations*

Definition 13.8. ([9]) *Given a binary relation ρ on U; an element $x \in U$ with $x\rho = \emptyset$ is called a* **solitary element** *with respect to ρ. The set of all solitary elements with respect to ρ is called the* **solitary set** *and is denoted by S_ρ.*

Remember from Def. 4.13 that $[x]_\rho$ is just a special notation for foresets $x\rho$, in case ρ is an equivalence relation. This gives a direct path to the following generalization inspired by formulas (13.1) and (13.2):

Definition 13.9.

(a) *Let ρ be a relation on U. The pair $\langle U, \rho\rangle$ is called a* **generalized approximation space** *(GA-space).*

(b) *Let $\langle U, \rho\rangle$ be a GA space, and $V \subseteq U$. The lower ρ-approximation of V is the set obtained by taking those elements whose foresets $x\rho$ are included in the set V:*

$$V_\rho^\perp = \{x \in U : x\rho \subseteq V\} = \{x \in U : \forall y \in U[x\rho y \rightarrow y \in V]\}.$$

(c) *Let $\langle U, \rho\rangle$ be a GA-space, and $V \subseteq U$. The* **upper ρ-approximation** *of V is the set obtained by taking those elements whose foresets $x\rho$ have a non-empty intersection with the set V:*

$$V_\rho^\top = \{x \in U : x\rho \cap V \neq \emptyset\} = \{x \in U : \exists y \in U[x\rho y \wedge y \in V]\}.$$

Note that, by definition, we have $x \in \{y\}_\rho^\top \iff x\rho \cap \{y\} \neq \emptyset \iff y \in x\rho$. The binary relation ρ can thus be reconstructed from upper approximations of singleton subsets of U: $x\rho = \{y : x \in \{y\}_\rho^\top\}$.

There is a wide range from an arbitrary binary relation with no restrictions imposed at all to an equivalence relation as used in Pawlak RST. It is interesting to see what properties of approximations are induced by which properties of relations. We follow the results given in [7, 15, 16, 21]. Numberings like (2U) etc. used below refer to Th. 13.6.

Proposition 13.10. *For a GA-space $\langle U, \rho\rangle$ the following properties hold:*

(1L)	$U^{\perp}_{\rho} = U$		$U^{\top}_{\rho} = \overline{\mathsf{S}_{\rho}}$
	$\emptyset^{\perp}_{\rho} = \mathsf{S}_{\rho}$	(2U)	$\emptyset^{\top}_{\rho} = \emptyset$
(4L)	$(X \cap Y)^{\perp}_{\rho} = X^{\perp}_{\rho} \cap Y^{\perp}_{\rho}$	(4U)	$(X \cup Y)^{\top}_{\rho} = X^{\top}_{\rho} \cup Y^{\top}_{\rho}$
(6L)	$X \subseteq Y \rightarrow X^{\perp}_{\rho} \subseteq Y^{\perp}_{\rho}$	(6U)	$X \subseteq Y \rightarrow X^{\top}_{\rho} \subseteq Y^{\top}_{\rho}$
(8L)	$\overline{X}^{\perp}_{\rho} = \overline{X^{\top}_{\rho}}$	(8U)	$\overline{X}^{\top}_{\rho} = \overline{X^{\perp}_{\rho}}$
(9L)	$(X \cup Y)^{\perp}_{\rho} \supseteq X^{\perp}_{\rho} \cup Y^{\perp}_{\rho}$	(9U)	$(X \cap Y)^{\top}_{\rho} \subseteq X^{\top}_{\rho} \cap Y^{\top}_{\rho}$

Proof. (1L): Let $x \in U$; then we have $x\rho \subseteq U$, whence $x \in U^{\perp}_{\rho}$. Thus $U \subseteq U^{\perp}_{\rho}$. But by definition also $U^{\perp}_{\rho} \subseteq U$.

$$
\begin{aligned}
\text{(4L):}\quad (X \cap Y)^{\perp}_{\rho} &= \{x \in U : x\rho \subseteq X \cap Y\} \\
&= \{x \in U : x\rho \subseteq X \wedge x\rho \subseteq Y\} \\
&= \{x \in U : x\rho \subseteq X\} \cap \{x \in U : x\rho \subseteq Y\} \\
&= X^{\perp}_{\rho} \cap Y^{\perp}_{\rho}.
\end{aligned}
$$

(6L): Let $X \subseteq Y$ and $x \in X^{\perp}_{\rho}$; then $x\rho \subseteq X$, whence $x\rho \subseteq Y$. Thus $x \in Y^{\perp}_{\rho}$, and finally $X^{\perp}_{\rho} \subseteq Y^{\perp}_{\rho}$.

$$
\begin{aligned}
\text{(8L):}\quad \overline{\overline{X}^{\top}_{\rho}} &= \overline{\{x \in U : x\rho \cap \overline{X} \neq \emptyset\}} \\
&= \{x \in U : x\rho \cap \overline{X} = \emptyset\} \\
&= \{x \in U : x\rho \subseteq X\} \\
&= X^{\perp}_{\rho}.
\end{aligned}
$$

(9L): From $X \subseteq X \cup Y$ and $Y \subseteq X \cup Y$ we get $X^{\perp}_{\rho} \subseteq (X \cup Y)^{\perp}_{\rho}$ and $Y^{\perp}_{\rho} \subseteq (X \cup Y)^{\perp}_{\rho}$; thus $X^{\perp}_{\rho} \cup Y^{\perp}_{\rho} \subseteq (X \cup Y)^{\perp}_{\rho}$. □

We take the following simple example from [21] to show that arbitrary relations may fail to satisfy elementary properties requested for approximation operators. Let $U = \{a, b, c\}$ and $\rho = \{\langle a, b\rangle, \langle b, c\rangle\}$:

(3L) $\emptyset^{\perp}_{\rho} = \{c\} = \mathsf{S}_{\rho} \supset \emptyset$;
(5L) $(\{a, c\}^{\perp}_{\rho})^{\perp}_{\rho} = \{b, c\}^{\perp}_{\rho} = U \neq \{a, c\}^{\perp}_{\rho} = \{b, c\}$;
(10LU) $\emptyset^{\perp}_{\rho} = \{c\} \supset \emptyset = \emptyset^{\top}_{\rho}$.

By duality, also properties (3U) and (5U) may fail. So taking an arbitrary relation is not of any real interest, but only mathematically instructive. Let us now have a look at relation properties that make a GA-space a proper approximation space.

13.2.2.2 *Using Special Relations*

Proposition 13.11. *Let $\langle U, \rho\rangle$ be a GA-space; then following statements are equivalent:*

(1) ρ is serial;
(2) $X^{\perp}_{\rho} \subseteq X^{\top}_{\rho}$ for all $X \subseteq U$, i.e. (10LU);
(3) $\emptyset^{\perp}_{\rho} = \emptyset$, i.e. (2L);
(4) $U^{\top}_{\rho} = U$, i.e. (1U).

Proof. (1) ⇒ (2): Let $x \in X_\rho^\perp$; then $x\rho \subseteq X$, whence $x\rho \cap X = x\rho \neq \emptyset$. Thus $x \in X_\rho^\top$.
(2) ⇒ (1): Assume that ρ is not serial, i.e. $\exists x \in U[x\rho = \emptyset]$. This means $x \in X_\rho^\perp$, but $x \notin X_\rho^\top$ for any $X \subseteq U$, a contradiction to $X_\rho^\perp \subseteq X_\rho^\top$ for all $X \subseteq U$.
(2) $\Longleftrightarrow$ (3): Using Prop. 13.10 we have

$$\begin{aligned} X_\rho^\perp \subseteq X_\rho^\top &\leftrightarrow X_\rho^\perp \cap \overline{X_\rho^\top} = \emptyset \\ &\leftrightarrow X_\rho^\perp \cap \overline{X}_\rho^\perp = \emptyset \\ &\leftrightarrow (X \cap \overline{X})_\rho^\perp = \emptyset \\ &\leftrightarrow \emptyset_\rho^\perp = \emptyset \end{aligned}$$

(3) $\Longleftrightarrow$ (4): Using Prop. 13.10(8U) we get $\emptyset_\rho^\perp = \emptyset$ iff $\overline{\emptyset_\rho^\perp} = \overline{\emptyset}$ iff $U_\rho^\top = U$. □

Proposition 13.12. ([7, Proposition 137]) *Let $\langle U, \rho\rangle$ be a GA-space; then the following statements are equivalent:*

(1) ρ is reflexive;
(2) $X \subseteq X_\rho^\top$ for all $X \subseteq U$, i.e. (3U);
(3) $X_\rho^\perp \subseteq X$ for all $X \subseteq U$, i.e. (3L);

Proof. (1) ⇒ (2): Reflexivity of ρ implies $\forall x \in U[x \in x\rho]$. Hence $x \in X$ implies $x \in x\rho \cap X \neq \emptyset$, whence $x \in X_\rho^\top$.

(2) ⇒ (3): We have $\overline{X} \subseteq \overline{X}_\rho^\top$ and by Prop. 13.10(8U) $\overline{X}_\rho^\top = \overline{X_\rho^\perp}$. Hence $\overline{X} \subseteq \overline{X_\rho^\perp}$, which is equivalent to $X_\rho^\perp \subseteq X$.
(3) ⇒ (1): Assume ρ to be not reflexive, i.e. there is an x such that $\langle x, x\rangle \notin \rho$. Let $X = U - \{x\}$; since $X \subseteq U$ we have $X_\rho^\perp \subseteq X$ by assumption; but, by definition of X, we also have $\forall y \in U[x\rho y \rightarrow y \in X]$, i.e. $x\rho \subseteq X$. Thus, $x \in X_\rho^\perp$ by Def. 13.9 and $x \notin X$, a contradiction to $X_\rho^\perp \subseteq X$. □

Proposition 13.13. ([7, Proposition 139]) *Let $\langle U, \rho\rangle$ be a GA-space; then the following statements are equivalent:*

(1) ρ is transitive;
(2) $(X_\rho^\top)_\rho^\top \subseteq X_\rho^\top$ for all $X \subseteq U$, i.e. (5Ua);
(3) $X_\rho^\perp \subseteq (X_\rho^\perp)_\rho^\perp$ for all $X \subseteq U$, i.e. (5La);

Proof. (1) ⇒ (2): Let $x \in (X_\rho^\top)_\rho^\top$; then, by Def. 13.9, there is a $y \in X_\rho^\top$ with $x\rho y$. And since $y \in X_\rho^\top$, there is a $z \in X$ with $y\rho z$, whence $x\rho z$ by the transitivity of ρ. Hence $x \in X_\rho^\top$.
(2) ⇒ (3): By Prop. 13.10(8U) we get $\overline{(X_\rho^\perp)_\rho^\perp} = (\overline{X}_\rho^\top)_\rho^\top$, and $(\overline{X}_\rho^\top)_\rho^\top \subseteq \overline{X}_\rho^\top = \overline{X_\rho^\perp}$. Thus, $X_\rho^\perp \subseteq (X_\rho^\perp)_\rho^\perp$.
(3) ⇒ (1): Assume that ρ is not transitive, i.e. there exist $x, y, z \in U$ such that $x\rho y$, $y\rho z$, but not $x\rho z$. Consider the set $X = U - \{z\}$; then, by definition of X, $\forall t \in U[x\rho t \rightarrow t \in X]$, i.e. $x\rho \subseteq X$, whence $x \in X_\rho^\perp$. But since $z \notin X$, $y \notin X_\rho^\perp$ and hence $x \notin (X_\rho^\perp)_\rho^\perp$, a contradiction to $X_\rho^\perp \subseteq (X_\rho^\perp)_\rho^\perp$. □

From Prop. 13.10 - 13.13 we get the following

Corollary 13.14. *Let $\langle U, \rho\rangle$ be a GA-space; then the following statements are equivalent:*

(1) ρ is a preorder;
(2) The map $X \mapsto X_\rho^\top$ is a closure operator;
(3) The map $X \mapsto X_\rho^\perp$ is an interior operator.

Proof. We get idempotency by substituting $X_\rho^\top$ and $X_\rho^\perp$, respectively, for X in Prop. 13.12. □

In Table 13.3 we summarize the relationships between relation properties and the properties of the corresponding relation-based approximations for most items listed in Th. 13.6. See [1] and [4]. The # is a placeholder for $\perp$ or $\top$ with obvious meaning, and a • means that an operator has the corresponding property. It happens that always both of (nL) and (nU) are valid, so simply (n) suffices to denote both properties.

TABLE 13.3: Basic properties of relation-based approximations.

Property of ρ	(1)	(2)	(3)	(4)	(5a)	(6)	(8)	(9)	(10)	(11)
$X^\#_{arbitrary}$	•			•		•	•	•		
$X^\#_{reflexive}$	•	•	•	•		•	•	•	•	
$X^\#_{tolerance}$	•	•	•	•		•	•	•	•	•
$X^\#_{preorder}$	•	•	•	•	•	•	•	•	•	
$X^\#_{equivalence}$	•	•	•	•	•	•	•	•	•	•

References

[1] Allam, A. A., Bakeir, M. Y., Abo-Tabl, E. A.: New approach for basic rough set concepts. In Ślęzak, D., Wang, G. et al(Eds): *Rough Sets, Fuzzy Sets, Data Mining, and Granular Computing*, LNAI 3641, 64-73, Springer-Verlag Berlin Heidelberg (2005).

[2] Banerjee, M.: Algebraic Studies in Pawlak's Rough Set Theory, In *Reminiscing Ideas and Interactions*, Calcutta Logic Circle 2011, 13-31 (2011).

[3] Bian, X., Wang, P., Yu, Z., Bai, X., Cehn, B: Characterizations of coverings for upper approximation operators being closure operators, Information Sciences **314**, 41-54 (2015).

[4] Bloch, I.: On links between mathematical morphology and rough sets, Pattern Recognition **33**, 1487-1496 (2000).

[5] Demri, S. and Orłowska, E.: *Incomplete Information Structure, Inference, Complexity*, Springer-Verlag Berlin Heidelberg (2002).

[6] Ganter, B., Stumme, G. and Wille, R. (Eds): *Formal Concept Analysis*, Springer-Verlag Berlin Heidelberg (2005).

[7] Järvinen, J.: Lattice theory for rough sets. In: Peter, J.F., Skowron, A. et al.(Eds): *Transactions on Rough Sets VI*, LNCS 4374, 400-498, Springer-Verlag Berlin Heidelberg (2007).

[8] Khan, M. A., Banerjee, M.: Algebras for Information Systems. In: Skowron, A. and Suraj, Z.(Eds): *Rough Sets and Intelligent Systems-Professor Zdzisław Pawlak in Memoriam*, 381-407, Springer-Verlag Berlin Heidelberg (2013).

[9] Liu, G.,Zhu, W.: The algebraic structures of generalized rough set theory, Information Sciences **178**, 4105-4113 (2008).

[10] Orłowska, E. (Ed): *Incomplete Information: Rough Set Analysis*, Springer-Verlag Berlin Heidelberg (1998).

[11] Pawlak, Z.: Information system, theoretical foundations, Information Systems **6**, 205-218 (1981).

[12] Pawlak, Z.: Rough sets, Int. J. of Computer and Information Sciences **11**, 341-356 (1982).

[13] Pawlak, Z.: *Rough Sets-Theoretical Aspects of Reasoning about Data*, Springer Science+Business Media, Dordrecht (1991).

[14] Polkowski, L.: *Rough Sets-Mathematical Foundations*, Springer-Verlag Berlin Heidelberg (2002).

[15] Yang, L., Xu, L.: Algebraic aspects of generalized approximation spaces, Int. J. of Approximate Reasoning **51**, 151-161 (2009).

[16] Yao, Y.: Generalized rough set model. In: Polkowski, L., Skowron A. (Eds.), *Rough Sets in Knowledge Discovery 1. Methodology and Applications*, 286-318, Springer-Verlag Berlin Heidelberg (1998).

[17] Yao, Y. ,Yao, B.: Covering based rough set approximations, Information Sciences **200**, 91-107 (2012).

[18] Zadeh, L.A.: Fuzzy sets, Information and Control **8**, 338-353 (1965).

[19] Zadeh, L.A.: Fuzzy logic=computing with words, IEEE Transactions on Fuzzy Systems **4**, 103-111 (1996).

[20] Zhang, Y.-L., Luo, M.-K.: Relationships between covering-based rough sets and relation-based rough sets, Information Sciences **225**, 55-71 (2013).

[21] Zhu, W.: Generalized rough sets based on relations, Information Sciences **177**, 4997-5011 (2007).

[22] Zhu, W.: Relationship between generalized rough sets based on binary relation and covering, Information Sciences **179**, 210-225 (2009).

[23] Zhu, W.: Covering rough sets based on neighborhood: An approach without using neighborhoods , Int. J. of Approximate Reasoning **52**, 461-472 (2011).

[24] Zhu, W., Wang, F.-Y.: Reduction and axiomization of covering generalized rough sets, Information Sciences **152**, 217-230 (2003).

Chapter 14

Formal Concept Analysis

From a psychological and philosophical point of view, *concepts* can be seen "*as the basic units of thought formed in dynamic processes within social and cultural environments*" ([5, p. 2]). And how detailed and sophisticated formal models we may design, at the bottom of such a model we must somehow account for the old philosophical tradition (mentioned already at the beginning of Section 13.1), that a concept is constituted by its extension, comprising all objects for which the concept can correctly be asserted, and its intension, including all attributes (properties, meanings) which apply to all objects of the extension.

As we have already seen in Chapter 12, concepts live in relationships with many other concepts. The *subconcept-superconcept-relation* plays a prominent role. Being a subconcept of a superconcept means that the extension of the subconcept is contained in the extension of the superconcept; this is equivalent to the relationship that the intension of the subconcept contains the intension of the superconcept. *Formal Concept Analysis* (FCA) provides a mathematical model that allows to speak about *objects, attributes,* and about objects *having* certain attributes. Moreover it allows to deal with concept hierarchies.

14.1 The Basic Notions and Algebra of FCA

FCA is considered an important formalism for knowledge representation, extraction and analysis, which assumes that there are given a finite set of objects, a finite set of yes-or-no attributes, and a binary relation between these two sets. The triplet made up of the two sets and the relation is called a *formal context* and is one of the basic notions in FCA. A formal context represents the data to be analyzed and can, at least in principle, be depicted by a table describing a relationship between objects (represented by table rows) and attributes (represented by table columns).

The primary aim of FCA is to extract from the input data (1) all relevant *formal concepts,* along with the subconcept-superconcept hierarchy (called the *concept lattice*) and (2) a set of particular attribute dependencies (called *attribute implications*). In our presentation we follow [1, 2, 6, 9]. Basic for both aims is the notion of a *formal context* as given in

Definition 14.1. *Let G and M be two disjoint finite sets, and $I \subseteq G \times M$. Each element of G is called an object (in German:* Gegenstand*), and each element of M is called an attribute (in German:* Merkmal*); gIm is read as "object g has attribute m". The triple $\mathbb{C} = \langle G, M, I\rangle$ is called a* **formal context**. *The relation I is referred to as the* **incidence relation** *of the context.*

Every formal context $\mathbb{C}$ induces a pair of so-called *derivation operators*, which are used for defining the formal concepts of $\mathbb{C}$:

Definition 14.2. *Let $\mathbb{C} = \langle G, M, I\rangle$ be a formal context. Two operators $^{\uparrow} : \mathbf{2}^G \longrightarrow \mathbf{2}^M$ and $^{\downarrow} : \mathbf{2}^M \longrightarrow \mathbf{2}^G$ are defined for every $A \subseteq G$ and $B \subseteq M$ by*

(a) $A^{\uparrow} = \{m \in M : \forall g \in A[gIm]\}$;
(b) $B^{\downarrow} = \{g \in G : \forall m \in B[gIm]\}$;
(c) A **formal concept** *in $\mathbb{C}$ is a pair $\langle A, B\rangle$, such that $A^{\uparrow} = B$ and $B^{\downarrow} = A$.*

For a formal concept $\langle A, B\rangle$ in $\langle G, M, I\rangle$, A and B are called the **extent** and **intent** of $\langle A, B\rangle$, respectively. Put verbally, a pair $\langle A, B\rangle$ is a formal concept of a formal context$\langle G, M, I\rangle$ if and only if A contains the objects sharing all attributes from B and B contains the attributes shared by all objects from A. (As a mnemonic trick you might conceive of $^{\uparrow}$ as pointing from the world of solid real objects up to the sphere of mental constructs, and for $^{\downarrow}$ the other way round.)

The derivation operators have several natural properties, among them: $A_1 \subseteq A_2 \rightarrow A_2{}^{\uparrow} \subseteq A_1{}^{\uparrow}$ (this is just the well-known property that the more objects there are, the fewer attributes they have in common), and the dual counterpart $B_1 \subseteq B_2 \rightarrow B_2{}^{\downarrow} \subseteq B_1{}^{\downarrow}$. From the foregoing it is clear that formal concepts can be seen as models of concepts in the classical sense: a concept is given by a collection of properties, and the concept applies to an object if and only if the object satisfies every property in the collection.

(Note: To make things more convenient, we will usually drop the word "*formal*" from now on since it will always be clear if we speak about a formal or informal context or concept.)

TABLE 14.1: An example illustrating FCA concepts.

Obj./Attr.	m_1	m_2	m_3	m_4
o_1	1	1	1	1
o_2	1		1	1
o_3		1	1	1
o_4		1	1	1
o_5	1			

Table 14.1 (borrowed from [1, Example 2.6]) shows an example in form of a so-called **cross table** to illustrate the concepts introduced so far. (Note that cross tables are nothing else than specific representations of Boolean matrices; so we use 1's in stead of ×

symbols.) We have:

- $\{o_2\}^{\uparrow} = \{m_1, m_3, m_4\}$, $\{o_2, o_3\}^{\uparrow} = \{m_3, m_4\}$,
- $\{o_1, o_4, o_5\}^{\uparrow} = \emptyset$,
- $X^{\uparrow} = \emptyset$, $\emptyset^{\uparrow} = M$,
- $\{m_1\}^{\downarrow} = \{o_1, o_2, o_5\}$, $\{m_1, m_2\}^{\downarrow} = \{o_1\}$,
- $\{m_2, m_3\}^{\downarrow} = \{o_1, o_3, o_4\}$, $\{m_2, m_3, m_4\}^{\downarrow} = \{o_1, o_3, o_4\}$,
- $\emptyset^{\downarrow} = G$, $M^{\downarrow} = \{o_1\}$.
- The shaded rectangle with the boldface **1**'s represents the concept $\langle A_1, B_1\rangle = \langle\{o_1, o_2, o_3, o_4\}, \{m_3, m_4\}\rangle$.

Here are three more formal concepts occuring in the above table:

G/M	m_1	m_2	m_3	m_4
o_1	1	**1**	**1**	**1**
o_2	1		1	1
o_3		**1**	**1**	**1**
o_4		**1**	**1**	**1**
o_5	1			

G/M	m_1	m_2	m_3	m_4
o_1	**1**	1	**1**	**1**
o_2	**1**		**1**	**1**
o_3		1	1	1
o_4		1	1	1
o_5	1			

G/M	m_1	m_2	m_3	m_4
o_1	**1**	1	1	1
o_2	**1**		1	1
o_3		1	1	1
o_4		1	1	1
o_5	**1**			

As you probably do suspect from the given examples, formal concepts correspond to maximal rectangles in the input data table filled with **1** (more precisely, they correspond to rectangles up to a permutation of rows and columns of the input data table).

As can be seen from Def. 14.2 the extent A and the intent B of a concept $\langle A, B\rangle$ are closely connected by the incidence relation I. Prop. 14.3 gives some simple properties of this relationship.

Proposition 14.3. *Let* $\mathbb{C} = \langle G, M, I\rangle$ *be a context,* $A, A_1, A_2 \subseteq G$ *be sets of objects and* $B, B_1, B_2 \subseteq M$ *be sets of attributes. Then the following holds:*

(1G)	$A_1 \subseteq A_2 \rightarrow A_2{}^{\uparrow} \subseteq A_1{}^{\uparrow}$	(1M)	$B_1 \subseteq B_2 \rightarrow B_2{}^{\downarrow} \subseteq B_1{}^{\downarrow}$
(2G)	$A \subseteq A^{\uparrow\downarrow}$	(2M)	$B \subseteq B^{\downarrow\uparrow}$
(3G)	$A^{\uparrow} = A^{\uparrow\downarrow\uparrow}$	(3M)	$B^{\downarrow} = B^{\downarrow\uparrow\downarrow}$

(4GM) $A \subseteq B^{\downarrow} \leftrightarrow B \subseteq A^{\uparrow} \leftrightarrow A \times B \subseteq I.$

Proof. (1G): If $m \in A_2{}^{\uparrow}$, then $\forall g \in A_2[gIm]$, i.e. in particular $\forall g \in A_1[gIm]$, since $A_1 \subseteq A_2$, whence $m \in A_1{}^{\uparrow}$.
(2G): For $g \in A$ we have $\forall m \in A^{\uparrow}[gIm]$, whence $g \in A^{\uparrow\downarrow}$.
(3G): From (2G) we have $A \subseteq A^{\uparrow\downarrow}$, which together with (1G) gives $A^{\uparrow} \supseteq A^{\uparrow\downarrow\uparrow}$. The reverse inclusion follows directly from (2M) with $B := A^{\uparrow}$. □

Prop. 14.3 shows that the derivation operators establish a Galois connection between $(\mathbf{2}^G, \subseteq)$ and $(\mathbf{2}^M, \subseteq)$ (see Def. 2.81(a)). The following proposition is thus an immediate consequence of Prop. 2.84.

Proposition 14.4. ([6, Proposition 11]) *Let $A_j \subseteq G$ be sets of objects for all $j \in J$. Then the following holds:*

$$(\bigcup_{j\in J} A_j)^{\uparrow} = \bigcap_{j\in J} A_j{}^{\uparrow}$$

Proof. We have $m \in \left(\bigcup_{j\in J} A_j\right)^{\uparrow} \iff gIm$ for all $g \in \bigcup_{j\in J} A_j \iff gIm$ for all $g \in A_j$ for all $j \in J \iff m \in A_j{}^{\uparrow}$ for all $j \in J \iff m \in \bigcap_{j\in J} A_j{}^{\uparrow}$. □

Definition 14.5. *Let $\mathbb{C} = \langle G, M, I\rangle$ be a formal context, and $\langle A_1, B_1\rangle$ and $\langle A_2, B_2\rangle$ be two concepts in $\mathbb{C}$.*

(a) *$\langle A_1, B_1\rangle$ is called a* **subconcept** *of $\langle A_2, B_2\rangle$ (and accordingly $\langle A_2, B_2\rangle$ a* **superconcept** *of $\langle A_1, B_1\rangle$), if $A_1 \subseteq A_2$ (which is equivalent to $B_2 \subseteq B_1$). This will be denoted by $\langle A_1, B_1\rangle \leq \langle A_2, B_2\rangle$.*

(b) *The set of all formal concepts in $\mathbb{C}$ ordered by $\leq$ is denoted by $\mathfrak{B}(G, M, I)$ (from the German noun $\mathfrak{B}$egriff), or in short by $\mathfrak{B}(\mathbb{C})$, and is called the* **concept lattice** *of $\mathbb{C}$:*

$$\mathfrak{B}(G, M, I) = \{\langle A, B\rangle \in \mathbf{2}^G \times \mathbf{2}^M : A^{\uparrow} = B \wedge B^{\downarrow} = A\}.$$

(c) *We denote the set of all extents and intents, respectively, of the context $\mathbb{C} = \langle G, M, I\rangle$ by*

$$\mathfrak{E}(G, M, I) = \mathfrak{E}(\mathbb{C}) = \{A \in \mathbf{2}^G : \langle A, B\rangle \in \mathfrak{B}(G, M, I)\},$$
$$\mathfrak{I}(G, M, I) = \mathfrak{I}(\mathbb{C}) = \{B \in \mathbf{2}^M : \langle A, B\rangle \in \mathfrak{B}(G, M, I)\}.$$

Def. 14.2 and Def. 14.5 together form a mathematical model of the well-known notion of a concept which goes back to Port-Royal logic: A concept is determined by a collection of objects (extent) which fall under the concept and a collection of attributes (intent) covered by the concept. Concepts are naturally ordered by a subconcept-superconcept relation which in turn is based on inclusion on the sets of objects and attributes. So $\langle A_1, B_1\rangle \leq \langle A_2, B_2\rangle$ means that $\langle A_1, B_1\rangle$ is more specific than $\langle A_2, B_2\rangle$, or that $\langle A_2, B_2\rangle$ is more general, e.g. CAT $\leq$ MAMMAL, or BIRD $\leq$ VERTEBRATE.

For the formulation of the fundamental result on concept lattices we need the following notions: A set of elements $S \subseteq L$ of a complete lattice L is called *supremum-dense* in L if every lattice element is a supremum of elements from this set, i.e. $L = \{\bigvee T \mid T \subseteq S\}$. Dually, a set $S \subseteq L$ is called *infimum-dense* in L if $L = \{\bigwedge T \mid T \subseteq S\}$.

Theorem 14.6. ([6, Theorem 3], [8, p. 449]) (The Basic Theorem on Concept Lattices)

(1) *$\mathfrak{B}(G, M, I)$ is a complete lattice with infima and suprema given by*

$$\bigwedge_{j\in J} \langle A_j, B_j\rangle = \langle \bigcap_{j\in J} A_j, (\bigcup_{j\in J} B_j)^{\downarrow\uparrow}\rangle \tag{14.1}$$

$$\bigvee_{j\in J} \langle A_j, B_j\rangle = \langle (\bigcup_{j\in J} A_j)^{\uparrow\downarrow}, \bigcap_{j\in J} B_j\rangle. \tag{14.2}$$

(2) *Conversely, a complete lattice $(L, \leq)$ is isomorphic to $\mathfrak{B}(G, M, I)$ iff there are mappings $\varphi : G \longrightarrow L$ and $\psi : M \longrightarrow L$ such that*

(a) $\varphi(G)$ *is supremum-dense in* L, *and* $\psi(M)$ *is infimum-dense in* L;
(b) $gIm \leftrightarrow \varphi(g) \leq \psi(m)$ *for all* $g \in G$ *and* $m \in M$.

The first part of Th. 14.6 gives the precise formulation for the infimum and supremum of arbitrary sets of formal concepts, and the second says that concept lattices do not have any special properties: every complete lattice is (isomorphic to) a concept lattice. In other words, for every complete lattice $(L, \leq)$ we can find a a set G of objects, a set M of attributes, and a suitable relation I, such that the given lattice is isomorphic to $\mathfrak{B}(G, M, I)$.

14.2 Attribute Implications

Above we have said that discovering attribute implications is a second important goal of FCA, since they represent data dependencies such as: "*Every number divisible by 2 and 3 is divisible by 6.*" or "*Every patient which exhibits a set of symptoms S, also has symptoms* s_1 *and* s_5.". In this section, we will give a very short overview only in order to give an idea about this aspect, and we begin by formally introducing implications in contexts.

Definition 14.7. *Let* $\mathbb{C} = \langle G, M, I\rangle$ *be a formal context.*

(a) *An* **attribute implication** *over* M *is an ordered pair written as* $A \Rightarrow B$, *with* $A, B \subseteq M$.
(b) $T \subseteq M$ *is an* **attribute model** *of* $A \Rightarrow B$, *denoted by* $T \models A \Rightarrow B$, *if* $A \not\subseteq T$ *or* $B \subseteq T$, *and is a model of a set of implications* $L = \{l_1, l_2, \ldots, l_n\}$, $T \models L$, *if* $T \models l_i$, *for all* $i \in \underline{n}$.
(c) *An attribute implication* $A \Rightarrow B$ *over* M *holds in* $\mathbb{C} = \langle G, M, I\rangle$ *if* $\{g\}^{\uparrow} \models A \Rightarrow B$, *for all* $g \in G$, *i.e. if it holds in the set of the objects' intents. In this case, we say that* $A \Rightarrow B$ *is an implication of* $\mathbb{C}$.

As an example consider the following context, dealing with some animals, together with the attribute implications that hold in it:

	feathers	hairy	fly	hunt	run
hawk	1		1	1	
eagle	1		1	1	
fox		1		1	1
dog		1			1

{fly} ⇒ {feathers, hunt}
{hairy} ⇒ {run}
{hunt, fly} ⇒ {feathers}

Proposition 14.8. ([7, 3.3.3 Proposition]) *An attribute implication* $A \Rightarrow B$ *over* M *holds in* $\mathbb{C} = \langle G, M, I\rangle$ *iff* $B \subseteq A^{\downarrow\uparrow}$. *It then also holds in* $\mathfrak{I}(G, M, I)$.

Proof. Let $A \Rightarrow B$ be an implication of the context $\langle G, M, I\rangle$. According to Prop. 14.4

$$A^{\downarrow\uparrow} = (A^{\downarrow})^{\uparrow} = \left(\bigcup_{g\in A^{\downarrow}}\{g\}\right)^{\uparrow} = \bigcap_{g\in A^{\downarrow}} g^{\uparrow}.$$

So $A \subseteq g^{\uparrow}$ for all $g \in A^{\downarrow}$ by Prop. 14.3(2M). Then, since $A \Rightarrow B$ holds in $\langle G, M, I\rangle$, $A \not\subseteq g^{\uparrow}$ or $B \subseteq g^{\uparrow}$, for all $g \in G$. In particular, this implies that $B \subseteq g^{\uparrow}$ for all $g \in A^{\downarrow}$, so $B \subseteq \bigcap_{g\in A^{\downarrow}} g^{\uparrow} = A^{\downarrow\uparrow}$. Conversely assume that $B \subseteq A^{\downarrow\uparrow}$. If $A \not\subseteq g^{\uparrow}$, we are done. Otherwise, $A \subseteq g^{\uparrow}$ implies $A^{\downarrow\uparrow} \subseteq g^{\uparrow\downarrow\uparrow} = g^{\uparrow}$ by Prop. 14.3(1M), (1G) and (3G), so $B \subseteq g^{\uparrow}$.

Now suppose that $A \Rightarrow B$ holds in $\langle G, M, I\rangle$. Then $B \subseteq A^{\downarrow\uparrow}$ by the first part of the proof. For every intent N of $\langle G, M, I\rangle$ either $A \not\subseteq N$ or $A \subseteq N$. If the latter holds, we have to show $B \subseteq N$: As an intent, $N = F^{\uparrow}$ for some $F \subseteq G$; and from $A \subseteq N$ we get $A^{\downarrow\uparrow} \subseteq N^{\downarrow\uparrow} = F^{\uparrow\downarrow\uparrow} = F^{\uparrow} = N$ using Prop. 14.3(1),(3G). Therefore $B \subseteq N$, whence $A \Rightarrow B$ holds in $\mathfrak{I}(G, M, I)$. □

By Prop. 14.3(4GM), the condition $B \subseteq (A^{\downarrow})^{\uparrow}$ of the above proposition is equivalent to $\{g \in G : \forall m \in A[gIm]\} = A^{\downarrow} \subseteq B^{\downarrow} = \{g \in G : \forall m \in B[gIm]\}$. This means that $A \Rightarrow B$ holds in a context $\langle G, M, I\rangle$ if and only if every object having all the attributes in A must also have all the attributes in B, which corresponds to the familiar notion of a valid implication.

An interesting question is for a procedure that allows to read off implications from a concept lattice. By the next lemma it is shown that it is sufficient to know a way for implications of the form $A \Rightarrow \{m\}$, henceforth simply written as $A \Rightarrow m$.

Lemma 14.9. *$A \Rightarrow B$ holds in $\mathfrak{I}(G, M, I)$ iff $A \Rightarrow m$ holds in $\mathfrak{I}(G, M, I)$ for each $m \in B$.*

Proof. By Prop. 14.8 we have that $A \Rightarrow B$ holds in $\mathfrak{I}(G, M, I)$ iff $B \subseteq A^{\downarrow\uparrow}$ iff $\forall m \in B[\{m\} \subseteq A^{\downarrow\uparrow}$ iff $\forall m \in B[A \Rightarrow m$ holds in $\mathfrak{I}(G, M, I)]$. □

We finally like to mention that one can show that the concept lattice can be inferred from the attribute logic i.e. the rules between certain sets of attributes. This is particularly convenient in the case of contexts with a large object and a relatively small attribute set: Instead of having to laboriously write down all the objects, we can simply infer the concept lattice from the rules between sets of attributes: Let $\mathsf{Md}(L)$ denote the set of all models of a set L of implications in M, i.e. $\mathsf{Md}(L) = \{X \subseteq M : X \models L\}$

Proposition 14.10. ([7, 3.3.8 Proposition]) *If $\mathsf{Imp}(G, M, I)$ denotes the set of all implications of a context $\langle G, M, I\rangle$, then $\mathsf{Md}(\mathsf{Imp}(G, M, I)) = \mathfrak{I}(G, M, I)$.*

Proof. By Prop. 14.3(2M) we have $B \subseteq B^{\downarrow\uparrow}$ for all $B \subseteq M$, so by Prop. 14.8 we get $\mathfrak{I}(G, M, I) \subseteq \mathsf{Md}(\mathsf{Imp}(G, M, I))$. We will now show that every $X \in \mathsf{Md}(\mathsf{Imp}(G, M, I))$ is an intent of $\langle G, M, I\rangle$. From $X^{\downarrow\uparrow} \subseteq X^{\downarrow\uparrow}$ we get that $X \Rightarrow X^{\downarrow\uparrow}$ holds in $\langle G, M, I\rangle$ by Prop. 14.8, so $X \Rightarrow X^{\downarrow\uparrow} \in \mathsf{Imp}(G, M, I)$. Therefore, $X \models X \Rightarrow X^{\downarrow\uparrow}$, whence, by definition of $\models$, $X \not\subseteq X$ or $X^{\downarrow\uparrow} \subseteq X$, so $X^{\downarrow\uparrow} \subseteq X$, which together with Prop. 14.3(2M) yields $X = X^{\downarrow\uparrow}$. Thus by Prop. 14.8 $X \Rightarrow X$ holds in $\langle G, M, I\rangle$, hence also in $\mathfrak{I}(G, M, I)$. So we have shown $\mathsf{Md}(\mathsf{Imp}(G, M, I)) \subseteq \mathfrak{I}(G, M, I)$. □

14.3 FCA and Relations

If you have made it up to here, you will hopefully say "*Quite interesting!*", and probably you will ask "*But where is the FCA-specific relation stuff?*". I could then say "*Didn't you notice the incidence relation, not to speak about subsethood and equality, or the partial orders involved in the definition of the Galois connection?*". Of course, that is obviously not all what should be expected from this chapter. So I hope to provide a slightly more satisfactory answer to the reader's question below.

First of all note that concepts in FCA are maximal rectangles of the incidence relation and can, therefore, be analyzed from this point of view; recall Section 6.2.1.3.

Another aspect is the following: $\mathfrak{B}(G,M,I)$ being a complete lattice, a natural question then is for complete sublattices of it. It turns out that they correspond to certain subrelations of the incidence relation I of the formal context $\langle G,M,I\rangle$; see [4, Section 8.4]. We call $J \subseteq I$ a *closed subrelation* of a context $\langle G,M,I\rangle$ if each concept in $\langle G,M,J\rangle$ is also a concept of $\langle G,M,I\rangle$. Without proof we report

Theorem 14.11. ([4, p. 106, Satz 26], [6, p. 112, Theorem 13]) *Let J be a closed subrelation of $\langle G,M,I\rangle$. Then $\langle G,M,J\rangle$ is a complete sublattice of $\langle G,M,I\rangle$. Conversely, if $\mathfrak{U}$ is a complete sublattice of $\mathfrak{B}(G,M,I)$, then $J := \bigcap\{A \times B : \langle A,B\rangle \in \mathfrak{U}\}$ is a closed subrelation of $\langle G,M,I\rangle$ and $\mathfrak{B}(G,M,J) = \mathfrak{U}$.*

This means that the closed relations are in a one-to-one correspondence to the complete sublattices.

A special case of concept lattices is induced by Ferrers relations: Assume that $F \subseteq G \times M$ is Ferrers; then for all $g,h \in G$ and all $m,n \in M$ we have

$$gFm \wedge hFn \rightarrow gFn \vee hFm.$$

Visualized in a cross table for the context $\langle G,M,F\rangle$ this means that

×	
	×

must not occur as a pattern, i.e. three of the four cells of such a square must be filled. This is equivalent to the property that such a cross table can be transformed into a stair-shaped form by rearranging the rows and columns. As immediate consequences of the above results we get

Corollary 14.12. ([6, p. 237, Proposition 103])

(1) *$F \subseteq G \times M$ is Ferrers iff $\mathfrak{B}(G,M,F)$ is a chain.*
(2) *The complete chains in the concept lattice of $\mathfrak{B}(G,M,I)$ correspond to the closed Ferrers relation in $\langle G,M,I\rangle$.*

Proof. (1): Recall from Def. 3.21(biorder) that for a Ferrers relation we have

$$\forall u,v,x,y[u\rho v \wedge \neg x\rho v \wedge x\rho y \rightarrow u\rho y],$$

and that $\langle A_1,B_1\rangle \leq \langle A_2,B_2\rangle$ if $A_1 \subseteq A_2$ (which is equivalent to $B_2 \subseteq B_1$); see Def. 14.5(a).

Now assume that F is Ferrers, and that $\langle A_1, B_1\rangle \not\leq \langle A_2, B_2\rangle$; then there are an object $g \in A_1$ (with $g \notin A_2$) and an attribute $n \in B_2$ (with $n \notin B_1$) such that $\neg gFn$. But, by definition, $\forall m \in B_1[gFm]$ and $\forall h \in A_2[hFn]$. Therefore, we get $hFn \wedge \neg gFn \wedge gFm$, whence hFm due to the Ferrers condition, for all $h \in A_2$ and $m \in B_1$. So, for $h \in A_2$ we get $\forall m \in B_1[hFm]$, that is $h \in B_1^{\downarrow} = A_1$. Thus we have shown $A_2 \subseteq A_1$, whence $\langle A_2, B_2\rangle \leq \langle A_1, B_1\rangle$, and this means that any two concepts in the concept lattice $\mathfrak{B}(G, M, F)$ are comparable which, therefore, is a chain.

Conversely, assume that $\mathfrak{B}(G, M, F)$ is a chain; then $\langle A_1, B_1\rangle \leq \langle A_2, B_2\rangle$ or $\langle A_2, B_2\rangle \leq \langle A_1, B_1\rangle$. As above let $g \in A_1$, $h \in A_2$, and $m \in B_1$, $n \in B_2$. Do further assume that F is not of Ferrers type, i.e. that

$$\text{(NF)} \qquad hFn \wedge \neg gFn \wedge gFm \wedge \neg hFm$$

holds. We first consider the case $A_1 \subseteq A_2$: then we get $g \in A_2$, hence $\forall n' \in B_2[gFn']$. But this is a contradiction to (NF), so this assumption cannot hold. And similar for the case $A_2 \subseteq A_1$. □

References

[1] Bělohlávek, R.: Introduction to formal context analysis. Tech. Rep., Dpt. of Computer Science, Palacký University, Olomouc (2008), available at `https://phoenix.inf.upol.cz/esf/ucebni/formal.pdf`.

[2] Bělohlávek, R.: Formal Concept Analysis: Classical and Fuzzy. In [3], 177-207 (2011).

[3] Bělohlávek, R., Klir, G.J. (Eds): *Concepts and Fuzzy Logic*, The MIT Press, Cambridge, Massachusetts (2011).

[4] Ganter, B.: *Diskrete Mathematik: Geordnete Mengen*, Springer-Verlag Berlin Heidelberg (2013).

[5] Ganter, B., Stumme, G. and Wille, R. (Eds): *Formal Concept Analysis*, Springer-Verlag Berlin Heidelberg, LNAI 3626 (2005).

[6] Ganter, B., Wille, R.: *Formal Concept Analysis-Mathematical Foundations*, Springer-Verlag Berlin Heidelberg (1999).

[7] Lambrechts, L.: *Formal Concept Analysis*, Master Thesis, Vrije Univ. Brussel (2012), available at `https://we.vub.ac.be/sites/default/files/files/thesis_lieve_lambrechts.pdf`.

[8] Wille, R.: Restructuring lattice theory: An approach based on hierarchy of concepts. In Rival, I. (Ed.): *Ordered sets*, 445-470, Reidel, Dordrecht-Boston (1982). (Reprinted in Ferré, S., Rudolph, S. (Eds.): *Formal Concept Analysis, ICFCA 2009*, 314-339, Springer-Verlag Berlin Heidelberg (2009).)

[9] Wille, R.: Formal Concept Analysis as Mathematical Theory of Concepts and Concept Hierarchies. In [5], 1-33 (2005).

Chapter 15

Mereology and Qualitative Spatial and Temporal Reasoning

Mereology is the theory that deals with the relations of part to whole and the relations of part to part within a whole. The importance of *Qualitative Reasoning* (QR) lies in its ability of abstracting from metrical details of the physical world. This enables computers to make predictions about relations, even when precise quantitative information is not available or irrelevant. Qualitative representations can also handle some forms of indefinite information by considering equivalence classes rather than values. Moreover, they provide a natural human-computer interface as people often think and communicate in terms of qualitative classifications. Two very important concepts of commonsense knowledge are time and space; accordingly, the two main directions in QR so far are spatial and temporal reasoning.

Mereological systems are another example of how a set together with a relation can be made a powerful tool for building sophisticated models for a certain domain of discourse. *Mereology* (from the Greek $\mu\epsilon\rho o\varsigma$, 'part') is the theory that deals with the relations of part to whole and the relations of part to part within a whole, from a formal point of view. Like with syllogisms, its roots can be traced back to the early days of philosophy, beginning with the Presocratics and continuing throughout the writings of ontologists all the time until it made its way into contemporary philosophy. As done for syllogisms, formal models have also been developed to capture mereological concepts, starting in the first half of the twentieth century with [20] and [19]. Since then mereology has become a topic of central interest for modern ontologists and logicians.

Qualitative reasoning is an approach to handle commonsense knowledge without using numbers. Instead, one tries to represent knowledge using the limited vocabulary of a suitably chosen small set of qualitative relations between given or assumed entities. An important motivation for using a qualitative approach is that it is considered to be closer to the way humans represent and reason about knowledge by basing the formal models on qualitative abstractions of temporal and spatial aspects of the background knowledge on which our human perception of physical reality is based. Another motivation is that it is possible to deal with incomplete knowledge, however different from fuzzy models.

As mentioned above, time and space are two very important concepts of commonsense knowledge. Qualitative temporal reasoning has emerged early, since time is a simple scalar concept, and has generated a lot of research effort and important results. In contrast, space is much more complex due to its multi-dimensionality which leads to a higher number of description possibilities of entities and relationships between them. This becomes obvious when enumerating natural language expressions involving space or time. While temporal expressions mainly describe order and duration (like "*before*", "*during*", "*long*", or "*a while*"), spatial expressions are manifold. They are used for describing direction ("*left, above*"), distance ("*far, near*"), size ("*large, tiny*"), shape ("*oval, convex*"), or topology ("*touch, inside*").

In the area of spatial reasoning, topological reasoning about regions [28], and reasoning about directions [12] can be distinguished. For temporal reasoning, either points [36] or intervals [1] are used as basic entities. Qualitative reasoning is an area that can be structured in the following three subfields: representation, decidability, and computation. The first deals with the different forms of spatial or temporal knowledge and how it can be formalized and represented in a computational framework, and the last one with methods and techniques for reasoning and developing efficient algorithms. By a (first-order) theory we mean a set of formulas that is closed under logical consequence. A theory Γ is *decidable* if there is an effective procedure that, given an arbitrary formula (in the signature of the theory) decides whether the formula is a member of the theory or not.

According to its introductory character, this chapter will have its focus on the first of the three subfields just mentioned. The goals of the considerations to come are not located within the philosophical or ontological realms (see e.g. the overview in [35]); therefore, the ontological issues related to the axioms postulated for the variants considered are discussed now and then and in a short manner only. That is to say, we will in general not present detailed arguments for why an axiom or some of its consequences should, or cannot, be accepted. Instead we will have a closer look on formal aspects and on what can be done by using relations in modeling mereological and spatial configurations. This will result in a brief overview about some common mereo(topo)logical and temporal systems and their interrelationships.

15.1 Mereology

15.1.1 *Basic Notions of Mereology*

More than anything else considered in this book, mereology has tight connections towards life itself. It is directly based on our perceptual equipment as it has developed during biological evolution that finally led to (what once has been called in a maybe very optimistic or, more probably, overbearing mood) *Homo sapiens sapiens*. Parthood is a relation of significant importance in both perception and conceptual modeling: "If we examine the world that surrounds us we can say that it comes to us in parts. We never experience the world as a whole, but only its bigger or smaller parts." ([14], p. 147).

Two prominent kinds of such parts are regions and solid objects, and how they are made up and how they are related to each other is a very important kind of knowledge. But already doing the first step into the topic of this chapter, one has to be cautious: That we are dealing with an epistemologically basic perceptual mechanism does in no case mean that we have to do with clearly defined notions, or uncontroversial concepts and mathematical models only. And even where there is agreement on vocabulary, such as the use of a binary *connection* predicate, it is not always interpreted in the same way. See e.g. the introductory sections of [34, 35] from which we will use some examples in the following considerations.

The first thing to observe is that the word "*part*" has many different meanings in ordinary English, not all of which correspond to the same relation. But mereology does not differentiate among the roles that parts play within the structure of an aggregate, i.e., it does perceive all parts as being of the same type. On the other hand, mereology assumes no ontological restriction on the field of "*part*" : It can be used to indicate any portion of a given entity, regardless of any internal distinctions, and the operands of the parthood relation can be material bodies, events, geometric entities, or spatio-temporal regions, as in (1)-(5) below, or abstract entities such as properties, propositions, types, or kinds, as in the last three examples:

(1) The handle is part of the mug.
(2) The remote control is part of the stereo system.
(3) The left half is your part of the cake.
(4) The cutlery is part of the tableware.
(5) The larval stage is part of some insect's life.
(6) The antecedent is the 'if' part of the conditional.
(7) Rationality is part of personhood.
(8) The letter 'm' is part of the word "*mereology*".

The portion indicated by "*part*" can itself have additional properties: It can be attached to the remainder, as in (1); or detached as in (2). It can be cognitively or functionally salient, as in (1) and (2), or arbitrarily demarcated, as in (3); self-connected, as in (1)-(3), or disconnected, as in (4), and so on. If you remember how syllogisms are structured or how well formed formulas of various logical systems are defined out of variables, connectives, terms and quantifiers, then (6) gives sense. But does such a kind of composition really apply to (7)? And if 'm' is part of "*mereology*" then clearly also 'e'. But this letter occurs twice.

Mereology is best understood as a general theory of the parthood relation, say P, regardless of whether the second operand of the relation may be said to qualify as a whole, and regardless of the role the operands play in an aggregate. Certainly, a hand is part of the arm it belongs to, and this arm is part of the body of say John Smith; moreover, also the hand is part of this body. So we get an undisputed example for transitivity: $hPa \wedge aPb \rightarrow hPb$. And if John Smith is member in or part of a team, can we say that transitivity still holds, i.e. that his hand or arm or body are also part of this team?

We will stop here, because no more examples are needed to demonstrate that part-whole discussions go far beyond what can be captured by a mathematical model with only two *ingredients* (this is the term originally used in [20]), viz. a set and a relation. Nevertheless, as we will show below, a lot of configurations and relationships can be built upon this simple basis.

15.1.2 *Ground Mereology*

Mereology is typically constructed as a theory of certain formal properties and relations that are exemplified across a wide range of domains, independent from the nature of the entities under consideration. In the following we adhere to the commonly used notation to avoid unnecessary and error-prone translations between formalisms, i.e. we will use uppercase letters that provide a kind of mnemonic support.

Virtually every mereological theory agrees on treating parthood as a partial ordering, i.e. as a reflexive, antisymmetric, transitive relation. Thus, the core set of axioms postulated in any standard mereological system is as follows:

Definition 15.1. *A pair* $\langle X, \mathsf{P}\rangle$ *consisting of a non-empty set of entities (objects,...)* X *and a relation* P *called 'part of' is said to be a* **Ground Mereology** [33], *denoted by* M, *if* P *is a partial order, i.e. if the following axioms hold in* X*:*

(P1) $\forall x \quad [x \,\mathsf{P}\, x]$
(P2) $\forall x, y \quad [x \,\mathsf{P}\, y \wedge y \,\mathsf{P}\, x \rightarrow x = y]$
(P3) $\forall x, y, z \, [x \,\mathsf{P}\, y \wedge y \,\mathsf{P}\, z \rightarrow x \,\mathsf{P}\, z]$

Formulated in colloquial terms this means: "*Everything is part of itself*", "*Two distinct things cannot be part of each other.*", and "*Any part of any part of a thing is itself part of that thing.*". Taking reflexivity as constitutive of the meaning of "*part*" amounts to accept identity as a limiting case of, or improper, parthood. Note that both reflexivity and transitivity of parthood have been argued against and therefore become a matter of discussion; for details see e.g. [17] and [34].

Based on the simple mathematical structure of M the following additional mereological relations can be introduced by definition (see [33, 34]):

Definition 15.2.

(a)	*(Equality)*	$x \,\mathsf{EQ}\, y$	$:= x \,\mathsf{P}\, y \wedge y \,\mathsf{P}\, x$
(b)	*(Proper Part)*	$x \,\mathsf{PP}\, y$	$:= x \,\mathsf{P}\, y \wedge \neg y \,\mathsf{P}\, x$
(c)	*(Proper Extension)*	$x \,\mathsf{PE}\, y$	$:= \neg x \,\mathsf{P}\, y \wedge y \,\mathsf{P}\, x$
(d)	*(Overlap)*	$x \,\mathsf{O}\, y$	$:= \exists z [z \,\mathsf{P}\, x \wedge z \,\mathsf{P}\, y]$
(e)	*(Underlap)*	$x \,\mathsf{U}\, y$	$:= \exists z [x \,\mathsf{P}\, z \wedge y \,\mathsf{P}\, z]$
(f)	*(Over-Crossing)*	$x \,\mathsf{OX}\, y$	$:= x \,\mathsf{O}\, y \wedge \neg x \,\mathsf{P}\, y$
(g)	*(Under-Crossing)*	$x \,\mathsf{UX}\, y$	$:= x \,\mathsf{U}\, y \wedge \neg y \,\mathsf{P}\, x$
(h)	*(Proper Overlap)*	$x \,\mathsf{PO}\, y$	$:= x \,\mathsf{OX}\, y \wedge y \,\mathsf{OX}\, x$
(i)	*(Proper Underlap)*	$x \,\mathsf{PU}\, y$	$:= x \,\mathsf{UX}\, y \wedge y \,\mathsf{UX}\, x$
(j)	*(Disjointness)*	$x \perp y$	$:= \neg x \,\mathsf{O}\, y$

It is easy to show that the following properties hold for these derived relations. We use this opportunity to give one more demonstration of how to use the rules we have learned about in Section 2.2.

Proposition 15.3.

(1) EQ *is an equivalence relation.*
(2) *Both* PP *and* PE *are irreflexive, asymmetric and transitive.*
(3) O *and* U *are reflexive and symmetric, but not transitive.*
(4) $x \mathsf{P} y \leftrightarrow x \mathsf{PP} y \vee x = y$.
(5) $x \mathsf{P} y \rightarrow \forall z[z \mathsf{O} x \rightarrow z \mathsf{O} y]$.
(6) $x \mathsf{P} y \rightarrow \forall z[y \mathsf{U} z \rightarrow x \mathsf{U} z]$.
(7) $x \mathsf{P} y \rightarrow x \mathsf{O} y$.

Proof.

(5): $x \mathsf{P} y \wedge z \mathsf{O} x$	$\rightarrow \exists z'[z' \mathsf{P} z \wedge z' \mathsf{P} x]$	Def. 15.2(d)
	$\rightarrow t \mathsf{P} z \wedge t \mathsf{P} x$	(EI), Th. 2.38(3)
	$\rightarrow t \mathsf{P} z \wedge t \mathsf{P} y$	Def. 15.1(P3)
	$\rightarrow \exists z'[z' \mathsf{P} z \wedge z' \mathsf{P} y]$	(EG), Th. 2.38(4)
	$\rightarrow z \mathsf{O} y$	Def. 15.2(d)
$x \mathsf{P} y$	$\rightarrow (z \mathsf{O} x \rightarrow z \mathsf{O} y)$	Th. 2.33(12)
	$\rightarrow \forall z'[z' \mathsf{O} x \rightarrow z' \mathsf{O} y]$	(UG), Th. 2.38(2), since z arbitrary

(7): $x \mathsf{P} y \rightarrow x \mathsf{P} x \wedge x \mathsf{P} y \rightarrow \exists z[z \mathsf{P} x \wedge z \mathsf{P} y]$, using Def. 15.1(P1) and Th. 2.38(4). □

Interpreting P as spatial inclusion, the four basic patterns of mereological relationship can be depicted as shown in Fig. 15.1 (adapted from Figure 1 of [33]). Note that for the leftmost situation the clauses below the dashed line hold only due to the presence of the rectangle as a larger entity z including both x and y.

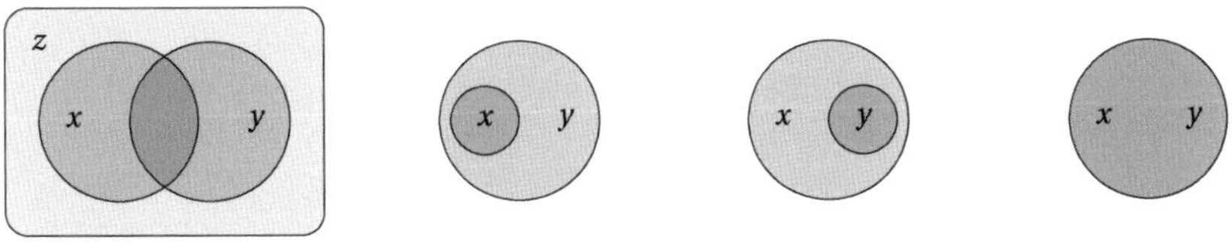

$x \mathsf{O} y$	$y \mathsf{O} x$	$x \mathsf{O} y$	$y \mathsf{O} x$	$x \mathsf{O} y$	$y \mathsf{O} x$	$x \mathsf{O} y$	$y \mathsf{O} x$
$x \mathsf{OX} y$	$y \mathsf{OX} x$	$x \mathsf{P} y$	$y \mathsf{OX} x$	$x \mathsf{OX} y$	$y \mathsf{P} x$	$x \mathsf{U} y$	$y \mathsf{U} x$
$x \mathsf{PO} y$	$y \mathsf{PO} x$	$x \mathsf{U} y$	$y \mathsf{U} x$	$x \mathsf{U} y$	$y \mathsf{U} x$	$x \mathsf{P} y$	$y \mathsf{P} x$
$x \mathsf{U} y$	$y \mathsf{U} x$	$x \mathsf{PP} y$	$y \mathsf{PE} x$	$x \mathsf{PE} y$	$y \mathsf{PP} x$	$x = y$	$y = x$
$x \mathsf{UX} y$	$y \mathsf{UX} x$	$x \mathsf{UX} y$			$y \mathsf{UX} x$		
$x \mathsf{PU} y$	$y \mathsf{PU} x$						

FIG. 15.1: The basic patterns of mereological relationship.

15.1.3 *Minimal and Extensional Mereology, Atoms*

As already mentioned above M is by default viewed as embodying the common core of any mereological theory. So we have to think about possible directions or strategies of how to extend this minimal mereological framework. Generally speaking, such extended systems result from M by adding principles that assert the existence of certain mereological items given the existence of other items.

One way to enrich M is by means of so-called *decomposition principles*, i.e. postulates dealing with the part structure of a given whole. One fundamental intuition here is that whenever something has a proper part, it has more than one, i.e. a whole cannot be decomposed into a single proper part. This intuition is captured in the (*Weak*) *Supplementation Principle*, WSP in short:

(P4w) $\forall x, y\ [x \mathsf{PP}\, y \rightarrow \exists z[z \mathsf{PP}\, y \wedge z \perp x]]$ (*Weak*) *Supplementation*

The mereological theory comprising (P1) - (P4w) is called **Minimal Mereology**, abbreviated by MM, perhaps because these four postulates are by some authors regarded as the minimal requirements to be imposed on a meaningful formalization of parthood. For a detailed discussion on variants of (weak) supplementation and their properties see section 3.1 of [35].

There is another, stronger way of expressing the supplementation intuition. It corresponds to the following axiom:

(P4) $\forall x, y\ [\neg y \mathsf{P}\, x \rightarrow \exists z[z \mathsf{P}\, y \wedge z \perp x]]$ *Strong Supplementation*

Intuitively, this says that if an object fails to include another among its parts, then there must be a remainder, something that makes up for the difference. It can be shown that in M (P4w) is equivalent to $x \mathsf{PP}\, y \rightarrow \exists z[z \mathsf{P}\, y \wedge z \perp x]$; so, (P4) differs from (P4w) in the antecedent only. The extension of M with (P4) is called **Extensional Mereology**, abbreviated by EM.

Furthermore, in M (P4) implies (P4w), so any mereological system violating (P4w) will *a fortiori* violate (P4). The converse, (P4w) $\rightarrow$ (P4) is not true as shown in the following example (adapted from [35]).

(P4w) holds for $u \mathsf{P}\, x$, $u \mathsf{P}\, y$, $v \mathsf{P}\, x$, and $v \mathsf{P}\, y$ since each proper part counts as a supplement of the other, but (P4) is false. EM is thus a proper extension of MM.

It is a theorem of EM that no composite objects with the same proper parts can be distinguished:

$$\mathsf{EM} \vdash (\exists z[z \mathsf{PP}\, x] \vee \exists z[z \mathsf{PP}\, y]) \rightarrow (x = y \leftrightarrow \forall z[z \mathsf{PP}\, x \leftrightarrow z \mathsf{PP}\, y]) \qquad (15.1)$$

The attribute *extensional* is justified by the exclusion of models where distinct objects decompose into the same proper parts like in the one given above. Thus, EM is a theory incorporating the view that an object is exhaustively defined by its constituent parts.

More generally, let Q be a reflexive and symmetric binary predicate; then the following generic definition captures extensionality for Q:

$$\forall x, y\big[\forall z[(Q(x,z) \leftrightarrow Q(y,z)) \rightarrow x = y]\big] \tag{15.2}$$

But sameness of parts does not necessarily mean identity: For instance, two objects can be distinct solely with respect to how their parts are arranged, and one can make the distinction (or not) between an object and the matter constituting it, like a vase and the clay. So, even in case of highly plausible and seemingly 'innocent' postulates sometimes unexpected obstacles occur. And as indicated at the beginning of this chapter, a lot of this kind of philosophical considerations and discussions can be witnessed in mereology throughout its development. For a detailed and general exposition of different axiom systems for mereologies, their logical interrelationships, and various arguments in favor of or against them, see [35].

Another important group of decomposition principles deals with the question of atomism. Mereologically, an atom is an entity with no proper parts, regardless of whether it is considered point-like or is meant to have spatial (and/or temporal) extension. Formally this gives

$$\mathsf{At}\,x := \neg\exists y[y\,\mathsf{PP}\,x].$$

By definition of PP, all atoms are pairwise disjoint and can only overlap things of which they are part. Concerning atoms there are two mutually incompatible possibilities for extending mereological systems X.

(AT1) $\forall x\exists y\,[\mathsf{At}\,y \wedge y\,\mathsf{P}\,x]$ *Atomicity*

(AT2) $\forall x\exists y\,[y\,\mathsf{PP}\,x]$ *Atomlessness*

Adding (AT1) to X yields a corresponding *atomistic* version AX, and adding (AT2) an *atomless* version $\bar{\mathsf{A}}\mathsf{X}$ of X.

15.1.4 *Closure and General Mereologies*

A second way of extending M is by adding one or more so-called *composition principles* to it that regiment the behavior of parthood P by some rules taking us from a parts to wholes. (In [5] they are called *closure principles* in accordance to the naming of the corresponding mereologies.) The idea behind is that a mereological domain must be closed under mereological operations of various sorts, like mereological sum, product, or difference.

Definition 15.4. ([33, Definition 3]) *The theories of* **Closed Mereology** CM, *and* **Closed Extensional Mereology** CEM, *respectively, are the extensions of* M *and* EM *obtained by adding the following axioms:*

(P5) $\forall x, y\,\big[x\,\mathsf{U}\,y \rightarrow \exists z\forall w[w\,\mathsf{O}\,z \leftrightarrow (w\,\mathsf{O}\,x \vee w\,\mathsf{O}\,y)]\big]$ *(Sum, Fusion)*

(P6) $\forall x, y\,\big[x\,\mathsf{O}\,y \rightarrow \exists z\forall w[w\,\mathsf{P}\,z \leftrightarrow (w\,\mathsf{P}\,x \wedge w\,\mathsf{P}\,y)]\big]$ *(Product)*

(P7) $\forall x, y\,\big[\exists z[z\,\mathsf{P}\,x \wedge \neg z\,\mathsf{O}\,y] \rightarrow \exists z\forall w[w\,\mathsf{P}\,z \leftrightarrow (w\,\mathsf{P}\,x \wedge \neg w\,\mathsf{O}\,y)]\big]$

(Be careful: Unfortunately, names differ (in e.g. [5] CM and CEM are called *Closure Mereology* and *Extensional Closure Mereology*), but acronyms are standard.)

In the presence of (P4), the entities whose conditional existence is asserted by (P5)-(P7) are unique by (15.1)[1], and one can therefore introduce a special notation for these entities: $x + y, x \times y$, and $x - y$. So $x + y := \iota z[\forall w[w \mathsf{O} z \leftrightarrow (w \mathsf{O} x \vee w \mathsf{O} y)]]$.

Among various proposals for defining the intuitive idea of a sum of two entities (see e.g. [34,35]), (P5) is a good compromise: Informally, it says that for any pair of U-related entities x and y there is something that overlaps exactly those things that overlap either x or y. (P5) is strong enough to rule out the model on the left, but weak enough to be compatible with the model on the right given in the figure below (taken from [35]).

As always when mathematicians start to formalize an intuitive appealing concept, they first try do it in a way that produces a structure with some neat properties; if this strategy fails, they have to invent new concepts themselves. In case of parts, sums, and products it can be shown that the definitions chosen behave as expected: Both sum and product are idempotent, commutative, associative, and well-behaved with respect to both parthood and distributivity. See again [35, Section 4.2] on sums.

The strongest fusion principle is as follows: Let $\phi(x)$ be a first-order formula in which x is a free variable, i.e. a formula describing a property or condition. Then the idea of unrestricted fusion requires that for every satisfied ϕ there is an entity consisting of all entities that satisfy $\phi(x)$.

Definition 15.5. *The theories of* **General Mereology** GM, *and* **General Extensional Mereology** GEM, *respectively, are the extensions of* CM *and* CEM *respectively, obtained by adding the following axiom scheme:*

(P8) $\exists x[\phi(x)] \rightarrow \exists z \forall w[w \mathsf{O} z \leftrightarrow \exists x[\phi(x) \wedge w \mathsf{O} x]]$ *(Unrestricted Fusion)*

15.2 Introducing Qualitative Models and Inferences

Qualitative spatial and temporal reasoning (QSTR) is a sub-field of knowledge representation involved with representations of spatial and temporal domains that are based on finite sets of so-called qualitative relations denoting qualitative relationships between temporal or spatial entities. The need for reasoning about time and space arises in

[1] If there is exactly one object with a certain property ϕ, then it is convenient to have an expression that names this object. A common notation for this purpose is $\iota x[\phi x]$, which may be read as "*the thing that is* ϕ" or more briefly as "the ϕ". An expression of the form $\iota x[\phi x]$ is called *definite description*; the *definite description operator* ι binds variables like the quantifiers $\forall$ and $\exists$.

many areas of AI like computer vision, natural language understanding, geographic information systems (GIS), scheduling, planning, diagnosis and genetics. Computationally, QSTR has become linked to constraint satisfaction problems (CSPs) over infinite domains where the qualitative relations serve as constraints. Typical domains for temporal representations are points and intervals and, for spatial representations, regions or orientations in the Euclidean plane.

Following the example given in the Introduction of [23], consider a boat race (or equally suitable, a race involving cars, bicycles, or horses), and suppose you get the following information about the state of the competition: (1) A is behind B, (2) C is in front of B, (3) D is behind B, and (4) E and D are at the same level. Furthermore assume you get an additional message from another source: (5) C is behind E.

No doubt, this may be an every-day situation with a natural and common way of communication: a parsimonious set of qualitative relations is used to transmit the relevant information on the context under consideration. This level of description can easily be abstracted in a framework consisting of a line representing the race course and three relations: *in front of, behind,* and *at the same level.* It is clear that these three relations are the only possible relations between two boats, and that they constitute what is known as a **JEPD set of relations** (*jointly exhaustive and pairwise disjoint*), i.e. each pair $\langle a, b\rangle$ over the domain under consideration is member of exactly one of these relations. Relations in a JEPD set are called *atomic, base,* or *basic relations.*

In mathematical terms we can use a system $\langle R, <, \mathrm{eq}, >\rangle$, where R represents the race course in form of a real interval, and $<$ and $>$ stand for "*is behind*" and "*is in front of*" and represent the relative position of two points in R. Clearly, $<, \mathrm{eq}$, and $>$ are the converses of $>, \mathrm{eq}$, and $<$, respectively. The information about the race can then be reformulated as: (1) $A < B$, (2) $C > B$, (3) $D < B$, and (4) $E \,\mathrm{eq}\, D$; the additional message is (5) $C < E$.

Given these pieces of information, one can derive others: e.g. $B < C$ from $C > B$, and $A < B$ together with $B < C$ gives $A < C$. From (1) and (3) we know that $A < B$ and $B > D$; what can be inferred on the position of D relative to A from this knowledge? Nothing, since either of the three relations can hold between them. We can express this by saying that *behind* $\circ$ *in front of* is the relation known as the universal relation, which is the union of the three base relations. This example shows a further characteristic of QSTR: The composition of two basic relations need not give a result that is again one of the basic relations, but can yield an indefinite information representable by the subset (or union) of those base relations that can possibly hold; and if there is no information at all, i.e. each of the base relations can hold, we use the universal (relative to the given JEPD set) relation which is the union of all base relations. Such subsets of basic relations are called *disjunctive relations.*

Typically, Qualitative Constraint Networks (QCNs) are used to express information on a spatial or temporal situation. Each constraint of a QCN is defined by a set of basic relations and represents a set of acceptable qualitative configurations between some temporal or spatial entities. This allows us to establish a connection to the domain of *constraint-based reasoning.* The data of our example, the five objects and the relations between them, can be represented by a directed graph having five vertices $A, ..., E$ and arcs labeled according to the relations provided by the informants. This is shown in Fig. 15.2, with the informations (1) - (4) drawn as solid arrows, and information (5) as a dashed one.

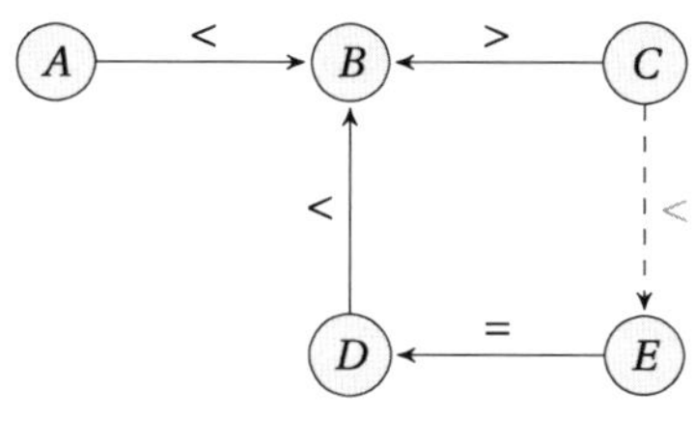

FIG. 15.2: A QCN for the boat race example.

The basic operations can be considered in terms of paths: Inversion corresponds to replacing an arc by the opposite one, and composition to combining two successive arcs, say $\langle x, y\rangle$ and $\langle y, z\rangle$, to get a path from x to z.

Let us now consider the situation of the recipient of the five messages given above. For better decoding of the information he wants to graphically represent the five boats as points on the directed line. But this is possible only if the information provided is *consistent*, i.e. if it represents a possible configuration. Obviously, this is not the case in our example: We know already that $B < C$ which, together with (3) $D < B$, gives $D < C$. Ont the other hand, $C < E$ and $D \,\mathrm{eq}\, E$ imply $C < D$. The information provided by the two informants taken together is thus contradictory. Leaving away the dashed arc corresponding to (5), however, yields a consistent network with three compatible configurations shown in Fig. 15.3.

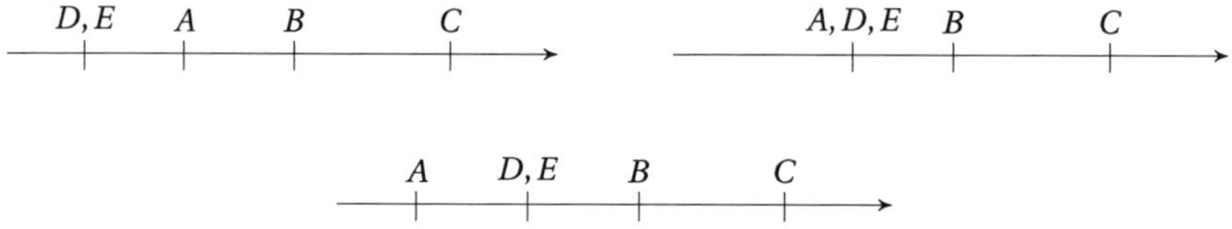

FIG. 15.3: Compatible configurations for the boat race example.

15.2.1 *Some Formal Background*

From a formal point of view, a binary qualitative calculus QC is based on a finite set $\mathcal{B} = \{\rho_1, \rho_2, ..., \rho_k\}$ of so-called *base relations* on a domain D chosen in a way that make $\mathcal{B}$ a JEPD set, i.e. $\mathcal{B}$ forms a partition of $\mathrm{D} \times \mathrm{D}$. A **disjunctive relation** is the union of two ore more base relations, commonly represented as the set of base relations contained in it. Hence all the $2^{|\mathcal{B}|}$ relations available in QC are given by $\mathbf{2}^{\mathcal{B}}$. Disjunctive relations allow us to represent indefinite knowledge: the constraint $r\{\rho_i, \rho_j, \rho_k\}s$ specifies that the relation between r and s is one of ρ_i, ρ_j, or ρ_k, where $\{\rho_i, \rho_j, \rho_k\} \subseteq \mathscr{B}$.

Once a JEPD set $\mathscr{B}$ has been chosen, we have fixed the granularity and the vocabulary for describing particular configurations in D. At this level of generality it may, however, occur that the composition of two basic relations is outside $\mathscr{B}$, e.g. for Allen's calculus. But having established that $r\rho s$ and $s\sigma t$ for some $r, s, t \in \mathsf{D}$ and $\rho, \sigma \in \mathscr{B}$ the natural question arises if we can say something about the relationship of r and t. So we need a new plausible concept, which has given the name *weak composition.*

The set $\mathbf{2}^{\mathscr{B}}$ is equipped with the usual set-theoretic operations of union and intersection, the converse operation (for every $\rho \in \mathscr{B}$ also $\rho^{\smile} \in \mathscr{B}$), and weak composition; moreover we assume that $I_{\mathsf{D}} \in \mathscr{B}$.

- The *converse of a disjunctive relation* R is the union of the converses of its base relations: Let $R = \{\rho_1, \rho_2, ..., \rho_k\}$, then $R^{\smile} = \{\rho_1{}^{\smile}, \rho_2{}^{\smile}, ..., \rho_k{}^{\smile}\}$.
- The **weak composition** *of two base relations* ρ and σ is

$$\rho \diamond \sigma := \{\tau \in \mathscr{B} : (\rho \circ \sigma) \cap \tau \neq \emptyset\}. \tag{15.3}$$

- The **weak composition** *of two disjunctive relations* R and S is

$$R \diamond S := \bigcup_{\rho \in R, \sigma \in S} \rho \diamond \sigma. \tag{15.4}$$

(Notice that $\diamond$ is only defined relative to a particular partition, and thus is not a purely relational concept like $\circ$.)

Intuitively, if restricted to the language provided by the partition induced by $\mathscr{B}$, weak composition is the best approximation (from above) we can get to the usual $\circ$-composition, i.e. it is the smallest relation $\tau \in \mathbf{2}^{\mathscr{B}}$ which contains $\rho \circ \sigma$. Thus, the given set of relations $\mathbf{2}^{\mathscr{B}}$ is closed with respect to the algebraic operations of weak composition, union, intersection, and converse. Note that in some qualitative algebras (e.g. the Point Algebra PA introduced in Section 15.3.1 below), $\rho \diamond \sigma$ and $\rho \circ \sigma$ coincide.

Proposition 15.6. *Let $\mathscr{B}$ be a JEPD set, and $\rho, \sigma, \tau \in \mathscr{B}$. Then*

(1) $\rho \circ \sigma \subseteq \rho \diamond \sigma$.
(2) $(\rho \diamond \sigma)^{\smile} = \sigma^{\smile} \diamond \rho^{\smile}$.
(3) $(\rho \diamond \sigma) \cap \tau = \emptyset$ *iff* $(\rho \circ \sigma) \cap \tau = \emptyset$.

Proof. (1) Any $\langle r, s\rangle \in \rho \circ \sigma$ is member of a unique $\tau \in \mathscr{B}$. Since this τ has an element in common with $\rho \circ \sigma$, it must belong to $\rho \diamond \sigma$.

(2): An easy consequence of the corresponding result for composition.

(3): Since $\rho \circ \sigma \subseteq \rho \diamond \sigma$ by (1), $(\rho \diamond \sigma) \cap \tau = \emptyset$ clearly implies $(\rho \circ \sigma) \cap \tau = \emptyset$. The other direction is shown by indirect proof: If $(\rho \diamond \sigma) \cap \tau \neq \emptyset$, then, since $\rho \diamond \sigma$ is a union of $\beta_i \in \mathscr{B}$, it contains τ; and by the definition of weak composition this means $(\rho \circ \sigma) \cap \tau \neq \emptyset$. □

In Table 15.1 we provide a summary of the various relational operations relevant for qualitative reasoning. The idea of what is now known as *composition table* has been introduced in [1]. From that time on the use of composition tables has become a key technique in providing efficient inference mechanisms in QSTR calculi. Given the set $\mathscr{B}$ of base relations as a fixed vocabulary, such a table enables one to answer the following question by a simple look-up: given two relational facts $r\rho s$ and $s\sigma t$, what are the possible relations in $\mathscr{B}$ that can hold between r and t? More details on composition tables can be found in Section 15.5.1 below.

TABLE 15.1: Relational Operations in $\mathscr{B} = \mathsf{D} \times \mathsf{D}$ and $\mathbf{2}^{\mathscr{B}}$.

Relational Operations	Set-theoretic definition $\rho, \sigma \in \mathsf{D} \times \mathsf{D}$	Algebraic definition $R, S \in \mathbf{2}^{\mathscr{B}}$
Intersection $\cap$	$\rho \cap \sigma :=$ as usual	$R \cap S := \{\rho : \rho \in R \wedge \rho \in S\}$
Converse $\breve{}$	$\rho^{\smile} :=$ as usual	$R^{\smile} := \{\rho^{\smile} : \rho \in R\}$
Composition $\circ$	$\rho \circ \sigma :=$ as usual	$R \circ S := \{\rho \circ \sigma : \rho \in R \wedge \sigma \in S\}$
Weak comp. $\diamond$	$\rho \diamond \sigma := \{\tau \in \mathscr{B} : (\rho \circ \sigma) \cap \tau \neq \emptyset\}$	$R \diamond S := \{\rho \diamond \sigma : \rho \in R \wedge \sigma \in S\}$

We close this section with [23, Example 11.1] using the linear order on $\mathsf{D} = \{a, b\}$. As JEPD set of relations on $\mathsf{D} \times \mathsf{D}$ we take $< := \{\langle a, b\rangle\}, \mathsf{eq} := \{\langle a, a\rangle, \langle b, b\rangle\}$, and $> := \{\langle b, a\rangle\}$. Then $< \circ < \, = \, > \circ > \, = \emptyset$, $< \circ > = \{\langle a, a\rangle\}$, and $> \circ < = \{\langle b, b\rangle\}$. Therefore $< \diamond < \, = \, > \diamond > \, = \emptyset$, and $< \diamond > \, = \, > \diamond < \, = \mathsf{eq}$.

15.3 Qualitative Temporal Calculi

Representing and reasoning about time play an important role in many areas of artificial intelligence as e.g. commonsense reasoning, plan recognition, scheduling, planning, and qualitative reasoning. Due to this increasing interest in automation of temporal reasoning various formal models have been proposed and explored during the last three decades.

We will consider two calculi: The first has as its domain a collection of time points forming a totally and strictly ordered set $\langle \mathsf{D}, < \rangle$, i.e. $<$ is an irreflexive, transitive and strongly connected relation on D, and is interpreted as "*before*" or "*precedes*" in time. Using this point-based model of time we then introduce an interval-based model which is able to represent how activities with temporal extent can relate to each other.

15.3.1 *Point Algebra*

The formal background for the considerations of Section 15.2 is the following: The concepts described are based on reasoning about relations between points that form a relation algebra, called the **Point Algebra** (PA). Note that points are in fact primitive, and although one can intuitively think of points as "points in time", there are no particular demands on them.

$<$ is commonly interpreted over $\mathbb{R}$, $\mathbb{Q}$, or $\mathbb{N}$ with the usual meaning. Any pair $\langle r, s\rangle \in \mathsf{D} \times \mathsf{D}$ is then contained in exactly one of the following three relations:

- $< := \{\langle r, s\rangle \in \mathsf{D} \times \mathsf{D} : r < s\}$,

- $> := \{\langle r, s\rangle \in \mathsf{D} \times \mathsf{D} : r > s\}$,
- $\mathsf{eq} := \{\langle r, s\rangle \in \mathsf{D} \times \mathsf{D} : r = s\}$.

Obviously, these three relations are a JEPD set P of base relations on D. P contains the identity relation eq, and is closed under the inverse operation $\breve{}$: $<\breve{} = >, >\breve{} = <$, and $\mathsf{eq}\breve{} = \mathsf{eq}$.

PA, one of the first and most prominent frameworks for representing qualitative temporal relations and reasoning about them, is a finite relation algebra with the three basic relations given above as atoms, and therefore 2^3 elements in $\mathbf{2}^{\mathsf{P}}$. Remember that an element $R \in \mathbf{2}^{\mathsf{P}}$ containing two or more base relations is called a *disjunctive relation* and is equal to the union of the base relations involved. Some of these relations have common names: e.g. $\{<, \mathsf{eq}\}$ we know as $\leq$, and $\{<, >\}$ as $\neq$; ? denotes the whole set of base relations $\mathsf{P} = \{<, \mathsf{eq}, >\}$ and corresponds to the "*nothing can be concluded*", "*no information available*", or "*everything is possible*" situation. The point relations then are

$$\mathsf{PA} := \big\{\emptyset, \{<\}, \{\mathsf{eq}\}, \{<, \mathsf{eq}\}, \{>\}, \{<, >\}, \{>, \mathsf{eq}\}, \{<, \mathsf{eq}, >\}\big\}$$

Usually, we will use the more sloppy, but common notations

$$\mathsf{PA} = \{\emptyset, <, \mathsf{eq}, \leq, >, \neq, \geq, ?\}.$$

TABLE 15.2: The composition table for PA (for dense domains).

∘	eq	<	>
eq	eq	<	>
<	<	<	?
>	>	?	>

The entries of the composition Table 15.2 for PA look straightforward, but let us have a closer look at it in

Proposition 15.7. ([23, PROPOSITION 1.1]) *Let* D *be dense, e.g.* $\mathbb{R}$ *or* $\mathbb{Q}$*; then* $< \circ > = > \circ < = \{<, \mathsf{eq}, >\}$.

Proof. Assume $r < s$ and $s > t$. This is consistent with $t < r$, $t = r$, and t between r and s, hence $t > r$. Thus $< \circ > \subseteq \{<, \mathsf{eq}, >\}$. Conversely, we show $< \subseteq < \circ >$: Assume $r < s$; then it is sufficient to take a t such that $t > s$, since we then have $r < t$ and $t > s$. In a similar way we get $> \subseteq < \circ >$ and $\mathsf{eq} \subseteq < \circ >$; thus $< \cup\, \mathsf{eq} \cup > = \{<, \mathsf{eq}, >\} \subseteq < \circ >$. □

In addition, we can verify that we not only have the inclusions of elements $< \circ < \subseteq <$ and $> \circ > \subseteq >$, which express the transitivity of $<$ and $>$, but also $< \circ < = <$ and $> \circ > = >$. Again, this follows from the density property which is the crucial point in the proofs: We can always find a point between two distinct points, and, given any point, there exists one point which precedes it and one which follows it.

If PA-relations are interpreted over a non-dense domain, e.g. $\mathbb{N}$ then PA is not closed under composition. For instance, $3 < 4$ but there exists no natural number y with $3 < y < 4$. Hence, according to the definition of relation composition, $\langle 3,4\rangle \notin < \circ <$. But $\langle 3,4\rangle \in < \diamond <$, since $< \circ < \cap < \neq \emptyset$, containing e.g. $\langle 3,5\rangle$.

15.3.2 *Interval Algebra*

Since in most cases it is advisable to give an introduction to a new topic in terms of the original intuitions, concepts, and formalizations, I will proceed in just this manner. The story of qualitative temporal and spatial reasoning begins with Allen's 1983 paper [1]. His **interval algebra** IA is the best known temporal logic within AI.

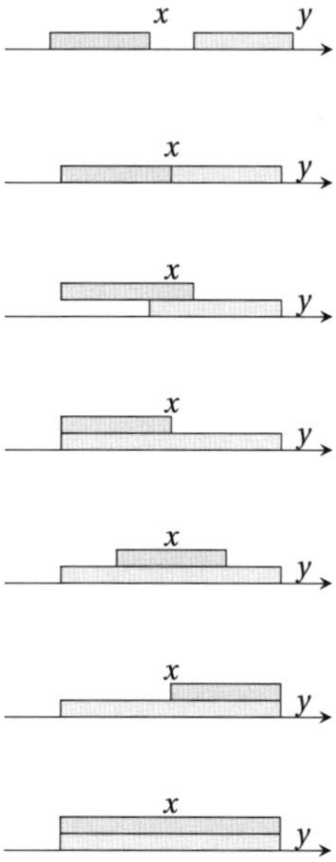

$x\mathsf{b}y$	"*x is before y*"	$x_l < x_r < y_l < y_r$
$y\mathsf{b}^{\smile}x$	"*y is after x*"	
$x\mathsf{m}y$	"*x meets y*"	$x_l < x_r = y_l < y_r$
$y\mathsf{m}^{\smile}x$	"*y is met by x*"	
$x\mathsf{o}y$	"*x overlaps y*"	$x_l < y_l < x_r < y_r$
$y\mathsf{o}^{\smile}x$	"*y is overlapped by x*"	
$x\mathsf{s}y$	"*x starts y*"	$x_l = y_l < x_r < y_r$
$y\mathsf{s}^{\smile}x$	"*y is started by x*"	
$x\mathsf{d}y$	"*x during y*"	$y_l < x_l < x_r < y_r$
$y\mathsf{d}^{\smile}x$	"*y contains x*"	
$x\mathsf{f}y$	"*x finishes y*"	$y_l < x_l < x_r = y_r$
$y\mathsf{f}^{\smile}x$	"*y is finished by x*"	
$x\mathsf{eq}y$	"*x equals y*"	$y_l = x_l < x_r = y_r$

FIG. 15.4: The base relations of Interval Algebra.

Allen presented a set of 13 qualitative relations describing all possible relationships between two (closed and finite) intervals say $x = [x_l, x_r]$ and $y = [y_l, y_r]$ based only on the relative orderings of their start and end points as shown in Fig. 15.4 (adapted from [23][2]).

These 13 basic relations constitute a partition of the set of all the possible relationships between two intervals, i.e. they have the JEPD property. As an example for a disjunctive IA relation consider $x\{\mathsf{s},\mathsf{s}^{\smile},\mathsf{d},\mathsf{d}^{\smile},\mathsf{f},\mathsf{f}^{\smile},\mathsf{o},\mathsf{o}^{\smile},\mathsf{eq}\}y$ expressing the fact that intervals x and y are not disjoint.

[2] The first two chapters of this book contain a detailed discussion of Allen's calculus together with the results obtained during the 25 years after its publication concerning its computational tractability.

As second example consider the following description of events (taken from [8, Example 12.2]):

> *"Fred was reading the paper while eating his breakfast. He put the paper down and drank the last of his coffee. After breakfast he went for a walk."*

The first sentence tells us that the interval of time over which Fred read the paper intersects with the interval of time over which Fred ate breakfast:

$$\mathit{Paper}\{\mathsf{s},\mathsf{s}^{\smile},\mathsf{d},\mathsf{d}^{\smile},\mathsf{f},\mathsf{f}^{\smile},\mathsf{o},\mathsf{o}^{\smile},\mathsf{eq}\}\mathit{Breakfast}.$$

The second information is:

$$\mathit{Paper}\{\mathsf{d},\mathsf{o},\mathsf{s}\}\mathit{Coffee}.$$

It is clear that drinking coffee occurs during breakfast and that walk is after breakfast:

$$\mathit{Coffee}\{\mathsf{d}\}\mathit{Breakfast} \quad \text{and} \quad \mathit{Walk}\{\mathsf{b}^{\smile}\}\mathit{Breakfast}.$$

Fig. 15.5 shows the temporal constraint network and a consistent scenario for this example.

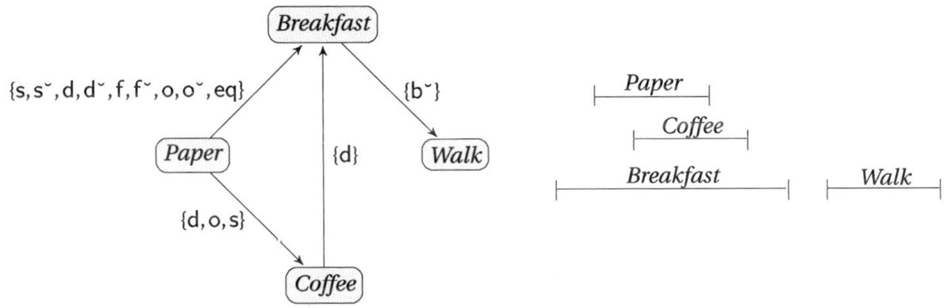

(a) network (b) consistent scenario

FIG. 15.5: The Breakfast example.

15.4 Qualitative Spatial Calculi: Mereotopology

In *Qualitative Spatial Reasoning* (QSR) one usually considers a particular spatial aspect such as topology, direction, or distance and develops a system of qualitative relationships between spatial entities which suitably cover the aspect under consideration, and which appears to be useful from an applicational point of view. For a recent overview see [6]. From the aspects just mentioned we will consider only what is subsumed under the heading of mereotopology; see e.g. [].

15.4.1 *Background and Ontological Aspects*

If one accepts regions as primitives, mereotopology is a mere combination of the mereological concept of parthood with the topological concept of *contact* or *connection*. And although this is a really simple combination, it allows for manifold options to build mereotopological theories.

From a formal point of view, mereotopology is a first-order theory, embodying mereological and topological concepts of the relations among wholes, parts, parts of parts, and the boundaries between parts. Enhancing mereology with topological aspects is due to the need of properly characterizing temporal and spatial relationships in many areas; consequently, temporal and spatial reasoning has become a sub-discipline in Artificial Intelligence during the 1990s with many applications in areas such as Geographical Information Systems (GIS) or spatial query languages.

The main reason for topology having become a second major ingredient of a theory of spatial reasoning is as simple as it is fundamental: A pure theory of parthood cannot tell us anything on *wholes*, i.e. self-connected pieces such as a stone, as opposed to a scattered entity made up of several disconnected parts, such as a bikini, or perhaps a symphony consisting of several movements. The connection relation is a topological relation that must be formally treated independent from mere mereological aspects. (But connectedness is not the only aspect that lies outside the modeling capabilities of pure mereological approaches: This also holds for e.g. the relationship between an object and its surface, or the description of being inside or outside, or touching something else; see [33]).

15.4.2 *Ground Mereotopology*

We have just given arguments for why we have to expand our language from a mereological to a mereotopological one, and we do so by adding a second predicate constant C to be intuitively understood as the relation of topological *connection* or *contact*. The minimal requirements imposed on C are that it be reflexive, symmetric, and monotonic with respect to P (whatever is connected to a part is also connected to the whole).

Definition 15.8. *A pair $\langle X, \mathsf{C}\rangle$ consisting of a non-empty set of entities (objects,...) X and a relation C called 'connected to' is said to be a* **Ground Topology** [33], *denoted by* T, *if the following axioms hold for C in X:*

$$\begin{array}{lll}(\mathrm{C1}) & \forall x & [x \mathsf{C} x]\\ (\mathrm{C2}) & \forall x,y & [x \mathsf{C} y \to y \mathsf{C} x]\\ (\mathrm{C3}) & \forall x,y,z & [z \mathsf{C} x \wedge x \mathsf{P} y \to z \mathsf{C} y]\end{array}$$

Using Th. 2.33(12),(13) and Th. 2.36(4), it is easy to show that (C3) is equivalent to

$$(\mathrm{C3}') \quad \forall x,y\big[x \mathsf{P} y \to \forall z[z \mathsf{C} x \to z \mathsf{C} y]\big]$$

a formulation that is also to be found in literature, e.g. in [33, Definition 5] or [34, p. 980/981].

There is no reason for discussing the intuitive adequacy of (C1) and (C2), whence they are part of all common mereotopologies. A theory $\mathfrak{T}$ based on only (C1) and (C2), however, is extremely weak and a lot will have to be added before we achieve an interesting qualitative topology. Obviously, by interpreting C as mereological overlap O, one gets a model of $\mathfrak{T}$; hence, what further principles to add to $\mathfrak{T}$ so as to distinguish

C from O is one of the questions depending on intuitive interpretation and ontological commitments concerning the notion of contact or connection. Following [34], we will give some reasons that should show why (C3) is a plausible candidate and does not just appear from nowhere.

Consider the following relation

$$x \mathsf{E} y := \forall z[z \mathsf{C} x \rightarrow z \mathsf{C} y] \qquad (\textit{Enclosure}).$$

Due to (C1) and (C2) E is reflexive and symmetric; if C is extensional (see (15.2)), then E is also antisymmetric, i.e. a partial order and we have the same properties for P and E. Then, for each mereological predicate using P one could now introduce a corresponding topological predicate using E instead. Probably, this is not what we are looking for. Also the simple option of just putting together mereological and topological axioms is of little interest. What one needs are rules for how P and C should behave together.

There is one sort of such a bridging principle that most theories accept: no matter how P and C are characterized, they must be related in such a way that a whole and its parts are tightly connected. Three possibilities for capturing this intuition are the following:

(C3a) $\forall x, y\ [x \mathsf{P} y \rightarrow x \mathsf{C} y]$
(C3b) $\forall x, y\ [x \mathsf{O} y \rightarrow x \mathsf{C} y]$
(C3′) $\forall x, y\ [x \mathsf{P} y \rightarrow x \mathsf{E} y]$,

where the third clause is just an equivalent reformulation of Def. 15.8(C3). The first principle is perhaps the most immediate: It simply asserts that everything must be connected to its constitutive parts. This is not to say that the parts must all be connected to one another: The two main parts of a bikini are not, but they are, in an obvious sense, connected to their sum. (C3a) is, however, a very weak requirement: It does not even capture the idea that if something is part of two things, then those things are connected because of that common part. This is made explicit in (C3b). (Note that (C3b) is stronger than (C3a): We have $x \mathsf{P} y \rightarrow x \mathsf{O} y$ by Prop. 15.3(7). Assuming $x \mathsf{O} y \rightarrow x \mathsf{C} y$ we get

$$(x \mathsf{P} y \rightarrow x \mathsf{O} y) \wedge (x \mathsf{O} y \rightarrow x \mathsf{C} y) \rightarrow (x \mathsf{P} y \rightarrow x \mathsf{C} y),$$

and hence

$$(x \mathsf{P} y \rightarrow x \mathsf{O} y) \rightarrow [(x \mathsf{O} y \rightarrow x \mathsf{C} y) \rightarrow (x \mathsf{P} y \rightarrow x \mathsf{C} y)],$$

using Th. 2.33(10) and (12).)

It is out of question that in mereotopology there should be a distinction between (topological) connection and (mereological) overlap. Indeed, (C3b) gives rise to define a notion of *external connection*, or *touching*, that can only hold between mereologically disjoint entities:

$$x \mathsf{EC} y := x \mathsf{C} y \wedge x \perp y \qquad (\textit{External Connection})$$

As Varzi [34] puts it: "*This is an important notion, which makes all the difference between mereology and mereotopology.*" Yet (C3b) is not strong enough to secure that *touching* a part, like overlapping, is sufficient for being connected: Surely someone can touch a mug just by touching its handle. Thus it is (C3) that finally proves to be a suitable requirement corresponding to intuition. (Note that (C3) is stronger than (C3b): Assume $x \mathsf{O} y$; then, by Def. 15.2(d), there is a z such that $z \mathsf{P} x \wedge z \mathsf{P} y$. So $\forall t[t \mathsf{C} z \rightarrow t \mathsf{C} x] \wedge \forall t[t \mathsf{C} z \rightarrow t \mathsf{C} y]$ by (C3) using the enclosure condition. Since $z \mathsf{C} z$ by the reflexivity of C we get $z \mathsf{C} x$ from the left implication, whence $x \mathsf{C} z$ by symmetry of C, and finally $x \mathsf{C} y$ from the right implication.)

Ground Topology T is still a very simple theory and yet it does already provide the basis for the definition of a number of important spatial relations which, like EC, cannot be distinguished within a purely mereological setting: We can, for example, express the difference between a proper part that lies entirely within the interior of the whole and a proper part that is connected with its exterior. Following [33, 34], the definition below presents a few of the newly available spatial relations.

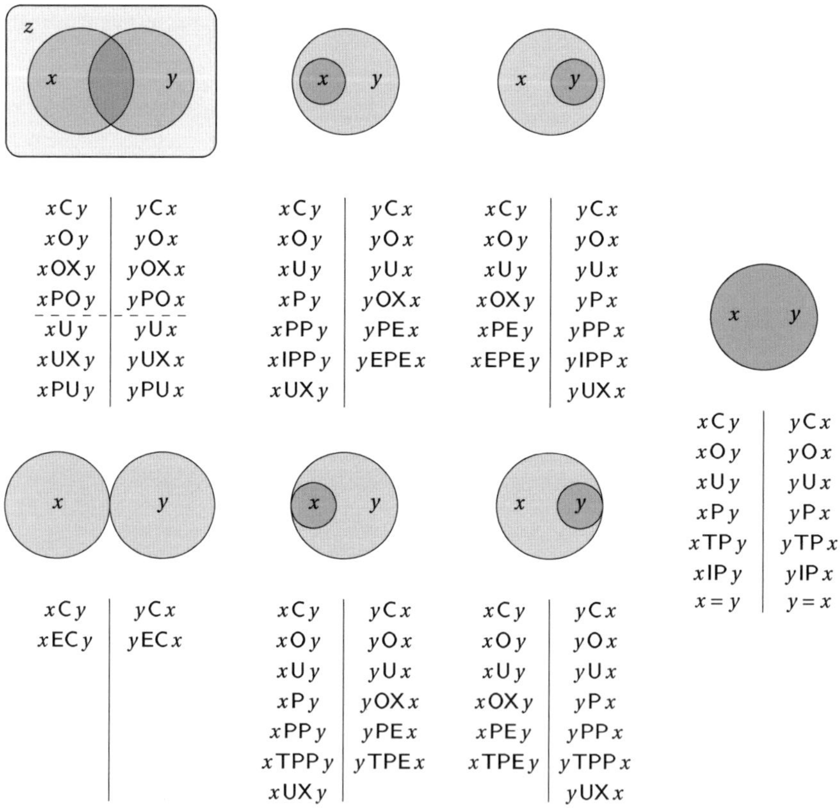

FIG. 15.6: The basic patterns of mereotopological relationship.

Definition 15.9.

(a) *(Tangential Part)* $x \mathsf{TP}\, y := x \mathsf{P}\, y \wedge \exists z[z \mathsf{EC}\, x \wedge z \mathsf{EC}\, y]]$
(b) *(Internal Part)* $x \mathsf{IP}\, y := x \mathsf{P}\, y \wedge \neg x \mathsf{TP}\, y$
(c) *(Internal Proper Part)* $x \mathsf{IPP}\, y := x \mathsf{PP}\, y \wedge \neg x \mathsf{TP}\, y$
(d) *(Tangential Proper Part)* $x \mathsf{TPP}\, y := x \mathsf{PP}\, y \wedge \neg x \mathsf{IPP}\, y$
(e) *(External Proper Extension)* $x \mathsf{EPE}\, y := x \mathsf{PE}\, y \wedge \exists z[z \mathsf{C}\, y \rightarrow z \mathsf{O}\, x]$
(f) *(Tangential Proper Extension)* $x \mathsf{TPE}\, y := x \mathsf{PE}\, y \wedge \neg x \mathsf{IPP}\, y$

Analogous to Fig. 15.1 we give an illustration of the basic patterns of the connection relations in Fig. 15.6 , again adapted from [33]. We now stop with introducing derived relations from M and/or $\mathfrak{T}$ because we think that the material presented above should provide sufficient information about what concepts are discussed and how formalizations are developed in the area of mereotopology. The final information we give is the definition of the so-called *Ground Mereotopology* together with an illustration of its basic patterns.

Definition 15.10. Ground Mereotopology [33], *denoted by* MT , *is the theory defined by the* M *-axioms (P1)-(P3) together with the* T *-axioms (C1)-(C3).*

Richer theories than MT result by adding T to stronger mereological bases than M: Let X be any mereological theory. The *Ground Mereotopology induced by* X, labelled XT, is the extension of X obtained by adding the T-axioms (C1)-(C3).

15.4.3 *The Region-Connection Calculus*

The Region Connection Calculus (RCC) is probably the best known QSR formalism in AI literature for representing and reasoning about spatial relations. RCC is a first order theory that was initially described in [28] and is based on a single primitive relation between spatial regions, the "connectedness" relation C. Using this relation various other binary relations can be defined, among them following set:

R-8 :={DC, EC, PO, EQ, TPP, NTPP, TPPi, NTPPi}.

(Note that we will retain our notation for inverses and write $\mathsf{R}^{\smile}$ instead of Ri which is commonly used in QSTR.) This set of relations is interesting for a number of reasons: It is the smallest set of base relations which allows topological distinctions rather than just mereological ones and which forms a relation algebra. Furthermore, the semantics of these relations can be described by using propositional rather than first-order logic, a property which allows to prove decidability ([2]).

The most popular variant of RCC is known as RCC-8 since it is based on the set of the eight base relations listed above. These and some other relations are defined as follows:

Definition 15.11. *Let* C *be a relation on a set X which is reflexive and symmetric, i.e. it obeys axioms (C1) and (C2) of the Ground Topology (see Def. 15.8).*

(a) $x \mathsf{DC}\, y := \neg x \mathsf{C}\, y$, "$x$ is disconnected from y";

(b) $x \mathsf{P} y$	$:= \forall z[z \mathsf{C} x \rightarrow z \mathsf{C} y]$,	"x is a part of y";
(c) $x \mathsf{PP} y$	$:= x \mathsf{P} y \wedge \neg y \mathsf{P} x$,	"x is a proper part of y";
(d) $x \mathsf{EQ} y$	$:= x \mathsf{P} y \wedge y \mathsf{P} x$,	"x is identical with y";
(e) $x \mathsf{O} y$	$:= \exists z[z \mathsf{P} x \wedge z \mathsf{P} y]$,	"x overlaps y";
(f) $x \mathsf{PO} y$	$:= x \mathsf{O} y \wedge \neg x \mathsf{P} y \wedge \neg y \mathsf{P} x$,	"x partially overlaps y";
(g) $x \mathsf{EC} y$	$:= x \mathsf{C} y \wedge \neg x \mathsf{O} y$,	"x is externally connected to y";
(h) $x \mathsf{DR} y$	$:= \neg x \mathsf{O} y$,	"x is discrete from y";
(i) $x \mathsf{TPP} y$	$:= x \mathsf{PP} y \wedge \exists z[z \mathsf{EC} x \wedge z \mathsf{EC} y]$,	"x is a tangential PP of y";
(j) $x \mathsf{NTPP} y$	$:= x \mathsf{PP} y \wedge \neg \exists z[z \mathsf{EC} x \wedge z \mathsf{EC} y]$,	"x is a non-tangential PP of y";

Pictures representing intuitive interpretations of the RCC-8 relations (without the obvious case DC) correspond to those shown in Fig. 15.6 for Ground Topology:

$$\mathsf{EC},\ \mathsf{PO},\ \mathsf{EQ},\ \mathsf{TPP},\ \mathsf{NTPP},\ \mathsf{TPP}^{\smile},\ \mathsf{NTPP}^{\smile}.$$

The composition table for RCC-8 is shown in Table 15.3 in the original form as given in [29]; the neutral element EQ has not been included. Sometimes one can find abbreviating versions using $\mathsf{DR} = \mathsf{DC} \vee \mathsf{EC}$ and $\mathsf{PP} = \mathsf{TPP} \vee \mathsf{NTPP}$ like e.g. in [24] and [13]. These relationships become clear from the "subsumption hierarchy" of possible relations between pairs of regions, all of which are definable purely in terms of C, shown in Fig. 15.7. Lines connect more restrictive relations (below) to more inclusive ones (above).

Take, for instance, $r \mathsf{TPP}\, s \wedge s \mathsf{EC}\, t$; then the following two kinds of relative location of r and t are possible, i .e. $r \mathsf{DC}\, t \vee r \mathsf{EC}\, t$ can be deduced:

15.5 Some Logical and Algebraic Considerations

As already stated at the beginning of this chapter, qualitative reasoning aims at describing some aspects of common-sense background knowledge on which our human perspective on physical reality is based. From the preceding sections we have then learned that, roughly speaking, qualitative calculi usually start from a finite set of so-called base relations, which is a family of relations that partitions the set of all pairs of domain elements $x \in \mathsf{D}$ which denote spatial or temporal entities. Qualitative knowledge is represented by specifying qualitative relations between these entities. It is then possible to apply qualitative reasoning methods for tasks such as checking consistency of the given information, deriving previously unknown information or answering queries.

Well-known qualitative calculi include Point Algebra PA and Interval Algebra IA for representing temporal relations, and the Region Connection Calculi RCC-5 and RCC-8,

TABLE 15.3: The composition table for RCC-8 (? specifies "*unknown*", i.e. R-8 $\times$ R-8).

$\diamond$	DC	EC	TPP	TPP˘	PO	NTPP	NTPP˘
DC	?	DC, EC, PO, TPP, NTPP	DC, EC, PO, TPP, NTPP	DC	DC, EC, PO, TPP, NTPP	DC, EC, PO, TPP, NTPP	DC
EC	DC, EC, PO, TPP˘, NTPP˘	DC, EC, PO, TPP, TPP˘, EQ	EC, PO, TPP, NTPP	DC, EC	DC, EC, PO, TPP, NTPP	PO, TPP, NTPP	DC
TPP	DC	DC, EC	TPP NTPP	DC, EC, PO, TPP, TPP˘, EQ	DC, EC, PO, TPP, NTPP	NTPP	DC, EC, PO, TPP˘, NTPP˘
TPP˘	DC, EC, PO, TPP˘, NTPP˘	EC, PO, TPP˘, NTPP˘	PO, EQ, TPP, TPP˘	TPP˘ NTPP˘	PO, TPP˘, NTPP˘	PO, TPP, NTPP	NTPP˘
PO	DC, EC, PO, TPP˘, NTPP˘	DC, EC, PO, TPP˘, NTPP˘	PO, TPP, NTPP	DC, EC, PO, TPP˘, NTPP˘	?	PO, TPP, NTPP	DC, EC, PO, TPP˘, NTPP˘
NTPP	DC	DC	NTPP	DC, EC, PO, TPP, NTPP	DC, EC, PO, TPP, NTPP	NTPP	?
NTPP˘	DC, EC, PO, TPP˘, NTPP˘	PO, TPP˘, NTPP˘	PO, TPP˘, NTPP˘	NTPP˘	PO, TPP˘, NTPP˘	PO, TPP˘, TPP, NTPP, NTPP˘, EQ	NTPP˘

Cardinal Relation Calculus CDC [21], and Rectangle Algebra RA [26], an extension of the interval algebra IA to a bidimensional space, for representing spatial relations.

In this section our focus shifts from concrete examples to more abstract algebraic properties and structures occurring in connection to qualitative spatio-temporal calculi. There exist a lot of algebraic results on such calculi, see e.g. [–11] for recent examples; but, as appropriate for an introductory text, we will confine ourselves to some basic and historically early considerations and results, beginning with what we have already met in Section 15.2.1 above.

Let D be a non-empty set, called the *universe* or *domain of interest*, which is the set of the entities that the qualitative formalism deals with (e.g. temporal entities such as points or intervals, or spatial entities such as points or regions). We start by formalizing the concept which forms the basis of most of the many spatial or temporal qualitative formalisms to be found in literature:

Definition 15.12. *Let* $D \neq \emptyset$ *be a domain and* $\mathcal{B} = \{\beta_1, \ldots, \beta_m\}$ *be a finite set of relations on* D. *The pair* $(D, \mathcal{B})$ *is called a* **partition scheme** *if it is JEPD,* $I_D \in \mathcal{B}$*, and for every* $\beta_i \in \mathcal{B}$ *also* $\beta_i{}^{\smile} \in \mathcal{B}$. *The* $\beta_i \in \mathcal{B}$ *are called* **base relations**, *and elements* $R \in \mathbf{2}^{\mathcal{B}}$ **disjunctive relations**, *denoting the relation* $\bigcup\{\beta : \beta \in R\} \subseteq D \times D$.

Once a partition scheme has been chosen, we have fixed the granularity and the vocabulary for describing particular configurations between elements in D. Recall that

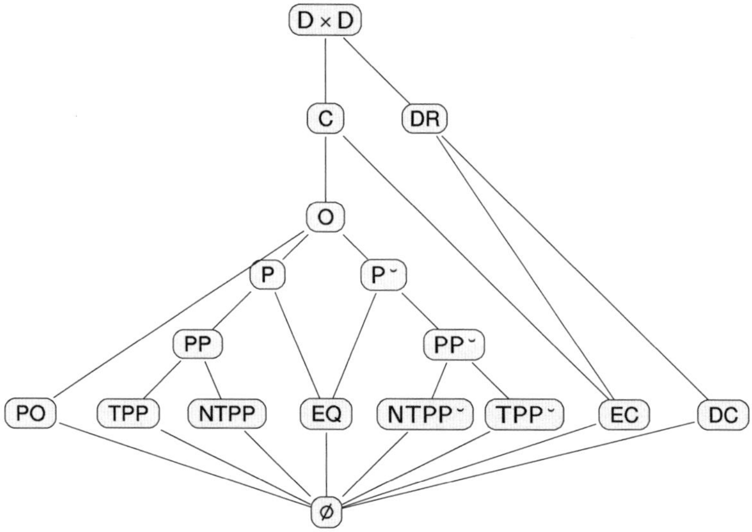

FIG. 15.7: A subsumption hierarchy for RCC relations.

each pair $\langle d, d' \rangle \in \mathsf{D} \times \mathsf{D}$ belongs to exactly one of the $\beta_i \in \mathscr{B}$, whence the β_i represent definite knowledge with respect to the given level of granularity. Indefinite knowledge can be specified using disjunctive relations, i.e. by unions of possible basic relations. We will see soon that the set of all possible unions of base relations can be made the carrier of a relation algebra.

15.5.1 *Composition Tables*

In Section 15.2.1 we have introduced the concept of weak composition for a JEPD set of base relations, which has been used for the computation of our examples for composition tables: Table 15.2 for the point algebra PA, and Table 15.3 for RCC-8. In this section we will go one step further towards a formal viewpoint turning to some algebraic aspects of composition tables and consider how reasoning about a given qualitative knowledge can be done.

The domains arising in spatial and temporal contexts are usually infinite: there are infinitely many time points or temporal intervals on the time line, and infinitely many regions in a two or three dimensional space. Therefore, it is not feasible to represent relations as sets of ordered pairs, and we have to look for new mechanisms for knowledge representation and for how to make use of it.

Since we are not dealing with explicit pairs anymore, we have to compute the algebraic operations for the base relations used. Union, complement, converse, and intersection of relations are the usual set-theoretic operations while composition is not as straightforward, as we have already mentioned above.

Constraints can be propagated using a suitable composition operation on relations.

The concept of *weak composition* has been introduced since for an arbitrary partition scheme $(\mathsf{D},\mathscr{B})$ the set $\mathscr{B}$ is not necessarily closed under usual relational composition (see (15.3) and (15.4)). Therefore weak composition $\diamond$ has been established as the best approximation (from above) that can be formulated in the language of $(\mathsf{D},\mathscr{B})$. (Recall that weak composition is defined relative to a particular partition scheme only, whence it is not to an absolute relational concept like that of usual composition.)

From a general formal point of view we start out with a set Rels of relation symbols (see e.g. [4, 24]) for which we assume the following structure: we have a symbol $eq = eq^{\smile} \in \mathsf{Rels}$ as well as pairs of symbols $r_i, r_i^{\smile}$ different from eq. A **composition table** (for Rels) is a mapping $\mathbf{CT}: \mathsf{Rels} \times \mathsf{Rels} \longrightarrow \mathbf{2}^{\mathsf{Rels}}$ that specifies for all $r, s \in \mathsf{Rels}$, their composition as a subset $\mathbf{CT}(r,s) = \{t_1, \dots, t_k\} \subseteq \mathsf{Rels}$. When Rels is finite with $|\mathsf{Rels}| = n$, **CT** can concisely be represented in an $n \times n$ matrix.

Following [24, p. 123] we say that a pair (D, v) is a **model of** (Rels, **CT**), where D is a set and v is a mapping $v: \mathsf{Rels} \longrightarrow \mathscr{B}_{\mathsf{D}}$ such that the set $\{v(r) : r \in \mathsf{Rels}\}$ is a partition of $\mathsf{D} \times \mathsf{D}$, and $v(r) \circ v(s) \subseteq \bigcup_{t \in \mathbf{CT}(r,s)} v(t)$ for all $r, s \in \mathsf{Rels}$. A model is called

- **consistent** if $t \in \mathbf{CT}(r,s) \leftrightarrow [v(r) \circ v(s)] \cap v(t) \neq \emptyset$ for all $r, s, t \in \mathsf{Rels}$, and
- **extensional** if it is consistent and $v(r) \circ v(s) = \bigcup_{t \in \mathbf{CT}(r,s)} v(t)$ for all $r, s \in \mathsf{Rels}$.

If, for sake of convenience, we put $v(r) = \rho$ etc., then in a consistent model of (Rels, **CT**)we have $t \in \mathbf{CT}(r,s)$ iff there are $a, b, c \in \mathsf{D}$ such that $a\rho b \wedge b\sigma c \wedge a\tau c$. And if the model is extensional and $t \in \mathbf{CT}(r,s)$ then whenever $a\tau c$ then there must exist some $b \in \mathsf{D}$ with $a\rho b \wedge b\sigma c$.

It is obvious that these definitions are inspired by and targeted towards the situations that we have considered earlier in this chapter as concrete examples in Section 15.2. And that symbols in Rels come in pairs clearly indicates that we allude to the concept of a relation algebra. Putting things together we introduce a new notion in

Definition 15.13. *Given a partition scheme* $(\mathsf{D},\mathscr{B})$*, then* $\mathbf{2}^{\mathscr{B}}$ *is a Boolean algebra with the usual set-theoretic operations of union, intersection, and complementation. Moreover the operation of* **weak composition** *is defined for the base relations as*

$$\rho \diamond \sigma := \{\tau \in \mathscr{B} : (\rho \circ \sigma) \cap \tau \neq \emptyset\} = \{\tau \in \mathscr{B} : \exists r, s, t \in \mathsf{D}[r\rho s \wedge s\sigma t \wedge r\tau t]\},$$

and for disjunctive relations $R, S \in \mathbf{2}^{\mathscr{B}}$ *as* $R \diamond S := \bigcup_{\rho \in R, \sigma \in S} \rho \diamond \sigma$. *Conversion is extended to disjunctive relations in the obvious way: Let* $R = \{\rho_1, \rho_2, \dots, \rho_k\}$*, then* $R^{\smile} = \{\rho_1^{\smile}, \rho_2^{\smile}, \dots, \rho_k^{\smile}\}$. $(\mathbf{2}^{\mathscr{B}}, \cup, ^{-}, \diamond, ^{\smile}, \{eq\})$ *is of type* $\langle 2,1,2,1,0 \rangle$*, and is called a* **qualitative algebra**.

Recalling Def. 10.29 one recognizes that qualitative algebras have the same type as relation algebras; so the question for the actual algebraic structure of the former immediately arises. We need the following result on relation algebras which we state without proof:

Lemma 15.14. ([16, Lemma 3.12]) *Let $\mathfrak{A} \in \mathsf{NA}$; then, given the validity of (R1) - (R3), (R5), (R7) - (R9), the Peircean law* $(x;y)\cdot z^{\smile}=0 \leftrightarrow (y;z)\cdot x^{\smile}=0$ *is equivalent to (R10).*

Proposition 15.15. *A qualitative algebra is (the representation of) a non-associative relation algebra.*

Proof. Clearly $\mathbf{2}^{\mathscr{B}}$ is a Boolean algebra under the usual set-theoretic operations, so (R1) - (R3) hold. Since $\{\rho\}\diamond\{\sigma\} = \rho\diamond\sigma$, it suffices to prove properties of weak composition for disjunctive relations. (R7) and (R8) are obvious, the remaining conditions can be shown by straightforward calculations:

(R5): $(R\cup S)\diamond T = \bigcup_{\gamma\in R\cup S,\tau\in T}\gamma\diamond\tau = \bigcup_{\gamma\in R,\tau\in T}\gamma\diamond\tau \cup \bigcup_{\gamma\in S,\tau\in T}\gamma\diamond\tau$
$= (R\diamond T)\cup(S\diamond T)$

(R6): $R\diamond\{eq\} = \bigcup_{\rho\in R}\rho\diamond eq = \bigcup_{\rho\in R}\{\tau\in\mathscr{B} : \rho\cap\tau\neq\emptyset\} = R$, since $\mathscr{B}$ is a JEPD set.

(R9): $(R\diamond S)^{\smile} = \{\tau^{\smile}\in\mathscr{B} : \tau\in R\diamond S\}$
$= \{\tau^{\smile}\in\mathscr{B} : \bigcup_{\rho\in R,\sigma\in S}\{\tau^{\smile}\in\mathscr{B} : \rho\circ\sigma\cap\tau\neq\emptyset\}\}$
$= \{\tau^{\smile}\in\mathscr{B} : \bigcup_{\rho\in R,\sigma\in S}\{\tau^{\smile}\in\mathscr{B} : \sigma^{\smile}\circ\rho^{\smile}\cap\tau^{\smile}\neq\emptyset\}\}$
$= \{\tau^{\smile}\in\mathscr{B} : \tau^{\smile}\in S^{\smile}\diamond R^{\smile}\}$

(R10): According to Lemma 15.14, we will show that for disjunctive relations R,S,T we have $(R\diamond S)\cap T^{\smile} = \emptyset$ iff $(S\diamond T)\cap R^{\smile} = \emptyset$; see also Cor. 10.40. We then get $(R\diamond S)\cap T^{\smile} = \bigcup_{\rho\in R,\sigma\in S,\tau\in T}\{\gamma\cap\tau^{\smile} : \rho\sigma\cap\gamma\neq\emptyset,\ \gamma\in R\diamond S\}$. Applying Cor. 10.40 and Prop. 15.6(3) yields the following equivalences:

$$(\rho\diamond\sigma)\cap\gamma\neq\emptyset \leftrightarrow (\rho\sigma)\cap\gamma\neq\emptyset \leftrightarrow (\sigma\gamma)\cap\rho^{\smile}\neq\emptyset \leftrightarrow (\sigma\diamond\gamma)\cap\rho^{\smile}\neq\emptyset.$$

Thus $(R\diamond S)\cap T^{\smile} = \bigcup_{\rho\in R,\sigma\in S,\tau\in T}\{(\sigma\diamond\gamma)\cap\rho^{\smile}\neq\emptyset,\ \gamma\in R\diamond S\} = (S\diamond T)\cap R^{\smile}$. □

From the classical calculi, all of the Point Algebra, Allen's interval calculus or RCC-8 are relation algebras. However, INDU, an extension of the interval calculus taking account of durations is not associative (see [27]), whereas the Cardinal Direction Calculus CDC is. For much more detailed information see the summary given in [11, Fig. 1].

This offers a good opportunity to insert some short (intuitive, not formal) remarks on different views of the boat race example from Fig. 15.3 where the Point Algebra PA has been used for modeling. We follow [22, Section 3.1] and will see how these views give rise to the mentioned qualitative calculi in a natural way.

Although the interval algebra was introduced for temporal representation and reasoning, there are also spatial calculi derived from it for applications that can require only a one-dimensional spatial representation. Traffic management is an example ([31, Section 7.3]): Roads and railway lines can be regarded as one-dimensional routes, and also air and sea traffic mainly operates on given routes. Such a route can be represented as a one-dimensional space and the vehicles on it as intervals. The main difference of vehicles on a route to IA is that vehicles and also routes have a direction.

In the case of our boat race example, each boat goes in the same direction; so we can forget about directions for the moment, and we thus work with Allen's original interval calculus. IA allows us to express finer distinctions of the relative position of two boats by viewing them as objects extending from prow to stern, as e.g. a boat overtaking another: the prow comes to the same level as the stern of the other, then it overlaps the other, etc.

Now forget about the race; then boats can go upstream or downstream. Thus we get various versions of overlapping: "*b overlaps b′ from behind going in the same direction*" or "*b overlaps b′ in the front going in the opposite direction*" etc. In sum there are 26 base relations in the Directed Intervals Algebra DIA as conceived in [30].

The models considered so far assume that each boat can move in an one-dimensional space only. If we assume that the boats are on a lake, or consider ships at sea, then the objects under consideration can now move in a two-dimensional space. Likewise you may want to look at e.g. some European cities. The relative positions can then be described using the cardinal directions North, South, East, West, and intermediate directions North West etc. So, together with equality, we get a total of nine basic relations: =, n, w, s, e, nw, sw, se, ne. See e.g. [21].

15.5.2 *Constraint-based Qualitative Reasoning*

What makes (natural or artificial) intelligent systems intelligent is their ability to reason about given knowledge. From the different reasoning tasks an intelligent system might have to perform we will look at the one of deriving new knowledge from, and on checking the consistency of some given information. For this to become possible, a suitable representation has to be used.

Knowledge about the relationships between spatial or temporal entities is often given in the form of *constraints*. When trying to place furniture in a room, you will e.g. have to take account for the room's length (unary), for the requirement that "*the desk should be placed in front of the window*" (binary), or that "*the table should be placed between the sofa and the armchair*" (ternary). The considerations below will be restricted to binary constraints and to those basic concepts which are needed for understanding the most fundamental principles of the modeling and propagation approaches.

Knowledge between different entities can be represented by using constraints, encoded as a relation ρ over a domain D: A binary constraint $x\rho y$ between two variables x and y restricts the possible instantiations of x and y to the pairs contained in the relation ρ. Spatio-temporal constellations can then be represented by qualitative constraint networks, which are defined as follows:

Definition 15.16. *Let* $P = (D, \mathscr{B})$ *be a partition scheme. A* P-**constraint network** *is a pair* $\Theta = (V, h)$ *consisting of a finite set* V *of* variables *representing the spatial or temporal entities, and a map* $h : V \times V \longrightarrow \mathbf{2}^{\mathscr{B}}$ *such that* $h(v, v') = h(v', v)^{\smile}$ *and* $h(v, v) = I_V$ *for all* $v, v' \in V$.

The notion of a composition table (CT), combined with constraint based techniques, has become a key tool used in QSR adopted from work in temporal reasoning [1]. Reasoning can be done by exploiting compositions of the base relations which are precomputed and stored in a composition table. For a fixed vocabulary of relations, say

$\mathscr{B}$, the composition table allows the answer of the following question by simple look-up: given two relational facts $r\rho s$ and $s\sigma t$, $\rho, \sigma \in \mathscr{B}$, what are the possible relations in $\mathscr{B}$ that can hold between r and t?

A constraint network $\Theta = (V, h)$ can be visualized as a directed graph, whose nodes are from V and whose edges are labeled by the corresponding $h(v, v')$. Note that due to the definition given above loops at a node labeled with I_V can be omitted, and it is sufficient to draw only one of the two directed edges between a pair of nodes, since the other is determined by the requirement $h(v, v') = h(v', v)^{\smile}$. Moreover, we can also do without edges between nodes v, v' labeled with $\mathscr{B}$, which corresponds to the idea that edges represent constraints: $\mathscr{B}$ has the meaning that there is none, since all is possible or allowed.

A fundamental problem in the domain of qualitative reasoning then is to determine whether a given constraint network is consistent, and if this is the case, what are the compatible configurations.

Definition 15.17. *Let $P = (D, \mathscr{B})$ be a partition scheme and $\Theta = (V, h)$ be a P-constraint network. A map $s: V \longrightarrow D$ is a* **solution** *of Θ if $\langle s(x), s(y)\rangle \in h(x, y)$ for all $x, y \in V$. Θ is* **consistent** *if it has a solution.*

An example for a qualitative constraint network has already been given above in Fig. 15.2. This QCN (without the dashed arrow) is consistent, since there are the three compatible configurations of Fig. 15.3.

The question whether a given set of constraints is consistent is a special case of a *constraint satisfaction problem* (CSP). A CSP can be viewed as a triple $(V, \mathrm{D}, \mathscr{C})$ composed of a finite set of variables V, a domain D with possible instantiations for each variable $v_i \in V$ and a finite set $\mathscr{C}$ of constraints between the variables of V. The constraint satisfaction problem consists in determining if there exists an assignment of values to these variables such that all the constraints of $\mathscr{C}$ are satisfied: A *solution* of a CSP is an instantiation of each variable $v_i \in V$ with a value $d_i \in \mathrm{D}$ such that for each constraint $v_i \rho v_j \in \mathscr{C}$ we have $d_i \rho d_j$. If a CSP has a solution, it is called *consistent* or *satisfiable.*

Given a consistent set of constraints, one does, of course, ask what additional information and relationships on the entities involved can be concluded. Reasoning about temporal or spatial qualitative constraint networks in the same kind as done in CSPs has proven a fruitful approach. Readers interested in constraint-based reasoning and its computational tractability might want to have a look to e.g. [8, 18]. For a short introductory overview on 'Reasoning Problems Associated with QCNs' you might also consult the equally named Section 3.3 of [32]; an introductory article of a more general character is [7].

Additional Reading

Of course, there have been developed formal representations that combine spatial and temporal knowledge into single, so-called *spatio-temporal* frameworks. Qualitative spatio-temporal reasoning is concerned with reasoning over time and space, in particular reasoning about spatial change. But these combined frameworks are beyond the scope of this chapter. As starting points we refer to [25], an older paper but including a good overview, or to [3] for a concrete example.

For a recent overview on spatial representations see [6]. We also have to mention an area we have not dealt with at all: the questions of tractability and complexity in connection with composition tables. For these aspects you may want to start with [31].

References

[1] Allen, J.F.: Maintaining knowledge about temporal intervals, Communications of the ACM **26**, 832-843 (1983).

[2] Bennett, B.: Modal Logics for Qualitative Spatial Reasoning, Logic J. of the IGPL **4**, 23-45 (1996).

[3] Bennett, B., Cohn, A. et al.: Multi-Dimensional Modal Logic as a Framework for Spatio-Temporal Reasoning, Applied Intelligence **17**, 239-251 (2002).

[4] Bennett, B., Isli, A., Cohn, A.: When does a composition table provide a complete and tractable proof procedure for a relational constraint language. In: Proceedings of the IJCAI-97 workshop on Spatial and Temporal Reasoning, Nagoya, Japan (1997).

[5] Casati, R., Varzi, A.C.: *Parts and Places-The Structures of Spatial Representation,* MIT Press, Cambridge, Massachusetts (1999).

[6] Chen, J., Cohn, A.G. et al.: A survey of qualitative spatial representations. The Knowledge Engineering Review, **30**, 106-136 (2015).

[7] Cohn, A.G., Renz, J.: Qualitative Spatial Representation and Reasoning. In: van Harmelen et. al., *Handbook of Knowledge Representation,* 551 596. Elsevier (2007).

[8] Dechter, R.: *Constraint Processing,* Morgan Kaufmann, San Francisco (2003).

[9] Düntsch, I.: Relation algebras and their application in temporal and spatial reasoning, Artificial Intelligence Review **23**, 315-357 (2005).

[10] Dylla, F., Lee, J. H., Mossakowski, T., Schneider, T. et al.: A survey of qualitative spatial and temporal calculi: algebraic and computational properties, ACM Computing Surveys **50**, 7 (2017).

[11] Dylla F., Mossakowski T., Schneider T., Wolter D.: Algebraic Properties of Qualitative Spatio-temporal Calculi. In Tenbrink T., Stell J., Galton A., Wood Z. (Eds): *Spatial Information Theory.* COSIT 2013. Lecture Notes in Computer Science, vol 8116. Springer (2013).

[12] Freksa, C.: Using Orientation Information for Qualitative Spatial Reasoning. In Frank, A.U., Campari, I., Formentini, U. (Eds.): *Theories and Methods of Spatial-Temporal in Geographic Space Reasoning,* 162-78, Springer-Verlag (1992).

[13] Gotts, N.M., Gooday, J.M., Cohn, A.G.: A Connection Based Approach to Common-Sense Topological Description and Reasoning, The Monist **79**, 51-75 (1996).

[14] Gruszczynski, R., Pietruszczak, A.: Space, points and mereology. On foundations of point-free Euclidean geometry, Logic and Logical Philosophy **18**, 145-188 (2009).

[15] Hahmann, T., Gruninger, M.: Region-based theories of space: Mereotopology and beyond. In Hazarika, S.M. et al. (Eds.): *Qualitative Spatio-Temporal Representation and Reasoning: Trends and Future Direction,* IGI Global, 1-62 (2012).

[16] Hirsch, R., Hodkinson, I.: *Relation Algebras by Games*, Elsevier Science B.V., Amsterdam (2002).

[17] Johansson, I.: On the transitivity of the parthood relations. In Hochberg, H. and Mulligan, K. (Eds.): *Relations and predicates*, Ontos Verlag, Frankfurt Lancaster, 161-181 (2004).

[18] Lecoutre, C.: *Constraint Networks-Techniques and Algorithms*, ISTE Ltd and John Wiley & Sons, London & Hoboken, NJ (2009).

[19] Leonard, H.S., Goodman, N.: The Calculus of Individuals and Its Uses, Journal of Symbolic Logic **5**, 45-55 (1940).

[20] Leśniewski, S.: *Podstawy ogólnej teoryi mnogości. I*, Moskow: Prace Polskiego Koła Naukowego w Moskwie, Sekcya matematyczno-przyrodnicza; Eng. trans. by D.I. Barnett: 'Foundations of the General Theory of Sets. I', in S. Leśniewski, *Collected Works* (ed. by S.J. Surma et al.), Dordrecht: Kluwer, 1992, Vol. 1, 129-173 (1916).

[21] Ligozat, G.: Reasoning about cardinal directions, J. of Visual Languages and Computing **9**, 23-44 (1998).

[22] Ligozat, G.: Categorical methods in qualitative reasoning: The case for weak representations. In Cohn, A.G., Mark, D.M. (Eds.): *COSIT 2005, International Conference on Spatial Information Theory*, Springer-Verlag Berlin Heidelberg, 265-282 (2005).

[23] Ligozat, G.: *Qualitative Spatial and Temporal Reasoning*, ISTE Ltd and John Wiley & Sons, Inc., London and Hoboken, NJ (2012).

[24] Li, S., Ying, M.: Region Connection Calculus-Its models and composition table, Artificial Intelligence **145**, 121-146 (2003).

[25] Muller, P.: Topological spatio-temporal reasoning and representation, Computational Intelligence **18**, 420-450 (2002).

[26] Navarrete, I., Morales, A., Sciavicco, G., & Cardenas-Viedma, M. A.: Spatial reasoning with rectangular cardinal relations, Annals of Mathematics and Artificial Intelligence **67**, 31-70 (2013).

[27] Pujari, A. K., Vijaya Kumari, G., and Sattar, A. : INDU: An Interval and Duration Network. In *Proceedings of the 12th Australian Joint Conference on Artificial Intelligence*, Springer-Verlag Berlin Heidelberg, 291-303 (1999).

[28] Randell, D.A., Cui, Z., Cohn, A.G.: A spatial logic based on regions and connection. In B. Nebel, W. Swartout, C. Rich (Eds.): *Proceedings of the 3rd International Conf Knowledge Representation and Reasoning*, Morgan Kaufmann, Los Allos, CA,165- 176 (1992).

[29] Randell, D.A., Cohn, A.G.: Exploiting lattices in a theory of space and time, Computers Math. Applic. **23**, 459-476 (1992).

[30] Renz, J.: Spatial Odyssey of the Interval Algebra: 1. Directed Intervals. In *Proceedings of IJCAI 2001*, 51-56 (2001).

[31] Renz, J., Nebel, B.: Qualitative spatial reasoning using constraint calculi. In Aiello, M., Pratt-Hartmann, I., van Benthem, J. (Eds.): *Handbook of Spatial Logics*, Springer, Dordrecht, The Netherlands, 161-215 (2007).

[32] Sioutis, M.: Algorithmic contributions to qualitative constraint-based spatial and temporal reasoning, PhD thesis, Université d'Artois (2017).

[33] Varzi, A.C.: Parts, Wholes, and Part-Whole Relations: The Prospects of Mereotopology, Data and Knowledge Engineering **20**, 259-286 (1996).

[34] Varzi, A.C.: Spatial reasoning and ontology-parts, wholes, and locations. In Aiello, M., Pratt-Hartmann, I., van Benthem, J. (Eds.): *Handbook of Spatial Logics*, Springer, Dordrecht, The Netherlands, 945-1038 (2007).

[35] Varzi, A.C.: "Mereology", The Stanford Encyclopedia of Philosophy (February 2016), (Working URL as of May 2020: `http://plato.stanford.edu/entries/mereology/`).

[36] Vilain, M.B., Kautz, H.A., van Beek, P.G.: Constraint propagation algorithms for temporal reasoning: A revised report. In Weld, D.S., De Kleer, J. (Eds.): *Readings in Qualitative Reasoning About Physical Systems,* 373-382 (1989).

Chapter 16

Argumentation Frameworks

16.1 Introduction

"Argumentation constitutes a major component of human intelligence. The ability to engage in arguments is essential for humans to understand new problems, to perform scientific reasoning, to express, clarify and defend their opinions in their daily lives." This is how Dung's seminal paper [18] begins, and nothing more has to be said to demonstrate the importance of this kind of finding a conclusion or decision based upon given data (a knowledge base) - some of them being facts, others only hypotheses, and some of them eventually conflicting with each other. Given that this kind of reasoning is ubiquitous, it is no surprise that argumentation itself has become a major focus of research in Artificial Intelligence (AI) since the mid nineties..

But, of course, how humans think and make their conclusions has been of interest long before: In Western Philosophy, the systematic study of reasoning can be traced back to the ancient Greek philosophers, who searched for the requirements that make an argument correct (remember the topics of Section 1.1.3). From their work to present day, the understanding of human reasoning processes has been slowly progressing and finally resulted in the many formalisms that have been created mainly during the development of modern mathematical logic, computer science and artificial intelligence.

As an example for human reasoning (taken from [13]), assume that you have to make an important decision like buying a house or accepting a job you have been offered. The conclusion you have to reach in such situations usually will not be achieved by deductive reasoning. Instead, most likely one will construct arguments in favor of the new job (higher salary, better prospects for the future) and arguments against it (farther away, bigger workload, and thus less time for your family). Finally, one has to weigh the pro and con arguments against each other to reach a decision.

Argumentation can be seen as a particular useful and intuitive paradigm for doing non-monotonic reasoning; it structures the reasoning process into modular and quite intuitive steps as follows:

(1) Construct a set of arguments in favor or against statements given a priori.
(2) Evaluate the strength of each argument according to some chosen weighting method.

(3) Determine the different interactions between these arguments (e.g. possibly existing conflicts).
(4) Select the most acceptable arguments according to a predefined set of required properties.
(5) Use these selected arguments to make your conclusions.

Roughly one can view the idea of argumentation-based reasoning as follows: Whether or not a rational agent believes in a statement depends on whether or not the arguments supporting this statement can be successfully defended against counterarguments. And an argument can be seen as a set of statements (propositions), made up of three parts, a conclusion, a set of premises, and an inference from the premises to the conclusion. The notion of argument commonly refers to the concepts of explanation, justification, or proof, so arguments aim to justify beliefs or decisions.

Before going into formal details, it is illustrative to elaborate on some characteristics of argumentation-based reasoning, among others the ones that cause it to differ considerably from the purely deductive approach of first order logic:

- Argumentation-based reasoning is inspired by human reasoning. It is performed by building arguments and examining their interactions, which is, to a large extent, similar to human reasoning. As a consequence, the argumentation approach can be given dialectical proof procedures that are quite close to the process by which humans would discuss an issue. See e.g. [21,22].
- An advantage of argumentation-based reasoning is its ability to perform *non-monotonic reasoning*, which is characterized by the fact that the introduction of new knowledge can invalidate previously derived conclusions; i.e., non-monotonic reasoning is able to revise previously reached conclusions when new knowledge is acquired. This property is useful because it allows to draw conclusions utilizing defaults and assumptions when complete knowledge is not available, and, if necessary, to make changes when new knowledge is acquired.
- Another advantage of argumentation-based reasoning is its ability to handle inconsistent information. Reasoning formalisms such as classical first order logic require knowledge bases to be consistent, because inconsistency leads to inappropriate reasoning outcomes: anything can be inferred from a contradiction.

This chapter is devoted to the presentation of the original concept of argumentation frameworks as introduced in 1995 by Dung [18], some of the extensions that have been propagated in the meantime, and its description in terms of relations and images.

16.2 Key Concepts of Argumentation Frameworks

In the just mentioned paper Dung introduced an abstract formalism for argumentation-based inference, called *Argumentation Framework*, which assumes as its components nothing else than a set of *arguments* A together with a binary *attack/defeat relation* ↱.

Though nothing is said on the structure of the arguments or the nature of the attack relation, a powerful and interesting theory has been developed on this simple basis. Recent years have witnessed intensive formal study, development and application of Dung's ideas in various directions. So Dung's formalism has become the most influential one, providing a basis for many extensions, generalizations, and instantiations.

In Dung's approach, arguments are considered as atomic items, focusing entirely on the relations between them. Consequently, it is assumed that these abstract arguments and the relevant relations between them have already been constructed, usually from a given background knowledge base. The obtained argumentation system is then evaluated on the abstract level, yielding possibly alternative sets of arguments which may be collectively accepted. Finally, these results are interpreted in terms of the original knowledge base.

Definition 16.1. *A* **Dung argumentation framework** *(DAF), also* **argumentation framework** *(AF), is a pair* $\mathscr{D} = \langle A, \Rsh \rangle$ *such that*

(1) $A \neq \emptyset$ *is a finite set of* arguments, *and*
(2) $\Rsh \subseteq A \times A$*, is a binary relation on* A*; if for* $a, b \in A$ *we have* $a \Rsh b$*, then we say that* a attacks *or* defeats b*, or equivalently, that* a *is a* counterargument *to* b.
(3) An argument $a \in A$ *is* attacked by a set of arguments $B \subseteq A$*, if there is a* $b \in B$ *such that* $b \Rsh a$*:* $B \Rsh a \;:\Leftrightarrow\; \exists b \in B[b \Rsh a]$.

A particular feature of abstract argumentation frameworks is their simple structure: abstract argumentation frameworks are just directed graphs with arguments as vertices, and edges indicating a conflict between the two connected arguments. Though Dung's concept is the most abstract, and hence most simple, formalism for abstract argumentation, it is quite powerful. Due to the great interest it has received, many semantics have been developed, defining the properties of sets of arguments $E \subseteq A$ that are regarded acceptable as a rational opinion.

Two main approaches to the definition of argumentation semantics are available in the literature: (a) The idea underlying the *labeling*-based approach is to give each argument exactly one label. A common choice for the set of labels is: `in`, `out` or `undec`, where the label `in` means that the argument is accepted, the label `out` means that the argument is rejected, and `undec` means that one abstains from an opinion on whether the argument is accepted or rejected. A specific labeling-based argumentation semantics provides a way to select 'reasonable' labellings among all the possible ones, according to some criterion embedded in its definition. (b) An *extension*-based argumentation semantics is defined by specifying the criteria for deriving, for a generic argumentation framework, sets of arguments which can survive the conflict together and thus represent collectively a reasonable position an autonomous reasoner might take. Such sets of acceptable arguments are called **extensions**.

While the adoption of the labeling or extension-based style is a matter of subjective preference, it is natural to ask about the expressiveness of the two styles. Obviously, any extension-based definition can be equivalently expressed in a simple labeling-based formulation using a set of two labels {`in`, `out`} with `in` corresponding to extension membership. On the other hand, an arbitrary labeling cannot in general be formulated in terms of extensions. However, labellings considered in the literature typically include a label called `in` which naturally corresponds to membership in an extension, while other labels correspond to a partition of other arguments. So equivalent extension-based formulations of labeling-based semantics are typically available. The following considerations we will be confined to extension-based semantics as originally introduced in [18].

Definition 16.2.

(a) *A* **semantics for argumentation frameworks** *is given via a function σ which assigns to each DAF $\mathscr{D} = \langle A, \vdash\rangle$ a set $\sigma(\mathscr{D}) \subseteq \mathbf{2}^A$ of extensions. A member $E \in \sigma(\mathscr{D})$ is called a σ-extension in $\mathscr{D}$.*

(b) *If a semantics σ satisfies the condition $|\sigma(\mathscr{D})| = 1$ for any argumentation framework, then σ is said to be a* **unique-status semantics**.

There is a general agreement that any extension E has to fulfill the following minimal requirements: E must be consistent, i.e. free of conflicts, and for each $a \in E$ we should be able to defend it. The basic idea is that an argument is acceptable with respect to a set S of arguments if it is defended by that set S against all its defeaters. This gives rise to the following formal concepts:

Definition 16.3. *Let $\mathscr{D} = \langle A, \vdash\rangle$ be a DAF.*

(a) *An argument $a \in A$ is* **acceptable w.r.t. a set** *$S \subseteq A$, if every attacker of a is attacked by S (it is also said that S* **defends an argument** *a (in $\mathscr{D}$)):*

$$S\,\mathsf{Def}\,a :\Leftrightarrow \forall b \in A[b \vdash a \rightarrow S \vdash b].$$

(b) *By $\mathsf{Def}(S) : \mathbf{2}^A \longrightarrow \mathbf{2}^A$ we denote the set of all $a \in A$ which are defended by S: $\mathsf{Def}(S) = \{a \in A : S\,\mathsf{Def}\,a\}$.*

(c) *A set $S \subseteq A$ is called* **conflict-free** *(in $\mathscr{D}$), denoted as $S \in cf(\mathscr{D})$, if no argument in S attacks another argument in S.*

(d) *Let $S \subseteq A$; the* **range of S**, *denoted as S^+, is the set $S \cup \{b : S \vdash b\}$.*

Based on these few simple concepts several semantics for acceptability have been defined in literature as follows (see e.g. [19]):

Definition 16.4. *Let $\mathscr{D} = \langle A, \vdash\rangle$ be a DAF, and $S \in cf(\mathscr{D})$. Then S is called a*

(a) **naive set** *in $\mathscr{D}$, denoted as $S \in naive(\mathscr{D})$, if there is no $T \in cf(\mathscr{D})$ with $T \supset S$, i.e. S is a $\subseteq$-maximal conflict-free set in $\mathscr{D}$;*

(b) **admissible set** *in $\mathscr{D}$, denoted as $S \in adm(\mathscr{D})$, if it is conflict-free and defends all its elements, i.e. $S \subseteq \mathsf{Def}(S)$;*

(c) **stable extension** *in $\mathscr{D}$, denoted as $S \in stb(\mathscr{D})$, if S is admissible and attacks all arguments outside it: $S \vdash = \overline{S}$ or, equivalently, $S^+ = A$;*

(d) **preferred extension** *in $\mathscr{D}$, denoted as $S \in prf(\mathscr{D})$, if there is no $T \in adm(\mathscr{D})$ with $T \supset S$, i.e. S is a $\subseteq$-maximal admissible set in $\mathscr{D}$;*

(e) **complete extension** *in $\mathscr{D}$, denoted as $S \in com(\mathscr{D})$, if it is admissible and already contains all arguments defended by S, i.e. $S = \mathsf{Def}(S)$;*

(f) **grounded extension** *in $\mathscr{D}$, denoted as $S \in grd(\mathscr{D})$, if there is no $T \in com(\mathscr{D})$ with $T \subset S$, i.e. S is a $\subseteq$-minimal complete extension;*

(Note that the definitions given above are not exactly the same as in [18], but are proven to be equivalent in [16].)

As already mentioned above, the idea underlying the extension-based approach is to identify sets of arguments which can survive the conflict together and thus represent collectively a rational position an autonomous reasoner might take: An *admissible set* of arguments is therefore conceived of as a conflict-free set $S \subseteq A$ such that the arguments in S defend themselves "collectively" against any external attack. A *complete extension* is a set of arguments that is conflict free and consists of all and only the arguments that it defends. Note that $\emptyset$ is always admissible, and that, contrary to the other notions defined above, stable extensions do not always exist. Every DAF has precisely one *grounded extension*, even if it is empty. The grounded extension contains all and only arguments whose defense can be traced back to any of the unattacked arguments.

We illustrate the definitions given so far with a running example in argumentation theory (often to be found in literature, e.g. in [9, 11, 13]) in Fig. 16.1:

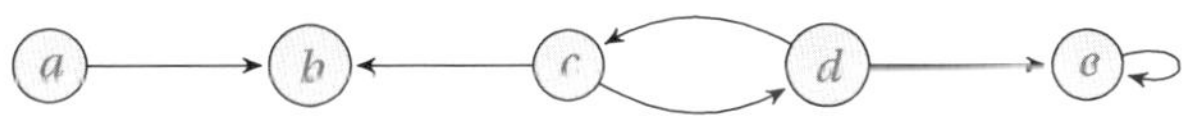

FIG. 16.1: A simple attack graph.

This attack graph represents the DAF $\mathscr{B} = \langle B, \upharpoonright\!\!\rightarrow \rangle$ with

$$B = \{a, b, c, d, e\} \text{ and } \upharpoonright\!\!\rightarrow = \{\langle a, b\rangle, \langle c, b\rangle, \langle c, d\rangle, \langle d, c\rangle, \langle d, e\rangle, \langle e, e\rangle\}.$$

We have $adm(\mathscr{B}) = \{\emptyset, \{a\}, \{c\}, \{d\}, \{a, c\}, \{a, d\}\}$, and according to the semantics given in Def. 16.4 we get the following acceptable sets:

$$\begin{aligned}
stb(\mathscr{B}) &= \{\{a, d\}\};\\
sem(\mathscr{B}) &= \{\{a, d\}\};\\
stage(\mathscr{B}) &= \{\{a, d\}\};\\
prf(\mathscr{B}) &= \{\{a, c\}, \{a, d\}\};\\
cf(\mathscr{B}) &= adm(\mathscr{B}) \cup \{\{b\}, \{b, d\}\}, \text{ thus}\\
naive(\mathscr{B}) &= prf(\mathscr{B}) \cup \{\{b, d\}\};\\
com(\mathscr{B}) &= \{\{a\}, \{a, c\}, \{a, d\}\},\\
grd(\mathscr{B}) &= \{\{a\}\}.
\end{aligned}$$

Some more extension-based semantics have been defined; among them the following are discussed frequently (see e.g. [7, 8], which give a much more detailed account on many aspects of argumentation semantics):

Definition 16.5. *Let $\mathscr{D} = \langle A, \Rsh \rangle$ be a DAF, and $S \in cf(\mathscr{D})$; S is called*

(a) a **stage extension** *in $\mathscr{D}$, denoted as $S \in stage(\mathscr{D})$, if there is no $T \in cf(\mathscr{D})$ with $T^+ \supset S^+$, i.e. S is conflict-free with $\subseteq$-maximal range;*
(b) a **semi-stable extension** *in $\mathscr{D}$, denoted as $S \in sem(\mathscr{D})$, if there is no $T \in adm(\mathscr{D})$ with $T^+ \supset S^+$, i.e. S is a complete extension with $\subseteq$-maximal range;*
(c) an **ideal extension** *in $\mathscr{D}$, if S is a maximal ideal set. (S is an* **ideal set** *if it is a subset of every preferred extension.)*

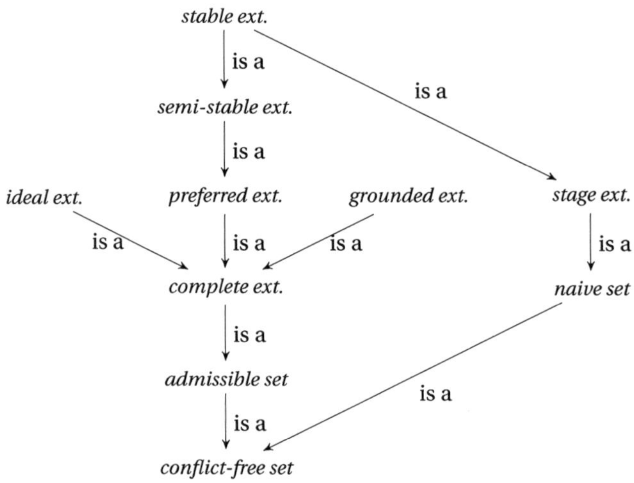

FIG. 16.2: Relationships of extension based argumentation semantics.

Some basic properties of these semantics are as follows: For any of the considered semantics σ except stable semantics we have that $\sigma(\mathscr{D}) \neq \emptyset$, i.e. these semantics always propose at least one extension; the grounded is a unique status semantics. Semi-stable semantics aims at guaranteeing the existence of extensions in any case (different from stable semantics) while coinciding with stable semantics (different from preferred semantics) when stable extensions exist. The intuition is that semi-stable extensions should be "close" to stable extensions, in case the latter do not exist. Ideal semantics (see [15]) provides an alternative unique-status approach which is less skeptical than grounded semantics, i.e. for any argumentation framework the (unique) ideal extension is a (sometimes strict) superset of the grounded extension. Furthermore it is known that the ideal extension is also complete.

For any DAF $\mathscr{D} = \langle A, \Rsh \rangle$, the following chains of subset relations hold:

$$stb(\mathscr{D}) \subseteq stage(\mathscr{D}) \subseteq naive(\mathscr{D}) \subseteq cf(\mathscr{D}),$$
$$stb(\mathscr{D}) \subseteq sem(\mathscr{D}) \subseteq prf(\mathscr{D}) \subseteq com(\mathscr{D}) \subseteq adm(\mathscr{D}) \subseteq cf(\mathscr{D}).$$

Taking all the results together we get the overall picture for the most commonly used extension semantics as shown in Fig. 16.2.

In addition to all the above concepts we can decide about how strict we like to be in acceptance of a calculated extension. This is expressed in

Definition 16.6. *Let $\mathscr{D} = \langle A, \Rsh \rangle$ be a DAF, $a \in A$ an argument, and σ be a semantics.*

*(a) a is σ-**credulously accepted** iff it is a member of at least one σ-extension in $\mathscr{D}$: $a \in \bigcup_{E \in \sigma(\mathscr{D})} E$;*

*(b) a is σ-**sceptically accepted** iff it is a member of every σ-extension in $\mathscr{D}$: $a \in \bigcap_{E \in \sigma(\mathscr{D})} E$.*

Although they present only some simple intuitive ideas, the above considerations show that argumentation frameworks are another example of how to create a whole new universe of interesting concepts out of a set and one arbitrary (!) relation: Nothing has been said about the attack relation $\Rsh$ beyond being one of two components in a DAF. A few hints at more advanced variants of DAFs can be found in Section 16.4 below.

16.3 DAFs from a Relational Viewpoint

After the short presentation of argumentation frameworks in intuitive terms given in the foregoing section, we will now recast some of the main concepts into a more technical notation as shown in Table 16.1, and will also report on some of their properties in terms of relations and images. We will also use Bandler-Kohout products and images introduced in Section 6.1.

TABLE 16.1: DAF concepts in relational formulation.

DAF concept	**relational form**	**FOL form**
a attacks b	$a \Rsh b$	-
S attacks b	$S \Rsh b$	$\exists s \in S\, [s \Rsh b]$
a is acceptable w.r.t. S S defends a	$S\, \mathsf{Def}\, a$	$\forall b \in A\, [b \Rsh a \rightarrow S \Rsh b]$
$S \in cf(\mathscr{D})$	$S \subseteq \overline{S \Rsh}$	$\neg \exists a, b \in S\, [a \Rsh b]$
$S \in stb(\mathscr{D})$	$S = \overline{S \Rsh}$	$\forall a \in \overline{S}\, \exists s \in S\, [s \Rsh a]$
$S \in adm(\mathscr{D})$	$S \subseteq \overline{S \Rsh} \wedge S \subseteq \overline{\overline{S \Rsh} \Rsh}$	-
$S \in com(\mathscr{D})$	$S \subseteq \overline{S \Rsh} \wedge S = \overline{\overline{S \Rsh} \Rsh}$	-

From the concepts introduced so far, admissible sets offer a good opportunity for comparing the usual DAF with an argumentation-free, i.e. purely syntactical notation. Let us begin with some elementary properties.

Remember that in Def. 16.4(b) above an admissible set of arguments in $\mathscr{D}$ has been defined as a conflict-free set $S \subseteq A$ (i.e. no argument in S attacks another argument in

S) such that S defends all its elements. (For the following recall that $S\Rsh = \Rsh[S]$. Do also note that we mix notation for sake of compactness.)

Lemma 16.7. *Let $\mathscr{D} = \langle A, \Rsh\rangle$ be a DAF, and $S \subseteq A$. Then the following holds:*

(1) $S \in cf(\mathscr{D}) \leftrightarrow S \subseteq \overline{S\Rsh}$, *and* $S \in cf(\mathscr{D}) \rightarrow \Rsh S \subseteq \overline{S}$.
(2) $\mathsf{Def}(S) = \{a \in A : \Rsh a \subseteq S\Rsh\}$.
(3) $\mathsf{Def}(S) = \Rsh^{\rhd}[S\Rsh]$.
(4) $\mathsf{Def}(S) = \overline{\overline{S\Rsh}\Rsh}$.
(5) $\mathsf{Def}(S \cap T) \subseteq \mathsf{Def}(S) \cap \mathsf{Def}(T)$.

Proof. (1): $S \in cf(\mathscr{D}) \;\leftrightarrow\; \neg\exists a, b \in S[a\Rsh b] \;\leftrightarrow\; S \subseteq \overline{S\Rsh}$. So, $\Rsh S = \{t \in A : \exists s \in S[t\Rsh s]\} \subseteq \overline{S}$.
(2): $\mathsf{Def}(S) = \{a \in A : S\,\mathsf{Def} a\} = \{a \in A : \forall b \in A[b\Rsh a \rightarrow S\Rsh b]\} = \{a \in A : \Rsh a \subseteq S\Rsh\}$.
(3): By (2), using Def. 6.2(c). We can exclude the trivial case $\Rsh a = \emptyset$: If a is not attacked by any argument there is no need to defend it; so we may set $\mathsf{Def}(S) = \{a \in A : \emptyset \neq \Rsh a \subseteq S\Rsh\}$
(4): Using (2) we get $\mathsf{Def}(S) = \{a \in A : \Rsh a \subseteq S\Rsh\} = \{a \in A : \Rsh a \cap \overline{S\Rsh} = \emptyset\} =$
$\overline{\{a \in A : \Rsh a \cap \overline{S\Rsh} \neq \emptyset\}} = \overline{\{a \in A : \exists b \in \overline{S\Rsh}\,[b\Rsh a]\}} = \overline{\overline{S\Rsh}\Rsh}$.
(For sake of comparison we reproduce the argumentation based proof as given in [9, Prop. 1]: By definition, all the attackers of an argument defended by S are attacked by S. Consequently, an argument defended by S is not attacked by an argument not attacked by S, that is $\mathsf{Def}(S) \subseteq \overline{\overline{S\Rsh}\Rsh}$. Let $a \in \overline{\overline{S\Rsh}\Rsh}$. Then no attacker of a belongs to $\overline{S\Rsh}$. Consequently, all the attackers of a are attacked by S, and hence $a \in \mathsf{Def}(S)$.)
(5): Using (2) and Prop. 3.9(6) for the two starting lines we get

$$\begin{aligned}
\mathsf{Def}(S \cap T) &= \{a \in A : \Rsh a \subseteq \Rsh[S \cap T]\} \\
&\subseteq \{a \in A : \Rsh a \subseteq \Rsh[S] \cap \Rsh[T]\} \\
&= \{a \in A : \Rsh a \subseteq \Rsh[S]\} \cap \{a \in A : \Rsh a \subseteq \Rsh[T]\} \\
&= \mathsf{Def}(S) \cap \mathsf{Def}(T).
\end{aligned}$$

□

Proposition 16.8. ([9, Theorem 5]) *Let $\mathscr{D} = \langle A, \Rsh\rangle$ be a DAF, and $S \in cf(\mathscr{D})$.*

$S \subseteq A$ is an admissible set iff (a) $S \subseteq \mathsf{Def}(S) \cap \overline{S\Rsh}$
iff (b) $S \subseteq \mathsf{Def}(S \cap \overline{\Rsh S})$
iff (c) $S \subseteq \mathsf{Def}(S) \cap \overline{\Rsh S}$
iff (d) $S \subseteq \mathsf{Def}(S \cap \overline{S\Rsh})$.

The proofs given in [9] are based on the concepts of argumentation theory. Clearly, Prop. 16.8 induces that the four sets on right hand sides of the set inequalities must be equal. In the following corollary we will present an 'argument-free' demonstration of this fact, based on relations and their properties.

Corollary 16.9. *Let (A), …, (D) denote the right hand sides of the set inequalities in Prop. 16.8(a), …, (d), respectively:*

$$(A) := \mathsf{Def}(S) \cap \overline{S\Rsh}, \qquad (B) := \mathsf{Def}(S \cap \overline{\Rsh S}),$$
$$(C) := \mathsf{Def}(S) \cap \overline{\Rsh S}\text{, and} \quad (D) := \mathsf{Def}(S \cap \overline{S\Rsh}).$$

Let $\mathscr{D} = \langle A, \Rsh \rangle$ be a DAF, and $S \in cf(\mathscr{D})$. Then $S \subseteq A$ is an admissible set iff $S \subseteq (A) = (B) = (C) = (D)$.

Proof. Assume $S \in adm(\mathscr{D})$; then $S \subseteq \overline{S\Rsh}$ by Lemma 16.7(1), $S \subseteq \mathsf{Def}(S)$ by Def. 16.4(b), and $\mathsf{Def}(S) = \overline{\overline{S\Rsh}\Rsh}$ by Lemma 16.7(4). $S \subseteq \overline{S\Rsh}$ implies $S\Rsh \subseteq \overline{S\Rsh}\Rsh$ using Prop. 3.9(4). Thus $\overline{\overline{S\Rsh}\Rsh} = \mathsf{Def}(S) \subseteq \overline{S\Rsh}$, whence $(A) := \mathsf{Def}(S) \cap \overline{S\Rsh} = \mathsf{Def}(S) = (D) := \mathsf{Def}(S \cap \overline{S\Rsh})$. Moreover, $\Rsh S \subseteq \overline{S}$, by Lemma 16.7(1). Hence $S \subseteq \overline{\Rsh S}$, so $(B) = \mathsf{Def}(S)$. Thus, $(B) = (A) = (D) = \mathsf{Def}(S)$.

Therefore, $(C) \subseteq (B)$. By Lemma 16.7(5) we have $(B) := \mathsf{Def}(S \cap \overline{\Rsh S}) \subseteq \mathsf{Def}(S) \cap \mathsf{Def}(\overline{\Rsh S})$. From Lemma 16.7(1), $\Rsh S \subseteq \overline{S}$, we get $S \subseteq \overline{\Rsh S}$. Moreover, it is easy to see that $\mathsf{Def}(S) \subseteq \mathsf{Def}(T)$ for $S \subseteq T$. Thus, $(B) \subseteq \mathsf{Def}(S) \cap \mathsf{Def}(\overline{\Rsh S}) = \mathsf{Def}(S) = (C)$.

Conversely, assume $S \subseteq (A) = (B) = (C) = (D)$ for $S \in cf(\mathscr{D})$. Then $S \subseteq \overline{S\Rsh}$ by Lemma 16.7(1), so $S \subseteq (D) = \mathsf{Def}(S)$. □

16.4 Generalizations and Variants

Even though DAFs provide a clear and precise approach to formalizing aspects of argumentation and are open for a long list of additional concepts, two restrictions can be noticed at once:

- Dung's abstract argumentation theory leaves open the question of how arguments actually look like. Arguments are treated as atomic: they do not have an internal structure or a content, and so are all treated as equal.
- DAFs allow for only one binary relation between arguments (attack, defeat, conflict), whence their expressive abilities are limited.

Furthermore, Dung's abstract argumentation theory leaves open the question of how arguments are constructed from a knowledge base and the conditions under which one argument defeats the other. These properties give first hints in which direction DAFs can be generalized and enriched.

In the following I will not deal with the logical aspects of DAFs (see e.g. [1, 2, 6, 23]) but instead will present some generalizations (following [13]) that are based on enhancing the modeling abilities by changes of the relational part. In the following only a short characterization of the various enhancements will be given; for more details and properties readers will have to consult the pertinent literature.

16.4.1 *Values and Preferences*

In a decision making situation we are usually confronted with pro and contra arguments, and we make our decisions on basis of our preferences among these arguments. So it is quite natural to include preference handling techniques in abstract argumentation. Considering information on preferences not only allows to model the problem more accurately, it can also reduce the number of extensions obtained.

Several proposals have extended Dung's framework in order to take into account the strength of arguments. Preference based argumentation frameworks (PAFs) are defined as follows (see [3–5]):

Definition 16.10. *Let $\mathscr{D} = \langle A, \Rsh\rangle$ be a DAF. A* **preference based argumentation framework** *(PAF) is a triple $\mathscr{P} = \langle A, \Rsh, \geq\rangle$ where $\geq \subseteq A \times A$ is a (partial or total) preorder representing some preference information on the arguments.*

The basic idea is to ignore an attack if the attacked argument is stronger than (or preferred to) its attacker, and to apply the given semantics only to the remaining attacks: Whenever a strictly less preferred argument a attacks a more preferred argument b, i.e. $a \Rsh b \wedge b > a$, then $a \Rsh b$ is simply disregarded. This means that from a given $\mathscr{P} = \langle A, \Rsh, \geq\rangle$ a standard DAF $\mathscr{D} = \langle A, \Rsh'\rangle$ is generated such that $\Rsh' := \Rsh - \{\langle a, b\rangle : b > a\}$.

Definition 16.11. *Let $\mathscr{D} = \langle A, \Rsh\rangle$ be a DAF. A* **value based argumentation framework** *(VAF) is a 5-tuple $\mathscr{V} = \langle A, \Rsh, V, val, P\rangle$ where V is a nonempty set of* values, *$val : A \longrightarrow V$, and $P \subseteq V \times V$ is a (transitive, irreflexive and asymmetric)* preference relation.

As for PAFs, the evaluation of a VAF is based on the generation of a DAF. Here attacks are disregarded whenever the attacked argument promotes a more preferred value than the attacker: $\mathscr{D} = \langle A, \Rsh'\rangle$ is built such that $\Rsh' = \Rsh \wedge \langle v(b), v(a)\rangle \notin P$.

16.4.2 *Bipolar Frameworks*

DAFs, in their original version, limit the interactions between arguments to conflicts described with the binary attack relation. An extended version which is capable of modeling a kind of positive interaction between arguments by additionally using a support relation, are the so-called *bipolar frameworks.* Both relations, attack and support, are assumed to be independent, i.e., the support relation is not defined in terms of the attack relation, and vice versa. The term 'bipolarity' refers to the usage of two independent kinds of information with a diametrically opposed nature which represent repellent forces.

Definition 16.12. *Let $\mathscr{D} = \langle A, \Rsh\rangle$ be a DAF. A* **bipolar argumentation framework** *(BAF) is a triple $\mathscr{B} = \langle A, \Rsh, \smile\rangle$ where $\smile \subseteq A \times A$ represents* support relation. *It is required that $\Rsh \cap \smile = \emptyset$.*

The combination of attack and support leads to various kinds of indirect attacks. There is a

- *supported attack* from a_1 to b if there is a sequence of support links from a_1 to a_n and an attack from a_n to b.
- *secondary attack* from b to a_n if there is a sequence of support links from a_1 to a_n and an attack from b to a_1.

- *mediated attack* from b to a_1 if there is a sequence of support links from a_1 to a_n and an attack from b to a_n.

For the generalization of Dung-style semantics to BAFs these indirect, complex notions of attack must be taken into account adequately.

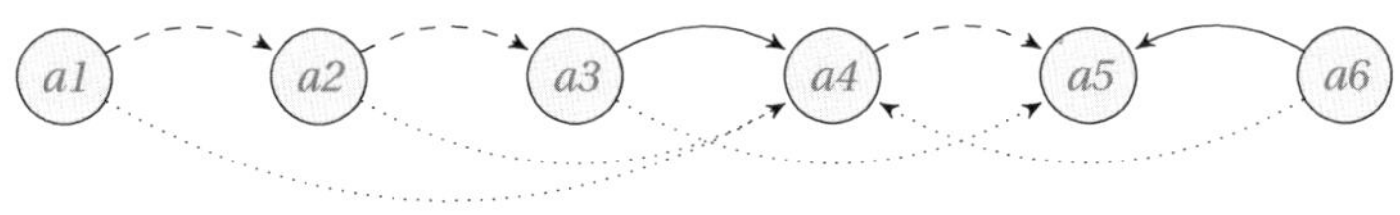

FIG. 16.3: A simple bipolar framework.

The example in Fig. 16.3, borrowed from [13], shows some of the indirect attacks (the normal attacks are depicted with normal, supports with dashed, and mediated attacks with dotted arrows): $\langle a_1, a_4\rangle$ and $\langle a_2, a_4\rangle$ are supported attacks, $\langle a_3, a_5\rangle$ is a secondary, and $\langle a_6, a_4\rangle$ a mediated attack.

For more and newer results on bipolar frameworks a good starting point is [17]. From the literature it can be seen that, on the one hand, there is no definite single interpretation of support and that, on the other hand, bipolar frameworks use the hidden assumption that conflict is stronger than support: No matter how strong the support for an argument, it still has to be defended from incoming attacks. So it is a nearby idea to specify for each argument, in what way support and attack interact. This is exactly the functionality provided by abstract dialectical frameworks to be described below.

16.4.3 *Abstract Dialectical Frameworks*

The concept of abstract dialectical frameworks (ADFs) has been proposed in [14] as another generalization of Dung-style AFs in the following way:

Definition 16.13. *An* **abstract dialectical framework** *is a triple* $\mathcal{L} = \langle S, \lambda, C\rangle$ *where* S *is a set of* statements, $\lambda \subseteq S \times S$ *is a set of* links, *and* $C = \{C_s\}_{s\in S}$ *is a set of total functions* $C_s : \mathbf{2}^{\lambda s} \longrightarrow \{\mathsf{in}, \mathsf{out}\}$. C_s *is called the* acceptance condition *for* s, *and members in* λs, *the set of nodes with a direct link to* s, *are called the* parents of s.

An ADF is a directed graph whose nodes represent arguments, statements or positions which can be accepted or not. The links represent dependencies: the status of a node s only depends on the status of its parents λs. C_s is a function that assigns to each subset of s's parents one of the values $\{\mathsf{in}, \mathsf{out}\}$. Intuitively, if for some $P \subseteq \lambda s$ we have $C_s(P) = \mathsf{in}$, then s will be accepted provided the nodes in P are accepted and those in $\lambda s - P$ are not.

Alternatively, since the acceptance conditions are Boolean functions, one can also use their propositional formula representation. An ADF can thus equivalently be represented as a graph such that each node s is annotated with a propositional formula $\varphi(s)$

built from the parent nodes of s: All and only parents of an argument appear as atoms in the acceptance condition of this argument. For details see [20, Chapter 3].

We now show how DAFs (have attacking links only and a single type of nodes) can easily be captured as an ADF: Let $\mathscr{D} = \langle A, \Rsh \rangle$ be a Dung-style framework. Its associated ADF is $\mathscr{L}_{\mathscr{D}} = \langle S, \lambda, C \rangle$, where $S = A$ and $\lambda = \Rsh$. Thus, the two components of $\mathscr{D}$ are retained; note however, that λ loses the meaning of $\Rsh$: it now represents the connections only, without any information as to their nature. The acceptance conditions are as follows: The acceptance condition C_a is defined as $\neg a_1 \wedge \ldots \wedge \neg a_n$, where $a_1, \ldots, a_n$ are all attackers of a. Formally, $C_a := \bigwedge_{b \in \Rsh a} \neg b$; in case $\Rsh a = \emptyset$ we get the empty conjunction which is defined in logic to always be true, which we denote by $\top$. Similarly, the empty disjunction is always false, denoted by $\bot$.

The associated ADF of the framework from Fig. 16.1 is shown in Fig. 16.4. Argument a is unattacked and can always be accepted, hence its acceptance condition is $\top$.

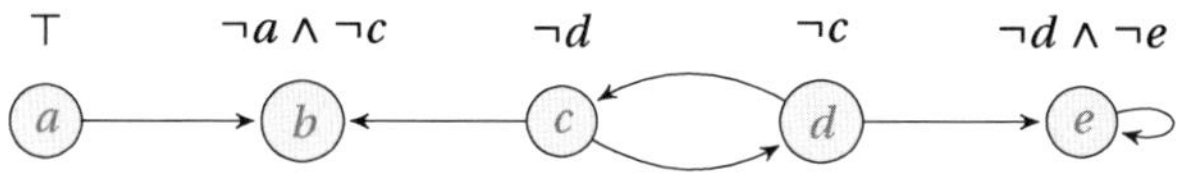

FIG. 16.4: A simple abstract dialectical framework.

Acceptance conditions allow us to introduce different node and link types. The influence a node may have on another node is entirely specified by its acceptance condition. So, in contrast to Dung frameworks, ADFs allow a variety of different dependencies to be represented in a flexible way. Consider the ADF shown in Fig. 16.5. The relation between b and its parents is an example of a group support: accepting either a or c is sufficient for b, one does not have to commit to both. An attack by a set of arguments, i.e. where more arguments are needed for a successful attack, is carried out by c and e on d since $\neg c \vee \neg e$ is equivalent to $\neg(c \wedge e)$.

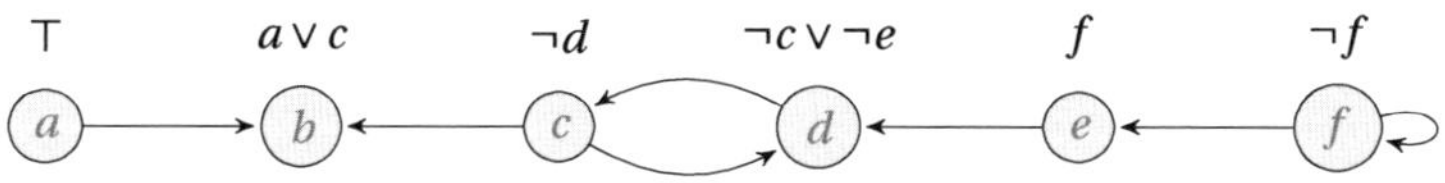

FIG. 16.5: An abstract dialectical framework example.

Additional Reading

Concerning logical aspects of argumentation you may want to start with [10, 12].

References

[1] Amgoud, L.: Postulates for logic-based argumentation systems, Int. J. of Approximate Reasoning **55**, 2028-2048 (2014).

[2] Amgoud, L., Besnard, P.: Logical limits of abstract argumentation frameworks, J. of Applied Non-Classical Logics **23**, 229-267 (2013).

[3] Amgoud, L., Cayrol, C.: A reasoning model based on the production of acceptable arguments, Annals of Mathematics and Artificial Intelligence **34**, 197-216 (2002).

[4] Amgoud, L., Vesic, S.: A new approach for preference-based argumentation framework, Annals of Mathematics and Artificial Intelligence **63**, 149-183 (2011).

[5] Amgoud, L., Vesic, S.: Rich preference-based argumentation framework, Int. J. of Approximate Reasoning **55**, 585-606 (2014).

[6] Arieli, O.: CLIMA XIV A Sequent-Based Representation of Logical Argumentation. In Leite, J. et al. (Eds.): *CLIMA XIV*, 69-85, Springer-Verlag Berlin Heidelberg (2013).

[7] Baroni, P., Giacomin, M.: Semantics of abstract argument systems. In Rahwan, I., Simari, G.(Eds.): *Argumentation in Artificial Intelligence*, 25-44, Springer Science+Business Media (2009).

[8] Baroni, P., Caminada, M., Giacomin, M.: An introduction to argumentation semantics, The Knowledge Engineering Review, **26**, 365-410 (2011).

[9] Besnard, P., Doutre, S.: Characterization of Semantics for Argument Systems, KR 2004, 183-193 (2004).

[10] Besnard, P., Hunter, A.: A logic-based theory of deductive arguments, Artificial Intelligence **128**, 203-235 (2001).

[11] Besnard, P., Hunter, A.: *Elements of Argumentation*, MIT Press, Cambridge, Massachusetts (2008).

[12] Besnard, P., Hunter, A.: Argumentation based on classical logic. In Rahwan, I., Simari, G.(Eds.): *Argumentation in Artificial Intelligence*, 133-152, Springer Science+Business Media (2009).

[13] Brewka, G., Polberg, S. , Woltran, S.: Generalizations of Dung frameworks and their role in formal argumentation, IEEE Intelligent Systems **29**, 30-38 (2014).

[14] Brewka, G., Woltran, S.: Abstract dialectical frameworks. In Lin, F., Sattler, U. and Truszczyński, M. (Eds.): *Principles of Knowledge Representation and Reasoning, Proceedings of the Twelfth International Conference* KR 2010, 102-111. AAAI Press, (2010).

[15] Caminada, M. W. A.: Comparing two unique extension semantics for formal argumentation: ideal and eager. In Mehdi Dastani, M. and de Jong, E. (Eds.): *Proceedings of the 19th Belgian–Dutch conference on artificial intelligence (BNAIC 2007)*, pp. 81–87 (2007).

[16] Caminada, M.W., Carnielli, W.A., Dunne, P.E.: Semi-stable semantics, J. Logic Computat. **22**, 1207-1254 (2011).

[17] Cayrol, C. Lagasquie-Schiex, M.-C.: Bipolarity in argumentation graphs: Towards a better understanding, Int. J. of Approximate Reasoning **54**, 876-899 (2013).

[18] Dung, P.M.: On the acceptability of arguments and its fundamental role in nonmonotonic reasoning, logic programming, and n-person games, Artificial Intelligence **77**, 321-357 (1995).

[19] Dvořák, W., Spanring, C.: Comparing the Expressiveness of Argumentation Semantics, DBAI-TR-2015-90 (2015).

[20] Ellmauthaler, S.: *Abstract Dialectical Frameworks: Properties, Complexity, and Implementation.* Master's thesis, Vienna University of Technology (2012).

[21] Grossi, D.: Abstract argument games via modal logic, Synthese **190**, 5-29 (2013).

[22] Modgil, S., Caminada, M.: Proof theories and algorithms for abstract argumentation frameworks. In Rahwan, I., Simari, G.(Eds.): *Argumentation in Artificial Intelligence*, 105-129,

Springer Science+Business Media (2009).

[23] Prakken, H.: An abstract framework for argumentation with structured arguments, Argument & Computation **1**, 93-124 (2010).

Chapter 17

Describing Graphs and Transition Systems

In the first part of this chapter a general concept of directed graphs is introduced, its properties are investigated and we show how they can be characterized with the help of relations. Due to their intimate connections to relations and graphs, transition systems play a major role in computer science (in a wide sense), especially as a suitable semantic model of reactive systems. Various notions of behavioral equivalence for such systems are presented in the second part of this chapter.

17.1 Graphs and Relations

Remark: Most of this section is based on small parts in chapters 5-7 of [15]. In this book you can find much more and much deeper concepts and results on graphs and relations.

17.1.1 *Basic Concepts*

I did already mention directed graphs and some of their properties in Section 5.2. This chapter takes a closer look at graphs and related concepts, eventually with some repetitions to make it self-contained. Following [10, p. 28] we begin with a more general conception of a directed graph than the one given in Def. 5.4; our 'official' definition of a digraph is presented in Def. 17.2 below.

A **directed graph** (or **digraph** for short) is a pair $G = (V, E)$ of disjoint sets together with two maps $tail\colon E \longrightarrow V$ and $head\colon E \longrightarrow V$. An element $v \in V$ is called a **vertex** (or **node**), and $e \in E$ an **edge** (or **arc**) of G. $v = tail(e)$ is called the **tail** (or **initial vertex**), and $w = head(e)$ the **head** (or **target vertex**) of e.

The terms "*head*" and "*tail*" clearly come from the arrows used to draw directed graphs. Note that the above definition allows **multiple edges**, i.e. edges having the same endpoints. This makes, of course, sense, e.g. when modeling a network of streets where usually there is more than only a single way from, say, A to B. When a digraph models a relation, each ordered pair $\langle tail, head\rangle \in V \times V$ is assigned to at most one edge of G; in this case, G is called **simple**, and we are back at Def. 5.4. An edge e is called **loop** if $tail(e) = head(e)$.

Definition 17.1. *Let $G = (V, E)$ be a directed graph, where $|V| = n$ and $|E| = m$.*

(a) *The* **adjacency matrix** *of G is the $n \times n$ matrix $A^G = [a_{ij}]$ with*
$$a_{ij} = |\{e : tail(e) = v_i \wedge head(e) = v_j\}|,$$
i.e. a_{ij} is the number of edges from v_i to v_j.

(b) *Let G be a loopless digraph; then the* **incidence matrix** *of G is the $n \times m$ matrix $M^G = [m_{ij}]$ with*
$$m_{ij} = \begin{cases} 1, & \textit{if } tail(e_j) = v_i, \\ -1, & \textit{if } head(e_j) = v_i, \\ 0, & \textit{otherwise.} \end{cases}$$

For considerations on a lot of more general relationships between matrices and graphs you may want to have a look at e.g. the classic [7] or [3, 13]. Our focus in this chapter is on the usual concept of two-valued relations ρ, i.e. the ones with $M_\rho = (m_{ij}) \in \boldsymbol{B}_{m\times n}$ (see Def. 3.19). Obviously, as defined above, the adjacency matrix is not necessarily, and the incidence matrix is definitely not a member of some $\boldsymbol{B}_{m\times n}$.

To capture incidence of nodes and edges in terms of relations in a suitable way, we will change the formal description, but will keep the intuitive arrow-based terminology in what will become our official definition of directed graphs:

Definition 17.2. (following [15, Section 5.1])

(a) *A tuple $G = (V, E, \iota, o)$ is called a* **directed graph** *(or* **digraph** *for short), if V and E are two disjoint sets of* **nodes** *(or* **vertices***) and* **edges** *(or* **arrows***), respectively. $\iota, o \subseteq E \times V$ are two mappings called* **ingoing** *and* **outgoing incidence**, *and $\mu := \iota \cup o$ is called the* **incidence** *of G.*

(b) *A node $v \in V$ is called the* **head** *(or* **target node***) of edge $e \in E$ if $\langle e, v\rangle \in \iota$, and the* **tail** *(or* **initial node***) if $\langle e, v\rangle \in o$.*

(c) *Two edges e, e' are called* **parallel** *if their heads and tails coincide, i.e. if there are $h, t \in V$ such that $e\iota h \wedge e'\iota h$ and $eot \wedge e'ot$.*

The concept of a digraph as given above in Def. 5.4 turns out as a special case which from now on we will call a **1-graph**[1] $G = (V, B)$, with $B \subseteq V \times V$ its **associated relation** ([15, p. 6]). (Note that in contrast to [15, Definition 5.1.2] but in accordance to [16, Definition 3.2.4], we let every edge have both a head and a tail, i.e. edges do not have an existence of their own.)

It is easy to see that for simple digraphs the entries in an adjacency matrix have their values in $\{0, 1\}$. So, in the following, by the unqualified term "*digraph*" we shall mean a simple digraph, with V and E finite.

[1]Given $k \in \mathbb{N}$, a graph G is called a k-graph if there are no distinct $k+1$ parallel arcs between two nodes.

Definition 17.3.

(a) *A vertex $v \in V$ is said to be* **incident** *with edge e if v is the tail or head of e. The two vertices incident with an edge e are called its* **endnodes**, *and e is said to be an* edge from its tail to its head.

(b) *Two vertices $v \neq w$ of G are called* **adjacent**, *or* **neighbors**, *if they are endnodes of an edge e. Two edges $e \neq f$ are* **adjacent** *if they have an endnode in common.*

Using the ingoing and outgoing incidence of a directed graph G, we see that

- ov denotes the set of edges starting in node v,
- ιv denotes the set of edges ending in node v,
- eo denotes the tail of e, and
- $e\iota$ denotes the head of e.

$|ov| \geq 0$ is called the **outdegree**, and $|\iota v| \geq 0$ the **indegree** of v. The two simple concepts of *incidence* and *adjacency* give rise to a lot of results on graphs in terms of relations. According to their definition, the former is described by heterogeneous, and the latter by homogeneous relations.

Obviously, incidence and adjacency are not independent: two distinct nodes are said to be adjacent if there is an edge incident with both of them, and two distinct edges are adjacent if there is a node that is incident with both. So one can express both (node) **adjacency** γ and **edge adjacency** κ in a directed graph G by the product of the incidence and its transpose as follows:

$$\gamma := \overline{I_V} \cap \mu^{\smile}\mu \quad \text{and} \quad \kappa := \overline{I_E} \cap \mu\mu^{\smile}. \tag{17.1}$$

Digraphs and 1-graphs are similar when drawn as pictures; in fact, 1-graphs and certain digraphs have identical graphic representations, though their formal definitions differ considerably. This is illustrated by the example shown in Fig. 17.1: Without the dashed arrow both graphs have identical drawings, but different relational descriptions.

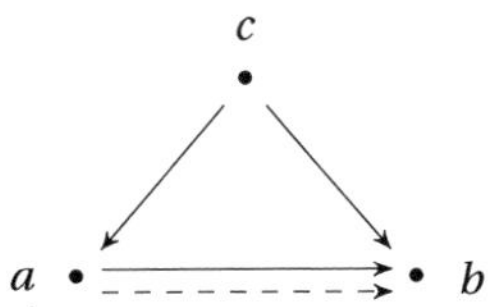

$$o = \begin{array}{c} \\ \swarrow \\ \searrow \\ \longrightarrow \\ \dashrightarrow \end{array} \begin{array}{c} \begin{array}{ccc} a & b & c \end{array} \\ \left[\begin{array}{ccc} 0 & 0 & 1 \\ 0 & 0 & 1 \\ 1 & 0 & 0 \\ \hline 1 & 0 & 0 \end{array}\right] \end{array}, \quad \iota = \begin{array}{c} \\ \swarrow \\ \searrow \\ \longrightarrow \\ \dashrightarrow \end{array} \begin{array}{c} \begin{array}{ccc} a & b & c \end{array} \\ \left[\begin{array}{ccc} 1 & 0 & 0 \\ 0 & 1 & 0 \\ 0 & 1 & 0 \\ \hline 0 & 1 & 0 \end{array}\right] \end{array}$$

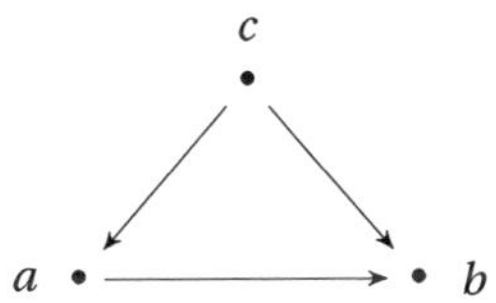

$$o^{\smile} = \left[\begin{array}{ccc|c} 0 & 0 & 1 & 1 \\ 0 & 0 & 0 & 0 \\ 1 & 1 & 0 & 0 \end{array}\right], \quad o^{\smile}\iota = \begin{array}{c} \\ a \\ b \\ c \end{array} \begin{array}{c} \begin{array}{ccc} a & b & c \end{array} \\ \left[\begin{array}{ccc} 0 & 1 & 0 \\ 0 & 0 & 0 \\ 1 & 1 & 0 \end{array}\right] \end{array}$$

FIG. 17.1: A digraph and its associated 1-graph.

Definition 17.4. *Let $G = (V, E, \iota, o)$ be a digraph; then $\alpha := o\,\breve{}\,\iota$ is called the* **associated relation**, *and (V, α) the* **associated 1-graph** *of G.*

An adjacency relation can be associated with a digraph $G = (V, E, \iota, o)$ in two ways: (i) by means of its incidence $\mu := \iota \cup o$ and $\gamma_\mu := \overline{I_V} \cap \mu\breve{}\,\mu$ (see Def. 17.2 and (17.1)); and (ii) via the associated relation $\alpha := o\breve{}\,\iota$ and $\gamma_\alpha := \overline{I_V} \cap (\alpha\breve{} \cup \alpha)$. By the following proposition γ_μ and γ_α are equal.

Proposition 17.5. ([15, Prop. 5.1.7]) *Let $G = (V, E, \iota, o)$ be a digraph with incidence $\mu := \iota \cup o$ and associated relation $\alpha := o\,\breve{}\,\iota$. Then $\overline{I_V} \cap \mu\,\breve{}\,\mu = \overline{I_V} \cap (\alpha\,\breve{} \cup \alpha)$.*

Proof. $\overline{I_V} \cap \mu\breve{}\,\mu = \overline{I_V} \cap (\iota\breve{} \cup o\breve{})(\iota \cup o) = \overline{I_V} \cap (\iota\breve{}\,\iota \cup \iota\breve{}\,o \cup o\breve{}\,\iota \cup o\breve{}\,o)$. But since ι and o are functional, by Prop. 4.2(1) we get $\iota\breve{}\,\iota \subseteq I_V$ and $o\breve{}\,o \subseteq I_V$, whence the last expression evaluates to $\overline{I_V} \cap (\iota\breve{}\,o \cup o\breve{}\,\iota) = \overline{I_V} \cap (\alpha\breve{} \cup \alpha)$. □

The example in Fig. 17.1 also shows that there is no one-to-one correspondence between digraphs and 1-graphs, since, obviously, the upper drawing yields two different digraphs, one with and the other without the dashed arrow, which both have the same associated relation and the associated 1-graph shown in the lower drawing. So given a directed graph G with its ingoing and outgoing incidence ι and o, one can easily calculate its associated relation α. But given α, ι and o are not uniquely determined, but only up to isomorphism (see [15, Theorem 7.2.1]).

17.1.2 *Describing a Graph by its Relations*

The relations introduced so far in the context of graphs can be seen as the ones that "define" or fix a certain exemplar. Now we will have a look at which kind of information they provide. For this purpose, to facilitate a proper understanding it is necessary to have a separate notation for nodes $x \in V$ when used as a singleton subset of V in form of relational point. Recall from the remarks to Fig. 6.4 that in this case we will write $\dot{x}$ for the latter, and from Prop. 6.29(1) that, for $\rho \in \mathscr{B}_X$ and $x, y \in X$, $\langle x, y\rangle \in \rho$ can equivalently be written as $\dot{x}\dot{y}\breve{} \subseteq \rho$ in a pure relational way.

The first thing that comes to mind is to consider a node x and its surrounding: There are the edges for which x is the head or the tail according to the ingoing and outgoing incidence relations. We start with a look on the relationship between 1-graphs and directed graphs in

Proposition 17.6. ([15, Prop. 5.2.4]) *Let $G = (V, E, \iota, o)$ be a digraph and put $\pi := oo\,\breve{} \cap \iota\iota\,\breve{} \subseteq E \times E$. Then*

(1) π is an equivalence on E.
(2) $d, e \in E$ are parallel iff $de\,\breve{} \subseteq \pi$.
(3) $\pi \subseteq I$ iff G has no parallel edges.

Proof. (1): $oo\breve{}$ is reflexive, since $e o t o\breve{} e$; symmetric, since $d o t o\breve{} e$ implies $e o t o\breve{} d$; and transitive, because from $c o t o\breve{} d \wedge d o t o\breve{} e$ we get $c o t o\breve{} e$ (note that t is uniquely determined since o is a mapping). Similar for $\iota^\smile$, whence π is an equivalence by Prop. 4.18(1).

(2): If two edges d, e are parallel, then, by Def. 17.2(c), they have the same heads and the same tails, i.e. $d\iota h \wedge e\iota h$ as well as $dot \wedge eot$. But then $\langle d, e\rangle \in \pi$, or equivalently $de^\smile \subseteq \pi$. The converse direction is also obvious.

(3): Assume $\pi \subseteq I$; then, using Prop. 6.28(2)(ii), from $de^\smile \subseteq \pi \subseteq I$ we get $d = e$. Conversely, assume $\pi \not\subseteq I$, i.e. $\pi \cap \overline{I} \neq \emptyset$. Then, by Prop. 6.29(1), there are two points $d, e \in \mathscr{B}_E$ such that $de^\smile \subseteq \pi \cap \overline{I}$. The edges d and e are distinct and parallel. □

We continue with the nodes which are tails of arrows ending in x, and the nodes which are heads of arrows starting in x. These two sets are called the **predecessors** and the **successors** of x. This gives rise to

Definition 17.7. *Let $G = (V, \alpha)$ be a digraph, with α its associated relation, and let $x \in V$ be a node of G. Then*

(a) $\alpha^\smile\dot{x}$ is the set of **successors**, *and $\alpha\dot{x}$ the set of* **predecessors** *of x.*

(b) $d^-(x) := |\alpha x|$ is the **predecessor degree**, *and $d^+(x) := |\alpha^\smile x|$ the* **successor degree** *of x.*

Note the subtle difference between the representations in (a) and (b): (a) uses a point-free, i.e. pure relational characterization with $\alpha\dot{x} \in \mathscr{B}_V$ a vector, whereas in (b) we use the concept of a *foreset* of x in α, i.e. $\alpha x \subseteq V$, whence we can determine a cardinality for it. The example shown in Fig. 17.2 illustrates the situation. It is easy to see that the following holds:

$\dot{x}\dot{y}^\smile \subseteq \alpha^\smile\alpha \iff$ x and y have a common predecessor,

$\dot{x}\dot{y}^\smile \subseteq \alpha\alpha^\smile \iff$ x and y have a common successor,

$\alpha\alpha^\smile \subseteq I \iff$ no two distinct elements have a common successor.

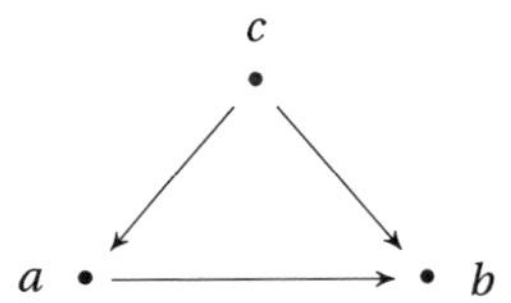

$$\alpha = \begin{array}{c} \\ a \\ b \\ c \end{array}\!\!\begin{array}{c} \begin{array}{ccc} a & b & c \end{array} \\ \begin{bmatrix} 0 & 1 & 0 \\ 0 & 0 & 0 \\ 1 & 1 & 0 \end{bmatrix} \end{array}, \quad \alpha^\smile = \begin{array}{c} \\ a \\ b \\ c \end{array}\!\!\begin{array}{c} \begin{array}{ccc} a & b & c \end{array} \\ \begin{bmatrix} 0 & 0 & 1 \\ 1 & 0 & 1 \\ 0 & 0 & 0 \end{bmatrix} \end{array}, \quad \begin{array}{l} c\alpha = \{a, b\} \\ \alpha b = \{a, c\} \end{array}$$

$$\dot{c} = \begin{array}{c} \\ a \\ b \\ c \end{array}\!\!\begin{array}{c} \begin{array}{ccc} a & b & c \end{array} \\ \begin{bmatrix} 0 & 0 & 0 \\ 0 & 0 & 0 \\ 1 & 1 & 1 \end{bmatrix} \end{array}, \quad \alpha^\smile\dot{c} = \begin{array}{c} \\ a \\ b \\ c \end{array}\!\!\begin{array}{c} \begin{array}{ccc} a & b & c \end{array} \\ \begin{bmatrix} 1 & 1 & 1 \\ 1 & 1 & 1 \\ 0 & 0 & 0 \end{bmatrix} \end{array}; \quad \dot{b} = \begin{array}{c} \\ a \\ b \\ c \end{array}\!\!\begin{array}{c} \begin{array}{ccc} a & b & c \end{array} \\ \begin{bmatrix} 0 & 0 & 0 \\ 1 & 1 & 1 \\ 0 & 0 & 0 \end{bmatrix} \end{array}, \quad \alpha\dot{b} = \begin{array}{c} \\ a \\ b \\ c \end{array}\!\!\begin{array}{c} \begin{array}{ccc} a & b & c \end{array} \\ \begin{bmatrix} 1 & 1 & 1 \\ 0 & 0 & 0 \\ 1 & 1 & 1 \end{bmatrix} \end{array}$$

FIG. 17.2: Predecessors and successors.

One can classify the points of a graph according to whether they have no, precisely one, or more than one predecessor:

Definition 17.8. *Let $G = (V, \alpha)$ be a digraph, with α its associated relation, and let $x \in V$ be a node of G. Then x is called*

(a) **initial** *(or* **source***), if $\alpha x = \emptyset$, i.e. if x has no predecessor;*
(b) **terminal** *(or* **sink***), if $\alpha^{\smile} x = \emptyset$, i.e. if x has no successor.*

So $\dot{x}$ is initial if $\alpha^{\smile}\dot{x} \subseteq \emptyset$; by the Schröder equivalence Th. 3.26(1)(a) we get $\alpha^{\smile}\dot{x} \subseteq \emptyset \leftrightarrow \alpha^{\smile}U \subseteq \overline{\dot{x}} \leftrightarrow \dot{x} \subseteq \overline{\alpha^{\smile}U}$. So $\overline{\alpha^{\smile}U}$ is the set of initial nodes, and similarly $\overline{\alpha U}$ is the set of terminal points: those from which one cannot proceed any further along α. An example is provided in Fig. 17.3.

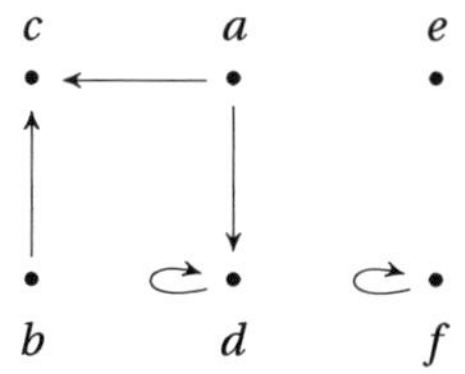

$$\alpha = \begin{array}{c} \\ a \\ b \\ c \\ d \\ e \\ f \end{array}\!\!\begin{array}{c} a\,b\,c\,d\,e\,f \\ \begin{bmatrix} 0&0&1&1&0&0 \\ 0&0&1&0&0&0 \\ 0&0&0&0&0&0 \\ 0&0&0&1&0&0 \\ 0&0&0&0&0&0 \\ 0&0&0&0&0&1 \end{bmatrix} \end{array}, \quad \overline{\alpha^{\smile}U} = \begin{bmatrix} 1 \\ 1 \\ 0 \\ 0 \\ 1 \\ 0 \end{bmatrix}, \quad \overline{\alpha U} = \begin{bmatrix} 0 \\ 0 \\ 1 \\ 0 \\ 1 \\ 0 \end{bmatrix}$$

FIG. 17.3: Initial and terminal nodes.

17.1.2.1 *Bipartite Graphs and Independence*

Given two different non-empty sets X and Y, a subset $\rho \subseteq X \times Y$ has been called a *heterogeneous* relation in Section 3.1. Heterogeneous relations can be represented in the usual way by (rectangular) Boolean matrices, or by so-called bipartite graphs.

Definition 17.9. *The quadruple $G = (U, W, \rho, \sigma)$ is called a* **bipartite graph** *if U and W are nonempty disjoint sets (of nodes), and $\rho \subseteq U \times W$ and $\sigma \subseteq W \times U$ are heterogeneous relations.*

So, each edge $e \in \rho \cup \sigma$ of a bipartite graph G connects nodes from the two partitioning sets, so there are no edges inside U or W, and therefore no loop in G. In other words, when drawing a bipartite graph it is possible to color the vertices using only two colors in such a way that adjacent vertices have different colors.

Obviously, the associated matrix α is a partitioned matrix with two blocks of zeros; see Section 5.1. Trees and even cycles are examples for bipartite graphs; odd cycles are not. An important example in computer science are Petri nets.

In Fig. 17.4 we show a bipartite graph $G = (l, r; \rho, \sigma)$, with $l = \{a, b, c\}$ and $r = \{d, e\}$ with the nodes typically drawn in two columns on the left and right side, and the edges between them.

In Fig. 17.5 we show another example of a bipartite (undirected) graph, usually denoted as $K_{3,3}$, which demonstrates how different drawings can be, even if they a represent the same relatively simple graph in a regular way.

As shown in the example given in Fig. 17.4, a bipartite graph gives rise to a 1-graph (V, E) with $V = l \cup r$ and E a block matrix of the form $\left(\begin{smallmatrix} O & \rho \\ \sigma & O \end{smallmatrix}\right)$. For the two vectors l and r

$$\rho = \begin{array}{c} \\ a \\ b \\ c \end{array}\begin{array}{c} d\ e \\ \left[\begin{array}{cc} 1 & 1 \\ 1 & 0 \\ 0 & 1 \end{array}\right] \end{array} \qquad \sigma = \begin{array}{c} \\ d \\ e \end{array}\begin{array}{c} a\ b\ c \\ \left[\begin{array}{ccc} 0 & 0 & 0 \\ 0 & 1 & 1 \end{array}\right] \end{array}$$

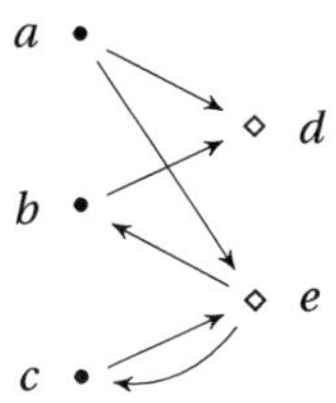

$$E = \begin{array}{c} \\ a \\ b \\ c \\ d \\ e \end{array}\begin{array}{c} a\ b\ c\ \ d\ e \\ \left[\begin{array}{ccc|cc} 0 & 0 & 0 & 1 & 1 \\ 0 & 0 & 0 & 1 & 0 \\ 0 & 0 & 0 & 0 & 1 \\ \hline 0 & 0 & 0 & 0 & 0 \\ 0 & 1 & 1 & 0 & 0 \end{array}\right] \end{array}, \; l = \begin{bmatrix} 1 \\ 1 \\ 1 \\ 0 \\ 0 \end{bmatrix}, \; r = \begin{bmatrix} 0 \\ 0 \\ 0 \\ 1 \\ 1 \end{bmatrix}$$

FIG. 17.4: A bipartite graph.

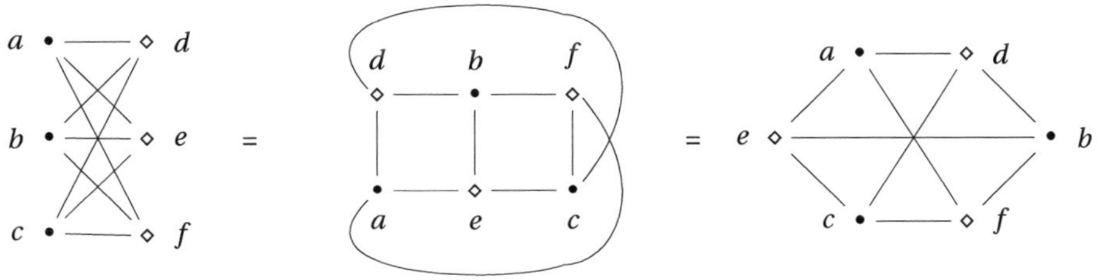

FIG. 17.5: A bipartite graph.

the following holds: $\overline{l} = r$, $l = \overline{r}$ (with respect to V), and $l^{\smile} l = U_V = r^{\smile} r$. Furthermore we have $l \cap r = \emptyset$, $l \cup r = V$, $l^{\smile} r = \emptyset = r^{\smile} l$, and $E \subseteq l r^{\smile} \cup r l^{\smile}$ (which specifies that edges of G either start in l and end in r or start in r and end in l).

A well-known theorem in graph theory says that a graph is bipartite if and only if it has no odd cycle, i.e. a cycle with an odd number of edges (see e.g. [17, Theorem 1.5.10]. Following [4] we will now present a proof of one direction using relation algebraic means. It is longer than the more direct graph-theoretical one, but this is not really surprising since we have to argue on a much more abstract level. We first give some preliminary results (for which it might be helpful to recall how to use vectors to describe subsets from Section 6.2.1.1).

Lemma 17.10. ([4, Lemma 2.1]) *Let u and v be vectors with $u \cap v = \emptyset$. Then the following holds:*

(1) $u^{\smile} v = \emptyset$ and $v^{\smile} u = \emptyset$.
(2) $uu^{\smile} \cup vv^{\smile}$ is transitive.
(3) $(uv^{\smile} \cup vu^{\smile})(uv^{\smile} \cup vu^{\smile}) \subseteq uu^{\smile} \cup vv^{\smile}$

Proof. (1): Using the modular law Th. 3.26(2)(b) and the vector property we get $u^{\smile} v = u^{\smile} v \cap U \subseteq u^{\smile}(v \cap u^{\smile\smile} U) = u^{\smile}(v \cap u) = \emptyset$.

(2): By straightforward calculation:

$$\begin{array}{lll} & (uu^{\smile} \cup vv^{\smile})(uu^{\smile} \cup vv^{\smile}) & \\ = & uu^{\smile}uu^{\smile} \cup uu^{\smile}vv^{\smile} \cup vv^{\smile}uu^{\smile} \cup vv^{\smile}vv^{\smile} & \\ = & uu^{\smile}uu^{\smile} \cup vv^{\smile}vv^{\smile} & \text{using (1)} \\ \subseteq & uUu^{\smile} \cup vUv^{\smile} & \text{by monotonicity} \end{array}$$

$= uu^{\smile} \cup vv^{\smile}$ using the vector property

(3) Similar to (2). □

Lemma 17.11. ([4, Theorem 3.1]) *Let ρ be a relation and u and v be vectors such that $u \cap v = \emptyset$. Then $\rho \subseteq uv^{\smile} \cup vu^{\smile}$ iff $\rho u \subseteq v$ and $\rho v \subseteq u$ and $\rho \subseteq (u \cup v)(u \cup v)^{\smile}$.*

Proof. [⇒] : Using Lemma 17.10(1) and that v is vector we get $\rho u \subseteq (uv^{\smile} \cup vu^{\smile})u = uv^{\smile}u \cup vu^{\smile}u = vu^{\smile}u \subseteq vU = v$, and similar for $\rho v \subseteq u$. Then $\rho \subseteq uv^{\smile} \cup vu^{\smile} \subseteq uu^{\smile} \cup uv^{\smile} \cup vu^{\smile} \cup vv^{\smile} = (u \cup v)(u \cup v)^{\smile}$.
[⇐] : Using the modular law Th. 3.26(2)(a) we first get $uu^{\smile} \cap \rho = (u \cap \rho u^{\smile\smile})u^{\smile} = (u \cap \rho u)u^{\smile} \subseteq (u \cap v)u^{\smile} = \emptyset$, and similar $\rho \cap vv^{\smile} = \emptyset$. Then we calculate as follows:

$$\begin{aligned}
\rho &= \rho \cap (u \cup v)(u \cup v)^{\smile} \\
&= \rho \cap (uu^{\smile} \cup uv^{\smile} \cup vu^{\smile} \cup vv^{\smile}) \\
&= (\rho \cap (uu^{\smile}) \cup (\rho \cap vv^{\smile}) \cup (\rho \cap (uv^{\smile} \cup vu^{\smile})) \\
&= \rho \cap (uv^{\smile} \cup vu^{\smile}),
\end{aligned}$$

so $\rho \subseteq uv^{\smile} \cup vu^{\smile}$. □

Definition 17.12. *Let tuple $G = (N, E, \iota, o)$ denote a digraph, and let γ and κ be the adjacency and edge adjacency as defined in (17.1) above: $\gamma := \overline{I_N} \cap \mu^{\smile}\mu$ and $\kappa := \overline{I_E} \cap \mu\mu^{\smile}$. Further, let $n \subseteq N$ and $e \subseteq E$ denote a set of nodes and edges, respectively.*

(a) *n is called* **independent** *iff no two nodes are adjacent*
(b) *e is called* **independent** *iff no two edges are adjacent*

It is not hard to see that independency for vertices and edges can be characterized by the following pairs of conditions:

$$\gamma n \subseteq \overline{n} \iff nn^{\smile} \subseteq \overline{\gamma} \quad \text{and} \quad \kappa e \subseteq \overline{e} \iff ee^{\smile} \subseteq \overline{\kappa} \tag{17.2}$$

It is obvious that any subset of an independent (node or edge) set is also independent, so the question arises for biggest possible independent sets. A set of independent edges is called a **matching**, and a set of independent nodes is called **stable**. If n is a stable set of nodes, then if an arc ends in n, it necessarily begins outside of n.

For the final step recall from the example given in Fig. 17.4 that a bipartite graph gives rise to a 1-graph (V, E) with $V = l \cup r$ and E a suitable block matrix. The vectors l and r fulfill $\overline{l} = r, l = \overline{r}$ (with respect to V), so $l \cap r = \emptyset, l \cup r = V$, and $E \subseteq lr^{\smile} \cup rl^{\smile}$ (which specifies that edges of G either start in l and end in r or start in r and end in l).

Now let ρ be the adjacency matrix of a digraph $G = (X, \rho)$, and let vectors v, w describe two subsets $V, W \subseteq X$. If we assume $v \cap w = \emptyset$ and $v \cup w = X$, then we get $w = \overline{v}$, whence $W = \overline{V}$ relative to X. So we get $\rho \subseteq (v \cup w)(v \cup w)^{\smile} = XX^{\smile}$, $\rho v \subseteq w \leftrightarrow \rho v \subseteq \overline{v}$, and $\rho w \subseteq v \leftrightarrow \rho\overline{v} \subseteq v$. Thus, Lemma 17.11 gives a relation-algebraic description of $(V, \overline{V})$ being a bipartition of G if and only if V and $\overline{V}$ are independent sets of G.

We are now ready to present the relation-algebraic proof that bipartite graphs do not have odd cycles in

Theorem 17.13. ([4, Theorem 3.2]) *Let ρ be a relation. If there exists a bipartition (u,v) of ρ, then $\rho(\rho\rho)^* \cap I = \emptyset$.*

Proof. We need two auxiliary results:

(i) $uv^{\smile} \cap I \subseteq u(v^{\smile} \cap u^{\smile} I) = u(u \cap v)^{\smile} = \emptyset$, and similar for $vu^{\smile} \cap I = \emptyset$, using the modular law.

(ii) For line (*) below note that $uu^{\smile} \cup vv^{\smile}$ is transitive by Lemma 17.10(2), that $\rho = \mathbf{t}(\rho)$ iff ρ is transitive by Prop. 3.60(3), and that $\mathbf{t}(\rho) = \rho^+$ by Th. 3.59(3).

Then

$$\begin{aligned}
\rho(\rho\rho)^* \cap I &\subseteq \rho((uv^{\smile} \cup vu^{\smile})(uv^{\smile} \cup vu^{\smile}))^* \cap I && \rho \text{ bipartition: } \rho \subseteq uv^{\smile} \cup vu^{\smile} \\
&\subseteq \rho(uu^{\smile} \cup vv^{\smile})^* \cap I && \text{Lemma 17.10(3)} \\
&= \rho(I \cup (uu^{\smile} \cup vv^{\smile})^+) \cap I && \text{Def. 3.57} \\
&= \rho(I \cup uu^{\smile} \cup vv^{\smile}) \cap I && \text{by (ii),} \qquad (*) \\
&= (\rho \cup \rho uu^{\smile} \cup \rho vv^{\smile}) \cap I && \\
&\subseteq (\rho \cup vu^{\smile} \cup uv^{\smile}) \cap I && \text{Lemma 17.11}[\Rightarrow]\text{: } \rho u \subseteq v, \rho v \subseteq u \\
&= (vu^{\smile} \cup uv^{\smile}) \cap I && \rho \text{ bipartition} \\
&= (vu^{\smile} \cap I) \cup (uv^{\smile} \cap I) && \\
&= \emptyset && \text{by (i)}
\end{aligned}$$

□

If ρ is the adjacency relation of a digraph $G = (V, \rho)$, then clearly $x\,\rho(\rho\rho)^*\,y$ if there is directed path from x to y with odd length. So $\rho(\rho\rho)^* \cap I = \emptyset$ means that there is no cycle in G with odd length. In [4, Section 4] the authors also give a relation-algebraic proof of the converse direction: Relations without odd cycles are bipartite.

17.1.3 *Paths and Reachability*

We will now turn our focus from nodes and pairs of nodes to sequences of nodes, or from local to global properties and structures in a graph, by presenting some basic concepts following [15, Chapter 6]. The question under consideration is for the possibility of getting from one node to some other by traversing a sequence of successive edges. From the viewpoint of applications, the graph-theoretical concept is called *reachability,* and it is not really surprising that we will have to look at the powers of the graph's associated relation for analyzing it.

Definition 17.14. *(Path as a sequence of points) Let $G = (V, \alpha)$ be a 1-graph, and $w = \langle n_0, \ldots, n_h \rangle$, $h \geq 0$, be a sequence of nodes.*

(a) w is called a **path** *from n_0 to n_h iff $n_{i-1} \subseteq \alpha n_i$, for $1 \leq i \leq h$;*
(b) w is called a **cycle** *iff w is a path from n_0 to n_h, $h \geq 1$, and $n_0 = n_h$.*

n_0 is called the **starting point** and n_h the **endpoint** of w. The **length** of the path $w = \langle n_0, \ldots, n_h \rangle$ is defined as $|w| = h$. A path or a circuit $w = \langle n_0, \ldots, n_h \rangle$ is called **elementary**, if no point occurs more than once (except $n_0 = n_h$ in the case of circuits). In the example shown in Fig. 17.6 w_0, w_1 are elementary paths, and w_3, w_4 elementary circuits.

(Recall from Fig. 17.2 that $\alpha\dot{y}$ is the set of predecessors of y; so $x \subseteq \alpha\dot{y}$ means that $\langle x, y\rangle \in \alpha$. From now on we will not be consequent in using the notation with a dot above a node x for its relational representation, i.e. each of $x \in \alpha y$, $x \subseteq \alpha y$, and $\langle x, y\rangle \in \alpha$ will express that x is a predecessor of y.)

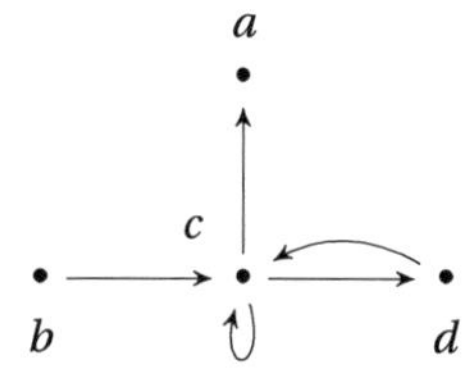

$w_0 = \langle a\rangle$ $\quad w_3 = \langle c, c\rangle$
$w_1 = \langle b, c, a\rangle$ $\quad w_4 = \langle c, d, c\rangle$
$w_2 = \langle b, c, c, d\rangle$ $\quad w_5 = \langle c, d, c, d, c\rangle$

FIG. 17.6: Paths and cycles.

Remembering Fig. 17.1 we see that the associated 1-graph of a digraph $G = (V, E, \iota, o)$ tells us if there is a path between two nodes x and y; but as soon as there are parallel edges e, f from x to y, we obviously cannot express that there are two different paths between the two nodes on basis of the path concept from Def. 17.14. So, if we want to be able to discriminate between parallel paths we need a more powerful concept including also the edges as additional information.

Definition 17.15. *(Path as a sequence of points and arcs) Let $G = (V, E, \iota, o)$ be a digraph and $w = \langle n_0, e_1, n_1 \ldots, e_h, n_h\rangle$, $h \geq 0$, an alternating sequence of nodes and edges.*

(a) *w is called a* **path** *rom starting point n_0 to endpoint n_h iff $n_0 = e_1 o$, $n_h = e_h\iota$, and $n_i = e_{i+1}o = e_i\iota$ for $i = 1, \ldots, h-1$. A path is called* **simple** *if no arc occurs more often than once in it. A sequence $\langle n_0\rangle$, i.e. $h = 0$, is the* **void path** *at node n_0.*
(b) *w is called a* **cycle** *iff w is a path from n_0 to n_h, $h \geq 1$, and $n_0 = n_h$.*

As mentioned above, to gain information about reachability in 1-graphs $G = (V, \alpha)$ the powers and the transitive closure of its associated relation α are of interest, since they are closely related to the existence of paths: If $\langle x = x_0, \ldots, x_h = y\rangle$ is a path from $\dot{x}$ to $\dot{y}$, then $\dot{x} \subseteq \alpha^h \dot{y}$.

Proposition 17.16. ([15, Proposition 6.1.3]) *Let $G = (V, \alpha)$ be a 1-graph and $x, y \in V$ be two arbitrary nodes, and $h \geq 0$. Then*

(1) $x \subseteq \alpha^h y$ *iff there is a path from x to y of length h;*
(2) $x \subseteq \alpha^* y$ *iff there is a path from x to y;*
(3) $x \subseteq \alpha^+ x$ *iff there is a cycle through x;*
(4) $\alpha^+ \subseteq \overline{I}$ *iff G has no cycles.*

Proof. (1): We have $x \subseteq \alpha^0 y = Iy = y \leftrightarrow x = y$ by Prop. 6.28(2)(ii), and $\langle x\rangle$ is a path of length 0 from x to y. Using induction on h, $x \subseteq \alpha^h y = \alpha^{h-1}\alpha y$ implies that there is a point z with $x \subseteq \alpha^{h-1}z$ and $z \subseteq \alpha y$ by Prop. 6.30. By the induction hypothesis there is a path $\langle x = x_0, x_1, \ldots, x_{h-1} = z\rangle$ of length $h-1$ from x to z, so $\langle x, \ldots, z, y\rangle$ is a path of length h from x to y.

(2): Assume $x \subseteq \alpha^* y$; by Prop. 6.27(1) this is equivalent to $xy^{\smile} \subseteq \alpha^*$. By Prop. 6.29(1) $xy^{\smile}$ is an atom, and so must be contained in one of the parts $\alpha^j, j \geq 0$ of α^*, say $xy^{\smile} \subseteq \alpha^k$. Again using Prop. 6.27(1), this is equivalent to $x \subseteq \alpha^k y$, whence there is a path of length k from x to y. Conversely, by (1) we get that $x \subseteq \alpha^j y$ for some j, whence $x \subseteq \alpha^* y$.

(3): α^+ represents paths of length ≥ 1. So the assertion follows by (2).

(4): Assume both $\alpha^+ \subseteq \overline{I}$ and $x \subseteq \alpha^+ x$. By Prop. 6.29(1) it follows that $xx^{\smile} \subseteq \alpha^+ \subseteq \overline{I}$; but this contradicts the injectivity of x, viz. $xx^{\smile} \subseteq I$. So $\alpha^+ \subseteq \overline{I}$ implies that G has no cycles. Conversely, assume $\alpha^+ \not\subseteq \overline{I}$. Then $\alpha^+ \cap I \neq \emptyset$, and by Prop. 6.29(2) there are points x, y such that $xy^{\smile} \subseteq \alpha^+ \cap I$. So, $x \subseteq (\alpha^+ \cap I)y$, whence $x \subseteq Iy = y$ and $x \subseteq \alpha^+ y$. The first inclusion implies $x = y$ by Prop. 6.28(2)(ii), and the second that there is a path of length ≥ 1 from x to y, i.e. a cycle through x. □

Definition 17.17. *Let $G = (V, \alpha)$ be a 1-graph.*

(a) G is called **strongly connected** *if*
$\alpha^* = U \iff I \subseteq \alpha^+ \iff$ *every point is reachable from any other.*

(b) The equivalence classes of $\alpha^* \cap \alpha^{*\smile}$ are called the **strongly connected components** *of G.*

A strongly connected component thus contains all nodes that can be reached from each other. We have to verify that $\alpha^* \cap \alpha^{*\smile}$ is indeed an equivalence: Symmetry follows from Prop. 3.52(5), and by Prop. 3.62(6) $\alpha^* = \mathbf{rt}(\alpha)$. So, using Prop. 3.25(1) and (6), $\alpha^{*\smile}$ is also reflexive and transitive, whence, finally, $\alpha^* \cap \alpha^{*\smile}$ is an equivalence by Prop. 3.52(2) and (6).

Proposition 17.18. ([15, Proposition 6.1.7]) *Let $G = (V, \alpha)$ be a 1-graph, and x, y be two points in G. Then*

(1) G is strongly connected $\iff \overline{I} \subseteq \alpha^+ \cap \alpha^{+\smile}$
$\iff$ *any two distinct points lie on a cycle.*

(2) x, y lie in different strongly connected components, i.e. $[x]_\alpha \cap [y]_\alpha = \emptyset$
$\iff$ *$x \neq y$, and there is no cycle through x, y.*

Proof. (1): By definition, $U = \alpha^* = I \cup \alpha^+ \leftrightarrow \overline{I} \subseteq \alpha^+ \iff \overline{I} \subseteq \alpha^{+\smile}$.

(2): $x \neq y$ implies $xy^{\smile} \subseteq \overline{I}$ by Prop. 6.28(2)(i). Thus, for the right hand side, we get $xy^{\smile} \subseteq \overline{I} \wedge (xy^{\smile} \not\subseteq \alpha^+ \vee xy^{\smile} \not\subseteq \alpha^{+\smile})$. And for the left hand side we have $xy^{\smile} \subseteq \overline{\alpha^* \cap \alpha^{*\smile}}$. These two expressions are equivalent, since by Prop. 6.29(1) $xy^{\smile}$ is an atom in $\mathscr{B}_V$. □

17.1.3.1 *Rooted Graphs and Rooted Trees*

A major role is played by graphs in which a particular node is distinguished as entry or starting point. Such graphs are used especially in computer science in connection with e.g. flowcharts, finite automata, or as binary trees and syntax diagrams.

Definition 17.19. *Let $G = (V, \alpha)$ be a 1-graph, and $a \in V$. a is said to be a* **root** *of G, iff all nodes can be reached from it.*

A graph can, of course, have several roots: in a strongly connected graph even every point is a root. In applications, typically one root is defined to be the entry point. So we call a pair (G, a) consisting of a graph G and a root a a **rooted graph**, and an injective and cycle-free rooted graph a **rooted tree**. Rooted trees have a simple structure: there are no parallel edges, no cycles, and each node, with exception of *the* root, has precisely one predecessor (there is just one root since a second would give rise to a cycle).

Corollary 17.20. *a is root of G* $\iff aU \subseteq \alpha^* \iff U \subseteq a^\smile\alpha^* \iff a \subseteq \overline{\overline{\alpha^*}U}$.

Proof. The first equivalence is obvious, because a is a vector, whence $aU = U \subseteq \alpha^*$ just says that all nodes a reachable from a. Using Prop. 6.27(1) we get $aU \subseteq \alpha^* \leftrightarrow (aU)^\smile = Ua^\smile \subseteq \alpha^{*\smile} \leftrightarrow U \subseteq \alpha^{*\smile}a \leftrightarrow U^\smile = U \subseteq a^\smile\alpha^*$, which proves the second equivalence. By the Schröder equivalence Th. 3.26(1)(b) we have $aU \subseteq \alpha^* \leftrightarrow \overline{\alpha^*}U \subseteq \overline{a} \leftrightarrow a \subseteq \overline{\overline{\alpha^*}U}$. □

Among the rooted graphs those with no parallel arcs and no circuits, and each point (with exception of the root) having precisely one predecessor are particularly simple: An injective and circuit-free rooted graph is called a **rooted tree** or **arborescence**. Thus, a rooted graph $G = (V, \alpha)$ is a rooted tree in case $\alpha\alpha^\smile \subseteq I$ and $\alpha^+ \subseteq \overline{I}$.

17.1.4 *Homomorphisms and Isomorphisms*

Remembering Section 10.2, we see that given any mathematical structure, the questions for substructures, homomorphisms and congruences are among the first that usually are considered in order to learn about properties that are equal to, or different from, the properties of other known structures. We will restrict our considerations to the simple case of 1-graphs.

Given two digraphs $G = (V, \alpha)$ and $H = (W, \beta)$, from a structure preserving map ϕ we expect that it sends pairs of points related by α to pairs which are related by β: So if $x, y \in V$ with images $\phi^\smile x$ and $\phi^\smile y$, then $xy^\smile \subseteq \alpha$ shall imply $\phi^\smile x(\phi^\smile y)^\smile \subseteq \beta$, where the latter is equivalent to $xy^\smile \subseteq \phi\beta\phi^\smile$ since ϕ is a mapping.

Definition 17.21. *Let $G = (V, \alpha)$ and $H = (W, \beta)$ be two digraphs, then $\phi : V \longrightarrow W$ is a* **homomorphism** *from G to H if ϕ is a mapping $\phi : V \longrightarrow W$ such that $\alpha \subseteq \phi\beta\phi^\smile$, i.e. if the following three inclusions hold:*

$$\phi^\smile\phi \subseteq I_W, \quad I_V \subseteq \phi\phi^\smile, \quad \alpha \subseteq \phi\beta\phi^\smile;$$

Such a homomorphism will also be written as $\phi : G \longrightarrow H$.

Using the first two inequalities, it is easy to see that the third is equivalent to each of the following three inclusions which arise in a cyclic way by suitable multiplications by ϕ or $\phi^\smile$. We have

$$\alpha\phi \subseteq \phi\beta \quad \leftrightarrow \quad \phi^{\smile}\alpha\phi \subseteq \beta \quad \leftrightarrow \quad \phi^{\smile}\alpha \subseteq \beta\phi^{\smile} \tag{17.3}$$

Composing two homomorphisms $\phi : G \longrightarrow H'$ and $\phi' : H' \longrightarrow H$ gives again a homomorphism: $\alpha \subseteq \phi\beta'\phi^{\smile} \subseteq \phi\phi'\beta\phi'^{\smile}\phi^{\smile} = (\phi\phi')\beta(\phi\phi')^{\smile}$. A nice example of homomorphic (symmetric, irreflexive) graphs is shown in Fig. 17.7 (following [14, p. 7]).

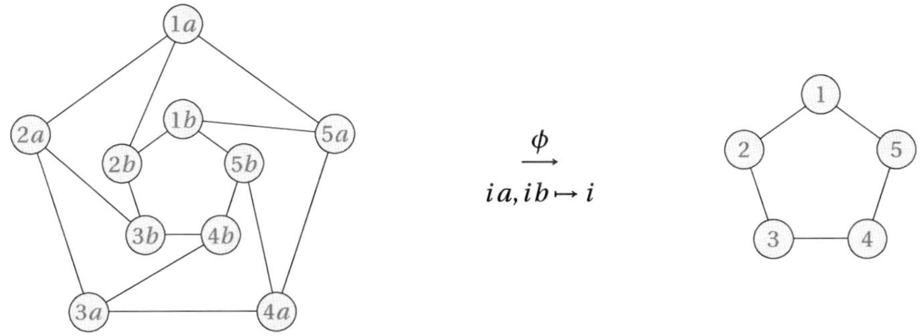

FIG. 17.7: Two homomorphic graphs.

In order to characterize systems that are "*essentially the same*" one uses the concept of isomorphism. It is well known that for algebraic structures as e.g. groups isomorphisms are just invertible homomorphisms, where the homomorphism condition is given by equations like $\phi(x + y) = \phi(x) +' \phi(y)$.

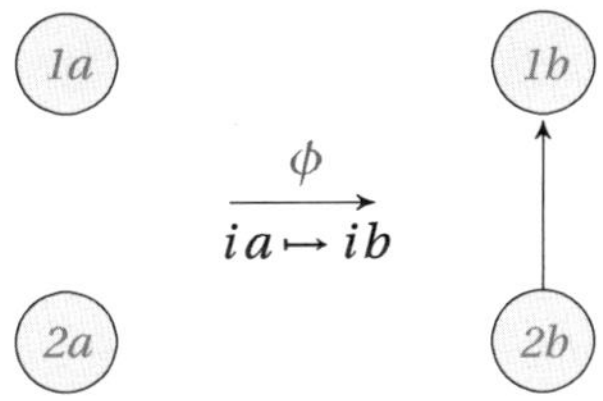

Apparently, this does not work fine for relational structures as illustrated in the figure on the left: ϕ certainly is a homomorphism from the left to the right digraph that is bijective. Nonetheless, no one would consider these two graphs to be isomorphic. So $\phi^{\smile}$ must be required to be a homomorphism too.

Definition 17.22. *Let $G = (V, \alpha)$ and $H = (W, \beta)$ be two digraphs, then $\phi : V \longrightarrow W$ is an* **isomorphism** *of G and H if both ϕ and $\phi^{\smile}$ are homomorphisms.*

So given graphs $G = (V, \alpha)$ and $H = (W, \beta)$, an isomorphism from G to H is a bijection $\phi : V \longrightarrow W$ such that $\phi(v)$ is adjacent to $\phi(v')$ in H if and only if v is adjacent to v' in G. By Prop. 4.2 an isomorphism is characterized by the following conditions:

$$\phi^{\smile}\phi = I_W, \quad I_V = \phi\phi^{\smile}, \quad \alpha\phi \subseteq \phi\beta, \quad \beta\phi^{\smile} \subseteq \phi^{\smile}\alpha.$$

Taking into account (17.3) each of the cyclically produced homomorphism conditions becomes an equality: E.g. $\phi^{\smile}\alpha = \beta\phi^{\smile}$ follows immediately from the two corresponding inequalities; moreover, multiplication of $\beta\phi^{\smile} \subseteq \phi^{\smile}\alpha$ gives $\beta \subseteq \phi^{\smile}\alpha\phi$. Thus

$$\alpha = \phi\beta\phi^{\smile}, \quad \alpha\phi = \phi\beta, \quad \phi^{\smile}\alpha\phi = \beta, \quad \phi^{\smile}\alpha = \beta\phi^{\smile}.$$

These four equations are no longer equivalent to one another if ϕ is not bijective as demonstrated by examples in [15, p. 147].

17.2 Transition Systems and Behavioral Equivalence

Due to their intimate connections to relations and graphs, transition systems play a major role in computer science (in a wide sense), especially as a suitable semantic model of reactive systems. Therefore, we hold it worthwhile to provide a very cursory overview on the topic and its basic notions; for more details see e.g. [11] or [12, Chapter 2] as starting points, the latter one a concise and easily comprehensible introduction.

A **process** is the behavior of a system over time, where a system can comprise such diverse things as an elementary particle, a simple vending machine, your smartphone, or an airplane; but also a bee, an ant colony, or a chess player etc. *Process theory* is the study of processes, i.e. the analysis and investigation of an actual or the possible behaviors of systems, with *modelling* and *verification* as two main activities. The former is the task of representing processes by suitable mathematical structures, and the latter that of proving that the actual behavior of a system is compatible with, or equal to, its intended and specified behavior.

Of course, for deciding whether a model is *valid*, i.e. really describes a given system in an adequate way, or whether two such descriptions are compatible, we need a criterion that tells us if two systems behave similarly. This concerns the semantics of the equality concept employed in a process theory. So, another task of process theory is to determine which semantics is suitable for which application.

Systems are viewed as being in interaction with an environment. For our considerations these interactions are taken to be ruled by a simple synchronization mechanism without exchange of values. And as always when building models of a given system, which aspects of its behavior are important to a certain user depends on the environment in which the system will be running, and on the particular interests of this user.

From the broad spectrum available in process theory we will present a few basic concepts only together with a small example. We restrict our considerations to **sequential systems**: Actions are assumed to be *atomic*, i.e. they are either performed completely or not at all, and also to be *instantaneous*, i.e. two actions cannot occur at the same time

The behaviour of a system is what we can observe by interacting with it: e.g. how a machine reacts to our inputs like pressing a button or inserting coins. Other things as the colour or the shape of the machine, are irrelevant; we are interested in what the machine does, not in what it looks like.

Transition systems are a standard class of models which are often used to describe the behavior and properties of natural and artificial objects under consideration. They are basically directed graphs where nodes represent *states*, and edges model *transitions*, i.e., state changes. A state describes some information about a system at a certain moment, whereas transitions specify how the system can evolve from one state to another. For instance, the state of a switch indicates its current position: on or off, or some other information like a certain temperature for a washing machine. State changes describe the reaction of a system to some external influence like pressing a button or inserting and turning a key.

Definition 17.23. (e.g. [6]) *A* **labelled transition system** *(LTS) is a quadruple* $\mathscr{L} = (S, A, \{\xrightarrow{a} : a \in A\}, s)$ *such that* $S \neq \emptyset$ *denotes a set of* states, $A \neq \emptyset$ *is a set of* labels *or* actions, *and* $\xrightarrow{a}$ *is a relation over* S *for each* $a \in A$, *called a* transition relation. $s \in S$ *is called the* initial state.

An LTS thus informs about the states in which it can be and, for each state, about the interactions that are possible from that state. An interaction is represented by a labelled arc; in the LTS terminology it is called a *transition.* Transition systems can e.g. be viewed as an abstract model of computation with S modelling the possible states of a computer, with the actions standing for programs, and $s \xrightarrow{p} t$ meaning that there is an execution of the program p that starts in s and terminates in state t. Indeed, LTSs can be thought of as models for practically any dynamic situation: each transition takes us from the current state to some next state of the system due to some external influence.

An LTS $\mathscr{L}$ is called **finite** if both S and A are finite sets, and **deterministic** if $\forall s \in S \forall a \in A [|s \xrightarrow{a}| \leq 1]$. By A^* we denote the set of all finite words over A, including the empty string ϵ. An element $t \in A^*$ is called a **trace**. We set $p \xrightarrow{\epsilon}{}^* p$ for all $p \in S$. If $t = a_1 a_2 \ldots a_n$ and if there is a sequence of transitions $w = p \xrightarrow{a_1} p_1 \xrightarrow{a_2} p_2 \cdots \xrightarrow{a_n} q$ we call t a trace of p and simply write $p \xrightarrow{t}{}^* q$ according to Def. 3.57; recall from Prop. 3.62 that $\rho^* = \mathbf{rt}(\rho)$. Furthermore we write $Tr(p)$ for the set of all traces of p. w itself is called a **path** or **execution fragment** of $\mathscr{L}$.

Given two LTSs $\mathscr{L}_1$ and $\mathscr{L}_2$, one of the foremost questions is how they can be compared with respect to their possible evolutions over time. *Behavioral equivalence* relations are used to abstract from unwanted details and to identify those LTSs that react "similarly" to external experiments; e.g. they characterize which system behaviors are sufficiently good approximations, or equivalent in a certain sense. Intuitively, two processes should be equivalent if they cannot be distinguished by an observer subjecting them to identical experiments, i.e. interacting with them using the same sequence of actions.

The problem is, however, not as simple as it sounds. As a first intuitive idea one might think that graph isomorphism would be a suitable candidate. But this idea is immediately refuted by the example shown in Fig. 17.8: Obviously both systems are able to perform an arbitrary number of a-actions, taking p and q as their initial, and p and q_1 as their final states. So they cannot be distinguished by interaction, but there is no isomorphism between them.

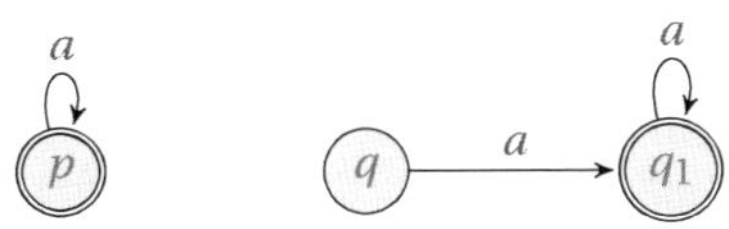

FIG. 17.8: Two indistinguishable LTSs.

Before we introduce some examples of equivalence concepts for LTSs, we present three different models for a very simple vending machine that delivers a cup of either coffee or tea after the insertion of two coins. Vending machines are running examples for LTSs; the ones shown in Fig. 17.9 are used in e.g. [9].

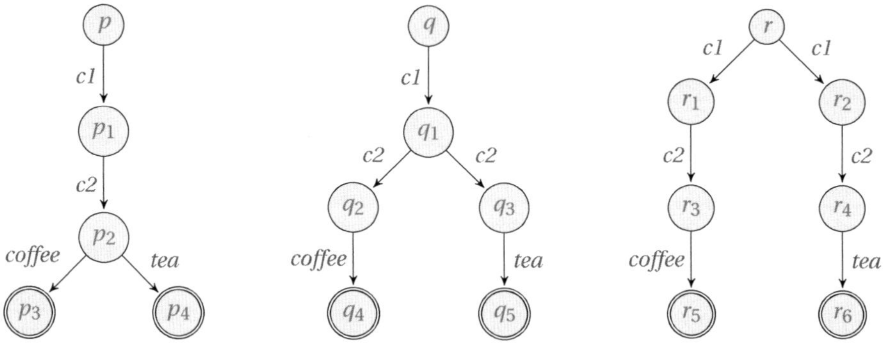

FIG. 17.9: Three vending machines.

As above, though they are different, it is easy to see that all three systems can perform the same sequences of actions. In the viewpoint of automata theory: all three do accept the same language (implicitly taking the roots p, q, and r as initial, and the leafs as final states): $\{c_1 \cdot c_2 \cdot \mathit{coffee},\ c_1 \cdot c_2 \cdot tea\}$.

17.2.1 *Trace Equivalences*

Given a state p, the set $Tr(p)$ has been defined above as the set of all possible finite sequences of interactions that may take place starting in p. So it is reasonable to require a notion of behavioral equivalence to relate only processes p and q with $Tr(p) = Tr(q)$.

Definition 17.24.

(a) *Two states p, q are called* **trace equivalent** *if $Tr(p) = Tr(q)$, denoted as $p \sim_T q$.*

(b) *Two labelled transition systems are called* **trace equivalent** *if their initial states are.*

Now, since $Tr(p) = Tr(q) = Tr(r)$, the three vending machines shown in Fig. 17.9 are trace equivalent. A string is considered 'accepted' if it gives us at least one path from the initial state to a final state. So the existence of other paths that fail, i.e. that end in non-final states where no further action is possible, is irrelevant. A non-final state q such that $\neg\exists a \in A\,[\exists q' \in S\,[q \xrightarrow{a} q']]$ is called a **deadlock**, denoted by $q \not\rightarrow$. This means that $\sim_T$ is not sensitive to deadlocks. However, the so-called *completed trace equivalence* is, where the set of completed traces of p is $CTr(p) = \{t \in A^* : \exists q \in S\,[p \xrightarrow{t} q \wedge q \not\rightarrow\}$. Two states p, q are called **completed trace equivalent** if $Tr(p) = Tr(q)$ and $CTr(p) = CTr(q)$, denoted as $p \sim_{CT} q$. Note that completed trace equivalence also does identify the three vending machines of Fig. 17.9, but it can differentiate between the two LTSs in Fig. 17.10 since, though $Tr(q_1) = Tr(q_5) = \{a \cdot b\}$, $CTr(q_1) = \{a\}$ and $CTr(q_5) = \emptyset$, whence $q_1 \not\sim_{CT} q_5$.

FIG. 17.10: Trace equivalent LTSs with different deadlock behavior.

It should by now have become clear that LTSs as models of reactive systems are looked at in a quite different manner than are automata when viewed as language recognizers: Analyzing our three vending machines from a customer's point of view, they exhibit significantly different behavior. People not indifferent between coffee and tea would certainly prefer the leftmost variant, since it is the only one that enables them to choose what (s)he really likes. So we will have to look for alternatives more suitable for processes.

17.2.2 *Simulation Equivalences*

Definition 17.25. *Let $\mathscr{L} = (S, A, \{\xrightarrow{a} : a \in A\}, s_0)$ be a labelled transition system. A relation $\sigma \subseteq S \times S$ is called a* **simulation** *if, whenever $\langle p, q\rangle \in \sigma$, for all $a \in A$ we have*

$$p \xrightarrow{a} p' \text{ implies } \exists q' \in S\,[q \xrightarrow{a} q' \wedge \langle p', q'\rangle \in \sigma].$$

$p \in S$ is simulated by a state $q \in S$, denoted $p \preceq q$, if there exists a simulation σ such that $\langle p, q\rangle \in \sigma$. p and q are called **simulation equivalent**, *denoted $p \sim_S q$, if $p \preceq q$ and $q \preceq p$.*

Simulation $\preceq$ thus requires that whenever $p \preceq q$, q can mimic all stepwise behavior of p: whatever action p performs, reaching p', q can respond with the same action reaching a state q' that is able to simulate p'. Simulation can be lifted to entire transition systems by comparing their initial states: LTS $\mathscr{L}_1 = (P, A, \{\xrightarrow{a}_1 : a \in A\}, p_0)$ is simulated by $\mathscr{L}_2 = (Q, A, \{\xrightarrow{a}_2 : a \in A\}, q_0)$ if there exists a simulation $\sigma \subseteq P \times Q$ such that $\langle p_0, q_0\rangle \in \sigma$ written $\mathscr{L}_1 \preceq \mathscr{L}_2$.

Note that two LTSs can be combined to one by their disjoint union; so it does not matter whether we relate states from only one or two LTSs. Let $\mathscr{L}_1 = (S_1, A_1, \longrightarrow_1, s_1)$ and $\mathscr{L}_2 = (S_2, A_2, \longrightarrow_2, s_2)$ be two LTSs, where $\longrightarrow_i := \{\xrightarrow{a}_i : a \in A_i\}$ and $S_1 \cap S_2 = \emptyset$ is assumed (otherwise one can use an isomorphic copy of one of them). Then we set $\mathscr{L} = (S, A, \{\xrightarrow{a} : a \in A\}, s)$ where

- $S := S_1 \cup S_2 \cup \{s\}$, with $s \notin S_1 \cup S_2$,
- $A := A_1 \cup A_2 \cup \{\alpha\}$ with $\alpha \notin A_1 \cup A_2$, and
- $\longrightarrow := \longrightarrow_1 \cup \longrightarrow_2 \cup \{s \xrightarrow{\alpha} s_1, s \xrightarrow{\alpha} s_2\}$.

With this definition, structural properties obviously keep maintained, since the only additions are the non-deterministic transitions from the new starting state $s \in \mathscr{L}$ to the starting states of $\mathscr{L}_1$ and $\mathscr{L}_2$, and nothing vanishes.

Proposition 17.26. *Let $\mathscr{L} = (S, A, \{\xrightarrow{a} : a \in A\}, s_0)$ be a labelled transition system. Then for simulation the following properties hold:*

(1) For any $p, q \in S$, $p \preceq q$ implies $Tr(p) \subseteq Tr(q)$.
(2) $p \sim_S q$ implies $p \sim_T q$.
(3) id_S is a simulation relation.
(4) The composition $\rho\sigma$ of two simulation relations ρ and σ is a simulation.
(5) The union $\bigcup_{i \in I} \rho_i$ of simulation relations ρ_i is a simulation.
(6) $\preceq \subseteq S \times S$ is the largest simulation relation.

Proof. (1): If $p \preceq q$, then $\langle p, q\rangle \in \rho$ for some simulation relation ρ. By routine induction on the length of traces $t \in A^*$ one proves the somewhat stronger fact that $p \xrightarrow{t}{}^* p'$ implies $q \xrightarrow{t}{}^* q'$ with $\langle p', q'\rangle \in \rho$, whence $Tr(p) \subseteq Tr(q)$.

(4): $p\rho\sigma q$ implies that $p\rho p'\sigma q$ for some p'. So, if $p \xrightarrow{a} p_1$ then $p' \xrightarrow{a} q'$ for some q' with $p_1 \rho q'$; and from $p'\sigma q$ and $p' \xrightarrow{a} q'$ we get $q \xrightarrow{a} q_1$ for some q_1 with $q'\sigma q_1$, i.e. $p_1\rho\sigma q_1$. Thus we have shown that for any pair $\langle p, q\rangle \in \rho\sigma$, if $p \xrightarrow{a} p_1$, then there is a q_1 such that $q \xrightarrow{a} q_1$ and $p_1\rho\sigma q_1$, as required.

(5): Assuming $\langle p, q\rangle \in \bigcup_{i\in I} \rho_i$ means that $\langle p, q\rangle \in \rho_j$ for some $j \in I$. Hence, if $p \xrightarrow{a} p'$, then there is a q' with $q \xrightarrow{a} q'$ and $\langle p', q'\rangle \in \rho_j \subseteq \bigcup_{i\in I} \rho_i$.

(6): By definition, $p \preceq q$ if there exists a simulation relation ρ with $\langle p, q\rangle \in \rho$, i.e. $\preceq = \bigcup\{\rho \subseteq S \times S : \rho$ is a simulation relation$\}$. By (5) $\preceq$ is a simulation relation. □

Simulation relations are often used for verification purposes to show that one system correctly implements another, more abstract system, its specification, sometimes ranging over a few steps. This is possible due to the facts stated in the next proposition which justifies the choice of the symbol $\preceq$ and the notion of simulation equivalence.

Proposition 17.27. *Let $\mathscr{L} = (S, A, \{\xrightarrow{a} : a \in A\}, s_0)$ be a labelled transition system. Then $\preceq$ is a preorder and $\sim_S$ is an equivalence relation.*

Proof. By Prop. 17.26(3) id_S is a simulation relation and, since $id_S \subseteq \preceq$, the latter is reflexive. Now assume $p \preceq r \preceq q$; then there are simulations ρ, σ with $p\rho r\sigma q$, so $p\rho\sigma q$. By Prop. 17.26(4), $\rho\sigma$ is a simulation relation, whence $\rho\sigma \subseteq \preceq$; therefore, $p \preceq q$.

From Prop. 4.8 we get that both $\preceq^\smallsmile$ and $\preceq \cap \preceq^\smallsmile = \sim_S$ are preorders, and since the latter is symmetric by Prop. 3.52(5), $\sim_S \in \mathsf{Eq}_S$. □

Fig. 17.10 demonstrates that simulation equivalence is not sensitive to deadlock. The two LTSs are simulation equivalent since $\rho_1 := \{\langle q_5, q_1\rangle, \langle q_6, q_2\rangle, \langle q_7, q_4\rangle\}$ and $\rho_2 := \{\langle q_1, q_5\rangle, \langle q_2, q_6\rangle, \langle q_3, q_6\rangle, \langle q_4, q_7\rangle\}$ are simulation relations between them.

As the last concept of process theory we present bisimulation in a variant that uses two LTSs in its definition (see e.g. [8, p. 341] or [2, Definition 7.1]):

Definition 17.28. *Let $\mathscr{L}_1 = (P, A, \{\xrightarrow{a}_1 : a \in A\}, p_0)$ and $\mathscr{L}_2 = (Q, A, \{\xrightarrow{a}_2 : a \in A\}, q_0)$ be two labelled transition systems.*

(a) *A relation* $\beta \subseteq P \times Q$ *is called a* (**strong**) **bisimulation** *for* $\mathscr{L}_1$ *and* $\mathscr{L}_2$ *if, whenever* $\langle p,q\rangle \in \beta$, *for all* $a \in A$ *we have*

1. $\langle p_0, q_0\rangle \in \beta$ *and* $\langle q_0, p_0\rangle \in \beta$;
2. $p \xrightarrow{a}_1 p'$ *implies* $\exists q' \in Q\,[q \xrightarrow{a}_2 q' \wedge \langle p', q'\rangle \in \beta]$;
3. $q \xrightarrow{a}_2 q'$ *implies* $\exists p' \in P\,[p \xrightarrow{a}_1 p' \wedge \langle p', q'\rangle \in \beta]$.

(b) $\mathscr{L}_1$ *and* $\mathscr{L}_2$ *are called* **bisimilar**, *denoted as* $\mathscr{L}_1 \sim_B \mathscr{L}_2$, *in case there is a bisimulation for* $\mathscr{L}_1$ *and* $\mathscr{L}_2$.

Two states p, q are called bisimilar, $p \sim_B q$, if there is a bisimulation that relates them. An example for bisimulation equivalent LTSs is given in Fig. 17.8 since $p \sim_B q$.

Trace equivalence and bisimulation based equivalence are at the opposite ends of a whole spectrum of behavior equivalence concepts (see [11]), among them also one, called decorated-traces equivalence, that discriminates between the left one and the other two, which seems to be an appropriate grouping for the above example. Bisimulation distinguishes all of the three vending machines shown in Fig. 17.9, because its basic idea is to consider two states equivalent if by performing the same sequences of actions from these states it is possible to reach equivalent states. The first and the second machine are distinguished because after receiving two coins the first machine still offers the user the possibility of choosing between having coffee or tea while the second does not; and the second and third machine are distinguished because already after the insertion of one coin the user looses control of the third machine.

Based on equality, $\sim_T$ is obviously an equivalence. Among other important properties, this is shown for $\sim_B$ in the next theorem.

Theorem 17.29. ([1, Theorem 3.1], [2, Lemma 7.4])
Let $\mathscr{L}_i = (S_i, A, \{\xrightarrow{a}_i : a \in A\}, s_i)$, $i = 1,2$, *be labelled transition systems; a bisimulation relation* $\sim_B$

(1) *is an equivalence relation;*
(2) *is the largest strong bisimulation;*
(3) *satisfies the following property:* $p \sim_B q$ *iff, for each* $a \in A$,
(a) $p \xrightarrow{a} p'$ *implies* $\exists q' \in S_2\,[q \xrightarrow{a} q' \wedge p' \sim_B q']$, *and*
(b) $q \xrightarrow{a} q'$ *implies* $\exists p' \in S_1\,[p \xrightarrow{a} p' \wedge p' \sim_B q']$.

Proof. (1): id_S is a bisimulation, whence $s \sim_B s$ for each $s \in S$, so $\mathscr{L} \sim_B \mathscr{L}$.
Let $\beta \subseteq S_1 \times S_2$ be a bisimulation from $\mathscr{L}_1$ to $\mathscr{L}_2$. For the starting states we then have $\langle s_1, s_2\rangle \in \beta$ and $\langle s_2, s_1\rangle \in \beta$; thus both pairs are in $\beta^{\smile}$, whence (a).1 of Def. 17.28 obviously holds. Since $\beta^{\smile}$ is obtained by swapping the states of each pair in β, by symmetry of (a).2 and (a).3 these conditions also hold for $\beta^{\smile}$, so it a bisimulation from $\mathscr{L}_2$ to $\mathscr{L}_1$. Therefore, $\mathscr{L}_1 \sim_B \mathscr{L}_2$ implies $\mathscr{L}_2 \sim_B \mathscr{L}_1$.
To show transitivity of $\sim_B$, assume that β_1 is a bisimulation from $\mathscr{L}_1$ to $\mathscr{L}_2$, and that β_2 is a bisimulation from $\mathscr{L}_2$ to $\mathscr{L}_3$. Set $\beta := \beta_1\beta_2 \subseteq S_1 \times S_3$. Since $\langle s_1, s_2\rangle \in \beta_1$ and $\langle s_2, s_3\rangle \in \beta_2$, we get $\langle s_1, s_3\rangle \in \beta$, and similarly $\langle s_3, s_1\rangle \in \beta$. So (a).1 holds for β. Now assume that $\langle r, t\rangle \in \beta$; then $\langle r, s\rangle \in \beta_1 \wedge \langle s, t\rangle \in \beta_2$ for some $s \in S_2$. We now show (a).2; (a).3

has an analogous proof. Assume $\langle r,t\rangle \in \beta$ and $\langle r,s\rangle \in \beta_1$; then, since be_1 is a bisimulation relation, we get $r \xrightarrow{a}_1 r' \to \exists s' \in S_2\,[s \xrightarrow{a}_2 s' \wedge \langle r',s'\rangle \in \beta_1]$. $\langle s,t\rangle \in \beta_2$ and $s \xrightarrow{a}_2 s'$ implies $\exists t' \in S_3\,[t \xrightarrow{a}_3 t' \wedge \langle s',t'\rangle \in \beta_2]$. Together this gives that for $\langle r,t\rangle \in \beta$ we have $r \xrightarrow{a}_1 r' \to \exists t' \in S_3\,[t \xrightarrow{a}_3 t' \wedge \langle r',t'\rangle \in \beta]$. So $\mathcal{L}_1 \sim_B \mathcal{L}_3$.

(2): Similar to Prop. 17.26(5) and (6).

(3): The implication from left to right is immediate since by (2) $\sim_B$ is a bisimulation. Conversely, suppose that p and q are two states such that

$$(*)\quad \begin{cases} \text{for all } a \in A: \\ p \xrightarrow{a} p' \text{ implies } q \xrightarrow{a} q' \text{ such that } p' \sim_B q',\text{ for some } q' \in S, \text{ and} \\ q \xrightarrow{a} q' \text{ implies } p \xrightarrow{a} p' \text{ such that } p' \sim_B q',\text{ for some } p' \in S. \end{cases}$$

We now have to construct a bisimulation ρ with $\langle p,q\rangle \in \rho$. Furthermore, to each transition $p \xrightarrow{a} p'$ we must have a transition $q \xrightarrow{a} q'$ with $\langle p',q'\rangle \in \rho$. We set $\rho := \{\langle p,q\rangle, \langle q,p\rangle\} \cup \sim_B$. Then $\langle p,q\rangle \in \rho$ by construction, and from $(*)$ it follows that $p \xrightarrow{a} p'$ implies $q \xrightarrow{a} q'$ such that $\langle p',q'\rangle \in \rho$, for some $q' \in S$. It is straightforward to show that ρ is a bisimulation. □

We now rephrase Def. 6.75 in a general form: Let $\equiv\, \subseteq X \times X$ be an equivalence relation; then $\equiv$ is said to be a **regular equivalence** with respect to $\rho \subseteq X \times X$, if for all $a,b,c \in X$, $a \equiv b$ implies

$$(a\rho c \to \exists d \in X\,[b\rho d \wedge c \equiv d]) \quad \wedge \quad (c\rho a \to \exists d \in X\,[d\rho b \wedge c \equiv d]).$$

That two nodes $a,b \in X$ are *regular equivalent,* is denoted by $a \equiv_{re} b$.

Corollary 17.30. *A relation $\beta \subseteq S \times S$ is a bisimulation iff, for all $a \in A$,*

$$\beta \circ \xrightarrow{a} \;=\; \xrightarrow{a} \circ \beta \quad \textit{and} \quad \beta^{\smile} \circ \xrightarrow{a} \;=\; \xrightarrow{a} \circ \beta^{\smile}. \qquad (*)$$

Proof. By Th. 17.29(3) we see that bisimulation $\sim_B$ is a regular equivalence w.r.t. the transition relations $\xrightarrow{a}$. It thus follows from Prop. 6.76 that $(*)$ holds for all $a \in A$. Conversely, assume that $(*)$ holds for all $a \in A$, and that $\langle p,q\rangle \in \beta$ and $q \xrightarrow{a} q'$. Then $p\beta \cdot \xrightarrow{a} q'$, and therefore $p \xrightarrow{a} \cdot \beta q'$. So $p \xrightarrow{a} p'\beta q'$ for some p'. Together this gives that, for $\langle p,q\rangle \in \beta$, $q \xrightarrow{a} q'$ implies $\exists p' \in S\,[p \xrightarrow{a} p' \wedge p'\beta q']$. Analogous arguments for the second case complete the proof. □

Some results analogous to those for simulations also hold for bisimulations:

Proposition 17.31. *Let $\mathcal{L} = (S, A, \{\xrightarrow{a} : a \in A\}, s_0)$ be a labelled transition system. Then for bisimulation the following properties hold:*

(1) *id_S is a bisimulation relation.*
(2) *The composition $\rho\sigma$ of two bisimulation relations ρ and σ is a bisimulation.*
(3) *The union $\bigcup_{i\in I} \rho_i$ of bisimulation relations ρ_i is a bisimulation.*
(4) *$\sim_B \subseteq S \times S$ is the largest bisimulation relation.*

A relation-algebraic analysis of bisimulations can be found in [18] and [5]. As a final remark on bisimulations we mention that they do not need to be symmetric relations. On the other hand, since $\rho = \rho^{\smile}$ for symmetric relations, a symmetric simulation relation is always a bisimulation.

We conclude this section with information about the relationships of the behavioral equivalence concepts mentioned (due to the existence of counterexamples all implications are strict):

$$\text{Isomorphism} \longrightarrow \sim_B \longrightarrow \sim_S, \sim_{CT} \longrightarrow \sim_T .$$

Trace equivalence and bisimulation equivalence coincide over deterministic LTSs.

References

[1] Aceto, L., Ingólfsdóttir, A. et al.: *Reactive Systems-Modelling, Specification and Verification*, Cambridge University Press (2007).

[2] Baier, C., Katoen, J.-P.: *Principles of Model Checking*, The MIT Press, Cambridge, Massachusetts (2008).

[3] Bapat, R.B.: *Graphs and Matrices*, 2nd ed., Springer-Verlag, London (2014).

[4] Berghammer, R., Stucke, I. Winter, M.: Investigating and computing bipartitions with algebraic means. In Kahl, W. et.al (Eds): *RAMiCS 2015*, 257-274, Springer International Publishing Switzerland (2015).

[5] Berghammer, R., Winter, M.: Embedding mappings and splittings with applications, Acta Informatica **47**, 77-110 (2010).

[6] Blackburn, P., Rijke, M., Venema, Y.: *Modal Logic*, Cambridge University Press (2001).

[7] Brualdi, R.A., Ryser, H.J.: *Combinatorial Matrix Theory*, Cambridge University Press (1991).

[8] Chen, X.J., De Nicola, R.: Algebraic characterizations of trace and decorated trace equivalences over tree-like structures, Theoretical Computer Science **254**, 337-361 (2001).

[9] De Nicola, R.:. Behavioral equivalences. In Padua, D. (Ed): *Encyclopedia of Parallel Computing*, 120-127, Springer Science+Business Media (2011).

[10] Diestel, R.: *Graph Theory*, 3rd ed., Springer-Verlag Heidelberg, New York (2005).

[11] van Glabbeek, R.J.: The Linear Time - Branching Time Spectrum I: The Semantics of Concrete, Sequential Processes. In Bergstra, J.A., Ponse, A., and Smolka, S.A.(Eds.): *Handbook of Process Algebra*, 3-99. Elsevier, Amsterdam (2001).

[12] Gorrieri, R., Versari, C.: *Introduction to Concurrency Theory*, Springer International Publishing Switzerland (2015).

[13] Liu, B., Lai, H.-J.: *Matrices in Combinatorics and Graph Theory*, Kluwer Academic Publishers, Dordrecht (2000).

[14] Roberson, D.E.: Variations on a theme: Graph homomorphisms, PhD Thesis, University of Waterloo (2013).

[15] Schmidt, G., Ströhlein, T.: *Relations and Graphs*, Springer-Verlag Berlin Heidelberg (1993).

[16] Schmidt, G.: *Relational mathematics*, Cambridge University Press (2011).

[17] West, D.B.: *Introduction to Graph Theory*, 2nd ed., Pearson Education, Singapore (2001).

[18] Winter, M.: A relation-algebraic theory of bisimulations, Fundamenta Informaticae **83**, 1-21 (2008).

Chapter 18

Order Relations and Preference Modelling

In this chapter we cover another important area for both theoretical developments and various applications of binary relations: Its focus is on the question of what kind of relations contribute to the study of preferences and measurement theory. This is done in two parts: In Section 18.2 we give some more information on order relations that have a leading part in modeling preferences, and on their algebraic properties; and Section 18.3 contains considerations specific for various aspects of preference, utility and measurement. But before we deal with the topics just mentioned, we have to learn about one of the basic ingredients in Section 18.1, viz. on Ferrers relations and their properties.

As for almost all topics dealt with in the book, there is a rich literature also for the current one. So, throughout this chapter it is important to keep in mind that we have to restrict our presentation to some classical concepts and results, which means to some early and relatively simple formalisms to model and measure individual preference. Even within this limitation proofs will often be omitted in case they seem to be too involved to be presented in an introductory overview. However, some rather elementary example proofs will be included in order to provide an idea of the kind of argumentation needed. This way the reader should get a rough idea of some of the main concepts and techniques; references will be supplied for those interested in more details.

Only deterministic binary models, which have been the historically first and basic approach to preference measurements, are introduced. Of course, they suffer from the weakness that a person's preferences (and indifferences) may not remain invariant during an experiment. Readers interested in probabilistic or fuzzy extensions are referred to corresponding literature.

18.1 Ferrers Relations

As we shall see below, Ferrers relations are a basic concept for the conception of preference models. So this is a good opportunity to have a closer look at some of their properties. Recall from Def. 3.21 that a relation ρ is said to be a *Ferrers relation*, or of *Ferrers type*, if

$$\forall x, y, z, t \in X[x\rho y \wedge z\rho t \rightarrow x\rho t \vee z\rho y], \quad \text{or equivalently,} \quad \rho\rho^d\rho \subseteq \rho.$$

In terms of graphs this means: Given any two arrows, there exists an arrow from one of the starting points leading to the ending point of the other arrow; see Fig. 18.1. A fact that has already been mentioned in Section 14.3 in our discussion of Formal Concept Analysis is that visualized in matrix representation the definition means that $\begin{pmatrix} 1 & 0 \\ 0 & 1 \end{pmatrix}$ must not occur as a pattern, i.e. three of the four cells of such a square must be filled. This is equivalent to the property that such a matrix can be transformed into a stair-shaped block form by rearranging its rows and columns. Ferrers relations thus have the interesting property that they can be decomposed in rectangular blocks of either 0's or 1's.

Proposition 18.1.

(1) *Either none or all of $\rho, \rho^{\smile}$, and $\overline{\rho}$ are of Ferrers type.*
(2) *ρ is of Ferrers type iff $\rho^{\smile}\overline{\rho}$ is.*
(3) *ρ is of Ferrers type iff $\overline{\rho}\rho^{\smile}$ is.*

Proof. (1): From Prop. 3.13(6) we get $\rho\rho^d\rho \subseteq \rho$ iff $\rho^{\smile}\overline{\rho}\rho^{\smile} \subseteq \rho^{\smile}$; and since $\overline{\rho} = (\rho^{\smile})^d$, by Cor. 3.31(3), $\rho^{\smile}$ is Ferrers iff ρ is. From the first Schröder equivalence Th. 3.26(1)(a) we get $(\rho\rho^d)\rho \subseteq \rho \leftrightarrow ((\rho^d)^{\smile}\rho^{\smile})\overline{\rho} \subseteq \overline{\rho}$. By Cor. 3.31 we then have $(\rho^d)^{\smile} = \overline{\rho}$ and $\rho^{\smile} = \overline{\rho}^d$, which gives the desired result.

(2): For each relation ρ, from $\rho^{\smile}\overline{\rho} = \rho^{\smile}\overline{\rho}$ we have $\rho\overline{\rho^{\smile}\overline{\rho}} \subseteq \rho$ by Th. 3.26(1)(a), and $(\rho^{\smile}\overline{\rho})^{\smile} = \rho^d\rho$. Applying these two results and Prop. 3.13(6) we get

$$\rho^{\smile}\overline{\rho}\cdot(\overline{\rho^{\smile}\overline{\rho}})^{\smile}\cdot\rho^{\smile}\overline{\rho} \subseteq \rho^{\smile}\overline{\rho} \quad \text{iff}$$
$$\rho^d\rho\cdot(\overline{\rho^{\smile}\overline{\rho}})\cdot\rho^d\rho \subseteq \rho^d\rho$$

But this last inequality holds since $\rho^d[\rho\cdot(\overline{\rho^{\smile}\overline{\rho}})]\cdot\rho^d\rho \subseteq \rho^d[\rho]\cdot\rho^d\rho \subseteq \rho^d\rho$ due to the assumption that ρ is of Ferrers type. So the first inequality also holds, i.e. $\rho^{\smile}\overline{\rho}$ is a Ferrers relation. □

Proposition 18.2.

(1) *A reflexive Ferrers relation is strongly complete.*
(2) *An irreflexive Ferrers relation is asymmetric and transitive.*
(3) *An irreflexive Ferrers relation is negatively transitive.*
(4) *If ρ is a Ferrers relation on a set X then $\rho\rho^d$ is a weak order.*

Proof. (1): $x\rho x \wedge y\rho y \rightarrow x\rho y \vee y\rho x$.

(2): $x\rho y \wedge y\rho z \rightarrow x\rho z \vee y\rho y$, so ρ is transitive. Thus, by Prop. 3.24(1) it is also asymmetric.

(3): By (2) and Prop. 3.24(7).

(4): According to Prop. 3.32(3) it suffices to show that $\rho\rho^d$ is asymmetric. Assume this to fail; then there exist $x, y \in X$ such that $x\rho\rho^d z$ and $z\rho\rho^d x$. This implies $x\rho y \wedge z\overline{\rho} y \wedge z\rho t \wedge x\overline{\rho} t$ for some $t, y \in X$, contradicting the definition of a Ferrers relation. □

18.2 More on Order Relations

We did already have a look at order relations in Section 4.2 where we have presented some of their algebraic properties. Below we will elaborate on some kinds of relations that play a central role in the context of describing and analyzing preferences. For this we need some new concepts, and we begin by listing the corresponding formal definitions, together with some replications for easier look-up:

Definition 18.3. *Let ρ be a relation on a set X. Then ρ is called*

(a) **negatively transitive** *(1) if $\forall x, y, z[x\rho z \rightarrow x\rho y \vee y\rho z]$;*
(b) **negatively transitive** *(2) if $\forall x, y, z[\neg x\rho y \wedge \neg y\rho z \rightarrow \neg x\rho z]$;*
(c) **semitransitive** *if $\forall x, y, z, t \in X[x\rho y \wedge y\rho z \rightarrow x\rho t \vee t\rho z]$, or $\rho\rho\rho^d \subseteq \rho$.*

Lemma 18.4. *Let ρ be an asymmetric relation on X. Then*

(1) $\rho \subseteq \rho^d$.
(2) Let $\asymp_\rho := \overline{\rho} \cap \rho^d$; then $\rho \cup \asymp_\rho = \rho^d$.

Proof. (1): $x\rho y \rightarrow \neg y\rho x \rightarrow \overline{x\rho^\smile y} \rightarrow x\rho^d y$.
(2): $\rho \cup \asymp_\rho = (\rho \cup \overline{\rho}) \cap (\rho \cup \rho^d) = U \cap \rho^d = \rho^d$, using (1). □

Definition 18.5. *Let ρ be a relation on a set X. Then ρ is called*

(a) a **strict partial order** *if it is irreflexive and transitive;*
(b) a **weak order** *if it is asymmetric and negatively transitive;*
(c) an **interval order** *if it is an irreflexive Ferrers relation;*
(d) a **semiorder** *if it is a semitransitive interval order;*
(e) a **total preorder** *if it is a complete preorder.*

Be careful concerning the notion of a *weak order*, since it sometimes is used for what we call a *total preorder* (e.g. in [10, p. 69] or [11, p. 14]). In the same way as we do, this notion is used in e.g. [2, 5, 8]. For sake of completeness we list some easy consequences of the above definition in

Corollary 18.6.

(1) An interval order is asymmetric and transitive.
(2) A semiorder is asymmetric and transitive.
(3) Interval orders, and hence also semiorders, are weak orders.
(4) A weak order is acyclic.
(5) A complete weak order is linear.

Proof. (1): By Prop. 18.2(2).
(2): By (1), since semiorders are interval orders.
(3): By (1),(2) and Prop. 3.24(7).
(4): By definition, a weak order is asymmetric and negatively transitive: Thus it is also transitive by Prop. 3.24(6) and irreflexive by Prop. 3.24(3), so finally acyclic by

Prop. 3.24(2).
(5): By Prop. 3.24(3),(6). □

The intuition behind the definition of an interval order $<$ is that four elements w, x, y, z behave as if they were the end-points of two intervals $[x, y]$ and $[z, w]$ in $(\mathbb{R}, \leq)$, for which clearly the Ferrers property $x < y \wedge z < w \rightarrow x < w \vee z < y$ holds. This justifies the name for this kind of order on an arbitrary set.

Two main classes of relations are used to formalize preferences leading to two different kinds of models (see [2, p 21 ff]): The first one uses 'strict' preference, i.e. an individual that has to choose between a and b will definitely prefer one of them, and the second way describes a 'weak' preference, in the sense that an individual prefers one of a and b, or is indifferent between them. In the first case the preference relation is taken to be asymmetric, and in the other it must be reflexive and generally is assumed to be also complete.

These two ways to formalize preferences are equivalent in the following sense: Let $<$ denote an asymmetric relation; then "*a is indifferent to b*" if neither a is preferred to b, nor b is preferred to a. Then the **indifference relation** $\sim_<$ associated to $<$ is

$$a \sim_< b :\Longleftrightarrow \neg(a < b \vee b < a), \text{ i.e. } \sim_< = \overline{< \cup <^{\smile}} = \overline{<} \cap <^d, \tag{18.1}$$

where ρ^d is the dual of ρ as defined in Def. 3.30. The associated weak preference can be taken as $\leq := < \cup \sim_<$, and according to Lemma 18.4(2) we have $\leq = <^d$. The usual way of reading $a < b$ is "*b is better than a*", and that of $a \leq b$ is "*b is at least as good as a*".

Indifference might arise in several ways (see [8, p. 12]): First, an individual might feel that there is no real difference between a and b. Secondly, an individual is uncertain as to his preference between the alternatives. He might find the comparison difficult and may decline to commit himself to a strict preference judgment while not being sure that he regards a and b as equally desirable (or undesirable). Thirdly, $a \sim_< b$ might arise when the individual considers a and b incomparable in some sense.

Using strict preference, asymmetry can be viewed as an obvious consistency criterion: If you prefer a to b, you should not simultaneously prefer b to a. And transitivity seems like a reasonable criterion of coherence for an individual's preferences: If you prefer a to b and also b to c, common sense suggests that you should also prefer a to c.

In Prop. 18.9(2) it is shown that weak orders are transitive, so they are an intuitively plausible candidate for modeling preferences. But we must not overlook the important fact that the associated indifference is also transitive, and actually an equivalence. This certainly is a fine thing from a mathematical point of view, and also a kind of consistency requirement that we expect a rational decision maker to follow: It may be seen to be of a normative nature and fits well if we are concerned with how individuals *should* make their decisions. However, as is known from many experiments conducted by psychologists and behavioral scientists, indifference is often observed to be intransitive. So there is the equally valid concern to model the way how people *actually do* make decisions. This is where interval and semiorders will enter the picture. We will say more on this in Section 18.3.2 below.

Definition 18.7. *Two ordered sets $P = (X, \leq_P)$ and $Q = (Y, \leq_Q)$ are said to be* **isomorphic**, *(***dually isomorphic***) if there exists a bijection $f : X \longrightarrow Y$ such that for all $x, y \in X$*

$$x \leq_P y \leftrightarrow f(x) \leq_Q f(y), \qquad (x \leq_P y \leftrightarrow f(x) \geq_Q f(y)).$$

Concerning the relationship between asymmetric and total binary relations we have the following result:

Proposition 18.8. ([6, Proposition 7.17]) *The sets of asymmetric and of strongly complete binary relations on a set X and ordered by inclusion are dual semilattices via the coduality map $\rho \mapsto \rho^d$.*

Proof. Clearly, the intersection of two asymmetric relations on X is asymmetric, and the union of two strongly complete relations is strongly complete; so the set of asymmetric (strongly complete) relations on a set X is a meet (join) semilattice. By Cor. 3.31 we have $\rho \subseteq \sigma \leftrightarrow \sigma^d \subseteq \rho^d$ and $(\rho^d)^d = \rho$. Furthermore, ρ is asymmetric $\leftrightarrow \rho \cap \rho^{\smile} = \emptyset \leftrightarrow \rho \subseteq \rho^d$, and ρ is strongly complete $\leftrightarrow \rho \cup \rho^{\smile} = U \leftrightarrow \overline{\rho} \cap \rho^d = \emptyset \leftrightarrow \rho^d \subseteq \rho$. It then follows that the coduality map is a dual isomorphism between the two semilattices. □

This duality between the classes of asymmetric and of strongly complete relations allows us to immediately translate any result obtained for one class to a result for the other one. It is thus equivalent to work either with asymmetric relations formalizing strict preferences or with strongly complete relations formalizing weak preferences. It is, however, more economical to work with the first group: they contain a smaller number of ordered pairs than complete relations. We now proceed with some algebraic properties of the types of orders mentioned above.

Proposition 18.9. ([8, Theorem 2.1]) *Let $\prec$ be a weak order on a set X, $\sim$ its associated indifference relation, and $\preceq := \prec \cup \sim$. Then*

(1) *exactly one of $x \prec y$, $y \prec x$ or $x \sim y$ holds for all $x, y \in X$;*
(2) *$\prec$ is transitive, i.e. a strict partial order;*
(3) *$\sim$ is an equivalence;*
(4) *$x \prec y \wedge y \sim z \to x \prec z$ and $x \sim y \wedge y \prec z \to x \prec z$;*
(5) *$\preceq$ is transitive and strongly complete;*
(6) *$\prec'$ on $X/\sim$, with $a \prec' b$ iff $x \prec y$ for some $x \in a$ and $y \in b$ is a linear order.*

Proof. (1): Follows from asymmetry and the definition of $\sim$.

(2): By Prop. 3.24(6).

(3): Assume that $x \sim y \wedge y \sim z$ but not $x \sim z$. Then, by (1), either $x \prec z$ or $z \prec x$; thus, by definition of negative transitivity, we get that either $x \prec y \vee y \prec z$ or $z \prec y \vee y \prec x$ must hold, a contradiction to $x \sim y$, $y \sim z$ and (1). So we have shown $x \sim z$.

(4): Assume $x \prec y \wedge y \sim z$; then, by Def. 18.3(a), $x \prec z \vee z \prec y$, whence $x \prec z$ by (1).

(5): Transitivity follows from (2), (3), and (4). Assuming $\neg x \preceq y \wedge \neg y \preceq x$ gives $\neg x \prec y \wedge \neg x \sim y \wedge \neg y \prec x$, violating (1)

(6): (i) $\prec'$ is asymmetric: Assume that $a \prec' b \wedge b \prec' a$. Then we have $x \prec y \wedge y' \prec x'$ for some $x, x' \in a$ and $y, y' \in b$, with $x \sim x'$ and $y \sim y'$. So, by (4), we first get $x' \prec y$, and then $x' \prec y'$ a contradiction to $y' \prec x'$. (ii) $\prec'$ is negatively transitive: If $a \prec' c$, with $x \in a, z \in c$ and $x \prec z$, then, for any $b \in X/\sim$ and any $y \in b$, $x \prec y \vee y \prec z$, i.e. $a \prec' b \vee b \prec' c$. (iii) $\prec'$ is complete: Assume $a, b \in X/\sim$ and $a \neq b$. Then $a \cap b = \emptyset$, whence for all $x \in a, y \in b$ $\neg x \sim y$. Si by (1) either $x \prec y$ or $y \prec x$, i.e. either $a \prec' b$ or $b \prec' a$. So we have shown that $\prec'$ is a complete weak order, and the assertion now follows from Cor. 18.6(5). □

An analogous result for strict partial orders is given in the next proposition. Note that the indifference $\sim$ as defined in (18.1) but induced by a strict partial order $\prec$ need not be transitive; so if $x \sim y$ and $y \sim z$, we may still have $x \prec z$. So $\sim$ is not necessarily an equivalence relation. However, the relation

$$x \approx_{\prec} y :\Longleftrightarrow \forall z \in X[x \sim_{\prec} z \leftrightarrow y \sim_{\prec} z] \tag{18.2}$$

is, as shown in Prop. 18.10(2) below.

Proposition 18.10. ([8, Theorem 2.3]) *Let $\prec$ be a strict partial order on a set X. Then*

(1) exactly one of $x \prec y, y \prec x$, $x \approx y$, or $(x \sim y \wedge \neg x \approx y)$ holds for all $x, y \in X$;
(2) $\approx$ is an equivalence;
(3) $x \approx y$ iff $(x \prec z \leftrightarrow y \prec z) \wedge (z \prec x \leftrightarrow z \prec y)$, for all $z \in X$;
(4) $x \prec y \wedge y \approx z \rightarrow x \prec z$ and $x \approx y \wedge y \prec z \rightarrow x \prec z$;
(5) $\preceq$ is transitive and strongly complete;
(6) $\prec^*$ on $X/\approx$, with $a \prec^* b$ iff $x \prec y$ for some $x \in a$ and $y \in b$, is a strict partial order.

Corollary 18.11. ([2, p. 23, Theorem 2.1]) *Let $\prec$ be a strict partial order; then the following conditions are equivalent:*

(1) $\prec$ is a weak order;
(2) $\sim$ is an equivalence;
(3) $\prec^d$ is a strongly complete preorder.

Proof. (1) ⇒ (2): By Prop. 18.9(3).

(2) ⇒ (1): By Prop. 3.24(1) $\prec$ is asymmetric. Now assume that $x \bar{\prec} y \wedge y \bar{\prec} z$, but $x \prec z$. Then $y \prec x$ is impossible, since it would imply $y \prec z$ by the transitivity of $\prec$; so $y \bar{\prec} x$, i.e. $x \prec^d y$. Since by hypothesis $x \bar{\prec} y$, we get $x \sim y$. A similar reasoning shows $y \sim z$, whence $x \sim z$, i.e. $x \bar{\prec} z$, a contradiction to our assumption $x \prec z$.

(1) ⇔ (3): We have to show that ($\prec^d$ is reflexive and transitive and strongly complete) $\Longleftrightarrow$ ($\prec$ is asymmetric and negatively transitive). By Prop. 18.8, $\prec^d$ is strongly complete iff $\prec$ is asymmetric. Now assume the right hand condition; then $\prec$ is transitive by Prop. 3.24(6), so $\breve{\prec}$ is transitive and asymmetric by Prop. 3.25(4),(6). Therefore, $\breve{\prec}$ is irreflexive by Prop. 3.24(1), whence $\prec^d$ is reflexive Prop. 3.25(2). Furthermore, using Prop. 18.9(1) and (4), we get that $x \prec^d y \wedge y \prec^d z$, i.e. $\neg y \prec x \wedge \neg z \prec y$, implies $(x \prec y \vee x \sim y) \wedge (y \prec z \vee y \sim z)$, and therefore $x \prec z$, which implies $\neg z \prec x$. Thus $\neg x \breve{\prec} z$, i.e. $x \prec^d z$.

For the converse direction it remains to deduce that $<$ is negatively transitive from the left hand condition. We did already mention that, by Prop. 18.8, $<^d$ is strongly complete iff $<$ is asymmetric. So we can apply Prop. 18.9(1): Assume $x \overline{<} y \wedge y \overline{<} z$ then $(y < x \vee y \sim x) \wedge (z < y \vee z \sim y)$. So $z < x \vee z \sim x$, whence $x \overline{<} z$. □

For the further discussion of orders in the context of preferences it is illustrative to have a picture of some important configurations which appear as forbidden substructures. We let $\mathbf{2}+\mathbf{2} := \{\langle x, y\rangle, \langle z, t\rangle\}$, and $\mathbf{1}+\mathbf{3} := \{\langle x, y\rangle, \langle y, z\rangle\}$ denote two relations on the set $\{x, y, z, t\}$. In Fig. 18.1 the dashed arrows between two points x and y stand for $x\overline{\rho}y$, i.e. the last but one configuration cannot be part of a Ferrers, and the right one not that of a semitransitive relation, since one of the two dashed arrows must occur in them. So clearly $\mathbf{2}+\mathbf{2}$ and $\mathbf{1}+\mathbf{3}$ become forbidden sub-configurations in interval and semiorders.

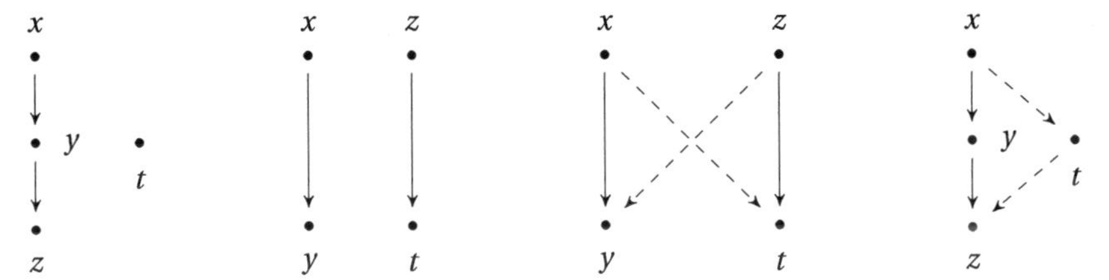

1+3 **2+2** Ferrers condition semitransitivity

FIG. 18.1: Some characteristic configurations in order relations.

It has been said above that indifference has been found in various experiments to be intransitive. So, in addition to the beautiful world of equivalences, we need additional mathematical concepts to come to grips with this fact. It is here, where interval and semiorders come into play. In the propositions below we give algebraic characterizations of these kinds of order relations, but will present or reproduce only parts of the proofs. First of all we mention the following

Lemma 18.12. ([6, p. 198, Lemma 7.10,4.]) *Let ρ be a strict partial order on X, and $\sim = \overline{\rho} \cap \rho^d$. Then $\rho^d = \rho \cup \sim$, $\rho \cdot \sim = \rho\rho^d$, and $\sim \cdot \rho = \rho^d \rho$.*

Proof. For an asymmetric relation we have $\rho \cap \rho^{\smile} = \emptyset$, so $\rho \subseteq \rho^d$. Thus, $\rho^d = \rho^d \cap (\rho \cup \overline{\rho}) = (\rho^d \cap \rho) \cup (\rho^d \cap \overline{\rho}) = \rho \cup \sim$.
Since, for any $\rho \in \mathscr{B}_X$, $\rho \cdot \sim = \rho(\overline{\rho} \cap \rho^d) \subseteq \rho\overline{\rho} \cap \rho\rho^d \subseteq \rho\rho^d$, we must prove the inverse inequality $\rho\rho^d \subseteq \rho \cdot \sim$. Assume $x\rho z \rho^d y$ for some z, then $y\overline{\rho}z$. Given $z\overline{\rho}y$ then $y\rho^d z$, so $y \sim z$, whence $x\rho \cdot \sim z$; on the other hand, if $z\rho y$, then $x\rho y$ by the transitivity of ρ, whence $x\rho y \sim y$. Thus, $\rho\rho^d \subseteq \rho \cdot \sim$. □

Proposition 18.13. ([2, p. 59, Lemma 3.1]) *Let ρ be a binary relation on a set X and $\sim = \overline{\rho} \cap \rho^d$. Then the following assertions are equivalent:*

(1) ρ is an interval order.
(2) ρ is a strict partial order without $\mathbf{2}+\mathbf{2}$ as substructure.
(3) ρ is irreflexive and $\rho\cdot\sim\cdot\rho\subseteq\rho$.
(4) ρ is irreflexive and $\rho\cdot\sim$ and $\sim\cdot\rho$ are weak orders.
(5) ρ is asymmetric and every cycle of ρ^d contains at least two consecutive pairs of $\sim$.

Proof. (1) $\Rightarrow$ (2): ρ is irreflexive by definition, and transitive by Cor. 18.6(1). It is obvious from Fig. 18.1 that interval orders cannot contain $\mathbf{2}+\mathbf{2}$ as a substructure.

(2) $\Rightarrow$ (3): Assume $x\rho y, y\sim z$ and $z\rho t$. Since $\mathbf{2}+\mathbf{2} := \{\langle x,y\rangle,\langle z,t\rangle\}$ is not a substructure of ρ, $x\sim t := x\overline{\rho}t \wedge t\overline{\rho}x$ is impossible (see Fig. 18.1), whence $x\rho t\vee t\rho x$ must hold. $t\rho x$ is impossible since otherwise we would get $z\rho y$ by the transitivity of ρ, a contradiction to $y\sim z$. So we must have $x\rho t$, i.e. $\rho\cdot\sim\cdot\rho\subseteq\rho$.

(3) $\Rightarrow$ (4): $\rho\cdot\sim\cdot\rho\subseteq\rho$ implies $(\rho\cdot\sim)^2\subseteq\rho\cdot\sim$, so $\rho\cdot\sim$ is transitive. It is also irreflexive, since $x\rho z\sim x$ would imply $z\overline{\rho}x\wedge x\overline{\rho}z$, contradicting $x\rho z$. Thus, $\rho\cdot\sim$ is asymmetric by Prop. 3.24(1), and hence also negatively transitive by Prop. 3.24(7), i.e. a weak order. A similar proof works for $\sim\cdot\rho$.

(4) $\Rightarrow$ (5): ρ is irreflexive by assumption, and $\rho\cdot\sim$ acyclic by Cor. 18.6(4). Assume that $x\rho y\wedge y\rho z$; since both $z\rho x$ and $x\sim z := \neg z\rho x\wedge\neg x\rho z$ would give a contradiction, we must conclude that $x\rho z$, i.e. that ρ is transitive. (The contradictions arise as follows: Since ρ is irreflexive, $x\sim x$, for all x, and by definition $\sim$ is symmetric. Thus $x\rho y\wedge y\rho z\wedge z\rho x$ would imply the cycle $x\rho y\sim y\rho z\sim z\rho x\sim x$ in $\rho\cdot\sim$; and $x\rho y\wedge y\rho z\wedge x\sim z$ gives rise to $x\rho y\sim y\rho z\sim x$.) We have shown that ρ is a strict partial order, so ρ is asymmetric by Prop. 3.24(1).
Let $C=\langle x_1,\dots,x_i,\dots,x_m=x_1\rangle$ be a cycle of ρ^d which, by Lemma 18.12, equals the disjoint union $\rho\cup\sim$ (obviously $\rho\cap\sim=\emptyset$). Since ρ is transitive, any sequence $x_i\rho\dots\rho x_k$ can be replaced by $x_i\rho x_k$, and thus induces another cycle C' leaving the number of pairs $x\sim y$ unchanged. If C' does not contain a sequence of the form $x\sim y\sim z$, then it would consist of a sequence of alternating pairs of ρ and $\sim$, and we would have a cycle $(\rho\cdot\sim)^j$, for some $j\geq 2$, a contradiction to Prop. 3.24(2). So C', whence also C, contains two consecutive pairs of $\sim$.

(5) $\Rightarrow$ (1): We have to show that ρ is an irreflexive Ferrers relation. Since ρ is irreflexive by Prop. 3.24(3), it remains to verify that $\forall x,y,z,t\in X[x\rho y\wedge z\rho t\rightarrow x\rho t\vee z\rho y]$. Assume that $x\rho y\wedge z\rho t$ but $x\overline{\rho}t\wedge z\overline{\rho}y$. This would result in the cycle $x\rho y\rho^d z\rho t\rho^d x$. Since $\sim$ is reflexive this yields $x\rho y\sim y\rho^d z\rho t\sim t\rho^d x$. By Lemma 18.12, $\rho^d=\rho\cup\sim$; so we get four possibilities

(1) $x\rho y\sim y\rho z\rho t\sim t\rho x$, whence $x\rho y\sim y\rho z\sim z\rho t\sim t\rho x\sim x$, i.e. $x(\rho\cdot\sim)^4x$, a contradiction to the acyclicity of $\rho\cdot\sim$.
(2) $x\rho y\sim y\rho z\rho t\sim t\sim x$, whence $x\rho y\sim y\rho z\sim z\rho t\sim x$, again a contradiction.
(3) $x\rho y\sim y\sim z\rho t\sim t\rho x$, whence $x\rho y\sim z\rho t\sim t\rho x\sim x$, again a contradiction.
(4) $x\rho y\sim y\sim z\rho t\sim t\sim x$, whence $x\rho y\sim z\rho t\sim x$, again a contradiction.

So we have shown that ρ is a Ferrers relation, and thus an interval order. □

Without proof we note a similar characterization of semiorders in

Proposition 18.14. ([2, p. 62, Lemma 3.2]) *Let ρ be a binary relation on a set X. Then the following assertions are equivalent:*

(1) *ρ is a semiorder.*
(2) *ρ is a strict partial order without* **2** + **2** *and* **1** + **3** *as substructures.*
(3) *ρ is irreflexive and $\rho \cdot \sim \cdot \rho \subseteq \rho$ and $\rho^2 \cdot \sim \subseteq \rho$.*
(4) *ρ is irreflexive and $\rho \cdot \sim \cup \sim \cdot \rho$ is a weak order.*
(5) *ρ is asymmetric and for every cycle γ of ρ^d, $|\sim \cap \gamma| > |\rho \cap \gamma|$.*

The above theorem finishes our formal algebraic exposition on various kinds of order relations. In the following we will turn to more intuitive considerations that deal with how preferences can be modeled by numbers in a plausible and consistent way.

18.3 Preferences and their Measurement

This section gives an overview on basic concepts and results concerning an *individual's* preferences: how they can be modeled by various kinds of order relations, what are plausible consistency requirements, and how they can be measured. As main sources I have used Fishburn [8, Chapter 2], Roberts [16, Sections 2.1 and 3.1], and Krantz, Luce et al. [11, Chapter 1], all of them classical textbooks.

As stated in [18], to develop preferences is an inevitable step in nearly each human activity and is necessary for making one's decision when buying a car or a house, or only some pair of shoes, or fruits, or choosing a film on TV etc. Consequently, to elicit and model preferences has been the goal in various fields like economy, psychology, sociology, operational research, and decision theory. Thus, in connection with preferences, we do not only have to think about various kinds of order relations and their algebraic properties from a purely formal point of view as done in Section 18.2. We also have to investigate how these relations can be used to evaluate preferences, and what are their properties with respect to this task. In a sense, we should change our viewpoint from a formal mathematical to a more behavioral and operational perspective. Of course, if the two viewpoints fit together, we should meet concepts and results of the preceding section again, however with new clothes.

18.3.1 *A Short Sketch of Measurement Theory*

We begin by locating **measurement theory** in the network of sciences, and will also consider its most basic aspects: According to [10, p. 324], measurement theory is a body of concepts lying somewhere between mathematics and *philosophy of science*; see e.g. [23, Chapters 1-4]. In the simplest (and usual) case, when measuring some attribute of a class of objects, we associate numbers with the objects in such a way that the relations holding in reality for a property amongst the objects are faithfully represented as numerical properties; see [11, p.1]. Technically spoken, we are looking for an homomorphism from an observed, i.e. empirical, relational system $\mathfrak{A}$ to some numerical relational system $\mathfrak{B}$.[1]

Following [16, p. 50], consider as examples the measurement of temperature or mass: We are after an assignment of numbers that preserve the observed relations "*warmer than*" or "*heavier than*": Let A be a set of objects, and W, H be two relations on A, where aWb holds iff you judge a to be warmer than b, and aHb holds iff "*a is heavier than b*". Then, for all $a, b \in A$, we want to assign real numbers $f(a), f(b) \in \mathbb{R}$ such that

$$aWb \iff f(a) > f(b) \qquad \text{and} \qquad aHb \iff f(a) > f(b). \tag{18.3}$$

However, in the case of mass, we want our assignment to be additive in the sense that the mass of the combination of two objects is the sum of their masses (think of packing a precious thing into a massive box). More formally, when given an operation $\oplus$ on A representing a combination of two objects in a way that both their masses become relevant, then we want a real-valued function f on A that not only satisfies condition (18.3) but also preserves $\oplus$: for all $a, b \in A$, $f(a \oplus b) = f(a) + f(b)$.

The above are two typical kinds of problems which are tackled by various ideas and techniques developed in the field of measurement theory. From the wide range of concepts we will present only the most basic (and historically early) ones, since they provide fine examples for our view on relations in this book. But, as is the case for all other application areas mentioned, more sophisticated and complex extensions have been invented in the course of time.

Measurement theory has to face two basic problems, the first one being called the **representation problem** ([16, p. 54]): Given a numerical relational system $\mathfrak{B}$, one has to find conditions on an observed relational system $\mathfrak{A}$ that are sufficient for the existence of a homomorphism f from $\mathfrak{A}$ into $\mathfrak{B}$. These conditions are usually called axioms for the representation, and the theorem stating their sufficiency is usually called a **representation theorem**. The triple $\langle \mathfrak{A}, \mathfrak{B}, f \rangle$, or f in short, is called a **scale**.

From this point of view, measurement may be regarded as the construction of homomorphisms from empirical relational structures of interest into numerical relational structures that are useful. It is of great practical value if there is a constructive proof of a representation theorem that not only shows that a representation is possible, but also gives advice of how to construct it.

As an example, assume that we seek a homomorphism $f : (A, \rho) \longrightarrow (\mathbb{R}, >)$. If such a homomorphism f exists, then ρ must be transitive, since $a\rho b \wedge b\rho c$ implies $f(a) > f(b) \wedge f(b) > f(c)$, so $f(a) > f(c)$ and $a\rho c$ follow. Thus, the transitivity of the (empirical relation) ρ is a typical measurement axiom. Such axioms are called *necessary* in [11, p.

[1] Of what we give a short overview only below is known as *representational measurement theory* (RMT). Though being the dominant view on measurement since the middle of the 20th century, it is not an undisputed bible and has been under critical discussion for quite a while; see e.g. [20] or [21], to name one early and one recent contribution. But these methodological considerations belong to the area of philosophy of science (where there is a huge amount of literature on this topic), and are not our concern here: We are only looking at the interesting relationships between formal properties of relations and the kind of numerical representations which they induce, and what are the ideas behind the various concepts. For a most recent general discussion you might want to have a look at [13].

21] in the sense that they are a consequence of the existence of the homomorphism one is trying to establish. In order to illustrate the representation problem we consider the simple case of the existence of an order homomorphism on a finite weakly ordered set:

Theorem 18.15. ([5, (1.2.1) Theorem]) *Let $\prec$ be a binary relation on a finite nonempty set X; then the following assertions are equivalent:*

(1) $\prec$ is a weak order.
(2) There exists a function $f: X \longrightarrow \mathbb{R}$, such that $x \prec y \leftrightarrow f(x) < f(y)$.

Proof. Let $\prec$ be a weak order on X. We define a real-valued function f on X by $f(x) := |\prec x|$, i.e. $f(x)$ is the number of elements $y \in X$ with $y \prec x$. If $y \prec x$, then $\prec y \subseteq \prec x$, since, by the transitivity of $\prec$, $z \prec y \prec x$ implies $z \prec x$. Thus, $f(y) \le f(x)$. We have $y \in \prec x$, but $y \notin \prec y$ since $\prec$ is irreflexive; so $f(y) < f(x)$. Conversely, assume $f(y) < f(x)$, but $\neg y \prec x$. Then, by the negative transitivity of $\prec$, we get $\neg(z \prec y) \rightarrow \neg(z \prec x)$, whence $z \prec x \rightarrow z \prec y$ from Th. 2.33(8). So $\prec x \subseteq \prec y$, which gives $f(x) \le f(y)$, a contradiction to our assumption $f(y) < f(x)$; therefore $y \prec x$.

Let f be a real-valued function f on X, such that $x \prec y \leftrightarrow f(x) < f(y)$. Then, since $y \prec x$ implies $f(y) < f(x)$, $\prec$ is asymmetric. Moreover, if $\neg(y \prec x)$ and $\neg(z \prec y)$ then $f(x) \le f(y)$ and $f(y) \le f(z)$, so $f(x) \le f(z)$, whence $\neg(z \prec x)$, i.e. $\prec$ is negatively transitive. □

The second basic problem of measurement theory is the **uniqueness problem**: In general, a whole class of scales $\mathsf{S}(\mathfrak{A}, \mathfrak{B})$ exists that homomorphically map a given empirical relational system $\mathfrak{A}$ in a numerical relational system $\mathfrak{B}$. So a natural question is, if all these homomorphisms are of the same kind. The answer to this question is given by so-called **uniqueness theorems**; they tell us what kind of scale $f \in \mathsf{S}(\mathfrak{A}, \mathfrak{B})$ is, and give rise to a theory of meaningfulness of statements involving scales. In particular, uniqueness theorems put restrictions on the mathematical operations that can be performed on the numbers arising as scale values.

Before we can proceed, we first need the concept of an *admissible transformation* which becomes clear from the viewpoint of homomorphisms between relational structures: A transformation $\phi: f \longrightarrow f'$ of a scale f is **admissible** if and only if $f': \mathfrak{A} \longrightarrow \mathfrak{B}$ also is a homomorphism from $\mathfrak{A}$ into $\mathfrak{B}$. Thus, if f is order preserving and additive (as required above for mass measurement), then clearly also each $f'_{\mathsf{c}} := \phi_{\mathsf{c}} \circ f = \mathsf{c} \cdot f$, with $\mathsf{c} > 0$.

According to [16, Sections 2.2 and 2.3] a scale $f \in \mathsf{S}(\mathfrak{A}, \mathfrak{B})$ is called **regular** if for any other scale $g \in \mathsf{S}(\mathfrak{A}, \mathfrak{B})$ there is a transformation $\phi: f(A) \longrightarrow B$ such that $g = \phi \circ f$. If every homomorphism $f: \mathfrak{A} \longrightarrow \mathfrak{B}$ is regular, we call $\mathfrak{A} \longrightarrow \mathfrak{B}$ a **regular representation**. So a representation $\mathfrak{A} \longrightarrow \mathfrak{B}$ is regular if, given any two scales $f, g \in \mathsf{S}(\mathfrak{A}, \mathfrak{B})$, one can map each into the other by an admissible transformation. And if this is the case, then the class of admissible transformations defines the *scale type*. What is typically shown in a **uniqueness theorem** is the fact that all admissible transformations are of the same

TABLE 18.1: Some common scale types.

Scale Type	Admissible Transformation	Properties and Examples
nominal	any one-to-one ϕ	numbers are labels only, without any mathematical meaning, i.e. they have none of the properties of real numbers (especially no inherent order); thus it is the least informative form of measurement qualitative, categorical Blood groups, Religion, Phone numbers, Social Security numbers
ordinal	strictly monotone increasing transformation: $x < y \leftrightarrow \phi(x) < \phi(y)$	scores reflect order in meaning, i.e. a higher score value means that more of a property is present; the distance between numerical scores still has no meaning qualitative, categorical Mohs scale of hardness, Beaufort wind scale, *"low/medium/ high"*
interval	positive linear transformation: $\phi(x) := ax + b$, $a > 0$	scores are ordered, and the magnitude of the dis- tance between scores has meaning; a distance d means the same everywhere, but a score of $2s$ does not mean doubling s; we may change both the unit and the zero point quantitative, numerical temperature (Fahrenheit, centigrade)
ratio	similarity transformation: $\phi(x) := ax$, $a > 0$	have all properties of the real numbers: order, magnitude, and, unlike the other three scale types, a meaningful origin; due to the last fact, only multiplication and division (i.e. changes in the unit of measurement via a scaling factor) maintain the structure of the scale and meaning of the scores; only unit can be changed quantitative, numerical temperature (Kelvin; due to absolute zero), time intervals (second, hour, week), length, mass

form. Now we are ready to state as a counterpart to Th. 18.15 the following uniqueness theorem. We first need the following

Lemma 18.16. ([16, Theorem 2.1]) *$(\mathfrak{A}, \mathfrak{B}, f)$ is regular iff $f(a) = f(b)$ implies $g(a) = g(b)$, for every other homomorphism g from $\mathfrak{A}$ into $\mathfrak{B}$ and all $a, b \in A$.*

Proof. If f is regular, then there is a function $\phi: f(A) \longrightarrow B$ such that $g = \phi \circ f$. Thus $f(a) = f(b)$ implies $g(a) = \phi(f(a)) = \phi(f(b)) = g(b)$. Conversely, given g, we define $\phi(f(a)) = g(a)$. Thus, since $f(a) = f(b)$ implies $g(a) = g(b)$, ϕ is well defined, and $g = \phi \circ f$, whence f is regular. □

Theorem 18.17. ([16, Theorem 3.2]) *Let $\prec$ be a binary relation on a finite nonempty set A, and f be a a real-valued function on A satisfying*

$$a \prec b \leftrightarrow f(a) < f(b). \tag{18.4}$$

Taking $\mathfrak{A} = (A, \prec)$ and $\mathfrak{B} = (\mathbb{R}, <)$, $\langle \mathfrak{A}, \mathfrak{B}, f \rangle$ is an ordinal scale and $\mathfrak{A} \longrightarrow \mathfrak{B}$ is a regular representation.

Proof. We first show that every function f satisfying (18.4) is a regular scale: Suppose g is another function satisfying (18.4); then $f(a) = f(b)$ implies $\neg a \prec b \wedge \neg b \prec a$, so $g(a) = g(b)$. Therefore, f is regular by Lemma 18.16. Moreover, $(\mathfrak{A}, \mathfrak{B}, f)$ is ordinal: If $\phi: f(A) \longrightarrow B$ is any strictly monotone increasing function, then $\phi \circ f$ satisfies (18.4) whenever f does, since $(\phi \circ f)(a) < (\phi \circ f)(b) \leftrightarrow f(a) < f(b) \leftrightarrow a \prec b$. So we have shown that the admissible transformations are the strictly monotone increasing functions, whence $(\mathfrak{A}, \mathfrak{B}, f)$ is ordinal. □

A final question is for what can be called a *meaningful statement* involving numerical scales; see e.g. [17]. As we have seen above, uniqueness theorems identify the class of admissible transformation for a scale. From this point of view we say that a statement is *meaningful* if and only if its truth (or falsity) remains unchanged under all admissible transformations of all the scales involved.

As the last point in this section, we will now illustrate some of the concepts introduced above. First, consider the statement S:="*a weighs twice as much as b.*". Then S can be written as

$$f(a) = 2f(b), \tag{18.5}$$

which is meaningful if, for every admissible transformation ϕ of f, (18.5) holds iff

$$\phi \circ f(a) = 2\phi \circ f(b). \tag{18.6}$$

Since weight is a ratio scale, ϕ has the form $\phi(x) = Cx$, for some $C \in \mathbb{R}_+$. So (18.6) becomes $Cf(a) = 2Cf(b)$, which is equivalent to (18.5) which, therefore, is meaningful. Denoting by a any child, and by b any elephant, statement S is still meaningful, but is false whatever unit we take to measure weight, e.g. pounds or kilograms. Thus, being meaningful is not the same as being true: Meaningfulness is concerned with whether or not an assertion based on scales of measurement is an accident of the particular scales used, or is independent of the scales and thus says something significant about the fundamental relations among the objects being measured.

Measurement on ratio scales is the most important for scientific applications, but not the only one. As mentioned above in Table 18.1, temperature, for example, if measured in Celsius or Fahrenheit degrees, is on an interval scale, and we have the well known transformation $t_C = \frac{5}{9}(t_F - 32) = \frac{5}{9}t_F - \frac{160}{9}$. Now consider the statement S:="*a is twice as warm as b.*". This can again be written as (18.5), but with the admissible transformations for interval scales, $\phi(x) := Ax + B,\ A > 0$, we now get as analogue to (18.6)

$$\phi \circ f(a) = Af(a) + B = 2[Af(b) + B] = 2\phi \circ f(b),$$

and it is quite possible that (18.5) holds, but the last equation does not: a can be twice as warm as b in Fahrenheit, but not in Centigrade.

In case you wonder where the names come from: Assume that f_p denotes the weight in pounds, and f_k the weight in kilograms. Then we have $f_p(a)/f_p(b) = f_k(a)/f_k(b)$, i.e. the quotient is independent from the unit we use for measurement. So it makes sense to compare ratios. Similarly, on ordinal scales one can compare size: $f(a) > f(b)$ is meaningful, and on interval scale differences, i.e. interval lengths: $f(a) - f(b) < f(c) - f(d)$.

18.3.2 *Semiorders and Interval Orders*

We continue our chapter on preferences with a short overview on the algebraic properties and numerical representations characteristic for those kinds of order relations that allow for intransitive indifference, and take the result of Th. 18.15 as a starting point for our considerations. Though interval orders and semiorders still have their problems they are interesting because the strict preference relation is transitive, whereas the indifference relation may be intransitive, and because they give a beautiful example for the straightforward evolution (with respect to both syntax and semantics, so to speak) of an initial nearby concept.

Recall that we have already argued on page 496 that the transitivity of the associated indifference relation of a weak order is much more a normative consistency requirement than a hard fact in real life. This can be easily made plausible as follows: Suppose that we have three objects x, y and z such that both x and y and y and z are "close" to each other, so that we are indifferent between them. Nevertheless, x and z may be far enough apart, whence we prefer z to x.

Returning to the problem of nontransitive indifference, from real life and psychological experiments it is a well known fact that perceived or measured orderings of elements (e.g. bodies, alternatives) according to some property (e.g. mass, preference) do not always faithfully mirror reality when a sensory or instrumental threshold prevents the detection of "small differences" in the relevant property between distinct elements. This idea, that indifference corresponds to closeness, suggests using the notion of a "threshold" or "*just noticeable difference*" (jnd), and to construct numerical representations in a way that indifference between two objects occurs if and only if the corresponding numbers are close, i.e. within some fixed positive distance d which measures the threshold. It will turn out that in the case of interval orders the threshold may vary from element to element, whereas for semiorders the threshold is uniform (constant jnd). To see that for an interval order $\prec$ the associated relations $\preceq$ and $\sim$ may fail to be transitive, consider $\prec$ defined on $\mathbb{R}$ as $x \prec y :\leftrightarrow\ x + 1 < y$.

So much on the reasons why these order structures have found considerable interest during the last few decades. We now turn to their mathematical aspects. As already remarked at the begin of this chapter, we will scratch on some early classical results only (for newer developments see e.g. [1,3] or [7]).

Theorem 18.18. ([6, Proposition 7.13]) *Let ρ be a binary relation on a finite set X. The following conditions are equivalent:*

(1) ρ is a Ferrers relation.
(2) There exist two functions $f, g : X \longrightarrow \mathbb{R}$ such that $x\rho y \leftrightarrow f(x) < g(y)$.
(3) There exist two functions $u, d : X \longrightarrow \mathbb{R}$ such that $x\rho y \leftrightarrow u(x) + d(x) < u(y)$.

Proof. (1) ⇒ (2): By Prop. 18.2(4), $\rho\rho^d$ is a weak order, so, by Th. 18.15, there is a function $f : X \longrightarrow \mathbb{R}$ with $x\rho\rho^d z$ iff $f(x) < f(z)$. We define $g : X \longrightarrow \mathbb{R}$ as follows:

$$g(y) = \begin{cases} \max\{f(z) : z \in X\} + 1 & \text{if } y\rho^d = \emptyset \\ \min\{f(z) : y\rho^d z\} & \text{otherwise.} \end{cases}$$

We first show that $x\rho y$ implies $f(x) < g(y)$: Assume $x\rho y$ for some $x, y \in X$; then, if $y\rho^d = \emptyset$, we get $f(x) \le \max\{f(z) : z \in X\} < g(y)$. And if $y\rho^d \neq \emptyset$, then $y\rho^d z$ for some $z \in X$; so $x\rho\rho^d z$, using the assumption $x\rho y$. Therefore, $f(x) < f(z)$ for each $z \in y\rho^d$, whence $f(x) < \min\{f(z) : y\rho^d z\} = g(y)$.

For the converse direction, assume $f(x) < g(y)$. If $y\rho^d = \emptyset$, then $\overline{y\rho^d}x$, i.e. $y\overline{\overline{\rho}}^{\smile}x$ for all x; so we have $x\rho y$. Otherwise, since $f(x) < \min\{f(z) : y\rho^d z\}$, $x \notin y\rho^d$, so again $\overline{y\rho^d}x$.

(2) ⇒ (3): For every $t \in X$ let $u(t) := g(t)$ and $d(t) := f(t) - u(t)$. Then $x\rho y$ iff $f(x) = u(x) + d(x) < u(y)$.

(3) ⇒ (1): Assume that the Ferrers condition does not hold, i.e. $x\rho y \wedge z\overline{\rho}y \wedge z\rho t \wedge x\overline{\rho}t$ for some x, y, z, t. Then, since clearly $z\overline{\rho}y \leftrightarrow u(y) \le u(z) + d(z)$, we get $u(x) + d(x) < u(y) \le u(z) + d(z) < u(t) \le u(x) + d(x)$, a contradiction. □

Theorem 18.19. ([6, Proposition 7.14]) *Let ρ be a binary relation on a finite set X. The following conditions are equivalent:*

(1) ρ is an interval order.
(2) There exist two functions $u : X \longrightarrow \mathbb{R}$ and $d : X \longrightarrow \mathbb{R}_+$ such that

$$x\rho y \leftrightarrow u(x) + d(x) < u(y).$$

(3) There exists a function F from X to the set of intervals of $\mathbb{R}$ such that

$$x\rho y \leftrightarrow F(x) < F(y),$$

where $F(x) < F(y)$ if, for all $a \in F(x), b \in F(y)$, we have $a < b$.

Proof. (1) ⇔ (2): Since an interval order is an (irreflexive) Ferrers relation by definition, this equivalence follows directly from Th. 18.18 taking into account that ρ is irreflexive. So, for all x, $x\overline{\rho}x$, whence $u(x) + d(x) \ge u(x)$, i.e. $d(x) \ge 0$.

(2) ⇒ (3): For every $x \in X$ we set $F(x) := [u(x), u(x) + d(x)]$.

(3) ⇒ (1): Since, for all x, $\neg[F(x) < F(x)]$, we get that ρ is irreflexive. Now assume that

$x\rho y \wedge z\rho t \wedge x\overline{\rho}t$. The first two imply $F(x) < F(y)$ and $F(z) < F(t)$, and the third $F(x) \geq F(t)$, whence $F(z) < F(y)$. Thus, $z\rho y$. So ρ is an irreflexive Ferrers relation, i.e. an interval order. □

We continue with the theorem on the numerical representation of semiorders, the so-called Scott-Suppes Theorem due to [22], this time omitting the proof.

Theorem 18.20. ([16, Theorem 6.1]) *Let ρ be a binary relation on a finite set X and $d \in \mathbb{R}_+$. The following conditions are equivalent:*

(1) ρ is an semiorder.
(2) There is a function $f : X \longrightarrow \mathbb{R}$ such that

$$x\rho y \leftrightarrow f(x) + d < f(y). \tag{18.7}$$

Corollary 18.21. ([16, p. 251, Corollary]) *If a binary relation ρ on a finite set X is representable in the form (18.7), then it is representable in this form for any $\delta \in \mathbb{R}_+$. In particular, it is representable as*

$$x\rho y \leftrightarrow f(x) + 1 < f(y). \tag{18.8}$$

Proof. If f satisfies (18.7), then so does $(d'/d) \cdot f$ with d' in place of d. □

That the Scott-Suppes Theorem is false without the assumption that X is finite can be seen as follows: Let $r \notin \mathbb{N}_+$, with $\mathbb{N}_+$ the set of positive integers, and take $X := \mathbb{N}_+ \cup \{r\}$. With ρ defined on X by

$$\begin{cases} x\rho y \leftrightarrow x+1 < y & \text{for } x, y \in \mathbb{N}_+ \\ x\rho r & \text{for all } x \in \mathbb{N}_+. \end{cases}$$

Then ρ is a semiorder on X, but there is no $f : X \longrightarrow \mathbb{R}$ such that (18.8) holds. Note that, by definition of ρ, we have $1\,\rho\,3\,\rho\,5\,\rho \ldots \rho\,[2(n-1)+1]\,\rho\,(2n+1)\,\rho \ldots \rho\,r$. So $f(1)+1 < f(3)$ which gives us $f(1) + n < f(2n+1)$, for $n = 1$, as the base clause for an induction proof as follows: Given the induction hypotheses for $n = k$, i.e. $f(1) + k < f(2k+1)$, then, since $(2k+1)\,\rho\,[2(k+1)+1]$, we get $f(2k+1)+1 < f(2(k+1)+1)$. But from the induction hypotheses we get $f(1)+k+1 < f(2(k+1)+1)$. Thus, $f(1)+n < f(2n+1)$, for all $n \in \mathbb{N}_+$. Moreover, since $n\,\rho\,r$ for all $n \in \mathbb{N}_+$, $f(1)+n < f(2n+1) < f(r)$, i.e $f(r)$ is larger than any positive integer, a contradiction.

We still are in debt of showing that ρ is in fact a semiorder: For all $x \in \mathbb{N}_+$ we have $\neg x+1 < x \wedge x \neq r$, so ρ is irreflexive. Assume that the Ferrers condition $x\rho y \wedge z\rho t \rightarrow x\rho t \vee z\rho y$ fails to hold, i.e. $x+1 < y \wedge z+1 < t \wedge x+1 \geq t \wedge z+1 \geq y$ for some $x, y, z, t \in \mathbb{N}_+$. This implies $t \leq x+1 < y \leq z+1 < t$, a contradiction. And if $y = r$ or $t = r$, the condition obviously holds. Next assume that ρ is not semitransitive, i.e. $x\rho y \wedge y\rho z \wedge x\overline{\rho}t \wedge t\overline{\rho}z$ for some $x, y, z, t \in \mathbb{N}_+$. Then we get $x\rho z$ by Prop. 18.2(2) and $x\overline{\rho}z$ by Prop. 18.2(3), a contradiction. If $z = r$, the semitransitivity condition obviously holds, and the case $y = r$ is excluded since we cannot have $r\rho z$.

18.4 Preference Structures

In the last section of this chapter we will have a short look on what is called a *preference structure*. This concept has been used for the original definition of a semiorder in Luce [12]. (Let us note that the concept of a semiorder is usually attributed to this paper, but, as a matter of fact, it is older and dates back under a different name to [24] of which a "translation" to modern notation can be found in [9].)

As stated in [18], and already mentioned above, to develop preferences is an inevitable step in a lot of human activities and is necessary for making one's decisions in daily life. In a sense, the conception of preference structures starts out from a behavioral point of view by considering the reaction possibilities for decision makers (individuals, agents) if they have to judge elements from a set A of alternatives (actions, candidates, objects, ...).

Following [4,14,18], let us consider a set A of alternatives. When asked about a pair of distinct elements in A, individuals are supposed to respond with one of the two answers "*yes*" or "*no*" to the question "*Is object a at least as good as object b?*". Asking this question for all ordered pairs of $A \times A$ results in the definition of a binary relation $\preceq$ on A, and we say $a \preceq b$ iff a is confirmed to be at least as good as b by a "*yes*" to the above question. You may think of $a \preceq b$ meaning "*a comes before or is equally good as b*"; it is then natural to consider $\preceq$ as reflexive. Moreover, we suppose that individuals are consistent in their judgements, so that we presuppose $\preceq$ to be transitive too, i.e. a preorder.

Based on two "*yes*" |"*no*" answers for every two objects $a, b \in A$ one of the following four patterns applies:

(1) $a \preceq b \wedge \neg b \preceq a$, denoted by aPb, and interpreted as "*a is strictly preferred to b*";
(2) $\neg a \preceq b \wedge b \preceq a$, denoted by bPa, and interpreted as "*b is strictly preferred to a*";
(3) $a \preceq b \wedge b \preceq a$, denoted by aIb, and interpreted as "*a is indifferent to b*";
(4) $\neg a \preceq b \wedge \neg b \preceq a$, denoted by aJb, and interpreted as "*a is incomparable to b*".

So one, and only one, of the following three situations occurs: An individual

- clearly prefers an object a over another object b, or
- feels indifferent about them, or
- is unable to compare the two alternatives under consideration.

It is thus a triplet of binary relations that contains all the possible answers of any decision maker in a pairwise comparison experiment. According to these cases, the following offers itself as a suitable framework since, by construction, P is asymmetric, I and J are symmetric, and from the reflexivity of $\preceq$ we get that I is also reflexive, and that J is irreflexive. Given that all pairs are presented to a decision maker, we get three subsets of the set of all ordered pairs in $A \times A$ as follows:

(1) $P := \{\langle a, b\rangle : a \textit{ is preferred to } b\}$,
(2) $I := \{\langle a, b\rangle : a \textit{ is indifferent to } b\}$, and

(3) $J := \{\langle a,b\rangle : a \text{ and } b \text{ cannot be compared }\}$.

Taking I as the complement of $P \cup P^{\smile}$ with respect to $A \times A - J$, gives us the following

Definition 18.22. ([14, Section 3.6]) *A* **preference structure** *on a set A is a triple* $\langle P, I, J\rangle$ *of relations on A such that:*

(1) P is asymmetric;
(2) I is reflexive and symmetric, i.e. a tolerance;
(3) J is irreflexive and symmetric;
(4) $P \cup P^{\smile} \cup I \cup J = A^2$;
(5) P, I, and J are pairwise disjoint.

Note that, due to the last two requirements, $P, P^{\smile}, I$ and J form a partition of $A \times A$, since the asymmetry of P implies $P \cap P^{\smile} = \emptyset$. Likewise, since I is symmetric, we cannot have $xP^{\smile}y \wedge xIy$, so $I \cap P^{\smile} = \emptyset$, and also $J \cap P^{\smile} = \emptyset$ by the same argument.

(Remark concerning the notation: Using $P, I, J, \ldots$ to denote relations we follow the usual habit in the field. Therefore, in order to avoid misunderstandings in the usage of I, we will use Δ_X for the diagonal $\{\langle x,x\rangle : x \in X\}$ if necessary.)

Any preference structure (P, I, J) on a set A can be completely characterized by a single relation, called its **characteristic relation**, defined as $R := P \cup I$. R clearly is reflexive, since $\Delta_A \subseteq I$, and one has $R^d = P \cup J$. Fig. 18.2, shaped similar to Figure 1 in [19], visualizes the above mentioned partition, and also the characteristic relation R; note that $R^{\smile} = P^{\smile} \cup I$, since I is symmetric. Moreover, using elementary calculations, it is not hard to check that the following holds:

$$P = R \cap R^d, \quad I = R \cap R^{\smile}, \text{ and } \quad J = R^{\smile} \cap R^d.$$

As an example we give a proof of the second equation:

$$\begin{aligned} R \cap R^{\smile} &= (P \cup I) \cap (P \cup I)^{\smile} \\ &= (P \cup I) \cap (P^{\smile} \cup I^{\smile}) \\ &= [(P \cup I) \cap P^{\smile}] \cup [(P \cup I) \cap I^{\smile})] \\ &= [(P \cap P^{\smile}) \cup (I \cap P^{\smile})] \cup [(P \cap I^{\smile}) \cup (I \cap I^{\smile})] \\ &= [\emptyset \cup \emptyset] \cup [\emptyset \cup I] \\ &= I. \end{aligned} \tag{18.9}$$

The definitions of preference structures which one can find in literature differ along three dimensions:

- They are defined by the properties of its characteristic relation, or by the properties of its constituting relations P, I, and J;
- they differ depending on whether individuals are assumed (urged in an experient /able) to judge for every two distinct elements $a, b \in A$ which one they prefer, or not. In the former case we have $J = \emptyset$ and will write the resulting preference structure as (P, I);

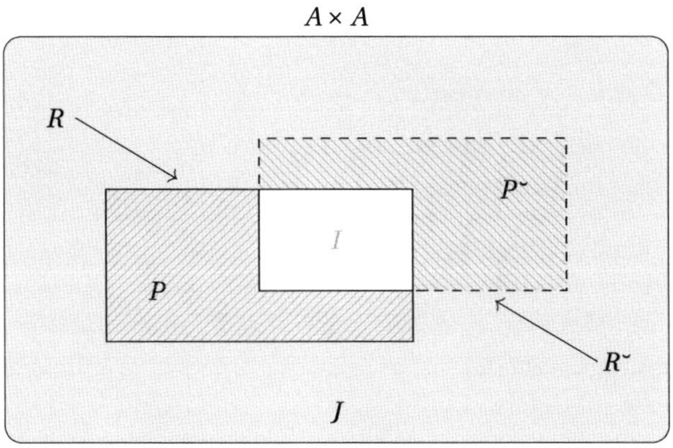

FIG. 18.2: The partition of a preference structure.

- they may differ in the assumptions made about the basic relation observed from an individual's judgements being irreflexive or reflexive, and which is usually assumed to fulfill the consistency requirement of transitivity.

We begin with a characterization of *total preorders,* which we did not consider up to now. We only know from Cor. 18.11 and Cor. 3.31(2) that the dual of a total preorder is a weak order. In fact, there is a bijective map between these two sets of order relations (see [15, Theorem 3.6]). The next theorem gives some properties of total preorders as a model of preference. In order to prove it we have to use the following results from the theory of valued graphs, which we present without proof:

Definition 18.23. *A* **valued relation** *on a set A is a pair (ρ, v) consisting of a relation $\rho \subseteq A \times A$ and a function $v : \rho \longrightarrow \mathbb{R}$. For $\sigma \subseteq \rho$ we define the value of σ by $v(\sigma) := \sum_{\langle a,b\rangle \in \sigma} v(\langle a,b\rangle)$. $g : A \longrightarrow \mathbb{R}$ is called a* **potential function** *of the valued relation (ρ, v) if $g(a) - g(b) \geq v(a,b)$ for every $\langle a,b\rangle \in \rho$.*

Corollary 18.24. ([2, Lemma 3.3]) *A valued relation (ρ, v) admits a potential function f iff for every cycle C of ρ we have $v(C) \leq 0$.*

Theorem 18.25. ([15, Theorem 3.4]) *Let R be the characteristic relation of a preference structure (P, I) on a finite set A. Then the following conditions are equivalent:*

(1) R is a total preorder.

(2) $\begin{cases} P \text{ is negatively transitive,} \\ R \text{ is reflexive and complete.} \end{cases}$

(3) $\begin{cases} P \text{ is transitive,} \\ I \text{ is transitive,} \\ R \text{ is reflexive and complete.} \end{cases}$

(4) $\begin{cases} P \text{ is transitive,} \\ PI \subseteq P \text{ and } IP \subseteq P, \\ R \text{ is reflexive and complete.} \end{cases}$

(5) $\begin{cases} \text{there is no cycle of length } \leq 3 \text{ in } R \text{ containing a } P, \\ R \text{ is reflexive and complete.} \end{cases}$

(6) $\begin{cases} \text{there is no cycle in } R \text{ containing a } P, \\ R \text{ is reflexive and complete.} \end{cases}$

(7) *There exists a function* $g : A \longrightarrow \mathbb{R}$ *such that, for all* $a, b \in A$,
$$\begin{cases} aPb \leftrightarrow g(a) > g(b), \\ aIb \leftrightarrow g(a) = g(b). \end{cases}$$

Proof. (1) ⇒ (2): Assume $\neg aPb \wedge \neg bPc$; then $\neg aIb \wedge \neg bIc$, whence $\neg aRb \wedge \neg bRc$. Since R is complete, we get $bRa \wedge cRb$, so cRa. Therefore, $cPa \vee cIa$, so $\neg aPc$, since $cIa \rightarrow aIc \rightarrow aRc \rightarrow aRb$, a contradiction to $\neg aRb$.

(2) ⇒ (3): P is transitive by Prop. 3.24(6). Moreover, since $aIb := a \preceq b \wedge b \preceq a$, it is easy to see from the transitivity of $\preceq$ that I is transitive too (and thus an equivalence relation),

(3) ⇒ (4): Assume $aPb \wedge bIc$; so also cIb. Since R is complete, we must have $aRc \vee cRa$. But $cIa \wedge bIc \rightarrow bIa \rightarrow aIb$ a contradiction to aPb, and $cPa \wedge aPb \rightarrow cPb$ contradicting cIb. So aRc. aIc would imply cIa, which we have just seen to yield a contradiction, so aPc, i.e. $PI \subseteq P$. An analogous proof shows $IP \subseteq P$.

(4) ⇒ (5): Since P is irreflexive by Prop. 3.24(3), $\neg aPa$, for all $a \in A$, so there is no cycle of length 1 containing P. Since $P^2 \subseteq P$, $IP \subseteq P$, and $PI \subseteq P$, there is also no cycle with length 2. Finally, we have $P^3 \subseteq P$, $PIP \subseteq P^2 \subseteq P$, $IPI \subseteq PI \subseteq P$, and $PIP \subseteq P^2 \subseteq P$; thus there is no cycle of length 3 in R containing P. (Remember: $P \cap I = \emptyset$.)

(5) ⇒ (6): Suppose R has a cycle with length ≥ 4 containing a P. Then there are elements $a, b, c \in A$ such that $aPbIc$, $aIbPc$, or $aPbPc$ are part of this cycle. We get a shorter cycle using aPc instead. Repeated application of this replacement proves the assertion.

(6) ⇒ (7): We let $v(\langle a,b\rangle) = 0$, for all $\langle a,b\rangle \in I$, and $v(\langle a,b\rangle) = d > 0$, for all $\langle a,b\rangle \in P$. Then we get a valued relation (R, v) without positively valued cycles. By Cor. 18.24 there is a potential function $g : A \longrightarrow \mathbb{R}$ such that
$$aPb \rightarrow g(a) - g(b) \geq d, \text{ so } g(a) > g(b).$$
Further, aIb implies bIa, so that $g(a) - g(b) \geq 0$ and $g(b) - g(a) \geq 0$. Therefore,
$$aIb \rightarrow g(a) = g(b).$$
For the converse direction note that R is complete, so we must have either $g(a) > g(b)$ or $g(a) = g(b)$, for all $a, b \in A$. Assume $g(a) = g(b) \rightarrow aPb$; since $\neg bPa$ we would get the contradiction $g(b) \neq g(a)$. And since $P \cap I = \emptyset$ we get $g(a) > g(b) \rightarrow aPb$ and $g(a) = g(b) \rightarrow aIb$.

(7) ⇒ (1): We have $aRb \leftrightarrow g(a) \geq g(b)$, for all $a, b \in A$. Thus R is reflexive, transitive, and complete. □

We conclude this chapter with a short look at semi- and total orders, and will describe them in a similar way as we did for total preorders. In order to obtain better comparability with total preorders, we use a a reflexive version of semiorders. As one should expect, ensuring nontransitive indifference makes things a little bit more complex. Since the proof uses the same kind of arguments it is omitted. Comparing the last clause of the theorem below with Th. 18.25(7) one can see that total preorders can formally be regarded as the limit case of semiorders with $q = 0$. But this is the decisive difference: the condition $q = 0$ shrinks the indifference interval to a point, making it thus transitive.

Theorem 18.26. ([15, Theorem 3.1]) *Let R be the characteristic relation of a preference structure (P, I) on a finite set A. Then the following conditions are equivalent:*

(1) *R is a reflexive semiorder.*

(2) $\begin{cases} R \cup R^{\smile} = A^2 & (R \text{ is reflexive and complete}), \\ RR^dR \subseteq R & (R \text{ is a Ferrers relation}), \\ RRR^d \subseteq R & (R \text{ is semitransitive}). \end{cases}$

(3) $\begin{cases} PIP \subseteq P, \\ P^2 \cap I^2 = \emptyset, \\ R \quad \text{is reflexive and complete.} \end{cases}$

(4) $\begin{cases} PIP \subseteq P, \\ P^2 I \subseteq P \text{ and } IP^2 \subseteq P, \\ R \quad \text{is reflexive and complete.} \end{cases}$

(5) $\begin{cases} \text{every cycle of length } \leq 4 \text{ in } R \text{ contains strictly more } I \text{ than } P, \\ R \text{ is reflexive and complete.} \end{cases}$

(6) $\begin{cases} \text{every cycle in } R \text{ contains strictly more } I \text{ than } P, \\ R \text{ is reflexive and complete.} \end{cases}$

(7) *There exists a functions* $g' : A \longrightarrow \mathbb{R}$ *and* $q' : A \longrightarrow \mathbb{R}_+$ *such that, for all* $a, b \in A$,

$$\begin{cases} aPb & \leftrightarrow g'(a) > g'(b) + q'(b), \\ aIb & \leftrightarrow \begin{cases} g'(a) \leq g'(b) + q'(b), \\ g'(b) \leq g'(a) + q'(a). \end{cases} \\ g'(a) > g'(b) & \rightarrow g'(a) + q'(a) \geq g'(b) + q'(b). \end{cases}$$

(8) *There exists a function* $g : A \longrightarrow \mathbb{R}$ *and a constant* $q \in \mathbb{R}_+$ *such that, for all* $a, b \in A$,

$$\begin{cases} aPb & \leftrightarrow \quad g(a) > g(b) + q, \\ aIb & \leftrightarrow \quad |g(a) - g(b)| \leq q. \end{cases}$$

Proof. (1) ⇔ (2): We take this equivalence as the definition for the reflexive variant of a semiorder. □

Proposition 18.27. ([14, Definition 6.3]) *Let R be the characteristic relation of a preference structure (P, I) on a set A. Then the following conditions are equivalent:*

(1) *R is a total order.*

(2) *R is reflexive, antisymmetric, complete and transitive.*

(3) $\begin{cases} I = \Delta_A, \\ P \text{ is transitive}, \\ R \text{ is reflexive and complete.} \end{cases}$

(4) $\begin{cases} P \text{ is transitive}, \\ PI \subseteq P \text{ and } IP \subseteq P, \\ R \text{ is reflexive and complete.} \end{cases}$

Proof. (1) $\Leftrightarrow$ (2): By Def. 4.7(d).

(2) $\Rightarrow$ (3): P is asymmetric, hence irreflexive by Prop. 3.24(3); but $R := P \cup I$ is reflexive, so $\Delta_A \subseteq I$. And since R is antisymmetric we must have $I = \Delta_A$. By Prop. 3.24(3) P is antisymmetric, whence transitive by Prop. 3.24(4). Finally, R is reflexive and complete by assumption.

(3) $\Rightarrow$ (4): Obvious.

(4) $\Rightarrow$ (2): To show the transitivity of R note that $P^2 \subseteq P$, and $IP \cup PI \subseteq P$ by hypothesis. Moreover, since $aIb := a \preceq b \wedge b \preceq a$, it is easy to see from the transitivity of $\preceq$ that I is transitive too (and thus an equivalence relation), so that $(P \cup I)^2 = P^2 \cup PI \cup IP \cup I^2 \subseteq P \cup I^2 = P \cup I$, since equivalence relations are idempotent by Prop. 4.18(2). Thus R is transitive.

For antisymmetry note that we get $\Delta_A \subseteq I$ as above which implies $P = P\Delta_A \subseteq PI$. But by hypothesis $PI \subseteq P$, so $\Delta_A = I$. Thus $(P \cup I) \cap (P \cup I)^\smile = (P \cup I) \cap (P^\smile \cup I^\smile) = P \cap P^\smile \cup I \cap P^\smile \cup P \cap I^\smile \cup I \cap I^\smile = \emptyset \cup \emptyset \cup P \cap I \cup I = I = \Delta_A$, noting that $I = I^\smile$. □

And as one would expect, the following numerical representation for total orders is possible:

Theorem 18.28. ([4, Theorem 2.1]) *Let (P, I) be a preference structure on a finite set A with characteristic relation R. Then the following conditions are equivalent:*

(1) *R is a total order.*

(2) *There exists a function $u : X \longrightarrow \mathbb{R}$ such that for all $a, b \in A$*

$$xRy \leftrightarrow u(a) \geq u(b) \quad \text{and} \quad u(a) = u(b) \rightarrow a = b.$$

References

[1] Abrísqueta, F.J., Campión, M.J., Catalán, R. et al.: New trends on the numerical representability of semiordered structures, Mathware & Soft Computing **19**, 25-37 (2012).

[2] Aleskerov, F., Bouyssou, D., Monjardet, B.: *Utility Maximization, Choice and Preference*, 2nd ed., Springer-Verlag Berlin Heidelberg (2007).

[3] Bosi, G., Estevan, A. et al.: Continuous representability of interval orders: the topological compatibility setting. Int. J. of Uncertainty, Fuzziness and Knowledge-Based Systems **23**, 345-365 (2015).

[4] Bouyssou, D., Vincke, P.: Binary relations and preference modeling. In D. Bouyssou, D. Dubois, M. Pirlot, H. Prade (Eds.): *Decision-Making Process: Concepts and Methods*, 49-84, ISTE Ltd and John Wiley & Sons, Inc. (2009).

[5] Bridges, D.S., Mehta, G.B.: *Representations of Preference Orderings*, Springer-Verlag Berlin Heidelberg (1995).

[6] Caspard, N., Leclerc, B., Monjardet, B.: *Finite Ordered Sets-Concepts, Results and Uses*, Cambridge University Press (2012).

[7] Estevan, A.: Some results on biordered structures, in particular distributed systems. J. of Mathematical Psychology **77**, 70-81 (2017).

[8] Fishburn, P.C.: *Utility Theory for Decision Making*, John Wiley & Sons (1970).

[9] Fishburn, P.C., Monjardet, B.: Norbert Wiener on the theory of measurement (1914 1915 1921), J. of Mathematical Psychology **36**, 165-184 (1992).

[10] French, S.: *Decision Theory - An Introduction to the Mathematics of Rationality*, Ellis Horwood, Chichester (1988).

[11] Krantz, D.H., Luce, R.D., Suppes, P., Tversky, A.: *Foundations of Measurement, Vol. I-Additive and Polynomial Representations*, Academic Press, New York and London (1971).

[12] Luce, R. D.: Semiorders and a theory of utility discrimination, Econometrica **24**, 178-191 (1956).

[13] Mari, L., Maul, A., Irribarra, D. T., Wilson, M.: Quantities, Quantification, and the Necessary and Sufficient Conditions for Measurement. Measurement **100** 115-121 (2017).

[14] Moretti, S., Öztürk, M., Tsoukiàs, A.: Preference Modelling. In Greco, S., Figueira, J., Ehrgott, M. (Eds.): *Multiple Criteria Decision Analysis: State of the Art Surveys*, 2nd ed., 43-95, Springer Science+Business Media New York (2016).

[15] Pirlot, M., Vincke, Ph.: *Semiorders-Properties, Representations, Applications*, Springer Science+Business Media Dordrecht (1997).

[16] Roberts, F.S.: *Measurement Theory-with Applications to Decisionmaking, Utility and the Social Sciences*, Cambridge University Press, Cambridge UK (1985).

[17] Roberts, F.S.: Meaningless Statements, Matching Experiments, and Colored Digraphs. In Roberts, F.S. (Ed): *Applications of Combinatorics and Graph Theory to the Biological and Social Sciences*, 277-294, Springer-Verlag New York Inc. (1989).

[18] Roubens, M., Vincke, P.: *Preference Modelling*, Springer-Verlag Berlin Heidelberg (1985).

[19] Rudas, I.J., Fodor, J.: Fundamentals of fuzzy preference modeling. In *Proc. of the 8th WSEAS International Conference on Applied Informatics and Communications (AIC'08)*, 134-145 (2008).

[20] Schwager, K.W.: The representational theory of measurement: An assessment, Psychological Bulletin **110**, 618-626 (1991).

[21] Saint-Mont, U.: What measurement is all about, Theory & Psychology **22**, 467-485 (2012).

[22] Scott, D., Suppes, P.: Foundational Aspects of Theories of Measurement, The Journal of Symbolic Logic **23**, 113-128 (1958).

[23] Suppes, P.: *Representation and Invariance of Scientific Structures*, CSLI Publications (2002).

[24] Wiener, N.: Contribution to the theory of relative position. Mathematical Proceedings of the Cambridge Philosophical Society **17**, 441-449 (1914).

Chapter 19

Relations and Syllogisms Again

In Chapter 1 we started with a look back into history to shed some light on how and when relations entered the mathematical stage as a formal concept. And in doing so we arrived at ARISTOTLE's system of syllogisms, one of the most important and influential analyses of reasoning rules, which, after more than two thousand years, is still a fascinating topic of investigation.

In the first section of the last chapter we will return to syllogisms, but taking the other way round by showing how relations can be used to analyze and describe them. As the final topic we will have a short look at the so-called Square of Opposition which has gained some interest in recent years.

19.1 Relational Analysis of Syllogisms

In Section 1.2 we have mentioned that $MaP, SaM \vdash SaP$ is a modern, self-explaining and comfortable notation for the '*Barbara*' or AAA-1 syllogism. Rewriting this as $A; A \subseteq A$ or as $\alpha\alpha \subseteq \alpha$ we arrive at the formal machinery for our relational analysis of syllogisms. For obvious reasons we will adhere to the notation $a \cdot a \subseteq a$, as is also done in [13] and [14, Chapter 1.2] on which this section is based.

We start out by postulating that a is a reflexive relation; in other words, we make "*all P are P*", or

$$PaP \tag{19.1}$$

a self-evident proposition. In the same way we take it for evident that "*from SaM and MaP 'we can conclude SaP*", or

$$SaM \text{ and } MaP \Rightarrow SaP. \tag{19.2}$$

So we use two axioms from which we develop our considerations below. In the next step we define the three additional relations as follows:

$$SiP \;:\leftrightarrow\; MaS \text{ and } MaP \text{ for some } M, \tag{19.3}$$

$$SeP \;:\leftrightarrow\; \neg SiP, \text{ and} \tag{19.4}$$

$$SoP \;:\leftrightarrow\; MaS \text{ and } MeP \text{ for some } M. \tag{19.5}$$

Note that the definition of I, $S(x) \wedge P(x)$, for some x, as given in Table 1.1, differs from the one of i above in that the latter talks about predicates instead of subjects. To make this definition meaningful, we must assume that only predicates of a so-called concept-pyramid be considered.

From the axioms and definitions we get the following properties: a is transitive and i are symmetric. Thus, by Prop. 3.25(3) e is also symmetric. So we have

$$a \cdot a \subseteq a, \qquad i = i^{\smile}, \qquad e = e^{\smile}.$$

It is instructive to visualize categorical propositions by circles which represent the extensions of the predicates involved. This way we get the following diagrams (which are *not* so-called Venn diagrams; see e.g. [12, Sections 4.3, 4.6 and 5.2]):

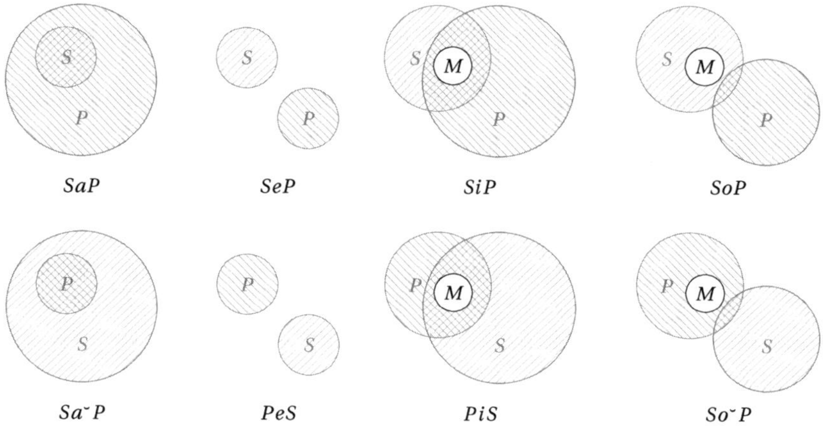

It is easy to see that a and o are not symmetric, whereas e and i are: If the circle for S is contained in (is a proper part of) the circle for P then this does not apply in the other direction. Instead, we have $P a^{\smile} S$.

Neither ARISTOTLE nor the scholastics knew about this concept of a converse relation. But as we will see in a moment, using converse relations will considerably simplify the investigation of what can be inferred as conclusion from two premises. Clearly, there are four possibilities to arrange two premises on two places, yielding the following inference forms in which the variables ρ, σ, τ stand for one of the relations a, e, i, o:

$$\begin{aligned} M\rho P \text{ and } S\sigma M \quad &\Rightarrow \quad S\tau P, \\ P\rho M \text{ and } S\sigma M \quad &\Rightarrow \quad S\tau P, \\ M\rho P \text{ and } M\sigma S \quad &\Rightarrow \quad S\tau P, \\ P\rho M \text{ and } M\sigma S \quad &\Rightarrow \quad S\tau P. \end{aligned}$$

Since we have at our disposal also the converses for each of the relations a, e, i, o, it suffices to consider only one of the above inference patterns, and we choose the first one:

$$S\rho M \text{ and } M\sigma P \quad \Rightarrow \quad S\tau P, \tag{19.6}$$

in which the variables ρ, σ, τ will be substituted by one of the relations in the set $\mathscr{S} := \{a, a^{\smile}, e, i, o, o^{\smile}\}$. Remember that $i = i^{\smile}$ and $e = e^{\smile}$ due to their symmetry, so $\mathscr{S}$ contains the four classical relations $\{a, e, i, o\}$ and also all their converses. Inferences of the form (19.6) are called **normed syllogisms**. So axiom (19.2), SaM and $MaP \Rightarrow SaP$, is a normed syllogism. Also definitions (19.3) and (19.5) give rise to such syllogisms, namely

$$Sa^{\smile}M \text{ and } MaP \Rightarrow SiP,$$
$$Sa^{\smile}M \text{ and } MeP \Rightarrow SoP.$$

But according to the definitions we even get

$$Sa^{\smile}M \text{ and } MaP \text{ for some } M \Leftrightarrow SiP, \tag{19.7}$$
$$Sa^{\smile}M \text{ and } MeP \text{ for some } M \Leftrightarrow SoP. \tag{19.8}$$

We can also strengthen axiom (19.2) to

$$SaM \text{ and } MaP \text{ for some } M \Leftrightarrow SaP, \tag{19.9}$$

since we have SaS by axiom (19.1): choosing S for M we get $SaP \Rightarrow SaS$ and SaP.

If we can find out which products of relations in $\mathscr{S}$ are again in this set, then we know which inferences are possible that conclude from two given sentences, the premises, a third one using just these given relations. From the equivalences (19.7) – (19.9) we thus get the following identities:

$$a \cdot a = a \quad a^{\smile} \cdot a = i \quad a^{\smile} \cdot e = o \tag{19.10}$$

And from the first two it follows that

$$i \cdot a = a^{\smile} \cdot a \cdot a = a^{\smile} \cdot a = i, \tag{19.11}$$

which means that SiM and $MaP \Rightarrow SiP$. By the SCHRÖDER rule Th. 3.26(1)(b) we get $ia \subseteq i \leftrightarrow \bar{i}a^{\smile} \subseteq \bar{i} \leftrightarrow ea^{\smile} \subseteq e$. Therefore, $ae^{\smile} \subseteq e^{\smile}$, whence $ae \subseteq e$ by the symmetry of e. Note that using Th. 3.26(1)(b) we have to be careful to get things right in terms of inferences: $\rho\sigma \subseteq \tau \leftrightarrow \bar{\tau}\sigma^{\smile} \subseteq \bar{\rho}$ applied to $i \cdot a = i$ means that

$$SiM \text{ and } MaP \Rightarrow SiP \leftrightarrow S\bar{i}P \text{ and } Pa^{\smile}M \Rightarrow S\bar{i}M.$$

Since $e = \bar{i}$ we thus get SeP and $Pa^{\smile}M \Rightarrow SeM$; therefore, by the conversion rule $\rho\sigma = \tau \leftrightarrow \sigma^{\smile}\rho^{\smile} = \tau^{\smile}$ and the symmetry of e, MaP and $PeS \Rightarrow MeS$. Using (19.1) with $P = M$ this can again be strengthened to MaP and PeS for some $P \Leftrightarrow MeS$. Therefore we get the identity

$$a \cdot e = e. \tag{19.12}$$

Using the conversion rule for relations, $\rho\sigma = \tau \leftrightarrow \sigma^{\smile}\rho^{\smile} = \tau^{\smile}$, and the symmetry of i and e, from the five identities (19.10) – (19.12) found so far we additionally get the following ones (the conversion of $a^{\smile} \cdot a = i$ replicates itself):

$$a^{\smile} \cdot o = a^{\smile}, \quad e \cdot a = o^{\smile}, \quad a^{\smile} \cdot i = a, \quad e \cdot a^{\smile} = e. \tag{19.13}$$

Furthermore, from the identities shown above, we get

$$a^{\smile} \cdot o = a^{\smile} \cdot a^{\smile} \cdot e = a^{\smile} \cdot e = o, \tag{19.14}$$
$$o \cdot a^{\smile} = a^{\smile} \cdot e \cdot a^{\smile} = a^{\smile} \cdot e = o, \tag{19.15}$$
$$e \cdot i = e \cdot a^{\smile} \cdot a = e \cdot a = o^{\smile}, \tag{19.16}$$

TABLE 19.1: Syllogistic composition table.

$\mathscr{S}$	$a^{\smile}$	a	i	e	o	$o^{\smile}$
a		a		e		$o^{\smile}$
$a^{\smile}$	$a^{\smile}$	i	i	o	o	
i		i		o		
e	e	$o^{\smile}$	$o^{\smile}$			
$o^{\smile}$		$o^{\smile}$				
o	o					

and from these, again by conversion,

$$o^{\smile}\cdot a = o^{\smile}, \quad a\cdot o^{\smile} = o^{\smile}, \quad i\cdot e = o. \tag{19.17}$$

Altogether we have the fifteen identities listed in (19.10) – (19.17) which can be arranged in a converse-symmetrical scheme as shown in Table 19.1. Each of these identities corresponds to a syllogism.

Using axiom 1 (19.1) and the second and third identity in (19.10) the following implications follow:

$$SaP \Rightarrow SiP, \quad Sa^{\smile}P \Rightarrow SiP, \quad SeP \Rightarrow SoP, \quad SeP \Rightarrow So^{\smile}P. \tag{19.18}$$

To prove e.g. $Sa^{\smile}P \Rightarrow SiP$, note that from identity $a^{\smile}\cdot a = i$ in (19.10) we have $Sa^{\smile}P$ and $PaP \Rightarrow SiP$. Therefore, if the conclusion of a syllogism is e.g. an e-sentence we obtain two other syllogisms with the same premises and o and $o^{\smile}$ as conclusions.

Looking at Table 19.1 we find that each of $a, a^{\smile}$ appears once, and that e appears twice as conclusion. Thus in addition to the fifteen *strong* syllogisms (for which implications hold in both direction) we get six *weak* ones, which are represented as relational inclusions:

$$\begin{aligned} a\cdot a &\subseteq i, & a^{\smile}\cdot a^{\smile} &\subseteq i, \\ a\cdot e &\subseteq o, & a\cdot e &\subseteq o^{\smile}, \\ e\cdot a^{\smile} &\subseteq o, & e\cdot a^{\smile} &\subseteq o^{\smile}. \end{aligned} \tag{19.19}$$

Altogether we have then found twenty-one normed syllogisms.

We can compare the traditional and the relational analysis of syllogisms in an adapted version of Table 1.3. First note that the traditional syllogistic distinguishes inferences according to the four figures and does not admit converses, whence it arrives at a different number of inferences. Ten out of the fifteen strong syllogisms listed in Table 19.1 have a, e, i or o, and four of the six weak syllogisms in (19.19) have i or o as a result. These fourteen inferences are shown in the first column of Table 19.2.

Considering the normed syllogisms, from the traditional viewpoint a syllogism with a premise containing e or i must be counted twice due to the symmetry of e and i. From the above mentioned fourteen inferences six do not contain e or i, seven contain e or i in one of the premises, and one inference ($i\cdot e = o$) contains both. So we get $6\cdot 1+7\cdot 2+1\cdot 4 =$

24 different valid syllogisms as shown in Table 19.2. Thus we could reconstruct *all* the traditionally valid syllogisms in the relational setting.

In order to prove this statement in the relational framework, we must show that none of the other twenty-one products for which no result appears in Table 19.1 does permit an inference of the form $\rho\cdot\sigma\subseteq\tau$; they are called *non-inferential* in [14]. Five of them lie in the main diagonal (the shaded cells):

$$a\cdot a^{\smile},\ i\cdot i,\ e\cdot e,\ o^{\smile}\cdot o \text{ and } o\cdot o^{\smile}. \tag{19.20}$$

From the remaining sixteen products it suffices to consider those lying above the diagonal, because the lower half consists of their converses:

$$a\cdot i,\ a\cdot o,\ a^{\smile}\cdot o^{\smile},\ i\cdot o,\ i\cdot o^{\smile},\ e\cdot o,\ e\cdot o^{\smile} \text{ and } o^{\smile}\cdot o^{\smile}. \tag{19.21}$$

TABLE 19.2: Relational form of the traditionally valid syllogisms.

	Figure 1	**Figure 2**	**Figure 3**	**Figure 4**
	Ⓜ P S Ⓜ	P Ⓜ S Ⓜ	Ⓜ P Ⓜ S	P Ⓜ Ⓜ S
$a\cdot a\ = a$	Barbara			
$a\cdot e\ = e$	Celarent	Cesare		
$e\cdot a^{\smile} = e$		Camestres		Calemes
$i\cdot a\ = i$	Darii		Datisi	
$a^{\smile}\cdot i = i$			Disamis	Dimatis
$i\cdot e\ = o$	Ferio	Festino	Ferison	Fresison
$o\cdot a^{\smile} = o$		Baroco		
$a^{\smile}\cdot o = o$			Bocardo	
$a^{\smile}\cdot e = o$			Felapton	Fesapo
$a^{\smile}\cdot a = i$			Darapti	
$a^{\smile}\cdot a^{\smile}\subseteq i$				Bamalip
$a\cdot a\ \subseteq i$	Barbari			
$a\cdot e\ \subseteq o$	Celaront	Cesaro		
$e\cdot a^{\smile}\subseteq o$		Camestros		Calemos

Lemma 19.1. *Given that the following four products*

$$a\cdot a^{\smile},\ a\cdot o,\ a^{\smile}\cdot o^{\smile},\ e\cdot e \tag{19.22}$$

are non-inferential then also all the other nine products mentioned in (19.20) and (19.21) are.

Proof. ***a·i, i·i***: Assume $a\cdot i\subseteq\tau$ for some τ, then also $a\cdot a^{\smile}\subseteq\tau$, since $a^{\smile}\subseteq i$ by (19.18) and monotonicity of product, a contradiction to the hypothesis that $a\cdot a^{\smile}$ is non-inferential. Analogous arguments apply to the case $i\cdot i$.

i·o: Next assume $i\cdot o\subseteq\tau$ for some τ, then also $a\cdot o\subseteq\tau$, since $a\subseteq i$ by (19.18), again a contradiction.

i·o$^{\smile}$: contradicts $a^{\smile}\cdot o^{\smile}$.

e·o, e·o$^{\smile}$***, o***$^{\smile}$***·o, o·o***$^{\smile}$***, o***$^{\smile}$***·o***$^{\smile}$: contradict $e\cdot e$. □

So there remain only the four products (19.22) which must be shown to be non-inferential. We can further simplify this task: If we had $\rho\sigma \subseteq a$ or $\rho\sigma \subseteq a^{\smile}$ for some product $\rho\sigma$, then also $\rho\sigma \subseteq i$ would be valid due to (19.18); and if $\rho\sigma \subseteq e$ would be valid then also $\rho\sigma \subseteq o$ or $\rho\sigma \subseteq o^{\smile}$. To prove that a product $\rho\sigma$ is non-inferential it is thus sufficient to show the *invalidity* of the following three inferences:

$$\rho\sigma \subseteq i,\ \rho\sigma \subseteq o,\ \rho\sigma \subseteq o^{\smile}. \tag{19.23}$$

A logical aside: (We did not include the following in Section 2.2 because it is needed only for the rest of the current section.) Without going into technical details we will sketch the proof strategy used for (19.23) in very rough terms. The notions you have to look for in case you want to get more information including the whole formal machinery are: consequence, interpretation, model, satisfiable, validity. As introductory books you may try [11] or [15].

Let $\varphi(x_1,\ldots,x_n)$ be a formula with variables $x_1,\ldots,x_n$. Then φ is said to be *satisfiable* if there exist n objects o_j out of some universe M such that $\varphi(o_1,\ldots,o_n)$ holds (becomes true). Sloppy speaking, we call the mapping $m: \{x_1,\ldots,x_n\} \longrightarrow M, x_i \mapsto o_i$, an *interpretation*, and say that m is a *model* of φ if $\mathbf{o}_n := (o_1,\ldots,o_n)$ satisfies φ. A formula φ is *valid* (or a *tautology*), if it holds under all interpretations; examples are $\varphi \vee \neg\varphi$ or $\exists x[x = x]$.

Let $\Phi := \{\varphi_1,\ldots,\varphi_m\}$ be a set of formulas; Φ is called *satisfiable*, if there is a model for all $\varphi_j \in \Phi$, or equivalently for $\varphi_1 \wedge \ldots \wedge \varphi_m$. φ is said to be a *logical consequence* of Φ, written $\Phi \models \varphi$, iff every model of $\varphi_1 \wedge \ldots \wedge \varphi_m$ is also a model of φ. From formal logic the following equivalences are well known:

$\Phi \models \varphi$
iff there is no interpretation which is a model of $\varphi_1 \wedge \ldots \wedge \varphi_m$ but not a model of φ
iff there is no interpretation which is a model of $\varphi_1 \wedge \ldots \wedge \varphi_m \wedge \neg\varphi$
iff $\varphi_1 \wedge \ldots \wedge \varphi_m \rightarrow \varphi$ is a tautology.

Thus, in order to prove that $\rho\sigma \subseteq \tau$ is invalid, we have to show that $\{x\rho y,\ y\sigma z\} \not\models x\tau z$, i.e. that there is a model for $x\rho y \wedge y\sigma z$ which is not a model for $x\tau z$.

Proposition 19.2. ([14, p. 16f]) *None of the three inequalities of (19.23),*

$$\rho\sigma \subseteq i,\ \rho\sigma \subseteq o,\ \rho\sigma \subseteq o^{\smile}$$

describes a valid inference.

Proof. $\rho\sigma \subseteq \boldsymbol{i}$: We must find predicates S, M and P such that $S\rho M$ and $M\sigma P$ are true but SiP is not, whence SeP is. In other words,

$$S\rho M \wedge M\sigma P \wedge SeP \tag{19.24}$$

must be satisfiable.

$\rho\sigma \subseteq \boldsymbol{o}$: We must find predicates S, M and P such that $S\rho M$ and $M\sigma P$ are true but SoP is not. In traditional syllogistic a and o are contradictory, i.e. exactly one of them is true (see Def. 19.3). So we may try to prove within the relational setting that if $S\rho M \wedge M\sigma P \wedge SaP$ is satisfiable then $S\rho M \wedge M\sigma P \wedge SoP$ is not. Suppose that $\rho\sigma \subseteq o$, and that $S\rho M \wedge M\sigma P \wedge SaP$ is satisfiable; then there are two predicates S and P with SaP and SoP. But since $o = a^{\smile} \cdot e$ (no. 9 of the strong syllogisms in Table 19.2), there is a predicate N with $Sa^{\smile}N$ and NeP. So we have NaS and SaP, which implies NaP using $a \cdot a = a$. By (19.18) we get NiP, a contradiction to NeP. Since

$$S\rho M \wedge M\sigma P \wedge SaP \tag{19.25}$$

will be shown to be satisfiable, $\rho\sigma \subseteq o$ cannot be valid.

$\rho\sigma \subseteq \boldsymbol{o}^{\smile}$: We must find predicates S, M and P such that $S\rho M$ and $M\sigma P$ are true but $So^{\smile}P$ is not. This can be shown in a similar way as above, using $o^{\smile} = e \cdot a$ and demonstrating that

$$S\rho M \wedge M\sigma P \wedge Sa^{\smile}P \tag{19.26}$$

is satisfiable.

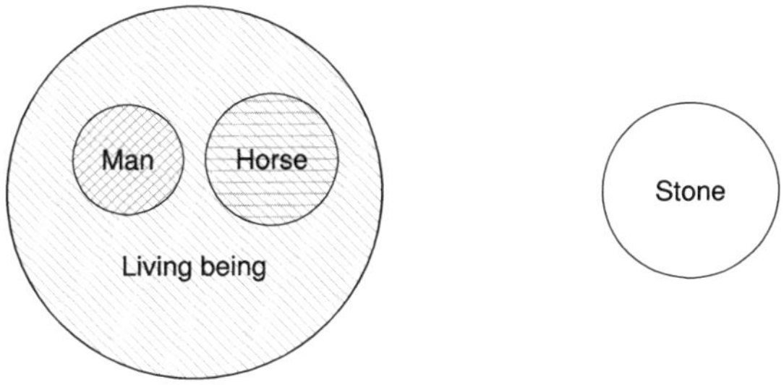

FIG. 19.1: Predicates for satisfiability.

Concerning the ten strong syllogisms, it remains to show that all three of the inference patterns $S\rho M \wedge M\sigma P \wedge SeP\,/\,SaP\,/\,Sa^{\smile}P$ of (19.24)-(19.26) are indeed satisfiable for the four products in (19.22). For this purpose four predicates P_1, P_2, P_3, P_4 are sufficient such that the following holds:

$$P_1\, a\, P_3, \quad P_2\, a\, P_3, \quad P_1\, e\, P_2 \quad \text{and} \quad P_3\, e\, P_4.$$

Aristotle uses as examples the predicates: $P_1 :=$ Man, $P_2 :=$ Horse, $P_3 :=$ Living being, and $P_4 :=$ Stone. Their relationships can be illustrated as shown in Fig. 19.1.

The satisfiability of the sentential patterns $S\rho M \wedge M\sigma P \wedge S\tau P$ by means of three predicates to be substituted for S, M and P, and where the product $\rho\sigma$ is substituted by one of the four products $a \cdot a^{\smile}$, $a \cdot o$, $a^{\smile} \cdot o^{\smile}$, $e \cdot e$ appearing in (19.22) and $\tau \in \{e, a, a^{\smile}\}$, can be seen from Table 19.3.

Let us consider a few examples listed in Table 19.3:

- *"All men are living beings"*, *"all horses are living beings"*, *"no men is a horse"* satisfies $SaM \wedge Pa^{\smile}M \wedge SeP$;
- *"All men are men"*, *"some men are not horses"*, *"no men is a horse"* satisfies $SaM \wedge MoP \wedge SeP$;
- *"All men are living beings"*, *"some living beings are men"*, *"all living beings are living beings"*, satisfies $Ma^{\smile}S \wedge Po^{\smile}M \wedge Pa^{\smile}S$.

Concerning the weak syllogisms, we have to show that (19.19) lists all the valid ones. This is possible by first disproving the following five inclusions:

$$e \subseteq i, \; a \subseteq o, \; a \subseteq o^{\smile}, \; a \subseteq a^{\smile}, \; o \subseteq o^{\smile}. \tag{19.27}$$

TABLE 19.3: Satisfying predicates.

ρ σ τ	S	M	P
a $a^{\smile}$ e	man	living being	horse
a $a^{\smile}$ a	stones	stones	stones
a $a^{\smile}$ $a^{\smile}$	stones	stones	stones
a o e	man	man	horse
a o a	man	living being	man
a o $a^{\smile}$	man	living being	man
$a^{\smile}$ $o^{\smile}$ e	man	man	horse
$a^{\smile}$ $o^{\smile}$ a	living being	man	living being
$a^{\smile}$ $o^{\smile}$ $a^{\smile}$	living being	man	living being
e e e	man	horse	stones
e e a	man	horse	man
e e $a^{\smile}$	man	horse	man

As above we have to find predicates S and P such that SeP holds but SiP does not etc. For the inclusions just mentioned this can be achieved by the following predicate pairs for S/P, in the order of their appearance in (19.27): horse/man, horse/living being, man/man, horse/living being, and living being/horse. The last one reads as: LoH, i.e. "*some living beings are not horses*", and $Lo^{\smile}H = HoL$, i.e. "*some horses are not living beings*" which certainly is false, so the former cannot imply the latter.

The invalidity of the remaining ones is shown by deducing contradictions, denoted by ※ (recall that $a \subseteq i,\ a^{\smile} \subseteq i,\ e \subseteq o,\ e \subseteq o^{\smile}$ by (19.18)):

$a \subseteq e$:	implies $a \subseteq o$, since $e \subseteq o$ ※;
$a^{\smile} \subseteq e$:	implies $a^{\smile\smile} \subseteq e^{\smile}$, i.e. $a \subseteq e$, but $a \subseteq i$ ※;
$a^{\smile} \subseteq o$:	implies $a^{\smile\smile} \subseteq o^{\smile}$, i.e. $a \subseteq o^{\smile}$ ※;
$a^{\smile} \subseteq o^{\smile}$:	implies $a \subseteq o$ ※;
$i \subseteq o$:	implies $a^{\smile} \subseteq i \subseteq o$, i.e. $a \subseteq o^{\smile}$ ※;
$i \subseteq o^{\smile}$:	implies $a \subseteq i \subseteq o^{\smile}$ ※.

Taking together the four valid inclusions from (19.18), the five invalid from (19.27) and the six just proven to be invalid, we get the fifteen pairs of relations above the main diagonal in the square matrix that arises out of $\mathscr{S} := \{a, a^{\smile}, e, i, o, o^{\smile}\}$ containing the 6×6 possible pairs. Altogether we have thus shown that the twenty-one inferences listed in Table 19.1 and (19.19) exhaust all normed syllogisms, and that therefore the twenty-four traditional syllogisms are all the valid inferences in the four figures. □

19.2 Relations and the Square of Opposition

19.2.1 *The Square of Opposition*

As its name suggests, a square of opposition is a structure consisting of four formulas and certain logically necessary relations between them. Our concern here is with the so-called **traditional square of opposition** which results by illustrating the relations among the four kinds of categorical propositions (see Section 1.1.1). For sake of convenience we shortly recapitulate the concepts needed in the considerations below, including parts of Table 1.1

TABLE 19.4: ARISTOTLE's proposition types.

Code	Proposition	Logic	Set theory
A	*All A are B*	$\forall x\,[A(x) \to B(x)]$	$A \subseteq B$
E	*All A are not B*	$\forall x\,[A(x) \to \neg B(x)]$	$A \subseteq \overline{B}$
I	*Some A are B*	$\exists x\,[A(x) \wedge B(x)]$	$A \cap B \neq \emptyset$
O	*Some A are not B*	$\exists x\,[A(x) \wedge \neg B(x)]$	$A \cap \overline{B} \neq \emptyset$

Categorical propositions express how two sets of things relate to each other and are constructed from (1) a grammatical *subject*, and (2) a grammatical *predicate*; they have two properties, viz. (a) *quality* (affirmative, negative) and (b) *quantity* (universal, particular). Do also remember that, contrary to the modern Boolean interpretation, the Aristotelian interpretation requires that a universal categorical sentence like "*All A are B*" implies that there are any *A*'s (i.e. $\emptyset \subset A$). Besides this problem of *existential import* (already discussed in Section 1.1.1), the following assumptions have to be made: Neither the **A** nor the **E** proposition must be necessarily true, because two propositions can only be contraries if they can both be false, but a necessarily true proposition cannot possibly be false. Similarly, **I** and **O** propositions must not necessarily be false. Propositions that are neither necessarily true nor necessarily false are called **contingent.**

This requirement for contingent propositions translates to the set theoretical context as follows (see Fig. 19.2): Let $A, B \subseteq X$ be two subsets of an underlying universe X. That neither **A**: $A \subseteq B$ nor **E**: $A \subseteq \overline{B}$ must be necessarily true implies $A \neq \emptyset$ and $\overline{B} \neq X$, i.e. $\emptyset \subset A, B$ (whence **A** entails **I**). And that **O**: $A \cap \overline{B} \neq \emptyset$ must not be necessarily false implies $\overline{B} \neq \emptyset$, i.e. $B \subset X$.

Definition 19.3. *The (traditional, classical, or Aristotelian)* **square of opposition** *is a set of four propositions (denoted by* **A**, **E**, **I**, *and* **O**, *according to their historical origin) with the following relationships between them:*

(a) ***A** and **O**, and also **E** and **I**, are negations of each other, i.e. exactly one of them is true:* $\boldsymbol{A} \equiv \neg\boldsymbol{O}$ *and* $\boldsymbol{E} \equiv \neg\boldsymbol{I}$;
(b) ***A** and **E** cannot both be true, i.e. at least one of them is false:* $\neg\boldsymbol{A} \vee \neg\boldsymbol{E}$;
(c) ***I** and **O** cannot both be false, i.e. at least one of them s true:* $\boldsymbol{I} \vee \boldsymbol{O}$;

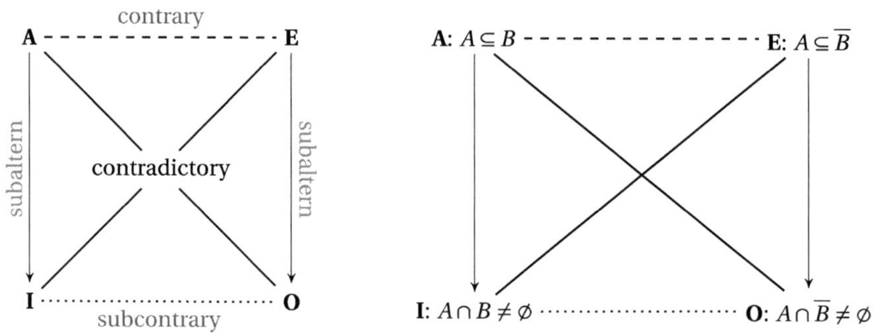

FIG. 19.2: The traditional square of opposition.

(d) ***A** entails **I**, and **E** entails **O**: $\neg$**A** $\vee$ **I** and $\neg$**E** $\vee$ **O**.*

Fig. 19.2 shows the traditional square of opposition both in its general form, and also in terms of set relationships corresponding to the categorical propositions as given in Table 19.4 above.

A considerable number of useful immediate inferences can be drawn from the information embedded in the traditional square of opposition, given the truth, or the falsehood, of any one of the four standard-form categorical propositions, as shown in Table 19.5. To give an example, if a certain **A** proposition is true, the corresponding **E** proposition is false (because at least one must be false); but if **A** is false, the corresponding **E** proposition could be either true or false, i.e. it is undetermined. In ordinary language we have e.g. if "*All cars are green*" is false, then clearly "*No car is green*" can be true or false.

TABLE 19.5: Inferences from the square of opposition.

		A	**E**	**I**	**O**
true	**A**	–	false	true	false
	E	false	–	false	true
	I	undetermined	false	–	undetermined
	O	false	undetermined	undetermined	–
false	**A**	–	undetermined	undetermined	true
	E	undetermined	–	true	undetermined
	I	false	true	–	true
	O	true	false	true	–

The logical relationships coded in the square of oppositions are not restricted to categorical propositions, which are its source of origin. In Table 19.6 we list some more examples, among them sets (repeated from Table 19.4) and relations (discussed in detail in Section 19.2.2 below). CPL denotes classical propositional logic, S5 the corresponding system of modal, and KD the one of deontic logic; see [17, p. 531].

TABLE 19.6: Instances of the Square.

Code	**CPL**	**S5**	**KD**	**Set theory**	**Rel-1**	**Rel-2**
A	$p \wedge q$	$\Box p$	$\mathrm{O}p$	$A \subseteq B$	$\overline{\overline{S}\rho}$	$\overline{\overline{S}\,\overline{\rho}}$
E	$\neg(p \vee q)$	$\Box\neg p$	$\mathrm{O}\neg p$	$A \cap B = \emptyset$	$\overline{S\rho}$	$\overline{S\overline{\rho}}$
I	$p \vee q$	$\Diamond p$	$\mathrm{P}p$	$A \cap B \neq \emptyset$	$S\rho$	$S\overline{\rho}$
O	$\neg(p \wedge q)$	$\Diamond\neg p$	$\mathrm{P}\neg p$	$A \not\subseteq B$	$\overline{S}\rho$	$\overline{S}\,\overline{\rho}$

One can construct a cube called the *cube of opposition* in [5], which did already appear with a slight variation in [16]) and might perhaps more aptly be called the *syllogistic cube*, as follows: Make all predications that are shown in Table 19.4 for A's for not-A's, saying e.g. **a**: "*every not-A is a not-B*", or **o**: "*some not-A is a B*". This results in the cube shown in Fig. 19.3. (The small letters **a**, **e**, **i**, and **o** have been used in [3, Section II].)

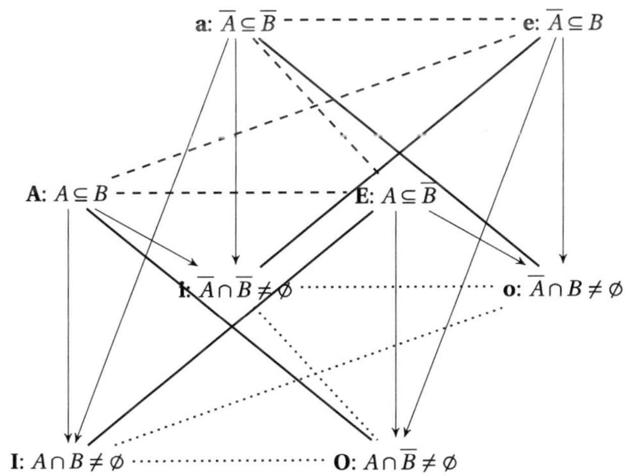

FIG. 19.3: The syllogistic cube.

Note that the assumptions we had to make in connection with Fig. 19.2, $\emptyset \subset A, B$ and $B \subset X$ for some underlying universe X, imply $\overline{A}, \overline{B} \subset X$ and $\overline{B} \neq \emptyset$. Thus **a** and **e** are not necessarily true. In order that **o** does not become necessarily false, we need $\overline{A} \neq \emptyset$, i.e. $A \subset X$, as an additional assumption (whence **a** entails **i**). One can easily check that the back face of Fig. 19.3 constitutes a traditional square of opposition. So the overall requirement $\emptyset \subset A, B \subset X$ ensures both the front and back face of the cube to constitute a square of opposition.

Moreover, with these assumptions it is not hard to show that the relationships indicated by the drawing are indeed valid for the top, bottom and side faces: For example, for the top face we have that **a** and **E** cannot both be true; for the bottom face that **I** and **o** cannot both be false, and for the side faces that **e** entails **O**.

19.2.2 *Opposition Structures Induced by Sets and Relations*

We now turn to another kind of squares of opposition, shown in Fig. 19.4, that arise when considering sets and relationships between them that are inspired by the relations existing between the logical statements in the Aristotelian square of opposition as given in Def. 19.3, taking a set $X \neq \emptyset$ as underlying universe:

Definition 19.4. *A* **set-based square of opposition** *is a collection of four sets A, E, I, O with the following relationships between them:*

(a) A and O, and also E and I, are complements of each other, i.e. $A \cap O = \emptyset = E \cap I$ and $A \cup O = X = E \cup I$;
(b) $A \cap E = \emptyset$ (but possibly $A \cup E \subset X$);
(c) $I \cup O = X$ (but possibly $I \cap O \neq \emptyset$);
(d) $A \subseteq I$, and $E \subseteq O$.

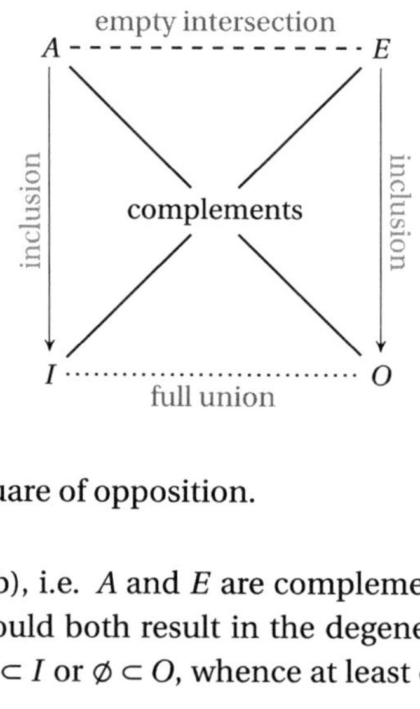

FIG. 19.4: The set-based square of opposition.

Note that $A \cup E = X$ in (b), i.e. A and E are complements, or $I \cap O = \emptyset$ in (c), i.e. I and O are complements, would both result in the degenerated case $A = I$ and $E = O$. Moreover, from (c) we get $\emptyset \subset I$ or $\emptyset \subset O$, whence at least one of the inclusions in (d) is non-trivial.

Following e.g. [2] instances of the set-based square of opposition can be constructed by taking a relation $\emptyset \subset \rho \subset X \times Y$ and a subset $\emptyset \subset S \subset X$: for the images of S and $\overline{S}$ under ρ and $\overline{\rho}$, respectively, the conditions of Def. 19.4 hold if both $\rho^{\smile}$ and $\overline{\rho}^{\smile}$ are total , i.e.

(a) $\overline{\overline{S}\rho}$ and $\overline{S}\rho$, and also $\overline{S\rho}$ and $S\rho$, are complements of each other;
(b) $\overline{\overline{S}\rho} \cap \overline{S\rho} = \emptyset$ (but possibly $\overline{\overline{S}\rho} \cup \overline{S\rho} \subset Y$);
(c) $S\rho \cup \overline{S}\rho = Y$ (but possibly $S\rho \cap \overline{S}\rho \neq \emptyset$);

(d) $\overline{\overline{S}\rho} \subseteq S\rho$, and $\overline{S\rho} \subseteq \overline{S}\rho$.

Note that the restrictions on $\rho \subset X \times Y$ and $S \subset X$ ensure that we also have $\emptyset \subset \overline{\rho}$ and $\emptyset \subset \overline{S}$ which allows us to substitute the latter for the former, which will be needed to construct a relational cube below. Since (a) obviously holds, it remains to show (b) - (c), which is done in the next two lemmas for both ρ and $\overline{\rho}$.

To make such a square work as desired, we further have to assume some requirements analogous to those for the non-emptiness of classes in categorical propositions. For sake of conciseness, let us call a relation ρ **admissible** for the remainder of this section if (i) $\emptyset \subset \rho \subset X \times Y$, with X, Y finite sets, and (ii) both $\rho^{\smile}$ and $\overline{\rho}^{\smile}$ are total, i.e. $\rho y \neq \emptyset$ and $\overline{\rho} y \neq \emptyset$, for all $y \in Y$.

Lemma 19.5. *Let ρ be an admissible relation on $X \times Y$ and $\emptyset \subset S \subset X$. Then*

(1) $\overline{S\rho} \subseteq \overline{S}\rho$;
(2) $\overline{\overline{S}\rho} \subseteq S\rho$;
(3) $\overline{S\overline{\rho}} \subseteq \overline{S}\overline{\rho}$;
(4) $\overline{\overline{S}\overline{\rho}} \subseteq S\overline{\rho}$;

Proof. (1): We have $S\rho = \{y \in Y : \exists s \in S[s\rho y]\} = \{y \in Y : S \cap \rho y \neq \emptyset\}$. So [1]

$$\begin{aligned}\overline{S\rho} &= \{y \in Y : S \cap \rho y = \emptyset\} = \{y \in Y : S \subseteq \overline{\rho y}\} = \{y \in Y : \rho y \subseteq \overline{S}\} \\ &\subseteq \{y \subset Y : \overline{S} \cap \rho y \neq \emptyset\} = \overline{S}\rho\end{aligned}$$

(2): From (1) with the substitution $S := \overline{S}$.

(3): From (2) with $S := \overline{S}$ and $\rho := \overline{\rho}$.

(4): From (3) with $S := \overline{S}$. □

Lemma 19.6. *Let ρ be an admissible relation on $X \times Y$ and $\emptyset \subset S \subset X$. Then*

(1) $\overline{\overline{S}\rho} \cap \overline{S\rho} = \emptyset$;
(2) $S\rho \cup \overline{S}\rho = Y$.
(3) $\overline{\overline{S}\overline{\rho}} \cap \overline{S\overline{\rho}} = \emptyset$;
(4) $S\overline{\rho} \cup \overline{S}\overline{\rho} = Y$.

Proof. (1): We have

$$S\rho = \{y \in Y : \exists s \in S[s\rho y]\} = \textstyle\bigcup_{s \in S} s\rho, \quad \overline{S}\rho = \{y \in Y : \exists s \in \overline{S}[s\rho y]\} = \textstyle\bigcup_{s \in \overline{S}} s\rho.$$

So their complements are given by

$$\overline{S\rho} = \overline{\textstyle\bigcup_{s \in S} s\rho} = \textstyle\bigcap_{s \in S} \overline{s\rho} = \textstyle\bigcap_{s \in S} s\overline{\rho}, \quad \text{and} \quad \overline{\overline{S}\rho} = \textstyle\bigcap_{s \in \overline{S}} s\overline{\rho}.$$

Therefore, $\overline{S\rho} \cap \overline{\overline{S}\rho} = \bigcap_{s \in X} s\overline{\rho} = \emptyset$, since assuming $\bigcap_{s \in X} s\overline{\rho} \neq \emptyset$ implies that there is a $y_0 \in Y$ such that $\forall s \in X[\neg s\rho y_0]$, i.e. $\rho y_0 = \emptyset$, contradicting the assumption $\forall y \in Y[\rho y \neq \emptyset]$.

[1] Note that it is here where the requirement for the totality of $\rho^{\smile}$, i.e. $\rho y \neq \emptyset$, for all $y \in Y$, comes into play: Assume $\rho y_0 = \emptyset$, then certainly $\rho y_0 \subseteq \overline{S} = \{y \in Y : S \cap \rho y = \emptyset\}$, whence $y_0 \in \overline{S\rho}$, for all $S \subseteq X$; but clearly $y_0 \notin \overline{S}\rho = \{y \in Y : \overline{S} \cap \rho y \neq \emptyset\}$ for any $S \subseteq X$, and the desired inclusion would not hold.

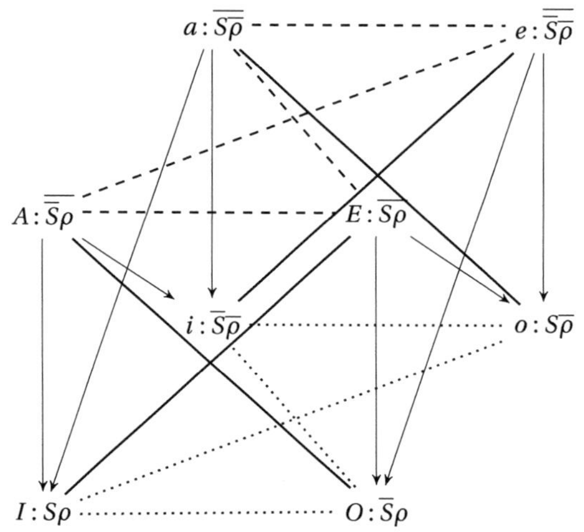

FIG. 19.5: The relational cube.

(2): Using (1) we get $Y=\overline{\emptyset}=\overline{\overline{S\rho}\cap\overline{\overline{S}\rho}}=\overline{\overline{S\rho}}\cup\overline{\overline{\overline{S}\rho}}=S\rho\cup\overline{S}\rho$.

(3),(4): By substituting $\overline{\rho}$ for ρ in (1) and (2). □

Using these two lemmas we have indeed demonstrated that the last two columns of Table 19.6 form set-based squares of opposition:

Proposition 19.7. *Let ρ be an admissible relation on $X \times Y$ and $\emptyset \subset S \subset X$. Then the front and the back face of the cube shown in Fig. 19.5 are set-based squares of opposition. For the bottom, top, and side faces the following remarkable properties hold:*

- *all the eight arrows in the side faces characterize set inclusions of the form tail $\subseteq$ head;*
- *the dotted lines in the bottom face stand for full unions, and*
- *the dashed lines in the top face for empty intersections.*

Proof. In the light of Lemma 19.5 and Lemma 19.6, we have to show the corresponding properties for diagonals only: We have e.g. $\overline{S\rho}=\{y\in Y:\forall s\in S[\neg s\rho y\}\subseteq\{y\in Y:\exists s\in S[\neg s\rho y\}=S\overline{\rho}$ for the right face; or $S\overline{\rho}\cup S\rho=\{y\in Y:\exists s\in S[\neg s\rho y]\}\cup\{y\in Y:\exists s\in S[s\rho y]\}=Y$ for the bottom, and finally for the top face $\overline{S\overline{\rho}}=\{y\in Y:S\cap\overline{\rho}y=\emptyset\}=\{y\in Y:S\subseteq\rho y\}$ and similarly $\overline{S\rho}=\{y\in Y:S\subseteq\overline{\rho y}\}$, whence $\overline{S\overline{\rho}}\cap\overline{S\rho}=\emptyset$. □

Note that the front and the back face of the *relational cube* are the only two set-based squares of opposition in it since in all other faces diagonals do not link complements.

We will now consider two instances of the relational cube belonging to the contexts of rough sets and formal concepts, two areas that have been heavily discussed during the last few decades. In our presentation we follow [1]. (But be careful when reading this paper, because the authors use the notation $\rho(S)$ to denote what is usually called the pre-image or inverse image and is written as in Def. 3.8 as $\rho^{-1}(S) = \{x \in U : \exists y \in S[x\rho y]\} = \rho S = \rho^{\smile}[S] = S\rho^{\smile}$, whereas their notation for foresets ρy and aftersets $x\rho$ is the same as ours.) Some more examples for the occurrence of the cube of opposition are given in [7,8].

Recall from Section 13.2.2.1 that a pair $\langle U, \rho\rangle$, with U a finite set and ρ a relation on U, has been called a *generalized approximation space*. Then the *lower* and *upper* ρ-*approximations* of V have been defined as the sets

$$V^{\perp}_{\rho} = \{x \in U : x\rho \subseteq V\} \quad \text{and} \quad V^{\top}_{\rho} = \{x \in U : x\rho \cap V \neq \emptyset\},$$

and the *positive (negative) region* of V comprises all $x \in U$ that definitely do (not) belong to V:

$$pos^{V}_{\rho} = V^{\perp}_{\rho} \qquad (neg^{V}_{\rho} = U - V^{\top}_{\rho} = \overline{V^{\top}_{\rho}}).$$

Some simple calculations ensure that the front face of the RST cube in Fig. 19.6 is indeed a square of opposition:

- A: $V^{\perp}_{\rho} = \{x \in U : x\rho \subseteq V\} = \{x \in U : \rho^{\smile}x \subseteq V\} = \{x \in U : \overline{V} \subseteq \overline{\rho^{\smile}x}\}$
  $= \{x \in U : \overline{V} \cap \rho^{\smile}x = \emptyset\} = \overline{\overline{V}\rho^{\smile}}$;
- I: $V^{\top}_{\rho} = \{x \in U : x\rho \cap V \neq \emptyset\} = \{x \in U : \rho^{\smile}x \cap V \neq \emptyset\} = V\rho^{\smile}$;
- E: $\overline{V^{\top}_{\rho}} = \overline{V\rho^{\smile}}$;
- O: $\overline{V}^{\top}_{\rho} = \overline{V^{\perp}_{\rho}}$ by Prop. 13.10(8U), and $\overline{V^{\perp}_{\rho}} = \overline{V}\rho^{\smile}$;
- Requiring $\rho^{\smile}$ to be admissible means that $\rho^{\smile\smile} = \rho$ is total, whence $V^{\perp}_{\rho} \subseteq V^{\top}_{\rho}$ by Prop. 13.11.

To capture also the corners involving the negation of $\rho^{\smile}$, we have to consider other operators one can find in the rough set literature: The so-called **sufficiency operator**[2] is defined as

$$[\![V]\!]_{\rho} := \{x \in U : V \subseteq x\rho\},$$

and its dual as

$$\langle\!\langle V\rangle\!\rangle_{\rho} := \{x \in U : \overline{V} \cap x\overline{\rho} \neq \emptyset\}.$$

We get

- a: $[\![V]\!]_{\rho} = \{x \in U : V \subseteq \overline{x\overline{\rho}}\} = \{x \in U : V \subseteq \overline{\overline{\rho}^{\smile}x}\} = \{x \in U : V \cap \overline{\rho^{\smile}}x = \emptyset\}$
  $= \overline{V\overline{\rho^{\smile}}}$;
- e: $\overline{\langle\!\langle V\rangle\!\rangle_{\rho}} = \{x \in U : \overline{V} \cap x\overline{\rho} = \emptyset\} = \{x \in U : \overline{V} \subseteq \overline{x\overline{\rho}} = x\rho\} = [\![\overline{V}]\!]_{\rho} = \overline{\overline{V}\,\overline{\rho^{\smile}}}$;
- i: $\langle\!\langle V\rangle\!\rangle_{\rho} = \overline{V}\,\overline{\rho^{\smile}}$;
- o: $\overline{[\![V]\!]_{\rho}} = \{x \in U : V \cap \overline{\rho^{\smile}}x \neq \emptyset\} = \{x \in U : V \cap x\overline{\rho} \neq \emptyset\} = \langle\!\langle \overline{V}\rangle\!\rangle_{\rho} = V\overline{\rho^{\smile}}$.

So we have shown that the left cube of Fig. 19.6 is a relational cube of opposition.

[2] Both $[\![\cdot]\!]_{\rho}$ and $\langle\!\langle\cdot\rangle\!\rangle_{\rho}$ are examples of **modal operators**. For more details we refer to e.g. [9] or [10]. (Note that $x \in [\![V]\!]_{\rho}$ iff $\forall y[y \in V \to x\rho y]$, i.e. $y \in V$ is sufficient for $x\rho y$.)

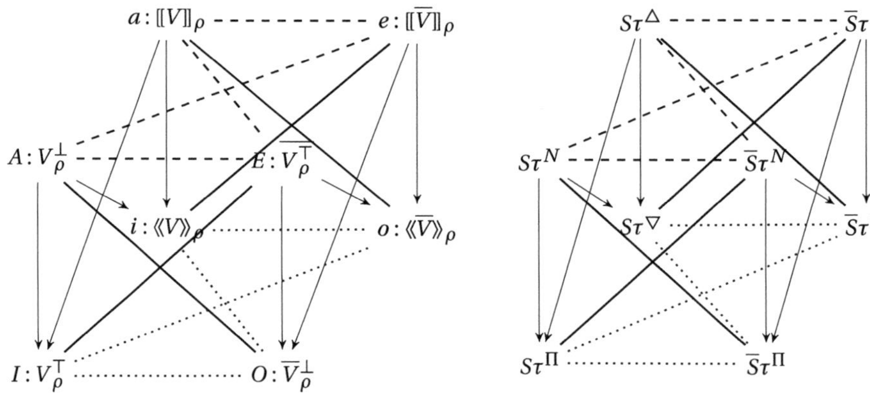

FIG. 19.6: The relational RST and FCA cubes.

Recall from Section 14.1 that a triple $\langle O, P, \tau\rangle$ with O and P two disjoint finite sets and $\tau \subseteq O \times P$, has been called a *formal context*, and that $o\tau p$ is read as "*object o has property p*". Two operators $^{\uparrow} : \mathbf{2}^O \longrightarrow \mathbf{2}^P$ and $^{\downarrow} : \mathbf{2}^P \longrightarrow \mathbf{2}^O$ have been defined as $A^{\uparrow} = \{p \in P : \forall o \in A[o\tau p]\}$ and $B^{\downarrow} = \{o \in O : \forall p \in B[o\tau p]\}$ for every $A \subseteq O$ and $B \subseteq P$. A **formal concept** is a pair $\langle A, B\rangle$, such that $A^{\uparrow} = B$ and $B^{\downarrow} = A$, where A and B are called the *extent* and *intent* of $\langle A, B\rangle$, respectively.

So $o\tau$ denotes the set of properties of object o, and $\tau p = p\tau^{\smile}$ is the set of objects with property p. The requirements we had to impose on the relation on the base of which the cube is to be constructed now have the following meaning: $\emptyset \subset p\tau^{\smile}$ says that at least one object has property p, for all $p \in P$ (otherwise p does not contribute any information about the set S of objects under consideration); and $p\tau^{\smile} \subset S$ that it is not the case that all objects have property p (which again would make p useless since it cannot discriminate objects in S). Of course, we also assume $\emptyset \subset S \subset O$. Following [4] - [6] one can define the following four sets:

- $S\tau^{\Pi} = \{p \in P : \tau y \cap S \neq \emptyset\} = \bigcup_{o \in S} o\tau$ is the set of properties that are possessed by *at least one* object in S.
- $S\tau^{N} = \{p \in P : \tau y \subseteq S\} = \bigcap_{o \notin S} \overline{o\tau}$ is the set of properties such that any object that satisfies *one* of them is necessarily in S. Thus, having any $p \in S\tau^{N}$ is sufficient for $o \in S$.
- $S\tau^{\triangle} = \{p \in P : \tau y \supseteq S\} = \bigcap_{o \in S} o\tau$ is the set of properties shared by *all* objects in S. In other words, satisfying all properties in $S\tau^{\triangle}$ is a necessary for $o \in S$.
- $S\tau^{\triangledown} = \{p \in P : \tau y \cup S \neq O\} = \bigcup_{o \notin S} \overline{o\tau}$ is the set of properties that are not satisfied by at least one object in $\overline{S}$. So for any $p \in S\tau^{\triangle}$ there is an object outside S that misses it.

Similar to the RST case it is not hard to check that the following holds: $S\tau^{\Pi} = S\tau$, $S\tau^{N} = \overline{\overline{S}\tau}$, $S\tau^{\triangle} = \overline{S\overline{\tau}}$, and $S\tau^{\triangledown} = \overline{S}\overline{\tau}$. Comparing this with Fig. 19.5 we see that the four

FCA operators form the left side face of the relational cube. In fact, standard FCA uses only $\tau^{\triangle}$: For a formal concept $\langle A, B\rangle$ we have $B = A^{\uparrow} = S\tau^{\triangle}$ and $A = B^{\uparrow} = S\tau^{\smile\triangle}$. (Note that we also have $S\tau^{\triangle} = [\![S]\!]_{\tau}$.) To complete the cube we have to consider the corresponding information on the set $\overline{S}$ and get $\overline{S}\tau^{\Pi} = \overline{S}\tau$, $\overline{S}\tau^{N} = \overline{S\tau}$, $\overline{S}\tau^{\triangle} = \overline{\overline{S}\overline{\tau}}$, and $\overline{S}\tau^{\nabla} = S\overline{\tau}$.

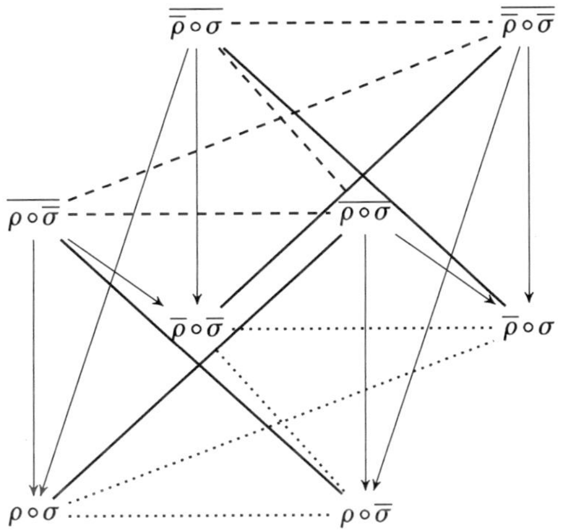

FIG. 19.7: The relational cube with composition.

The last example of a relational cube we would like to mention is from [2, p. 368] and involves relational composition. Given the usual provisions (i.e. that both ρ and σ and their converses are total), we use the following four sets to form the front square of the cube (recall that $x\overline{\rho} = \overline{x\rho}$ and $\overline{\rho}y = \overline{\rho y}$ from Prop. 3.12(1)):

$$
\begin{array}{lll}
(I) & \rho\circ\sigma = \{\langle x,z\rangle : \exists y[x\rho y \wedge y\sigma z]\} & = \{\langle x,z\rangle : x\rho \cap \sigma z \neq \emptyset\} \\
(E) & \overline{\rho\circ\sigma} = \{\langle x,z\rangle : \neg\exists y[x\rho y \wedge y\sigma z]\} & = \{\langle x,z\rangle : \forall y[\neg x\rho y \vee \neg y\sigma z]\} \\
 & \quad = \{\langle x,z\rangle : \forall y[x\rho y \rightarrow \neg y\sigma z]\} & = \{\langle x,z\rangle : x\rho \subseteq \overline{\sigma z}\} \\
(A) & \overline{\rho\circ\overline{\sigma}} = \{\langle x,z\rangle : \forall y[x\rho y \rightarrow y\sigma z]\} & = \{\langle x,z\rangle : x\rho \subseteq \sigma z\} \\
(O) & \rho\circ\overline{\sigma} = \{\langle x,z\rangle : \forall y[x\rho y \wedge \neg y\sigma z]\} & = \{\langle x,z\rangle : x\rho \cap \overline{\sigma z} \neq \emptyset\}
\end{array}
$$

It is straightforward to see that all the relationships required for a set-based square of opposition hold; see Fig. 19.4. For the bottom line we have $(\rho\circ\sigma)\cup(\rho\circ\overline{\sigma}) = \rho\circ(\sigma\cup\overline{\sigma}) = \rho\circ U_{Y\times Z}$, and since ρ is assumed to be total we get $\rho\circ U_{Y\times Z} = X\times Z$.

Similarly, and with the normalization conditions applied to $\overline{\rho}$ and $\overline{\sigma}$, the back square arises from

$$
\begin{array}{lll}
(o) & \overline{\rho}\circ\sigma = \{\langle x,z\rangle : \exists y[\neg x\rho y \wedge y\sigma z]\} & = \{\langle x,z\rangle : \overline{x\rho} \cap \sigma z \neq \emptyset\} \\
(a) & \overline{\overline{\rho}\circ\sigma} = \{\langle x,z\rangle : \neg\exists y[\neg x\rho y \wedge y\sigma z]\} & = \{\langle x,z\rangle : \forall y[x\rho y \vee \neg y\sigma z]\} \\
 & \quad = \{\langle x,z\rangle : \forall y[\neg x\rho y \rightarrow \neg y\sigma z]\} & = \{\langle x,z\rangle : \overline{x\rho} \subseteq \overline{\sigma z}\} \\
(e) & \overline{\overline{\rho}\circ\overline{\sigma}} = \{\langle x,z\rangle : \forall y[\neg x\rho y \rightarrow y\sigma z]\} & = \{\langle x,z\rangle : \forall y[x\rho y \vee y\sigma z]\} \\
 & \quad = \{\langle x,z\rangle : x\rho \cup \sigma z = Y\} & \\
(i) & \overline{\rho}\circ\overline{\sigma} = \{\langle x,z\rangle : \forall y[x\rho y \wedge \neg y\sigma z]\} & = \{\langle x,z\rangle : \overline{x\rho} \cap \overline{\sigma z} \neq \emptyset\}
\end{array}
$$

Recall that we have assumed ρ to be an *admissible* relation above, so both $\rho^{\smile}$ and $\overline{\rho}^{\smile}$ are total, i.e. $y\rho \neq \emptyset$ and $y\overline{\rho} \neq \emptyset$, for all y. This means that (A) corresponds to DeBaets-Kerre's version of the subdirect product $\rho \lhd \sigma$, and (a) to subdirect product $\overline{\rho} \lhd \overline{\sigma}$.

Moreover, we can easily see that in fact $A = \overline{\rho \circ \overline{\sigma}} \subseteq \rho \circ \sigma = I$. It is equally straightforward to check the relationships on the other faces of the cube: E.g. we have $A = \overline{\rho \circ \overline{\sigma}} = \{\langle x,z\rangle : \neg\exists y[x\rho y \wedge y\overline{\sigma}z]\} = \{\langle x,z\rangle : \forall y[x\overline{\rho}y \vee y\sigma z]\} = \{\langle x,z\rangle : \forall y[x\rho y \rightarrow y\sigma z]\}$; and $e = \overline{\overline{\rho} \circ \overline{\sigma}} = \{\langle x,z\rangle : \neg\exists y[x\overline{\rho}y \wedge y\overline{\sigma}z]\} = \{\langle x,z\rangle : \forall y[x\rho y \vee y\sigma z]\} = \{\langle x,z\rangle : \forall y[x\overline{\rho}y \rightarrow y\sigma z]\}$. So $A \cap e = \emptyset$, whence $A \subseteq \overline{e} = i$.

References

[1] Ciucci, D., Dubois, D., Prade, H.: The structure of oppositions in rough set theory and formal concept analysis - Toward a new bridge between the two settings. In: Beierle, C., Meghini, C. (Eds.): *Proceedings FoIKS 2014*, 154-173, LNCS Vol. 8367, Springer International Publishing Switzerland (2014).

[2] Ciucci, D., Dubois, D., Prade, H.: Structures of opposition induced by relations - The Boolean and the gradual cases, Annals of Mathematics and Artificial Intelligence **76**, 351-373 (2016).

[3] De Morgan, A.: On the structure of the syllogism, Trans. Cambridge Philos. Soc. **8**, 1379-1408 (1859).

[4] Dubois, D., de Saint-Cyr, F.D., Prade, H.: A possibility-theoretic view of formal concept analysis, Fundamenta Informaticae **75**, 195-213 (2007).

[5] Dubois, D., Prade, H.: From Blanch´e's hexagonal organization of concepts to formal concept analysis and possibility theory, Logica Universalis **6**, 149-169 (2012).

[6] Dubois, D., Prade, H.: Formal concept analysis from the standpoint of possibility theory. In J. Baixeries et al. (Eds.): *Proceedings ICFCA 2015*, 21-38, LNAI 9113, Springer International Publishing Switzerland (2015).

[7] Dubois, D., Prade, H., Rico, A.: The Cube of Opposition: A Structure Underlying Many Knowledge Representation Formalisms. In *Proceedings IJCAI 2015*, 2933-2939 (2015).

[8] Dubois, D., Prade, H., Rico, A.: Graded cubes of opposition and possibility theory with fuzzy events, Int. J. of Approximate Reasoning **84**, 168-185 (2017).

[9] Düntsch, I., Gediga, G.: Approximation operators in qualitative data analysis. In H. de Swart et al. (Eds.): *Theory and Applications of Relational Structures as Knowledge Instruments*, 214-230, Springer-Verlag Berlin Heidelberg (2003).

[10] Düntsch, I., Orłowska, E.: Boolean algebras arising from information systems, Annals of Pure and Applied Logic **127**, 77-98 (2004).

[11] Ebbinghaus, H.-D., Flum, J., Thomas, W.: *Mathematical Logic*, 2nd ed., Springer Science+Business Media, New York (1994).

[12] Hurley, P.J.: *A Concise Introduction to Logic*, 11th ed., Wadsworth, Belmont (2012).

[13] Lorenzen, P.: Über die Syllogismen als Relationenmultiplikationen. Archiv für mathematische Logik und Grundlagenforschung **3**, 112-116 (1957).

[14] Lorenzen, P.: *Formal Logic*, D. Reidel Publishing Company, Dordrecht (1965).

[15] Rautenberg, W.: *A Concise Introduction to Mathematical Logic*, 3rd ed., Springer Science+Business Media, New York (2010).

[16] Reichenbach, H.: The syllogism revised. Philosophy of Science **19**, 1-16 (1952).

[17] Smessaert, H., Demey, L.: Logical Geometries and Information in the Square of Oppositions, J Log Lang Inf **23**, 527-565 (2014).

Index

L-fuzzy relation, 231
L-fuzzy set, 202, 222
L-interpretation, 305
L-structure, 304
L-valued matrix, 232
Ω-algebra, 44
α-cut, 224
α-level set, 224
α-restriction, 224
ρ-boundary, 411
ρ-definable set, 411
g-inverse, 173
n-ary operation on A, 44
n-ary relation, 70
t-covering, 126
$\mathscr{A}$-assignment, 305
1-graph, 472
2-category, 378
  locally posetal, 379
2-morphism, 378

Abelian semigroup, 43
absorption laws, 50
abstract class, 305
abstract dialectical framework, 467
acceptable w.r.t. a set $S \subseteq A$, 460
accuracy of the ρ-approximation, 413
acyclic, 81, 98, 136
additive operation, 306
adjacency, 473
adjacency matrix, 136, 472
adjacent edges, 473
adjacent vertices, 473
adjoint, 142
  left, 62
  lower, 62
  right, 62
  upper, 62
adjoint couple, 213
adjointess property, 213
admissible relation, 529
admissible set, 460
admissible transformation, 503
afterset, 74
  in a fuzzy relation, 231
algebra
  Boolean, 56
  degenerate, 44
  finite, 44
  qualitative, 449
  similar, 44
  simple, 298
  subdirectly irreducible, 302
  trivial, 44
algebra of binary relations, 296
  full, 296
  square, 297
algebra of finite type, 45
algebra of type θ, 44
allegory, 379
  distributive, 385
  division, 386
  pretabular, 383
  tabular, 383
  unitary, 383
ambiguity, 399
annihilator
  left, 62
  right, 62
antichain, 49

antiring, 48
antisymmetric, 81, 98
antitone mapping, 49
approximate reasoning, 260
approximation space
  covering, 414
  generalized, 414
  Pawlak, 410
arborescence, 482
arc, 136, 471
argument, 7
  deductive, 7
  valid, 7
argumentation framework, 459
  bipolar, 466
  preference based, 466
  value based, 466
Aristotelian interpretation, 7
Aristotelian Logic, 5
Aristotelian syllogism, 8
arity, 44
arrow
  converse, 379
  epic, 342
  map, 381
  monic, 342
  total, 381
  univalent, 381
arrow category, 388
arrow product, 352
arrows, 336, 472
associated 1-graph, 474
associated relation, 472, 474
associative
  binary operation, 43
asymmetric, 81, 98
atom, 58
atomic
  Boolean algebra, 58
atomless
  Boolean algebra, 58
attribute
  deterministic, 408
  nondeterministic, 408
attribute implication, 423
attribute model, 423
Automorphic equivalence, 189
automorphic equivalence, 189
automorphism, 46
Baer semigroup, 281
balanced category, 343
band, 47
Bandler-Kohout products, 153
BAO, 306
base of $\mathfrak{Re}$, 296
base relation, 447
basis, 285
  of a vector space, 138
belongs to
  X-element b to m, 376
bicategory, 378
bidirectional residual, 96
bijective, 33, 81, 98, 113
binary matrix, 79
binary operation on S, 42
binary relation, 70
biorder, 81, 98
bipartite graph, 476
bipolar argumentation framework, 466
biresiduation, 215
bisimilar, 489
bisimulation, 489
  strong, 489
BK products, 153
BL-algebra, 213
block
  of a matrix, 133
  of a partition, 30
  of a relation, 124
block diagonal matrix, 134
Boolean algebra, 54, 56
  atomic, 58
  atomless, 58
  binary, 54
Boolean algebra with operators, 306
Boolean congruence, 325
Boolean homomorphism, 325
Boolean interpretation, 7
Boolean matrix, 79
Boolean relation algebra, 328
Boolean ring, 55
Boolean vector space of dimension, 137
border conjunctor, 250
border disjunctor, 250
border implicator, 250
bound variables, 39
bounded, 49
bounded from above, 49
bounded from below, 49

bounded lattice, 53
Brouwerian lattice, 211

canonical homomorphism, 299
cardinality, 28
carrier, 44
cartesian category, 358
Cartesian product, 31
  of fuzzy sets, 230
categorical proposition, 5
categorical statement, 5
categorical syllogism, 9
category, 336
  E/M-structured, 364
  balanced, 343
  cartesian, 358
  concrete, 338
  Dedekind, 386
  dual, 339
  finitely complete, 358
  large, 337
  locally small, 337
  opposite, 339
  order enriched, 378
  regular, 362
  Schröder, 386
  slice, 339
  small, 337
Cayley table, 43
chain, 49
characteristic function, 201
characteristic relation, 510
circle product, 153
circular, 81, 98
class
  abstract, 305
class operators, 303
classificational system, 402
Closed Extensional Mereology, 433
Closed Mereology, 433
closed set, 36, 37
closed with respect to f, 297
closure
  reflexive, 103
  symmetric, 103
  transitive, 103
closure of S, 36
closure of a relation ρ, 103
closure operator, 37
closure system, 36
  topological, 36
coarse, 403
cocone for D, 357
codomain
  of f, 336
coequalizer, 345
colimit, 357
collection, 28
column basis, 285
column consistent relation, 175
column diagonally dominant, 149
column rank, 285
column reduced, 285
column space, 285
commutative
  binary operation, 43
  quantale, 212
commutative semigroup, 43
commuting diagram, 337
commuting relations, 180
comparable, 48
compatibility
  degree of, 246
compatible, 275
  left, 275
  on $\mathfrak{A}$, 298
  right, 275
  with n-ary operation, 298
complement, 53
  fuzzy set, 224
  of a set, 28
  relative, 53
complement of ρ, 70
complemented lattice, 53
complete, 81, 98
complete extension, 461
complete lattice, 50
completely additive operation, 306
completely join-distributive, 255
composition, 74
  of functions, 33
composition operator, 336
composition table, 449
compositional rule of inference, 261
compound, 101
concatenation, 118
concept
  fuzzy, 404
concept lattice, 422
conclusion, 7, 9

concrete, 338
conditionally valid, 12
cone for D, 356
conflict-free, 460
conformable matrices, 132
congruence, 275, 298
  Boolean, 325
  left, 275
  right, 275
conjugate of f, 307
conjunctor, 250
connected, 82, 98
constants, 44
constraint network, 451
  consistent, 452
constraint satisfaction problem, 452
contingent proposition, 525
control theory, 262
convergent element, 139
converse, 232
converse arrow, 379
converse relation, 74, 371
coproduct, 351
core of an L-fuzzy, 224
coreflexive, 82, 98
cotransitive, 82, 98
cover, 360
covering, 29
covering approximation space, 414
covers, 49, 194
credulously accepted, 463
crisp set, 222
cycle, 136, 479, 480
cyclic semigroup, 139

De Morgan algebra, 54
De Morgan triple, 227
deadlock, 486
Dedekind category, 386
Dedekind rule, 86, 236, 380
deductive argument, 7
defends an argument, 460
degenerate algebra, 44
degree of compatibility, 246
degree of equality, 246
denotation, 396
dense, 82, 98
dependent vector, 138
designation, 396
deterministic, 82, 98
deterministic attribute, 408
deterministic information system, 408
deterministic property, 408
diagonal matrix, 134
diagonally dominant, 150
  column, 149
  row, 149
  weakly, 149
diagram, 337
  commuting, 337
  in $\mathscr{C}$, 356
difunctional, 82, 98
difunctional relation, 178
digraph, 136, 471, 472
  simple, 471
direct image, 155
direct product, 300
directed, 49
directed from above, 49
directed from below, 49
directed graph, 136, 471, 472
directly indecomposable, 302
disjoint, 29
disjoint, pairwise, 29
disjunctive relation, 436, 447
disjunctor, 250
distributive allegory, 385
distributive lattice, 50
distributivity, 29
diversity, 78, 310
division allegory, 386
domain, 33
  of f, 336
  of a fuzzy relation, 232
  of a relation, 71, 75
down-set, 205
dual category, 339
dual Galois connection, 210
dual isomorphism, 497
dual poset, 204
dual relation, 89
dual residual of f, 209
Duality Principle, 204
dually adjoint, 62
dually residuated
  mapping, 206
Dung argumentation framework, 459

edge, 136, 471
edge adjacency, 473

edges, 472
  adjacent, 473
  parallel, 472
element
  X-element of $\mathscr{C}$-object B, 376
  domain, 319
  equivalence, 319
  functional, 319
  left ideal, 319
  of $\mathscr{C}$-object X, 347
  range, 319
  reflexive, 319
  right ideal, 319
  subidentity, 319
  symmetric, 319
  transitive, 319
element, of a set, 27
elementary path, 479
embedding, 46
empty relation, 70
empty set, 28
endomorphism, 46
epi-mono factorization structure, 364
epic arrow, 342
epimorphism, 46, 342
  extremal, 360
  regular, 360
  split, 343
  strong, 360
equality
  degree of, 246
equalizer, 345
equational class, 305
equational theory, 305
equivalence
  automorphic, 189
  regular, 189, 190
  structural, 188, 189
equivalence class, 32, 119
  fuzzy, 242
equivalence element, 319
equivalence relation, 32, 118
  fuzzy, 241
  independent, 181
  orthogonal, 181
  partial, 118, 128
Euclidean, 82
execution fragment, 485
Existential generalization, 41
existential import, 7
Existential instantiation, 41
expansion, 298
extension, 398, 459
  complete, 461
  grounded, 461
  ideal, 462
  preferred, 460
  semi-stable, 462
  stable, 460
  stage, 462
extensional, 82, 98, 244
Extensional Mereology, 432
extent
  of a concept, 420
extremal epimorphism, 360

factor congruence, 301
factorization structure
  epi-mono, 364
Ferrers relation, 82, 98, 493
fiber product, 352
field
  of a relation, 71
  of sets, 54
figure of a syllogism, 10
filter
  of a poset, 205
  principal, 205
finer, 30
finite algebra, 44
finite limits, $\mathscr{C}$ has all, 358
finitely complete category, 358
foreset, 74
  in a fuzzy relation, 231
formal concept, 169, 420, 532
formal context, 420
frame, 212
framework
  abstract dialectical, 467
free variables, 39
fringe, 172
Frobenius normal form, 135
full algebra of binary relations, 296
full subcategory, 339
full transformation semigroup, 282
function, 32, 113
  bijective, 33
  characteristic, 201
  identity, 33
  injective, 33

one-to-one, 33
onto, 33
partial, 32
surjective, 33
total, 32
functional, 82, 98, 113
functional element, 319
functor, 340
functors, 341
fuzzy concept, 404
fuzzy equality, 241
fuzzy equivalence
class, 242
relation, 241
fuzzy partition, 246
fuzzy point, 244
fuzzy relation
ϵ-reflexive, 238
reflexive, 238
strictly ϵ-reflexive, 238
strongly reflexive, 238
symmetric, 238
transitive, 238
weakly reflexive, 238
fuzzy semi-partition, 246
fuzzy set, 202, 222, 400
normal, 224
fuzzy sets
Cartesian product, 230
fuzzy subset, 223

g.l.b., 49
Gödel algebra, 213
GA-space, 414
Galois connection, 61
dual, 210
General Extensional Mereology, 434
General Mereology, 434
generalized approximation space, 414
generalized inverse, 135, 173
generators
of a semigroup, 139
graph
bipartite, 476
directed, 136, 471, 472
strongly connected, 481
greatest element, 49
greatest lower bound, 49
Green's Lemma, 277
Green's relations, 274
Ground Mereology, 430
Ground Mereotopology, 445
Ground Topology, 442
grounded extension, 461
group, 47
groupoid, 43
simple, 273

Hall matrix, 145
has images, 350
head, 471, 472
height of an L-fuzzy, 224
heterogeneous relation, 70
Heyting algebra, 211, 217
Heyting lattice, 211
hom-set, 336
homogeneous relation, 70
homomorphic image, 46
homomorphism, 45, 255
Boolean, 325
canonical, 299
from ρ to σ, 108
of digraphs, 482
horizontal composition, 378
hybrid structure, 45
hypothetical syllogisms, 260

ideal
in a Boolean algebra, 325
in a relation algebra, 325
in a ring, 326
of a groupoid, 273
left, 273
right, 273
two-sided, 273
of a poset, 205
principal, 205
ideal element, 319
ideal extension, 462
ideal set, 462
idempotent
binary operation, 43
element, 47
quantale, 212
semigroup, 47
identity, 43
identity function, 33
identity on B, 336
identity relation, 70
image

of A under ρ, 72
of S under f, 33
of a morphism, 350
under a fuzzy relation, 232
image factorization, 367
implicative lattice, 211
implicator, 250
incidence, 472
ingoing, 472
outgoing, 472
incidence matrix, 472
incidence relation, 420
incident vertex, 473
incline, 141
0-1-, 141
incomparable, 48
indecomposable
directly, 302
indegree, 473
independent
Boolean vectors, 138
edge set, 478
equivalence relations, 181
set, 285
set of axioms, 311
vertex set, 478
index
of element a, 139
indifference relation, 496
indiscernibility, 409
indiscernible, 244
indistinguishable, 244
inf-morphism, 255
inference rule, 260
valid, 260
infimum, 49
information system, 408
complete, 408
deterministic, 408
incomplete, 408
nondeterministic, 408
information table, 408
ingoing incidence, 472
initial node, 472, 476
initial object, 347
initial vertex, 136, 471
injective, 33, 82, 98, 113
integral
quantale, 212
intension, 398
intent
of a concept, 420
interpretation, 305
intersection
of fuzzy sets, 223
interval
of a lattice, 53
interval algebra, 440
interval order, 495
intransitive, 82, 98
inverse, 43
generalized, 173
Moore-Penrose, 173
of a function, 35
of a relation, 173
sub-, 173
inverse elements, 272
inverse image of B under ρ, 72
inverse matrix, 135, 145
inverse semigroup, 272
invertible matrix, 135, 145
involution semigroup, 272
irreducible matrix, 135
irreflexive, 82, 98
iso arrow, 343
isomorphic
sets, 301
spans, 368
isomorphism, 46
between ρ and σ, 109
dual, 497
in a category, 343
of digraphs, 483
isotone mapping, 49

JEPD set of relations, 435
join decomposition, 254
irredundant, 255
join semilattice, 50
join-decomposable, 254
join-irreducible, 254
jointly monic, 368

kernel
of a function, 33
kernel pair, 354
Kronecker symbol, 132

l.u.b., 49
labelled transition system, 485

large category, 337
lattice, 50
  relatively complemented, 53
  bounded, 53
  Brouwerian, 211
  complemented, 53
  complete, 50
  distributive, 50
  implicative, 211
  pseudocomplemented, 211
  relatively pseudocomplemented, 210
  residuated, 213
least element, 49
least upper bound, 49
left adjoint, 62
left cancellative, 34
left compatible, 275
left congruence, 275
left Euclidean, 82, 98
left ideal
  element, 319
  of a groupoid, 273
  principal, 273
left ideal of S generated by A, 273
left identity, 43
left inverse, 43, 145
  of a function, 34
  of a relation, 173
left residual, 91, 386
left residuated, 64
left total, 82, 98
left translation, 64
left unique, 82, 98
left zero, 272
level set of A, 224
lexicographic order, 118
limit, 356
linear element, 390
linear order, 116
linearly ordered set, 49
locale, 212
locally posetal, 379
locally reflexive, 82, 98
locally small category, 337
loop, 471
lower ρ-approximation, 411
lower adjoint, 62
lower bound, 49
lower triangular, 134
LTS, 485
Lukasiewicz algebra, 213

major premise, 9
major term, 9
many-to-one, 82, 98
map, 82, 98
  see mapping, 32
map arrow, 381
mapping, 32, 82, 98, 113
  antitone, 49
  dually residuated, 206
  isotone, 49
  left cancellative, 34
  residuated, 62, 206
  right cancellative, 34
maps S into T, 33
maps S onto T, 33
matching, 478
matrices
  conformable, 132
matrix
  L-valued, 232
  binary, 79
  block diagonal, 134
  block lower triangular, 134
  block upper triangular, 134
  Boolean, 79
  column diagonally dominant, 149
  diagonal, 134
  diagonally dominant, 150
  inverse, 135, 145
  invertible, 135, 145
  irreducible, 135
  left invertible, 145
  lower triangular, 134
  nilpotent , 135
  nonsingular, 135
  partitioned, 133
  permutation, 133
  primitive, 135
  reducible, 134
  reflexive, 145
  right invertible, 145
  row diagonally dominant, 149
  symmetric, 132
  transitive, 145
  upper triangular, 134
  weakly diagonally dominant, 149
matrix product, 132
maximal, 49

maximal rectangle, 169
maximally surjective, 256
maximum, 49
meaning triangle, 396
measurement theory, 501
mediate, 82, 98
meet semilattice, 50
membership degree, 222
membership function, 222
mereology, 427
  closed, 433
  closed extensional, 433
  extensional, 432
  general, 434
  general extensional, 434
  ground, 430
  minimal, 432
mereotopology
  ground, 445
metric, 123
middle term, 9
minimal, 49
Minimal Mereology, 432
minimum, 49
minor, 142
minor premise, 9
minor term, 9
modal, 224
modal operator, 531
model, 305
  of a composition table, 449
model class, 305
Modular laws, 86, 236
modus ponens, 203, 260
modus tollens, 260
monic
  arrow, 342
  jointly, 368
monoid, 47
  integral, 226
monomorphism, 46, 342
  split, 343
monotone Galois connection, 62
mood, 9
Moore family, 36
Moore-Penrose inverse, 173
morphisms, 336
multiple edges, 471
multiplicative unit, 47
MV-algebra, 214

naive set, 460
natural mapping, 299
natural partial order, 279
natural transformation, 341
negation, 213
negative region, 411
negatively transitive, 82, 98, 495
negator, 227
  involutive, 227
  strong, 227
neighbor vertices, 473
neighborhood of x
  in τ, 124
neutral element, 43
nilpotent matrix, 135
node, 136, 471
  initial, 476
  terminal, 476
nodes, 472
non-enlargeable rectangle, 169
nonassociative relation algebra, 312
nondeterministic attribute, 408
nondeterministic property, 408
nonsingular matrix, 135
normal
  fuzzy set, 224
  operation, 306
normed syllogisms, 519

objects, 336
observable, 244
offshoots, 256
one, 47
one-to-many, 98
one-to-one, 33, 113
onto, 33, 82, 98, 113
operation
  additive, 306
  completely additive, 306
  normal, 306
operational structure, 45
operator, 306
  closure, 37
  extensive, 37
  isotone, 37
  normal, 306
opposite category, 339
order
  strict partial, 116
  interval, 495

lexicographic, 118
linear, 116
of a semigroup, 139
of an element, 139
partial, 48, 116
total, 116
weak, 495
order enriched category, 378
ordered pair, 31
ordered semigroup, 64
orthodox semigroup, 47
orthogonal
arrows in $\mathscr{C}$, 359
equivalence relations, 181
morphism classes, 359
relation, 175
oscillating element, 139
outdegree, 473
outgoing incidence, 472

PA-space, 410
pair
residuated, 206
pairwise disjoint, 29
parallel edges, 472
partial equivalence relation, 118, 128
partial function, 32
partial identity, 382
partial map, 113, 282
partial order, 48, 116
partial partition, 29
partial transformation, 282
partially ordered semigroup, 64
partially ordered set, 48, 116
partition, 30
fuzzy, 246
partial, 29
restriction of, 182
partition scheme, 447
partitioned, 133
partitioned matrix, 133
path, 136, 479, 480
elementary, 479
simple, 480
void, 480
path algebra, 142
Pawlak approximation space, 410
period
of element a, 139
periodic semigroup, 139
permanent, 142
permutation, 282
permutation matrix, 133, 145
permutation relation, 175
Poincaré paradox, 243
point, 163
point algebra, 438
point axiom, 166
pomonoid, 64
posemigroup, 64
poset, 48, 116
dual, 204
positive region, 411
potential function, 511
power set, 29
powerset relation, 168
pre-block
of a relation, 124
pre-image of B under ρ, 72
pre-image of T under f, 33
predecessor, 475
predecessor degree, 475
predicate term, 6
preference based argumentation framework, 466
preference structure, 510
preferred extension, 460
prefix, 118
premise, 7, 9
preorder, 116
total, 495
pretabular allegory, 383
primitive matrix, 135
principal down-set, 205
principal filter, 205
principal ideal, 205
of a relation algebra, 325
principal left ideal, 273
principal right ideal, 273
principal two-sided ideal, 273
principal up-set, 205
process, 484
product, 74
circle, 153
direct, 300
in a category, 350
of Boolean matrices, 79
of fuzzy sets, 224, 234
of matrices, 132
square, 153

subdirect, 153, 302
superdirect, 153
product arrow, 351
product of L-valued matrices, 232
projection, 31
arrows, 350
proper fuzzy subset, 223
proper relation algebra, 296
property
deterministic, 408
nondeterministic, 408
pseudo-Boolean algebras, 211
pseudo-order, 194
pseudocomplement
relative, 210
pseudocomplemented, 211
pseudocomplemented lattice, 211
pseudoinverse
of a relation, 173
pullback, 352
Pullback Lemma, 355
pullback stable, 355
pushout, 353

qualitative algebra, 449
quality, 6
quantale, 212
commutative, 212
idempotent , 212
integral, 212
unital, 212
quantity, 6
quasi-order, 116
quotient algebra, 299
quotient set, 32, 242
of a tolerance, 124
of an equivalence, 119

range, 33
of a fuzzy relation, 232
of a relation, 71, 75
range of S, 460
rank
of a partial transformation, 283
of an operation, 44
rectangle, 170
containing ρ, 171
inside ρ, 171
maximal, 169
non-enlargeable, 169
of ρ, 169
of a relation algebra, 319
outside ρ, 171
rectangular
relation, 170
reduced relation, 285
reducible matrix, 134
reduct, 298
reference, 396
refinement, 403
partition, 30
refinement mapping, 402
class preserving, 402
proper, 402
reflexive, 82, 98
reflexive closure, 103
reflexive element, 319
reflexive matrix, 145
reflexive relation, 371
regular
$\mathscr{D}$-class, 278
category, 362
element, 271
epimorphism, 360
relation, 82, 98, 173
semigroup, 271
regular equivalence, 190
w.r.t. ρ, 490
regular equivalence, 189
regular representation, 503
regular scale, 503
relation, 32, 69
L-fuzzy, 231
n-ary, 70
acyclic, 81, 98
admissible, 529
antisymmetric, 81, 98
asymmetric, 81, 98
bijective, 81, 98, 113
binary, 32, 70
biorder, 81, 98
characteristic, 510
circular, 81, 98
column consistent, 175
column reduced, 285
complete, 81, 98
composition, 74, 232
connected, 82, 98
converse, 74, 371
coreflexive, 82, 98

cotransitive, 82, 98
dense, 82, 98
deterministic, 82, 98
difunctional, 82, 98, 178
disjunctive, 435, 436
dual, 89
empty, 70
equivalence, 32, 118
Euclidean, 82
extensional, 82, 98
Ferrers, 82, 98, 493
functional, 82, 98
heterogeneous, 70
homogeneous, 70
in a category, 369
indifference, 496
injective, 82, 98, 113
intransitive, 82, 98
inverse, 173
irreflexive, 82, 98
left Euclidean, 82, 98
left inverse, 173
left total, 82, 98
left unique, 82, 98
locally reflexive, 82, 98
many-to-one, 82, 98
mediate, 82, 98
negatively transitive, 82, 98, 495
one-to-many, 98
one-to-one, 113
onto, 82, 98, 113
orthogonal, 175
partial equivalence, 118, 128
powerset, 168
product, 74, 232
pseudoinverse, 173
reduced, 285
reflexive, 32, 82, 98, 371
regular, 82, 98, 173
right Euclidean, 82, 98
right inverse, 173
right unique, 82, 98
row consistent, 175
row reduced, 285
semitransitive, 82, 98, 495
separating, 82, 98
serial, 82, 98
strongly complete, 82, 98
strongly connected, 82, 98
subidentity, 82, 98
surjective, 82, 99, 113
symmetric, 32, 82, 99, 371
ternary, 70
tolerance, 118
total, 82, 99
totally connected, 82, 99
transitive, 32, 82, 99, 371
unary, 70
univalent, 82, 99
universal, 70
valued, 511
relation algebra, 310–312
atomic, 327
Boolean, 327, 328
commutative, 327
integral, 327
nonassociative, 312
proper, 296
representable, 328
semiassociative, 312
simple, 328
symmetric, 327
trivial, 327
weakly associative, 312
relation from S to T, 70
relation on S, 70
relational cube, 530
relational structure of finite type, 45
relational structure of type θ, 45
relations
commuting, 180
relative complement, 53
relative pseudocomplement, 210
relative sum, 19
relatively complemented lattice, 53
relatively pseudocomplemented lattice, 210
representable, 328
representation
of ρ, 79
of a relation algebra, 328
regular, 503
subdirect, 302
representation of ρ, 232
representation problem, 502
representation theorem, 328, 502
residual, 62, 90
bidirectional, 96
left, 91, 386
right, 90, 387
residual of f, 206

residual pair, 209
residuated
  lattice, 213
  left, 64
  mapping, 206
  pair, 206
  right, 64
  semigroup, 64
residuated mapping, 62
residuation, 213
restriction
  of f to S, 33
  of a partition, 182
restriction of a relation, 70
retraction, 343
right adjoint, 62
right cancellative, 34
right compatible, 275
right congruence, 275
right Euclidean, 82, 98
right ideal
  element, 319
  of a groupoid, 273
  principal, 273
right identity, 43
right inverse, 43, 145
  of a function, 34
  of a relation, 173
right residual, 90, 387
right residuated, 64
right translation, 64
right unique, 82, 98
right zero, 272
ring, 47
  Boolean, 55
role assignment, 188
role graph, 188
root, 482
root system, 255
rooted graph, 482
rooted tree, 482
rough set, 411, 412
row basis, 285
row consistent relation, 175
row diagonally dominant, 149
row rank, 285
row reduced, 285
row space, 174, 285
rule of inference, 260
s-crisp relation (in a Dedekind category), 389
scalar, 387
scalar multiplication, 138
scalar relation, 388
scale, 502
  regular, 503
sceptically accepted, 463
Schröder category, 386
Schröder equivalences, 22, 85, 236
section, 343
semantically independent, 185, 187
semantics
  for argumentation frameworks, 460
  unique-status, 460
semi-partition
  fuzzy, 246
semi-stable extension, 462
semiassociative relation algebra, 312
semigroup, 43, 46
  Abelian, 43, 47
  commutative, 43, 47
  cyclic, 139
  full transformation, 282
  idempotent, 47
  inverse, 272
  involution, 272
  ordered, 64
  orthodox, 47
  partially ordered, 64
  periodic, 139
  regular, 271
  residuated, 64
  with zero, 62
  zero, 272
semilattice, 50
  join, 50
  meet, 50
semiorder, 495
semiring, 47
  additively idempotent, 47
  commutative, 47
  zerosumfree, 48
semitransitive, 82, 98
separating, 82, 98
sequential system, 484
serial, 82, 98
set, 27
  L-fuzzy, 202, 222
  ρ-definable, 411
  ρ-rough, 411

closed, 297
difference, 28
fuzzy, 202, 222
intersection, 28
linearly ordered, 49
partially ordered, 48
union, 28
sets
isomorphic, 301
signature, 44
similarity, 409
similarity relation, 241
similarity relations, 241
similarity type, 44
simple algebra, 298
simple digraph, 471
simple groupoid, 273
simulation, 487
sink, 476
slice category, 339
small category, 337
solitary element, 414
solitary set, 414
source, 476
of f, 336
span, 138
in $\mathscr{C}$, 368
spans
composition of, 370
isomorphic, 368
split epimorphism, 343
split monomorphism, 343
square, 319
square algebra of binary relations, 297
square image, 155
square of opposition, 525
set-based, 528
square product, 153
stable
set of nodes, 478
stable extension, 460
stable under pullback, 355
stage extension, 462
standard categorical syllogism, 9
standard form, 9
star, 103
stem, 256
Stone representation theorem, 328
strict partial order, 49, 116, 495
strong components, 136
strong epimorphism, 360
strongly complete, 82, 98
strongly connected, 82, 98
components, 481
graph, 136, 481
nodes, 136
structural equivalence, 188, 189
structure
operational, 45
subalgebra, 297
subcategory, 339
full, 339
subconcept, 422
subdirect image, 155
subdirect product, 153, 302
subdirect representation, 302
subdirectly irreducible, 302
subidentity, 82, 98
subidentity element, 319
subinverse, 173
subject term, 6
subobject, 347, 348
subrelation, 371
subsemigroup, 139
subset, 28
fuzzy, 223
proper, 28
proper fuzzy, 223
subspace, 138
substitution property, 298
subuniverse, 297
successor, 475
successor degree, 475
sufficiency operator, 531
suffix, 118
sup-morphism, 255
superconcept, 422
superdirect image, 155
superdirect product, 153
support of an L-fuzzy, 224
supremum, 49
surjective, 33, 82, 99, 113
syllogism
categorical, 9
hypothetical, 260
normed, 519
standard categorical, 9
symmetric, 82, 99
symmetric closure, 103
symmetric difference, 30

symmetric element, 319
symmetric group, 282
symmetric inverse semigroup, 282
symmetric matrix, 132
symmetric quotient, 93
symmetric relation, 371
system over A, 29

t-conorm, 226
t-norm, 226
tabular allegory, 383
tabulation of ρ, 370
tabulation of r, 383
tail, 471, 472
target
  of f, 336
target node, 472
target vertex, 136, 471
Tarski rule, 102
terminal node, 476
terminal object, 347
ternary relation, 70
Theorem K, 15, 87
tolerance class, 124
tolerance relation, 118
  fuzzy, 241
tolerance space, 124
total, 82, 99, 113
total arrow, 381
total order, 116
total preorder, 495
totally connected, 82, 99
trace, 485
trace equivalent
  completed, 486
  LTSs, 486
  states, 486
traditional square of opposition, 525
transformation, 282
  partial, 282
transitive, 82, 99
transitive closure, 103
transitive element, 319
transitive matrix, 145
transitive relation, 371
translation
  left, 64
  right, 64
transpose, 132
tree
  rooted, 482
triangular conorm, 226
triangular norm, 226
trivial algebra, 44
two-sided ideal
  of a groupoid, 273
  principal, 273
type, 44, 45

unary relation, 70
unconditionally valid, 12
union
  of fuzzy sets, 223
unique-status semantics, 460
uniqueness problem, 503
uniqueness theorem, 503
uniqueness theorems, 503
unit
  in an allegory, 383
unit interval, 202
unital
  quantale, 212
unitary allegory, 383
univalent, 82, 99
univalent arrow, 381
Universal generalization, 41
Universal instantiation, 41
universal relation, 70
universe of discourse, 222
up-set, 205
upper ρ-approximation, 411, 414
upper adjoint, 62
upper bound, 49
upper triangular, 134

vagueness, 401
valid argument, 7
value based argumentation framework, 466
valued relation, 511
variables
  bound, 39
  free, 39
variety, 303
vector, 163
vector space, 138
  Boolean, 137
vertex, 136, 471
  incident, 473
vertical composition, 378
vertices, 472

adjacent, 473
neighbor, 473
void path, 480

weak
composition of relations, 437
order, 495
weak composition, 449
weakly associative relation algebra, 312
weakly diagonally dominant, 149

zero, 47
zero element, 62, 272
zero semigroup, 272
Zorn's Lemma, 125